Basic and Applied Aspects of Biotechnology

Varsha Gupta • Manjistha Sengupta
Jaya Prakash • Baishnab Charan Tripathy

Basic and Applied Aspects of Biotechnology

Springer

Varsha Gupta
Institute of Biosciences and Biotechnology
Chhatrapati Shahu Ji Maharaj University
Kanpur, UP, India

Jaya Prakash
Orthopaedics Unit
Community Health Centre
Kanpur, UP, India

Manjistha Sengupta
George Washington University
Washington DC, USA

Baishnab Charan Tripathy
School of Life sciences
Jawaharlal Nehru University
New Delhi, India

ISBN 978-981-10-0873-3 ISBN 978-981-10-0875-7 (eBook)
DOI 10.1007/978-981-10-0875-7

Library of Congress Control Number: 2016937500

Printed on acid-free paper

This Springer imprint is published by Springer Nature
The registered company is Springer Science+Business Media Singapore Pte Ltd.

Dedicated to teaching fraternity and students

Acknowledgments

We would like to express our heartfelt gratitude to everyone, who has contributed selflessly to the growth and development of science and kept the spirit of knowledge alive. Their quest for knowledge and the reflections of it in the form of writings, publications, books, and other documents have immensely helped us to live that spirit. That immortal spirit of investigation has always been an inspiration for us. Through this book, we hope to pass on that spirit to its readers.

We would like to thank Abhishek Gupta (managing director, i3 Consulting), who has provided his valuable inputs in the chapter on drug design. We would like to acknowledge the contribution of Dr. Kajal Biswas (staff scientist, Mouse Cancer Genetics Program, National Cancer Institute, Frederick, Maryland, USA) for his contribution in the chapter on animal cell culture. We would also thank Meghana and Agastya for spending their time and helping us with the writings and diagrams that could have been well spent on playing and watching cartoon movies. We would also like to thank Neerja, Vishal, Sipahee, Vivek, Gunjan, Sovon, Suruchi, and Meetu for their continued encouragement, support, and cooperation throughout.

The acknowledgment cannot be complete without expressing our gratitude to Dr. Mamta Kapila, Ms. Raman Shukla, Ms. Akanksha, Mr. Magesh Kaarthick, and to the entire Springer team for their cooperation and support.

Last but not the least, through this acknowledgment, we would like to express our gratitude to our family and friends for their immense patience, unfailing cooperation, love and support, and being a constant source of inspiration during the preparation of this project work.

We would like to close by thanking the Almighty for helping us through the entire journey and making this experience enjoyable.

Contents

About the Authors

Varsha Gupta, Ph.D. is working as Assistant Professor in Biotechnology Department, Chhatrapati Shahu Ji Maharaj University, Kanpur. She is specialized in genomics and proteomics techniques. Her research group is presently focusing on molecular diagnostics of autoimmune and orthopaedics disorders like rheumatoid arthritis, psoriatic arthritic, diabetes mellitus and osteoarthritis.

Her publications are:

1. Patel SL, Kumar V, Mishra R, Chandra V, Negi MPS, Tripathy BC, Prakash J and Gupta Varsha (2014) Effectiveness of methotrexate therapy with occasional corticosteroid in rheumatoid arthritis. Current Orthopaedics Practice.
2. Prakash J, Singh H, *Gupta Varsha* (2012) Evaluating arthroscopic debridement as a surgical option for patients with differing grades of knee osteoarthritis. Current Orthopaedics Practice. 23: 218-220.
3. Mishra R, Singh A, Chandra V, Negi MPS, Tripathy BC, Prakash J, and Gupta Varsha (2011) A comparative analysis of serological parameters and oxidative stress in Osteoarthritis and Rheumatoid arthritis. Accepted for publication in Rheumatology International (Clinical and Experimental Investigations) DOI: 10.1007/s00296-011-1964-1.
4. Gaur R, Sethy NK, Choudhary S, Shokeen B, *Gupta Varsha*, Bhatia S. (2011) Advancing the STMS genomic resources for defining new locations on the intra specific genetic linkage map of chickpea (Cicerarietinum L.). BMC Genomics. Feb 17;12:117.

Manjistha Sengupta, Ph.D. research interests broadly include the study of pathogenesis of infectious and autoimmune diseases. Her research encompasses several aspects of human health such as establishment of infection by pathogenic micro-organisms, mechanism of disease progression, multidrug resistance and study of virulence factors associated with clinically relevant pathogens, for example Candida albicans and Vibrio cholerae. Her studies include mechanisms of low copy virulence plasmid inheritance and maintenance in E.coli. She is currently studying autoimmune neuromuscular disorder myasthenia gravis at George Washington University, Washington, DC.

1. Manjistha Sengupta, Amrita Cheema, Henry Kaminski, Linda Kusner and The Muscle Study Group (2014). Serum Metabolomic Response of

Myasthenia Gravis Patients to Chronic Prednisone Treatment. PLoS ONE. 9(7): e102635. doi:10.1371/journal.pone.0102635.
2. James A Sawitzke, Brenda Youngren, Lynn C. Thomason, Teresa Baker, Manjistha Sengupta, Donald Court and Stuart Austin (2012). The segregation of E.coliminichromosomes constructed in vivo by recombineering. Plasmid. 67: 148–54.
3. Manjistha Sengupta and Stuart Austin (2011). The prevalence and significance of plasmid maintenance functions in virulence plasmids of pathogenic bacteria. Infection and Immunity.79: 2502–9.
4. Swagata Ghosh, Hanumanta Rao, Manjistha Sengupta, Sujit Bhattacharya, AsisDatta (2011). Two gene clusters coordinate for a functional N-acetylglucosamine catabolic pathway in Vibrio cholerae. Molecular Microbiology. 80: 1549–60.

Jaya Prakash, M.B.B.S., D.Ortho, D.N.B. (sec) Orthopaedics is senior medical officer, orthopaedic surgeon working with Uttar Pradesh Health Services. He is specialized in trauma and arthroscopic surgery. He handles various aspects of orthopaedic surgeries and disorders including autoimmune diseases like rheumatoid and psoriatic arthritic. He has contributed immensely for welfare of patients and has been awarded Kannauj Ratna for working selflessly for society and people. His some of the contributions are:

1. Patel SL, Kumar V, Mishra R, Chandra V, Negi MPS, Tripathy BC, Prakash J and Gupta Varsha (2014) Effectiveness of methotrexate therapy with occasional corticosteroid in rheumatoid arthritis. Current Orthopaedics Practice.
2. Prakash J, Singh H, Gupta Varsha (2012) Evaluating arthroscopic debridement as a surgical option for patients with differing grades of knee osteoarthritis. Current Orthopaedics Practice. 23: 218–220.
3. Mishra R, Singh A, Chandra V, Negi MPS, Tripathy BC, Prakash J, and Gupta Varsha (2011) A comparative analysis of serological parameters and oxidative stress in Osteoarthritis and Rheumatoid arthritis. Accepted for publication in Rheumatology International (Clinical and Experimental Investigations) DOI: 10.1007/s00296-011-1964-1.

Baishnab Charan Tripathy, Ph.D. is presently Dean of School of Life Sciences, Jawaharlal Nehru University, New Delhi. He has worked as Vice-Chancellor, Ravenshaw University, Cuttack, Odisha, from November 25, 2011–November 24, 2014. He has received several honours and awards for his contributions in the field of photobiology. He is distinguished scientist who has worked for NASA's Space Biology Project. He is visiting scientist at many international universities and institutes. He has organized many national and international conferences and is member of several academic bodies of JNU. More than 20 students have got their PhD degree under his expert guidance and supervision. He has got a patent to his credit (US2009/0291476 A1) for overexpressing gene of protochlorophyllide oxidoreductase C for imparting resistance in the plants against biotic and abiotic stresses. He has served as reviewer to several prestigious national and international journals like

"The Plant cell," "Plant physiology" to name a few. He is the reviewer for proposals submitted to National Science Foundation, USA; Hungarian Academy of Sciences; and German Science Foundation. He has 80 publications in peer-reviewed high-impact journals and co-authored two books with Springer, Dordrecht, The Netherlands.

Some of his publications after 2010 are:

1. Leelavathi, S, Bhardwaj, A., Kumar, S., Dass, A, Pathak, R, Pandey, SS, Tripathy, BC, Padmalatha, KV, Dhandapani, G, Kanakachari, M, Ananda Kumar, P, Cella, R, Reddy, VS (2011) Genome-wide transcriptome and proteome analyses of tobacco psaA and psbA deletion mutants. Plant Mol Biol. 76: 407–423.
2. Pattanayak, G. K., Tripathy, BC (2011) Overexpression of protochlorophyllide oxidoreductase C regulates oxidative stress in Arabidopsis. Plos One (Published online October, 2011).
3. Biswal AK, Pattanayak GK, Pandey SS, Leelavathi S, Reddy VS, Govindjee, Tripathy BC (2012). Light intensity-dependent modulation of chlorophyll b biosynthesis and photosynthesis by overexpression of chlorophyllide a oxygenase in tobacco. Plant Physiol

An Introduction to Biotechnology 1

Abstract

Biotechnology is multidisciplinary field which has major impact on our lives. The technology is known since years which involves working with cells or cell-derived molecules for various applications. It has wide range of uses and is termed "technology of hope" which impact human health, well being of other life forms and our environment. It has revolutionized diagnostics and therapeutics; however, the major challenges to the human beings have been threats posed by deadly virus infections as avian flu, Chikungunya, Ebola, Influenza A, SARS, West Nile, and the latest Zika virus. Personalized medicine is increasingly recognized in healthcare system. In this chapter, the readers would understand the applications of biotechnology in human health care system. It has also impacted the environment which is loaded by toxic compounds due to human industrialization and urbanization. Bioremediation process utilizes use of natural or recombinant organisms for the cleanup of environmental toxic pollutants. The development of insect and pest resistant crops and herbicide tolerant crops has greatly reduced the environmental load of toxic insecticides and pesticides. The increase in crop productivity for solving world food and feed problem is addressed in agricultural biotechnology. The technological advancements have focused on development of alternate, renewable, and sustainable energy sources for production of biofuels. Marine biotechnology explores the products which can be obtained from aquatic organisms. As with every research area, the field of biotechnology is associated with many ethical issues and unseen fears. These are important in defining laws governing the feasibility and approval for the conduct of particular research.

V. Gupta et al., *Basic and Applied Aspects of Biotechnology*, DOI 10.1007/978-981-10-0875-7_1

1.1 Introduction

The term "biotechnology" was coined by a Hungarian engineer Karl Ereky, in 1919, to refer to the science and methods that permit products to be produced from raw materials with the aid of living organisms. Biotechnology is a diverse field which involves either working with living cells or using molecules derived from them for applications oriented toward human welfare using varied types of tools and technologies. It is an amalgamation of biological science with engineering whereby living organisms or cells or parts are used for production of products and services. The main subfields of biotechnology are medical (red) biotechnology, agricultural (green) biotechnology, industrial (white) biotechnology, marine (blue) biotechnology, food biotechnology, and environmental biotechnology (Fig. 1.1.). In this chapter the readers will understand the potential applications of biotechnology in several fields like production of medicines; diagnostics; therapeutics like monoclonal antibodies, stem cells, and gene therapy; agricultural biotechnology; pollution control (bioremediation); industrial and marine biotechnology; and biomaterials, as well as the ethical and safety issues associated with some of the products.

The biotechnology came into being centuries ago when plants and animals began to be selectively bred and microorganisms were used to make beer, wine, cheese, and bread. However, the field gradually evolved, and presently it is the use or manipulation of living organisms to produce beneficiary substances which may have medical, agricultural, and/or industrial utilization. Conventional biotechnology is referred to as the technique that makes use of living organism for specific purposes as bread/cheese making, whereas modern biotechnology deals with the technique that makes use of cellular molecules like DNA, monoclonal antibodies, biologics, etc. Before we go into technical advances of DNA and thus recombinant DNA technology, let us have the basic understanding about DNA and its function.

The foundation of biotechnology was laid down after the discovery of structure of DNA in

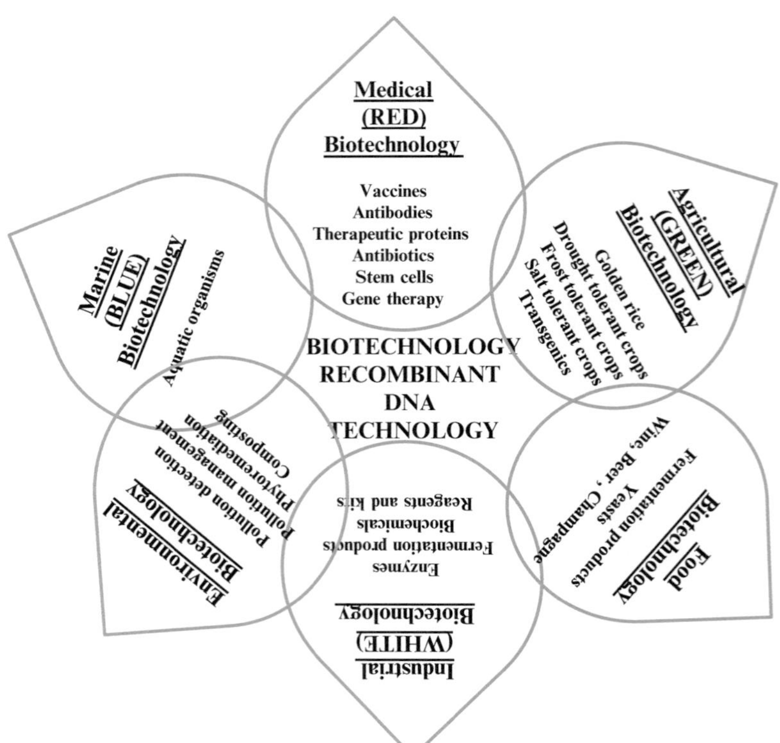

Fig. 1.1 Major applications of biotechnology in different areas and some of their important products

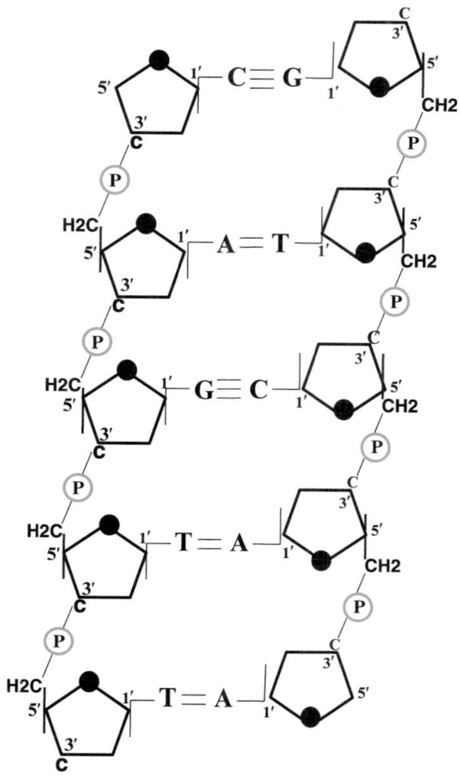

Fig. 1.2 The double helical structure of DNA where both strands are running in opposite direction. Elongation of the chain occurs due to formation of phosphodiester bond between phosphate at 5′ and hydroxyl group of sugar at 3′ of the adjacent sugar of the nucleotide in 5–3′ direction. The sugar is attached to the base. Bases are of four kinds: adenine (*A*), guanine (*G*) (purines), thymine (*T*), and cytosine (*C*) (pyrimidines). Adenine base pairs with two hydrogen bonds with thymine on the opposite antiparallel strand and guanine base pairs with three hydrogen bonds with cytosine present on the opposite antiparallel strand

the early 1950s. The hereditary material is deoxyribonucleic acid (DNA) which contains all the information that dictates each and every step of an individual's life. The DNA consists of deoxyribose sugar, phosphate, and four nitrogenous bases (adenine, guanine, cytosine, and thymine). The base and sugar collectively form nucleoside, while base, sugar, and phosphate form nucleotide (Fig. 1.2). These are arranged in particular orientation on DNA called order or sequence and contain information to express them in the form of protein. DNA has double helical structure, with two strands being complimentary and antiparallel to each other, in which A on one strand base pairs with T and G base pairs with C with two and three bonds, respectively. DNA is the long but compact molecule which is nicely packaged in our nucleus. The DNA is capable of making more copies like itself with the information present in it, as order or sequence of bases. This is called DNA replication. When the cell divides into two, the DNA also replicates and divides equally into two. The process of DNA replication is shown in Fig. 1.3, highlighting important steps.

DNA contains whole information for the working of the cell. The part of the DNA which has information to dictate the biosynthesis of a polypeptide is called a "gene." The arrangement or order of nucleotides determines the kind of proteins which we produce. Each gene is responsible for coding a functional polypeptide. The genes have the information to make a complimentary copy of mRNA. The information of DNA which makes RNA in turn helps cells to incorporate amino acids according to arrangement of letters for making many kinds of proteins. These letters are transcribed into mRNA in the form of triplet codon, where each codon specifies one particular amino acid. The polypeptide is thus made by adding respective amino acids according to the instructions present on RNA. Therefore, the arrangement of four bases (adenine, guanine, cytosine, and thymine) dictates the information to add any of the 20 amino acids to make all the proteins in all the living organisms. Few genes need to be expressed continuously, as their products are required by the cell, and these are known as "constitutive genes." They are responsible for basic housekeeping functions of the cells. However, depending upon the physiological demand and cell's requirement at a particular time, some genes are active and some are inactive, that is, they do not code for any protein. The information contained in the DNA is used to make mRNA in the process of "transcription" (factors shown in Table 1.1). The information of mRNA is used in the process of "translation" for production of protein. Transcription occurs in the nucleus and translation in the cytoplasm of the cell. In translation several initiation factors help in the assembly of mRNA with 40S ribosome and prevent binding of

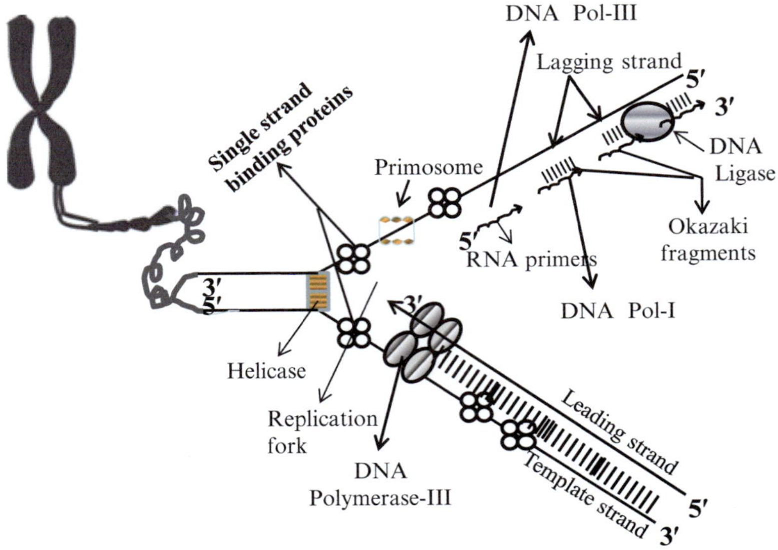

Fig. 1.3 The process of DNA replication. The DNA is densely packed and packaged in the chromosomes. The process requires the action of several factors and enzymes. DNA helicase unwinds the double helix. Topoisomerase relaxes DNA from its super coiled nature. Single-strand binding proteins bind to single-stranded open DNA and prevent its reannealing and maintains strand separation. DNA polymerase is an enzyme which builds a new complimentary DNA strand and has proofreading activity. DNA clamp is a protein which prevents dissociation of DNA polymerase. Primase provides a short RNA sequence for DNA polymerase to begin synthesis. DNA ligase reanneals and joins the Okazaki fragments of the lagging strand. DNA duplication follows semiconservative replication, where each strand serves as template which leads to the production of two complimentary strands. In the newly formed DNA, one strand is old and the other one is new (semiconservative replication). DNA polymerase can extend existing short DNA or RNA strand which is paired to template strand and is called primer. Primer is required as DNA polymerase cannot start the synthesis directly. DNA polymerase is capable of proofreading, that is, correction of wrongly incorporated nucleotide. One strand is replicated continuously with single primer, and it is called as leading strand. Other strand is discontinuous and requires the addition of several primers. The extension is done in the form of short fragments called as Okazaki fragments. The gaps are sealed by DNA ligase. Replication always occurs in 5′–3′ direction

Table 1.1 Factors involved in transcription process

Eukaryotic transcription		
Transcription factor (TF)	Functions	
TFIID	TATA binding	It recognizes
	Protein (TBP)	TATA box
	Subunit	
	TBP associated	Regulate DNA
	Factors	Binding by TBP
TFIIB	Recognizes TFIIB recognition elements (BRE); positions RNA polymerase (RNA pol)	
TFIIF	Stabilizes RNA pol; attracts TFIIE and TFIIF	
TFIIE	Regulates TFIIH	
TFIIH	Unwinds DNA at transcription start point; releases RNA polymerase from promoter	

both ribosomal subunits; they also associate with cap and poly(A) tail. Several elongation factors play an important role in chain elongation. Though each cell of the body has the same genetic makeup, but each is specialized to perform unique functions, the activation and expression of genes is different in each cell. Thus, one type of cells can express some of its genes at one time and may not express the same genes some other time. This is called "temporal regulation" of the gene. In the body different cells express different genes and thus different proteins. For example, liver cell and lymphocyte, would express different genes. This is known as spatial regulation of the gene. Therefore, in the cells of the body, the activation of genes is under spatial regulation (cells present at different locations and different organs produce different proteins) and temporal regulation (same cells produce different proteins at different times). The proteins are formed by the information contained in the DNA and perform a variety of cellular functions. The proteins may be structural (responsible for cell shape and size), or they may be functional like enzymes, signaling intermediates, regulatory proteins, and defense system proteins. However, any kind of genetic defect results in defective protein or alters protein folding which can compromise the functioning of the body and is responsible for the diseases. Figure 1.4 shows the outline of the process of transcription and translation with important steps.

The biotechnological tools are employed toward modification of the gene for gain of function or loss of function of the protein. The technique of removing, adding, or modifying genes in the genome or chromosomes of an organism to bring about the changes in the protein information is called genetic engineering or recombinant DNA technology (Fig. 1.5). DNA recombination made possible the sequencing of the human genome and laid the foundation for the nascent fields of bioinformatics, nanomedicine, and individualized therapy. Multicellular organisms like cows, goats, sheep, rats, corn, potato, and tobacco plants have been genetically engineered to produce substances medically useful to humans. Genetic engineering has revolutionized medicine, enabling mass production of safe, pure, more effective versions of biochemicals that the human body produces naturally [20–22].

The technological advancements have resulted in (1) many biopharmaceuticals and vaccines, (2) new and specific ways to diagnose, (3) increasing the productivity and introduction of quality traits in agricultural crops, (4) the ways to tackle the pollutants efficiently for sustainable environmental practices, (5) helped the forensic experts by DNA fingerprinting and profiling, (6) fermenta-

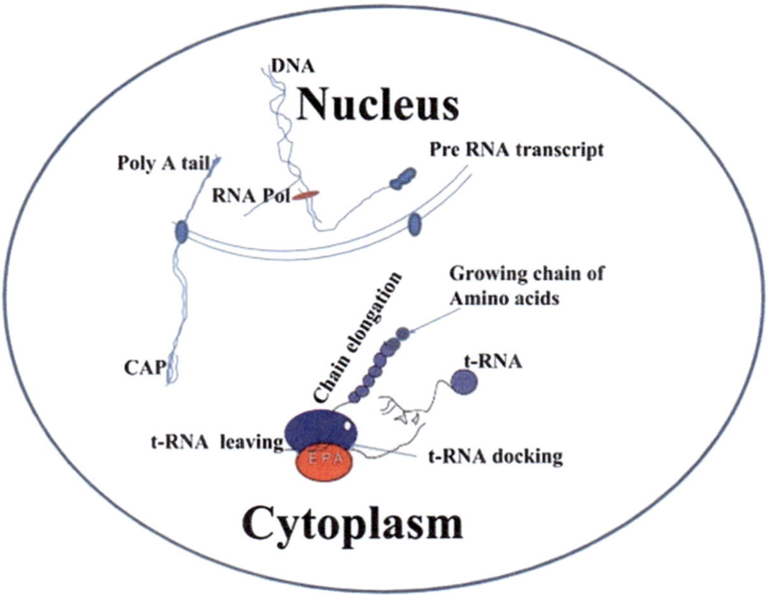

Fig. 1.4 It shows the process of transcription and translation. Transcription occurs in the nucleus and requires the usage of three polymerase enzymes. RNApol I for rRNA, pol II for mRNA, and pol III for both rRNA and tRNA. RNApol II initiates the process by associating itself with seven transcription factors, TFIIA, TFIIB, TFIID, TFIIE, TFIIH, and TFIIJ. After the synthesis, preRNA transcript undergoes processing to form mRNA by removal of introns by splicing and polyadenylation and capping. Protein synthesis is initiated by formation of ribosome and initiator tRNA complex to initiation codon (AUG) of mRNA and involves 11 factors. Chain elongation occurs after sequential addition of amino acids by formation of peptide bonds. Then polypeptide can fold or conjugate itself to other biomolecules and may undergo post-translational modifications as glycosylation or phosphorylation to perform its biological functions

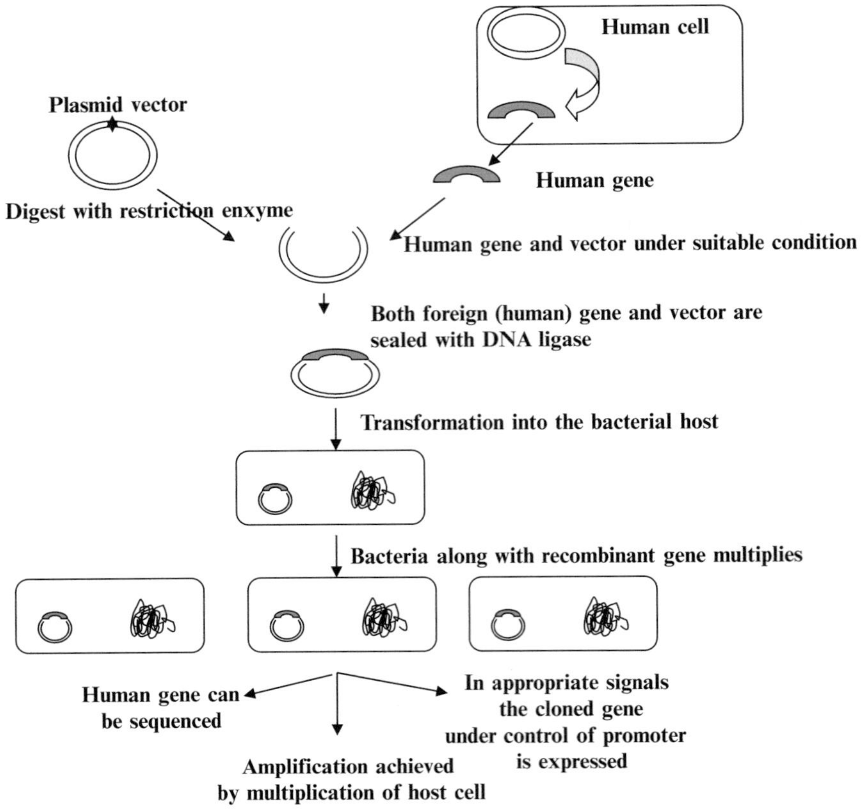

Fig. 1.5 The process of recombinant DNA technology. The gene of interest from human nucleus is isolated and cloned in a plasmid vector. The gene is ligated with the help of DNA ligase. The vector is transformed into a bacterial host. After appropriate selections, the gene is amplified when bacteria multiply or the gene can be sequenced or the gene can be expressed to produce protein

tion technology for production of industrially important products. The list is very long with tremendous advancements and products which have boosted the economy of biotechnology sector worldwide [16]. The biotechnology industry and the products are regulated by various government organizations such as the US Food and Drug Administration (FDA), the Environmental Protection Agency (EPA), and the US Department of Agriculture (USDA).

1.2 Medical Biotechnology

This field of biotechnology has many applications and is involved in production of recombinant pharmaceuticals, tissue engineering products, regenerative medicines such as stem cell and gene therapy, and many more biotechnology products for better human life (Fig. 1.6). Biotechnological tools produce purified bio-therapeutic agents on industrial scales. These include both novel agents and agents formerly available only in small quantities. Crude vaccines were used in antiquity in China, India, and Persia. For example, vaccination with scabs that contained the smallpox virus was a practice in the East for centuries. In 1798 English country doctor Edward Jenner demonstrated that inoculation with pus from sores due to infection by a related cowpox virus could prevent smallpox far less dangerously. It marked the beginning of vaccination. Humans have been benefited incalculably from the implementation of vaccination programs.

Tremendous progress has been made since the early recombinant DNA technology (RDT) experiments from which the lively—and highly profitable—biotechnology industry emerged.

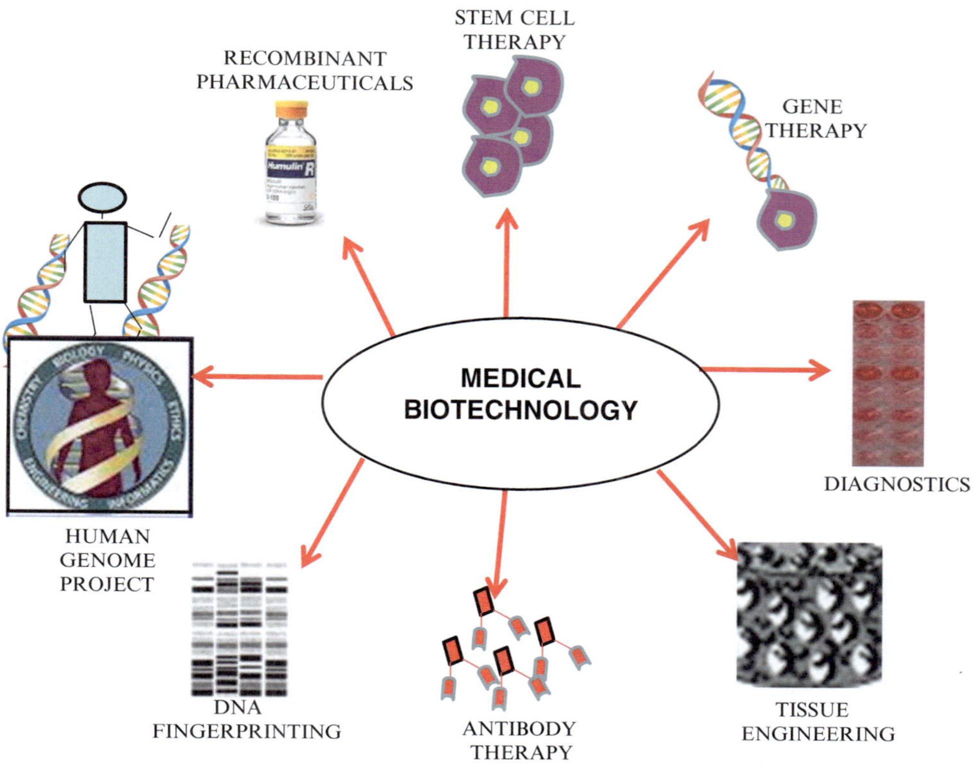

Fig. 1.6 Various applications of medical biotechnology

RDT has fomented multiple revolutions in medicine. Safe and improved drugs, accelerated drug discovery, better diagnostic and quick methods for detecting an infection either active or latent, development of new and safe vaccines, and completely novel classes of therapeutics and other medical applications are added feathers in its cap. The technology has revolutionized understanding of diseases as diverse as cystic fibrosis and cancer. Pharmaceutical biotechnology being one of the important sectors involves using animals or hybrids of tumor cells or leukocytes or cells (prokaryotic, mammalian) to produce safer, more efficacious, and cost-effective versions of conventionally produced biopharmaceuticals. The launch of the new biopharmaceutical or drug requires screening and development. Mice were widely used as research animals for screening. However, in the wake of animal protection, animal cell culture offers accurate and inexpensive source of cells for drug screening and efficacy testing. Pharmaceutical biotechnology's greatest potential lies in gene therapy and stem cell-based therapy. The underlying cause of defect of many inherited diseases is now located and characterized. Gene therapy is the insertion of the functional gene in place of defective gene into cells to prevent, control, or cure disease. It also involves addition of genes for pro-drug or cytokines to eliminate or suppress the growth of the tumors in cancer treatment.

But the progress so far is viewed by many scientists as only a beginning. They believe that, in the not-so-distant future, the refinement of "targeted therapies" should dramatically improve drug safety and efficacy. The development of predictive technologies may lead to a new era in disease prevention, particularly in some of the world's rapidly developing economies. Yet the risks cannot be ignored as new developments and discoveries pose new questions, particularly in areas as gene therapy, the ethics of stem cell research, and the misuse of genomic information.

Many bio-therapeutic agents in clinical use are biotech pharmaceuticals. Insulin was among the earliest recombinant drugs. Canadian physiologists

Frederick Banting and Charles Best discovered and isolated insulin in 1921. In that time many patients diagnosed with diabetes died within a few years. In the mid-1960s, several groups reported synthesizing the hormone.

The first "bioengineered" drug, a recombinant form of human insulin, was approved by the US Food and Drug Administration (FDA) in 1982. Until then, insulin was obtained from a limited supply of beef or pork pancreas tissue. By inserting the human gene for insulin into bacteria, scientists were able to achieve lifesaving insulin production in large quantities. In the near future, patients with diabetes may be able to inhale insulin, eliminating the need for injections. Inhaled insulin products like Exubera® were approved by the USFDA; however, it was pulled out and other products like Technosphere® insulin (Afrezza®) are under investigation. They may provide relief from prandial insulin. Afrezza consists of a pre-meal insulin powder loaded into a cartridge for oral inhalation.

Technosphere technology: The technology allows administration of therapeutics through pulmonary route which otherwise were required to be given as injections. These formulations have broad spectrum of physicochemical characteristics and are prepared with a diverse assortment of drugs with protein or small molecule which may be hydrobhobic or hydrophilic or anionic or cationic in nature. The technology can have its applicability not only through pulmonary route but also for other routes of administration including local lung delivery.

The first recombinant vaccine, approved in 1986, was produced by cloning a gene fragment from the hepatitis B virus into yeast (Merck's Recombivax HB). The fragment was translated by the yeast's genetic machinery into an antigenic protein. This was present on the surface of the virus that stimulates the immune response. This avoided the need to extract the antigen from the serum of people infected with hepatitis B.

The Food and Drug administration (FDA) approved more biotech drugs in 1997 than in the previous several years combined. The FDA has approved many recombinant drugs for human health conditions. These include AIDS, anemia, cancers (Kaposi's sarcoma, leukemia, and colorectal, kidney, and ovarian cancers), certain circulatory problems, certain hereditary disorders (cystic fibrosis, familial hypercholesterolemia, Gaucher's disease, hemophilia A, severe combined immunodeficiency disease, and Turner's syndrome), diabetic foot ulcers, diphtheria, genital warts, hepatitis B, hepatitis C, human growth hormone deficiency, and multiple sclerosis. Today there are more than 100 recombinant drugs and vaccines. Because of their efficiency, safety, and relatively low cost, molecular diagnostic tests and recombinant vaccines may have particular relevance for combating long-standing diseases of developing countries, including leishmaniasis (a tropical infection causing fever and lesions) and malaria.

Stem cell research is very promising and holds tremendous potential to treat neurodegenerative disorders, spinal cord injuries, and other conditions related to organ or tissue loss.

DNA analysis is another powerful technique which compares DNA pattern [14] after performing RFLP and probing it by minisatellite repeat sequence between two or more individuals. Its modification, DNA profiling (process of matching the DNA profiles for STS markers in two or more individuals; see chapter 18), which utilizes multilocus PCR analysis of DNA of suspect and victims, is very powerful and is useful in criminal investigation, paternity disputes, and so many other legal issues. The analysis is very useful in criminal investigations and involves evaluation of DNA from samples of the hair, body fluids, or skin at a crime scene and comparison of these with those obtained from the suspects.

1.2.1 Improved Diagnostic and Therapeutic Capabilities

The sequencing of the human genome in 2003, has given scientists an incredibly rich "parts list" with which to better understand why and how disease happens. It has given added power to gene expression profiling, a method of monitoring expression of thousands of genes simultaneously on a glass slide called a microarray. This technique can predict the aggressiveness of cancer.

The development of monoclonal antibodies in 1975 led to a medical revolution. The body normally produces a wide range of antibodies—the immune system proteins—that defend our body and eliminate microorganisms and other foreign invaders. By fusing antibody-producing cells with myeloma cells, scientists were able to generate antibodies that would, like "magic bullets," bind with specific targets including unique markers, called antigenic determinants (epitopes), on the surfaces of inflammatory cells. When tagged with radioisotopes or other contrast agents, monoclonal antibodies can help in detecting the location of cancer cells, thereby improving the precision of surgery and radiation therapy and showing—within 48 h—whether a tumor is responding to chemotherapy.

The polymerase chain reaction, a method for amplifying tiny bits of DNA first described in the mid-1980s, has been crucial to the development of blood tests that can quickly determine exposure to the human immunodeficiency virus (HIV). Genetic testing currently is available for many rare monogenic disorders, such as hemophilia, Duchenne muscular dystrophy, sickle cell anemia, thalassemia, etc.

Another rapidly developing field is proteomics, which deals with analysis and identification of proteins. The analysis is done by two-dimensional gel electrophoresis of the sample and then performing mass spectrometric analysis for each individual protein. The technique may be helpful in detecting the disease-associated protein in the biological sample. They may indicate early signs of disease, even before symptoms appear. One such marker is C-reactive protein, an indicator of inflammatory changes in blood vessel walls that presage atherosclerosis.

Nanomedicine is a rapidly moving field. Scientists are developing a wide variety of nanoparticles and nanodevices, scarcely a millionth of an inch in diameter, to improve detection of cancer, boost immune responses, repair damaged tissue, and thwart atherosclerosis. The FDA has approved a paclitaxel albumin-stabilized nanoparticle formulation (Abraxane® for injectable suspension, made by Abraxis BioScience) for the treatment of metastatic adenocarcinoma of the pancreas. Nanoparticles are being explored in heart patients in the USA as a way to keep their heart arteries open following angioplasty.

Therapeutic proteins are those, which can replace the patients naturally occurring proteins, when levels of the natural proteins are low or absent due to the disease. High-throughput screening, conducted with sophisticated robotic and computer technologies, enables scientists to test tens of thousands of small molecules in a single day for their ability to bind to or modulate the activity of a "target," such as a receptor for a neurotransmitter in the brain. The goal is to improve the speed and accuracy of therapeutic protein or potential drug discovery while lowering the cost and improving the safety of pharmaceuticals that make it to market.

Many of the molecules utilized for detection also have therapeutic potential too, for example, monoclonal antibodies. The monoclonal antibodies are approved for the treatment of many diseases as cancer, multiple sclerosis, and rheumatoid arthritis. They are currently being tested in patients as potential treatments for asthma, Crohn's disease, and muscular dystrophy. As the antibodies may be efficiently targeted against a particular cell surface marker, thus they are used to deliver a lethal dose of toxic drug to cancer cells, avoiding collateral damage to nearby normal tissues.

1.3 Agricultural Biotechnology

The man has made tremendous progress in crop improvement in terms of yield; still the ultimate production of crop is less than their full genetic potential. There are many reasons like environmental stresses (weather condition as rain, cold, frost), diseases, pests, and many other factors which reduce the ultimate desired yield affecting crop productivity. The efforts are going on to design crops which may be grown irrespective of their season or can be grown in frost or drought conditions for maximum utilization of land, which would ultimately affect crop productivity [24]. Agricultural biotechnology aims to introduce sustainable agricultural practices with best yield potential and minimal adverse effects on environment (Fig. 1.7). For example, combating pests was

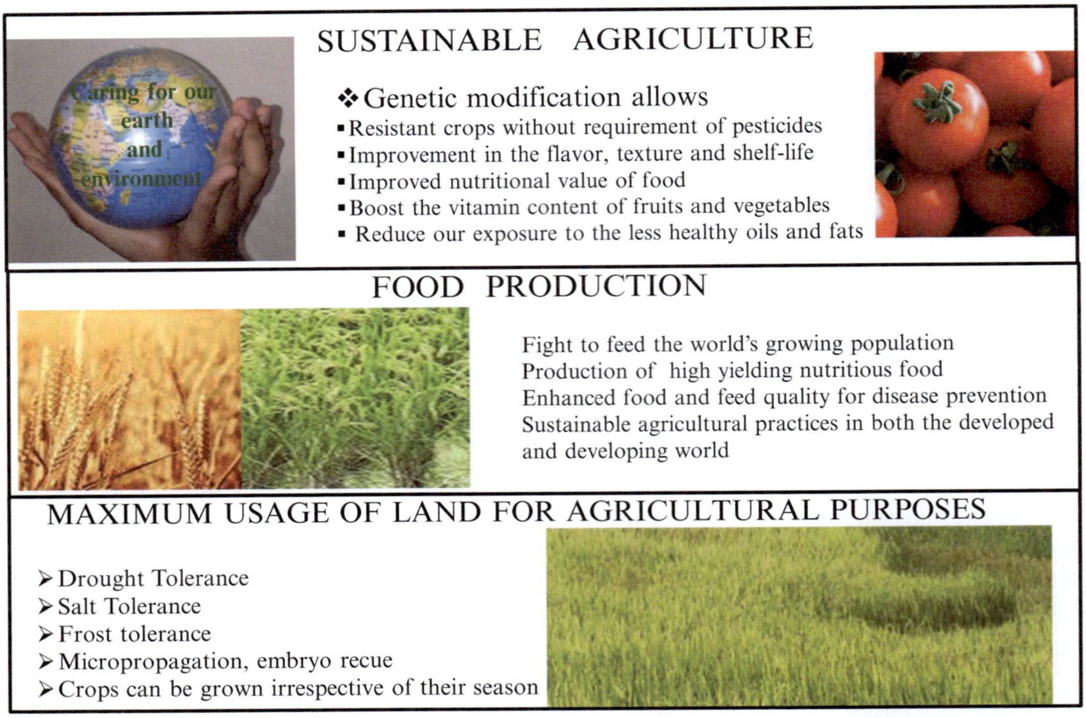

Fig. 1.7 Various applications of agricultural biotechnology

a major challenge. Thus, the gene from bacterium *Bacillus thuringiensis*, the Bt gene, that functions as insect-resistant gene when inserted into crop plants like cotton, corn, and soybean helps prevent the invasion of pathogen, and the tool is called *integrated pest management*. This management is helpful in reducing usage of potentially dangerous pesticides on the crop. Not only the minimal or low usage of pesticides is beneficial for the crop but also the load of the polluting pesticides on environment is greatly reduced [24].

Resistance to Infectious Agents Through Genetic Engineering

(a) Bt crops
- The gene comes from the soil bacterium *Bacillus thuringiensis*.
- The gene produces crystal proteins called Cry proteins. More than 100 different variants of the Bt toxins

(continued)

have been identified which have different specificity to target insect lepidoptera. For eg., CryIa for butterflies and CRYIII for beetles.
- These Cry proteins are toxic to larvae of insects like tobacco budworm, armyworm, and beetles.
- The Cry proteins exist as an inactive protoxins.
- These are converted into active toxin in alkaline pH of the gut upon solubilization when ingested by the insect.
- After the toxin is activated, it binds to the surface of epithelial cells of midgut and creates pores causing swelling and lysis of cells leading to the death of the insect (larva).
- The genes (cry genes) encoding this protein are isolated from the bacterium and incorporated into several

(continued)

crop plants like cotton, tomato, corn, rice, and soybean.

The proteins encoded by the following cry genes control the pest given against them:

- Cry I Ac and cry II Ab control cotton bollworms.
- Cry I Ab controls corn borer.
- Cry III Ab controls Colorado potato beetle.
- Cry III Bb controls corn rootworm.

(b) Protection against nematodes
- A nematode *Meloidogyne incognita* infects tobacco plants and reduces their yield.
- The specific genes (in the form of cDNA) from the parasite are introduced into the plant using *Agrobacterium*-mediated transformation.
- The genes are introduced in such a way that both sense/coding RNA and antisense RNA (complimentary to the sense/coding RNA) are produced.
- Since these two RNAs are complementary, they form a double-stranded RNA (ds RNA).
- This neutralizes the specific RNA of the nematode, by a process called RNA – interference.
- As a result, the parasite cannot multiply in the transgenic host, and the transgenic plant is protected from the pest.

These resistant crops result in reduced application of pesticides. The yield is high without the pathogen infestations and insecticides. This also helps to reduce load of these toxic chemicals in the environment.

The transformation techniques and their applications are being utilized to develop rice, cassava, and tomato, free of viral diseases by "International Laboratory for Tropical Agricultural Biotechnology" (ILTAB). ILTAB in 1995 reported the first transfer of a resistance gene from a wild-type species of rice to a susceptible cultivated rice variety. The transferred gene expressed and imparted resistance to crop-destroying bacterium *Xanthomonas oryzae*. The resistant gene was transferred into susceptible rice varieties that are cultivated on more than 24 million hectares around the world [6].

The recombinant DNA technology reduces the time between the identification of a particular gene to its application for betterment of crops by skipping the labor-intensive and time-consuming conventional breeding [3]. For example, the alteration of known gene in plant for the improvement of yield or tolerance to adverse environmental conditions or resistance to insect in one generation and its inheritance to further generations. Plant cell and tissue culture techniques are one of the applications where virus-free plants can be grown and multiplied irrespective of their season on large scale (micropropogation), raising haploids, or embryo rescue. It also opens an opportunity to cross two manipulated varieties or two incompatible varieties (protoplast culture) for obtaining best variety for cultivation.

With the help of technology, new, improved, and safe agricultural products may emerge which would be helpful for maintaining contamination-free environment. Biotechnology has the potential to produce:

- High crop yields [4]
- Crops are engineered to have desirable nutrients and better taste (e.g., tomatoes and other edible crops with increased levels of vitamin C, vitamin E, and/or beta-carotene protect against the risk of some prevalent chronic diseases and rice with increased iron levels protects against anemia)
- Insect- and disease-resistant plants
- Genetic modification avoids nonselective changes
- Longer shelf life of fruits and vegetables

The potential of biotechnology may contribute to increasing agricultural, food, and feed production, protecting the environment, mitigating pollution, sustaining agricultural practices, and improving human and animal health. Some agricultural crops as corn and marine organisms can be potential alternative for biofuel production. The by-products of the process may be processed to produce other chemical feedstocks for various products. It is estimated that the world's chemical and fuel demand could be supplied by such renewable resources in the first half of the next century [5].

1.3.1 Food Biotechnology

Food biotechnology is an emerging field, which can increase the production of food, improving its nutritional content and improving the taste of the food. The food is safe and beneficial as it needs fewer pesticides and insecticides. The technology aims to produce foods which have more flavors, contain more vitamins and minerals, and absorb less fat when cooked. Food biotechnology may remove allergens and toxic components from foods, for their better utility [6, 7].

1.4 Environmental Biotechnology

Environmental biotechnology grossly deals with maintenance of environment, which is pollution-free, the water is contamination-free, and the atmosphere is free of toxic gases. Thus, it deals with waste treatment, monitoring of environmental changes, and pollution prevention. Bioremediation in which utilization of higher living organisms (plants: phytoremediation) or certain microbial species for decontamination or conversion of harmful products is done is the main application of environmental biotechnology. The enzyme bioreactors are also being developed which would pretreat some industrial and food waste components and allow their removal through the sewage system rather than through solid waste disposal mechanisms. The production of biofuel from waste can solve the fuel crisis (biogas). Microbes may be engineered to produce enzymes required for conversion of plant and vegetable materials into building blocks for biodegradable plastics. In some cases, the by-products of the pollution-fighting microorganisms are themselves useful. For example, methane can be derived from a form of bacteria that degrades sulfur liquor, a waste product of paper manufacturing. This methane thus obtained is used as a fuel or in other industrial processes. Insect- and pest-resistant crops have reduced the use and environmental load of insecticides and pesticides. Insect-protected crops allow for less potential exposure of farmers and groundwater to chemical residues while providing farmers with season-long control.

1.5 Industrial Biotechnology

The utilization of biotechnological tools (bioprocessing) for the manufacturing of biotechnology-derived products (fuels, plastics, enzymes, chemicals, and many more compounds) on industrial scale is industrial biotechnology. The aim is to develop newer industrial manufacturing processes and products, which are economical and better than preexisting ones with minimal environmental impact. In industrial biotechnology, (1) microorganisms are being explored for producing material goods like fermentation products as cheese; (2) biorefineries where oils, sugars, and biomass may be converted into biofuels, bioplastics, and biopolymers; (3) and value-added chemicals from biomass. The utilization of modern techniques can improve the efficiency and reduces the environmental impacts of industrial processes like textile, paper, pulp, and chemical manufacturing. For example, development and usage of biocatalysts, such as enzymes, to synthesize chemicals and development of antibiotics and better tasting liquors and their usage in food industry have provided safe and effective processing for sustainable productions. Biotechnological tools in the textile industry are utilized for the finishing of fabrics and garments. Biotechnology also produces spider silk and biotech-derived cotton that is warmer and stronger and has improved dye uptake and retention, enhanced absorbency, and wrinkle and shrink resistance.

Biofuels may be derived from photosynthetic organisms, which capture solar energy, transform it in other products like carbohydrates and oils, and store them. Different plants can be used for fuel production:

1. Bioethanol can be obtained from sugar (as sugarcane or sugar beet) or starch (like corn or maize). These are fermented to produce ethanol, a liquid fuel commonly used for transportation.
2. Biodiesel can be obtained from natural oils from plants like oil palm, soybean, or algae. They can be burned directly in a diesel engine or a furnace, or blended with petroleum, to produce fuels such as biodiesel.
3. Wood and its by-products can be converted into liquid biofuels, such as methanol or ethanol, or into wood gas. Wood can also be burned as solid fuel, like the irewood.

In these kinds of biological reaction, there are many renewable chemicals of economic importance coproduced as side streams of bioenergy and biofuels as levulinic acid, itaconic acid, and sorbitol. These have tremendous economic potential and their fruitful usage would depend upon the collaboration for research and development between the government and the private sector.

1.5.1 Enzyme Production

The enzymes have big commercial and industrial significance. They have wide applications in food industry, leather industry, pharmaceuticals, chemicals, detergents, and research. In detergents the alkaline protease, subtilisin (from *Bacillus subtilis*), was used by Novo Industries, Denmark. The production of enzymes is an important industrial application with world market of approximately 5 billion dollars. The enzymes can be obtained from animals, plants, or microorganisms. The production from microorganisms is preferred as they are easy to maintain in culture with simple media requirements and easy scale-up. The important enzymes for the industrial applications are in food industry, human application, and research. A few animal enzymes are also important as a group of proteolytic enzymes, for example, plasminogen activators, which act on inactive plasminogen and activate it to plasmin, which destroys fibrin network of blood clot. Some of the plasminogen activators are urokinase and tissue plasminogen activators (t-PA). Urokinase (from urine) is difficult to obtain in ample quantity; thus, t-PA is obtained from cells grown in culture medium. Streptokinase (bacterial enzyme) is also a plasminogen activator but is nonspecific and immunogenic.

Enzyme engineering is also being tried where modifications of specific amino acid residue are done for improving the enzyme properties. One of the enzymes chymosin (rennin) coagulates milk for cheese manufacturing.

The enzymes can be produced by culturing cells, growing them with appropriate substrates in culture conditions. After optimum time the enzymes may be obtained by cell disruption (enzymatic/freeze–thaw/osmotic shock) followed by preparative steps (centrifugation, filtration), purification, and analysis. The product is then packaged and ultimately launched in the market.

After their production, they can be immobilized on large range of materials (agar, cellulose, porous glass, or porous alumina) for subsequent reuse. Some of the important industrial enzymes are α-amylase (used for starch hydrolysis), amyloglucosidase (dextrin hydrolysis), β-galactosidase (lactose hydrolysis), aminoacylase (hydrolysis of acylated L-amino acids), glucose oxidase (oxidation of glucose), and luciferase (bioluminescence). Some of the medically important enzymes are urokinase and t-PA for blood clot removal and L-asparaginase for removal of L-asparagine essential for tumor growth and thus used for cancer chemotherapy in leukemia.

1.5.2 Exploring Algae for Production of Biofuels

The energy requirement of present population is increasing and gradually fossil fuels are rapidly depleting. Thus, renewable energy sources like solar energy and wind-, hydro-, and biomass-based energy are being explored worldwide. One of the feedstocks may be microalgae, which are fast-growing, photosynthetic organisms requiring carbon dioxide, some nutrients, and water for its growth. They produce large amount of lipids and carbohydrates, which can be processed into different biofuels and commercially important coproducts. The production of biofuels using algal biomass is advantageous as they (1) can grow throughout the year and thus their productivity is higher than other oil seed crops, (2) have high tolerance to high carbon dioxide content, (3) utilize less water, (4) do not require herbicides or pesticides with high growth potential (waste water can be utilized for algal cultivation), (5) can sustain harsh atmospheric conditions, and (6) do not interfere with productivity of conventional crops as they do not require agricultural land. The production of various biofuels from algae is schematically represented in Fig. 1.8.

Algae can serve as potential source for biofuel production; however, biomass production is low. The production has certain limitations, as cultivation cost is high with requirement of high energy [1].

1.6 Marine or Aquatic Biotechnology

Marine or aquatic biotechnology also referred to as "blue biotechnology" deals with exploring and utilizing the marine resources of the world. Aquatic or marine life has been intriguing and a source of livelihood for many since years. As major part of earth is acquired by water, thus nearly 75–80 % types of life forms exist in oceans and aquatic systems. It studies the wide diversity found in the structure and physiology of marine organisms. They are unique in their own ways and lack their equivalent on land. These organisms have been explored and utilized for numerous applications as searching new treatment for cancer or exploring other marine resources, because of which the field is gradually gaining momentum and economic opportunities [19]. The global economic benefits are estimated to be very high. The field aims to:

1. Fulfill the increasing food supply needs
2. Identify and isolate important compounds which may benefit health of humans
3. Manipulate the existing traits in sea animals for their improvement
4. Protect marine ecosystem and gain knowledge about the geochemical processes occurring in oceans

Some of the major applications are discussed:

- Aquaculture: Aquaculture refers to the growth of aquatic organisms in culture condition for commercial purposes. These animals may be shellfish, finfish, and many others. Mariculture refers to the cultivation of marine animals. Their main applications are in food, food ingredients, pharmaceuticals, and fuels, the products are in high demand, and various industries are in aquaculture business, for example, crawfish farming (Louisiana), catfish industry (Alabama and Mississippi Delta), and trout farming (Idaho and West Virginia).
- Biotechnology discoveries and products.
  - Transgenic species of salmon with growth hormone gene has accelerated growth of salmons.
  - *Molt-inhibiting hormone* (MIH) from blue crabs leads to soft-shelled crab.
  - *Antifreeze proteins*: A novel protein antifreeze protein (AFP) was identified. AFPs were isolated from Northern cod (bottom-dwelling fish) living at the Eastern Canada coast and teleosts living in extremely cold weather of Antarctica. AFPs have been isolated from *Osmerus mordax* (smelt), *Clupea harengus* (herring), *Pleuronectes americanus* (winter flounder), and many others. Due to antifreeze properties (lowering the minimal freezing temperature by 2–3 °C), the gene has potential for raising plants which are cold tolerant (e.g., tomatoes).
  - *Green fluorescent protein*: A much popular green fluorescent protein (GFP) was obtained from jellyfish *Aequorea victoria*.

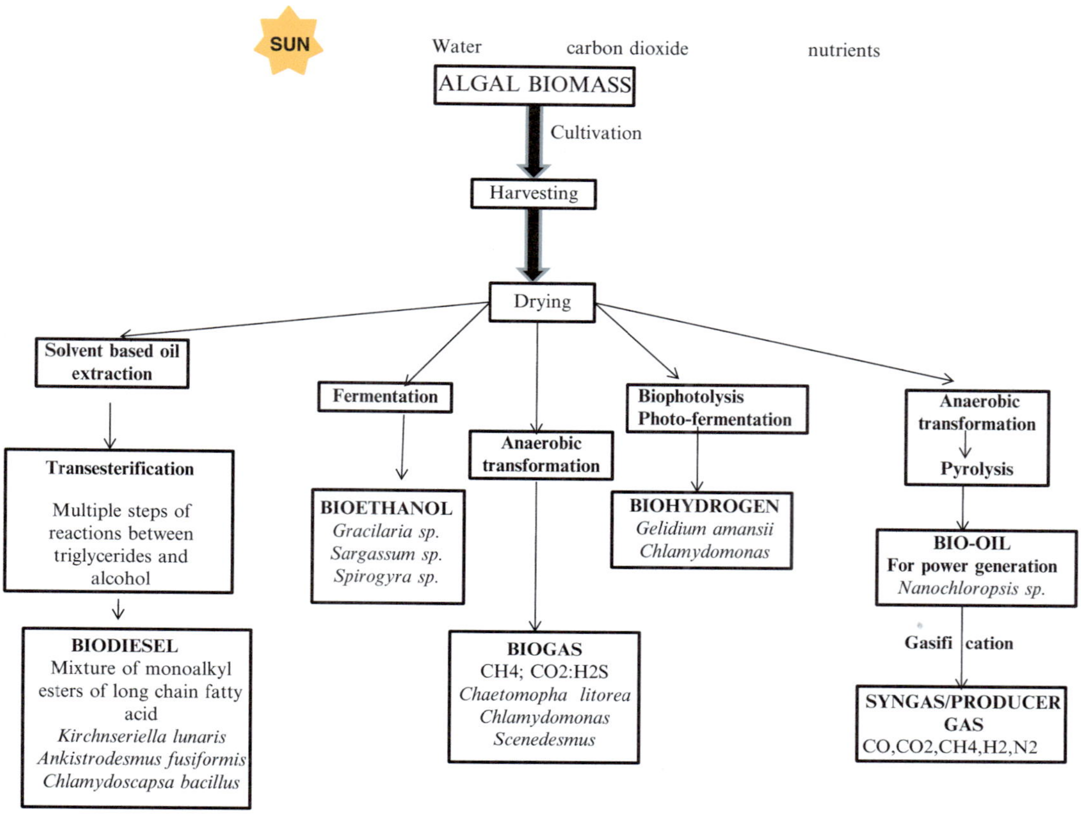

Fig. 1.8 Different biofuel productions by using microalgae. The algae use sunlight, CO_2, water, and some nutrients

It can fluoresce and thus glow in the dark. Many marine microorganisms have bioluminescent capability. GFP is widely used as reporter gene in experiments related to gene cloning, expression, and transgenics. A transgenic strain of zebra fish in the name of GloFish was created by Yorktown Industries, Texas, in 2004. This was with red fluorescent protein gene obtained from sea anemones, and it was the first genetically modified pet animal in the market.

Medicinal applications: For osteoporosis, salmon calcitonin (calcitonin is thyroid hormone promoting calcium uptake and bone calcification) with 20 times higher bioactivity is available as injection and nasal spray.

Hydroxyapatite (*HA*): Obtained from coral reefs and is an important component of bone and cartilage matrix. Its implants are prepared by Interpore Internationals which may be used for filling gaps in fractured bones.

Byssal fibers: Are protein-rich superadhesive which have elastic properties obtained from mussels (*Mytilus edulis*). Their isolation would not be very economical, but they can have wide applications in surgical sutures, artificial tendons, and ligament grafts.

Many anti-inflammatory, analgesic, anticancerous compounds have been identified from sea organisms which can have tremendous potential for human health.

Tetrodotoxin (TTX) is the most toxic poison (10,000 times more lethal than cyanide) produced by Japanese pufferfish or blowfish (*Fugu rubripes*). TTX is being used to study and understand its effect on sodium channels which can help guide the development of drugs with anesthetic and analgesic properties.

Other Products

1. Taq polymerase from *Thermus aquaticus* which is used in PCR reactions and obtained from hot spring Archaea.
2. Collagenase (protease) obtained from *Vibrio* is used in tissue engineering and culturing.

1.7 Transgenic Animals and Plants

In the early 1980s, inserting DNA from humans into mice and other animals became possible. The animals and plants which have foreign gene in each of their cells are referred to as transgenic organisms and the inserted gene as transgene. Expression of human genes in these transgenic animals can be useful in studies, as models for the development of diabetes, atherosclerosis, and Alzheimer's disease. They also can generate large quantities of potentially therapeutic human proteins. Transgenic plants also offer many economic, safe, and practical solutions for production of variety of biopharmaceuticals. The plants have been engineered to produce many blood products (human serum albumin, cytokines), human growth hormone, recombinant antibodies, and subunit vaccines.

The usage of transgenic plants for the production of recombinant pharmaceuticals might open new avenues in biotechnology. As plants can be grown inexpensively with minimal complicated requirements, thus they may have tremendous therapeutic potential. The plants have been engineered to produce more nutrients or better shelf life. The transgenic plants have been created which have genes for insect resistance (Bt cotton, soybean, corn). Now billion acres of land is used for cultivation of genetically engineered crops of cotton, corn, and soybean as they have higher yield and are pest resistant. However, due to social, ethical, and biosafety issues, they have received acceptance as well as rejections at many places and health and environment-related concerns in many parts of the world [8].

1.8 Response to Antibiotic Resistance

Antibiotics are one of the broadly used therapeutic molecules produced by certain classes of microorganisms (bacteria and fungi) which can be used in diverse clinical situations to eliminate bacteria, improve symptoms, and prevent number of infections. Antibiotics have various other applications apart from clinical aspects. They can be used for the treatment of tumors and treatment of meat, in cattles and livestocks, in basic biotechnological work. However, their effectiveness is a matter of concern as bacteria which are continuously exposed to certain antibiotics might become resistant to it due to accumulation of mutations. These days antibiotic-resistant bacteria have increased and some of them have developed multiple drug resistance. Thus, it has become very difficult to initiate therapy in diseases like tuberculosis and leprosy. Biotechnology is solving the urgent and growing problem of antibiotic resistance. With the help of bioinformatics—powerful computer programs capable of analyzing billions of bits of genomic sequence data—scientists are cracking the genetic codes of bacteria and discovering "weak spots" vulnerable to attack by compounds identified via high-throughput screening. This kind of work led in 2000 to the approval of Zyvox (linezolid), an antibiotic to reach the market in 35 years.

Lytic bacteriophage viruses that infect and kill bacteria may be another way to counter resistance. First used to treat infection in the 1920s, "phage therapy" was largely eclipsed by the development of antibiotics. However, researchers in the former Soviet Republic of Georgia reported that a biodegradable polymer impregnated with bacteriophages and the antibiotic Cipro successfully healed wounds infected with a drug-resistant bacterium.

Case Study
After exposure of strontium-90, three Georgian lumberjacks from village Lia had systemic effects, and two of them developed severe local radiation injuries which subsequently became infected with *Staphylococcus aureus*. Upon hospitalization, the patients were treated with various medications, including antibiotics and topical ointments; however, wound healing was only moderately successful, and their *S. aureus* infection could not be eliminated. Approximately 1 month after hospitalization, treatment with PhagoBioDerm (a wound-healing preparation consisting of a biodegradable polymer impregnated with ciprofloxacin and bacteriophages) was initiated. Purulent drainage stopped within 2–7 days. Clinical improvement was associated with rapid (7 days) elimination of the etiologic agent, and a strain of *S. aureus* responsible for infection was resistant to many antibiotics (including ciprofloxacin) but was susceptible to the bacteriophages contained in the PhagoBioDerm preparation [11].

1.9 The Challenges for the Technology

1.9.1 Gene Therapy

Some biotech approaches to better health have proven to be more challenging than others. An example is gene transfer, where the defective gene is replaced with a normally functioning one. The normal gene is delivered to target tissues in most cases by virus that is genetically altered to render it harmless. The first ex vivo gene transfer experiment, conducted in 1990 at the National Institutes of Health (NIH), on Ashanti DeSilva who was suffering from severe combined immunodeficiency (SCID) helped boost her immune response and successfully corrected an enzyme deficiency. However, treatment was required every few months. However, 9 years later, a major setback occurred in gene therapy trial after the death of 18-year-old Jesse Gelsinger suffering from ornithine transcarbamylase (OTC) deficiency due to intense inflammatory responses followed by gene therapy treatment. There were some positive experiences and some setbacks from gene therapy trials leading to stricter safety requirements in clinical trials.

1.9.2 Designer Babies

The fancy term designer baby was invented by media. Many people in society prefer embryos with better traits, intellect, and intelligence. They want to select embryo post germline engineering. This technique is still in infancy but is capable of creating lot of differences in the society thus requires appropriate guidelines.

1.9.3 Genetically Modified Food

Genetically modified crops harboring genes for insect resistance were grown on billion of acres of land. These crops became very popular due to high yield and pest resistance. However, some of the pests gradually developed resistance for a few of these transgenic crops posing resistant pest threat. The other technologies as "traitor" and "terminator" technologies pose serious risk on crop biodiversity and would impart negative characters in the crop (they were not released due to public outcry).

1.9.4 Pharmacogenomics

Scientists do not believe they will find a single gene for every disease. As a result, they are studying relationships between genes and probing populations for variations in the genetic code, called single nucleotide polymorphisms, or SNPs, that may increase one's risk for a particular disease or determine one's response to a given medication. This powerful ability to assign risk and response to genetic variations is fueling the movement toward "individualized medicine." The goal is prevention, earlier diagnosis, and more effective therapy by prescribing interventions that match patients' particular genetic characteristics.

1.9.5 Tissue Engineering

Tissue engineering is one of the emerging fields with tremendous potential to supply replacement tissue and organ option for many diseases. Lot is achieved, lot more need to be achieved.

1.10 Ethical Issues

The pursuit of cutting-edge research "brings us closer to our ultimate goal of eliminating disability and disease through the best care which modern medicine can provide." Understanding of the genetics of heart disease and cancer will aid the development of screening tools and interventions that can help prevent the spread of these devastating disorders into the world's most rapidly developing economies.

Biotechnology is a neutral tool; nevertheless, its capabilities raise troubling ethical questions. Should prospective parents be allowed to "engineer" the physical characteristics of their embryos? Should science tinker with the human germ line, or would that alter in profound and irrevocable ways what it means to be human?

More immediately, shouldn't researchers apply biotechnology—if they can—to eliminate health disparities among racial and ethnic groups? While genetic variation is one of many factors contributing to differences in health outcome (others include environment, socioeconomic status, health-care access, stress, and behavior), the growing ability to mine DNA databases from diverse populations should enable scientists to parse the roles these and other factors play.

Biotechnology along with supportive health-care infrastructure can solve complicated health problems. Accessibility to the new screening tests, vaccines, and medications and cultural, economic, and political barriers to change must be overcome. Research must include more people from disadvantaged groups, which will require overcoming long-held concerns, some of them have had about medical science.

Biotechnology has been a significant force which has improved the quality of lives and has incalculably benefitted human beings. However, technology does have prospects of doing harm also due to unanticipated consequences. Each technology is subjected to ethical assessment and requires a different ethical approach. Obviously the changes are necessary as technology can have major impact on the world; thus, a righteous approach should be followed. There is uncertainty in predicting consequences, as this powerful technology has potential to manipulate humans themselves. Ethical concerns are even more important as the future of humanity can change which require careful attention and consideration. Therefore, wisdom is required to articulate our responsibilities toward environment, animals, nature, and ourselves for the coming future generations. We need to differentiate what is important technologically rather that what technology can do. For an imperative question, that is, whether this can be achieved, the research must answer "Why should it be achieved"? Who would it benefit?

1.11 Issues Related to Safety

- As the new GM crops are entering the market, the issue regarding their consumption, whether they are safe, without any risk, is one of the important concerns [2]. Though the results related to safety and usage are well reported

(as compared to conventional crops), unknown fear from these products makes them non acceptable at many places [20].
- As insect- and pest-resistant varieties are being prepared and used as Bt genes in corn and cotton crops, there exists a risk of development of resistance insect population. Another important factor is that these resistant crops may harm other species like birds and butterfly.
- The development of more weeds may occur as cross-pollination might result in production of weeds with herbicide resistance which would be difficult to control.
- The gene transfers might cross the natural species boundary and affect biological diversity.
- The judgment of their usage would depend upon the clear understanding of risks associated with safety of these products in determining the impact of these on environment, other crops, and other animal species.

1.12 Future of the Technology

With the understanding of science, we should understand that genetic transfers have been occurring in animals and plant systems; thus, the risk of the biotechnology-derived products is similar as conventional crops [12].

The biotechnology products would be acceptable to many if they are beneficial and safe. People are willing to buy crops free of pesticides and insecticides. Nowadays people are also accepting crops grown without the usage of chemical fertilizers or pesticides, which are high in nutritive values.

The labeling of the product is also an ethical issue as some believe that labeling any product as biotechnology product might be taken by consumer as warning signs; however, others believe that labeling should be done as consumer has every right to know what he is consuming [9]. The products may be acceptable if consumers can accept the food derived from biotechnology weighing all pros and cons and, if the price is right, has more nutritive values, is good in taste, and is safe to consume [10].

Biotechnology is at the crossroads in terms of fears and thus public acceptance [15]. Surprisingly the therapeutic products are all accepted and find major place in biopharmaceutical industry, but food crops are still facing problems in worldwide acceptance. The future of the world food supply depends upon how well scientists, government, and the food industry are able to communicate with consumers about the benefits and safety of the technology [13, 16]. Several major initiatives are under way to strengthen the regulatory process and to communicate more effectively with consumers by conducting educational programs [18, 23].

1.13 Chapter End Summary

- The advantages of biotechnology are so broad that it is finding its place in virtually every industry. It has applications in areas as diverse as pharmaceuticals, diagnostics, textiles, aquaculture, forestry, chemicals, household products, environmental cleanup, food processing, and forensics to name a few.
- Biotechnology is enabling these industries to make new or better products, often with greater speed, efficiency, and flexibility.
- With the applications of recombinant DNA technology, more safer and therapeutic drugs are produced. These recombinant products do not elicit unwanted immunological response which is observed when the product is obtained from other live or dead sources. Many of these therapeutics are approved for human usage, and many of them are in the phase of development.
- Immunological and DNA-based techniques like PCR (polymerase chain reaction) are used for early diagnosis of disorders. PCR and NAAT with microarray can be utilized for the diagnosis of many diseases, and it can detect mutations in gene.
- The technology holds promise through stem cell research and gene therapy and holds applications in forensic medicine.
- The technique may be helpful in developing useful and beneficial plants. It overcomes the limitations of traditional plant breeding. The

techniques of plant tissue culture, transgenics, and marker-assisted selections are largely used for selecting better yielding varieties and imparting quality traits in plants.

- It is also helpful in maintaining environment by bioremediation and other treatment. The areas where it finds applications are:
  - Food industries. Production of single-cell protein, spirulina, enzymes, and solid-state fermentations
  - Increase and improvement of agricultural production
  - Production of therapeutic pharmaceuticals
  - Production of vaccines and monoclonal antibodies
  - Cultivation of virus for vaccine production

Multiple Choice Questions

1. Which abiotic stress can be tolerated by genetically modified crops?
   (a) Insects
   (b) Pests
   (c) Drought
   (d) All of the above
2. The golden rice is a crop having high nutritive value in:
   (a) Vitamin A
   (b) Vitamin B
   (c) Vitamin C
   (d) Vitamin D and calcium
3. Bt toxin gene which encodes cry protein is:
   (a) bccryI
   (b) dbcryII
   (c) cryIAc
   (d) cryIdb
4. The first recombinant product to reach the market was:
   (a) Growth hormone
   (b) Tissue plasminogen activator
   (c) Factor VIII
   (d) Insulin
5. Biotechnology deals with:
   (a) Genetically modifying organism
   (b) Production of therapeutics
   (c) Production of better diagnosis
   (d) All of the above
6. Green revolution is:
   (a) Increase in yield of crops
   (b) Improved crop varieties
   (c) Lesser fertilizers and agrochemicals
   (d) All of these
7. Insecticidal protein cry does not kill bacillus because:
   (a) It is resistant to it.
   (b) The toxin is enclosed in vesicle.
   (c) The toxin is present in inactive form.
   (d) None of these.
8. DNA defects may be solved by:
   (a) Gene therapy
   (b) Replacement protein therapy
   (c) Stem cell therapy
   (d) All of these
9. The use of insect resistant crop would be:
   (a) The productivity would improve.
   (b) The usage of chemical agent would be reduced.
   (c) The environment and crop would be insecticide free.
   (d) All of the above.
10. Bioremediation can be helpful in:
    (a) Detoxifying waste material
    (b) Burying waste material
    (c) Burning waste material
    (d) None of these
11. Which of the following statements are true?
    (1) In all the cells of our body, all the genes are active.
    (2) In different cells of our body, different genes are active.
    (3) Gene expression is spatially and temporally regulated.
      (a) All 1, 2, and 3 are correct.
      (b) 1 and 2 are correct.
      (c) 1 and 3 are correct.
      (d) 2 and 3 are correct.
12. In a classic experiment, Dr. Edward Jenner demonstrated that:
    (a) Inoculation with monoclonal antibody was able to prevent small pox.
    (b) Inoculation with pus from sores due to cowpox could prevent small pox.
    (c) Attenuated vaccine was able to prevent small pox.
    (d) None of the above.

Answers

1. (c); 2. (a); 3. (c); 4. (d); 5. (d); 6. (d); 7. (c); 8. (a); 9. (d); 10. (a); 11. (d); 12. (b)

Review Questions

Q1. What are cry proteins? What is their importance?

Q2. Give some applications of biotechnology in agriculture.

Q3. What is your opinion about labeling of biotechnology-based food product as rDNA technology derived product?

Q4. What are applications of biotechnology in maintaining environment?

Q5. What is medical biotechnology?

Q6. What are the challenges faced by biotechnology industry?

Q7. What do you think about GM crops?

References

1. Behera et al. (2015) Scope of algae as third generation biofuels. Front Bioeng Biotechnol 2: doi:10.3389/fbioe.2014.00090
2. Bruhn CM (1992) Consumer concerns and educational strategies: focus on biotechnology. Food Technol 46:80–102
3. Council for Agricultural Science and Technology: "Applications of Biotechnology to Crops: Benefits and Risks", Issue Paper, Number 12, Dec. 1999
4. Definition of Biotechnology-Economic Research Service at United States Department of Agriculture
5. Erickson B, Nelson JE, Winters P (2012) Perspective on opportunities in industrial biotechnology in renewable chemicals. Biotechnol J 7:176–185
6. Food biotechnology-benefits for the developing countries. http://ificinfo.health.org/insight/janfeb99/foodbiotechnology.htm
7. Food biotechnology: health and harvest for our times. http://ificinfo.health.org/brochure/biobroch.htm
8. Genetically engineered foods, fears and facts: FDA. Consumer 27(1). January/February 1993, pp. 11–14. http://www.fda.gov/fdac/100_toc.html
9. Hoban T, Kendall P (1994) Consumer attitudes about food biotechnology: final project report. North Carolina Cooperative Extension Service, North Carolina State University, Raleigh
10. IFIC Foundation- Americans remain positive on food biotechnology. http://ificinfo.health.org/press/positivebio.htm
11. Jikia et al (2005) The use of a novel biodegradable preparation capable of the sustained release of bacteriophages and ciprofloxacin, in the complex treatment of multidrug-resistant Staphylococcus aureus-infected local radiation injuries caused by exposure to Sr90. Clin Exp Dermatol 30:23–6
12. Myths and facts about food biotechnology, food insight, September/October 1999, pp. 2–3
13. National and international policy making in biotechnology. http://www.biotechknowledge.com/showlib.php3?194
14. New York Times Editorial, titled: "Food….people who would have the most to lose", 19 Nov 1999
15. North Carolina Biotechnology Center. "About Biotech". http://www.ncbiotech.org/aboutbt/main.cfm
16. Principles of biotechnology. http://www.nal.usda.gov/bic/Education_res/iastate.info/bio1.html
17. Public perceptions of biotechnology. A summary of research by Dr. Thomas Hoban at North Carolina State University. http://www4.ncsu.edu/~hobantj/biotech.htm
18. The Council for Biotechnology Information. http://www.whybiotech.com
19. Thieman WJ, Palladino MA (2004) Introduction to biotechnology, 2nd edn. Pearson Publications, USA
20. U.S. Food and Drug Administration Center for food Safety and Applied Nutrition: October-November 1997. http://vm.cfsan.fda.gov/~tdms/pubalgry.html
21. U.S. Food and Drug Administration Center for food Safety and Applied Nutrition: Q & A Sheet: June 1992. http://vm.cfsan.fda.gov/~lrd/bioqa.html
22. United States Department of Agriculture "Agricultural Biotechnology Concepts and Definitions". http://www.biotechknowledge.com/showlib.php3?1739
23. van Beuzekom B, Arundel A (2006) OECD biotechnology statistics 2006. OECD, Paris
24. What the experts say about food biotechnology http://ificinfo.health.org/foodbiotech/whatexpertssay.htm

Some Related Resources

http://ificinfo.health.org/backgrnd/BKGR14.htm
http://www.bio.org/aboutbio/guide1.html
http://www.bio.org/aboutbio/guide2000/guide00_toc.html
http://www.bio.org/aboutbio/guide3.html
http://www.bio.org/aboutbio/guide4.html
http://www.dec.ny.gov/energy/44157.html
http://www.ers.usda.gov/whatsnew/issues/biotech/define.htm
http://www.nal.usda.gov/bic/bio21
http://www.nature.com/nbt/press_release/nbt1199.html
www.angelfire.com/scary/intern/links.html
www.bio-link.org/library.htm
www.biospace.com
www.dnai.org
www.fiercebiotech.com
www.iastate.edu
www.icgeb.trieste.it
www.ncbi.nlm.nih.gov

Fundamentals of Recombinant DNA Technology

2

Abstract

The primary aim of biologists is to understand the mechanisms of life processes and implement the knowledge for betterment of human health and quality of life. To achieve this goal, it is important to identify and characterize genes involved in biological pathways. The knowledge gained from basic research is then implemented to manipulate genes that are of importance in medicine, industries, agriculture, and science. Recombinant DNA technology has provided us with set of techniques to achieve this objective. This involves recombining genes from different sources in a new combination and propagating or expressing them in a host, which is usually *E. coli*. It involves generation of mutants and knockouts to study reverse genetics. It is a highly evolving field of study with improved and new techniques being discovered rapidly and steadily. In this chapter, we are going to discuss the basic techniques of gene cloning and the recent advances in this direction. The chapter takes the reader to restriction endonucleases, cloning vectors, DNA sequencing techniques, protein expression in the host along with other advance tools, and their applications.

2.1 Introduction

The primary aim of biologists is to understand the mechanisms of life processes and implement the knowledge for betterment of human health and quality of life. To achieve this goal, it is important to identify and characterize genes involved in biological pathways. The knowledge gained from basic research is then implemented to manipulate genes that are of importance in medicine, industries, agriculture, and science. Recombinant DNA technology has provided us with set of techniques to achieve this objective. This involves recombining genes from different sources in a new combination and propagating or expressing them in a host which is usually *E. coli*. It involves generation of mutants and knockouts to study reverse genetics. It is a highly evolving field of study with improved and new techniques being discovered rapidly and steadily. In this chapter, we are going to discuss the basic techniques of gene cloning and some of the recent advances in this direction.

V. Gupta et al., *Basic and Applied Aspects of Biotechnology*, DOI 10.1007/978-981-10-0875-7_2

2.2 Gene Cloning or Molecular Cloning

Molecular cloning refers to a set of experiments that is used to construct a recombinant DNA molecule that can replicate in the host [8]. Clone is referred to a group of cells that have originated from the same progenitor and are identical to each other. This is the basic step in recombinant DNA technology. It starts with isolating the gene of interest (GOI) from an organism, amplifying the fragment by using polymerase chain reaction (PCR), and ligating it to a vector. The construct is then propagated in a host strain and later characterized and isolated for specific use. The tools required are vectors, restriction enzymes, DNA ligase enzyme, and host strain. Several options are available depending on the requirement of the experiment. We will discuss the major tools for cloning and then the steps involved to construct a clone.

2.3 Restriction Endonucleases (RE)

Restriction endonucleases are enzymes that are produced by bacteria as a defense mechanism against invading foreign genome such as those from viruses. These enzymes recognize specific foreign DNA sequences and digest them to small fragments hence protecting the host. The bacterial genome is protected from their own restriction enzymes due to their DNA being methylated. Most RE does not recognize host methylated sequences. The RE isolated from different bacterial species has been used in genetic engineering. They act as molecular scissors. They identify specific palindromic sequences and make either a staggered cut in the double-stranded DNA producing a 5′ or 3′ overhang or a blunt cut producing a blunt end (Fig. 2.1a, b). The 5′ and 3′ overhangs so produced can be efficiently stitched with another fragment of DNA that has been digested with the same enzyme with the help of DNA ligase enzyme. There are several REs that are useful to cut DNA at specific sites and are routinely used for cloning (Fig. 2.1c). They use specific reaction buffer and temperature to digest DNA in vitro.

2.4 Cloning Vectors

Vectors are vehicles that are used to transfer and stably maintain recombinant DNA fragments in host organisms. There are several choices for cloning vectors depending on the fragment length to be inserted and specific use. However, all the vectors have common features such as origin of replication (OriC), multiple cloning sites (MCS) for insertion of foreign DNA, and resistance drug markers for selection of colonies that are transformed (Fig. 2.2). Some vectors are used for expressing the cloned genes. These expression vectors have promoters and regulators for tight regulation of expression. Some of the vector types are discussed below.

2.4.1 Plasmids

Plasmids are autonomously replicating, circular, extrachromosomal DNA fragment present in bacteria and yeast. Bacterial plasmids are vector of choice for molecular cloning as they are easy to grow having simple growth requirements, present in multiple copies, and have high yield. Host–vector systems consisting of bacterium *E. coli* K-12 and its vector plasmids have proven of immense value for the cloning, amplification, and analysis of DNA fragments from a wide variety of organisms. Commonly used bacterial plasmids for cloning are pUC19 (Fig. 2.2), pBS, pACYC, and pBR322. They have origin of replication or *OriC*. The host replication machineries bind to *OriC* region and help to replicate. They have multiple cloning sites (MCS) consisting of several unique restriction sites. These are used for inserting foreign DNA fragments. They carry antibiotic resistance genes such as ampicillin or kanamycin resistance gene. These help in selection of the plasmids on agar plates containing specific antibiotics. Some plasmids carry reporter genes for selection of clones that carry the fragment of interest. The most commonly used reporter system is *lacZ* alpha complementation for blue-white selection (discussed below Sect. 2.5). Plasmids can carry around 15 kb DNA. Some plasmids contain bacteriophage M13 origin of replication and can

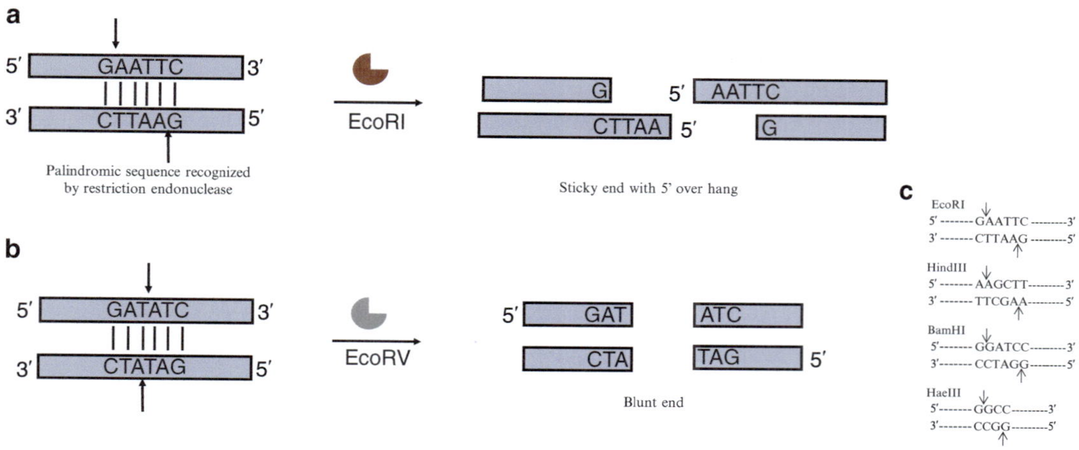

Fig. 2.1 Restriction endonucleases are enzymes isolated from bacteria that can recognize and digest DNA sequence specifically. They can form sticky ends or cohesive ends with 5′ or 3′ overhangs (**a**). The ends so formed are easy to ligate. The blunt ends are formed when there is no overhang (**b**). (**c**) Shows recognition sites for some commonly used restriction enzymes

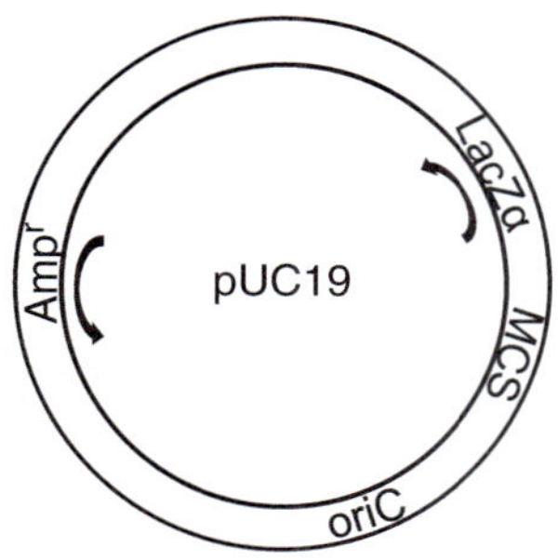

Fig. 2.2 Cloning vector. pUC19 is a prototype cloning vector. It has an origin of replication OriC, ampicillin-resistant gene, LacZα for blue-white selection of positive clones, and a multiple cloning site (MCS) for insert insertion. Since the MCS lies within the LacZα, cloned DNA fragment disrupts it and the colonies appear white on X-gal/IPTG plate

produce single-stranded DNA that can be used for DNA sequencing. Cloning in *E. coli* vector is relatively simple, but it is constrained as it can only accept small fragments of DNA.

2.4.2 Bacteriophage

For cloning large chunk of DNA, bacteriophage λ and M13 is used. Bacteriophage λ has a head that carries the 50 Kb lambda genome and a tail that is used to infect *E. coli*. Once infected it can either undergo lytic cycle where it produces millions of virion and lyses the *E. c*oli cell. Or it can integrate to the bacterial chromosome and exist as a temperate phage. For cloning vector, the region of the λ genome not required for lytic cycle is replaced by foreign DNA. A maximum of 25 kb can be cloned. λ vectors replicate as lytic viruses, killing the host cell and packaging the DNA into virions. They are useful to construct genomic DNA libraries.

2.4.3 Cosmid

These are plasmids that incorporate a segment of bacteriophage λ DNA that has the cohesive end site (cos) and contains elements required for packaging DNA into λ particles. It is a derived plasmid. It has origin of replication, MCS, and selectable markers. It is used to clone large DNA fragments between 30 and 45 Kb.

2.4.4 Bacterial Artificial Chromosome (BAC)

It is an engineered DNA molecule used to clone in bacterial cells. It utilizes the F-plasmid or fertility plasmid for transforming and cloning in

E. coli. It is maintained as an independent chromosome within the host. It can carry insert size of up to 350 kb. BACs are maintained in *E. coli* with a copy number of only 1 per cell. They are useful to manipulate mouse and human genome.

2.4.5 Yeast Artificial Chromosome (YAC)

It is an engineered DNA molecule to clone in yeast cells. Insert of up to 3000 kb may be carried by yeast artificial chromosome.

2.4.6 Human Artificial Chromosome (HAC)

Human artificial chromosome is a microchromosome that can be maintained independently in human cell as the 47th chromosome. It can carry around 6–10 Mb DNA fragment. HAC carries a functional kinetochore and is stably maintained. It is a promising system for gene delivery and expression with a potential to overcome many problems caused by the use of viral-based gene transfer systems such as limited cloning capacity, lack of copy number control, and insertional mutagenesis due to integration into host chromosomes [7].

2.5 *lac* Operon and Blue-White Screening

Operon is a cluster of genes that functions under the regulation of a single promoter. The genes are transcribed together into an mRNA. Bacterial *lac operon* is one of the most well-studied systems (see Fig. 2.3). It consists of three genes *lacZ*,

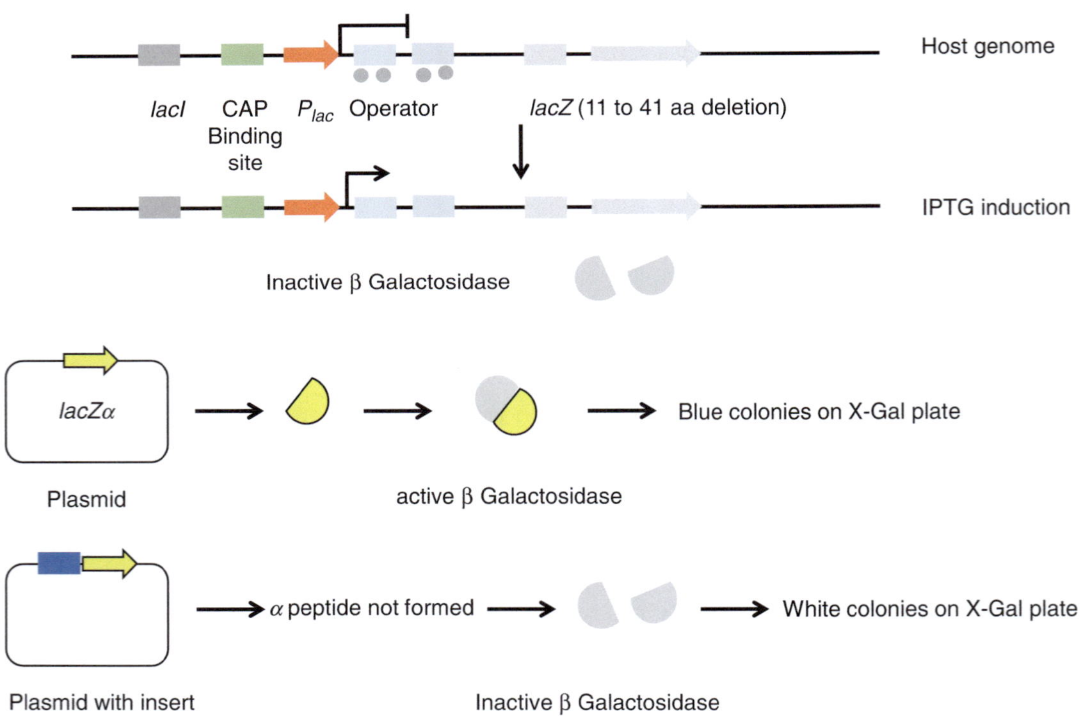

Fig. 2.3 α complementation and blue-white selection. The host *E. coli* strain carries the *lacZ* deletion mutant encoding the ω-peptide. The cloning vectors carry *lacZα* sequence encoding the α-peptide. Neither is functional by itself. However, when the two peptides are expressed together, they form a functional β-galactosidase enzyme and the host cells form blue colonies in X-gal/IPTG plates. When a DNA fragment is successfully cloned into the MCS of the plasmid, the *lacZα* is disrupted and functional enzyme is not formed. These cells form white colonies. This screening method is used for selecting clones with insert

lacY, and *lacA* responsible for lactose metabolism in absence of glucose. The gene product of *lacZ* is β-galactosidase that cleaves lactose to glucose and galactose. *LacY* encodes lactose permease that transports lactose into the cell. *LacA* encodes galactoside O-acetyltransferase. The lac operon uses a two-part control mechanism to ensure that the lac operon is activated only when necessary. In the absence of lactose, the lac repressor halts production of the enzymes encoded by the lac operon. In the presence of glucose, the catabolite activator protein (CAP), required for production of the enzymes, remains inactive, and EIIAGlc shuts down lactose permease to prevent transport of lactose into the cell. This dual control mechanism causes the sequential utilization of glucose and lactose in two distinct growth phases.

This system is utilized in blue-white screening during cloning where bacterial cells carrying plasmids with insert grow as white colonies when plated on X-gal plates whereas those without insert grow as blue colonies. β-galactosidase encoded by the *lacZ* exists as a homotetramer in its active state. A mutant *lacZ* derived from the M15 strain of *E. coli* produces ω-peptide, with N-terminal deletion (11–41 amino acid residues) that is unable to form a tetramer and is inactive. When N-terminal fragment of the protein, the α-peptide, is provided in trans, a functional protein is formed. The rescue of function of the mutant β-galactosidase by the α-peptide is called α-complementation. In blue-white screening, the host *E. coli* strain carries the *lacZ* deletion mutant which contains the ω-peptide, while the plasmids used for cloning carry the lacZα sequence which encodes the first 59 residues of β-galactosidase, the α-peptide. When the two peptides are expressed together, as when a plasmid containing the lacZα sequence is transformed into a lacZΔM15 cells, they form a functional β-galactosidase enzyme. The plasmid carries multiple cloning site (MCS) within the lacZα sequence. When a foreign DNA is integrated in the MCS, lacZα gene is disrupted. Hence the cells are unable to produce functional β-galactosidase (Fig. 2.3).

X-gal is a colorless analog of lactose that is cleaved by β-galactosidase to form 5-bromo-4-chloro-indoxyl, which then spontaneously dimerizes and oxidizes to form a bright blue insoluble pigment 5,5′-dibromo-4,4′-dichloro-indigo. This results in a characteristic blue color in cells containing a functional β-galactosidase. Isopropyl β-D-1-thiogalactopyranoside (IPTG) functions as the inducer of the lac operon. Blue colonies formed on X-Gal-IPTG plate show that they contain a vector with an uninterrupted lacZα, while white colonies indicate the presence of an insert in lacZα which disrupts the formation of an active β-galactosidase. The plasmids such as pBluescript and pUC19 contain the lacZα and are used for blue-white selection with *E. coli* cell with mutant *lacZ* gene such as JM109, DH5α, and XL1-Blue. On rare occasions, blue-white selection could be misleading. If the insert is "in frame" with the LacZα gene and a STOP codon is absent, it could lead to a fusion protein.

2.6 Polymerase Chain Reaction (PCR)

Polymerase chain reaction (PCR) is a method to amplify a DNA fragment in vitro. It is based on the ability of DNA polymerase enzyme to synthesize new strand of DNA complementary to the offered template strand. PCR involves repetitive cycles of template denaturation followed by DNA synthesis (Fig. 2.4). In this process as little as a single copy of a gene can be amplified to several million copies in a short period of time. This is a very powerful technique that has been developed by Kary B. Mullis in 1980s. He was awarded the Nobel Prize for Chemistry in 1993. PCR has a broad range of use in gene cloning, sequencing, diagnosis of bacterial and viral diseases, diagnosis of genetic disorder, and study of gene expression, among others.

PCR reaction requires template DNA, a pair of sequence-specific primers, heat-stable DNA polymerase enzyme, nucleotides, reaction buffer, and $MgCl_2$. The reaction is completed in a thermal cycler that can be programmed to alter the

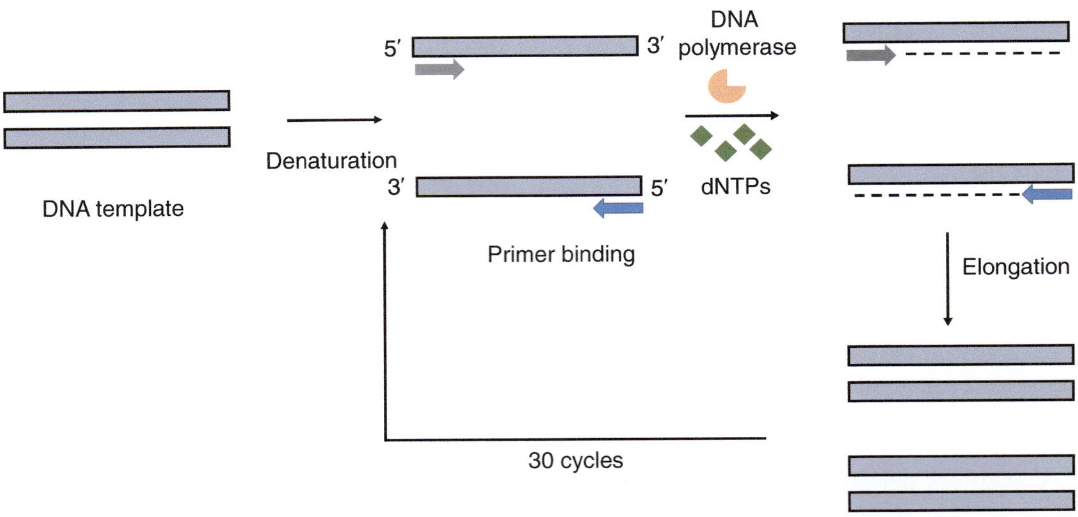

Fig. 2.4 PCR reaction. The double-stranded DNA template is denatured at high temperature. It is followed by primer binding. The DNA polymerase extends the strands from 5′ to 3′. After the elongation step, double-stranded DNA is formed with one old strand and a newly synthesized strand. The cycle is repeated 30 times to form large number of DNA

temperature of the reaction every few minutes to allow DNA denaturing and synthesis. The double-stranded DNA template is initially heat denatured to yield two single-stranded DNAs. This is followed by primer annealing. Since DNA polymerase can add a nucleotide only onto a preexisting 3′-OH group, a primer is required to which the first nucleotide can be added. These are short (20–50 bp) oligonucleotides that can bind to target DNA sequence specifically. This requirement makes it possible to delineate a specific region of template sequence to amplify. This is followed by the elongation step where a high-temperature DNA polymerase enzyme elongates the two strands using the original strands as templates. The most commonly used enzyme is Taq DNA polymerase isolated from *Thermus aquaticus*. Pfu DNA polymerase isolated from *Pyrococcus furiosus* is used widely because of its higher fidelity when copying DNA. Both of them are heat resistant. This process results in the duplication of the original DNA, with each of the new molecules containing one old and one new strand of DNA. Then each of these strands can be used to create two new copies, and so on. The cycle of denaturing and synthesizing new DNA is repeated as many as 30 or 40 times, leading to more than one billion exact copies of the original DNA segment.

2.7 Cloning Procedure

The cloning procedure usually begins with amplification of the target DNA by PCR. The target DNA or the insert is purified and digested with restriction enzymes. The vector selected is also digested by the same set of enzymes. The vector and the insert are stitched together with the help of DNA ligase enzyme (Fig. 2.5). The DNA ligase enzyme recognizes and acts on the DNA ends and covalently links them. The ligated insert and vector is then transferred to the host organism. In most cases it is a bacterial strain. The bacterial host is chemically treated to make them competent so the DNA molecules can be taken up in a process known as transformation. The transformed bacteria are plated on agar plates containing specific antibiotics. The cells that are transformed with DNA can form colonies. Only a single recombinant molecule will enter a bacterial cell. Thus the transformed colony may carry the vector alone, or the vector with the insert. They are then selected. Electroporation is another method for transform-

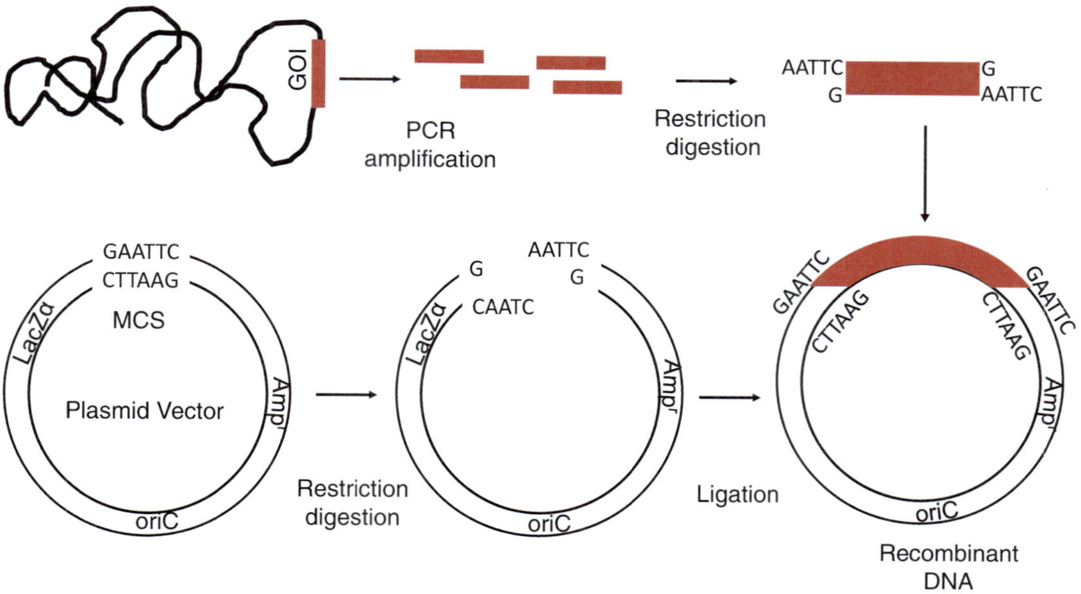

Fig. 2.5 Cloning. The gene of interest (GOI) is amplified and digested with RE. The vector plasmid is also digested by the same enzyme. The vector and the purified insert is ligated using DNA ligase. The sticky ends join to form construct of interest. The recombinant molecule is transformed into a bacterial host for selection and amplification

ing bacterial cells where high-voltage electrical pulses are used to translocate DNA into the cell. The efficiency of transformation is much higher in this case. When mammalian cells are used for transfer of DNA, the process is known as transfection. Sometimes viral particles are used to package the DNA, a process known as transduction. Once the cloning is done, the appropriate clone is selected by recovering the construct from the host and digesting it with the RE to check for the presence of insert. Then the construct is sequenced to check for any mutation that might have accumulated in this process.

After thorough evaluation, the clone is ready for functional analysis.

2.8 Library Construction

2.8.1 Genomic DNA Library

It is a collection of genomic DNA fragments cloned into vectors. The fragment of interest can be identified, isolated, and further manipulated with ease. Genomic DNA libraries contain large fragments of DNA in either bacteriophages or bacterial or P1-derived artificial chromosomes (BACs and PACs). The method to prepare a genomic DNA library is to isolate the genomic DNA from an organism. The DNA is fragmented to smaller pieces with rare cutter RE. The vector is digested with the same enzyme. The vector and the insert are ligated and transformed to the host [9].

2.8.2 cDNA Library

cDNA libraries are constructed with reverse-transcribed mRNA and therefore lack DNA sequences corresponding to genomic regions that are not expressed, such as introns and 5′- and 3′-noncoding regions. cDNA libraries generally contain much smaller fragments than genomic DNA libraries and are usually cloned into plasmid vectors. For library construction, mRNA is isolated and converted to complementary DNA or cDNA using reverse transcriptase enzyme. The cDNA thus obtained is cloned to vectors. cDNA library contains the expression profile of an organism under a specific condition.

2.9 Site-Directed Mutagenesis (SDM)

SDM is a process whereby a specific, intentional change is incorporated into a cloned DNA fragment. It can be used to introduce point mutation, deletion, or insertion in a gene. This process was first reported by Kunkel in 1983. According to Kunkel's method, the DNA fragment to be mutagenized is inserted into a phagemid and transferred to an *E. coli* strain that is deficient in dUTPase (dut) and uracil deglycosidase (ung) and incorporates dUTP instead of dTTP. The phage DNA replicated in this strain contains dUTP instead of dTTP, resulting in single-strand DNA that contains uracils (ssUDNA). The ssUDNA is used as template for mutagenesis. An oligonucleotide containing the desired mutation is used for primer extension. The heteroduplex DNA, that forms, consists of one parental non-mutated strand containing dUTP and a mutated strand containing dTTP. The DNA is then transformed into an *E. coli* strain carrying the wild-type *dut* and *ung* genes. Here, the uracil-containing parental DNA strand is degraded, so that nearly the entire resulting DNA consists of the mutated strand.

Currently most widely used methods do not require any modifications or unique strains and incorporate mutations into the plasmid by inverse PCR with standard primers (Fig. 2.6). For these methods, primers can be designed in either an overlapping or a back-to-back orientation. Overlapping primer design results in a product that will re-circularize to form a doubly nicked plasmid. Despite the presence of these nicks, this circular product can be directly transformed into *E. coli*, at a lower efficiency than non-nicked plasmids. Back-to-back primer design methods not only have the advantage of transforming non-nicked plasmids but also allow exponential amplification to generate significantly more of the desired product. In addition, because the primers do not overlap each other, deletions sizes are only limited by the plasmid and insertions are only limited by the constraints of modern primer synthesis. Currently, by splitting the insertion between the two primers, insertions up to 100 bp can routinely be created in one step using this method. The template DNA isolated from wild-type *E. coli* is naturally methylated. The PCR product containing specific mutation is digested with DpnI restriction endonuclease that digests methylated template

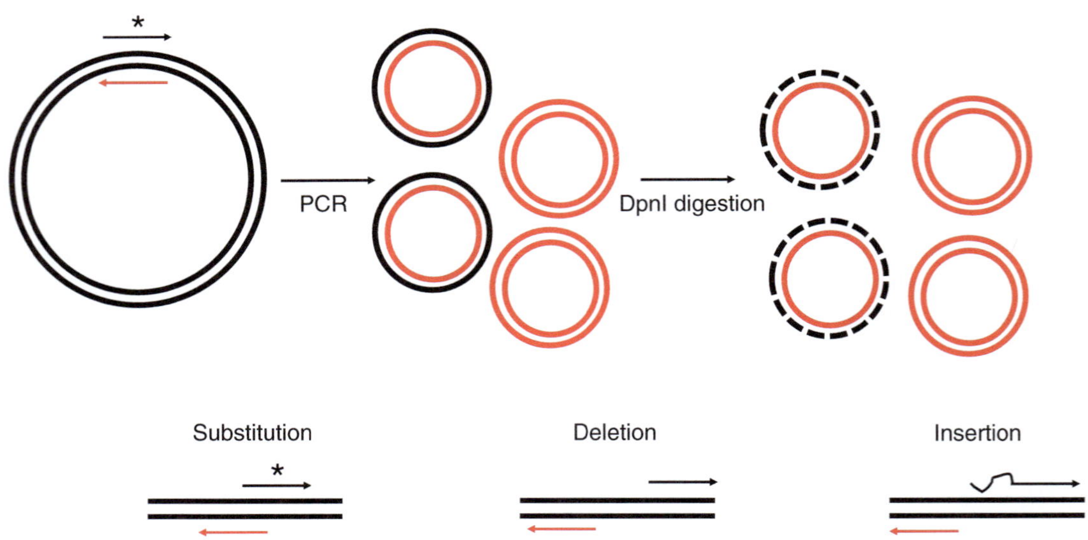

Fig. 2.6 Site-directed mutagenesis. The fragment of interest is cloned into a vector. The primers are designed to create desired mutations. Overlapping primers are used for substitution, back-to-back to primers to create deletion and insertion mutation. After PCR, the product is digested by DpnI that digests template DNA. The product is transformed in *E. coli* and selected

DNA strands. Hence the original template DNA is fragmented. Only the mutated DNA is present that is transformed to a different host.

Substitutions, deletions, and insertions are incorporated into plasmid DNA through the use of specifically designed forward (black) and reverse (red) primers (Fig. 2.6). Substitutions are created by incorporating the desired nucleotide change(s) in the center of the forward primer, including at least ten complementary nucleotides on the 3′ side of the mutation(s). The reverse primer is designed so that the 5′ ends of the two primers anneal back-to-back. Deletions are engineered by designing standard, non-mutagenic forward and reverse primers that flank the region to be deleted. Insertions less than or equal to six nucleotides are incorporated into the 5′ end of the forward primer, while the reverse primer anneals back-to-back with the 5′ end of the complementary region of the forward primer. Larger insertions can be created by incorporating half of the desired insertion into the 5′ ends of both primers. The maximum size of the insertion is largely dictated by oligonucleotide synthesis limitations.

2.10 DNA Sequencing

DNA sequencing enables determination of the exact nucleotide sequence of a DNA fragment. This technique has a very important implication in molecular biology as knowledge of DNA sequence has numerous applications such as determination of regulatory region and coding region of a gene, mutation analysis, splice variants, and homologous sequence, to name a few. This technique was developed independently by two groups. Maxam and Gilbert invented DNA sequencing by chemical cleavage method, and Sanger developed dideoxy chain termination method for sequencing.

2.10.1 Dideoxy Chain Termination Method or Sanger's Method

This method requires DNA synthesis in presence of dideoxy nucleotides (ddNTP) along with regular deoxyribonucleic acid (dNTPs) [14]. Dideoxynucleotides contain a hydrogen group on the 3′ carbon instead of a hydroxyl group (OH) that is present in nucleotides. When a ddNTP is incorporated, it stalls elongation as there is no free 3′ OH group for the next incoming nucleotide (Fig. 2.7). Individual reactions are set up for four nucleotides. Fragments of different length are created due to the insertion of the ddNTPs. These fragments are visualized on a gel and the length of each fragment indicates the position of the particular nucleotide in the template sequence.

The template DNA is denatured and annealed to a primer. The primer is usually radiolabeled to be able to visualize the fragments on a gel. For

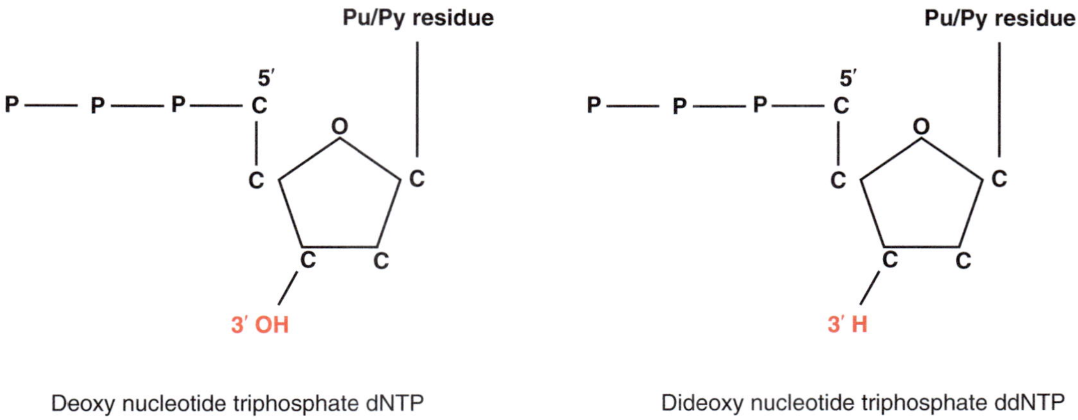

Fig. 2.7 Dideoxy chain termination sequencing. Dideoxy nucleotides have H attached to 3′ C instead of OH. DNA polymerase adds dNTP to 3′ OH group. Hence when it encounters a ddNTP, it is unable to elongate. In Sanger's method, ddNTPs are randomly incorporated forming fragments of template DNA

each reaction, all the dNTPs and a specific ddNTP (e.g., ddATP for A tube) are added. Thus, four reaction tubes are set for each template denoting A, T, G, and C. DNA polymerase is added and a regular PCR reaction is performed. During the course of reaction, the ddNTPs are randomly incorporated along with the dNTPs, and different sizes of fragments are synthesized. These DNA fragments are then run on a gel to read the sequences. Each fragment in a reaction tube denotes the position of the particular nucleotide in the template (Fig. 2.8).

There has been lot of advances in this technique. Currently only one reaction is set up for each template and individual ddNTPs are fluorescently labeled. Thus, four fluorochromes are used. After the PCR is done, the reaction is run on a capillary gel and passed through a laser coupled with a detector. As each fragment passes through the gel according to their molecular size, the laser excites the fluor attached to the terminal ddNTP and a light is emitted. This emitted light is detected by the detector and a peak is generated. So a chromatogram is created that tells the position of the nucleotides.

2.10.2 Maxam–Gilbert Reaction

Maxam–Gilbert reaction involves a two-step catalytic reaction to degrade the template DNA. In this method, the template DNA is treated with dimethyl sulfate and hydrazine to modify the purines and pyrimidines, respectively. Piperidine is used to catalyze phosphodiester bond cleavage where the base has been displaced. The DNA fragment to be sequenced is labeled at 5′ end. Chemical treatment generates breaks at a small proportion of one or two of the four nucleotide bases in each of four reactions (G, A+G, C, and C+ T). Formic acid depurinates the template DNA and causes break at A+G. The guanines are methylated by dimethyl sulfate in G reaction. The pyrimidines (C+T) are hydrolyzed using hydrazine. The addition of sodium chloride to the hydrazine protects thymine for the C-only reaction. The modified DNA is cleaved by hot piperidine at the position of the modified base. The concentration of the modifying chemicals is controlled to introduce on average one modification per DNA molecule. Thus a series of labeled fragments is generated, from the radiolabeled end to the first "cut" site in each molecule. The

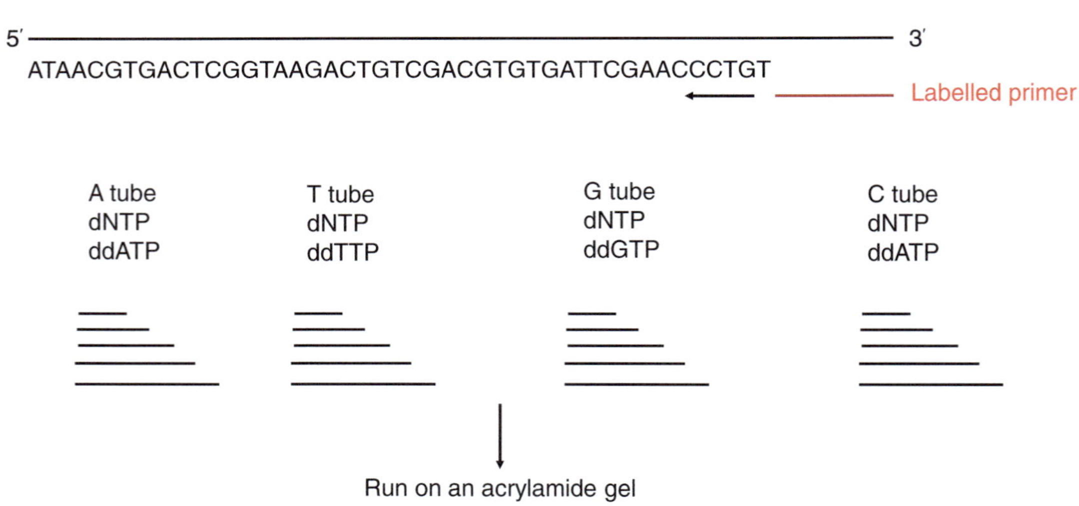

Fig. 2.8 Dideoxy chain termination method. Four reactions are set up using dNTPs and one of the ddNTPs. The ddNTPs are randomly integrated and the elongation is stalled. After the PCR is over, each reaction yields fragments of different lengths. They are run on an acrylamide gel and visualized. Each fragment shows the position of the particular nucleotide in the original template

reactions are loaded on polyacrylamide gels and the fragments resolved by electrophoresis.

2.10.3 Next-Generation Sequencing (NGS)

Next-generation sequencing or massive parallel sequencing refers to non-Sanger-based high-throughput DNA sequencing technologies [10]. In this method, millions or billions of DNA strands are sequenced in parallel, yielding more throughputs and minimizing the need for the fragment-cloning methods that are often used in Sanger sequencing of genomes. The major steps in NGS involve generation of fragment libraries from isolated cDNA or genomic DNA. These fragments are ligated to 3 and 5′ oligo sequences known as adapters and sequestered to beads or chips depending on the sequencing platforms. Millions of fragments are sequenced in parallel and short reads are generated. The information is matched with reference library and analyzed using sophisticated bioinformatics programs. Since cloning of DNA fragments are not involved in this process, cloning bias of genome can be avoided. The method has greatly reduced the time and cost involved in sequencing. The capillary-based sequencing instruments developed by Applied Biosystems were used by NIH-led and Celera-led Human Genome Project. The first human sequence published jointly in Science and Nature in 2001 required about 15 years to complete and close to 3 billion dollars. The NGS concept revolutionized the sequencing so much that presently 45 human genome can be sequenced in a day and the cost has been reduced to $1000 per genome. This has given rise to personal genome medicine.

There are several NGS platform. We will touch the principle and process for two of those in this chapter.

2.10.4 Pyrosequencing or 454 Sequencing

In this method, the pyrophosphate molecule released during nucleotide incorporation by DNA polymerase is used for a downstream reaction which produces light from the cleavage of oxyluciferin by luciferase. The steps include:

(a) Library preparation: The double-stranded DNA is fragmented to small pieces about 400–600 bp in size. The DNA fragments are linked to adapters and are then separated to single strands. Thus, the DNA library consists of randomly fragmented single-stranded templates attached to oligonucleotide adapters.
(b) Loading of DNA library to beads and emulsion PCR: The DNA fragments are attached to agarose beads via the adapters and these are used for emulsion PCR. Emulsion PCR uses a vigorously mixed oil and aqueous mixture containing PCR reactants, where isolated individual agarose beads containing a single unique DNA fragment is amplified.
(c) Sequencing: The amplified DNA hybridized to beads is loaded onto PicoTiterPlate™ for sequencing. These plates contain small wells. Each well can load only one bead. The plate is loaded to the instrument. The fluid system flows through the plate providing A, C, G, and T nucleotides sequentially. When these nucleotides are incorporated onto the DNA strands, a chemiluminescent reaction takes place producing light. The light is recorded by a CCD camera and the intensity is proportional to the number of nucleotide incorporated. The millions of sequence data so generated are aligned against the source reference data. This has advantage of speeding up time and deep sequencing.

2.10.5 Illumina Genome Analyzer

Illumina uses sequencing by synthesis (SBS) technique. Here the polymerase extends the template and a fluorescently labeled dNTP is added to it. The nucleotide is identified during each cycle by the specific fluorophore excitation. In this method, millions of strands are sequenced at the same time. It provides accurate, high yielding, short reads that are compared to reference library. The steps involved are:

(a) Library preparation: cDNA or DNA samples are fragmented randomly and are ligated to specific adaptors at both 3 and 5′ ends. These fragments are amplified by PCR and purified.
(b) Cluster Generation: The library is loaded on to a flow cell. The flow cell contains a lawn of oligos that are complementary to the adapters present in the library. Each fragment is amplified to clonal clusters. This is used for sequencing.
(c) Sequencing: Illumina SBS uses reversible termination-based sequencing where every base is recognized after being incorporated. All the four terminator-bound dNTPs are present in each cycle. This results in base-by-base sequencing.
(d) Data Analysis: The sequence reads are matched to reference genome. After alignment, various analyses are done.

2.10.6 Applications

NGS platform is used for answering questions about the genome, transcriptome, and epigenetic mechanisms in organisms. Some of the applications are:

(a) Genome studies:
   1. Whole genome sequencing (WGS): Genome-wide association studies (GWAS) are done to identify disease association across the genome. Study of single nucleotide changes that are associated with a disease is studied to evaluate their association with a particular disease state. This can be done very efficiently by using WGS approach where the entire human genome of several subjects can be evaluated in very less time. WGS is also used to study drug-resistant bacterial strains and their transmission.
   2. Exome sequencing: The protein-coding region of the genome is captured and sequenced to identify variations and mutations. This is a cheaper option than WGS.
   3. De novo sequencing: Sequencing of novel genome where no reference genome is available is done by de novo sequencing. Sequence reads are assembled as contigs. Deep sequencing allows filling out the gaps.
   4. Targeted sequencing: In this method, subset of genes is sequenced to identify variants in genes of interest. It is cheaper than whole-genome sequencing.
(b) Expression or transcriptome studies: RNA sequencing or RNA-Seq method required isolation of total RNA. Ribosomal RNA is removed from the sample and cDNA is constructed via reverse transcription. The cDNA is used for library construction.
   1. Total RNA and mRNA sequencing: RNA sequencing allows the study of expression profile of the whole genome.
   2. Targeted RNA sequencing: Expression of transcripts of interests is studied by this approach. Alternate splicing, allelic variation, isoforms, and differential expression can be studied.
   3. Small RNA and noncoding RNA sequencing: These are small noncoding RNAs, 18–22 bp, and play an important role in gene expression. Their expression profile can be studied by NGS.
(c) Epigenetic studies: Epigenetics is study of heritable or acquired alteration of DNA sequences. The major mechanisms are DNA methylation, histone modification, small RNA-mediated regulation, and DNA protein interaction.
   1. Methylation: Methylation of DNA changes the expression profile of different genes. Cytosine methylation regulates gene expression significantly. This is studied by two methods: whole-genome bisulfite sequencing (WGBS), where non-methylated cytosine is changed to uracil which is further converted to thymine during sequencing. Other method is reduced representation bisulfite sequencing (RRBS). Here DNA is digested with MspI, a restriction enzyme that remains unaffected by methylation. CpG and promoter-containing fragments are isolated to prepare library.
   2. ChIP: Protein–DNA and protein–RNA interactions are studied by combining chromatin immunoprecipitation and NGS.

3. Ribosome profiling: Deep sequencing of ribosome-protected mRNA fragments. All the ribosomes active at a particular time are studied.

2.11 Genome Editing

Targeted genome editing produces specifically designed changes in the genome of organisms and cell line. It is required to produce mutants to study reverse genetics, perform epitope tagging, and construct gene fusion with fluorescent protein to study expression. Several techniques have been developed over the years to produce deletion, insertional, and point mutations in bacteria, yeast, mouse, cell lines, and host of other models to study biological processes and disease models. Some of these methods are time consuming and are useful only for a particular organism. Most of all they leave behind scars such as selectable markers or sequences such as "loxP" sites in the host. Hence, it is important to develop new techniques that are efficient, universal, and time-saving. Most of the genome editing methods utilize the principles of DNA repair mechanisms to manipulate the genome. Hence it is of interest to review the main DNA repair pathways present in cells. DNA damage continually takes place in the cell due to environmental factors such as radiation and chemical assaults and also due to normal metabolism. DNA damage gives rise to single- or double-strand breakage. Hence the cells have developed several mechanisms to repair the damages. If the damages are not repaired, the cells either attend senescence or undergo programmed cell death or apoptosis. Imperfect repair mechanism is a major cause for uncontrolled cell proliferation resulting in cancer.

2.11.1 Double-Stranded DNA Repair Mechanisms

Double-strand breakage (DSB) has severe consequences as it can lead to genome rearrangement. There are two major mechanisms that exist for double-strand break repair, homologous recombination, and nonhomologous end joining.

2.11.1.1 Homologous Recombination (HR)

Homologous recombination ensures an accurate repair by using the undamaged sister chromatid or homologous chromosome as a template. It takes place in cells that are in S phase or in G2 phase of cell cycle where sister chromatids are in close vicinity and may provide template for sequence-driven DNA repair. This method is very effective to check mutations, deletions, and duplications in cells. Although homologous recombination varies widely among different organisms and cell types, most forms involve the same basic steps. After a double-strand break occurs, sections of DNA around the 5′ ends of the break are cut away in a process called resection. It is followed by strand invasion where an overhanging 3′ end of the broken DNA molecule "invades" a similar or identical DNA molecule that is not broken. After strand invasion, the further sequence of events may follow either of two main pathways: the DSBR (double-strand break repair) pathway or the SDSA (synthesis-dependent strand annealing) pathway. The proteins that are involved in mammalian systems are RPA, which coat the ssDNA. It is followed by co-localization of Rad51. Rad52, in conjunction with Rad51 and BRCA2, displaces RPA. Rad51 and Rad54 catalyze strand invasion and homology search with the undamaged template. Following DNA synthesis via polymerases, the resulting Holliday junctions are resolved. Homologous recombination that occurs during DNA repair tends to result in non-crossover products, in effect restoring the damaged DNA molecule as it existed before the double-strand break.

2.11.1.2 Nonhomologous End Joining

Nonhomologous end joining (NHEJ) is a pathway that repairs double-strand breaks in DNA in the absence of a template strand. It is an error-prone repair mechanism. In bacteria, Ku and LigD are the two proteins that are involved in NHEJ. The Ku protein homodimer binds to the broken DNA ends and recruits multifunctional LigD protein. LigD acts as a nuclease, polymerase, and ligase for end joining. In eukaryotes, there are several steps involving different proteins. End binding and tethering occurs early during DSB where the ends are identified and brought together. In yeast, the Mre11-Rad50-Xrs2 (MRX) complex is recruited to DSBs and is thought to promote bridging of the DNA ends. The corresponding mammalian complex of Mre11-Rad50-Nbs1 (MRN) is also involved in NHEJ. DNA-PKcs is also thought to participate in end bridging during mammalian NHEJ. Eukaryotic Ku is a heterodimer consisting of Ku70 and Ku80 and forms a complex with DNA-PKcs, which is present in mammals but absent in yeast. Ku may function as a docking site for other NHEJ proteins and is known to interact with the DNA ligase IV complex and XLF. End processing involves removal of damaged or mismatch nucleotides and resynthesis. This is followed by ligation of the ends using ligase IV complex [4].

2.11.2 Recombineering

Recombineering is an in vivo genetic engineering technology that exploits the homologous recombination machinery of the temperate bacteriophage λ to engineer DNA molecules such as chromosomes, plasmids, and episomes using *E. coli* as a host [16]. It is a simple, fast, and efficient system that does not require restriction sites or enzymes for DNA manipulation. It requires short 50 bp 5′ and 3′ homology sequence of the donor DNA and lambda recombination machinery [17].

Recombineering is based on homologous recombination and exploits bacteriophage λ Red system that includes the phage recombination genes *gam*, *bet*, and *exo*. The *gam* gene product, Gam, prevents *E. coli* nuclease, RecBCD, from degrading linear DNA fragments used as templates allowing preservation of transformed linear DNA in vivo. The *bet* gene product, Beta, is an ssDNA-binding protein that promotes annealing of two complementary DNA molecules and the *exo* gene product, Exo, has a 5–3′ dsDNA exonuclease activity. Exo and Beta together insert linear DNA at the desired target, creating genetic recombinants. For dsDNA, Red Exo is thought to degrade from both 5′ ends, exposing ssDNA that is bound by Red Beta (Fig. 2.9).

Single-stranded synthetic oligo or double-stranded PCR product with 50 bp homology on the either end is used for genetic manipulation. The fragments are electroporated in *E. coli* carrying the λ Red system. The λ Red system is tightly regulated by phage repressor system so that the proteins are expressed only under desired condition. This is achieved by the use of temperature-sensitive repressor expressed from the cI857 gene. λ repressor binds cooperatively at the three operator sites present at both the pL and pR promoters. These two sets of repressor-bound operators interact with each other by protein–protein-mediated looping between pL and pR to generate a handcuff of 12 repressor proteins. The λ Red genes are expressed from pL promoter and are controlled by the operators. At low temperature, the repressor is inactive. After electroporating the DNA fragments, the cells are transiently transferred to 42 °C, to express the proteins.

Targets are bacterial artificial chromosomes, bacterial chromosome, episomal DNA such as low copy P1, and F-plasmid derivatives that carry artificial chromosomes known as P1 artificial chromosome or PAC. This method is used for generating mutants, transgenic reporter constructs, and fusion tags.

The main advantages of recombineering over conventional DNA manipulation technologies is that it is free of DNA sequence-imposed limitations, and large fragments of DNA can be manipulated. With recombineering, multistep cloning strategies, many steps of which may damage DNA or be error prone, can be completely avoided. It is not restricted by presence or absence of existing restriction sites. Moreover, since recombineering is carried out in vivo, genetic changes can be made on large, complex DNA molecules that are currently difficult to manipulate in vitro, like human, mouse, bacterial, and phage chromosomes.

2.11.3 CRISPR/Cas9

Clustered regularly interspaced short palindromic repeats (CRISPR) and CRISPR-associated (Cas) genes are essential in adaptive immunity in select

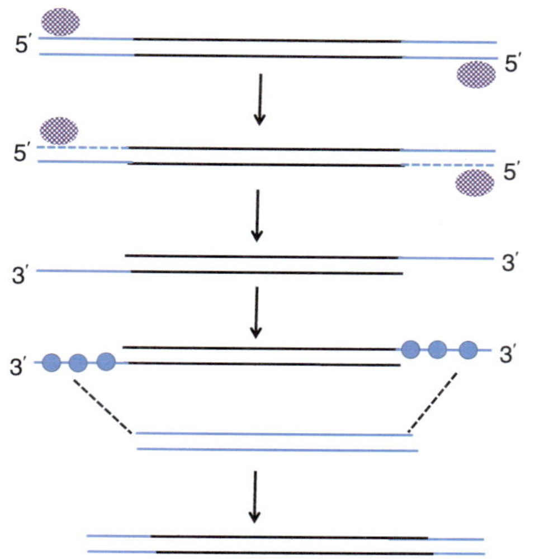

Fig. 2.9 Recombineering: Double-stranded DNA or an oligo containing short homology ends to the target is transformed to *E. coli* host having λ Red system. Exo binds to 5′ end and chews to form a 3′ overhang. Beta binds and protects the ssDNA end. Homologous recombination takes place. The changes are introduced to the target

bacteria and archaea. A novel technique based on a bacterial CRISPR-associated protein-9 nuclease (Cas9) from *Streptococcus pyogenes* has provided a RNA-guided system for genome editing.

CRISPR provides immunity to bacteria against invading hosts. It was initially discovered in *E. coli* in 1980, but its significance and function were appreciated much later when Barrangou et al. demonstrated in *S. thermophilus* that integration of a genome fragment of an infectious virus into its CRISPR locus provided acquired resistance against a bacteriophage [1].

There are three types of CRISPR mechanism, out of which type II has been studied in details. CRISPR locus consists of short 20 bp repeats. The invading DNA from virus and bacteria are cut into small fragments and are integrated into the CRISPR locus. They are transcribed and the primary transcripts are processed to form small crRNA or CRISPR RNA with the help of tracrRNA or trRNA that is complementary to crRNA. The matured crRNA-trRNA guides Cas9 nuclease to the invading DNA. They bind to the DNA sequence specifically and destroy them by cutting them down (Fig. 2.10). The double-stranded endonuclease activity of Cas9 also requires that the 3′ of the crRNA complementary sequence is followed by a short conserved sequence, (2–5 nts) known as protospacer associated motif (PAM). In type II system, only one protein, Cas9, is responsible for both crRNA generation and target DNA destruction. This property has been used by the scientists to develop CRISPR/Cas9 genome editing system.

CRISPR/Cas9 was developed as a genetic tool for editing sequences by Doudna and Charpentier labs in 2012. They developed a simplified two-component system by combining trRNA and crRNA into a single synthetic single-guide RNA (sgRNA). sgRNA-associated Cas9 was shown to be effective in guiding targeted gene alterations. Wild-type Cas9 can cleave double-stranded DNA site specifically resulting in the activation of the double-strand break (DSB) repair machinery. DSBs are repaired by the cellular nonhomologous end joining (NHEJ) pathway resulting in insertions and/or deletions (indels) which disrupt the targeted locus. Alternatively, if a donor template with homology to the targeted locus is supplied, the DSB may be repaired by the homology-directed repair (HDR) pathway allowing for precise replacement mutations to be made.

Several variations of Cas9 system have been developed [5]. A nuclease-deficient Cas9 (dCas9) harbors a mutation that inactivates cleavage activity, but retains DNA-binding activity. This variant is used to sequence-specifically target any region of the genome without cleavage. By fusing various effector domains, dCas9 can be used either as a gene silencing or activation tool.

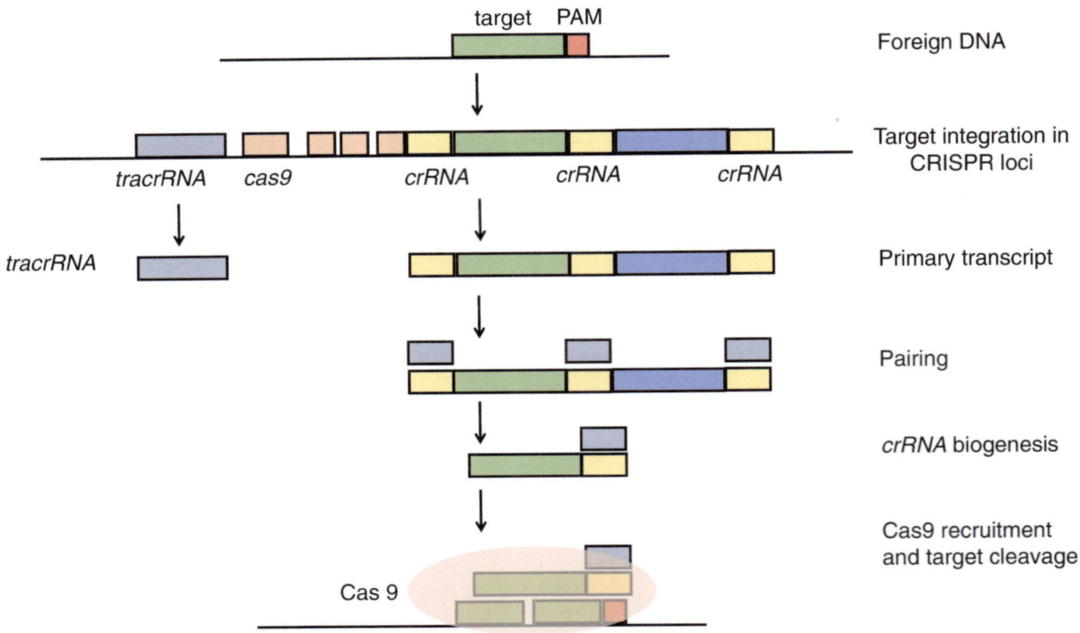

Fig. 2.10 CRISPR/Cas9 system in bacterial immunity. The invading foreign DNA integrates to the CRISPR loci of bacterial host and is expressed as primary transcript. *trcrRNA* is transcribed and pairs with *crRNA* region of the primary transcript. *crRNA* is formed after processing of primary transcript and contains A hybrid molecule with target-crRNA-trcrRNA. The crRNA recognizes the target and guides the Cas9 nuclease. The target is cleaved

Furthermore, it can be used as a visualization tool when fused to GFP protein (Fig. 2.11).

The targeting efficiency of this system is very high and it does not require drug marker for selection. This has been used to modulate zebra fish, frogs, mouse, human cell line, and plant genome very efficiently [13].

2.12 Gene Expression and Quantitation

2.12.1 Real-Time Quantitative PCR (q-PCR)

Nucleic acid detection and accurate quantitation has become integral in all areas of scientific research. It has a wide range of applications in basic science, biotechnology, medicine, diagnostics, and forensic science. q-PCR is a valuable technique for detection and quantification of DNA and RNA. Unlike endpoint PCR where the amplicon is quantitated and detected at the end of the reaction, in real-time PCR, the amount of DNA is measured after each cycle based on increasing yield of fluorescent signal produced by the accumulation of fluorescent dye. The signal is directly proportion to the number of molecules generated. Data collected in the exponential phase of the reaction yield quantitative information on the starting quantity of the target. The change in fluorescence is measured by a thermal cycler that has fluorescent dye scanning capability. By plotting fluorescence against the cycle number, it generates an amplification plot that represents the accumulation of product over the duration of the entire PCR reaction.

The steps in q-PCR are similar to that of conventional PCR. It includes denaturation of the DNA template, followed by primer annealing and extension. In case of RNA quantitation, a reverse transcription step is done before PCR where the RNA is reverse transcribed to complementary DNA (cDNA). The cDNA is used as a template for q-PCR. After the completion of every cycle, the fluorescence signal is measured and plotted

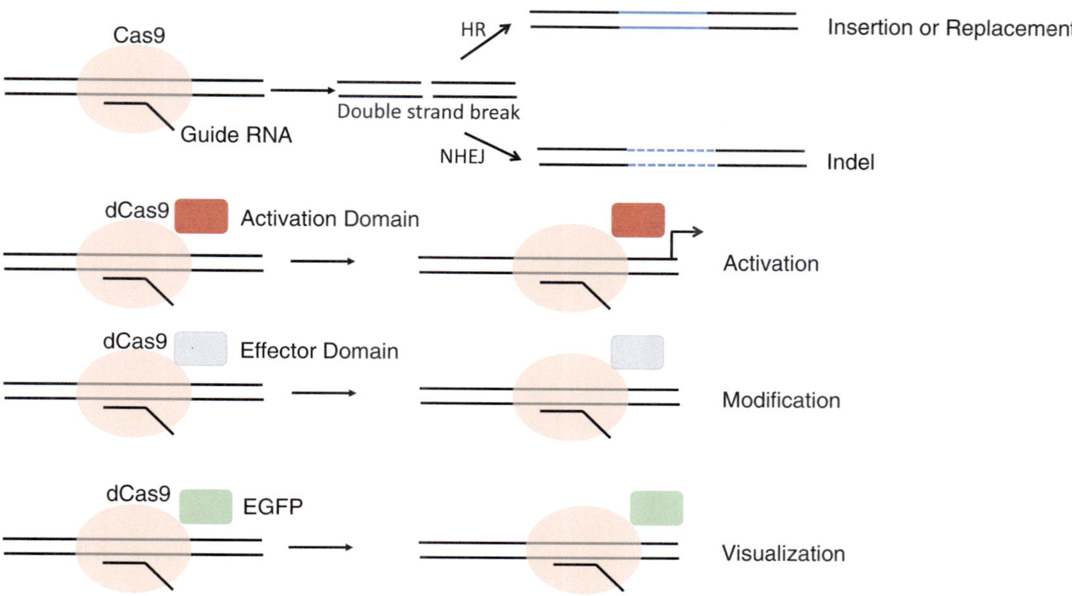

Fig. 2.11 CRISPR/Cas9 system for genome editing. A synthetic oligo with fused crRNA and trRNA is used as a guide RNA for genome modification. The guide RNA localizes Cas9 nucleases to the target sequence. This is used for construction of nuclease-mediated insertion, replacement, and deletion mutations. Mutated dCas9 can bind to DNA sequence specifically but is unable to cleave it. This is used for gene activation or silencing studies when tagged with other proteins. When tagged with GFP, it can be used for visualization

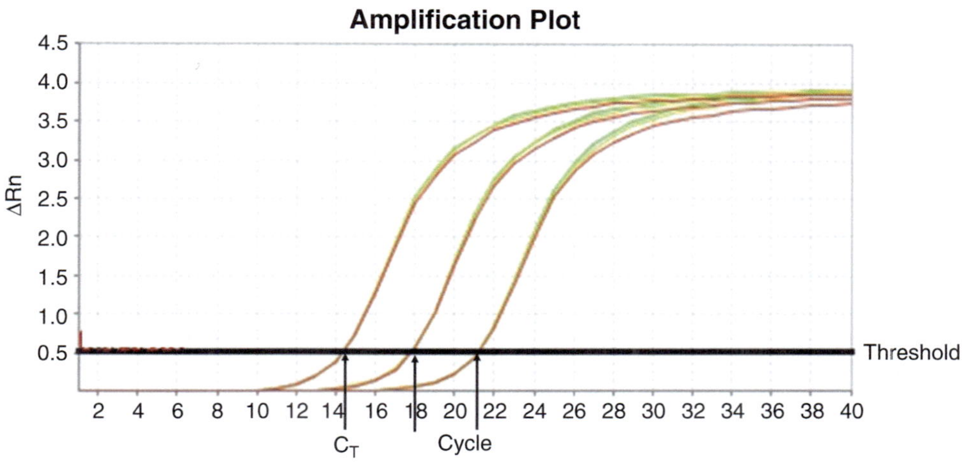

Fig. 2.12 Amplification plot of q-PCR. Fluorescence signal is plotted against cycle numbers. The threshold cycle (C_T) is calculated as the cycle at which the fluorescence signal reaches the threshold. C_T value is inversely proportional to the template amount

against the cycle number to provide an amplification curve (Fig. 2.12). Amplification curve has two phases: exponential and non-exponential phase. In exponential phase, the PCR product doubles after every cycle. Hence the fluorescence signal increases exponentially. The cycle number at which the fluorescence signal reaches detectable limit is called as threshold cycle or C_T. Since C_T is calculated at exponential phase when the reagents are not limiting, it can be used to calcu-

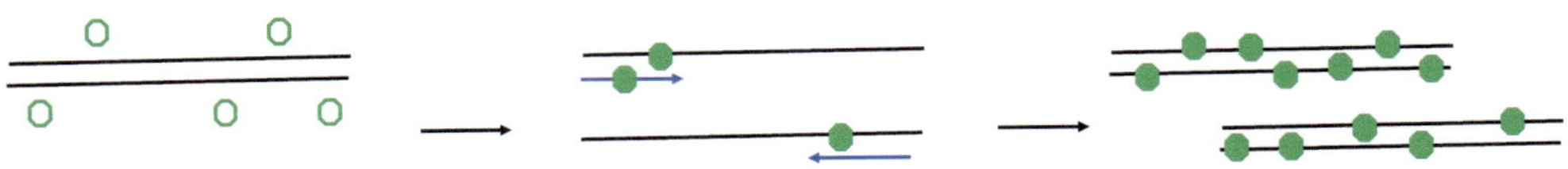

Fig. 2.13 q-PCR with SYBR Green. Unbound SYBR Green produces little fluorescence. During the course of PCR when double-stranded DNA is formed, SYBR Green binds to it and produces fluorescence. After every cycle, the signal is captured and plotted to give an amplification curve. The signal produced is related to the initial template concentration

late the initial amount of template accurately and reliably. C_T value is inversely related to the starting quantity of the template. The more the amount of the template, the earlier it will reach the C_T.

q-PCR can be used for exact quantitation or relative quantitation of a target. For exact quantitation, it is important to create a standard curve with known amount of the target. This can be done by performing q-PCR reaction with serial dilution of the purified target. Log of known concentration of the target template is plotted against the corresponding C_T value to prepare the standard curve. This is used for determining the exact quantity of the target template in the experimental sample and also to calculate the efficiency of the reaction. Most of the time, relative quantitation is done where the expression of the target template is calculated as fold changes with regard to an internal control that could either be a single-copy gene or a housekeeping gene that is expressed steadily in all tissue in every condition.

There are two types of PCR chemistry that is widely used for q-PCR: SYBR® Green dye-based assay and 5′ nuclease assay or TaqMan® Assay.

2.12.1.1 SYBR® Green Assay

SYBR® Green is a fluorescent dye that binds to the minor groove of any double-stranded DNA. The DNA-bound dye produces strong fluorescent signal when excited. This assay involves a pair of primers and follows conventional PCR protocol. After every cycle, there is an increase in double-stranded product and an increase in fluorescence signal that is recorded (Fig. 2.13). Eventually the signal surpasses the threshold value. SYBR Green can bind to any amplified product and hence has low specificity. The fluorescence signal is generated from both target and nonspecific amplification. Hence it is important to do a dissociation curve analysis after the final extension to determine the specificity of the reaction. Every amplicon has a unique melting temperature depending on sequence and fragment length. This is utilized during dissociation curve or melting curve analysis. Gradual dissociation of the PCR product causes a decrease in fluorescence. The first-order negative derivative of the dissociation curve is plotted as a peak. Usually this peak gives the Tm of the product. The melting temperature (Tm) is defined as the temperature at which half of the double-stranded DNA fragment is dissociated. Tm depends on the length of the DNA molecule and its specific nucleotide sequence. When there is specific product, it will form a sharp peak that will coincide with the calculated Tm of the product. It is necessary to do this specificity assay after every reaction when using SYBR Green.

2.12.1.2 TaqMan Assay or 5′ Nuclease Assay

This assay is based on the 5′ nuclease activity of Taq DNA polymerase that degrades DNA bound to the template downstream of synthesis. This includes two PCR primers and a TaqMan probe. The probe is an oligonucleotide that binds specifically to the sequence that is being amplified. The probe is tagged with two fluorescent dyes, a reporter and a quencher at the two ends. In unbound state, the probe folds in a natural posi-

Fluorophores

The technique of fluorescence labeling was developed and used since the 1980s. The technique involves usage of fluorophore. The fluorophores are chemical groups which absorb energy of specific wavelength and go to excited energy state; after a very short time, they come back to their ground state with emission of energy of longer wavelength. The nucleotide labeling with fluorophores may be done by incorporating 2′ deoxyuridine 5′ triphosphate (modified nucleotide) conjugated with appropriate fluorophore. In indirect labeling, fluorophore may be attached to any one of the highly interacting and affinity molecules as biotin and streptavidin.

The various fluorophores used are:

Fluorophores	Color	Excitation wavelength (nm)	Emission wavelength (nm)
Aminomethylcoumarin (AMCA)	Blue	350	450
4′,6-diamidino-2-phenylindole (DAPI)	Blue	358	461
Fluorescein isothiocyanate (FITC)	Green	492	520
Fluorescein	Green	494	523
Indocarbocyanine (CY3)	Red	550	570
Tetramethylrhodamine isothiocyanate (TRITC)	Red	554	575
Rhodamine	Red	570	590
Texas red	Red	596	620
Indodicarbocyanine (CY5)	Red	650	670

For detection of these fluorophores, either laser beam along with detectors or fluorescence microscopy is used

tion where the reporter and the quencher are as close enough and hence the fluorescence of the reporter is quenched. Once bound to the DNA template, the reporter and the quencher are close enough and hence light is not emitted. When the PCR reaction takes place, the TaqMan cleaves the probe and the reporter is freed. Hence fluorescence light is emitted. Thus only when a specific template is amplified, fluorescence signal is generated (Fig. 2.14). This increases the specificity of the assay.

The main advantage of q-PCR and RT-q-PCR is that by these methods very accurate quantitation of initial template concentration can be assayed over a dynamic range. There is no need to do gel electrophoresis of the PCR product. It is in a closed tube system with less chances of contamination.

2.12.2 Microarray

Microarray is a technique where nucleic acid targets are hybridized to a large array of oligonucleotides or probes immobilized on a solid surface. DNA microarray enables investigators to analyze expression of several genes in a single reaction, identify genomic variation associated with diseases, and allow gene mapping. Typically it uses a solid surface such as a glass slide or a silicon film known as chip. The chip is coated with several thousands of DNA fragments from a particular organism. The genetic identity of each of the probe is known. The complexity of these chips is very high and effectively covers the entire genome of an organism. The chip is hybridized to labeled cDNA or genomic DNA obtained from the same organism. After hybridization and wash-

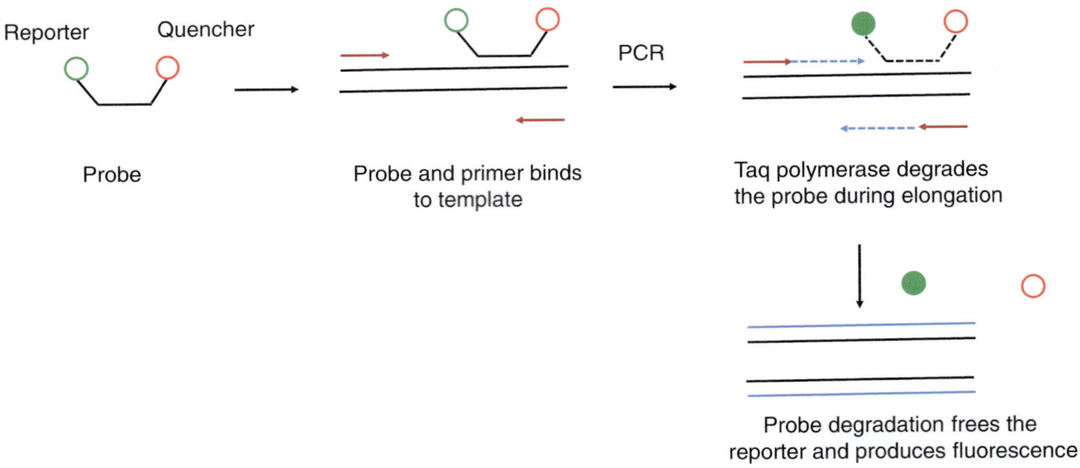

Fig. 2.14 TaqMan q-PCR principle. A gene-specific probe with a reporter and quencher binds along with the primer. Since the reporter is close to the quencher, it does not produce fluorescence. During PCR reaction, Taq DNA polymerase degrades the probe due to 5′ nuclease activity. The released reporter produces fluorescence signal that is captured after each cycle

ing, the chip is analyzed to study the intensity of probe binding. High intensity of a particular probe signifies more expression or greater abundance of a sequence. It enables high-throughput screening and detection. DNA microarray has increasing number of applications:

(a) Expression analysis: In this method, the DNA of an organism is spotted on a chip and hybridized to cDNA obtained from mRNA of the same organism. The intensity signifies the abundance of a particular mRNA in the sample, signifying the expression level of the gene under a given condition. This is the most commonly used technique.
(b) Mutation analysis: The genomic DNA is spotted on a chip and hybridized to DNA isolated from different individuals. Single nucleotide polymorphism can be identified by this method. It is important to analyze variation in a particular gene.
(c) Comparative genomic hybridization: Increase or decrease of chromosomal fragments associated with a disease state. Prenatal chromosomal aberrations can be studied.

2.13 RNA Interference or RNAi

RNAi is a process by which double-stranded RNA (dsRNA) can direct sequence-specific degradation of mRNA, thereby causing gene silencing. This mechanism has been adopted by organisms from plants to mammals as a defense against invading viruses and transposons. It is also used for regulation of gene expression. RNAi mechanism is used to shut down the expression of target genes in virtually all organisms and has recently evolved as a powerful tool to study loss of function as well as in drug development and target validation. Since RNAi targets the mRNA, the gene remains unaltered.

RNAi involves mainly two types of RNA, short interfering RNA (siRNA) and microRNA (miRNA). The double-stranded precursor molecules are cut into small fragments by an enzyme called Dicer. These fragments are about 20 bp long. The double-stranded RNA (dsRNA) is unwound and one of the strands is degraded. The other strand is incorporated to RNA-induced silencing complex or RISC complex. After integrating to the RNA-induced silencing complex (RISC) complex, the siRNA binds to the

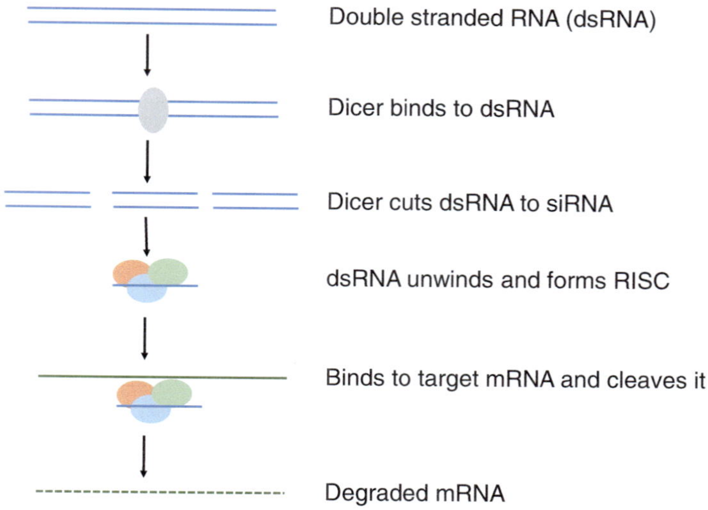

Fig. 2.15 RNAi method for gene silencing. Dicer binds to dsRNA and cleaves it to small fragments. The fragments unwind and form RISC complex. The RISC complex binds to target mRNA and cleaves its sequence specifically

target mRNA sequence specifically and degrades it (Fig. 2.15).

MicroRNAs (miRNAs) are small noncoding RNAs that are about 22 nucleotides long. They mediate posttranscriptional silencing of target genes. miRNAs are complementary to 3′ untranslated region (UTR) of target genes and cause silencing by translational inhibition, degradation of mRNA, or both. They are synthesized in the nucleus as pri-miRNA by RNA polymerase II. Long pri-miRNAs are cleaved by Drosha-DGCR8 to hairpin-like pre-miRNAs. These pre-miRNAs are exported to the cytoplasm by Exportin-5-Ran-GTP complex where they are processed by Dicer enzyme to form short RNA duplex. One strand of the duplex is incorporated in RNA-induced silencing complex (RISC). The miRNA loaded into RISC complex binds to the target mRNA causing repression of translation or mRNA degradation.

RNAi technique is widely used in biotechnology to suppress gene expression. It can target and silence any mRNA. It is used for library screening and is very useful to detect drug targets. This technology has been successfully used to create mutants. They are being developed as new class of drugs.

2.14 Recombinant Protein Expression and Purification

Recombinant protein expression and purification refers to a set of techniques by which a protein of interest is produced in a functional form in sufficient quantity in a host organism by using host protein synthesis machineries. The recombinant protein expressed is purified to study its structure, functions, modifications, localization, and interaction. In practical sense it involves cloning the gene of interest in an expression vector and expressing it in a suitable host that enables purification of the recombinant protein in a functional form in sufficient amount.

2.14.1 Protein Expression Systems

Both prokaryotic and eukaryotic systems are used for protein expression depending on the property of protein, the requirements for functional activity, and yield. The important expression systems have been discussed below. The advantages and disadvantages of each of them have been tabulated in Table 2.1 (for further details, refer to Chap. 4).

Table 2.1 Recombinant protein expression system

Systems	Merits	Problems
Bacterial system	Fast growth rate	Posttranscriptional modification absent
	Simple growth requirements	Posttranslational modification absent
	Easy genetic manipulation	Protein folding options limited
		Codon optimization required
		Formation of inclusion bodies
Yeast system	Fast growth rate	Limited size
	Simple growth requirements	Codon optimization required
	Easy genetic manipulation	
	Posttranscriptional modification	
	Posttranslational modification	
Insect cells with baculovirus expression system	Large size of expression protein	Expensive
	Posttranscriptional modification	
	Posttranslational modification	Needs technical expertise
Cell-free system	Not restricted by host	Expensive
	Expression of toxic proteins	Not suited for large-scale production
Mammalian system	Suitable for mammalian protein	Expensive
	Expression of mAb	

2.14.1.1 Bacterial System

E. coli by far is the choice of organism for protein purification as it has a fast growth rate with a doubling time of 20 min [12]. It can grow to a very high cell density which would increase expression of recombinant protein. The growth requirements are simple and inexpensive. Finally, there are a large number of tools to genetically manipulate *E. coli* and it is easily transformed [3].

The T7 system is the most popular approach for producing proteins in *E. coli*. Here an expression vector containing a gene of interest cloned downstream of the T7 promoter is introduced into a T7 expression host. The host carries a chromosomal copy of the phage T7 RNA polymerase gene under inducible promoter. When inducer is added, T7 RNA polymerase is expressed and becomes dedicated to transcription of the gene of interest. This mechanism has been discussed in details in the promoters section below (Sect. 2.14.2.4).

However, while expressing heterologous genes in *E. coli*, it is important to consider codon usage and optimization. All the mRNA codons are not equally used in *E. coli*. The major codons are those that occur in highly expressed proteins, whereas the minor or rare codons tend to be in genes expressed at a low level. Usually, the frequency of the codon usage reflects the abundance of their cognate t-RNAs. Therefore, when the codon usage of the target protein differs significantly from the average codon usage of the expression host, this could cause problems during expression. The problems due to different codon usages are (1) decrease in mRNA stability causing slowing down of protein expression; (2) premature termination of transcription or translation producing truncated protein; (3) frameshift, deletions, and mis-incorporation of amino acids; and (4) inhibition of protein synthesis and growth of cells. Codon optimization is replacing codons that are rarely found in highly expressed *E. coli* genes with more favorable codons throughout the whole gene.

Despite several advantages, there are certain limitations to protein expression in bacteria. Multi-domain eukaryotic proteins expressed in bacteria often are nonfunctional because the cells are not equipped to accomplish the required posttranslational modifications or molecular folding. Also, many proteins become insoluble as inclusion bodies that are very difficult to recover without harsh denaturants and subsequent cumbersome protein-refolding procedures.

2.14.1.2 Fungal System

Expression of proteins in yeast is a common alternative to prokaryotic and higher eukaryotic expression. Yeast cells offer many of the advantages of producing proteins in microbes such as fast growth rate, easy genetic manipulation, and simple growth conditions. It has some of the properties of higher eukaryotic systems such as posttranslational modifications and secretory expression. Several yeast protein expression systems exist in organisms such as *Saccharomyces*, *Pichia*, *Kluyveromyces*, *Hansenula*, and *Yarrowia*. The yeast expression vectors contain yeast promoter, terminator, and selectable marker. Many vectors have a secretory sequence that secretes the expressed protein in the medium. Most of these vectors can be maintained in *E. coli* as well and are known as shuttle vectors. Thus, the gene of interest can be cloned in *E. coli* and expressed in yeast. Some of the yeast vectors can be integrated into the yeast chromosome. They are stably maintained along with the host chromosome.

2.14.1.3 Insect Cells and Baculovirus Expression System

The baculovirus expression vector system (BEVS) is one of the most powerful and versatile eukaryotic expression systems available which has been used to express heterologous genes from different sources in insect cells. *Baculoviruses* (family *Baculoviridae*) belong to a diverse group of large double-stranded DNA viruses that infect different species of insects as their natural hosts. They are highly species specific and do not propagate in any non-invertebrate host. The baculovirus genome is replicated and transcribed in the nuclei of infected host cells where the large baculovirus DNA (between 80 and 200 kb) is packaged into rod-shaped nucleocapsids. Since the size of these nucleocapsids is flexible, recombinant baculovirus particles can accommodate large amounts of foreign DNA.

In BEVS, several nonessential baculovirus genes are replaced by heterologous genes. Since the baculovirus genome is generally too large to easily insert foreign genes, heterologous genes are cloned into transfer vectors. Co-transfection of the transfer vector and baculovirus DNA into insect cell line, *Spodoptera frugiperda* (Sf), allows recombination between homologous sites, transferring the heterologous gene from the vector. Baculovirus infection of Sf cells results in the shutoff of host gene expression allowing for a high rate of recombinant mRNA and protein production. Recombinant proteins can be produced at levels ranging between 0.1 % and 50 % of the total insect cell protein.

There are several advantages to use this system. Functional activities of the recombinant proteins are retained as insect cells are capable of post-transcriptional (splicing) and posttranslational modifications (phosphorylation, glycosylation, and acylation). The size of expressed protein is much larger. It is relatively a simple system.

2.14.1.4 Cell-Free System

Cell-free protein expression is performed without the use of living cells. All the components needed for transcription and translation such as ribosomes, t-RNAs, enzymes, cofactors, and amino acids are provided in vitro. Such solutions are obtained through making a cell lysate from a desired cell type. Cell-free mixtures have been made from both bacterial and eukaryotic cells. However they are not useful for large-scale protein expression. They are suited for a number of applications where the rapid generation of a smaller amount of recombinant protein is desirable. Cell-free expression is suitable for high-throughput screening of truncated proteins for structural or functional studies. It is used to express proteins that are toxic to expression hosts in vivo. It is also used for expression of proteins with modified amino acids, posttranslational modifications, or studies on protein folding.

2.14.1.5 Mammalian System

In order to produce functional mammalian protein, sometimes it is required to express the gene in mammalian cell line. The most widely used host mammalian cells are Chinese hamster ovary (CHO) cells, HEK293T, and mouse myeloma cells, including NS0 and Sp2/0 cells. The vectors used for expression in mammalian cell lines are usually derived from mammalian viruses such as adenovirus, vaccinia virus, retrovirus, and baculovirus. Mammalian cells are currently the main hosts for commercial production of therapeutic proteins, including monoclonal antibody (mAbs). However, it is an expensive procedure and requires technological skills [6].

2.14.2 Promoters

Several promoters are present that have their own advantages and limitations with regards to heterologous protein expression. Often combinations of promoters and regulators are used to obtain a favorable expression level. We will touch upon the commonly used promoter systems for protein expression.

2.14.2.1 P_{lac}/P_{lacUV5}

Most well-studied and commonly used promoter in *E. coli* system is the *lac* promoter. It is the key component of the *lac* operon and is induced by lactose or its non-hydrolysable analog isopropyl β-D-1-thiogalactopyranoside (IPTG). However, it is repressed by the presence of glucose (Fig. 2.16) by means of a process known as carbon catabolite repression. In absence or low level of glucose, cyclic adenosine monophosphate (cAMP) is produced, which is necessary for complete activation of the *lac* operon. To overcome catabolite repression, a mutant *lac* promoter P_{lacUV5} is used that has reduced sensitivity to catabolite repression and is expressed in presence of glucose (see Fig. 2.16). Lac promoter is negatively regulated by LacI which is a suppressor of lac operon. *lacI*^{Q} is a mutation of *lacI* gene that produces very high levels of the *lac* repressor. This provides a tight regulation of *lac* promoter and stops leaky expression in the absence of inducer. P_{lacUV5} in conjunction with *lacI*^{Q} is present in several expression vectors and provides stable expression in the presence of glucose and suppresses leaky expression in the absence of inducer. However, *lac*/*lacUV5* promoters are weak.

2.14.2.2 *tac*/*trc*

The *tac* promoter is a synthetic hybrid promoter. The promoter consists of the −35 region of the *trp* (tryptophan) promoter and the −10 region of the *lac* promoter. This is stronger than *lacUV5* promoter and is used in pMAL series of vectors.

2.14.2.3 T7

T7 promoter system is extremely popular for protein expression and is present in pET vectors (for figures and further details, refer to pET vectors in Sect. 2.14.5.1). It requires T7 RNA polymerase (T7 RNAP) that is expressed by λDE3 prophage present in the host. Expression of T7 RNAP is controlled by P_{lacUV5} promoter and is induced by IPTG. Basal level of T7 RNAP expression is controlled by *lacI*^{Q} and also by the expression of T7

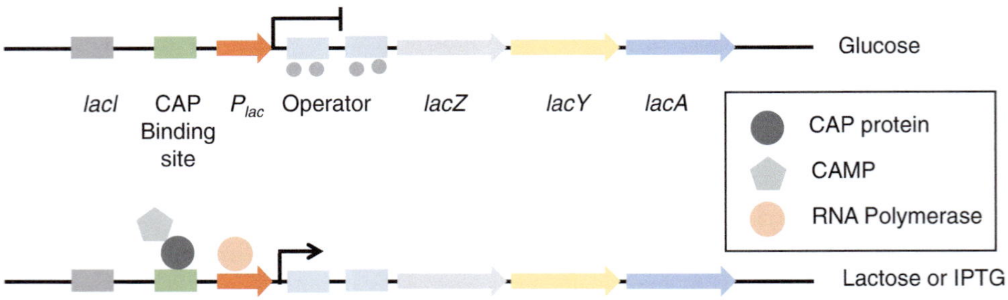

Fig. 2.16 The *lac* operon and its regulation. The *lac* operon consists of *lacZ*, *lacY*, and *lacA* genes under the control of *lac* promoter (P_{lac}) and operator. It is suppressed by lacI that binds to the operator sequences in presence of glucose. In presence of lactose or IPTG, lacI is degraded allowing the transcription of the operon by RNA polymerase. Under inducing condition, there is an increase in cAMP and CAP protein that allows further upregulation of the lac operon. P_{lacUV5} is a variant *lac* promoter that is used in several expression vectors. RNAP is recruited to this promoter more effectively resulting in higher rate of transcription. Further, it works independently of activator proteins and other cis regulatory elements other than the basal promoter. LacI^{Q} is a mutant LacI repressor protein that tightly regulates P_{lac} and P_{lacUV5} promoters

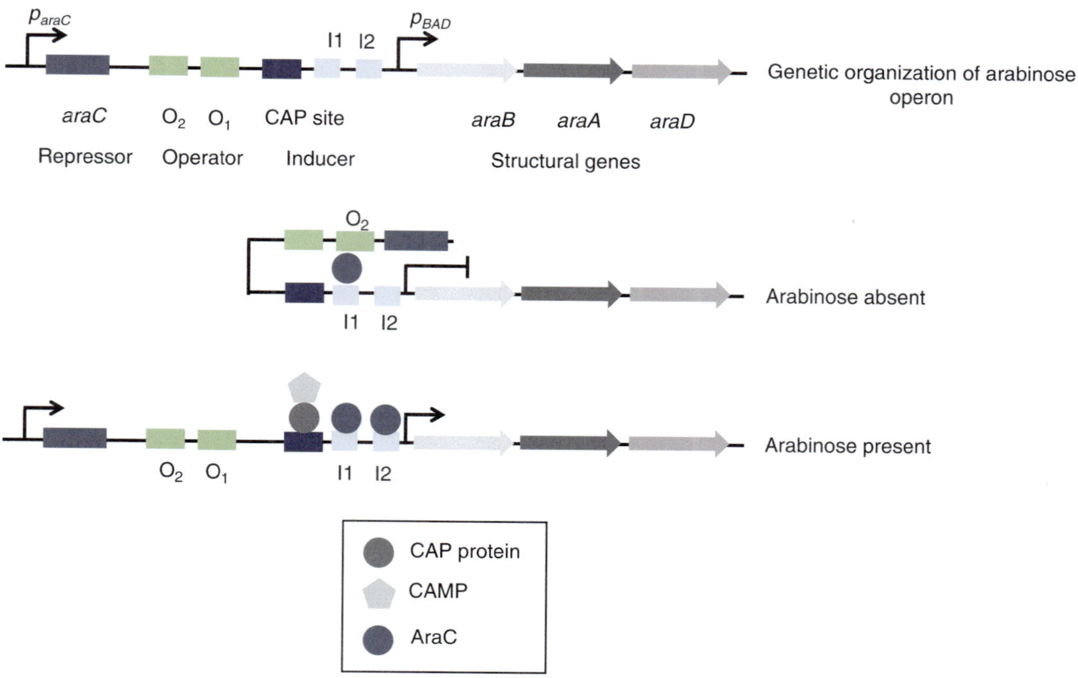

Fig. 2.17 Regulation of the arabinose operon. The arabinose operon consists of three structural genes, *araB*, *araA*, and *araD*. The *araC* gene encodes a transcriptional repressor. In the absence of arabinose, the AraC repressor protein binds the operator site, O2, and the inducer site I1, causing the DNA in this region to adopt a looped conformation preventing transcription. In the presence of arabinose, the AraC repressor changes shape and binds the adjacent inducer sites I1 and I2, causing the loop to be released so that transcription may occur. Binding of the CAP site by cAMP-CAP protein (CRP) leads to full induction of expression

lysozyme expressed from a separate plasmid. T7 lysozyme binds to T7 RNAP and inhibits leaky expression. After induction, excess T7 RNAP is produced that are engaged in transcription of the recombinant gene.

A hybrid T7/*lac* promoter has *lacO* operator downstream of the T7 promoter to control basal level of expression.

2.14.2.4 *ara*p$_{BAD}$

This expression system uses *ara* promoter of arabinose operon and the dual repressor/activator AraC protein (Fig. 2.17). In the absence of arabinose inducer, AraC represses translation by binding to two sites in the bacterial DNA. The protein–DNA complex forms a loop, preventing RNA polymerase from binding to the promoter. Upon addition of the arabinose, AraC switches into "activation mode" and promotes transcription from the *ara* promoter. In this way, arabinose is absolutely needed for induction.

2.14.2.5 pL Promoter

pL of phage lambda is expressed during early lytic phase. It is tightly repressed by the λcI repressor protein, which binds to the operator sequences during lysogenic growth. When the host SOS response is triggered by DNA damage, the expression of the protein RecA is stimulated, which in turn catalyzes the self-cleavage of λcI, allowing transcription of pL-controlled genes (Fig. 2.18). This mechanism is used in expression vectors containing the pL promoter. Addition of nalidixic acid, a DNA gyrase inhibitor, induces SOS response and recombinant protein expression. Alternatively, λcI production can be controlled by *lac* or *trp* promoters and can be induced by IPTG or tryptophan, respectively. A mutant

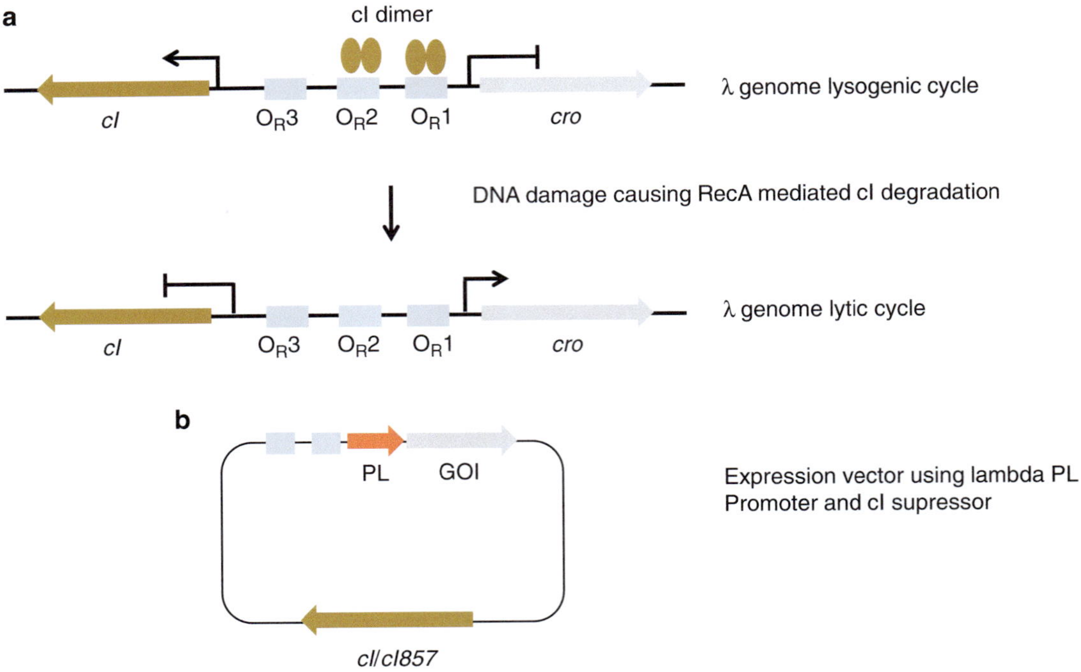

Fig. 2.18 (**a**) Regulation of lambda repressor (cI). The life cycle of λ phage is controlled by cI and Cro proteins. It remains in the lysogenic state when cI proteins predominate, but switches to lytic cycle when Cro proteins predominate. Transcription of the two proteins is regulated by the cI protein and the operator sites. When cI dimer binds to operators O_R1 and O_R2, Cro is repressed. When the host DNA is damaged, RecA protease is expressed that cleaves the cI protein. Cleaved cI proteins cannot bind to the operators. Thus, the Cro proteins are produced that transform the λ phage into the lytic cycle. (**b**) Lambda repressor protein cI and operator sites are frequently used in expression vectors for regulating gene expression. The gene of interest is cloned under the control of phage promoter (PL) and operator sites. The gene encoding cI repressor is expressed from the same construct and represses leaky expression. When induced, the cI is degraded causing the expression of the recombinant protein. λcI857 is temperature-sensitive cI repressor that is unstable at higher temperature. The host strain is transferred to 42 °C for the induction of gene expression. The cI and its derivatives ensure stringent control over protein expression

λcI repressor protein (λcI857) is temperature sensitive and is unstable at temperatures higher than 37 °C. *E. coli* host strains containing the λcI857 protein when shifted to higher temperature acts as an inducer.

2.14.3 Selection Markers

They are usually antibiotic-resistant genes that stop the growth of *E. coli* that do not contain the plasmid. Resistance to ampicillin is conferred by the *bla* gene whose product β-lactamase is a periplasmic enzyme that inactivates the β-lactam ring of β-lactam antibiotics. However, it degrades the ampicillin after sometime and then the cells lacking the plasmids can also grow in the medium. Tetracycline has been shown to be highly stable during cultivation, because resistance is based on active efflux of the antibiotic from resistant cells.

Antibiotics are expensive and are major causes of development and spread of antibiotic resistance. Hence alternative approaches such as plasmid addiction phenomenon are being used. Here an essential gene is supplied by the plasmid that is lacking in the host. Hence the host that loses the plasmid is unable to survive. Different subtypes of plasmid addiction systems exist such as toxin-/antitoxin-based systems, metabolism-based systems, and operator–repressor titration systems. For example, dihydrofolate reductase

(DHFR) or glutamine synthase (GS) gene synthesizes essential metabolite. In medium lacking the metabolites, the transformed cells have selective advantage due to presence of DHFR or GS genes resulting in their growth.

2.14.4 Affinity Tags and Affinity Purification

Affinity tags allow means for easy detection and purification of recombinant protein from *E. coli*. These are a stretch of amino acids (peptide tag) or a large polypeptide (fusion partner) that are expressed *in tandem* with the desired protein to form a chimeric protein. Tags are used to increase solubility of the protein as well. Small peptide tags can form N-terminal or C-terminal fusions and generally do not interfere with the functioning of the proteins. They are purified by using affinity columns where the tagged protein binds specifically and are later eluted. The common peptide tags are the poly-Arg, FLAG, poly-His, c-Myc, S, and Strep II tags. Commercial antibodies are available for their detection by Western blot analysis.

The 6xHis affinity tag is commonly used for protein purification [11]. It is small, less immunogenic, and uncharged at pH 8.0. It does not generally affect secretion, compartmentalization, or folding of the fusion protein within the cell. In most cases, the 6xHis tag does not interfere with the structure or function of the purified protein as demonstrated for a wide variety of proteins, including enzymes, transcription factors, and vaccines. It allows the immobilization of the protein on metal-chelating surfaces and simplifies many types of protein interaction studies. Anti-His antibodies are available for detection by Western blot or ELISA. C- or N-terminus fusion with the protein of interest facilitates binding to Ni^{2+} or Co^{2+} columns.

Non-peptide fusion partners increase solubility of the protein of interest by working as a chaperon. The most popular fusion tags are the maltose-binding protein (MBP), N-utilization substance protein A (NusA), and glutathione *S*-transferase (GST). However the fusion partners are required to be cleaved off eventually. Hence a cleavage recognition site is usually incorporated in the vector. In the case of tag removal by enzyme digestion, expression vectors possess sequences that encode for protease cleavage sites downstream of the gene coding for the tag. Some of the commonly used cleavage sites are enterokinase, thrombin, factor Xa, and the tobacco etch virus (TEV) protease that have all been successfully used for the removal of peptide tags and fusion partners.

Fusion proteins with specific affinity tags simplify their purification by employing affinity chromatography methods. Immobilized-metal affinity chromatography (IMAC) was first used to purify proteins in 1975 by Porath et al. using the chelating ligand iminodiacetic acid (IDA). IDA was charged with metal ions such as Zn^{2+}, Cu^{2+}, or Ni^{2+} and was used to purify a variety of different proteins and peptides. However, IDA has only three metal-chelating sites and cannot bind metal ions tightly. This results in low yields, impure products, and metal-ion contamination of isolated proteins. Nitrilotriacetic acid (NTA) column developed by QIAGEN for His-tagged protein purification binds metal ions more stably and retains the ions under a wide variety of conditions, especially under stringent wash conditions. NTA matrices can bind 6xHis-tagged proteins more tightly than IDA matrices, allowing one-step purification of proteins (Fig. 2.19). The basic steps of IMAC have been explained in Fig. 2.20. The His-tagged fusion protein is expressed in *E. coli*. The cells are lysed to obtain the protein either in native condition in the presence of phosphate or tris buffer or in denaturing condition in the presence of urea or guanidine HCl. The lysate is loaded on to the column for binding. The imidazole rings in the histidine residues of the 6xHis tag bind to the nickel ions immobilized by the NTA groups on the matrix. Subsequently, the column is washed several times with wash buffer containing low concentration of imidazole to remove impurities and nonspecific protein binding. Imidazole binds to the nickel ions and disrupts the binding of histidine residues in non-tagged background proteins. At low imidazole concentrations, nonspecific, low-

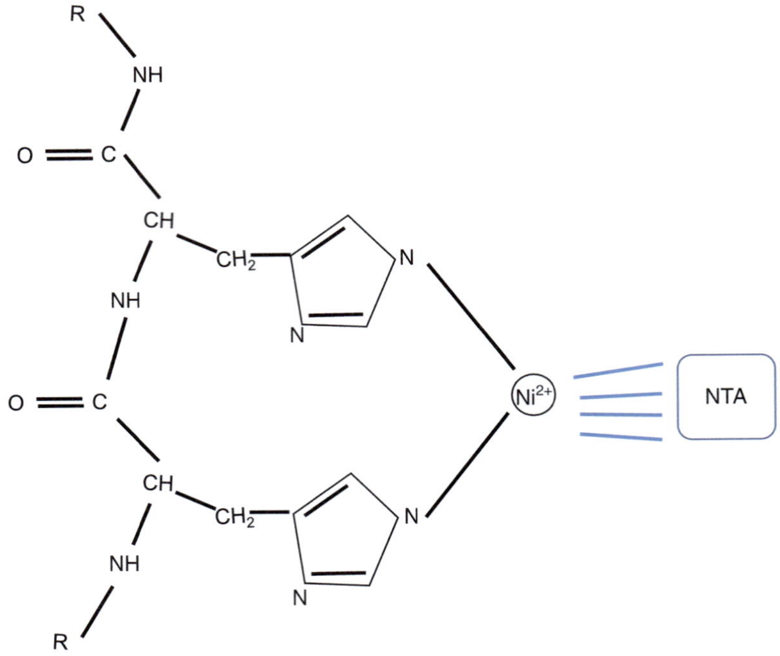

Fig. 2.19 Interaction of His molecules with Ni^{2+} attached to NTA matrix

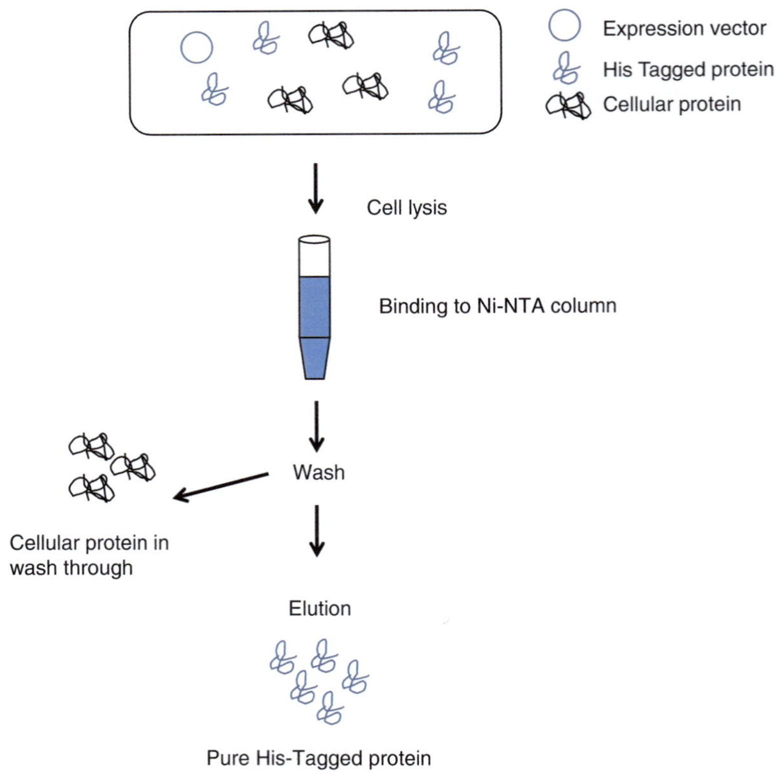

Fig. 2.20 Affinity purification of His-tagged protein using Ni-NTA column. The expression vector is induced in the host. Recombinant protein is expressed. The cells are lysed and the lysate is loaded on to Ni-NTA column for binding. The column is washed to remove cellular protein and impurities. The tagged protein is eluted in presence of high concentration of imidazole.

affinity binding of background proteins is prevented, while 6xHis-tagged proteins still bind strongly to the matrix. Therefore, adding imidazole to the lysis buffer and wash buffer leads to greater purity. The protein of interest is eluted using high concentration of imidazole in the elution buffer, which has structural resemblance with histidine, and thus aids in elution of His-tagged proteins (Fig. 2.21). 6xHis-tagged proteins dissociate since they can no longer compete for binding sites on the Ni-NTA resin.

2.14.5 Expression Vectors

Expression vectors have all the necessary elements for optimal gene expression. There are several expression vectors available depending on the host system, type of protein to be expressed, and the nature of the experiment. The important factors to consider while choosing a vector have been tabulated in Table 2.2. We will discuss essential features of *E. coli* expression vectors.

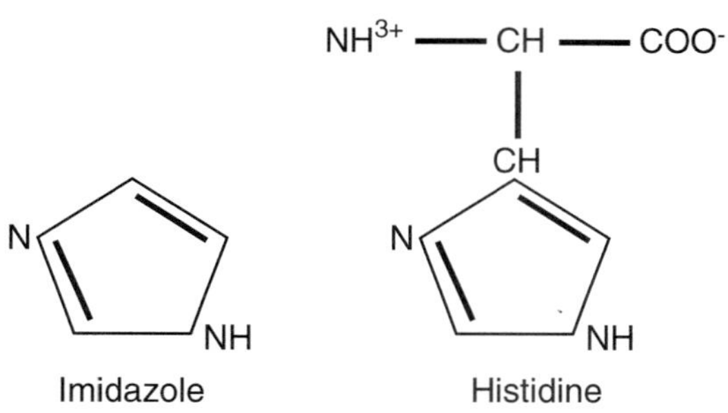

Fig. 2.21 Structure of imidazole and histidine. Due to similarity in structure, imidazole competes with histidine for binding to the matrix. It is used for removal of nonspecific binding and elution of His-tagged proteins during affinity purification

Plasmid copy number of expression vectors is an important consideration while choosing a vector. It refers to number of plasmids that are stably maintained in a host. It depends upon the origin of replication of the plasmid. Usually high copy number plasmids corelate with more expression of the recombinant protein. Commonly used expression vectors of pET series use pMB1 origin and have 15–60 copies per cell. pQE vectors from QIAGEN uses ColE1 origin and has 15–20 copies per cell. pBAD plasmids use p15A origin and are maintained at 10–12 copies per cell. However, if the protein is toxic or imposes metabolic burden on the host, it might cause a reduction in the growth rate and exhibit plasmid instability or loss of plasmids. Hence moderate to low copy number plasmid is preferred for expression. The pSC101 is a low copy number plasmid with less than five copies per cell. It is used when high expression is disastrous for cell growth. When more than one plasmid types are used in a same host cell, they should have different origins of replication; otherwise, plasmid incompatibility would lead to loss of one of the plasmids.

Expression vectors are equipped with strong promoter, regulatory elements, appropriate initiation,

Table 2.2 Expression vector properties

Ori (copy number)	Promoter	Affinity tag	Tag removal	Selection marker
pMB101 (15–60)	*lac*/*lac*UV5	*Peptide affinity tags*	Thrombin	*Antibiotic resistance*
ColE1 (15–20)	*tac*/*trc*	poly-Arg	TEV	Ampicillin
pUC (~500)	T7	poly-His	Factor Xa	Kanamycin
p15A (10–12)	*ara*p_{BAD}	FLAG	Enterokinase	Tetracycline
pSC101 (3–5)		c-Myc		Chloramphenicol
		Fusion partner		*Plasmid addiction system*
		MBP		Toxin–antitoxin
		NusA		Metabolic markers
		GST		Operator–repressor
		Ubiquitine		
		SUMO		

Plasmids

Plasmid ColE1: ColE1 is a plasmid found in bacteria. Its name derives as it carries a gene for colicin E1 (the *cea* gene). It also codes for immunity from this product with the *imm* gene. In addition the plasmid has a series of mobility (*mob*) genes. They are maintained in high copy number in the cell and are used for cloning and recombinant protein expression. Commonly used plasmids derived from ColE1 are pACYC, pUC18, pUC19, pBluescript, pBR322, and derivatives [2].

Plasmid pMB1: pMB1 belongs to ColE1 family of plasmids and is maintained as 15–20 copies per cell. pMB1 origin is used in cloning vector pBR322.

Plasmid p15A: Plasmid pl5A was first detected as one of three plasmids in *Escherichia coli* strain 15T. It is about 2.2 kb in size and one of the smallest naturally occurring plasmids known. It has been used to construct a cloning vector, pACYC184, which is useful because it is readily maintained by bacteria also carrying a ColEl derivative [15].

and termination sites and multiple cloning sites, epitopes for protein detection by Western blot or ELISA, and tags for affinity purification. Some of the commonly used expression vectors are described below.

2.14.5.1 pET

The pET vectors (Fig. 2.22) are developed for the cloning and expression of recombinant proteins in *E. coli*. The target genes cloned in pET plasmids are under the control of strong bacteriophage T7 transcription signal and are expressed only when T7 RNA polymerase is provided by the host cell. The T7 RNA polymerase is very selective and active. Once induced, almost all of the cell's resources are converted to target gene expression and the desired product can comprise more than 50 % of the total cell protein.

The host *E. coli* strains are lysogen of bacteriophage DE3 and carry a DNA fragment containing the *lacI* gene, the P_{lacUV5} promoter, and the gene for T7 RNA polymerase (T7 RNAP). The working of system is explained in Fig. 2.23. This fragment is inserted into the *int* gene, preventing DE3 from integrating into or excising from the chromosome without a helper phage. In a DE3 lysogen T7 RNAP gene is transcribed from the *lacUV5* promoter, which is inducible by isopropyl-b-D-thiogalactopyranoside (IPTG). Addition of IPTG to a growing culture of the lysogen induces T7 RNAP, which in turn transcribes

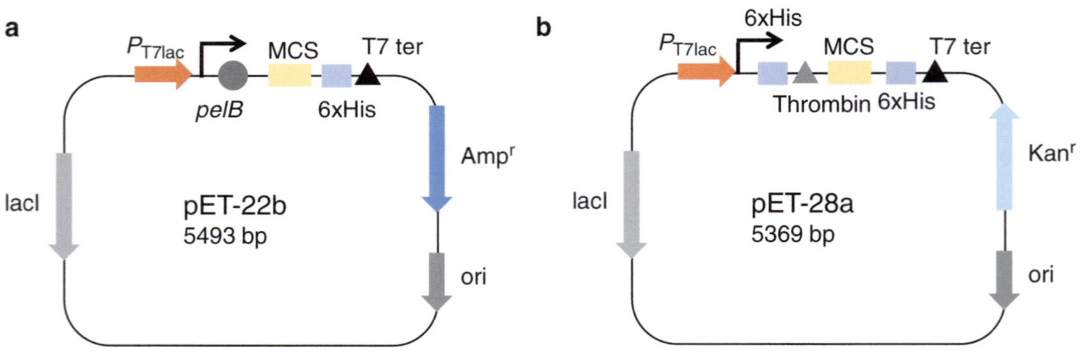

Fig. 2.22 Schematic diagram of pET vectors. Transcription is initiated by T7 promoter, regulated by *lac* operator/lacI, and terminated by T7 terminator. They carry pBR322 origin of replication and multiple cloning sites (MCS). (**a**) pET-22b carries N-terminal *pelB* signal sequence for periplasmic localization of the fusion protein and C-terminal His-tag sequence for purification. This plasmid carries ampicillin resistance marker. (**b**) pET-28a carry N-terminal His tag, thrombin cleavage site, and optional C-terminal His tag. This plasmid carries kanamycin resistance marker

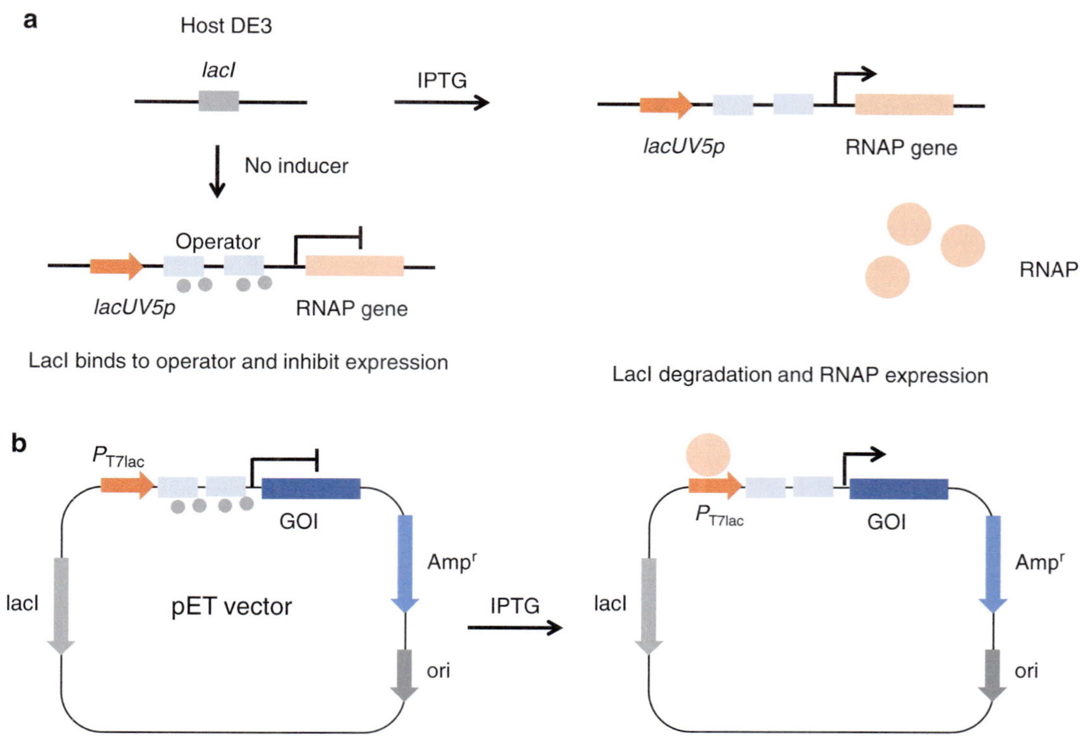

Fig. 2.23 Molecular mechanism of pET vector expression in DE3 lysogen. (**a**) The host cell carries RNAP gene under the control of P_{lacUV5} and *lac* operator. In absence of IPTG, LacI suppressor binds to the operator sequences and inhibits RNAP expression. In presence of IPTG, LacI is degraded and RNAP gene is expressed. (**b**) The RNA polymerase binds to the P_{T7lac} promoter present in the pET vector and induces transcription of the fusion protein P_{T7lac}. It is negatively regulated by LacI suppressor. In absence of inducer, LacI binds to the operator sequences and inhibits leaky expression of the fusion protein

the target DNA cloned in the pET plasmid. In uninduced state, the target genes remain transcriptionally silent. Target genes are initially cloned using hosts that do not contain the T7 RNAP gene, thus eliminating plasmid instability due to the production of proteins potentially toxic to the host cell. Once established, the plasmids are transferred into expression hosts and are induced by the addition of IPTG. Two types of T7 promoter (*T7* and $T7_{lac}$) and several hosts that differ in their stringency of suppressing basal expression levels are available, providing great flexibility and the ability to optimize the expression of a wide variety of target genes. The pET vectors are designed to be able to read the target protein in all the three reading frames and are designated with suffix a, b, and c depending on the translational start site with respect to the multiple cloning site. They have highly efficient ribosome binding site from the phage T7 major capsid protein. There are several vectors in this family which differ in selection markers such as ampicillin and kanamycin resistance. pET vectors also contain different sequences adjacent to the cloning sites that encode a number of peptide "tags," which perform various functions when fused with the target protein. Some of the fusion tags facilitate detection and purification of the target protein such as His tag, S tag, GST tag, and T7 tag. Others increase the probability of biological activity by affecting solubility in the cytoplasm (Trx tag) or export to the periplasm (PelB/OmpT). N- and C-terminal His tag and GST tags are mostly used for affinity purification. They carry protease cleavage sites such as thrombin, enterokinase, and factor Xa peptide cleavage site for removal of tags if required.

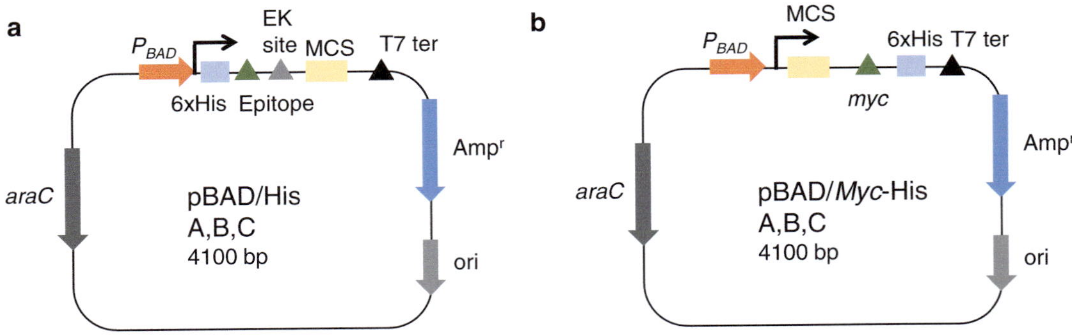

Fig. 2.24 Schematic diagram of pBAD vectors. Transcription is initiated by P_{BAD} promoter, regulated by AraC, and terminated by T7 terminator. They carry pBR322 origin of replication, multiple cloning sites (MCS), and ampicillin resistance marker. (**a**) pBAD/His carries N-terminal His-tag, epitope sequence followed by enterokinase recognition site. (**b**) pBAD/*Myc*-His carries myc epitope

2.14.5.2 pBAD

The pBAD expression vectors (Fig. 2.24) are derived from low copy pBR322 plasmid and are designed for regulated, dose-dependent recombinant protein expression and purification in *E. coli*. They utilize *ara*BAD promoter (PBAD) from *E. coli* for optimum levels of soluble, recombinant protein expression. PBAD is turned on in the presence of L-arabinose and turned off in presence of glucose. Glucose reduces the levels of 3′, 5′-cyclic AMP and cAMP activator protein (CAP) that is required for PBAD expression. By varying the concentration of L-arabinose, protein expression levels can be optimized to ensure maximum expression of soluble protein. The promoter is tightly regulated AraC which is present in pBAD plasmid. AraC forms a complex with arabinose and controls transcription. Hence tight regulation of PBAD by AraC is useful for expression of potentially toxic or essential genes. There are two types of pBAD plasmids, pBAD/His and pBAD/*Myc*-His. pBAD/His plasmid has N-terminal polyhistidine tag for affinity purification and act as epitope for antibody detection. pBAD/*Myc*-His has C-terminal polyhistidine tag and *c-myc* epitope. The epitopes are used for detection of the protein by Western blot using labeled secondary antibodies against them. The tags and epitopes can be removed conveniently by enterokinase digestion due to the presence of enterokinase cleavage site. The pBAD plasmids carry optimized ribosome binding site for increased efficiency of recombinant fusion protein expression, initiation ATG for translational initiation, multiple cloning sites for insertion of gene of interest, *rrn*B transcription termination region for efficient transcription termination, and ampicillin resistance gene (β-lactamase) for selection of the plasmid in *E. coli*. pBAD plasmids are available as A, B, and C for expressing the fusion protein in all three reading frames.

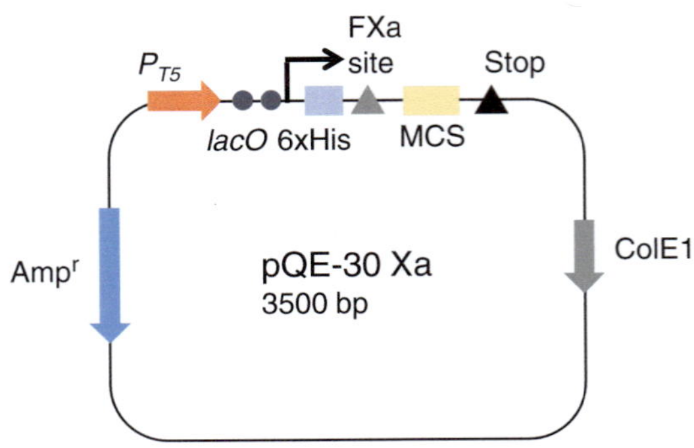

Fig. 2.25 Schematic diagram of pQE vector. It carries phage T5 promoter followed by two *lac* operator sequences for controlled gene expression. It has N-terminal His tag followed by factor Xa protease recognition sequence. It has two strong transcriptional terminators. pQE carries ColE1 origin of replication and ampicillin resistance gene

2.14.5.3 pQE

pQE vectors are shown in Fig. 2.25. The vectors from QIAGEN comprise a family of vectors that is used for expression of 6xHis-tagged recombinant proteins in bacterial, baculovirus, and mammalian expression systems. The pQE vectors contain an optimized promoter–operator element consisting of phage T5 promoter that is recognized by *E. coli* RNA polymerase and two *lac* operator sequences which increase *lac* repressor binding and ensure efficient repression of the powerful T5 promoter when not induced. They carry synthetic ribosomal binding site, RBSII, for high translation rates, 6xHis-tag coding sequence either 5′ or 3′ to the cloning region, multiple cloning site, and translational stop codons in all reading frames. The pQE vectors have two strong transcriptional terminators: *t0* from phage lambda and T1 from the *rrnB* operon of *E. coli*, to prevent read-through transcription and ensure stability of the expression construct. They have β-lactamase gene (*bla*) conferring resistance to ampicillin and ColE1 origin of replication. The 6X His tag can be added to N-terminal or C-terminal end of the protein. The fusion protein tagged with six consecutive histidine residues can bind to nickel-nitrilotriacetic acid (Ni-NTA) metal-affinity chromatography matrices and can be affinity purified.

2.14.5.4 pGEX

pGEX vectors are shown in Fig. 2.26. They are used for inducible, high-level expression and purification of recombinant fusion protein with glutathione S-transferase (GST) gene of *Schistosoma japonicum*. Expression in *E. coli* yields fusion proteins with the GST moiety at the amino terminus and the protein of interest at the carboxyl terminus. Expression is under the control of the *tac* promoter, which is induced by the lactose analog isopropyl β-D thiogalactoside (IPTG). The pGEX vectors carry *lac*Iq gene. The *lac*Iq gene product is a repressor protein that binds to the operator region of the *tac* promoter, preventing expression until induction by IPTG, thus maintaining tight control over expression of the insert. The fusion protein accumulates within the cytoplasm. GST fusion proteins are purified from bacterial lysates by affinity chromatography using immobilized glutathione. Glutathione is attached to Sepharose and the structure of glutathione is complementary to the glutathione S-transferase binding site. GST fusion proteins are captured by the affinity medium, and impurities are removed by washing. Fusion proteins are eluted under mild, non-denaturing conditions using reduced glutathione. The purification process preserves protein antigenicity and function. If desired, cleavage of the protein from GST can be achieved using a site-specific protease such as thrombin or factor Xa, whose recognition sequence is located immediately upstream from the multiple cloning site on the pGEX plasmid. The pGEX vector series consists of pGEX-P, pGEX-T, and pGEX-X that provide all three translational reading frames.

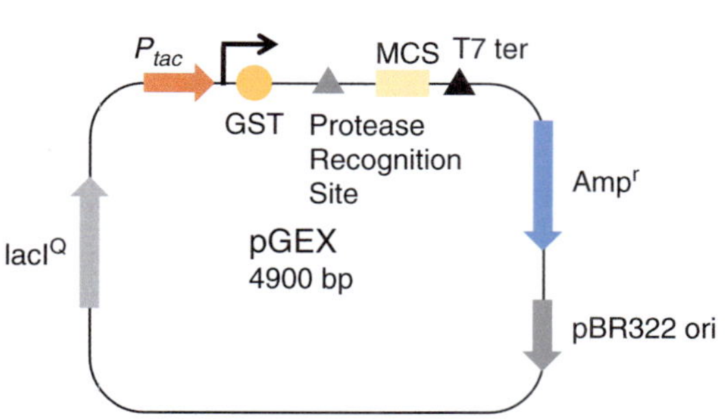

Fig. 2.26 Schematic diagram of pGEX vector. It carries P_{tac} promoter followed by glutathione S-transferase and protease recognition sequence. N-terminal GST fusion protein is expressed. pGEX carries pBR322 origin of replication and ampicillin resistance gene

2.15 Chapter End Summary

- Molecular cloning refers to a set of experiments that is used to construct a recombinant DNA molecule that can be replicate it in a host. It includes amplification of the gene of interest by PCR, restriction digestion of the insert and the vector, ligation of the fragments, and transformation into an appropriate host. Several kinds of cloning vectors are available depending upon the fragment size and the need of the experiment. They are plasmid, bacteriophage, cosmid, BAC, YAC, and HAC. Bacterial host and plasmid systems are by far the most popular. The bacterial lac operon is modified and used for selection of clones carrying insert by alpha complementation and blue-white selection.

- Genomic DNA and cDNA libraries are collection of clones carrying genomic DNA fragments or cDNA from organisms.
- Site-directed mutagenesis is a process to insert point mutation, deletion, and insertion in a DNA fragment that is cloned into a vector.
- DNA sequencing enables determination of the exact nucleotide sequence of a DNA fragment. Next-generation sequencing has widened the scope of sequencing for various applications such as genomic, proteomic, and expression studies. It has also reduced the time required for whole-genome sequencing.
- Targeted genome editing produces specifically designed changes in the genome of organisms and cell line. Recombineering uses the lambda recombinase system to induce changes in the genome. Recently CRISPR/Cas9 system is being used to produce changes in the genome. These methods do not depend on the availability of restriction sites and has wide applications.
- Real-time PCR enables detection and quantitation of DNA and RNA in a sample. It can accurately measure the initial template quantity. It is used to determine relative expression of a gene with respect to a housekeeping gene. There are two types of PCR chemistry that is widely used for q-PCR: SYBR® Green dye-based assay and 5′ nuclease assay or TaqMan® Assay.
- Microarray is a study of expression profile of an organism
- RNAi is a process by which double-stranded RNA (dsRNA) can direct sequence-specific degradation of mRNA, thereby causing gene silencing. It can be done via miRNA or shRNA-mediated silencing.
- Recombinant protein expression and purification refers to a set of techniques by which a protein of interest is produced in a functional form in sufficient quantity in a host organism by using host protein synthesis machineries. There are several systems for protein expression such as bacterial, yeast, insect cell, and baculovirus system; cell-free system; and mammalian system.

Multiple Choice Questions

1. Molecular cloning requires
   (a) Competent host cells
   (b) Vectors
   (c) Restriction endonucleases
   (d) All of the above
2. Restriction Endonucleases are
   (a) Endogenous bacterial enzymes
   (b) Produced by plants for self-defense
   (c) Man-made enzymes
   (d) Nutrients for bacteria
3. Plasmids are used to clone fragments
   (a) That are 35–50 kb in length
   (b) Less than 15 kb
   (c) 100 kb
   (d) 350 Mb
4. Blue-white selection is used to
   (a) Identify plasmids
   (b) Select plasmids with inserts
   (c) Propagate plasmids
   (d) Provide resistance
5. cDNA library contains
   (a) Reverse-transcribed RNA
   (b) Genomic DNA fragments
   (c) Exons
   (d) 3′ UTRs
6. Sanger's method of sequencing is based on
   (a) Incorporation of modified nucleotides
   (b) DNA synthesis
   (c) Unavailability of 3′ OH
   (d) All of the above
7. Pyrosequencing requires
   (a) Dideoxynucleotides
   (b) Cloning of DNA fragments
   (c) Chemical modification of pyrimidines and purines
   (d) Oligonucleotide adapters
8. Nonhomologous end joining
   (a) Requires template
   (b) Takes place during G2 or S phase
   (c) Error prone
   (d) Involves strand invasion
9. Recombineering is a method of gene editing based on
   (a) Bacterial recombination system
   (b) Bacteriophage recombination system
   (c) Bacterial DNA repair system
   (d) Bacterial replication system

10. CRISPR is
    (a) A RNA-guided system
    (b) Bacterial immunity system
    (c) Involves nuclease-mediated DNA restriction
    (d) All of the above
11. Q-PCR is a method where
    (a) Amount of DNA and RNA can be quantitated accurately
    (b) We can clone small amount of DNA
    (c) Only mRNA can be used as template
    (d) Product is quantitated at the end of the reaction
12. TaqMan Assay
    (a) Is very specific
    (b) Requires dissociation curve
    (c) Less expensive than SYBR Green
    (d) Requires less template
13. RNAi is process whereby the expression of a gene is suppressed by
    (a) Genetic modification
    (b) Epigenetic modification
    (c) Posttranslational modification
    (d) mRNA degradation
14. Dr. X needs to express and purify monoclonal antibody in large scale. He needs to express it in
    (a) Bacteria
    (b) Yeast
    (c) Insect
    (d) Mammalian cell culture

Answers

1.(d) 2. (a) 3. (b) 4. (b) 5. (a) 6. (d) 7. (d) 8. (c) 9. (b) 10. (d) 11. (a) 12. (a) 13. (d) 14. (d)

Review Questions

Q1. What are the different cloning vectors available and explain their significance in cloning?

Q2. Point out the main differences between Kunkel's method and currently used methods for SDM.

Q3. What are the applications of next-generation sequencing?

Q4. Explain how bacterial CRISPR/Cas9 is being manipulated to produce specific mutations in genome.

Q5. Differentiate between the basic principles of SYBR Green assay and TaqMan Assay of q-PCR.

Q6. Explain the use of RNAi to regulate gene expression.

References

1. Barrangou R et al (2007) CRISPR provides acquired resistance against viruses in prokaryotes. Science 315:1709–1712
2. Bhagwat AS, Person S (1981) Structure and properties of the region of homology between plasmids pMB1 and ColE1. Mol Gen Genet MGG 182:505–507
3. François B (1999) Recombinant protein expression in Escherichia coli. Curr Opin Biotechnol 10:411–421
4. Hollingworth R, Grand RJ (2015) Modulation of DNA damage and repair pathways by human tumour viruses. Viruses 7:2542–2591. doi:10.3390/v7052542
5. Jinek M et al (2012) A programmable dual-RNA–guided DNA endonuclease in adaptive bacterial immunity. Science 337:816–821
6. Kishwar HK (2013) Gene expression in mammalian cells and its applications. Adv Pharm Bull 3:257–263
7. Kouprina N, Earnshaw WC, Masumoto H, Larionov V (2013) A new generation of human artificial chromosomes for functional genomics and gene therapy. Cell Mol Life Sci 70:1135–1148. doi:10.1007/s00018-012-1113-3
8. Lodish H, Berk A, Zipursky SL et al (2000a) Molecular cell biology, 4th edn, Section 7.1, DNA Cloning with Plasmid Vectors. W. H. Freeman, New York, Available from: http://www.ncbi.nlm.nih.gov/books/NBK21498/
9. Lodish H, Berk A, Zipursky SL et al (2000b) Molecular cell biology, 4th edn, Section 7.2, Constructing DNA Libraries with λ Phage and Other Cloning Vectors. W. H. Freeman, New York, Available from: http://www.ncbi.nlm.nih.gov/books/NBK21696/
10. Mardis ER (2008) The impact of next-generation sequencing technology on genetics. Trends Genet 24:131–141
11. Porath J, Carlsson J, Olsson I, Belfrage G (1975) Metal chelate affinity chromatography, a new approach to protein fractionation. Nature 258:598–599
12. Rosano GL, Ceccarelli EA (2014) Recombinant protein expression in *Escherichia coli*: advances and challenges. Front Microbiol 5:1–17
13. Sander JD, Joung JK (2014) CRISPR-Cas9 systems for editing, regulating and targeting genomes. Nat Biotechnol 32:347–355

14. Sanger F, Coulson AR (1975) A rapid method for determining sequences in DNA by primed synthesis with DNA polymerase. J Mol Biol 94:441–448
15. Seizer G, Som T, Itoht T, Tomizawa JI (1983) The origin of replication of plasmid p15A and comparative studies on the nucleotide sequences around the origin of related plasmids. Cell 32:119–129
16. Sharan SK, Thomason LC, Kuznetsov SG, Court DL (2009) Recombineering: a homologous recombination-based method of genetic engineering. Nat Protoc 4:206–223
17. Thayalan V (2002) Recombineering: homologous recombination-mediated DNA engineering in *E. coli* 21. Curr Protoc Mol Biol 106:1.16.1–1.16.39

Related Sources

http://bitesizebio.com/21466/article-series-e-coli-plasmid-origins-of-replication-the-origin/
http://richsingiser.com/4402/Novagen/pET/system/manual.pdf
http://www.454.com/downloads/news-events/how-genome-sequencing-is-done_FINAL.pdf
http://www.bio-rad.com/en-us/applications-technologies/qpcr-real-time-pcr
http://www.britannica.com/science/recombinant-DNA-technology
http://www.gelifesciences.com/file_source/GELS/Service_and_Support/Documents_and_Downloads/Handbooks/pdfs/GST_gene_fusion_system_handbook.pdf
http://www.illumina.com/technology/next-generation-sequencing.html
http://www.ncbi.nlm.nih.gov/probe/docs/techmicroarray/
http://www.premierbiosoft.com/tech_notes/microarray.html
https://en.wikipedia.org/wiki/Blue_white_screen
https://en.wikipedia.org/wiki/Recombinant_DNA
https://tools.thermofisher.com/content/sfs/manuals/pbad_man.pdf
https://www.bdbiosciences.com/documents/Baculovirus_vector_system_manual.pdf
https://www.embl.de/pepcore/pepcore_services/protein_expression/ecoli/optimisation_expression_levels/
https://www.genome.gov/10000207
https://www.genome.gov/10000533
https://www.idtdna.com/pages/docs/educational-resources/dna-sequencing.pdf?sfvrsn=5
https://www.lifetechnologies.com/us/en/home/life-science/pcr/real-time-pcr/qpcr-education.html
https://www.neb.com/applications/cloning-and-synthetic-biology/site-directed-mutagenesis
https://www.neb.com/applications/protein-expression-and-purification/protein-expression-approaches/cell-free-protein-expression
https://www.neb.com/applications/protein-expression-and-purification/protein-expression-approaches/yeast-protein-expression
https://www.neb.com/tools-and-resources/feature-articles/crispr-cas9-and-targeted-genome-editing-a-new-era-in-molecular-biology
https://www.qiagen.com/us/products/genesandpathways/pathwaydetails/?pwid=143
https://www.qiagen.com/us/resources/resourcedetail?id=79ca2f7d-42fe-4d62-8676-4cfa948c9435&lang=en

3 Animal Cell Culture and Cryopreservation

Abstract

Cell culture is a process by which cells are grown under laboratory conditions outside their natural environment. The historical development of methods of cell culture is closely interrelated with tissue and organ culture. Animal cell culture has a long history of over 100 years, although major advancements have been accomplished in the last 30 years. It has become one of the major tools in life sciences. Almost 50 % of the biological products produced today or planned to be produced in the near future are based on animal cell culture. Therefore, there is an increasing interest in developing technologies for cultivation and maintenance of animal cells. Apart from developing new technologies for culturing and manipulating animal cells for producing biologics, researchers are also interested to look into developmental processes using animal cells as a model system. This chapter is designed to serve as a basic introduction to animal cell culture for the students and the laboratory workers who are interested to understand the key concepts and terminologies in this rapidly growing field.

3.1 Introduction

The technique of cell culture started in twentieth twentieth century to study the behavior of animal cells under laboratory conditions (in vitro). They have been closely related with tissue and organ culture. The technology involves selection of suitable cell and its source, isolation of the cell from its natural environment, and transfer to artificially controlled in vitro condition in suitable medium. These cells are standardized to grow in artificial conditions and have tremendous applications in research. Cell culture may be extremely useful in substituting animals for trials of drugs, they can provide model to study human pathogens and they are useful for large-scale culture and production of therapeutic compounds (Fig. 3.1). Today 70 % recombinant pharmaceuticals are produced by using mammalian cell culture. The technology has made possible the usage and preservation of stem cells for therapeutic purposes, future use, for gene therapy and starting material for tissue engineering. This chapter would help readers to get an insight about cell culture technique, their requirement, applications, and storage of cell culture.

V. Gupta et al., *Basic and Applied Aspects of Biotechnology*, DOI 10.1007/978-981-10-0875-7_3

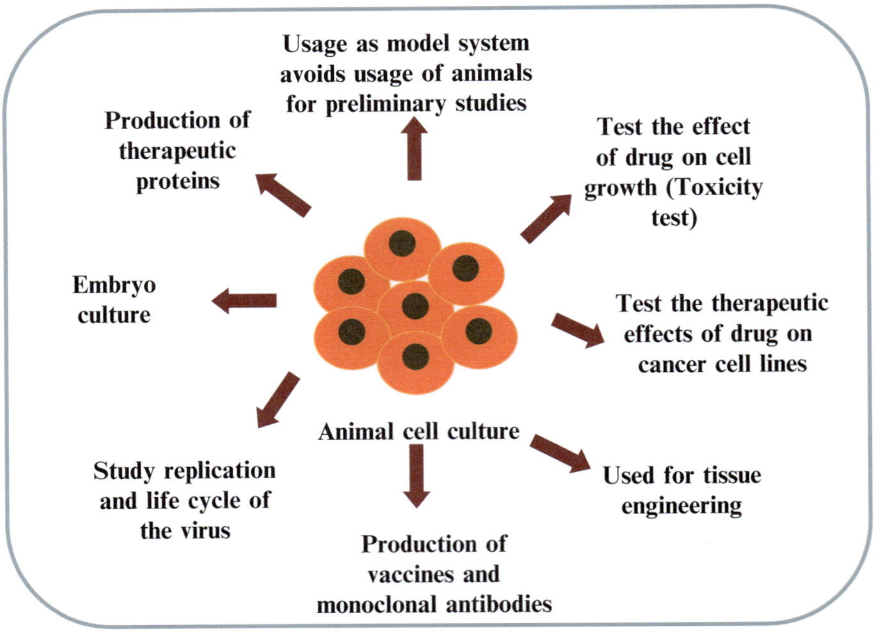

Fig. 3.1 The figure shows the important applications of animal cell culture. The animal cells are cultured for production of components of the media, production of therapeutic proteins, drug testing, and tissue engineering

3.2 Media Preparation and Sterilization

3.2.1 Cell Culture Media

The most important step for animal cell culture is selecting an appropriate growth medium for in vitro cultivation [3]. The selection of media is based on the cell type and the purpose of cell culture. Media used for cell cultivation mostly include two major parts.

1. Essential basal ingredients: This part of the media fulfills basic cellular requirements for nutrients and is known as the basal growth medium.
2. Supplemental ingredients: A set of supplements that satisfies specific cellular growth requirements and enables a particular cell type to grow.

Major components for growing animal cells in laboratory conditions include the following:

1. An energy source
2. Nitrogen source
3. Vitamins
4. Fats, fatty acids, and fat-soluble components
5. Inorganic salts
6. Nucleic acid
7. Antibiotics
8. Oxygen
9. pH-buffering systems
10. Hormones
11. Growth factor serum

There are two types of animal cell culture medium: completely natural media and artificial media with some natural products.

Natural Media: In the early days of animal cell culture, for in vitro cultivation of cells, natural media that were obtained from biological sources were used. But in recent years those are either used for specific purposes like toxicity or sterility testing or they are produced commercially as culture media. Some of the examples of natural media are:

(a) Body fluids such as plasma, serum, lymph, and amniotic fluid
(b) Tissue extracts like those from the liver, spleen, or bone marrow or from chick embryo
(c) Bovine embryo extract

Artificial Media: The basic criterion to choose artificial media is that they should provide all the required nutrients to the cell and maintain physiological pH (pH7) with the help of a buffering system, sterile and isotonic to the cultured cells.

Various artificial media developed for cell cultures may be grouped into following three classes:

(a) Serum-containing media

Serum is the most important component in animal cell culture media. Although the actual components and functions are still unknown, serum is considered to provide nutrients, hormones, and growth factors to the medium. Besides those, it provides the following:

1. Binds essential nutrients that are toxic when present in excessive amounts and releases them slowly in a controlled manner
2. Modulates the physical and chemical properties of the medium (viscosity, rate of diffusion)—protects cells in agitated culture
3. Has a pH-buffering function

Despite these advantages, there are several disadvantages associated with the use of serum:

1. Serum is the most expensive component and the actual composition is still not known. It is difficult to culture the cells that need defined media to grow.
2. There is the possibility of the presence of contaminants, for example, *Mycoplasma* and viruses.

(b) Serum-free media

A number of serum substituents are used in animal cell culture to reduce contamination from serum-borne pathogens such as:

1. Bovine pituitary extract (BPE)
2. Bovine or porcine tissue hydrolysates (peptones)
3. Soybean hydrolysates (peptones)

(c) Chemically defined media

In chemically defined media, every chemical component is precisely defined and quantified. This type of media provides minimum risk of bio-contamination and maximum process. But such media types are available in limited varieties and most cell lines grow very slowly.

3.2.2 Sterilization of Cell Culture Media

Cell culture media must be sterilized before use to prevent the growth of adventitious bacteria or other organisms. Autoclaving is not permitted for sterilization since exposure to high heat degrades the media components. Sterilization is generally accomplished by filtration with 0.2- or 0.1-μm-sterilizing-grade microporous membrane. Filtration (0.1 μm pore size) is the main method for eliminating mycoplasma contamination and gamma-irradiation is used to destroy viruses. Mycoplasma is the generic name assigned to microorganisms in the class molecules. They lack cell wall and are bound together by a trilaminar plasma membrane. Mycoplasma is the smallest bacterium to thrive in cell culture and cannot be retained by 0.2-μm-sterilizing-grade filters, due to their size and deformability. To reduce viruses and prions, pasteurization (short duration) is sometimes used prior to sterile filtration. Pasteurization processes do not cause any substantial changes in media quality and filterability.

Water used for culture media should be pyrogen-free. It is recommended to use fresh ultrapure water and not stored water, since some materials from plastic or glass may leech out and dissolve in the water, thus contaminating it.

Chemicals used for animal cell culture should be of the highest purity. Commercial chemicals, although pure, should be tested and certified for cell culture purpose since they inevitably contain traces of contaminants. Some of the traces may be toxic (like mercury). With regard to stability of media ingredients, inorganic chemicals are indefinitely stable. Vitamins, hormones, antibiotics, and growth factors are less stable and are recommended to be stored frozen (–20 °C) or refrigerated (0–4 °C). Most factors that affect the shelf life of media are the following:

- Natural decay rates of unstable compounds
- pH
- Moisture
- Storage temperature
- Access of oxygen
- Exposure to near-ultraviolet, daylight, or in fluorescence light

Most media should be stored at 4 °C and in a dark place. Freezing of media may cause loss of some purely soluble ingredients. Powdered media are stable for several years at room temperature (25° C).

The basal media (DMEM, medium 199, SGM, nutrient mixture Ham's F14, Basal Medium Eagle) used in the cell culture laboratory are purchased as gamma ray-sterilized solutions and certified as endotoxin-free. They can also be prepared in the laboratory using basic components and filter sterilized using 0.22-μm-pore size filter. The basic media are purchased without L-glutamine for increased stability since glutamine is one of the unstable components of cell culture. L-Glutamine, purchased as a sterile endotoxin-free solution, is added immediately prior to use.

The fetal bovine serum used in cell culture media should be certified as virus (BVD-IBR-PI3) and mycoplasma-tested and, since 1988, also bovine spongiform encephalopathy (BSE)-tested. Antibiotics, growth factors, or other supplements should be purchased as sterile-filtered solutions and with certified endotoxin-free or at a concentration less than 0.1 ng/μg.

Complete growth medium should be prepared under the laminar hood at sterile condition using disposable sterile plastic ware. After mixing all the components, it should be filtered using 500-ml Millipore Stericup (0.22 μm pore size). The cap and neck of the bottle containing the sterilized medium should be protected with Parafilm and stored at 4 °C.

Most complete media are stable at 4 °C for 1–2 months.

American-type culture collection (ATCC) established in 1914 helps in acquisition, authentication, production, preservation, development, and distribution of cell lines. The cell lines are maintained and verified by ATCC since 1962. It is the largest collection and one of the largest bioresources in the world and maintains cell lines of human, insect, fishes, and stem cells. Animal cell lines of about 150 different species and nearly 4,000 human cell lines including models for cancer (HeLa, OVCAR-3, and many more) are being maintained by ATCC.

Coriell Institute for Medical Research (ccr.coriell.org) and ATCC are nonprofit organizations which provide high-quality authentic cell lines.

3.3 Culturable Cells

Although theoretically, cells of any type can be cultured upon procurement at viable state from any organ or tissue, not all types of cells are capable of growing in an artificial environment. This is due to many reasons. Most importantly, the artificial environment fails to mimic the biochemical parameters of source environment. Some examples include the absence of growth regulators or cell to cell signal molecules. A cell line can be established by multiple passaging under optimal maintenance conditions until a pure-culture of specific cell type is obtained. The subculturing is repeatedly used to maintain a cell line. Using these procedures, several cancer cell lines have been established like HeLa, CHO, and so on.

Some Definitions

Primary culture: Tissue is obtained from the organisms, treated with digesting enzymes as trypsin or collagenase, and placed in suitable culture medium where they can divide and grow.

(continued)

Subculturing or passaging: When primary culture reaches the state of confluence (grows and fills up all the empty space), then they are dissociated enzymatically or by gentle scraping and some of them are transferred to fresh medium under sterile condition.

Monolayer culture: Monolayer refers to a layer of cells in which no cell is growing on top of another, but all are growing side by side and often touching each other on the same growth surface.

Suspension culture: It is a type of culture in which intact cells are maintained in suspension in the culture medium so that they are distributed evenly.

Anchorage-dependent cells: It is a cell that grows, survives, or maintains function only when attached to a suitable inert surface, such as glass or plastic.

Transformed cells: Eukaryotic cell line obtained in a quiescent or stationary phase which undergoes conversion to a state of unregulated growth in culture, resembling an in vitro tumor. It occurs spontaneously or through interaction with viruses, oncogenes, radiation, or drugs/chemicals.

Finite cell lines: The cells or cell lines which after a few generations of growth stop dividing and show signs of aging are finite lines.

Continuous cell lines: The cells which can continue to divide indefinitely or are immortal are continuous cell lines.

Neoplastically transformed: These cells are capable of forming tumors when injected into animals.

The cells which are cultured are characterized either on the basis of their morphology or their functional properties.

1. Epithelial-like cells: They are anchorage dependent, flat and polygonal in shape.
2. Lymphoblast-like cells: They are suspended and spherical in shape.
3. Fibroblast-like cells: These are anchorage dependent, elongated and bipolar and form swirls in confluent culture.

Culture conditions largely affect cell size and shape and cells in culture can attain multiple morphologies.

3.4 Development of Cell Lines

Cells for culturing in the media can be isolated from tissues in several ways.

(a) White blood cells can be easily purified from blood.
(b) By enzymatic digestion of soft tissues with enzymes such as collagenase, trypsin, or pronase. These enzymes break down the extracellular matrix and mononuclear cells are released.
(c) In explant culture, a piece of tissue is placed in growth media and the cells that grow out are available for culture.

Figure 3.2 shows the basic steps required for primary cell culture.

3.5 Primary and Established Cell Lines

Cells that are cultured directly from animal specimens are known as primary cells. Except some cell lines that are derived from tumors, most primary cell lines have limited life span. After a certain number of divisions, primary cell lines undergo senescence and stop dividing but they retain the viability.

An established or immortalized cell lines acquire the ability to proliferate indefinitely (Fig. 3.3). Establishment of cell lines can be achieved in several ways:

1. The original method for generating immortalized cell line is isolation from a naturally

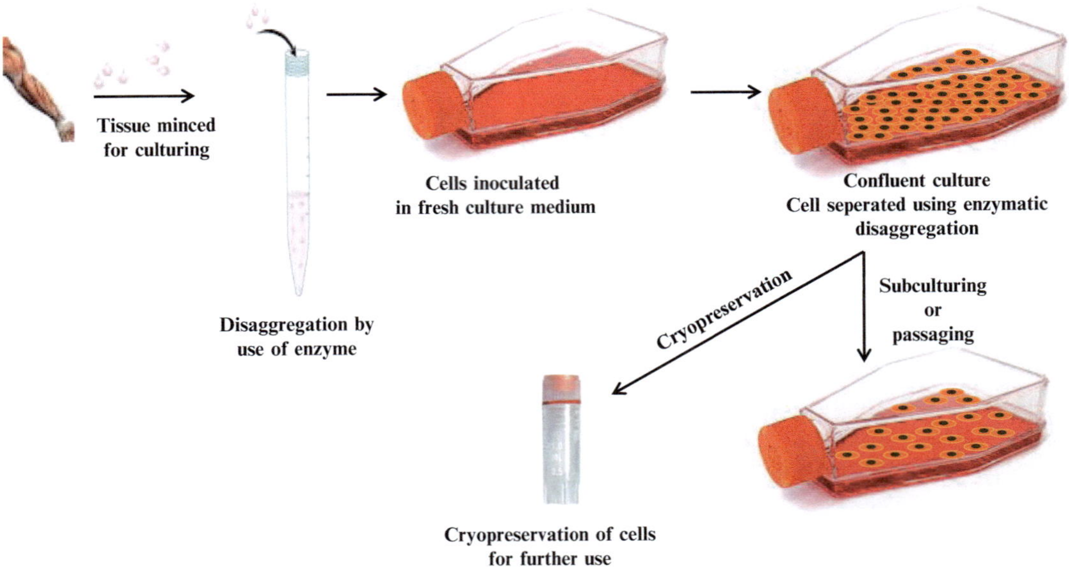

Fig. 3.2 The figure shows the culturing of cells. After confluent culture, the cells are passaged or cryopreserved for further usage

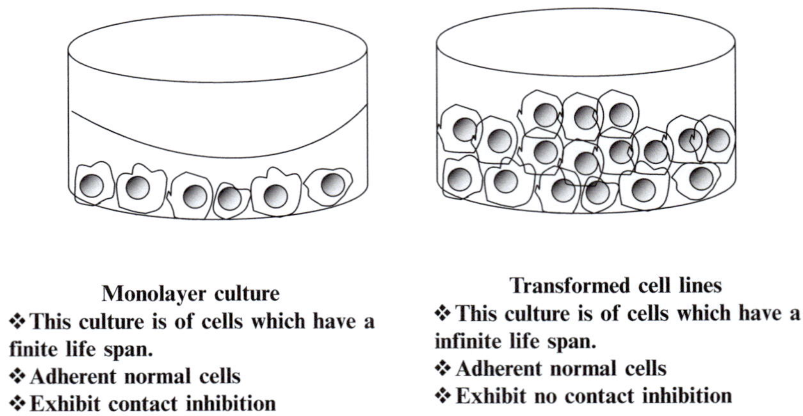

Fig. 3.3 This figure shows the properties of monolayer and transformed cells

occurring cancer. Major example includes human HeLa cells, obtained from a cervical cancer.

2. Random mutagenesis (spontaneous or induced) and selection for cells which are able to undergo division.
3. Introduction of some product into cells that will deregulate the cell cycle. Expression of viral gene (adenovirus E1) or key proteins required for immortality (telomerase) immortalize the cell lines.

3.6 Techniques of Mammalian Cell Culture

There are two types of mammalian cells based on their growth properties in cell culture media: adherent cells and suspension cells. Adherent cells attach on the surface of cell culture flasks and grow as monolayer. Suspension culture cells grow as suspension and they do not attach on the surface of cell culture flasks. Either in suspension or in adherent conditions, as cells grow, they must be subcultured or passaged. Failure to subculture produces confluent cells (in case of adherent cell lines) and results in reduced mitotic index and eventually cell death.

3.6.1 Subculturing (Passaging) of Cells

The first step in subculturing adherent cells is to detach them from the surface of the culture vessel. This can be done by trypsinization or by mechanical means. The resultant cell suspension is then divided and put into fresh culture. Secondary cultures are checked periodically for growth and fed and may subsequently be subcultured. The time to passage cells depends on the cell type and growth of the cells (Fig. 3.2).

Subculturing of suspension culture cells is less complicated than adherent culture as they are already in suspension and there is no need to disperse them enzymatically before passaging. However, before passaging, cells must reach confluency, i.e., they clump together and medium looks turbid when the flask is swirled.

3.6.2 Freezing Cells

It is always desirable to store cell lines for future study. To preserve cell lines for future use, they may be frozen with a cryoprotective agent. Freezing without cryoprotective agent will kill the cells in most cases. Generally dimethylsulfoxide (DMSO) is used as cryoprotective agent in conjunction with complete medium for preserving cells at –70 °C or lower. Gradual freezing reduces the risk of crystal ice formation and cell damage (Fig. 3.2). DMSO reduces freezing point and allows slower cooling rate. Freezing cells from suspension and adherent cultures are basically same except that the suspension cultures do not need detachment of cells (trypsinization).

3.6.3 Thawing and Recovering Cells

When frozen cells are needed for a study, they should be thawed rapidly and plated at high density to optimize recovery. DMSO from the freezing medium must be removed before putting into fresh medium; otherwise it will interfere with the growth of the cells. This can be done by spinning down the cells and removing the freezing medium.

3.6.4 Determining Viable Cell Numbers

It is important to determine the exact number of cells in a culture for standardizing culture conditions and also performing accurate quantitation in experiments. Cell numbers can be counted using a hemocytometer which is a thick glass slide with a central area designed as a counting chamber. The central portion of the slide has two counting platforms divided by a transverse grove. Each counting chamber consist a 3×3-mm grid. Each grid is further divided into nine 1×1-mm secondary squares. The four corner squares and the central square are used to determine cell count. The 4-corner squares are further subdivided into 16 and the central square into 25 tertiary squares. Each slide is accompanied with a thick, even-surfaced coverslip. Cell suspension applied to a defined area is used to count the cell density.

3.6.5 Preparing Cells for Transport

Both monolayer and suspension cultures can be shipped easily either in a tissue culture flask with media or in frozen conditions. For shipping in a flask with media, it is advisable to use fresh media and completely fill it to prevent drying. For shipping frozen cultures, cells are removed from liquid nitrogen and placed immediately on dry ice in an insulated container to prevent thawing during transport. Generally cultures are transported by same-day or overnight courier with a biohazard label outside the package.

3.7 Properties of Cell Lines

Primary cells have definite life span, whereas cell lines are capable of growing for indefinite time provided they have enough nutrients and space. All the cell lines are associated with genetic instability, immortal growth, and malignancy properties. Properties of cell lines can be described as follows:

Growth: immortal, loss of contact inhibition, high plating efficiency, shorter population-doubling time
Genetic: high spontaneous mutation rate, overexpressed or mutated oncogenes, gene and chromosomal translocations, aneuploidy
Structural: modified actin cytoskeleton and extracellular matrix, altered expression of cell adhesion molecules, disruption of cell polarity
Neoplastic: tumorigenic, angiogenic, invasive

3.8 Passaging

Passaging (also known as subculture or splitting cells) involves transferring a small number of cells into a fresh culture. Cell lines can be maintained for indefinite time if they are passaged regularly. This helps cells to get enough nutrients and space for growth and prevents senescence due to higher cell density. Adherent cultures are first detached from the surface using trypsin-EDTA. Passaging should be done when the surface is 80–90 % covered with cells (80–90 % confluent). Trypsin helps to digest the matrix and EDTA chelates calcium and magnesium that makes the matrix weaker. Mechanical scrapping of cells from the surface is also used for detachment of cells. A small number of detached cells can then be used to seed a new culture.

3.9 Measurement of Viability and Cytotoxicity

Viability of cells represents their capability to exist, survive, and develop. Cytotoxicity refers to the metabolic alterations of cells, including cell death due to exposure of some toxic chemicals or environment. To determine cytotoxicity of any chemicals or for screening a drug, cell viability and cytotoxicity assays are used. There are several methods of measuring cell viability and cytotoxicity based on various cell functions like enzyme activity, cell permeability, cell adherence, ATP production, coenzyme production, or nucleotide uptake activity. The outcome of the assay is determined by color change of a substrate, uptake of a radioactive substance, and counting the live and dead cells under the microscope or by their ability to form colonies. Some of those methods are described below.

3.9.1 Assays Based on Membrane Integrity

Most common method for determining cell viability is to determine membrane integrity. Due to cell disaggregation, cell separation, or freezing and thawing, cell membrane gets damaged. This can be measured by uptake of dyes, retention of dyes inside the cells, or release of radiolabeled chromium from cells. These methods are rapid but they can't distinguish the healthy cells from the cells that are losing functions but still alive.

3.9.1.1 Dye Exclusion Assay

The basis of this assay is that the viable cells are impermeable to several dyes such as trypan blue, eosin Y, nigrosin blue, erythrosine B, etc. Live cell membrane is not permeable to those dyes and is not stained. It stains only the dead cells. After staining, stained cells are counted using hemocytometer [6].

3.9.1.2 Dye Uptake Assay

The viable cells take up dyes and due to enzymatic activities they are converted to a substance that is impermeable to cell membrane. Dead cells are unable to convert them into cell impermeable substance. For example, diacetyl fluorescein is converted to fluorescein by hydrolysis inside the live cells and then the live cells emit fluorescent green color. This can be easily detected by fluorescent microscope or measured using a fluorimeter [1].

3.9.1.3 Chromium Release Assay

Radioisotope of chromium (^{51} Cr) binds to intracellular proteins as it binds to basic amino acids. Due to cell membrane damage, the labeled proteins leak out of the cell and the leakage is proportional to cell damage. This is a very sensitive method however; handling, storage, and disposal of radioisotopes are problematic and need special permission.

3.9.1.4 Enzyme Release Assay

When cell membranes are damaged, cellular enzymes are released from the cells. This can be assayed by measuring the activities of the released enzymes. Lactate dehydrogenase (LDH) assay is the most commonly used assay as this enzyme is stable during cell death and provides more reliable data.

3.9.2 Assays Based on Radioisotope Incorporation

By using radiolabeled substrates or metabolites that are incorporated into cell during cell growth and division, the level of incorporation is measured using scintillation counter. This method is used to determine drug toxicity. It is very sensitive but requires radioisotopes which cause various safety concerns.

3.9.2.1 Incorporation of Labeled Nucleotides

Tritium-labeled (^{3}H) thymidine or uridine is used in this method. Labeled ^{3}H- thymidine or ^{3}H-uridine are incorporated into cell nucleus during cell growth due to their incorporation into newly synthesized DNA and RNA. The amount of incorporated tritium can be measured using scintillation counter.

3.9.2.2 Release of Labeled Phosphates

Cells are labeled with ^{32}P-phosphates. Due to cell damage, they release labeled phosphates that can be measured.

3.9.3 Colorimetric Assays

These methods are based on the production of colored substance by live cells' activities. There is a good correlation between cell numbers and color production. This method is far superior to previously described methods as it is easy to use and safe and has higher reproducibility. Protein content, DNA, lysosomal activity, Golgi body activity, or other enzyme activities are used to determine cell number. One example is XTT assay in which conversion of the water-soluble XTT(2,3-bis-(2-methoxy-4-nitro-5-sulfophenyl)-2H-tetrazolium-5-carboxanilide) reagent to an orange formazan product by actively respiring cells is measured. The amount of water-soluble product generated from XTT is proportional to the number of living cells in the sample and can be quantified by measuring absorbance at wavelength of 490 nm.

3.9.4 Luminescence Assay

This method determines the level of ATP. It is based on the reaction between ATP and luciferin in the presence of oxygen and luciferase enzyme. The reaction produces AMP, 2Pi, carbon dioxide, and light. The production of light is measured by a luminometer. It can quantify the cell number as low as 20/ml.

3.9.5 Apoptosis Assay

Apoptosis can be used to determine cytotoxicity. It can be measured by detecting cellular morphology changes, DNA laddering, or detection of phosphatidyl serine in the membrane by using annexin V conjugated to fluorescein isothiocyanate or biotin.

3.10 Cell Lines and Maintenance

Cell line is an established culture of cells that are immortalized and has ability to grow indefinitely given appropriate nutrients (medium) and space. There are several cell lines established from human and other mammalian species. There is a summary of some of the commonly used cell lines and their origin and species in Table 3.1.

Table 3.1 Commonly used animal cell lines

Cell line	Organism	Origin tissue	Type of culture
293-T	Human	Kidney (embryonic)	Adherent
3 T3 cells	Mouse	Embryonic fibroblast	Adherent
BXPC3	Human	Pancreatic adenocarcinoma	Adherent
BHK-21	Hamster	Kidney	Adherent
CHO	Hamster	Ovary	Adherent
CMT	Dog	Mammary gland	Adherent
D17	Canine	Osteosarcoma	Adherent
EL4	Mouse	T-cell leukemia	Suspension
DU145	Human	Prostate	Adherent
HeLa	Human	Cervical cancer	Adherent
Jurkat	Human	T-cell leukemia	Suspension
JY	Human	Lymphoblastoid	Suspension
MCF-7	Human	Mammary gland	Adherent
Sf21	Insect	Ovary	Adherent
THP cell line	Human	Monocyte	Adherent
Vero cells	African green monkey	Kidney epithelium	Adherent
GH3	Rat	Pituitary tumor	Adherent

Animal cell lines are grown and maintained at an appropriate temperature and gas mixture (typically, 37 °C, 5 % CO_2 for mammalian cells) in an incubator (Fig. 3.5). Culture conditions and medium vary from one cell line to another and variation from that can yield a different phenotype or loss of properties. Cell culture media maintain appropriate pH and provides proper glucose concentration, growth factors, or other nutrients. The growth factors are often provided by the use of serum in media. But one complication of using blood-derived material is potential contamination of the culture with viruses or prions. But there is no better alternative to serum. The cell lines should always be maintained in aseptic conditions. To prevent bacterial growth, antibiotics are used in the media. All procedures should be done under the laminar flow hood that maintains aseptic condition and all the equipment should be sterilized properly [2].

Another critical factor during maintenance of cell line is to use proper seeding density (number of cells per volume of culture medium). This can vary from cell line to cell line. Difference in the seeding density may alter the properties of cell lines in some cell types. For example, a lower plating density makes granulose cells exhibit estrogen production.

Cell lines can be grown either in suspension or adherent cultures. Cells that exist in the blood stream naturally live in suspension and they do not need a surface to grow, whereas adherent cells require a surface that may be coated with extracellular matrix components to increase adhesion properties and provide other signals needed for growth and differentiation. Most cell lines derived from solid tissues are adherent. Another type of adherent culture is three-dimensional culture which involves growing cells in three-dimensional environment. This culture system is more similar to in vivo tissue both biochemically and physiologically, but it is technically challenging due to many factors (e.g., diffusion).

3.11 Bioreactors for Large-Scale Production of Animal Cells

Suspension culture of animal cells is in extensive use for the production of monoclonal antibodies, recombinant proteins, and vaccines. A fully automated bioreactor maintains physicochemical and biological factors to optimum level and maintains freely suspended cells [5]. Suspension culture

Apoptosis or Programmed Cell Death

The number of cells in the body of an organism is determined and balanced by rate of cell division and cell death. The rapid growth of cells is regulated by intracellular factors and by extracellular signals. In multicellular organisms, cell division occurs whenever there is requirement of more cells. The transition stages of the cell cycle are under strict regulation for cell division to occur.

Programmed cell death or apoptosis (apoptosis is derived from Greek word meaning falling off) is a very important process and is used to remove the cells which are either defective or not needed or inappropriate for the system or for the removal of infected cells (see Fig. 3.4). Apoptosis is mediated by caspase (cysteine-aspartic proteases or cysteine-dependent aspartate-directed proteases) family of proteases which have cysteine at their active site. They cleave their substrates on the C-terminal side of aspartate residue. They exist in cells in the form of procaspases (inactive precursors). The process is summarized here:

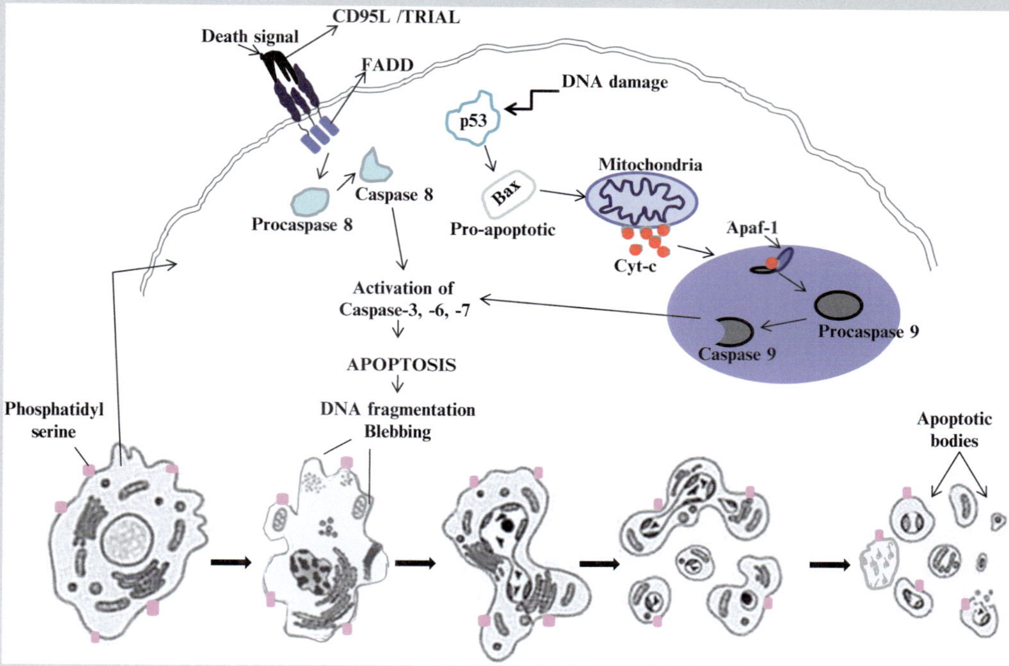

Fig. 3.4 In the cells, the programmed cell death is initiated when the system senses any abnormality or infection. The membrane receptor is bound to death signal as CD95L/FasL and TRAIL. After their binding, FADD (Fas-associated death domain) is clustered and activated at the cytoplasmic side. This causes activation of caspase-8, which then activates caspase-3, caspase-6, and caspase-7 for inducing apoptotic death. DNA damage or cellular abnormalities can activate p53 and Bax. These induce mitochondria to release cytochrome-c. Cytochrome-c binds to cytoplasmic protein Apaf1 and activates caspase-9, which further activates caspase-3, caspase-6, and caspase-7. These effector caspases along with other proteins start DNA fragmentation, membrane blebbing, cell fragmentation, phagocytosis, and removal of the target cell. Phosphatidyl serine (PS) is an important signal present on apoptotic cell and conveys "eat-me" signal to phagocytes for eating and removing the target cell

(continued)

- The process is initiated by autoactivation of caspase-8 and caspase-9.
- These cleave short domains from caspase-3, caspase-6, and caspase-7 (effector caspases).
- Effector caspases cleave many proteins as nuclear lamins leading to breakdown of nuclear envelope, cytoskeletal proteins, activates DNase which causes fragmentation of cellular DNA.

As the process of proteolysis is irreversible, therefore once apoptosis starts, it does not revert back. The activation of the caspase cascade may occur due to:

- Tumor necrosis factor (TNF) family members: TNF member family binds with death receptor, for example, Fas binds with its ligand (FasL) and forms trimers. The death receptor is also called as Fas receptor (FasR) or apoptosis antigen 1 (APO-1 or APT) or cluster of differentiation 95 (CD95) or TNF receptor superfamily member. It trimerizes Fas receptor, causing death domain to cluster at the cytoplasmic phase of the receptor. The clustered domains provide binding sites to adapter protein Fas-associated death domain (FADD), which activates and recruits procaspase-8 to initiate the caspase cascade.
- Inappropriate changes in the cell also trigger apoptosis. Activation of mitochondrial apoptotic pathway occurs after activation of Bax (acts downstream of p53). Binding of Bax to mitochondrial membrane forms oligomers, triggering *cytochrome-c* release, which activates procaspase-9.
- The caspase-8 or caspase-9 activates effector caspases and brings about DNA fragmentation, membrane blebbing, and fragmentation of the target cell.
- In the process of apoptosis, phosphatidylserine (PS) is externalized (normally present in inner leaflet of cell membrane) on the surface of cells undergoing apoptosis. This gives "eat-me" signal to phagocytic cells for phagocytosis of the target cell.

Blebbing in plasma membrane is required for generation of smaller fragments of dying cells which are called as apoptotic bodies. Bleb is produced due to intracellular pressure due to cell cortex and either rupture of the cortex or its detachment from the plasma membrane. The process is very important to keep a check on cell number and control. In cancer, the apoptotic cell death is lost, which results in increased and uncontrolled proliferation of cells (refer to Chap. 10 for cancer-related details).

must be stirred continuously to prevent cellular aggregation and cell death. Animal cells grow slower than bacterial cells and thus even small changes in culture bring unfavorable metabolic state to the cells. The main carbon and energy sources are glucose and glutamine and major metabolic products are lactate and ammonia. Therefore online monitoring for these products must be carried out during growth of the cells in bioreactors. It can be done by online flow injection analysis by using gas chromatography (GC) and high-performance liquid chromatography (HPLC). Mammalian cell bioreactors [4] can be categorized into two types: well-mixed continuous stirred tank (CSTR) and plug flow (tubular) reactors (Fig. 3.6). In the case of plug flow reactor (PFR), nutrient concentration will decrease from inlet to the distal end of the reactor, while metabolite concentration increases. In CSTR, there is no concentration gradient and it follows that the stream exiting from the reactor will have the same composition as the well-mixed fluid in the reactor. In considering the selection of bioreactors, the mixing characteristics and their relationship to scale up should be kept in mind. In case of PFR, it is intrinsically more difficult to scale up than mixing vessels, as the concentration gradient of essential nutrient becomes limiting in the downstream region.

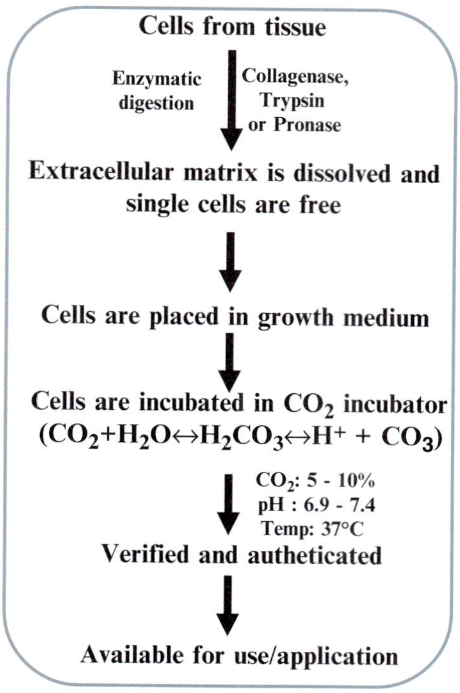

Fig. 3.5 The figure outlines the basic steps of animal cell culture

3.12 Applications

Cell culture represents standard in vitro model for biomedical research and studying complex physiological and biochemical properties without using laboratory animals. The application of cell culture ranges from basic research to the production of vaccines and antibodies in biotechnological and pharmaceutical industries.

3.12.1 Cancer Research

Animal cell lines are widely used in cancer research for studying cancer cell properties, screening new drugs, and generating xenograft model in mice for animal experiments.

3.12.2 Model System

Cell cultures are also used to study basic biology of the cells, metabolomics, and interaction of

Fig. 3.6 Schematic diagram of bioreactors used for large-scale animal cell culture

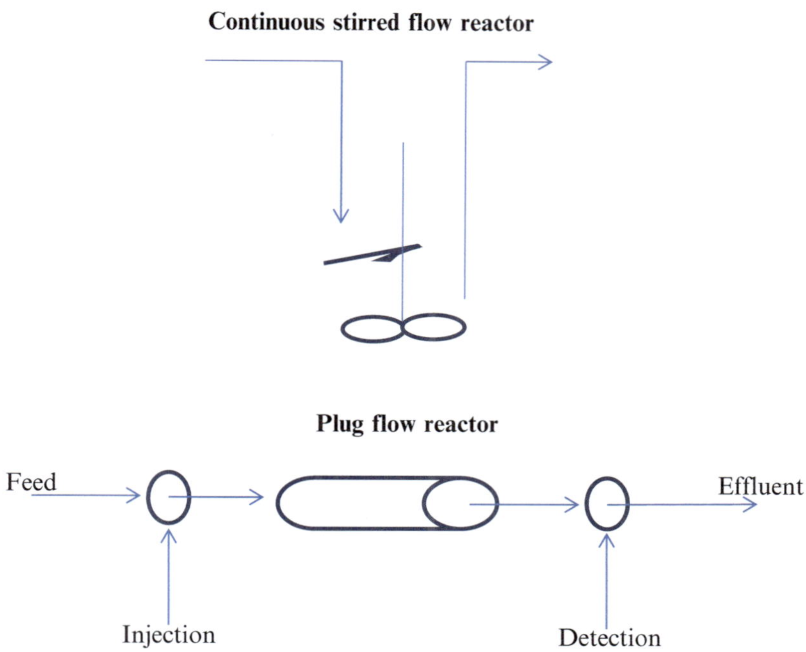

cells with pathogens like bacteria or fungus, in the study of aging.

3.12.3 Production of Antibodies, Vaccines, and Recombinant Proteins

Cell lines are widely used in biotechnological and pharmaceutical industries for the production of antibodies, vaccines, and recombinant proteins. It is also used to produce monoclonal antibodies, hormones, and insulin through genetic engineering of the cells. For example, vaccines for deadly diseases like polio, rabies, hepatitis, and chicken pox are produced using cell culture.

3.12.4 Virology

Animal cell culture is broadly used in virology studies to replicate the virus, studying the nature and mode of infection of virus and growth and development cycle of virus.

3.12.5 Drug Screening and Development and Cytotoxicity Test

Cell lines are used to find out effective and safe dosage of any new drug. Cell culture is used to test also the cytotoxicity of any toxic chemicals.

3.12.6 Gene Therapy

Cultured cells can be genetically altered and used for gene therapy. First cells are removed from patient and the defective gene is replaced by functional gene using genetic engineering or viral gene delivery. Those corrected cells are then put back into patient to correct the disease.

3.12.7 Replacement of Tissue or Organ

Animal cell culture can be used to generate tissue or organ for replacement. For example, cell culture system can generate skin for the patients of burns or ulcers. Research is going on to develop new organs like the liver or kidney using both embryonic and adult stem cells. Those cells have the ability to differentiate into many different types of cells.

3.12.8 Genetic Counseling

Fetal cell culture from pregnant women can be used to abnormalities of chromosomes or genes and determine fetal disorder.

3.13 Cryopreservation

Cryopreservation is a process where cells or tissues are preserved by cooling to sub-zero temperature (−196 °C). At low temperature, any biological activities including those reactions that could cause cell death are effectively stopped. The survival rate on freezing and thawing varies from cell line to cell line and depends on the ability of the cells to withstand the stresses during freezing and thawing. Freezing of the cells must be done using a cryoprotectant like glycerol or dimethyl sulfoxide (DMSO) to prevent formation of ice crystals inside the cells. Each cell type also has an optimum cooling rate at which survival rate is maximal. To get maximum survival, cells should be cooled at slow and steady rate of cooling.

Cryopreservation media generally consists of a base medium, cryopreservative, and a protein source. The cryopreservative and protein protects the cells from the stresses during freezing and thawing. Most common cryopreservatives are 10 % glycerol or 10 % DMSO.

3.13.1 Risks of Cryopreservation

Solute effects: When ice crystals form in freezing water, solutes are excluded and become concentrated. This can be very damaging to the cells.

Extracellular ice formation: As water migrates from inside the cells to outside during slow cooling, ice forms at extracellular spaces. This can cause mechanical damage to the cells.

Dehydration: Due to migration of water from inside of cells, cellular dehydration can occur.

Intracellular ice formation: Formation of intracellular ice is most detrimental and fatal to cells.

3.13.2 Methods to Avoid Risks

Controlled-rate and slow-freezing is the widely used method nowadays to prevent intracellular ice formation. If cooling is slow enough to permit sufficient water to leave the cell during freezing of extracellular fluid, intracellular ice crystal formation can be avoided. The rate differs from cells to cells but a typical 1 °C/min is appropriate for many mammalian cells in the presence of cryoprotectants.

3.13.3 Freezable Tissues

Generally cryopreservation is used for a thin section of tissues or a small clump of individual cells as it requires lesser doses of toxic cryoprotectants. Therefore, cryopreservation of larger tissues like the liver or kidney for transplant is still impractical.

Suitable cryopreservation has been achieved for the following biological materials:

Semen
Blood: for special cells
Stem cells
Umbilical cord blood
Oocytes
Embryos in 2-, 4-, or 8-cell stage
Ovarian tissue
Tissue samples from tumors and other histological cross sections

3.14 Chapter End Summary

- Animal cell culture includes processes for isolating, propagating, manipulating, culturing, and freezing mammalian cells in laboratory conditions using appropriate media, supplements, gas, temperature, and pressure.
- The cell lines are of two types: primary cell lines (directly obtained from animal tissue) and established cell lines (transformed cell lines).
- The cells are maintained in the laboratory in suitable culture condition in medium. Earlier body fluids were used for culture media, but due to potential risk of transmission of infection and variability in each lot, these are replaced by either compound obtained by chemical synthesis or by the use of recombinant DNA technology.
- The cells are maintained by regular passaging; however, prolonged passaging may bring about changes in the cells; thus when not required, they are cryopreserved with appropriate cryoprotectants.
- The cells in culture are periodically monitored for contamination. Cell viability is monitored and various assays are used for viability monitoring usually with effects of drugs. The cells used for production of therapeutic compounds are grown in bioreactor.
- Animal cell culture has revolutionized biomedical research and has a wide range of applications: (1) testing the effects of chemical agents and drugs on the growth of the cells, (2) the study of expression of genes in response to the drug, (3) used to check the efficacy and safety of a chemical compound on cell growth and metabolism (testing for cancer therapy or other diseases), and (4) used for basic research.
- The usage of cell culture has bypassed the initial usage of animals for each and every testing and has minimized animal usage as well as expedited the research process.

Multiple Choice Questions

1. Cryopreservation of cells is important because:

(a) The cells may be used when required.
(b) It is difficult to maintain cells in culture for a very long time.
(c) It is easier to transport them when preserved.
(d) All of the above.

2. Nowadays, serum-free and blood-free medium is preferred for cell culture because:
(a) They are very expensive.
(b) It is undefined media; thus, all the constituents are not known.
(c) There is significant variation with each lot of media.
(d) Risk of transmission of potential pathogen is reduced.
(e) All of the above.

3. The natural media is:
(a) Naturally occurring in environment
(b) Media based upon biological fluids
(c) Synthetic and simple chemical compounds
(d) None of these

4. The sterilization is important for the growth of cells in culture condition; the preferred method for sterilization is:
(a) Autoclaving
(b) Dry heating
(c) Filtration
(d) None of the above

5. The following condition is most critical and needs to be maintained during growth of cells in bioreactor:
(a) Oxygen
(b) Contamination
(c) Media ingredients
(d) None of these

6. The property of totipotency is present in:
(a) Hematopoietic stem cell
(b) Cells of mesoderm
(c) Blastula stage cells
(d) None of these

7. An established or transformed cell line is marked by:
(a) It is stem cell.
(b) It is undifferentiated cell.
(c) It is cancerous cell.
(d) It is redifferentiated cell.

8. Subculturing of adherent cells require trypsinization because:
(a) Cleaves the membrane for the release of product of choice.
(b) Detaches cells from culture vessel.
(c) Trypsin is provided as enzyme for digestion of food.
(d) None of these.

9. Common method of accessing cell viability is:
(a) Dye exclusion assays
(b) Dye uptake assays
(c) Chromium release assay
(d) All of these

10. Culturing of animal cells is important because:
(a) It helps to check cytotoxicity of the drug.
(b) It helps to see the effect of drug on genes.
(c) It helps in the production of therapeutic proteins.
(d) All of the above.

11. The important constituents of animal cell culture are:
(a) Growth factors
(b) Cytokines
(c) Glucose and glutamine
(d) None of these

12. During maintenance of animal cell culture, lactic acid is accumulated in culture fluid due to:
(a) Excessive oxygen
(b) Depleted oxygen
(c) Inhibition of glycolysis
(d) Production of ethyl alcohol

13. The example of stable primary cell line is:
(a) HeLa cells
(b) CHO cells
(c) Fibroblast cell
(d) None of these

14. Transformation in mammalian cell culture means:
(a) Phenotypic modification of cells
(b) Insertion of foreign DNA
(c) Viral infection
(d) None of these

Answers

1. (d); 2. (e); 3. (b); 4. (c); 5. (a); 6. (d); 7. (c); 8. (b); 9. (d); 10. (d); 11. (c); 12. (b); 13. (c); 14. (a)

Review Questions

Q1. What is the importance of animal cell culture?

Q2. Why is sterilization required for culturing of cells?

Q3. Write a note on medium required for animal cell culture.

Q4. Why is viability of cells monitored?

Q5. What are the methods of monitoring the viability of the cells?

Q6. What is cryopreservation?

References

1. Altman SA, Randers L, Rao G (1993) Comparison of trypan blue dye exclusion and fluorometric assays for mammalian cell viability determinations. Biotechnol Prog 9:671–674
2. Phelan MC (2007) Chapter 1:Unit 1.1. Basic techniques in mammalian cell tissue culture. Curr Protoc Cell Biol 36:1.1:1.1.1–1.1.18.
3. Vallier L (2011) Serum-free and feeder-free culture conditions for human embryonic stem cells. Methods Mol Biol 690:57–66
4. Wang D, Liu W, Han B, Xu R (2005) The bioreactor: a powerful tool for large-scale culture of animal cells. Curr Pharm Biotechnol 6:397–403
5. Warnock JN, Al-Rubeai M (2006) Bioreactor systems for the production of biopharmaceuticals from animal cells. Biotechnol Appl Biochem 45:1–12
6. Weisenthal LM, Dill PL, Kurnick NB, Lippman ME (1983) Comparison of dye exclusion assays with a clonogenic assay in the determination of drug-induced cytotoxicity. Cancer Res 43:258–264

Some Related Resources

http://www.opsdiagnostics.com/notes/ranpri/rpcryoprescell.htm

http://www.sigmaaldrich.com/technical-documents/protocols/biology/cryopreservation-and.html

https://www.atcc.org/~/.../Culture%20Guides/AnimCellCulture_Guide.pd

https://www.hpacultures.org.uk/collections.aspx

www.level.com.tw/html/ezcatfiles/.../img/.../intro_animal_cell_culture.pd

www.ncbi.nlm.nih.gov › NCBI › Literature › Bookshelf

Production of Recombinant Pharmaceutical Proteins

4

Abstract

The proteins produced in the body control and mediate the metabolic processes and help in its routine functioning. Any kind of impairment in protein production, such as production of mutated protein, or misfolded protein, leads to disruption of the pathway controlled by that protein. This may manifest in the form of the disease. However, these diseases can be treated, by supplying the protein from outside or exogenously. The supply of active exogenous protein requires its production on large scale to fulfill the growing demand. The process is complex, requiring higher protein expression, purification, and processing. Each product needs unique settings or standardizations for large-scale production and purification. As only large-scale production can fulfill the growing demand, thus it needs to be cost-effective. The tools of genetic engineering are utilized to produce the proteins of human origin in bacteria, fungi, insect, or mammalian host. Usage of recombinant DNA technology for large-scale production of proteins requires ample amount of time, labor, and resources, but it also offers many opportunities for economic growth. After reading this chapter, readers would be able to understand the basics about production of recombinant proteins in various hosts along with the advantages and limitations of each host system and properties and production of some of the important pharmaceutical compounds and growth factors.

4.1 Introduction

The proteins produced in the body control and mediate the metabolic processes and help in its routine functioning. Any kind of impairment in protein production, such as production of mutated protein or misfolded protein, leads to disruption of the pathway controlled by that protein. This may manifest in the form of the disease. However, these diseases can be treated, by supplying the protein from outside or exogenously. The supply of active exogenous protein requires its production on large scale to fulfill the growing demand. The process is complex, requiring higher protein expression, purification, and processing. Each product needs unique settings or standardizations for large-scale production and purification. As only large-scale production can fulfill the growing

V. Gupta et al., *Basic and Applied Aspects of Biotechnology*, DOI 10.1007/978-981-10-0875-7_4

demand, thus it needs to be cost-effective. The tools of genetic engineering are utilized to produce the proteins of human origin in bacteria, fungi, insect, or mammalian host (Fig. 4.1). Usage of recombinant DNA technology for large-scale production of proteins requires ample amount of time, labor, and resources, but it also offers many opportunities for economic growth. After reading this chapter, readers would be able to understand the basics about production of recombinant proteins in various hosts along with the advantages and limitations of each host system and properties and production of some of the important pharmaceutical compounds and growth factors.

4.2 Expression of Foreign Gene

In all living cells, the expression of gene occurs where genetic information contained in DNA is passed on to RNA in the process of transcription and from RNA to protein in the process of translation. Thus, the body synthesizes RNA and then proteins according to the instructions from the DNA. DNA-dependent RNA polymerase or RNA polymerase carries the transcription. Eukaryotes have three RNA polymerases: polymerase I transcribes ribosomal RNA genes (18S,5.8S and 28SrRNA), polymerase II transcribes all protein-coding genes (mRNA and small RNA), and polymerase III transcribes the genes for 5SrRNA and tRNA. In prokaryotic cells (*E. coli*), RNA polymerase consists of five subunits—two identical α subunits and one subunit each of β, β', and σ subunit. The σ subunit dissociates after polymerization ensues. Thus, the term "holoenzyme" is used for complete enzyme and "core enzyme" is used without σ subunit. The process of transcription is initiated by binding of RNA polymerase to DNA molecule at a very specific site called "promoters." The promoters are critical for the start of the transcription. The promoter sites are nearly 40 bp long and are mostly located before the first base, which is copied into RNA (Fig. 4.2a). This first base is called as start point of transcription and is denoted by +1.

4.2.1 Promoters

Promoters for RNApol II allow differential expression of genes and determine the rate at which the genes are transcribed. There are some

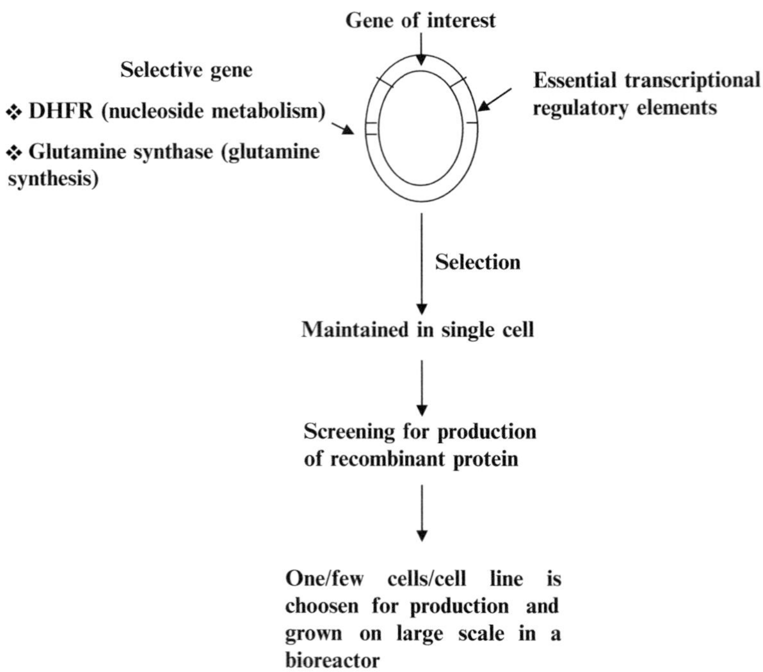

Fig. 4.1 The gene of interest is cloned in suitable vector. For expression of the cloned gene, the gene is with all the essential regulatory elements required for transcription of the gene. The gene is attached to the selective gene, which helps in the selection of the clones with gene of interest. The cells are screened for the synthesis of desirable product and then processed for large-scale production in the bioreactor with optimum condition for high yields

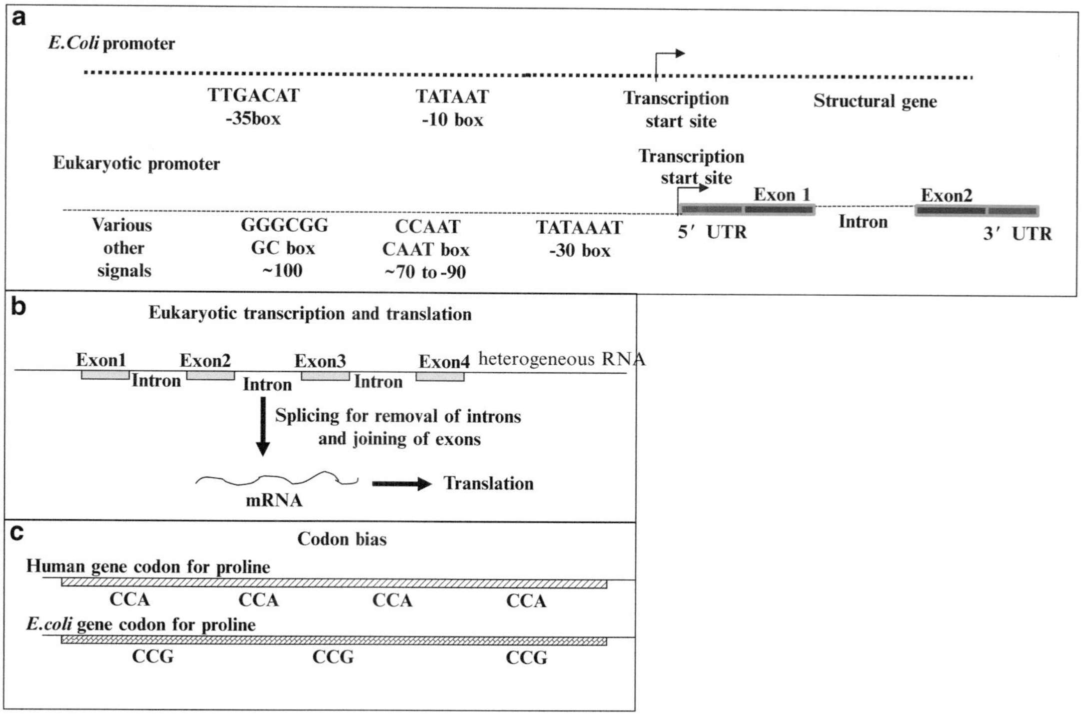

Fig. 4.2 The figure shows the problems encountered by *E. coli* due to sequence of foreign gene, when it is cloned and expressed in it. (**a**) It shows the promoter for *E. coli* and eukaryotes. As there are differences in the promoter of *E. coli* and eukaryotes, thus eukaryotic promoter might not work in *E. coli*, and the gene is placed under the control of *E. coli* promoter. (**b**) In the eukaryotes the splicing machinery removes introns from the target gene. Whereas the *E. coli* does not have any such system, therefore, the intronless version of gene is used in *E. coli*. (**c**) shows codon bias in bacterial and human system. As one amino acid is encoded by more than one codon, for example, proline, where preferential codons in humans are CCA whereas *E. coli* prefers CCG, if the gene containing the preferential codons for humans is used in *E. coli*, then it might result in inefficient translation. Thus, these problems need to be taken care of while using *E. coli* as host

promoters that cause the inserted genes to be expressed all the time; in all parts of the system, they are known as "constitutive" promoters. Others allow expression only at certain stages/certain tissue/organ of individuals and at certain time points. Gene expression is under temporal and spatial regulation.

In prokaryotes the position before start site at −10 and −35 can interact with σ subunit of holoenzyme of RNA polymerase. In eukaryotes the sequence (analogous to −10—consensus sequence of prokaryotes), TATAAAA, is present at −30 position in the promoter region. Eukaryotic promoter is shown (Fig. 4.2a) with TATA box forming the core promoter at −30 position (from −30 to −100), upstream of transcription start site. CAAT box and GC box are at approximately −70–90 and −100, respectively. The location of promoter is always on the same DNA molecule which they regulate. They are referred as cis-acting elements. The spacing of various elements is more important and much is dependent on locus-specific activators, either at core promoter or at distant sites. Various other signals as enhancers are also involved which are far apart from the target gene. They exert stimulatory effects on promoter activity and can be upstream, downstream or in the middle of the gene. Promoters of housekeeping genes or genes with complex patterns of expression have CpG islands rather than TATA box. The gene to be transcribed has 5′-untranslated region (5′ UTR), exons (coding region) separated by introns (noncoding region). The end has 3′-untranslated region (3′ UTR). In the cloning for gene expression, usually intronless version of the gene is used. In eukaryotes, the transcription is further enhanced by enhancers, which may be 2,000–3,000 bases away from the promoter region but are able to affect the rate of transcription.

4.2.2 General Considerations for Protein Production

The simplest host for the work of recombinant DNA technology is prokaryotic bacterial system. In the early 1980s, the first recombinant FDA-approved pharmaceutical, the human insulin (Humulin-US/Humuline-EU), was obtained from genetically engineered *Escherichia coli* (*E. coli*) for treatment of diabetes.

Due to increasing demand, many strains of microbial species are being designed with increased throughput and better recovery of the therapeutic protein from large-scale culture. The recombinant proteins approved by FDA are obtained either from *Escherichia coli* or other prokaryotes; from *Saccharomyces cerevisiae* or other fungal species; from insect cells, mammalian cells, or human cells; or from transgenic plants or animals. Cloning and production of protein in a particular host system are dependent upon a property of host to clone and express the desirable size of protein-encoding genes; production of correctly modified, folded, and functional protein; high yield of the protein; and low-cost requirements. The choice of host systems requires best system, which can fulfill the requirements [17].

The advantages of producing proteins using recombinant DNA technology are:

- As human gene may be cloned and expressed, it minimizes the risk of immune reaction and the specific activity of the protein is high.
- The therapeutic protein can be produced efficiently, maintaining its cost-effectiveness.
- It minimizes the risk of transmission of unknown pathogens present in animal and human sources.
- Appropriate modification with higher specificity, increased half-life, and improved functionality.
- Allow to create critical changes for better specificity and activity.

4.2.3 High Protein Expression in the Host

Expression and production of eukaryotic protein require cloning of its cDNA in an expression vector and subsequent transfer of the vector in suitable host. DNA is modified, cloned, and expressed in other host for the production of the protein; thus, optimum production conditions are required in each host. However, there are yield variations in different expression systems but high-level expression of the protein may be achieved by considering the following points:

- The recombinant gene should be with all the necessary elements for effective transcription initiation.
- Use is made of strong viral or cellular promoter/enhancer for efficient driving of transcription. The usage of viral T7 promoter in bacteria can result in higher yield of recombinant proteins. Transgene may be expressed under the control of either polyhedrin promoter of baculovirus, E1 promoter of adenovirus, or p7.5 promoter of *Vaccinia virus*. These are suitable for wide range of cell types. Mammalian cells may use promoter and enhancer of SV40 or the long-terminal-repeat promoter and enhancer of the *Rous sarcoma virus* or early promoter of the human cytomegalovirus (refer to Chap. 2 for promoters and expression vectors).
- Polyadenylation signals are helpful in eukaryotic genes. These terminators are required for defined 3′ end to the mRNA which extends by addition of poly A tail ultimately increasing the stability of RNA and facilitating its export to the cytoplasm.
- Removal of 3′ and 5′ UTRs (untranslated repeats) may influence gene expression. They may interfere with initiation of translation and

their secondary structure prevents efficient translation.

- Kozak's consensus 5′-CCRCCAUGG-3′ with purine at –3 position and guanine at +4 position affects transgene expression.
- Inclusion of one intron sequence, which is located between the promoter and cDNA coding sequence, gives better yield; thus, most expression vectors include at least one intron sequence. Intron presence is of importance when transgene needs to be expressed in mammalian cells.
- For eliminating the slow rates of translation due to codon biasing, the gene may be converted to a high-expressing gene by changing the t-RNA codons to the most abundant ones.
- The integration site has a major effect on the rate of the transcription of the recombinant gene (position effect).
- Some selective genes like dihydrofolate reductase (DHFR) or glutamine synthetase (GS) gene are incorporated. The genes are involved in nucleotide biosynthesis and glutamine synthetase, respectively; the selection occurs when the appropriate metabolite is missing preventing the growth of non-transformed cells.
- For increasing the yield of the recombinant protein, the protein is often fused with endogenous protein sequence (see Chap. 5 for selectable markers, reporters).
- To aid purification, the protein-encoding DNA contains coding DNA for specific protein or peptide that can be a target for affinity chromatography.
- For affinity purification, two systems are readily employed, (1) glutathione S-transferase (GST)–glutathione affinity and (2) polyhistidine–nickel ion affinity. GST has high affinity for glutathione and protein with side chains of histidine has high affinity for nickel ions.
- Stable transfection gives high yield of protein. After recombinant DNA is transfected into animal cells, it either can be integrated (stable expression) into the host genome or maintained in episomal form (transient expression). In stable expression systems, the foreign gene is passed on to the next descendants, and the expression is maintained generation after generation.

4.3 Microbial System for Production of Therapeutic Protein

Though various host systems are available for production of recombinant proteins, microbial hosts offer several advantages over other systems, as production is fast, cheap, and economic:

1. The molecular biology and physiology is well characterized and documented.
2. Easy to maintain and manipulate.
3. Utilizes inexpensive nutrition sources.
4. Rapid growth and biomass accumulation to achieve high cell densities.
5. Scale-up is easy and convenient.
6. Their expression machinery can be with variety of strong inducible promoters.

Inducible Promoters

The inducible promoter may require an inducer, or the depletion or addition of a specific nutrient, or pH change or changes in physicochemical factors in order to initiate the process of gene expression. The inducible systems suffer from the disadvantage that chemical inducers may be expensive and toxic and would require elimination during downstream processing when the product is intended for human usage. Thus, usage of thermoregulated systems has been used for production of recombinant pharmaceutical proteins as expression is dependent upon strong heat-regulated promoter minimizing the risks of any addition of chemical agent.

As with the cellular structure of bacteria, it can rapidly adapt to culturing conditions with very short replication time (20 min). The media requirement of bacterial cell is simple and consists of simple carbon and nitrogen source. Thus, the overall inputs in bacteria are 90 % lower than for mammalian cells. Some of the approved bacteria-derived (by either the European Union or FDA, USA) therapeutics include hormones (human insulin and insulin analogs, calcitonin, human growth hormone, glucagons, parathyroid hormone, somatropin, and insulin-like growth factor 1), interferons (alfa-1, alfa-2a, alfa-2b, and gamma-1b), interleukins 11 and 2, light and heavy chains raised against vascular endothelial growth factor-A, tumor necrosis factor alpha, cholera B subunit protein, granulocyte colony-stimulating factor, and plasminogen activator [3].

The microbe of first choice for production of recombinant protein is enterobacterium *E. coli*. The system offers quick and easy modifications, ease of growing in manageable environmental conditions and short life cycle. The bacterial cell can tolerate and adapt to changes in the environment rapidly, thus scale-up is easier. However, the system suffers from some of the disadvantages [11,16].

1. Human or mammalian genes cloned in bacteria cannot undergo splicing due to lack of splicing machinery, thus intron-less version of the gene is cloned for optimum results (Fig. 4.2b).
2. The signals involved in transcription of genes may vary; thus, the gene of interest is usually fused with bacterial gene under the control of its promoter, and the protein is obtained as a fusion product, which can be later cleaved, purified, and used.
3. Lac promoter is one of the most popular bacterial promoters. However, for high-level expression, T7 promoter is also preferred (present in pET vectors). It can drive the target protein expression to nearly 50 % of total cell protein. Gene of interest can be placed under the control of regulated promoter of phage.
4. For translation, abundance of t-RNA is related to the frequency of appearance of different codons (codon biasing). Therefore, codons rare for *E. coli* may cause amino acid misincorporation or premature termination affecting the yield of the therapeutic protein. This can be solved by either site-directed replacement of rare codons to the codons (for the same amino acid) preferred by *E. coli*. Another approach may be co-expression of rare t-RNAs in *E. coli* (strains of *E. coli*, BL21 codon plus, and Rosetta were designed for this purpose). For addition of amino acids during the process of translation, as there are more than one codon for several amino acids, thus, codon biasing occurs which is the preference of a particular codon of amino acid in a particular species (Fig. 4.2c).
5. Complexity: Eukaryotic cells have the advantage of producing fully functional and properly folded proteins. The antibodies with four subunits may be secreted by eukaryotic cells in fully functional form. On the other hand, it is very difficult to obtain multidomain protein from *E. coli*. Even if protein is obtained, the renaturation and folding in laboratory condition may either be very expensive or protein may lack its activity.
6. Lack of posttranslational modifications (PTMs) is the problem, which cannot be solved, and is mandatory for activity of many therapeutic proteins. The glycosylation is the most common modifications, and others are phosphorylation and formation of disulfide bond, which are essentially required for the full functional capability of many human proteins. PTMs play an important role in proper protein folding, processing, stability, tissue targeting, activity, immune reactivity, and half-life of the protein. Lack of these results in insoluble, unstable, or inactive product.

   However, the N-linked glycosylation system of *Campylobacter jejuni* has been successfully transferred to *E. coli*, thus opening a possibility for the production of glycosylated protein in it. Certain mutant *E. coli* are being developed to promote disulfide bond formation (AD494, Origami, Rosetta-gami) with reduced protease activity (BLZ1).
7. Overproduction of recombinant protein in bacteria might result in the loss of solubility and deposition of many protein species as protein aggregates or inclusion bodies. Alteration in growth conditions might render the product in insoluble form. Many eukaryotic proteins are found trapped in inclusion bodies with resistance to further processing. Success has

been obtained in purification of insulin and betaferon from inclusion bodies. The retrieval of proteins using denaturating condition with subsequent refolding and renaturation might not always be easy and prove to be extremely expensive.

8. With the *E. coli* host, it is very difficult to obtain protein larger than 60 kDa in soluble form.

Due to certain limitations for production of proteins in *E. coli*, other host systems are being discussed for production of proteins.

Glycosylation

Covalent attachment of carbohydrate group to the protein to form glycoprotein is called glycosylation. In glycoproteins, proteins constitute a major fraction. These play important roles in various physiological processes and are components of cell membranes.

The carbohydrates commonly attached to proteins may be fucose (Fuc), galactose (Gal), N-acetylgalactosamine (GalNAc), glucose (Glc), N-acetylglucosamine (GlcNAc), mannose (Man), and sialic acid (Sia).

These sugar moieties may associate through amide nitrogen atom of side chain of asparagine (Asn) termed as N-linked glycosylation or to the oxygen atom in the side chain of serine (Ser) or threonine (Thr) termed as O-linked glycosylation.

Not all the asparagine (Asn) present in polypeptide can accept the carbohydrate moiety. The residues with the sequence Asn-X-Ser or Asn-X-Thr, where X is any amino acid except proline, are targets for glycosylation. Not only the residual sequence but other aspects of the structure of the protein and cell type determine the glycosylation site.

All the N-linked sugar residues have a common core of pentasaccharides. These pentasaccharide consists of three mannose and two N-acetylglucosamine residues. The core may attach to different oligosaccharides to form different glycoproteins.

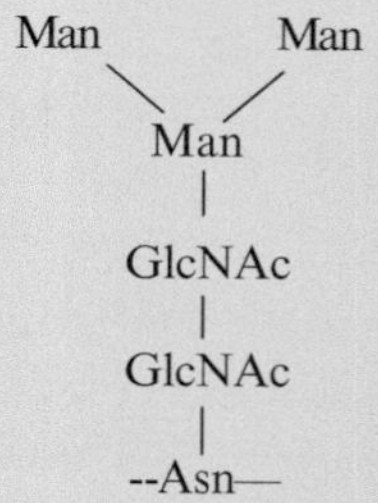

Pentasaccharide core

Glycosylation is one of the important posttranslational modifications, which occurs inside the lumen of the endoplasmic reticulum (ER) and in Golgi complex. The ER and Golgi complex are important in protein targeting and transport. N-linked glycosylation starts in the endoplasmic reticulum and continues in the Golgi complex after the polypeptide is synthesized on ribosomes. However, O-linked glycosylation exclusively occurs in the Golgi complex.

In the process, oligosaccharide to be attached to the protein associates with a specialized lipid present in ER, dolichol phosphate, which consists of about 20 isoprene (C5) units. Through phosphate of dolichol phosphate, oligosaccharide is transferred to specific asparagine residue of polypeptide chain on ribosomes. The enzyme responsible for glycosylating protein and activated oligosaccharide are located on the lumen side of ER.

Then these are transported to the Golgi complex, where the carbohydrate units are altered and finalized. Golgi complex is responsible for O-linked sugar attachment and modification of N-linked sugar. Then the proteins are targeted and transported to their destination.

(continued)

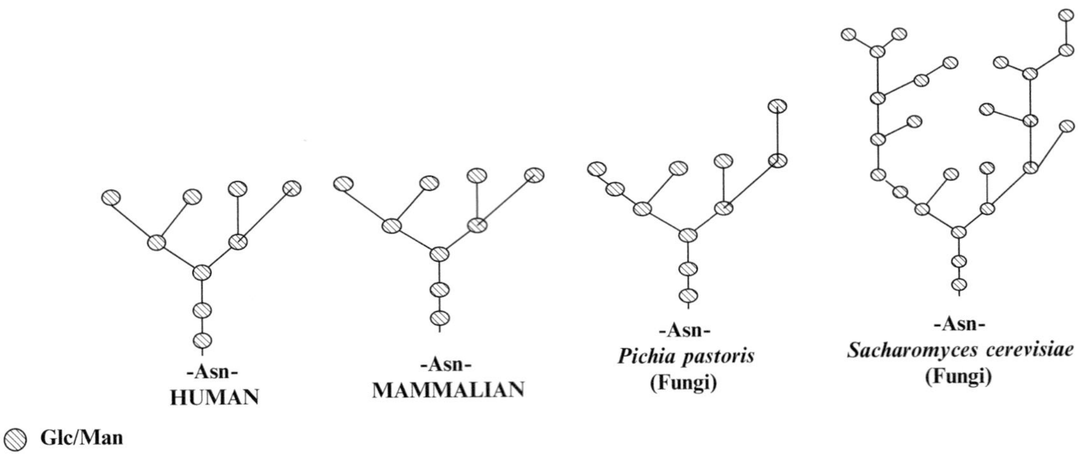

Fig. 4.3 Many of the human proteins are glycosylated (O-linked or N-linked glycosylation). Glycosylation is one of the important posttranslational modifications. The *E. coli* is unable to perform glycosylation. Fungi are simplest eukaryotic systems which can perform glycosylation. The use of fungal host (*Pichia pastoris* and *Saccharomyces cerevisiae*) for production of recombinant proteins results in hyper-glycosylation. The figure shows glycosylation pattern in human, mammalian, *Pichia*, and *Saccharomyces* systems. The hyper-glycosylations or abnormal glycosylations can make the protein highly immunogenic making it unsuitable for therapeutic purposes

4.4 Production of Recombinant Protein in Fungal Hosts

Due to the problems encountered in *E. coli* for production of larger proteins or modified proteins, the next cost-effective, fast, high-density, and easy to handle system is of fungi. *Saccharomyces cerevisiae* (yeast) was the system of choice when it was difficult to obtain therapeutic protein in soluble form and with appropriate posttranslational modifications in bacterial host. In yeast, mutants are available which can give high yield. The approved products obtained from yeast are hormones, vaccines, recombinant granulocyte macrophage colony-stimulating factors (GM-CSF), albumin, hirudin, and platelet-derived growth factor (PDGF).

The advantages of *S. cerevisiae* are the following: (1) it secretes recombinant protein in the culture, (2) protein is properly folded, and (3) it performs most posttranslational modifications. With the yeast system, high amounts of recombinant protein are obtained, and yeast is also capable of performing posttranslational modifications as O-linked glycosylation, phosphorylation, acetylation, and acylation but differs drastically in patterns of N-linked glycosylation (Fig. 4.3).

Protein Folding and Molecular Chaperons

Chaperones are family of highly conserved different proteins. The important functions of chaperones are:

- Prevention of aggregation and misfolding of newly synthesized polypeptide chain.
- They prevent irreversible aggregation of nonnative conformation and maintain the protein on the productive folding pathway.
- They prevent nonproductive interactions with other components of the cell.
- They help and guide the direct assembly of multisubunit protein complexes and larger proteins.
- The chaperones involved in folding recognize nonnative substrate proteins mainly via their exposed hydrophobic residues.

The major classes of molecular chaperones are:

Heat shock proteins are present in a variety of systems and prevent damage to the proteins under high heat.

(continued)

HSP60

- It is tetradecameric mitochondrial chaperonin.
- It is implicated in protein import and macromolecular assembly.
- Required for folding of precursor polypeptides in ATP-dependent manner.
- Prevents aggregation and mediates refolding of protein after heat shock.

HSP70

- They are central components of the cellular network of folding catalysts and molecular chaperones.
- They assist in different types of processes of protein folding in the cell by transient association of their substrate binding domain with short hydrophobic peptide.
- They bind and release their substrate by switching to low-affinity ATP-bound state and the high-affinity ADP-bound state.
- They form complex network of folding machines [15].

HSP90

- It is highly abundant chaperone.
- It plays an important role in many cellular processes, for example, cell cycle control, cell survival, and hormone and other signaling pathways.
- It is a key player in maintaining cellular homeostasis during stress.
- Has ATPase activity, whose binding and hydrolysis affects conformational dynamics of the protein.
- It has become a major therapeutic target for cancer, and its role is being explored in neurodegenerative disorders and infectious diseases [10].

CCT: Chaperonin Containing TCP1

- This is eukaryotic chaperonin tailless complex polypeptide 1 (TCP1) ring complex (TRiC).
- It facilitates the proper folding of many cellular proteins.

(continued)

GroEL

- It is a bacterial chaperone.
- It binds with partially folded and misfolded proteins.
- For its functionality GroEL requires its cofactor GroES.

Differences in N-glycosylation in yeast are with high or hypermannose which is highly immunogenic. Unmodified proteins are suitable for production in yeast.

Other members of fungi are *Pichia pastoris*, *Pichia methanolica*, *Candida boidinii*, and *Pichia angusta*, which are facultative methylotrophic yeast having great potential. *Pichia pastoris* is favored as high cell densities can be obtained; protein is secreted in high concentration (1 g/l), less hypermannosylation as compared to yeast and thus less immunogenic.

However, the disadvantage is that it requires methanol to induce gene expression as transgene is under the control of the promoter of alcohol oxidase 1 (AOX1) gene. Methanol may be flammable and is toxic to cells and humans if not thoroughly removed. Because of hypermannose-type glycosylation, the fungi are also unsuitable for production of many recombinant proteins.

4.5 Production of Recombinant Protein in Insect Cell

Insect cells: Insect cell can be infected with baculoviruses which are double-stranded circular DNA viruses with arthropods as host. Baculovirus-mediated gene expression in insects is a method of choice and is cost-effective, giving the much higher yield of recombinant protein compared to other systems. It is possible to produce large protein resulting in production of correctly processed and biologically active protein.

A baculovirus *Autographa californica nuclear polyhedrosis virus* (AcMNPV) is used as a cloning vector for insect cell lines. In this viral polyhedron protein is used, which is required in its normal habitat and exhibits high rate of transcription, but is not needed in cell culture. Thus, the coding sequence of the gene is replaced with foreign DNA. The gene is transcribed under the control of powerful polyhedron promoter with high yields (~30 % of total cell protein). The observed yield may be variable due to the course of virus infection and viral titer.

The production of recombinant protein in insect cell is time consuming (as compared to bacterial system) as cell growth is slow and the cost of medium is high. Every time fresh cells are required, viral infection is lethal for cells. It also has limitations in performing posttranslational modifications as it performs non-syalated N-linked glycosylation. All the other optimizations need to be perfect as yield depends upon the virus titer and time taken from infection to expression. Insect cells are preferred when active protein is difficult to obtain in *E. coli* system.

Genetic engineering has been used to select MIMIC™ (Invitrogen) and SfSWT-3, which are transgenic cell lines expressing all necessary enzymes to obtain humanized, complex N-linked glycosylation pattern. The system has been extensively used for structural studies as correctly folded eukaryotic proteins may be obtained in secreted form simplifying purification protocols. Some of the approved biopharmaceuticals from infected insect cell line Hi Five are Cervarix (recombinant papillomavirus C-terminal truncated major capsid protein L1 types 16 and 18, used as cancer vaccine).

Glycosylation is a problem which is encountered when insect cells are used for production of recombinant human glycosylated proteins. Lots of genetic engineering is required to produce human-like glycosylation in insect cell. Thus, the preferred system for therapeutic human protein production is mammalian system (Chinese hamster ovary cell line). Due to time and difficultly in maintaining insect cells, the mammalian cells were explored for production of recombinant protein. Mammalian cells, because of their properties of protein folding, assembly, and posttranslational modifications, have become the preferred system for protein production and are now accounting for major recombinant protein production.

4.6 Production of Recombinant Protein in Mammalian Cell

Mammalian Cells: Production of complex proteins requires extensive processing and posttranslational modifications. Mammalian cells have the advantage of performing PTMs (Fig. 4.3) correctly; they secrete recombinant protein into the medium in their natural form, thus skipping the critical steps of renaturation and refolding which sometimes leads to inactive proteins. Therefore, major therapeutic proteins (60–70 %) are produced in mammalian cells primarily Chinese hamster ovary (CHO) cells and baby hamster kidney cells (BHK). CHO cells are relatively easy to manipulate and their properties favor large-scale production in them [24]. The proteins produced are safe to use in humans with no adverse reactions because of similar glycosylation pattern. Chinese hamster ovary (CHO) and baby hamster kidney (BHK) cells are the prominent producers of recombinant proteins (Fig. 4.4).

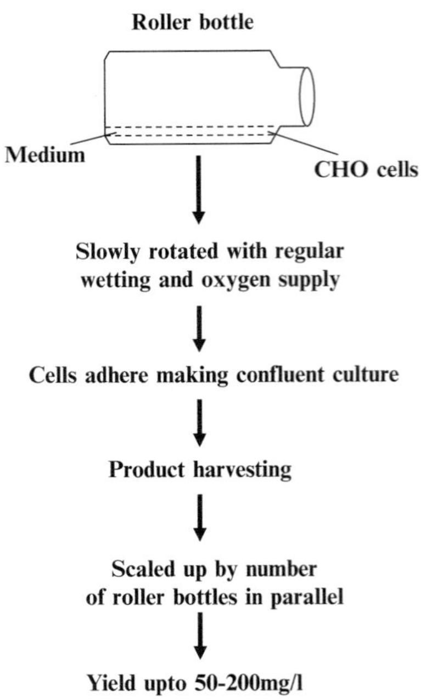

Fig. 4.4 The figure shows the culturing of mammalian cells using roller bottles. These cells are maintained in number of roller bottles. For adherent cells, the microcarrier beads are used. The cells adhere to beads and the beads are maintained in suspension culture

In mammalian cells, genes can be expressed either transiently or stably. Obtaining stably transformed cell lines requires the usage of some selectable marker. Another major advantage of these cells is they can be grown in suspension, in serum-free (SF), protein-free, and chemically defined media. The product is safe without the risk of prions of bovine spongiform encephalopathy (BSE) from bovine serum albumin and infections of variant Creutzfeldt–Jakob disease (vCJD) from the human serum.

Presently due to virus and prions in the donor plasma or blood samples, the manufacturers are opting for plasma-/serum-free growth medium for culturing the cell lines. Cells require some of these components (albumin, transferring) for their growth. Nowadays recombinant human albumin is available like human insulin (produced in *E. coli* and yeast) and yeast-derived animal-free recombinant human transferrin is available. These support plasma-free mammalian cell culture. Else, CHO cells may be engineered to produce its own transferring or insulin-like growth factor. The therapeutic protein obtained by mammalian cell is treated and tested for the presence of any pathogenic agents or viruses as contaminant. The therapeutic products obtained during processing are treated with various agents to remove inactive virus, but the presence of asymptomatic virus poses a serious risk. Common virus inactivation technique is using solvent/detergent, which is effective against lipid-enveloped viruses such as *human immunodeficiency virus* (HIV), *hepatitis B and C virus* (HBV and HCV), *human T-cell lymphotropic virus* (HTLV), and *West Nile virus* (WNV). Non-lipid-enveloped viruses such as parvoviruses, enteroviruses, and circoviruses are resistant to inactivation via solvent detergent treatment. Human herpesvirus 8, responsible for Kaposi's sarcoma, is shown to be transmitted through blood and blood product. Following outbreaks, strict requirements were imposed on manufacturers of biologics and medical devices. Such concerns prompted manufacturers to switch to sugar-based final formulations and develop recombinant plasma-free albumin produced in yeast for usage in biopharmaceutical manufacturing.

Plasma-free manufacturing involves the elimination of plasma derivatives in every step (like cell line development, upstream processing, downstream processing, and final formulations) of the process with appropriate postproduction checks. The manufacturers shifted from the use of serum to serum-free cell culture media with animal product-free media and ultimately to protein-free, completely synthetic chemically defined media. The media consists of protein hydrolysates derived from yeast, soy, and wheat with amino acids, peptides, carbohydrates, vitamins, and essential elements, which are ultrafiltered to remove any unwanted contaminants [8].

Case Study

The recombinant therapeutic product used for clinical applications was produced from mammalian cells. Mammalian cells were maintained in serum-/blood-/plasma-based medium; therefore, the presence of any infectious agent in the product might be deleterious if not properly removed. Infections may range from HIV, *coronavirus* (severe acute respiratory syndrome (SARS), non-lipid-enveloped (NLE) viruses as circoviruses (Torque tenovirus (TTV) and Torque tenominivirus (TTMV)), HBV, HCV, HTLV (human T-cell lymphotropic virus), to West Nile virus. Prions that are self-replicating infectious proteins may also be present which may lead to variant Creutzfeldt–Jakob disease (vCJD).

Pathogen transmission was a major concern in the manufacture of blood-derived coagulation factor. In the early 1980s, the factor replacement products derived from plasma, which were used to treat hemophilia, were found to be contaminated with HIV, HBV, and HCV viruses. In the year 1984, up to 78 % of US-based hemophilic patients were infected with HIV and 74–90 % were infected with HCV. Parvoviruses, B19 (B19V) and PARV4, were present as con-

(continued)

taminant in plasma-derived factor VIII. Therefore, the production urgently required regulatory measures.

Then came first recombinant factor VIII, Advate (Baxter), in the USA in 2003. Advate was produced in CHO cells grown in serum-free and protein-free medium with ultrafiltered soybean peptides with subsequent purification by immunoaffinity chromatography. The usage of Advate helped in eliminating the risk of transmitting emerging blood pathogens.

Prions pose a serious risk, as they are highly resistant to physical/chemical inactivation.

Early stage of prion infection is almost impossible to detect in plasma donor. Iatrogenic transmission of prions has occurred in patients who received human-derived pituitary hormones as human growth hormone (hGH) and gonadotropins. CJD was transmitted to over 160 recipients of cadaveric pituitary hGH before its withdrawal. Cadaveric pituitary-derived gonadotropins for infertility were associated with iatrogenic transmission of CJD. Later on cadaveric pituitary hGH and gonadotropins have been replaced with recombinant GH (produced in microbial system) and recombinant gonadotropins (produced in CHO cell lines).

4.7 Using Human Cells for Protein Production

Human cell lines: Human cell line-derived Dynepoerithropoietin (erythropoietin with increased shelf life), Elaprase-irudonate-2-sulfatase (lysosomal enzyme), and Replagal-alfa-galactosidase A (lysosomal hydrolase) have been approved by the European Union (EU) or Food and Drug Administration (FDA, USA). As these products are fully glycosylated when expressed in human cell lines and used as therapeutics in human beings.

4.8 Transgenics for Protein Production

4.8.1 Transgenic Animals

The transgenic animals are successfully used for production of recombinant proteins (for details refer to Chap. 5). Protein production poses great risk in terms of safety as transmission of infectious agents, allergic responses, immune reactivity, and autoimmune responses might occur.

ATryn was the only approved (approved in 2006 by European medical agency and in 2009 by FDA) recombinant biopharmaceutical using transgenic animals. It contains human antithrombin (432AA) with 15 % glycosylated moieties and is secreted into milk of transgenic goats.

Rhucin intended for acute attacks of angioedema in patients with congenital C1 inhibitor activity deficiency, obtained from transgenic rabbit, was denied approval. Transgenic plants are being explored as recombinant protein producers for research and diagnostic uses.

4.8.2 Transgenic Plants

Obtaining therapeutic proteins from their natural source poses threat for spread of diseases. Therefore, alternative systems for the production of therapeutic agents have their own benefits. Molecular farming in plants has been widely explored for production of recombinant pharmaceutical proteins. Their advantages are low cost, high mass production, scale-up, lack of human pathogens, and addition of eukaryotic PTMs. The first recombinant protein obtained in 1986 from tobacco plants was human growth hormone. However, sometimes plant-specific PTMs might result in adverse immune reactivity.

Production of recombinant heterologous proteins in plants is simple and is used for production of non-naturally occurring proteins as single chain Fv fragments (ScFvs). High yield of recombinant protein is the main goal of production system in transgenic plants. Therefore appropriate expression vectors and constructs are designed to achieve high yields of the engineered gene products [7, 12].

Protein Production in Plant System

The transgene in expression construct is chimeric structure as it is surrounded by various active regulatory elements. Polyadenylation sites play an important role for the high level of expression of transgene. *Cauliflower mosaic virus* (CaMV) 35S promoter works well with dicots. It is a strong constitutive promoter that is made more active by duplicating the enhancer region. However, in monocots maize ubiquitin-1 promoter is the preferred promoter. The presence of an intron in the 5′-untranslated region (5′-UTR) enhances transcription in monocots.

For obtaining high yield of the protein, several factors may be appropriately considered:

- The incorporation of polyadenylation sites may be from CaMV 35S transcripts or the *Agrobacterium tumefaciens* nos gene or the pea ssu gene.
- The yield can be controlled by placing the gene under the control of the promoter which is active in a particular tissue or developmental stage or particular environment (e.g., rice glutelin, pea legumin).
- Usage of inducible promoter (e.g., tomato hydroxyl-3-methylglutaryl CoA reductase-2 (HMGR2)) which has mechanical gene activation system developed by Cramer (Crop Tech Corp., Virginia, USA). Transcription starts when harvested tobacco leaves are sheared during processing.
- Codon bias in the host plant may be overcome by engineering of transgene at positions, which might lead to truncation and/or misincorporation, or slowing the process.

(continued)

- Subcellular targeting is also very important factor which affects the process of folding, assembly, and posttranslational modification and can be efficiently achieved by inclusion of an N-terminal signal peptide.
- Position of transgene integration.
- Structure of transgene locus.
- Gene copy number.
- Presence of truncated or rearranged transgene copies.
- Affinity tags as His or the FLAG epitopes can be used to ease the process of purification; however, these modifications not only affect the primary structure but also the properties of the protein.

Figure 4.5 shows the general vector pCAMBIA (it is small in size (7–12 kb) maintained in high copy number with pVS1 replicon that imparts chloramphenicol or kanamycin resistance and high stability in *Agrobacterium*) that is used for transgene expression in the plants. The modified vector has shown success in insulin production.

Nowadays plant system is efficiently engineered to produce human growth hormone, human serum albumin, erythropoietin, α-interferon, antibodies and ScFvs, toxins, subunit vaccines, and insulin [20].

4.9 Challenges of Production of Therapeutic Proteins

With increasing demand from the consumer, the companies are trying to increase the productivity [17]. Few challenges with large-scale production of proteins are:

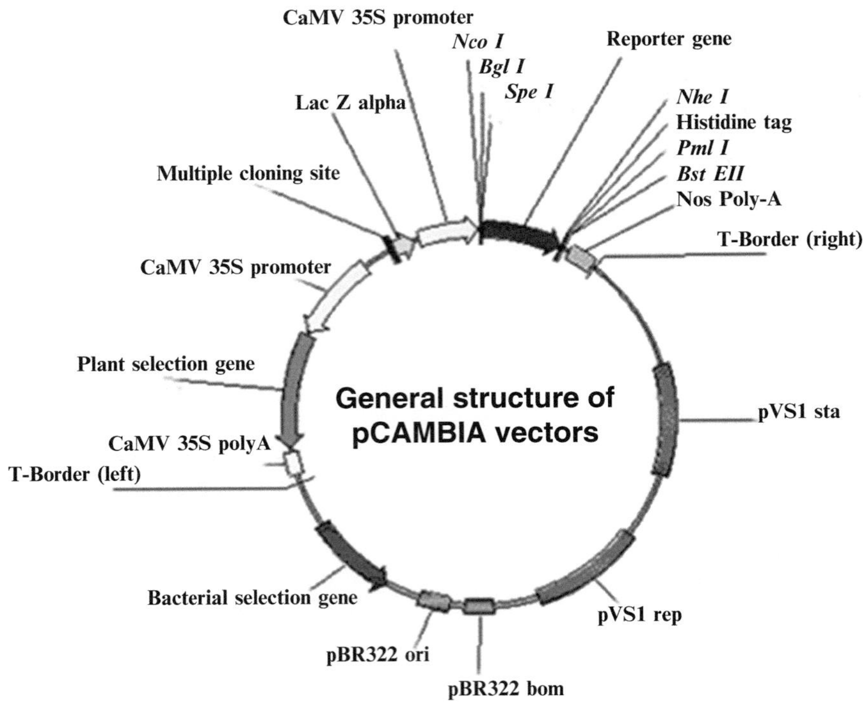

Fig. 4.5 The figure shows the structure of pCAMBIA vector (cambia.org) used for plant transformation. The vector has CaMV35S promoter, multiple cloning site, and reporter gene (GUS or GFP may be used). The vector can be modified to express genes for insulin (tomato) or Hep-B surface antigen (HBsAg) for recombinant therapeutics

1. Loss of expression: For high output, it is very important that gene of interest should give adequate protein production. However expression may be lost due to many factors, like, if there are structural changes in the recombinant gene or inactivation or disappearance of the gene from host cell. Other factors influencing yield may be increase in the copy number of insert, maintenance of optimum temperature, and toxicity of the expressed protein to the host. Chromosomal integration of the foreign gene might overcome the problem of expression stability, but in plasmid-based system, high copy number leads to increased yield.
2. Posttranslational processing: Protein folding requires foldases (accelerates protein folding) and chaperones (prevents protein formation of nonnative insoluble folding intermediates). Glycosylation is complex PTM requiring consecutive steps and enzymes. Glycosylation is important as it determines protein stability, solubility, antigenicity, folding, localization, biological activity, and circulation half-life. Getting correctly glycosylated and folded protein is required for therapeutic usage. In prokaryotes, with the discovery of N-glycosylation system in *Campylobacter jejuni*, several other systems of O-glycosylation were unraveled in both pathogenic and symbiotic bacteria. The production of recombinant proteins is commonly done by using *E. coli*, yeast, or cell lines derived from insects (SF9), mice (SP2/0), or CHO, but obtaining fully human PTMs is a challenging task. The commonly used mammalian cell lines of rodent origin (such as SP2/0, CHO, or BHK) are able to perform humanlike glycosylation. However, some human components are missing (such as the 2,6-linked sialylation), and a number of nonhuman components have been found to significantly increase (terminal galactose linked to another galactose or terminal sialic acids) which increases the high risk of immunogenic reactions. For this reason, human cell lines

providing a human glycosylation pattern have increased attention, and the efforts are being made for the development of novel glyco-engineered cell lines for production of fully glycosylated protein therapeutic.

3. Overexpression of therapeutic protein might result in the formation of inclusion bodies in prokaryotic system. Rapid intracellular protein accumulation and expression of large proteins increases the probability of aggregation. Aggregation protects proteins from proteolysis and can facilitate protein recovery. When the expressed protein is toxic to the host, the presence of protein in the inclusion body tends to protect it.

Precautions Recombinant protein production requires some precaution resulting in a loss of yield and/or product:

1. Contamination: At any stage, contamination might occur from any source. It poses big challenge and may be adverse when contaminated material is put for human usage.
2. Immune response: For the therapeutic agent, the body can mount the immune reactions which lead to deposition of immune complexes in various tissues, and condition of anaphylactic shock might occur (e.g., when some essential agent is lost since birth like factor VIII, then the patient might raise antibody response against the treatment). This also occurs when antibodies are used as therapeutic agents in the treatment of variety of cancers.
3. Protein aggregation: Any nonfunctional condition might result in aggregation or loss of activity of recombinant protein.
4. Posttranslational modifications and folding: After purification, the proper modifications and refolding are required for therapeutics.
5. Disulfide bond formation: It stabilizes protein structure; thus, strategies for specific extracellular excretion pathway or overexpression of chaperones is required for optimum production.
6. Degradation: Sometimes proteins, which are active in host cells like proteases or protein-modifying chemicals, might degrade the recombinant protein (e.g., PEG interferon is modified to have polyethylene glycol with prolonged presence and reduction in enzymatic degradation and renal clearance, thus extending its presence with lesser immunogenicity) [17].

4.10 Some Important Biopharmaceuticals

4.10.1 Tissue Plasminogen Activator (tPA)

Ischemic stroke and myocardial infarction are one of the leading causes of cardiovascular morbidity and mortality in the world. Important thrombolytic agents are urokinase (obtained from urine), tissue plasminogen activator (tPA), and streptokinase (obtained from bacteria). Among these, t-PA is largely used commercially. Plasminogen activators are serine proteases, which are responsible for conversion of inactive proenzyme plasminogen (Plg-a single chain glycoprotein) to serine protease called plasmin (Plm). The plasmin degrades the network of fibrin of the blood clots (Fig. 4.6a). There are two immunologically unrelated groups of plasminogen activators, the 55 kDa urokinase-type-PA (u-PA) and 72 kDa tissue-type PA (t-PA) (EC 3.4.21.68). The t-PA is the physiological vascular activator consisting of single polypeptide chain of 72 kDa consisting of 527 amino acids. It shows strong activity in the presence of fibrin [23].

The potential inhibitors of the thrombolytic cascade are type I plasminogen activator inhibitor (PAI-1 or serpin E1) and PAI-2 (secreted by placenta and present in significant amount during pregnancy). They act by competing with t-PA for binding sites on fibrin thus preventing the fibrinolytic cascade. PAI-1 complexes with t-PA for binding to fibrin. Thus, truncated t-PA in which the residues responsible for interacting with PAI-1 (296–299) are replaced with four alanine amino acids and three domains (finger, epidermal growth factor (EGF), and kringle 1 domains)) are deleted, and chimeric tetrapeptide Gly-His-Arg-Pro (GHRP) with high affinity to fibrin was added. Reteplase is the deletion mutant with a

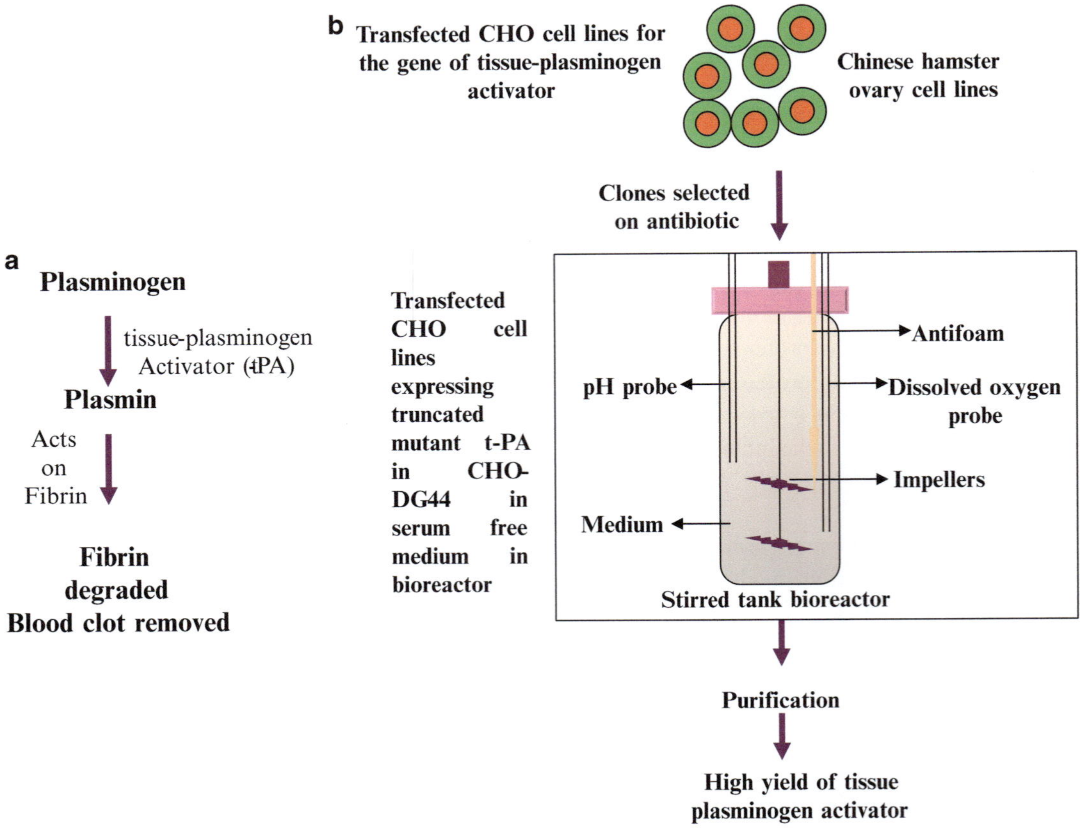

Fig. 4.6 The figure shows the function and production of tissue plasminogen activator. (**a**) This figure shows the function of t-PA that it helps in degradation of blood clot by degrading fibrin by activation of plasminogen into plasmin. (**b**) This figure shows the production of truncated t-PA in CHO DG44 cell line. The mutated form of t-PA is resistant to inhibition by plasminogen activator inhibitor and has better effectiveness in clinical usage

prolonged half-life, in which the finger, EGF, and kringle 1 domains of the full-length molecule are all deleted; thus, it is not inhibited.

Deficiency of PAI leads to over fibrinolysis and hemorrhagic diathesis (like deficiency of clotting factors). Tiplaxtinin is the inhibitor and is used for remodeling of blood vessels [2].

4.10.1.1 Production

Plasminogen activator has great clinical relevance for the management of stroke and myocardial infarctions. For production of t-PA, *E. coli* and yeast system did not work properly due to lack of posttranslational modification and over glycosylation, respectively.

A novel truncated form of t-PA with an improved fibrin affinity and an increased resistance to PAI-1 was expressed in a CHO DG44 expression system. Therapeutic protein was produced in stably transfected CHO DG44 cell lines. These cell lines were maintained in serum-free medium, with glutamine, hypoxanthine, and thymine in stirred tank bioreactor. The cells were grown at 37 °C, 140 rpm with 5 % CO_2 and 85 % humidity. The protein was then purified, with higher yield (Fig. 4.6b).

4.10.2 Factor VIII

Factor VIII is one of the important factors of all blood clotting factors. Deficiency of factor VIII causes bleeding disorder called hemophilia A. The hemophilia may be mild to severe depending upon factor VIII concentration in the body. In moderate and severe factor VIII deficiency, there can be spontaneous bleeding episodes in the joints. Hemophilia A affects 1 in 5,000–10,000 males. Replacement therapy is the treatment option for hemophiliacs either with human plasma-derived factor VIII (pdFVIII) or recombinant FVIII (rFVIII) [21].

Transfection of HEK 293 cell cultures in serum-free suspension is being tried for optimal yield. Recombinant factor VIII (rFVIII) is produced by culturing mammalian cells as baby hamster kidney (BHK) or Chinese hamster ovary cells (CHO), using large-scale bioreactors. Standardizations are done to maximize yields.

4.10.3 Insulin

Insulin is a peptide hormone consisting of 51 amino acids. It is secreted by β cells of islets of Langerhans of pancreas. The hormone is responsible for maintaining normal blood glucose level in blood. Insulin is stored in the form of proinsulin which contains two polypeptide chains, A and B, and is connected with a third peptide C-chain), which before secretion is cleaved with production of insulin and C-chain. The cleavage results in the removal of C-chain, and the A (21 amino acids) and B chain (30 amino acids) are linked by disulfide linkage to form mature insulin.

In the beginning, efforts were made to isolate mRNA for pre- and proinsulin from rat islets of Langerhans of pancreas and to synthesize cDNA. Thereafter, it was inserted into a plasmid. The recombinant plasmids were transferred into the *E. coli* cells, which secreted proinsulin [4, 5, 6, 19].

Scientists have chemically synthesized DNA sequences for two chains, A and B, of insulin and separately inserted into two pBR322 plasmids by the side of β-galactosidase gene. The recombinant plasmids were separately transferred into *E. coli* cells which secreted fused β-galactosidase-A chain and β-galactosidase-B chain separately. These chains were isolated in pure form by detaching from β-galactosidase with yields of about 10 mg/24 g of healthy and transformed cells. Production of recombinant insulin is shown in (Fig. 4.7a, b).

4.10.4 Human Growth Hormone (HGH)

Growth hormone is produced by the anterior lobe of the pituitary gland and is released in multiple pulses. The hGH is encoded by GHN gene cluster (an array of five closely related genes), which is localized on chromosome 17. It belongs to diverse gene family that has evolved by gene duplication events and has lots of structural similarity and some common functions. Of the various forms, the predominant form of hGH is 22 kDa protein of 191 amino acids with two disulfide bonds.

GH does not control the functions directly, but acts on certain hormones or somatomedins for its activity, for example, insulin-like growth factor 1 (IGF-1). It has a wide spectrum of roles to play as promotion of long bone growth, promotion of normal sex organ development through puberty, regulation of metabolism, stimulation of tissue growth and repair, anabolic/anti-catabolic effect via improved nitrogen retention, modulation of bone mineral density and metabolism throughout life, proliferation of some cell types of the immune system, appetite stimulation, and breakdown of fat (lipolysis). In clinical conditions, the therapy of growth hormone is given in the treatment of dwarfism, bone fractures, skin burns, bleeding ulcers, and AIDS.

Recombinant human growth hormone (rhGH) is 22 kDa consisting of long chain of amino acids. It is used in deficiency disorders of growth hormone. In children, it is used for growth abnormality as short stature and is used in chronic renal insufficiency. The therapy of growth hormone is also approved for adult growth hormone deficiency. GH is one of the most widely used hormones with the estimated market of more than 1.7 billion USD. Long experience in its administration has proven the therapy as safe and effective in various conditions of growth abnormality.

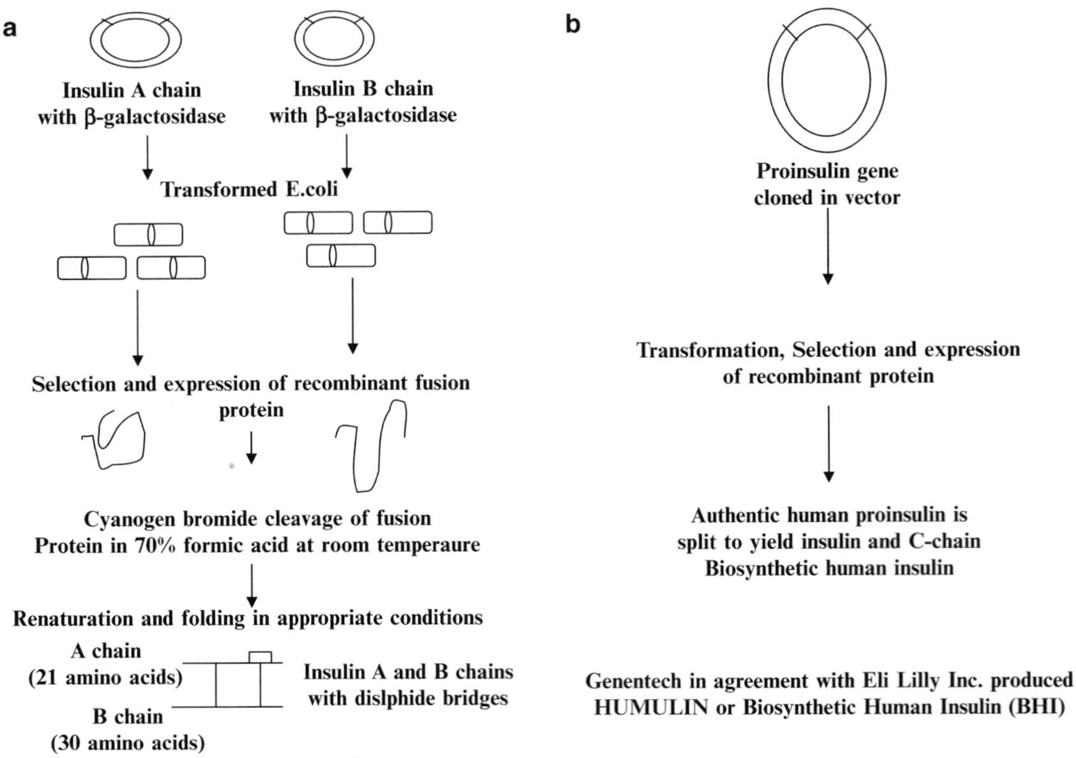

Fig. 4.7 The figure shows the production of insulin. (**a**) The A and B chains of insulin are cloned separately in the vector. The transformation is done and after selection, the recombinant protein is obtained for both A- and B-chain genes. Cleavage by using cyanogen bromide removes the bacterial protein. The protein is renatured and provided suitable conditions for folding. (**b**) This figure shows the production of insulin from proinsulin gene which yields insulin and C-peptide. This is preferred method for production of biosynthetic human insulin

Earlier pituitary-derived hGH was used but later on it was prohibited when found associated with Creutzfeldt–Jakob disease. Because of recombinant DNA technology, safe and abundant recombinant hGH was produced in various heterologous systems. As the non-glycosylated human growth hormone was biologically active, thus, the preferred system for its production is *E. coli*, which allows its rapid and economical production in large amounts [14].

Recombinant hGH (rhGH) is now used to treat:

- GH-deficient (GHD) short-stature children.
- Acceleration of wound healing.
- Increase in insulin-like growth factor (IGF)-1 levels.
- rhGH increases IGF-1, osteocalcin, type I procollagen pro-peptide (PICP), and bone density, when administered to children with GHD.

For optimal productivity, strong inducible promoters are preferred as IPL, IPR, trc, and T7 in *E. coli*. They are advantageous as they drive overproduction of recombinant proteins. Apart from *E. coli*, human somatotropin (hST) expression was tried in a biologically active, disulfide-bonded form in tobacco chloroplasts. The hormone is used for the treatment of hypopituitary dwarfism in children; additional indications are in treatment of Turner syndrome, chronic renal failure, HIV wasting syndrome, and possibly treatment of the elderly. Growth hormone deficiency in human occurs both in children and adults [18, 22].

4.10.5 Interferons

Interferons are a group of proteins which are secreted in response to viral infections. Resistance imparted by INFs is short lived and does not last forever. They are family of naturally occurring proteins that are made and secreted by all the cells (INF-α and INF-β) and lymphocytes (INF-γ). All these modulate the response of cells and immune system to viruses, bacteria, and cancer.

Interferons are produced by either an established cell line (lymphoblastoid) or fresh cells isolated from blood. The production involves induction with virus and priming (incubation with some interferon) with interferon, resulting in better yield. The affinity chromatography with monoclonal antibodies packed in the column has helped in purification of interferons. But before the clinical usage, the removal or inactivation of virus is very important. The interferon therapy is used for cancers and viral infections (INF-α), multiple sclerosis (INF-β), and chronic granulomatous disease (INF-γ). Multiferon is natural interferon-α, which consists of several subtypes. In some of the cancers like Merkel cell carcinoma, type I interferons (multiferon, which is a mix of various INF-α subtypes and INF-β) are highly effective. In Chap. 11, interferons are mentioned in protein therapeutics. Interferon is marketed as Roferon-A, Infergen, Intron A, and so on.

4.10.6 Erythropoietin

Hematopoietic growth factors consist of cytokines and protein hormones produced by the body which govern the production and maturation of the various cells produced during the process of hematopoiesis in the bone marrow from hematopoietic stem cell. The precursor cells in the presence of a particular growth factor differentiate and become a specialized kind of cell as monocyte, macrophage, lymphocytes, or red blood cell.

One of the important growth factor is erythropoietin which is a protein hormone produced by a specific type of cells in the kidney. In the presence of erythropoietin, progenitor cells are stimulated in the bone marrow to form mature erythrocytes (red blood cells). Thus the patients with chronic kidney disease are unable to maintain adequate amount of erythropoietin for normal development of erythrocytes in blood, resulting in low numbers of red blood cells and subsequent anemia. These patients either require blood transfusion or erythropoietin from outside. As supply is limited from the natural source that is kidney cells, thus a recombinant human erythropoietin EPOGEN® which is Amgen's trade name for epoetin alfa is marketed for anemic condition involving erythropoietin. The human gene encoding erythropoietin was cloned into the Chinese hamster ovary cell line for production of the human protein. This cell line continues to be used today for the production of EPOGEN®.

The half-life of erythropoietin can be increased by incorporating the glycosylation of the protein growth factor. Thus, Darbepoetin-α is an analog which is engineered for two extra amino acids which are substrates for glycosylation. Thus, production is done in CHO cell lines; the product has five N-linked sugar chains and has almost three times longer life than erythropoietin.

4.10.7 Platelet-Derived Growth Factor (PDGF)

PDGF regulates cell growth and division and plays a significant role in blood vessel formation (angiogenesis), the growth of blood vessels from already existing blood vessel in tissue, and may act in autocrine and paracrine stimulation of cell growth in vivo. PDGF plays the role in development, cell proliferation, cell migration, and angiogenesis and has been linked to atherosclerosis, fibrosis, and malignant diseases.

PDGF has five different isoforms: PDGFA, PDGFB, PDGFC, PDGFD, and AB heterodimer. PDGF-A and PDGF-B have 60 % similarity in amino acid sequence, but experiments suggest different biological functions for the two chains and different locations of these under different transcriptional controls. PGDF-AA is released in the medium, and PDGF-BB are insufficiently secreted and remain attached to the plasma membrane, and of these PGDF-BB and PDGF-AB are strong mitogens and are probably responsible for biological roles of PDGF. PDGF receptor (PDGFR) is receptor tyrosine kinase (RTK) (alpha and beta type). Upon activation by PDGF, these receptors dimerize leading to autophosphorylation of several sites on their cytosolic domain. PDGF being a mitogen promotes the proliferation of fibroblasts and smooth muscle cells in vitro.

PDGF shows considerable heterogeneity with sizes of 27–31 kDa; however, purified PDGF is cationic protein of 30 kDa. Recombinant human platelet-derived growth factor (rh-PDGF) was the first recombinant protein to be approved by the US Food and Drug Administration for treatment of chronic foot ulcers in diabetic patients (Regranex, Ethicon Inc. Somerville, NJ). It has the potential for use in bone regeneration and increasing bone density in long bones and spine. PDGF is commercially produced by using *E. coli* and mammalian cells.

4.10.8 Epidermal Growth Factor (EGF)

Human EGF protein has 53AA and three intracellular disulfide bonds and plays an important role in the regulation of cell growth and proliferation. It shows strong sequential and functional homology with human type- alpha transforming growth factor (hTGF alpha), which is a competitor for EGF receptor site. EGF acts by binding with high affinity to EGFR on cell surface and stimulates the intrinsic protein tyrosine kinase activity. EGF has many biological activities. Initial observations were centered around their proliferative effects on fibroblasts, keratinocytes, and epithelial cells. EGF modulates luteinizing hormone and thyroid hormone. EGF is produced commercially by engineered *E. coli*. The other systems are also being explored for optimum EGF production [9].

4.10.9 Fibroblast Growth Factor (FGF)

FGFs are commonly mitogens with multifunctional proteins with a wide variety of regulatory, morphological, and endocrine effects. There are 18 mammalian FGFs (1–10 and 16–23) which affect growth and functions of a wide variety of mesenchymal, endocrine, and nerve cells. The functions of FGFs in developmental processes include mesoderm induction, anteroposterior patterning, limb formation, neural tube induction, and brain development and in mature tissue/systems angiogenesis, keratinocyte organization, and wound healing processes. FGF is very important during normal development of both vertebrates and invertebrates, and any irregularities in their function lead to a range of developmental defects [1].

FGF1 and FGF2 show strong angiogenic properties with the promotion of endothelial cell proliferation and the physical organization of endothelial cells into tubelike structures and the growth of new blood vessels from the preexisting vasculature. FGF7 and FGF10 (also known as keratinocyte growth factors (KGF) and KGF2, respectively) stimulate the repair of injured skin and mucosal tissues by stimulating the proliferation, migration, and differentiation of epithelial cells, and they have direct chemotactic effects on tissue remodeling. Most FGFs are secreted proteins that bind heparan sulfates and can therefore be caught up in the extracellular matrix of tissues that contain heparan sulfate proteoglycans. This allows them to act locally in a paracrine fashion. However, the FGF19 subfamily (including FGF19, FGF21, and FGF23) which binds less tightly to heparan sulfates can act in an endocrine fashion on far away tissues, such as the intestine, liver, kidney, adipose, and bone. For example, FGF19 is produced by intestinal cells but acts on FGFR4-expressing liver cells to downregulate key genes in the bile acid synthase pathway; FGF23 is produced by the bone but acts on FGFR1-expressing kidney cells to regulate the synthesis of vitamin D and in turn affect calcium homeostasis. FGF may be synthesized using *E. coli* as host system.

4.10.9.1 Therapeutic Potential of FGFRs

FGF is involved in stimulating collateral vascularization and recovery from ischemia as well as enhancing wound healing, nerve regeneration, and repair of cartilage and has been alternately referred to as "pluripotent" (capable of developing into more than one cell type or tissue) growth factors and as "promiscuous" (biochemistry and pharmacology concept of how a variety of molecules can bind to and elicit a response from single receptor) growth factors due to its multiple actions on multiple cell types. The FGFs and small-molecule FGF receptor kinase inhibitor are used in the treatment of cancer and cardiovascular disease and have potential in the treatment of metabolic syndrome and hypophosphatemic diseases:

- Receptor tyrosine kinase inhibitor (Sunitinib) is approved for indications in renal cell carcinoma and gastrointestinal stromal tumors.
- Small FGFR inhibitors, SU5402, PD173074, and nordihydroguaiaretic acid, are effective in multiple myeloma cell lines.
- PD173074 can induce cell cycle arrest in endometrial cancer cells with mutated FGFR2.
- Antibody against FGFR3 has been shown to effectively cause apoptosis in mouse models of multiple myeloma and bladder cancer.

Thus FGFR inhibition can be very effective in the treatment of cancer.

4.10.10 Nerve Growth Factor (NGF)

The nerve growth factor (NGF) is an important member of the family of neurotrophins. Five protein nerve growth factors of the neurotrophin family are important. They regulate the development of the nervous system and play an important role in maintaining the structure, plasticity, and repair of the adult nervous system. All the neurotrophins are basic proteins of about 120AA, share 50% sequence homology, and are highly conserved in mammalian species.

Nerve growth factor (NGF) is a small secreted protein which induces the differentiation and survival of particular target neurons (nerve cells). Little is known of the biological action of neurotrophin apart from NGF. The nucleotide sequence of cDNA predicts that NGF is synthesized as pre-pro-NGF. Upon removal of hydrophobic signal, either 34 kDa or 27 kDa pro-NGF is generated depending on the size of the transcript. However, processing of the precursors in different tissues is not well understood. They are essential for normal development, growth, and differentiation of the sympathetic and sensory neurons and are also essential to maintain the normal function of these cells in adults. Thus it is important for maturation and survival of neurons and prevents degeneration of adult neurons. Apart from its important role in the nervous system, it has been shown to possess protective action of human pressure ulcer, corneal ulcer, and glaucoma. Reduced sensation may be observed in leprosy, wound healing, nerve injury, and diabetes. NGF may help to regulate the sensory fiber sensitivity and function directly or indirectly by stimulating other effectors. Administration of recombinant NGF may improve sensation and pain [13].

Cholinergic neurons of the basal forebrain show receptors for NGF; specific mRNAs for various NGFs have been identified in different areas of the brain. Cholinergic neuron loss is a cardinal feature of Alzheimer's disease. Nerve growth factor (NGF) stimulates cholinergic function, improves memory, and prevents cholinergic degeneration in animal models of injury, amyloid overexpression, and aging. NGF acts on intracellular calcium through tyrosine kinase receptor mechanism. Nerve growth factor enhances early regeneration of severed exons and is also important in maintaining the biochemical and morphological phenotype of mature basal forebrain cholinergic neurons (BFCNLs) after lesions or injury of the central nervous system (CNS). Thus, NGF may provide therapeutic option for preventing death of cholinergic neurons and other clinical conditions and is produced using *E. coli*.

4.10.11 Transforming Growth Factor Alpha (TGF-α)

TGF-α exerts its function in an autocrine and endocrine fashion for various cell types of ectodermal origin, including most epithelial cells. It is normally a transmembrane protein and functions in cell communication through its ability to activate a receptor tyrosine kinase. Ectodomain of TGF-α is cleaved in a highly regulated manner, releasing soluble TGF-α which activates paracrine signaling. The receptor is EGF/TGF alpha receptor; therefore, the focus is on understanding the important roles of TGF-α and EGF receptor signaling in carcinoma development.

4.10.12 Transforming Growth Factor Beta (TGF-β)

TGF-β is a large group of related proteins. Some of its family members include bone morphogenetic protein (BMPs), growth, and differentiation factors (GDP). It affects tissue remodeling, wound repair, hematopoiesis, morphogenesis, embryonic development, adult stem cell differentiation, immune regulation (is switch factor for IgA), and inflammation. It exists as multiple forms as TGF-β1, TGF-β2, and TGF-β3. It acts through transmembrane serine/threonine receptor kinase leading to the activation of Smads. It acts as tumor suppressor during cancer initiation but promoter during tumor progression. It has a role in the control of embryonic development, cellular differentiation, hormone secretion, and immune function. Its role as mesenchymal differentiation factor, with focus on the muscle, fat, and bone cell, might provide insights into its deregulation in skeletal and developmental diseases and is the area of active research. It is produced using CHO cell lines.

4.11 Future Prospects

There is huge potential for future therapies using proteins as therapeutic agents. The recombinant proteins are not only beneficial, but the researchers can further engineer them to improve their activity and prolonged stay in the body, for example, engineering of monoclonal antibodies to have toxin or radioisotope or generation of bispecific antibody. Still the technology is struggling hard to make the diseases completely a text of books and having a society free of diseases and the pathogens.

4.12 Chapter End Summary

- The production of proteins for therapeutic purpose is very important not only because of their specific action with minimum side effects but also due to their unique form and functions.
- Nowadays therapeutics (pancreatic enzymes from hog and pig pancreas or a-1-proteinase inhibitor from pooled human plasma) are not extracted from their native sources (from humans, animals) due to risk of transmission of pathogens. Other problems faced for extraction from native sources were the scarce availability of animal tissue for production, high cost of purification with less yield, and immunological reactions in the recipients.
- Majority of the biological therapeutic agents are produced by using various advance tools and technologies as cloning, selection, purification, and stability monitoring. It is also very important to monitor risks and side effects of therapeutics (safety analysis) obtained from a wide range of cells.
- The various biological systems are available for production of recombinant protein as bacteria, fungal system (yeast), insect cell, mammalian cell, human cell, and transgenic plants and animals.
- The system of choice is dependent upon the total cost of production and obtaining fully functional and appropriately modified (PTMs) (glycosylation, phosphorylation, or properly folded) protein. All these are essential for the biological activity of the protein.
- The bacterial system is the simplest with short cell division times, easy maintenance, easy to modify, and cost-effective production. The bacteria have limitation in production of

large-sized protein and are unable to perform posttranslational modification as glycosylation. The glycosylation is very important as it affects the activity, half-life, and immunogenicity of the recombinant protein.

- Due to limitations, the other host systems were used for production which could give better product with minimal immune reactions and are cost-effective. Fungal and insect systems are utilized for protein production but sometimes result in inappropriate glycosylation. Improper glycosylation may result in nonfunctional and highly immunogenic protein. Mammalian cells are the most preferred system for production of these proteins. CHO system is the system of choice for therapeutic protein production. It accounts for nearly 70 % production of all products available in the market.
- The protein products are having important therapeutic role. They are approved for marketing and use by various government agencies like the Food and Drug Administration (FDA) and European Union (EU). The approved recombinant biotechnology medicines include replacement products, monoclonal antibodies, interferons, vaccines, hormones, modified natural products, and many others. Many of them are in usage and helping to alleviate the symptoms or cause of the diseases.

Multiple Choice Questions

1. The function of proteins in the body is:
   (a) Initiation of transcription
   (b) Mediators of metabolic processes
   (c) Acts as enzymes
   (d) All of the above
2. The purpose of recombinant DNA technology is:
   (a) Genes can be cloned in *E. coli*
   (b) The product is cost-effective
   (c) Mammalian cells may be used for production
   (d) None of the above
3. The advantages of using *E. coli* as host for production of recombinant protein are:
   (a) Easy scale-up and simple media requirement
   (b) Can easily perform splicing of foreign DNA
   (c) Produces processed and properly modified protein
   (d) All of the above
4. The limitation of *E. coli* which poses a problem for recombinant protein production is:
   (a) Codon biasing
   (b) Splicing
   (c) Posttranslational modification
   (d) None of these
5. The fungal being eukaryote offers tremendous advantages in recombinant protein production but it does:
   (a) Proper protein folding
   (b) Posttranslational modification
   (c) Hypermannosylation
   (d) Easy scale-up
6. Insect cells may be infected with which viral vector for production of recombinant protein?
   (a) Retrovirus
   (b) *Coronavirus*
   (c) Baculovirus
   (d) Parvovirus
7. Most commonly used mammalian cell line is:
   (a) BHK
   (b) CHO
   (c) HeLa
   (d) All of these
8. Why is there a need of serum- or plasma-free medium in the production of therapeutic protein?
   (a) As their cost is high, thus the cost of product is high
   (b) Risk of transmission of unknown pathogens leading to diseases
   (c) Variation in each lot results in different yield
   (d) None of these
9. The first recombinant protein released in the market was:
   (a) Activase
   (b) Humulin
   (c) Trastuzumab
   (d) All of these
10. Growth factor inhibitors are finding place in therapeutics because:
    (a) Growth factors are harmful
    (b) Growth factor deactivation leads to healthy body

(c) Growth factors are required for progression of tumors
(d) All of the above

11. Darbepoetin is used for the treatment of:
(a) Short stature
(b) Anemia
(c) Blood clotting
(d) None of these

Answers

1. (d); 2. (b); 3. (a); 4. (c); 5. (c); 6. (c); 7. (b); 8. (b); 9. (b); 10. (c); 11. (b)

Review Questions

Q1. What are the advantages and disadvantages of using *E. coli* as host for production of recombinant proteins?
Q2. What is the role of codon biasing in expression of foreign gene?
Q3. What are the posttranslational modifications?
Q4. Write a note on fungal host for production of proteins.
Q5. When are insect cells preferred for the recombinant protein production?
Q6. Write a note on (a) insulin, (b) factor VIII, (c) GH, and (d) tissue plasminogen activator.
Q7. Write the trade names of insulin, tPA, and GH.

References

1. Beenken A, Mohammadi M (2009) The FGF family: biology, pathophysiology and therapy Nature reviews. Drug Discov 8:235–253
2. Fatemeh D, Barkhordari F, Alebouyeh M, Adeli A, Mahboudi F (2011) Combined TGE-SGE expression of novel PAI-1-resistant t-PA in CHO DG44 cells using orbitally shaking disposable bioreactors. J Microbiol Biotechnol 21:1299–1305
3. Ferrer MN, Domingo EJ, Corchero JL, Vazquez E, Villaverde A (2009) Microbial factories for recombinant pharmaceuticals. Microb Cell Fact 8:17
4. Frank BH, Chance RE (1985) The preparation and characterization of human insulin of recombinant DNA origin. In: Joyeaux A, Leygue G, Morre M, Roncucci R, Schmelck PH (eds) Therapeutic agents produced by genetic engineering "Quo Vadis". Toulouse-Labege, Symposium Sanofi Group, Paris MEDSI, pp 137–46
5. Galloway JA (1988) Chemistry and clinical use of insulin. In: Galloway JA, Potwin JH, Shuman CR (eds) Diabetes Mellitus, 9th edn. Lilly, Indianapolis, pp 106–137
6. Galloway JA, Hooper SA, Spradlin CT, Howey DC, Frank BH, Bowsher RR, Anderson JH (1992) Biosynthetic human proinsulin: review of chemistry, in vitro and in vivo receptor binding, animal and human pharmacology studies, and clinical trial experience. Diabetes Care 15:666–692
7. Goldstein DA, Thomas JA (2004) Biopharmaceuticals derived from genetically modified plants. Q J Med 97:705–716
8. Grillberger L, Kreil TR, Nasr S, Reiter M (2009) Emerging trends in plasma-free manufacturing of recombinant protein therapeutics expressed in mammalian cells. Biotechnol J 4:186–201
9. Guglietta A, Sullivan PB (1995) Clinical applications of epidermal growth factor. Eur J Gastroenterol Hepatol 7:945–950
10. Jackson SE (2013) Hsp90: structure and function. Top Curr Chem 328:155–240
11. Kamionka M (2011) Engineering of therapeutic proteins production in Escherichia coli. Curr Pharm Biotechnol 12:268–274
12. Ma JK, Drake PMW, Christou P (2003) The production of recombinant pharmaceutical proteins in plants. Nat Rev Genet 4:794–805
13. Manni L, Rocco ML, Bianchi P, Soligo M, Guaragna M, Barbaro SP, Aloe L (2013) Nerve growth factor: basic studies and possible therapeutic applications. Growth Factors 31:115–122
14. Martínez JAA, Saldaña HAB (2012) Genetic engineering and biotechnology of growth hormones, genetic engineering – basics, new applications and responsibilities, Prof. Hugo A. Barrera-Saldaña (Ed.), ISBN: 978-953-307-790-1, InTech, Available from: http://www.intechopen.com/books/genetic-engineering-basics-new-applications-and-responsibilities/geneticengineering-and- biotechnology-of-growth-hormones
15. Mayer MP, Bukau B (2005) Hsp70 chaperones: cellular functions and molecular mechanism. Cell Mol Life Sci 62:670–684
16. Miralles (2009) Microbial factories for recombinant pharmaceuticals. Microb Cell Fact 8:17
17. Palomares LA, Mondaca SE, Ramirez OT (2004) Production of recombinant proteins. In: Balbas P, Lorence A (eds) From methods in molecular biology. Humana Press, Totowa 267:15–51.
18. Rezaei M, Esfahani SHZ (2012) Optimization of production of recombinant human growth hormone in Escherichia coli. J Res Med Sci 17:681–685

19. Riggs AD (1981) Bacterial production of human insulin. Diabetes Care 4:65–68
20. Soltanmohammadi B et al (2014) Cloning, transformation and expression of proinsulin gene in tomato (Lycopersicum esculentum Mill). Jun J Nat Pharma Prod 9:9–15
21. Swiech K, Kamen A et al (2011) Transient transfection of serum-free suspension HEK 293 cell culture for efficient production of human rFVIII. BMC Biotechnol 11:11–114
22. Tabandeh F, Shojaosadati SA, Zomorodipour A, Khodabandeh M, Sanati MH, Yakhchali B (2004) Heat-induced production of human growth hormone by high cell density cultivation of recombinant Escherichia coli. Biotechnol Lett 26:245–250
23. Tor NY, Fredrik E, Lund B (1984) The structure of the human tissue-type plasminogen activator gene: correlation of intron and exon structures to functional and structural domains. Proc Natl Acad Sci U S A 81:5355–5359
24. Wurm FM (2004) Production of recombinant protein therapeutics in cultivated mammalian cells. Nat Biotechnol 22:1393–1398

Some Selected Resources

www.ibclifesciences.com
www.link.springer.com
www.ncbi.nlm.nih.gov/pubmed
www.qjmed.oxfordjournals.org
www.sciencedaily.com

Transgenic Animals and Plants

5

Abstract

Man from the very beginning in his evolutionary history is found to be involved in the manipulation of the traits of the animals and plants according to his need or desire. Prior to the development of molecular genetics, the only way of studying the regulation and function of mammalian genes was through the observation of inherited characteristics or spontaneous mutations, and selective breeding was a common practice among farmers for the enhancement of chosen traits. As the revolution took place in applied molecular genetics, new techniques evolved to study gene expression. Recent development of transgenic organism is one of those techniques.

Transgenic organisms as a subset of genetically modified organisms (GMOs) are organisms which have inserted DNA of different species origin; the inserted DNA/gene is known as transgene. Transgene is introduced into the organism, using recombinant DNA technology, and it must be transmitted through the germline so that every cell, including germ cells, of the body contains the same modified genetic material. In this chapter, readers would understand the methods of generation of transgenic animals and plants and the factors which affect the expression of transgene.

5.1 Introduction

Human from the very beginning in his evolutionary history has been involved in the manipulation of the traits of the animals and plants according to his need or desire. Prior to the development of molecular genetics, the only way of studying the regulation and function of mammalian genes was through the observation of inherited characteristics or spontaneous mutations, and selective breeding was a common practice among farmers for the enhancement of chosen traits. As the revolution took place in applied molecular genetics, new techniques evolved to study gene expression. Recent development of transgenic organism is one of those techniques.

Transgenic organisms are genetically modified organisms (GMOs) which have inserted DNA of different species origin. The inserted DNA/gene is known as transgene. Transgene is introduced into the organism, using recombinant DNA technology. The gene must be transmitted through the germline so that every cell, including germ cells, of the body contains the same modified genetic

V. Gupta et al., *Basic and Applied Aspects of Biotechnology*, DOI 10.1007/978-981-10-0875-7_5

material. In this chapter, readers would understand the methods of generation of transgenic animals and plants and the factors which affect the expression of transgene.

5.1.1 Basic Requirements

In gene transfer, the knowledge of the biochemical and physiological mechanisms of its action, regulation of gene expression, and safety of transgene and its product is important. The process requires some basic steps for successful transfer:

- The extraction of nucleic acid (DNA/RNA). It is the first step in the genetic engineering process.
- After DNA extraction and selection of suitable gene, the next step is gene cloning. Gene cloning has four stages: (1) generation of DNA fragments, (2) ligation to a vector, (3) propagation in a host cell, and (4) selection of the transformed host.
- Gene is designed in such a way that it is under appropriate promoter and correct regulatory elements. Selectable marker gene may be incorporated for selection of the transformants from non-transformants.
- Incorporation of reporter gene allows detection of gene expression in transformed cells and enhancers are essential for efficient transcriptional control.
- The gene of interest (GOI) is ligated to selectable marker gene and reporter gene. It is helpful for screening, detection, and identification of the transgene containing host cell. Reported genes may be β-glucuronidase gene (GUS) or green fluorescent protein or luciferase gene. GUS produces blue product when it acts on its substrate, and thus cells appear blue, whereas GFP whereas GFP and luciferase transformed cells emit light when excited with appropriate wavelength of light.

Major problem in raising transgenics is "position effects" which have prevented the reproducible success and limited the initial expectations of this technique. To overcome such position effects, the progressive addition of regulatory elements belonging to the same or to a heterologous expression domain has been made. This expression domain contains all regulatory elements that are needed to specifically control the expression of a given gene in time and space.

Position Effect

These are the effects, which are observed on the expression of a gene in changed location on the chromosome. Variable level of transgene expression is observed in transgenic animals and plants with same expression construct. This is affected by:

- Site of transgene integration (position effect). This occurs due to the effect of local regulatory elements on the transgene and architecture of the chromatin.
- Chromatic architecture is usually repressive and nonspecific. This may be due to transgene integration in the heterochromatin region of the chromosomes. Its features like deacetylated histones and hypermethylated DNA may result in partial expression or complete inactivation of the transgene.
- The inclusion of locus control region (LCR) of the human β-globin gene cluster confers high-level and position-independent expression.
- The mammalian gene shows correct and high-level expression when its flanking regions are associated with it. Thus, the use of genomic constructs with introns and large amounts of flanking sequence from the source gene positively affects its expression. They may have enhancers, regulatory elements, and LCRs that protect transgene from position effects.
- Site-specific recombination systems may be helpful to eliminate position effects.

5.2 Preparation of Transgene Construct

As we know that although the basic coding system is the same in all organisms, the fine details of gene regulation often differ. A gene from prokaryotes will not work correctly if it is introduced unmodified into eukaryotic cells. That is why the gene or the DNA segment that has to be introduced in the other organism requires some modification so that it can express itself correctly. The transgene constructed in this manner is a combination of the gene and a strong promoter sequence immediately upstream of the gene that has critical roles in the temporal and tissue-specific regulation of gene expression. Active virus derived promoter elements results in high transgene expression. Nonviral promoters suffer from limited transgene expression. For the preparation of a transgene cassette, the gene of interest is isolated and is cut at specific sites by restriction endonucleases. The use of restriction enzymes facilitates recombination of different functional regions of genes obtained from different species. The ends of the completed construct may be modified by the addition of polylinker sequences containing several different restriction enzyme recognition sites. The polylinker permits the construct to be inserted into a variety of vectors for testing and cloning. This cut DNA is then ligated with carrier vector that may be a virus, plasmid, or YAC vectors with the help of ligases. The vector carries this gene construct into the organism for expression (Fig. 5.1).

There are so many definitions of transgenic animals. The Federation of European Laboratory Animal Association defines the term as an animal in which there has been a deliberate modification of its *genome*. The success of transgenics lies in the correct incorporation of and expression of transgene in the host. Sperm-mediated gene transfer in rabbits was a pioneering experiment for the production of transgenic animals. Although gene transfer was accomplished in that study, germline transmission of the transgene was not successfully reported until retroviral gene transfer vectors in mice were used in 1976. In 1980, scientists showed that pronuclear microinjection of DNA into one-cell mouse zygotes was a relatively efficient method to accomplish germline gene transfer. Despite the technical difficulties in their construction, "transgenic mice"

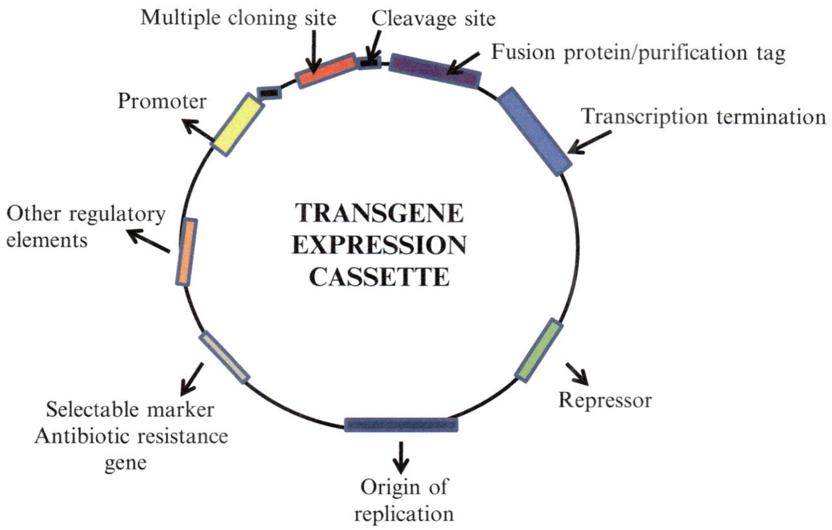

Fig. 5.1 Schematic diagram of the transgene expression cassette. Restriction enzyme recognition sites are clustered at either end of the cassette (i.e., upstream and downstream). The cassette has promoter for transcription; selectable marker gene, which would help in selection of transformants and fusion protein gene which would help in the purification of the recombinant protein

rapidly became part of the experimental biologists' arsenal, and after the development of transgenic mice, the development of other transgenic mammals and their use in molecular biology have become the standard tools for investigating gene expression, function, the disease pathology, and other experiments.

> The insertion of transgene can be random or targeted. Many methods result in random insertion where the transgene is placed anywhere in the genome. However if the DNA sequence, flanking the target loci, is known, then it is possible to place the gene at a specific site.

5.3 Production of Transgenic Animals

There are two basic strategies for producing transgenic animals. These are the so-called "gain of function" or "loss of function" transgenics. The "gain of function" relies on addition of a cloned fragment of DNA to an animal's genome to obtain new expression of a gene product that did not previously exist in that cell or tissue type. "Loss of function" is a more direct way to determine the function of a lost/inactive gene after its alteration (Fig. 5.2) and then observe the physical manifestation (phenotype) of that genetic trait, for example, the mouse is obese or diabetic. These types of mice are often called knockouts, and from several years now, it has been possible

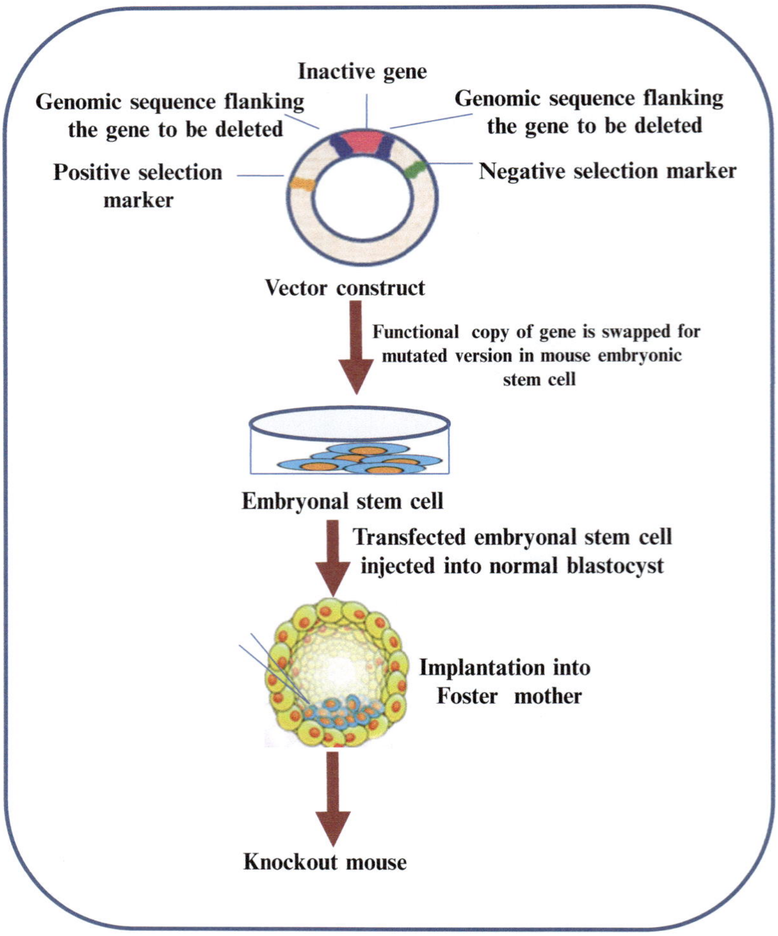

Fig. 5.2 The figure shows the loss of function cassette where inactive gene is incorporated for homologous recombination with genomic sequence flanking the target gene with positive and negative selection marker gene. The vector exchanges itself in mouse embryonic stem cell for functional copy resulting in the production of knockout mouse

to generate "knockout" mice, which carry specifically defined mutations in the gene of interest using gene targeting in mouse embryonic stem cells.

5.3.1 Transfer of Transgene in the Animal

The essence of this technology resides in successful gene transfer in the intended animal and its insertion in the genome. The insertion of gene in the genome can be random or targeted. There are several methods of producing transgenic animals: (1) DNA microinjection, (2) retrovirus-mediated gene transfer, (3) embryonic stem cell-mediated transfer, and (4) liposome-mediated DNA transfer into cells and embryos. A few modifications are also there for creating transgenics: (1) electroporation of DNA into the sperm, ova, or embryos and (2) nuclear transfer with somatic or embryonic cells.

5.3.2 Pronuclear Microinjection

Microinjection was the first technique used to generate transgenic mice by successful gene transfer using SV40 DNA directly into blastocoel cavity of preimplantation embryos and then their implantation in the foster mother [4, 8]. Eventually, the gene of interest may randomly integrate in the embryonic genome of some embryonic cells and ultimately contributes to germline. The technique was then extended to other animals (livestock). The method has poor efficiency and usually results in a high percentage of mosaics in which not all cells of the animal contained the transgene. The time and cost of screening for germline transmission from the major mosaic animals prohibits the generation of more transgenic animals. The method can also lead to high variability in transgene expression between animals due to mosaicism, variable efficiency of transgene integration, and chromosomal position effects. The major advantage of this method is its applicability to a wide variety of species. This method involves the direct injection of the gene construct into the cells to ensure guaranteed successful transfer of transgene to target cells [6]. Presently the established technique involves the injection of DNA into the pronuclei of the fertilized egg. The host cell is held in position with a suction pipette and DNA is transferred into the nucleus through a fine needle. Thereafter the embryos with the transgene are cultured in vitro up to the morula stage and then transferred to pseudopregnant foster mothers. The insertion of DNA is, however, a complete random process; thus, there is high probability that either the introduced gene might be inserted into the host genome or, less commonly, it may remain extrachromosomal for one or more cell divisions. Thus, the resulting animal may be transgenic or chimeric for transgene insertion. Five to 40 % of mice developing from manipulated eggs contain the transgene. Once the transgene is transmitted through the germline, it tends to be stably inherited over many generations. The transgene insertion is accompanied by deletions and rearrangements occurring in genomic DNA. The limitation of microinjection is that it only allows for the random addition of exogenous DNA rather than targeting to specific sites. DNA targeting is necessary in generating gene knockouts, for instance, to model human diseases. The success of pronuclear microinjection is evident in the generation of transgenic pigs, goats, sheep, and cattle; but its limitations have hindered the progress of transgenics in livestock.

5.3.3 DNA Transfer Through Virus

Gene transfer mediated by means of a carrier or vector, generally, a virus or a plasmid, provides increased probability of expression. Virus has natural ability to stably introduce DNA to the host upon infection; therefore recombinant retroviruses and replication defective adenoviruses can be efficiently used as vectors. Retrovirus is efficiently capable of infecting dividing cells. Initial attempts to produce transgenics by retroviral infection of early embryos invariably resulted in genetic mosaics caused by multiple insertion sites of the transgene. The short window of opportunity for the viral preintegration complexes to reach the embryonic chromatin during M-phase explains the delayed viral insertion, resulting in cell lineages with different insertion sites or no insertion at all. A significant advancement of this technique has been achieved recently by exposing metaphase II oocytes to transgene-containing retrovirus. Arrested oocytes are particularly appropriate because they have undergone nuclear envelope breakdown and remain at metaphase II for a longer period compared with the M-phase of somatic cells (Fig. 5.3).

This maximizes the probability of preintegration complexes gaining access to the oocyte chromatin. The infected embryos are implanted in the uterus of foster mother to get germline transfer of transgene. Adenovirus represents an alternative to retrovirus as vectors to insert recombinant DNA into the mammalian germline.

Adenoviruses Versus Retroviruses

There are some clear advantages of using adenovirus for transgenic purposes: it can infect a wide range of cell types, it can accommodate large pieces of exogenous DNA (>20 Kb), and high viral titer can be produced. The disadvantages of the adenoviruses are that it is inefficient mainly due to low integration frequencies and high toxicity.

However, the disadvantages of this technique are the possible interference of viral regulatory elements with the expression of genes and the susceptibility of virus to de novo methylation resulting in silencing of the genes. Though viruses are highly adapted to the process of gene transfer however, using virus as the vector may also produce health hazards as they may potentially arouse cancer/leukemia.

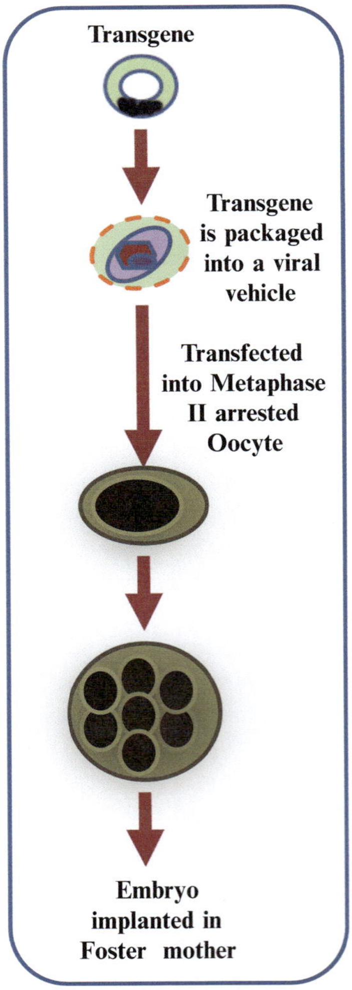

Fig. 5.3 Virus-mediated transgenic production where the virus along with transgene is allowed to transfect metaphase II arrested oocyte. After transfection of oocyte, it is implanted in foster mother resulting in production of transgenic animal

5.3.4 Embryonic Stem Cell Mediated Gene Transfer

Embryonic stem cells are obtained from inner cell mass of blastocyst. This method involves prior insertion of the desired DNA sequence by homologous recombination into an in vitro culture of embryonic stem (ES) cells, and the

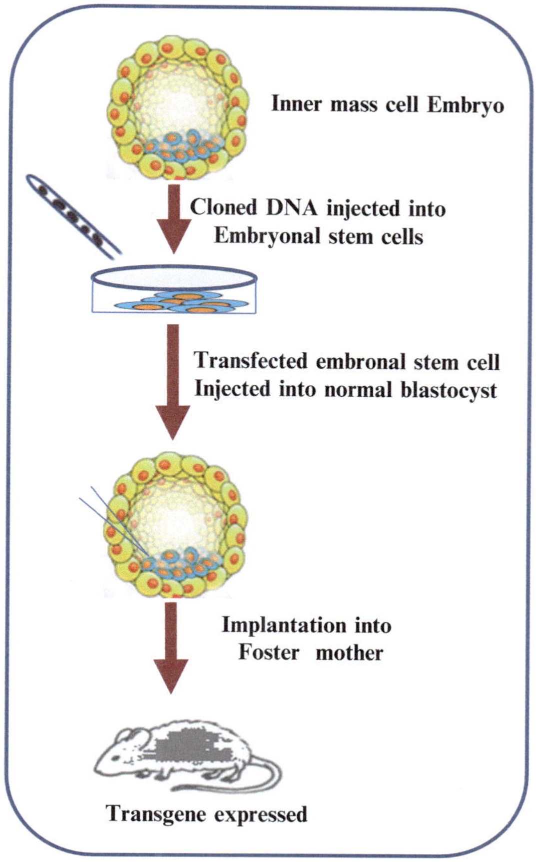

Fig. 5.4 Embryonic stem cell-mediated gene transfer. The transgene is injected into the embryonal stem cells. These stem cells with transgene are injected into normal blastocyst, which is then implanted into the foster mother

transformed cells can be selected using standard markers. Cells containing the desired DNA are incorporated into the blastocoel of host's embryo at the blastocyst stage, where they mix with inner cell mass (Fig. 5.4). Unlike the other two methods, which require live transgenic offspring to test for the presence of the desired transgene, this method allows testing for transgene at the cell stage. This is advantageous as there is no convenient way to select for eggs or embryos that have taken up foreign DNA.

The result is a true chimeric animal. ES cell-mediated gene transfer is the method of choice for gene inactivation, the so-called knockout method. This technique is of particular importance for the study of the genetic control of developmental processes. This technique works particularly well in mice. It has the advantage of allowing precise targeting of defined mutations in the gene via homologous recombination.

5.3.5 Sperm-Mediated Transgenesis

The ability of sperm cells to carry exogenous DNA spontaneously into the oocyte during fertilization was first reported by Brackett and coworkers in 1971 [3]. However, the concept of sperm-mediated transgenesis rested for 18 years until scientists reported the use of spermatozoa as DNA carriers to produce transgenic mice. The scientific community was challenged by an unsuccessful attempt to replicate the experiment met this work with skepticism. However afterward transgene delivery by sperm cells was used to produce transgenic animals in a wide variety of species, including cattle, pigs, rabbits, frogs, and zebra fish. Sperm-mediated transgenesis is the most straightforward technique to produce transgenic animals [10]. The spermatozoa are incubated with the DNA containing the gene of

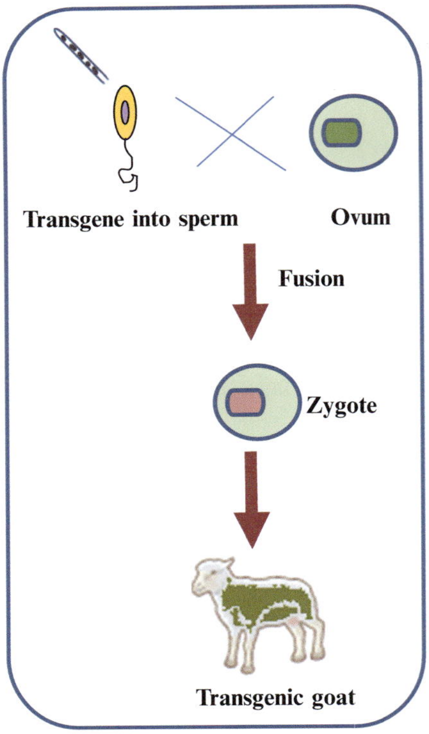

Fig. 5.5 Sperm-mediated gene transfer where transgene is inserted into male pronucleus which upon fusion with ovum results in the production of transgenic goat

interest followed by in vivo or in vitro insemination. DNA binds to the sperm's plasma membrane through specific DNA-binding proteins. After internalization, they are carried into the oocyte upon fertilization (Fig. 5.5). The most appealing characteristics of using sperm as vectors to produce transgenic animals are its simplicity (no embryo manipulation is required) and the possibility of performing mass production of genetically modified animals through in vivo or in vitro insemination of many oocytes. On the other hand, this technique also has limitations. Like pronuclear microinjection, no targeted modifications by homologous recombination could be achieved with this method due to the random integration of transgene.

Recently scientists reported successful generation of transgenic pigs carrying a human gene (human decay accelerating factor, hDAF). In this study, approximately 80 % of pigs had integrated the construct containing an hDAF minigene; from this 80 %, approximately 53 % expressed the foreign protein at different levels. There is no doubt that this success represents an important step toward production of humanized pig organs and tissues for human transplantation. If these findings are readily reproducible, sperm-mediated gene transfer may displace in short time the popular pronuclear microinjection method.

5.3.6 Somatic Cell Nuclear Transfer

Due to the absence of proven ES cells and the recent advances in nuclear transfer (NT), current emphasis for creating transgenic animals is on somatic cell nuclear transfer (SCNT). Nuclear transfer is a technique that is used to create a genetically identical copy, or a clone, of an animal. Nuclear transfer commonly involves the transfer or placement of a donor nucleus into the cytoplasm of an enucleated MII oocyte (Fig. 5.6). Donor cells can originate from a wide variety of cell types ranging from embryonic blastomeres all the way up to adult cells. Initial work on nuclear transfer was focused on using embryonic blastomeres as a donor source; the limited number of cells available in an early embryo hampered the process. Fetal or adult cells have been used successfully to clone all major livestock species, including sheep, cattle, goats, and swine. The ability to use cells where transgene can be introduced and which can be cultured increases the number of cells available to clone, thereby facilitating the ability to make transgenic animals [1, 2].

Success of SCNT: Dolly was the first mammal to be cloned using adult cell. Dolly was born after 277 attempts at Roslin Institute in Scotland (Fig. 5.7). Dolly, produced by the transfer of nucleus from a differentiated somatic mammary epithelial cell to an enucleated egg, had given new dimensions to animal cloning. In 1995 two lambs, Megan and Morag, were reported from nuclear transfer from cultured embryonic cells. The same group reported production of Dolly.

(continued)

Fig. 5.6 Somatic cell nuclear transfer where nucleus is removed from the egg cell and enucleated egg receives nucleus from donor somatic cell. The resultant egg with nucleus from somatic cell can be used for reproductive cloning or therapeutic cloning

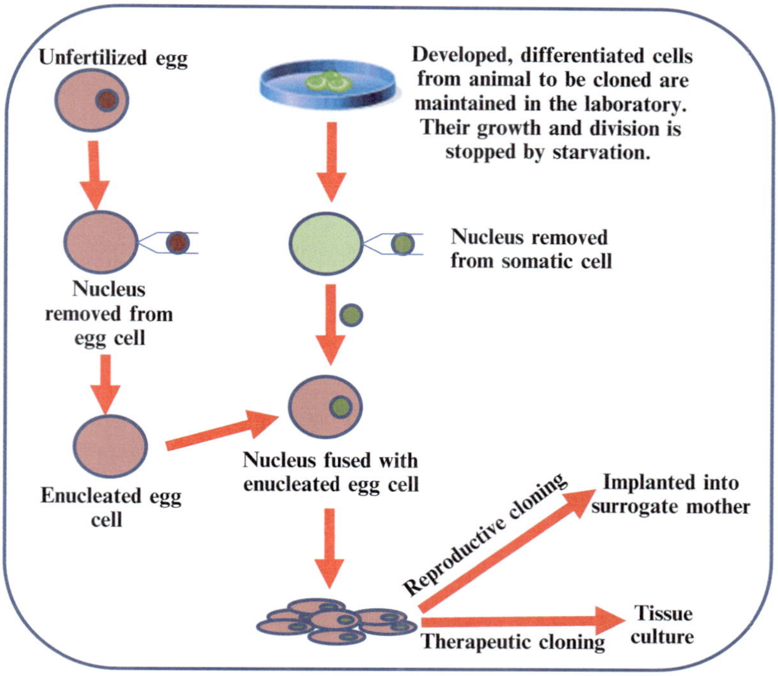

Fig. 5.7 The first cloned sheep Dolly is shown. Dolly was born by the technique of somatic cell nuclear transfer by using nuclei of adult somatic cell and enucleated egg cell

This opened up the routes for generation of transgenics and fueled debates and criticism for using it for human cloning.

Sheep Polly was a transgenic animal where nuclear transfer for human factor IX into fibroblasts of fetal sheep was done using the same nucleus for transfer to enucleated eggs. Polly produced the recombinant protein in her milk.

(continued)

Somatic cell nuclear transfer has facilitated the ability to make transgenic animals by circumventing most of the shortcomings of other transgenic techniques. First, the sex of the animal can be predetermined by choosing the donor material (i.e., male or female tissue). Second, the use of cell culture to propagate donor cells can lead to large numbers of transgenic cells that can be frozen and stored for long periods of time.

In conjunction with SCNT, these transgenic donor cells can eventually give rise to numerous cloned transgenic animals [5]. Transgene structure and expression can be tested by molecular techniques, such as PCR, Southern blot analysis, fluorescence in situ hybridization, and Western blot analysis, before initiating nuclear transfer and transferring the embryo to a recipient cow with a lengthy gestation time of 9 months. The proper use of SCNT also ensures that 100 % of animals produced are transgenic and that every cell of a cloned animal will have the transgene, thereby saving time and cost associated with recipient animals. The ability to use a clonal population of transgenic cells guarantees the same transgene insertion site for each clone, thus decreasing animal to animal variation in transgene expression levels. Further, transgene can be added to a specific genetic background. This technique can be used for the addition of DNA at random sites and targeted insertion of DNA by homologous recombination, which is vital in modulating specific gene expression as well as creating gene knockouts.

5.4 Transient and Stable Insertion of Transgene

When the foreign DNA is introduced in the cell of the host, it is called transfection, and if the gene is incorporated into the genome, it is called integration. When the foreign DNA fails to integrate into the genome of the host, it is maintained in the nucleus in extrachromosomal state. In the absence of origin of replication in this extrachromosomal DNA, the host maintains it for a very short time and subsequently it is degraded. This is known as "transient transfection." If the transgene is integrated in the genome of the host cell and is then passed on to all the descendants arising from the host cell, then it is known as "stable transfection." However the stably transformed cells need to be isolated from a large number of cells by selection.

Selection of stable transformants can be done by the use of certain enzymes or a drug-resistant gene:

- Thymidine kinase (TK): Any *TK*⁻ cell can be made wild-type *TK*⁺ by transfecting them with viral vector with wild-type *TK* gene.
- Hypoxanthine guanine phosphoribosyl transferase (*HGPRT*): Any *HGPRT*⁻ cell can be made wild type for the gene by transfecting wild-type *HGPRT* gene. The *TK*⁻ and *HGPRT*⁻ cells are grown in the presence of hypoxanthine–aminopterin–thymidine (HAT) medium.
- The presence of aminopterin blocks inosine monophosphate (IMP) and thymidine monophosphate (TMP) pathways for de novo DNA biosynthesis. In the absence of de novo DNA biosynthetic pathway, the cell is unable to synthesize DNA; therefore it would perish. The precursors of DNA biosynthesis (hypoxanthine and thymidine) provided in the medium are utilized by salvage pathway with the help of HGPRT and TK enzymes. Therefore the presence of HGPRT and TK in the cell resumes DNA biosynthesis pathway, but nontransfected cells without these genes (HGPRT and TK) are unable to survive in HAT medium.
- The markers used for selection are termed endogenous if they belong to a class of genes which are normally present in wild types, and they can be used only in the cells with nonfunctional version of these genes. Other endogenous selectable markers may be adenosine deaminase (ADA) and adenine phosphoribosyl transferase along with *TK* and *HGPRT*.
- Other markers may be drug resistant where their presence can impart cells to survive in the presence of drug. Cells lacking the resistant marker gene are susceptible to the particular drug thus are unable to survive.
- These include glycopeptide-binding protein, which can make the cells resistant to glycopeptide antibiotics, histidinol dehydrogenase can make the cells resistant for histidinol, neomycin transferase can give resistance to aminoglycoside antibiotics, and hygromycin phosphotransferase gives resistance in the presence of hygromycin-B.
- Methotrexate which is folic acid analog and competes for binding on the enzyme dihydrofolate reductase (DHFR) is used. Mutant DHFR is resistant to methotrexate-mediated effects.

5.5 Application of Transgenic Animals

In general the benefits derived from transgenic animals can be broadly categorized in three major groups:

1. Agriculture
2. Medicine
3. Industrial

5.5.1 Agricultural Applications

In agriculture cows and buffaloes are engineered to produce more milk, sheep to produce more wool, and large and fleshy animals with more meat for consumption as food [11, 13]. Of the few research reports describing the use of transgenic technologies in cattle, only one is directed toward a food production application in which they have worked on effect on milk production and composition.

The development of SCNT along with remarkable progress in gene mapping and genome sequencing endeavors in livestock will open a new set of possibilities for introduction of precise genetic modifications for agricultural applications. The host of possibilities includes progress in areas like milk production, growth rate, carcass composition, reproductive performance, and disease resistance.

We are now witnessing how some of these potential opportunities are being put into practice. One recent study used SCNT to create calves transgenic for two casein genes involved in milk protein production. When the resulting calves were induced to lactate, the levels of α- and β-casein protein in milk were altered, suggesting that the transgene did influence milk content [1, 2]. Another example of a practical application of SCNT combined with cell-based trangenesis is the production of four cloned calves carrying a genetic modification that would render them resistant to bovine spongiform encephalopathy (BSE).

5.5.2 Medical Applications

In medicine the transgenic animals may serve as disease models, knockouts for studies related to gene function, quality trait improvement, and pharmaceutical production [13].

5.5.2.1 Transgenic Animals as Disease Model

Certain transgenic animals are produced for specific traits and serve as disease model. Animals are genetically manipulated to exhibit disease symptoms; thus they can be used for studies related to causes of the disease and effective treatment. For example, Oncomouse® developed by Harvard scientists is genetically modified and carries a gene that promotes the development of various human cancers. Transgenic animals have enabled scientists to understand the role of genes in specific diseases. By either introducing or inactivating particular genes, researchers can discover the root causes of diseases associated with gene defects. Transgenic animals allow more effective treatments to be developed. Having found the genes implicated in a disease, scientists can then target these or design other therapies which act by influencing their expression [11].

Importance of Mice in Research and as Model System After the sequencing of mouse genome in 2002, it became animal of choice for most laboratory experiments. Mouse and humans show similarity in the organization of genes and their expression; they also suffer from some of the diseases as humans. Mice and humans along with other mammals including dogs, rabbits, monkeys, and apes have almost the same size of genome (nearly 3 billion base pairs). Although the genomes have great resemblance, some human genes have no counterpart in mouse and in a few diseases like HIV, there is no suitable animal model where disease can be explored well. Their reproduction is fast with short life span, easy to handle and manipulate, and inexpensive. Despite of the similarity in the size of the genome of human and mice, there are significant differences in the structure of genes and the encoded proteins. Complex gene–protein interactions, alternate splicing, and posttranslational modifications lead to many differences in the proteome:

- Knockout mice are the transgenic mice whose DNA is altered in such a way that the desired genes become nonfunctional.

(continued)

They serve as disease models for many diseases as diabetes, obesity, neurological disorders, etc.

- The transgenic animals make an ideal disease model for many of the diseases where wild-type animals do not serve as host for pathogenic microorganisms. For example, infection of poliovirus results in the onset of paralytic polio; however wild-type mice cannot be infected with poliovirus as they lack the cell surface molecule serving as receptor for the virus. Therefore transgenic mice who express human gene for receptor are created and subsequently used as disease model.
- Transgenic mice were created by overexpression of human mitochondrial transport protein, "uncoupling protein-3" in the skeletal muscle. Though the transgenic mice ate more than wild type, it remained leaner and lighter. They had an increased glucose clearance rate suggesting that this protein has potential in treating obesity. In the same vein, targeted removal of cyclooxygenase-2 (Cox2) which is implicated in inflammatory response prevented the development of autoimmune arthritis. This can be a potential target to treat inflammation in arthritis.
- Subsequently the models are used for Alzheimer's disease, AIDS, and Doogie mouse (show improved memory and capacity for learning). Apart from these, mouse model is being developed for human growth hormone (giant mouse). In the development of knockout mouse where a gene is inactivated, the researchers can determine the exact function of inactivated gene (fat mice, strong mice, and cold-tolerant mice).
- In nude mutation, the thymus fails to develop in mice (congenital birth defect in humans, DiGeorge's syndrome, and in nude mice: the condition results in loss of circulating T cells). This is due to mutation in the gene FoxN1 (forkhead box N1). FoxN1 (formerly known as Whn or Hfg11) encodes a transcriptional factor, located on chromosome 17 in mice, which is expressed on the thymic epithelium, epidermal keratinocytes, and hair follicles. This mutation in mice induces a hairless phenotype and a rudimentary thymus gland in mice (nude mouse). This has largely helped in studies involving FoxN1 gene function in the development of immune system and its diseases, in skin studies, and as cancer model.
- Antibody required for therapy may face immune rejection in human host as they are derived from mouse. Therefore genetic engineering approaches are used to place the specificity genes (complementarity determining regions, CDRs) from the mouse-derived hybridoma to corresponding regions of matched human immunoglobulin cDNA. The studies are also going on to develop mouse with large human immunoglobulin loci into mouse germline.
  - XenoMouse^R (Abgenix Inc., Fremont, CA).
  - HuMab Mouse^R (GenPharm, Medarex, San Jose, CA) carries both human V_H and V_L repertoire.
- Fusion with a myeloma cell can also be bypassed by using transgenic mice H-2kb-tsA58 (bacteriophage immunized), ImmortoMouse (Charles River Breeding Laboratory Wilmington, MA).
- Oncomouse®, developed by Harvard scientists, is genetically modified and carries a gene that promotes the development of various human cancers. Transgenic animals have enabled scientists to understand the role of genes in specific diseases.

(continued)

Transgenic animals help test the safety of new medicines and vaccines. Because transgenic models can highlight specific characteristics such as certain mechanisms involved in the formation of tumors, they can demonstrate more clearly the possible side effects of new therapies. Their use in early toxicity trials may also serve to prevent the subsequent use of a larger number of animals in the development phase.

5.5.2.2 Xenotransplantation

The gap between the demand for and the availability of human organs for transplantation is growing, a tendency that is not likely to reverse in the near future. The number of patients on waiting lists for organs has been growing every year, reflecting both an increased demand for organs for transplantation and a pronounced shortage of donors. Transgenic animals are being developed by some companies to provide new organs for transplantation such as kidneys, livers, and hearts. Pigs have long been considered as an alternative source of organs for xenotransplantation (i.e., transplantation of organs/tissues between different species, e.g., from animals to humans) to satisfy this increasing demand. Pigs seem to meet most of the requirements for an ideal animal donor: they are anatomically and physiologically similar to humans with organs of appropriate size, they are prolific, and they can be maintained under specific pathogen-free conditions. However, there are barriers to pig-to-human xenotransplantation; one of the main challenges is to avoid the hyperacute immunological rejection of the grafted tissue. When organs or tissues are transplanted between discordant species like pig to human, the host's immune system initiates a fast reaction known as hyperacute rejection (HAR).

Xenotransplants

Foreign tissue from pig has 1,3 galactose (Gal) residues. The xenotransplants obtained from pig are rejected by human host antibodies, which are against galactose (Gal) residues of pig followed by complement-mediated vascular damage. In most animals 1,3 galactosyl transferase is present which catalyzes the addition of galactose in the plasma membrane glycoproteins and glycolipids. However in humans 1,3 galactosyl transferase gene is inactive; therefore they neither have the enzyme nor the Gal. Hyperacute rejection (HAR) is initiated by antibody recognition of Gal; therefore to avoid this rejection, either this enzyme should be knocked off in pigs or immunosuppressive state especially of complement cascade be maintained in humans. Transgenic pig gave hope that it would be possible to generate "humanized" organs which would be transplanted and accepted in humans. Transgenic pigs with human MHC genes are being tried in the hope that their "humanized" organs will not be rejected by a host.

Pigs are being developed where one or both alleles for 1,3 galactosyl transferase were knocked out. Pigs bearing transgenes encoding major components of the complement regulatory pathway have been produced by pronuclear microinjection or sperm-mediated transgenesis. Expression of functional complement regulatory proteins by transgenic pig organs may be able to extend survival of xenotransplanted primates in the future.

5.5.3 Industrial Applications

As the products derived from transgenics should reach the market, the production is scaled up according to industry level as production of spider silk and pharmaceutical recombinant products. Transgenic animals can produce biological products. It may be possible to use transgenic animals to make rare biological products for medical treatment.

Human alpha-1-antitrypsin, a protein used to treat the rare genetic disorder of alpha-1-antitrypsin deficiency, is just one example. Another application for genetically modified cattle is the production of human therapeutic proteins. Human proteins that have been expressed in milk include human lactoferrin, human alpha lactalbumin, human serum albumin, and human bile salt-stimulated lipase [12]. The mammary gland in dairy cows is an excellent protein production factory. Large quantities of very complex proteins can be produced and collected at very low cost. Possibilities in biomedical applications include the production of important therapeutic proteins in milk such as antitrypsin for cystic fibrosis; blood clotting factors like antithrombin III, factor IX, and fibrinogen for bleeding disorders; and human serum albumin, which could be useful for treatment of burns. Although such possibilities originated when transgenic work was started using pronuclear microinjection, these ideas are now being realized through SCNT.

Pharmaceuticals produced in transgenic animal are less costly from pharmaceuticals produced in a bioreactor as providing suitable condition to culture is very costly. Besides this many complex proteins, coagulation factors are difficult to manufacture in their biologically active forms in the available cell lines. Since dairy animals can give high rates of heterologous protein secretion in the milk, and typically lactate for 10 months out of the year, production levels can easily be 100-fold higher than that currently achieved in cell culture.

Initially the transgenic models were being used to define the in vivo effector function(s) of genes by enhancing the expression of individual proteins in a specific organ or tissues. But nowadays the world of transgenic animals is continuously expanding. Its role in medical science and pharmaceutical production cannot be ignored. The function(s) of the majority of these genes is still virtually unknown. As a result, characterization of the effectors responses of these genes is the next major challenge for the scientific community. This functional analysis will require a variety of scientific approaches. In keeping with the concept of "in vivo veritas," the use of transgenic overexpression models has proven particularly useful in this regard. The mouse has become the target animal of choice in generating these models. This preference is the result of our extensive amount of knowledge regarding the murine genetic system. It is also due to the extensive investigations of the immune, injury, inflammatory, and healing responses in murine systems. This provides great convenience in designing and manipulating genes that one believes may be involved in human disease.

5.6 Transgenic Plants

Plants have provided us all the necessary things required to sustain on this earth. However now, they are being explored for production of recombinant heterologous proteins in simple and inexpensive system and for producing nonnatural proteins as single chain fragment variable (ScFvs).

The first pharmaceutical human growth hormone was obtained from transgenic tobacco in 1986. The gene encoding desirable trait is identified, selected, extracted, and transferred directly into another plant genome. The presence of the desired gene, controlling the trait, can be tested for at any stage of growth with the help of reliable marker.

The genetically engineered plants are widely used since their trail in 1990. The first GM crop was released in 1992. Though the area under cultivation of GM crops has increased widely, there is still strong opposition for this technology due to fears of contamination of the environment, cross-pollination and contamination of other crops, and effects of transgene. The plants have also being explored for production of recombinant pharmaceuticals and proteins. However, the yield of the recombinant product was the primary challenge in using plants as production system; thus, expression constructs are designed in a way to achieve high yields of the engineered gene. The transgene in expression constructs is chimeric structure as it is surrounded by various active regulatory elements. For driving the high level of expression of transgenic gene, the promoter and the polyadenylation sites are important. Affinity tags as His or the FLAG epitope can be used for the recombinant protein purification; however this modification affects not only the primary structure of the protein but also its properties.

Thus avoiding affinity tags and usage of specific technique for purification of the protein in its native structure would be more desirable. The cost of the bioprocessing is reduced when the product is more concentrated in the starting material. If conventional extraction from seeds is expensive, other strategies can be used to assist purification. One of the examples being the oleosin-fusion platform is developed by SemBioSys Genetics Inc., where recombinant protein is expressed in oilseed crops as a fusion with oleosin. The protein can be recovered from oil bodies using simple extraction procedure and separation from its fusion partner by endoprotease digestion.

CaMV (cauliflower mosaic virus) 35S promoter being strong and constitutive works well with dicots. The yield can also be controlled by placing the gene under the control of the promoter which is active in a particular tissue or developmental stage or particular environment, for example, rice gluten, pea legumin, and many others.

Genetic manipulation involves many diverse techniques; however their basic principle is similar and simple. The different techniques show effectiveness in different individuals thus the best suited and standardized technique is useful for creating recombinants. The transformation in the plant cells is discussed here by using particle bombardment or *Agrobacterium tumefaciens*-mediated gene transfer. Transformation: In this the gene in suitable vector is taken up by host cell.

5.6.1 Transgenic Production in Plants

Transformation methods: Though many methods are available in plant cell transformation, two methods of transformation are preferred, (1) *Agrobacterium* mediated and (2) particle bombardment.

5.6.1.1 *Agrobacterium tumefaciens*: Mediated Transformation

As DNA is shared among the living form naturally, thus *Agrobacterium tumefaciens*, a soil bacterium known as "nature's own genetic engineer," has the natural ability to genetically engineer plants. It causes crown gall disease in a wide range of broad-leaved plants, such as apple, pear, peach, cherry, almond, raspberry, and roses. The genes that cause the galls are removed and replaced with genes coding for desirable traits. Plant cells infected with the bacterium will not form galls but produce cells containing the desired gene, which when cultured in a special medium, will regenerate into plants and manifest the desired trait [7]:

- *Agrobacterium tumefaciens* is a soil pathogen which is an effective transformation tool widely used for most dicot species and is used for transformation of tobacco, alfalfa, pea, tomato, and potato.
- *A. tumefaciens* infects the dicot plants at the wounded site causing crown gall tumors.
- The bacterium has exceptional ability to transfer its particular segment of DNA (T-DNA), the part of tumor-inducing (Ti) plasmid stably into host genome.
- The Ti plasmid integrates in the host genome and is transcribed, causing the crown gall disease.
- The oncogenic genes present on T-DNA encode for enzymes involved in the biosynthesis of auxins and cytokinins, which are responsible for tumor formation (Fig. 5.8).

- T-DNA also has genes for the synthesis of opines. Other genes present on plasmid are responsible for the opine catabolism, mediating the process of T-DNA transfer from bacterium to the host, and bacterium-bacterium plasmid conjugative transfer.

- Virulent strains of *A. tumefaciens* and *A. rhizogenes*, when interacting with susceptible dicotyledonous plant cells, induce diseases known as crown gall and hairy roots, respectively, thus are called as Ti and Ri plasmids, respectively.
- The Ti plasmids are classified according to the opines which they release. The process of T-DNA transfer is mediated by the cooperative action of proteins encoded by virulence genes (vir genes) in the bacterial chromosome.
- The Ti plasmid also contains the genes for opine catabolism produced by the crown gall cells and regions for conjugative transfer

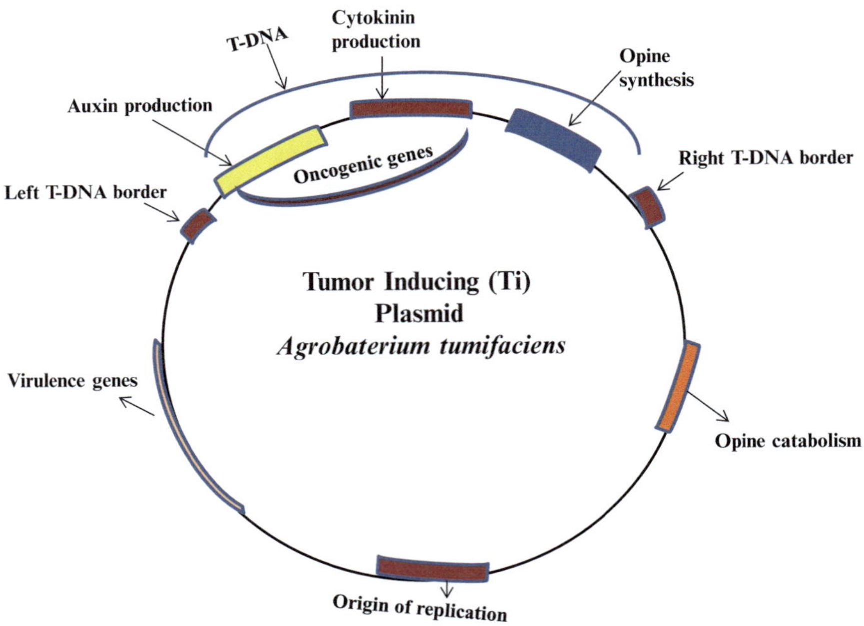

Fig. 5.8 The figure shows Ti plasmid of *Agrobacterium tumefaciens*

and for its own integrity and stability. The virulence (vir) region is organized in six operons that are essential for the T-DNA transfer (virA, virB, virD, and virG) or for the increase of transfer efficiency (virC and virE).

- VirA is a transmembrane dimeric sensor protein that detects signal molecules, mainly small phenolic compounds, released from wounded plants. The signals for virA activation include acidic pH; phenolic compounds, such as acetosyringone; and certain class of monosaccharides which acts sinergistically with phenolic compounds.
- The process of gene transfer from *A. tumefaciens* to plant cells (Fig. 5.9) implies several essential steps: (1) bacterial colonization, (2) induction of bacterial virulence system, (3) generation of T-DNA transfer complex, (4) T-DNA transfer, and (5) integration of T-DNA into plant genome.
- The T-DNA transfer to plant cells has three important steps:

1. The tumor formation is a transformation process of plant cells resulting from transfer, integration, and expression of T-DNA genes.
2. The T-DNA genes do not play any role in the transfer process and are transcribed only in plant cells.
3. The portion between left and right borders can include any foreign gene, which is required to be transferred in the plant.

The process has advantages as it reduces the copy number of the transgene, potentially leading to fewer problems with transgene cosuppression and instability. As only dicots are targets for *A. tumefaciens*, thus in other plants alternative transformation methods have been developed as particle bombardment, PEG mediated, microinjection, and protoplast fusion.

5.6.1.2 Particle Bombardment

It is one of the preferred transformation methods for legumes as soybean; cereals as rice, wheat, and maize; and plastid transformation for raising

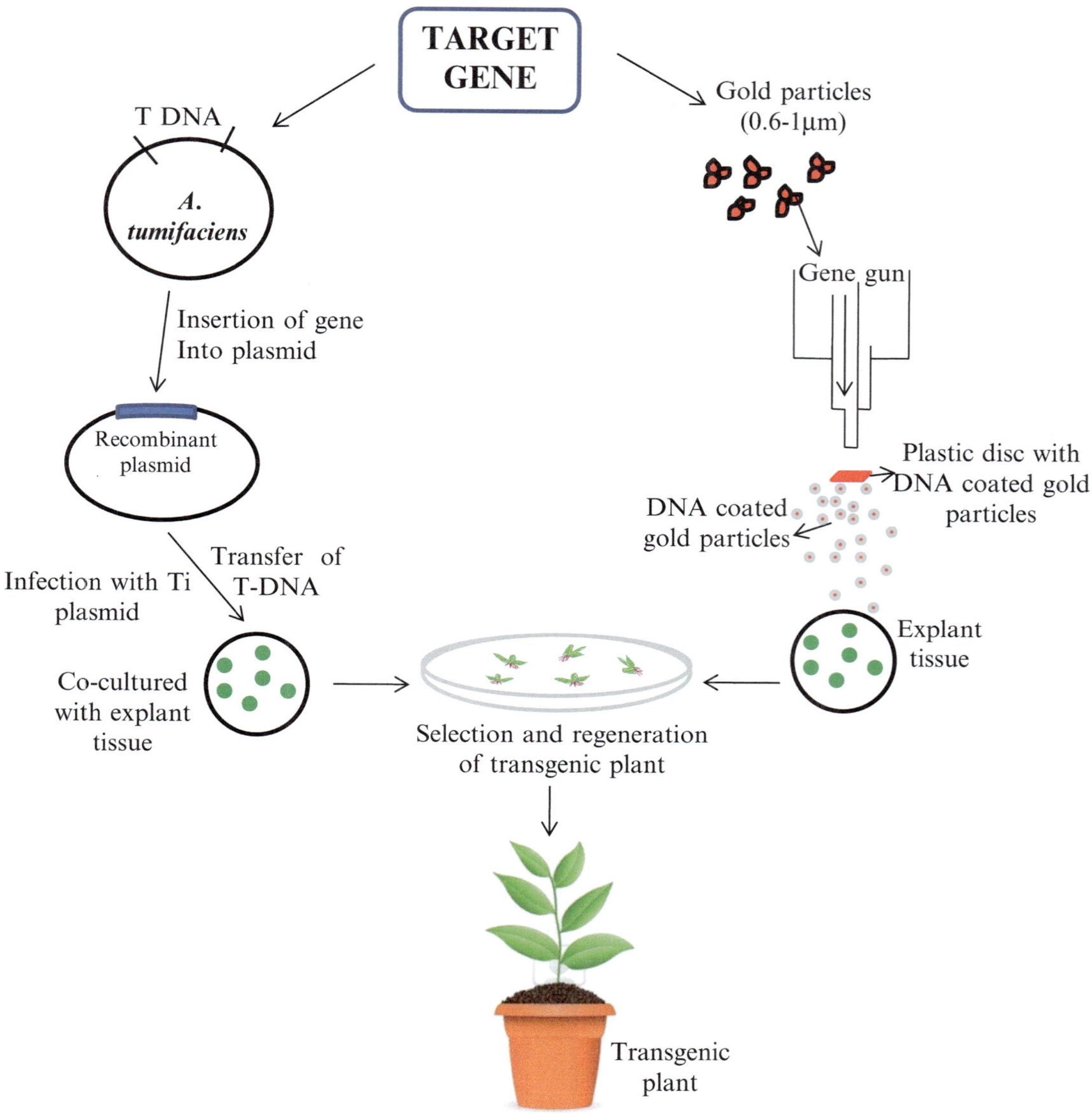

Fig. 5.9 The figure shows the process of plant transformation. *Agrobacterium*-mediated transformation is shown in the *left*. Particle bombardment is shown in the *right*

transplastomic (a transgenic plant in which the transgene is found in the plastid genome).

Particle Bombardment

- It is a mechanical method of introducing the desired gene.
- The desired gene is cloned into a suitable vector and introduced into the plant using the gene gun or particle gun.
- The gene gun uses minute particles of tungsten or gold as the bullet. These particles are coated with the DNA solution and fired to the plant cells through the force of the helium gas inside a vacuum-filled chamber.
- The DNA and the tungsten/gold particles get inside the cell, and within 12 h, the inserted DNA gets inside the nucleus and integrates with the plant DNA. The tungsten/gold particles are sequestered to the vacuole and eliminated later.
- Transformed cells are cultured in vitro and induced to form small plants (regeneration) that express the inserted gene (Fig. 5.9).

The main goal in any transformation procedure is to introduce the gene of interest into the nucleus of the cell without affecting the cell's ability to survive. If the introduced gene is functional, and the gene product is synthesized, then the plant is said to be transformed. Once the inserted gene is stable, inherited, and expressed in subsequent generations, then the plant is considered a transgenic. Detection of foreign gene can be done by PCR or other method. For the applications of transgenic plants, refer to Chap. 20.

5.7 Ethical Issues in Transgenic Production

The use of transgenic animals has sparked significant controversy in many areas. Some groups see the generation of genetically modified organisms as interfering with natural biological states or processes that have evolved over long periods of time, while some other groups are concerned with the limitation of modern science to fully comprehend the potential negative ramification and unforeseen possible effects of genetic manipulation.

Despite the importance of transgenic animals in biomedical research, there are some concerns raised about their use in research. Transgenic animals suffer more abnormalities than regular research animals. The introduction of DNA into an animal can be very complex and the possible side effects can be difficult to predict. Possible harm might arise from surgical techniques used to harvest and reimplant embryos, the collection of tissue from the tip of the tail for genotyping, and nonspecific effects caused by damage to genes adjoining the altered area. Transgenesis may also result in reduced fertility and/or oversize fetuses. In most cases the mutations affect highly specific metabolic processes or cell receptors without actually resulting in disease but causing discomfort, pain, or malformation in the animals.

Transgenic animals not expressing foreign DNA or not containing a particular gene modification are destroyed. Because transgenesis is a complex science, it is not 100% efficient. However, new methods are being developed to increase the accuracy in transgenesis. Again, it should be remembered that such genetic alteration can only be attempted if the authorities are persuaded that there is no other way to pursue important research. The potential risks of transgenics to animals, humans, and the environment are too great to justify their use. The Genetic Modification of Organisms regulations and the Environmental Protection Act (1990) address the risks to those working with animals and the impact on the environment of accidental or planned releases.

The intrinsic worth of animals may be devalued and their integrity violated by genetic modification. Transgenic animals have not chosen to have foreign DNA or other genetic modifications. However, this potential "cost" to the animals is routinely assessed under the ethical review of proposed procedures and weighed against the potential benefits. Medical researchers only employ this technology when no alternative research avenue exists. As the Royal Society concluded in its 2001 Report "The Use of Genetically Modified Animals," the use of transgenic animals is fundamentally little different from the use of other animals in biomedical research.

However, the technology has opened new frontiers in biomedical applications and has provided new opportunities for exploring the organization, biological pathway, regulation, and pathological function of molecular processes.

5.8 Chapter End Summary

- Transgenic animals as a subset of genetically modified organisms (GMOs) are organisms which have inserted DNA of different species origin; the inserted DNA/gene is known as transgene.
- Transgene is introduced into the animal, using recombinant DNA technology, and it must be transmitted through the germline so that every cell, including germ cells, of the animal contains the same modified genetic material.
- There are two basic strategies used when producing transgenic animals. These are the so-

called "gain of function" or "loss of function" transgenics.

- The "gain of function" relies on addition of a cloned fragment of DNA to an animal's genome to obtain new expression of a gene product that did not previously exist in that cell or tissue type. "Loss of function" is a more direct way to determine the function of a lost/inactive gene after its alteration and then observing its physical manifestation (phenotype).
- The transgene cassette is a combination of the gene under the control of a strong promoter sequence immediately upstream of the gene. Promoter has critical roles in the temporal and tissue-specific regulation of gene expression.
- There are several methods of producing transgenic animals: (1) DNA microinjection, (2) retrovirus-mediated gene transfer, (3) embryonic stem cell-mediated transfer, and (4) liposome-mediated DNA transfer into cells and embryos. A few modifications are also there for creating transgenics: (1) electroporation of DNA into sperm, ova, or embryos and (2) nuclear transfer with somatic or embryonic cells.
- In general the transgenic animals are having important applications in the field of agriculture, medicine, and industry.
- In agriculture cows and buffaloes are engineered to produce more milk, sheep to produce more wool, and large and fleshy animals with more meat for consumption as food. They are explored for their use in production of pharmaceuticals as production of therapeutic agent insulin in cow milk and sheep producing alpha-1-antitrypsin. In industrial applications the recombinant products derived from these are scaled up according to industry requirement, for example, production of spider silk.
- Transgenic plants are also being produced and are in use since 1992 as GMOs. The cotton, corn, soybean, etc., had insect resistance gene from *Bacillus thuringeniesis*. Nowadays many more plants are genetically modified for improving and increasing yield. The plants are also being explored for the production of recombinant pharmaceuticals like vaccines, antibodies, and other proteins.
- Ethical issues are very important concerns in transgenic production where the pros and cons of all experiments involving plant and animals are discussed, and approvals are granted according to their justifications for humans and other life forms.

Multiple Choice Questions

1. In transgenic animals foreign gene is ligated to the promoter:
   (a) Downstream to the promoter
   (b) Upstream to the promoter
   (c) Between promoter and regulator gene
   (d) None of these
2. Why is foreign gene generally injected into the male pronucleus?
   (a) Because male pronucleus is smaller than female
   (b) Because male pronucleus is larger than female
   (c) Because male pronucleus moves faster
   (d) Because male pronucleus has thin membrane
3. Embryonic stem cells (ESC) are used in transgenic animal production. These are derived from:
   (a) Outer cell mass of the blastocyst
   (b) Middle cell mass of the blastocyst
   (c) Inner cell mass of the blastocyst
   (d) Bone marrow
4. In transgenic animal production, the foreign gene is introduced into the cell via:
   (a) Viruses
   (b) Chemicals
   (c) Homologous recombination
   (d) All methods
5. Transgenic animals are generally used as:
   (a) Disease models
   (b) Xenoplanters
   (c) Transpharmers
   (d) Food sources and scientific models
   (e) All of the above

6. In which organ of the animals are transpharmers engineered to overexpress a particular gene?
   (a) Sweat glands
   (b) Liver
   (c) Mammary glands
   (d) Pancreas
   (e) None
7. Which of the following is a remarkable industrial application of transgenic animals to generate desired gene?
   (a) Antibody
   (b) Organ transplantation
   (c) Biosteel
   (d) Vitamin A
8. Why do we produce xenotransplanter organs in transplantation of the organs?
   (a) Xenotransplanters are histocompatible to the organisms.
   (b) Xenotransplanters are not histocompatible to the organisms.
   (c) Xenotransplanters produce antigens.
   (d) None.
9. Transgenic animals produce spider web protein which is used as industrial production of:
   (a) Monoclonal antibody
   (b) Plasminogen activator
   (c) Biosteel
   (d) Enzymes
10. Enviropig is a transgenic animal of:
   (a) Cat
   (b) Pig
   (c) Dog
   (d) Goat

Answers

1. (a); 2. (b); 3. (c); 4. (c); 5. (e); 6. (c); 7. (c); 8. (a); 9. (c); 10. (b)

Review Questions

Q1. What are transgenic animals?
Q2. What are the problems associated with transgene after it is transferred to host cell?
Q3. What is position effect?
Q4. How can one achieve loss of function phenotype?
Q5. Discuss the methods of generating transgenics.
Q6. What are applications of transgenics?

References

1. An X, Gou K, Zhu S, Guan H, Hou J, Lin A, Zeng S, Tian J, Chen Y (2002) Cloned calves produced by nuclear transfer from cultured cumulus cells. Sci China C Life Sci 45:201–210
2. Behboodi E, Groen W, Destrempes MM, Williams JL, Ohlrichs C, Gavin WG, Broek DM, Ziomek CA, Faber DC, Meade HM, Echelard Y (2001) Transgenic production from in vivo-derived embryos: effect on calf birth weight and sex ratio. Mol Reprod Dev 60:27–37
3. Brackett BG, Baranska W, Sawicki W, Koprowski H (1971) Uptake of heterologous genome by mammalian spermatozoa and its transfer to ova through fertilization. Proc Natl Acad Sci U S A 68:353–357
4. Capecchi MR (1980) High efficiency transformation by direct microinjection of DNA into cultured mammalian cells. Cell 22:479–488
5. Eyestone WH (1999) Production and breeding of transgenic cattle using in vitro embryo production technology. Theriogenology 51:509–517
6. Gordon JW, Ruddle FH (1981) Integration and stable germ line transmission of genes injected into mouse pronuclei. Science 214:1244–1246
7. Gustavo ADLR et al (1998) Agrobacterium tumefaciens: a natural tool for plant transformation. EJB Electron J Biotechnol 1:118–133
8. Jaenisch R, Mintz B (1974) Simian virus 40 DNA sequences in DNA of healthy adult mice derived from preimplantation blastocysts injected with viral DNA. Proc Natl Acad Sci U S A 71:1250–1254
9. Lacy E, Roberts S, Evans EP, Burtenshaw MD, Costantini FD (1983) A foreign beta-globin gene in transgenic mice: integration at abnormal chromosomal positions and expression in inappropriate tissues. Cell 34:343–358
10. Lavitrano M, Giovannoni R, Cerrito MG (2013) Methods for sperm-mediated gene transfer. Methods Mol Biol 927:519–529
11. Thomson AJ, McWhir J (2004) Biomedical and agricultural applications of animal transgenesis. Mol Biotechnol 27:231–244
12. Van Berkel PH, Welling MM, Geerts M, van Veen HA, Ravensbergen B, Salaheddine M, Pauwels EK, Pieper F, Nuijens JH, Nibbering PH (2002) Large scale production of recombinant human lactoferrin in the milk of transgenic cows. Nat Biotechnol 20:484–487
13. Yang X, Tian XC, Dai Y, Wang B (2000) Transgenic farm animals: applications in agriculture and biomedicine. Biotechnol Annu Rev 5:269–292

Some Selected Resources

www.actionbioscience.org/biotechnology/glenn.html
www.bbc.co.uk/ethics/animals/using/biotechnology_1.shtml
www.biomedcentral.com
www.bmg.gv.at
www.fda.gov/cvm/geanimals.htm
www.globalresearch.ca/
www.nhlbi.nih.gov

6 Genome Sequencing

Abstract

There is large unexplored world inside us which is complicated and interesting. Just a half-century ago, very little was known about the genetic factors that contribute to human diseases. In 1953, James Watson and Francis Crick described the double helix structure of deoxyribonucleic acid (DNA), the chemical compound that contains the genetic instructions for building, running, and maintaining living organisms. Methods to determine the order, or sequence, of the chemical letters in DNA were developed in the mid-1970s along with the advancement in technology of microscopy, molecular biology, and genetic engineering, because of which it was becoming possible to conduct a serious and systemic exploration of our internal world.

Each cell present in the body has two near identical chromosomes sets. The DNA is packaged in a tight but sophisticated manner in these chromosomes. Deciphering the order of every DNA base on each chromosome in a genome is genome sequence. A genome map identifies the landmarks, which help in navigating around the genome. In this chapter, the readers would understand the principles of human genome project, genome projects of model organisms, genomics of pathogens, and evolution of influenza, hepatitis B, and HIV genome.

6.1 Introduction

There is large unexplored world inside us which is complicated and interesting. Just a half-century ago, very little was known about the genetic factors that contribute to human diseases. In 1953, James Watson and Francis Crick described the double helix structure of deoxyribonucleic acid (DNA), the chemical compound that contains the genetic instructions for building, running, and maintaining living organisms. Methods to determine the order, or sequence, of the chemical letters in DNA were developed in the mid-1970s along with the advancement in technology of microscopy, molecular biology, and genetic engineering, because of which it was becoming possible to conduct a serious and systemic exploration of our internal world.

Each cell present in the body has two near identical chromosomes. The DNA is packaged in

V. Gupta et al., *Basic and Applied Aspects of Biotechnology*, DOI 10.1007/978-981-10-0875-7_6

a tight but sophisticated manner in these chromosomes. Deciphering the order of every DNA base on each chromosome in a genome is genome sequence. A genome map identifies the landmarks, which help in navigating around the genome. In this chapter, the readers would understand the principles of human genome project, genome projects of model organisms, genomics of pathogenesis, and evolution of influenza, hepatitis B, and HIV genome.

Human Genome Project An international endeavor in the form of human genome project (HGP) was officially launched in 1990. It was initiated by the US Department of Energy and National Institute of Health with a projected time span of 15 years with international partners to sequence all three billion letters, or base pairs, in the human genome, which is the complete set of DNA in the human body. This concerted, public effort was the human genome project (HGP). A working draft of the genome was released in 2000 and a complete one in 2003, with further analysis still being published.

A parallel project was conducted outside of government by the private industrial group Celera Genomics, led by Dr. Craig Venter in 1998; they used the technique of whole genome shotgun sequencing, instead of dividing the genome into sections like the HGP. Whole genome shotgun approach involves breaking DNA into smaller fragments by enzymatic digestion or mechanical method and cloning them in suitable vector so that they can be individually sequenced. The sequences were reassembled by alignment based on partial overlaps. From shotgun sequencing the massive parallel sequencing for next-generation sequencing (NGS) was adapted. NGS read template DNAs randomly along the genome [26]. For achieving this, the genome is broken into small pieces and then these pieces are ligated to adapters (small pieces of DNA) so that these can be read during DNA synthesis (sequencing by synthesis). The NGSnextgeneration sequencing (NGS) read length is shorter (50–500 continuous bp reads) and is called massively parallel sequencing.

Most of the government-sponsored sequencing was performed in universities and research centers from the USA, the UK, Japan, France, Germany, China, India, Canada, and New Zealand. The mapping of human genes is an important step in the development of medicines and other aspects of health care. Unraveling the bases of DNA on each chromosome in the complete genome is genome sequencing. The studies have led to identification of many genes responsible for the diseases (Table 6.1).

However, there might be errors as there are regions with highly repetitive DNA like centromeres and telomeres. In the genome, the proportion of genes or coding DNA is small as compared to the complete genome. The noncoding DNA or earlier referred as junk DNA is now shown to have various regulatory roles. The primary goal of the project was knowing the precise chemical instructions which define living organisms.

6.2 Human Genome Organization (HUGO)

Genome is the collective name for the different DNA molecules found in the cells of particular species. In humans, the genome comprises 25 different DNA molecules: a single type of mitochondrial DNA and 24 different nuclear DNA molecules (22 pairs of autosomes + X chromosome + Y chromosome). Since the amount of DNA in the nucleus is very large, therefore, a set of nuclear DNA is referred to as nuclear genome, and mitochondrial DNA is referred to as mitochondrial genome [11].

In the international human genome project (HGP), the blood (female) or sperm (male) samples were collected from a large number of donors. However only a few samples were processed as DNA resources and the donor's identities were protected so nobody could know whose DNA was sequenced. DNA clones from many libraries were used in the overall project.

The goal of HGP was (1) to acquire fundamental information concerning our genetic makeup which would strengthen the basic understanding of human genetics and of the role of various

Table 6.1 Disease genes localized on chromosomes

S.No.	Chromosome number	Number of genes	Disease genes	
1.	Chromosome 1	Over 3000 genes	p	UROD (porphyria cutanea tarda)
			q	GBA (Gaucher disease)
				GLC1A (glaucoma)
				HPC1 (prostrate cancer)
				PS2 (Alzheimer's disease)
2.	Chromosome 2	Over 2500 genes	p	ETM2 (essential tremor)
				MSH2 (colon cancer)
				MSH6 (colon cancer)
			q	PAX3 (Waardenburg syndrome)
3.	Chromosome 3	Over 1900 genes	p	VHL (von Hippel-Lindau)
				SCLC1 (lung cancer)
				MLH1 (colon cancer)
			q	ETM1 (essential tremor)
4.	Chromosome 4	Over 1600 genes	p	EVC (Ellis-van Creveld)
				HD (Huntington disease)
				FGFR3 (achondroplasia)
			q	NRCLP (narcolepsy)
				SNCA (Parkinson's disease)
				FOP (fibrodysplasia ossificans progressiva)
5.	Chromosome 5	Over 1700 genes	p	SRD5A1 (steroid 5-alpha reductase 1)
			q	CKN1 (Cockayne syndrome)
				SMN1 (spinal muscular atrophy)
				Asthma
				DTD (diastrophic dysplasia)
6.	Chromosome 6	Over 1900 genes	p	SCA1 (spinocerebellar ataxia)
				IDDM1 (diabetes)
				HFE (hemochromatosis)
				CYP21A (congenital adrenal hyperplasia due to 21-hydroxylase deficiency)
			q	EPM2A (epilepsy)
7.	Chromosome 7	Over 1800 genes	p	GCK (diabetes)
			q	ELN (Williams syndrome)
				CFTR (cystic fibrosis)
				Pendrin (Pendred syndrome)
				OB (obesity)
8.	Chromosome 8	Over 1400 genes	p	WRN (Werner syndrome)
			q	MYC (Burkitt lymphoma)
9.	Chromosome 9	Over 1400 genes	p	CDKN2 (malignant melanoma)
			q	FRDA (Friedrich's ataxia)
				ABCI (Tangier disease)
				TSC1 (tuberous sclerosis)
				ABL (chronic myeloid leukemia)
10.	Chromosome 10	Over 1400 genes	p	PAHF (Refsum disease)
			q	OAT (gyrate atrophy)

(continued)

Table 6.1 (continued)

S.No.	Chromosome number	Number of genes	Disease genes	
11.	Chromosome 11	Over 2000 genes	p	HRAS (Harvey Ross oncogene)
				IDDM2 (diabetes)
				LQT (long QT syndrome)
			q	VMD2 (vest disease)
				MEN1 (multiple endocrine neoplasia)
				ATM (ataxia telangiectasia)
12.	Chromosome 12	Over 1600 genes	p	PXR1 (Zellweger syndrome)
			q	PAH (phenylketonuria)
13.	Chromosome 13	Over 800 genes	p	Myeloproliferative syndrome (translocation bet 13 and 8th chromosome)
			q	CX26 (autosomal recessive neurosensory deafness)
				BRCA2 (breast cancer)
				RB1 (retinoblastoma)
				14.ATP7B (Wilson disease)
14.	Chromosome 14	Over 1200 genes	p	–
			q	PS1 (AD3) Alzheimer's disease)
				SERPINA1 (alpha-1-antitrypsin deficiency)
15.	Chromosome 15	Over 1200 genes	p	–
			q	SNRPN (Prader–Willi syndrome)
				UBE3A (Angelman syndrome)
				FBN1 (Marfan syndrome)
				HEXA (Tay–Sach's disease)
16.	Chromosome 16	Over 1300 genes	p	HBA1, HBA2 (alpha thalassemia)
				FMF (familial Mediterranean fever)
				PKD1 (polycystic kidney disease)
			q	Crohn's disease
17.	Chromosome 17	Over 1600 genes	p	p53 (tumor suppressor protein)
				CMT1A (Charcot–Marie–Tooth syndrome)
			q	BRCA1 (breast cancer)
18.	Chromosome 18	Over 600 genes	p	–
			q	NPC1 (Niemann–Pick disease)
				DPC4 (pancreatic tumor)
19.	Chromosome 19	Over 1700 genes	p	Jak3 (severe combined immunodeficiency)
			q	BCKDHA (maple syrup urine disease)
				APOE (atherosclerosis)
				DMPK (myotonic dystrophy)
20.	Chromosome 20	Over 900 genes	p	–
			q	ADA (severe combined immunodeficiency)
21.	Chromosome 21	Over 400 genes	p	TPTE (tyrosine phosphatase)
			q	SOD1 (amyotrophic lateral sclerosis)
				APS1 (autoimmune polyglandular syndrome)
22.	Chromosome 22	Over 800 genes	p	RTN4R reticulon 4 receptor (schizophrenia)
			q	DGS (DiGeorge syndrome)
				BCR (chronic myeloid leukemia)
				SGLT1 (glucose–galactose malabsorption)
				NF2 (neurofibromatosis)

(continued)

Table 6.1 (continued)

S.No.	Chromosome number	Number of genes	Disease genes	
X	Chromosome X	Over 1400 genes	p	PIGA (paroxysmal nocturnal hemoglobinuria)
				DMD (Duchenne muscular dystrophy)
			q	ATP7A (Menkes syndrome)
				COL4A5 (Alport syndrome)
				IL2RG (A-linked SCID)
				TNFSF5 (immunodeficiency with hyper-IgM)
				HPRT1 (Lesch–Nyhan syndrome)
				FMR1 (fragile X syndrome)
				ALD (adrenoleukodystrophy)
				MECP2 (Rett syndrome)
				HEMA (hemophilia A)
	Chromosome Y	Over 200 genes	p	SRY (testes-determining factor)
			q	–

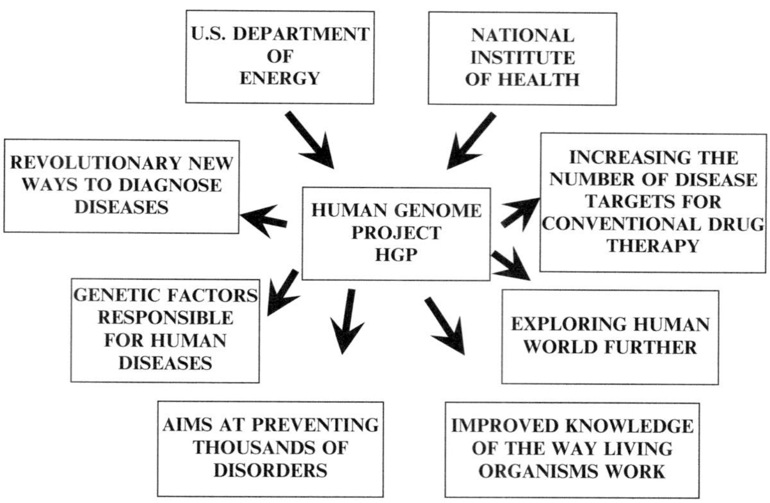

Fig. 6.1 The human genome project was initiated by DOE and NIH. The project was launched in 1990 with projected time span of 15 years, but was completed in 2003

genes in health and disease paving the way for new strategies for their diagnosis, treatment, and prevention (Fig. 6.1) and (2) to unravel the mystery of most evolved individual on earth, the humans, which required high-resolution human genetic maps, which could then be used as a framework for high-resolution physical maps with the complete sequence of human genome (Fig. 6.1).

To undertake this tough task, appropriate tools and techniques were developed and genome projects of other organism were undertaken. The human genome project supported an ethical, legal, and social implication research program to address many complex issues that might arise from this science. The dideoxy DNA sequencing method developed by Fred Sanger was the technique of choice utilized for sequencing. Automated fluorescent probe and gel-based DNA sequencers were used which employ multiple tiny (capillary) tubes to run standard electrophoretic separations. These separations are much faster because the tubes dissipate heat well and allow the use of much higher electric fields to complete sequencing in shorter times (Fig. 6.2).

There was an urgent need to manage and store huge amount of sequencing data that was being produced by the institutes working in collaboration.

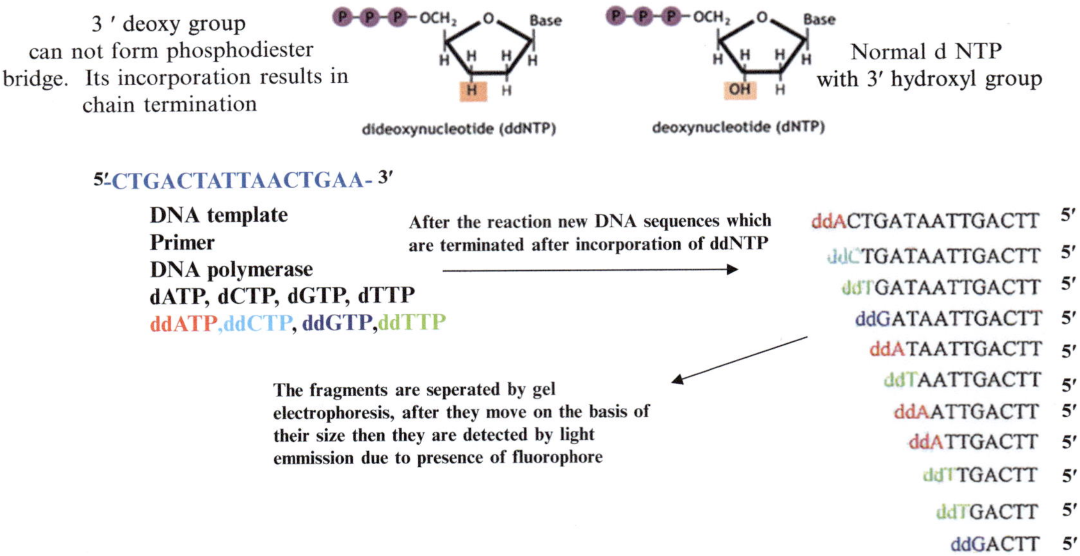

Fig. 6.2 Sanger's method of dideoxy sequencing. On the top is the structure of deoxyribose nucleic acid and dideoxyribose nucleic acid. The dideoxy is unable to participate in phosphodiester linkage and its incorporation results in premature chain termination. The PCR reaction using the template (in *blue*) using fluorescently labeled dideoxy nucleotides is performed. The amplification and incorporation of these results in premature termination and these amplified but terminated chains are separated by capillary PAGE electrophoresis. The laser beam at the end results in recording of appropriate colored peak which is later converted by the computer in readable format

Secondly, there was a requirement of sharing of the data for transparency and communication between various institutes in network; therefore central repositories for storing sequence data, which was freely accessible on the Internet, was established [11].

6.2.1 Physical Mapping

Physical mapping of human genome was started with genomic and cDNA library. Genomic DNA contains DNA fragments representing the entire genome of an organism, and cDNA library contains only complementary DNA molecules synthesized from mRNA molecules in a particular cell (Fig. 6.3).

The genome was broken into pieces through restriction digestion, and after size separation, these pieces were then ligated into a type of vector known as "bacterial artificial chromosomes," or BACs, which were derived from bacterial chromosomes, which have been genetically engineered, or λ-phage. The vectors containing the genes were then inserted into bacteria where they are copied by the bacterial DNA replication machinery. Every piece of this type was then sequenced separately as a small "shotgun" project and then assembled. The larger 150,000 base pairs go together to create chromosomes. This is known as the "hierarchial shotgun" approach, as initially the genome is broken into relatively large chunks, which are then mapped to chromosomes before being selected for sequencing (Fig. 6.3).

The size of the human genome is very large as compared to the length of DNA which can be directly sequenced; therefore it was necessary to divide the genome into fragments. Then fragments were put into order by digesting a copy of each fragment into still smaller fragments using sequence-specific endonucleases and then determine where the large fragments overlapped one another by contig mapping (Fig. 6.4).

The goal of HGP was the isolation of the entire human genome in overlapping clones and the development of physical maps of the cloned DNA. Though many vectors were used, yeast artificial chromosomes (YAC) allowed the cloning

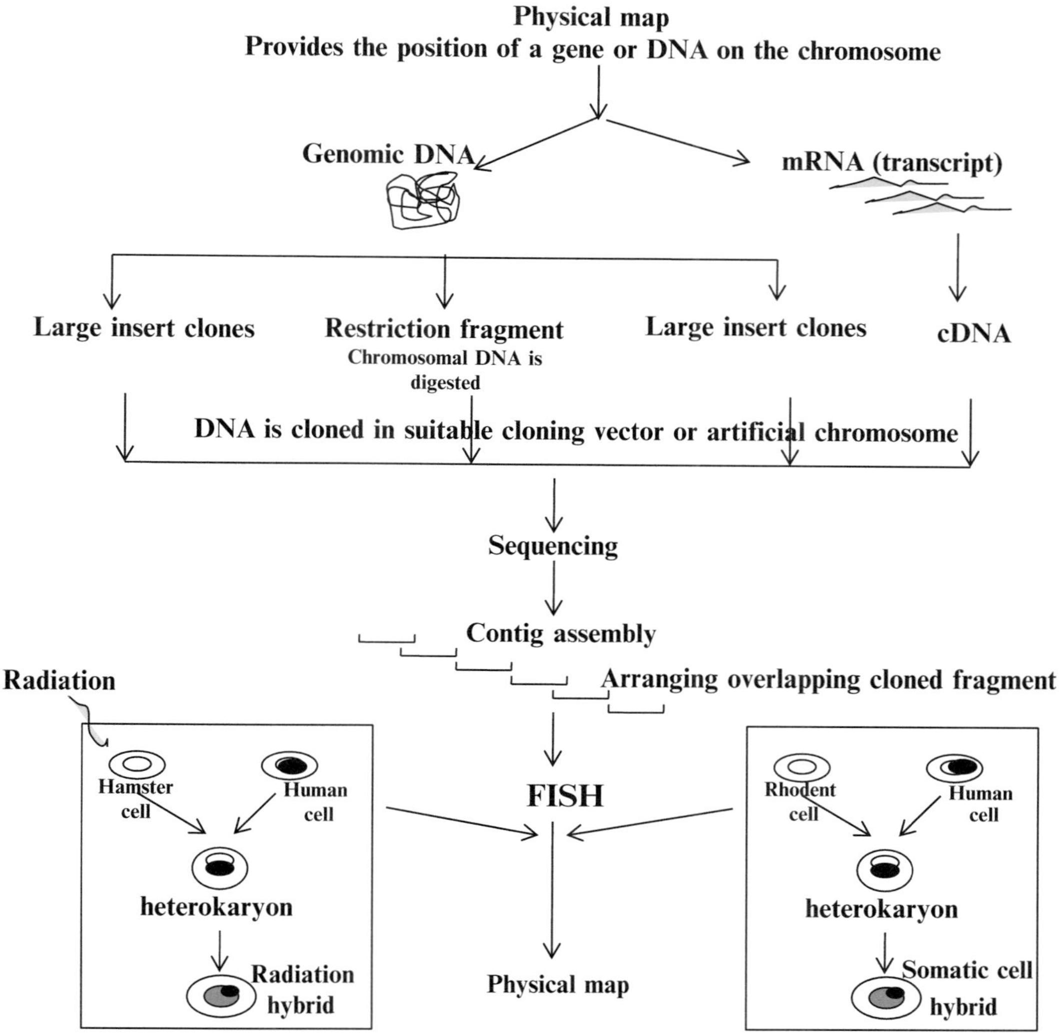

Fig. 6.3 The physical mapping of the genome by creating genomic and cDNA library. The inserts are cloned in the suitable vectors and sequenced. The clone contigs are prepared with the help of computer programs by looking at overlaps. The clone contigs are localized on genomic DNA using fluorescent in situ hybridization. For larger representation of the genomic/chromosomal DNA, somatic cell/radiation hybrids are used for creating physical maps

of large (100–1000 kb) DNA. The disadvantage with YAC cloning was the presence of clones containing two or more unrelated pieces of DNA (i.e., segments arising from different regions of the genome of origin).

Thus development of high-resolution physical maps required the assembly of overlapping clones into a contig, from which a map can be derived [19]. There were two strategies used to produce such maps:

1. Fingerprinting, in which a characteristic sequence-dependent pattern is generated for each randomly selected clone (e.g., by sizing restriction fragments), and clone overlaps are inferred based on shared parts of the pattern.
2. Chromosome walking, in which unique sequence probes are used to obtain clones, screening the clone library with an initial probe, and then, again with probes derived from the ends of clones obtained in the previ-

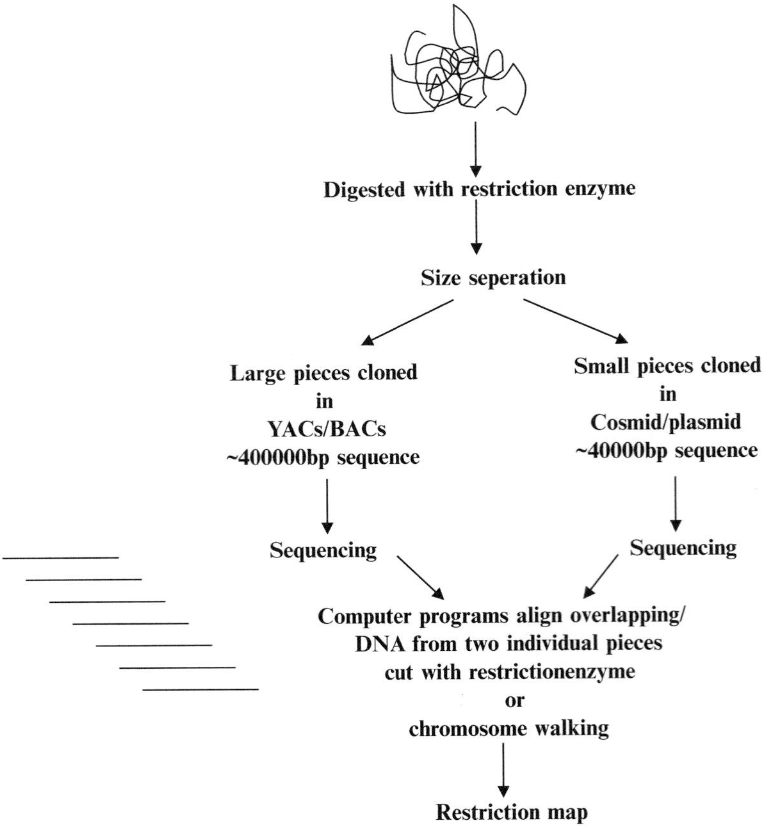

Fig. 6.4 The restriction mapping of the genomic DNA. The DNA is digested with restriction enzyme. The fragments are separated on the basis of size and cloned in suitable vector. The inserts are sequenced and contiguous clones are created by computer software by finding overlaps

ous screen, until the region of interest is fully isolated. Chromosome walking has the advantages of being targeted to the region of interest and of being able to detect small overlaps between clones, thereby allowing the construction of larger contigs. The readings in a contig can be summed to form a contiguous consensus sequence, and the length of this sequence is the length of each of the contig. Each contig is a genomic clone, usually in a cosmid or a BAC/YAC. "The contig may contain sequence gaps (within a clone), but it does not include gaps between clones [11]."

Random cDNA sequencing: cDNA may be synthesized from mRNA or by the use of cDNA representing UTR (expressed sequence tags EST) which may be randomly selected from the cDNA library. The advantage of EST is that they are rarely separated by introns; therefore PCR primer specific for a 3′-UTR amplifies the specific sequence on genomic DNA. After obtaining large number of human ESTs, they were placed on the physical maps either by typing YAC contigs or by screening radiation hybrid. Human gene map was constructed by placing ESTs on the physical maps.

In clonal populations of bacteria, each population maintaining a single artificial chromosome is stored in various laboratories around the world. The artificial chromosomes (BAC) can be grown, extracted, and labeled, in any lab. These fragments are on the order of 100,000 base pairs and are the basis for most fluorescence in situ hybridization (FISH) probes. FISH is used to localize the presence of specific DNA sequences on chromosomes (Fig. 6.5).

FISH uses fluorescent probes that bind to only those parts of the chromosome with which they show a high degree of sequence similarity. Fluorescence microscopy can be used to find out where the fluorescent probe is bound to the

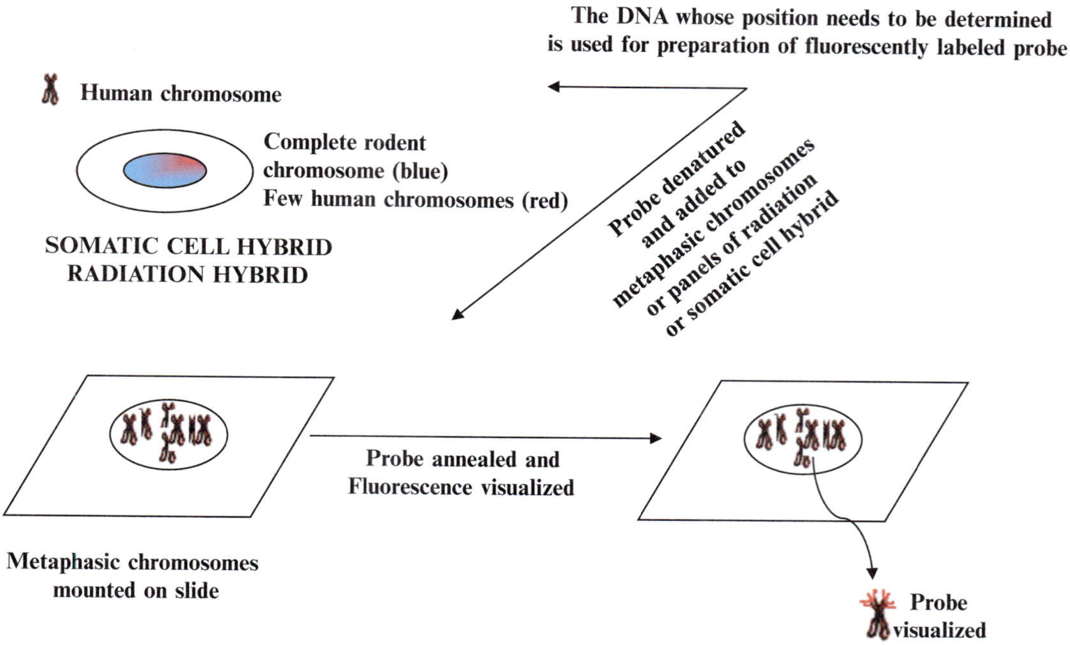

Fig. 6.5 The sequenced DNA fragment can be used for preparation of fluorescently labeled probe. The probe is hybridized to the panel of radiation hybrid or somatic cell hybrid or metaphasic chromosome mounted on glass slide. The fluorescent probe is allowed to anneal and bound probe is visualized by exposure to UV

chromosomes. Thus, FISH helps in localization of contig/radiation hybrid/somatic cell hybrid to a particular chromosome.

6.2.2 Genetic Mapping

The classical genetic maps based on crossing different mutants to determine the linkage between two loci were not possible in humans. Therefore, geneticists used polymorphic markers which were not disease related but followed the Mendelian pattern of inheritance; these were scored to obtain the genetic map. Here the markers are introduced for easy understanding [10, 11].

DNA markers or genetic marker or molecular markers are specific and heritable sequence of DNA which are capable of detecting polymorphism (existence of different forms (alleles) of the same gene in plants, animals, or their population) in the specific chromosomal region or are non-polymorphic associated with unique chromosomal locations or they may be random. They are phenotypically neutral and environmentally stable. They can be detected by hybridization-based techniques (Southern blotting) or by PCR. The differences between alleles are tracked as molecular markers. They can help in the identification of desired gene and its trait. The basic methodology is summarized:

- The sequencing of humans required construction of framework map for each chromosomal DNA molecule.
- These framework maps were based on contigs of clones which were organized in linear series of cloned overlapping DNA fragments, which together represented chromosomal DNA sequences.
- Assaying of these clones for the presence of certain DNA marker (STS) known to map in the approximate sub-chromosomal region by either typing a panel of artificially constructed somatic cell hybrid or radiation hybrid or by using FISH and assigning these markers their positions on the chromosomes.
- DNA marker maps can be built in different ways. Polymorphic markers may be assayed

in members of multigeneration pedigrees to identify groups of linked markers that must map to the same chromosome. If specific alleles from two or more markers segregated together in the family studies, the DNA marker could be allocated to a particular linkage group. Individual linkage groups may in turn be assigned to specific chromosomes by physical mapping of one or more of the constituent marker. For mapping studies overall likelihood of the pedigree is calculated with assumption that the loci are linked (θ) or not linked; recombination fraction is 0.5. The ratio of these two likelihoods gives the odds of linkage and logarithm of the odds is the LOD score (Z). If value of $Z=3$, then the linkage is accepted; $Z<2.0$, then linkage is rejected; and Z is between 2 and 3; the values are nonconclusive. In disease studies, families are typed for marker after marker until positive LODS are obtained [10, 11, 17].

- Non-polymorphic markers can be placed on marker maps by using PCR to assay the markers in panels of radiation hybrid cells or in panels of YAC clones or other genomic DNA clones with large inserts.
- Common markers are STS markers, derived from partly sequenced clones, including previously identified and sequenced DNA clones such as gene clones or other DNA clones of interest that have already been well characterized and mapped.

6.2.3 Molecular Markers and Mapping

Genetic polymorphism is occurrence of different alleles at a particular locus in the same population or different genotypes. If in the typed individual the alleles do not show any variation, they are said to be non-polymorphic. If typing shows the presence of different alleles of the same locus in different individuals, the marker is said to be polymorphic. Polymorphism can arise due to:

1. Every individual inherits one set of chromosomes from each of the parents.
2. Due to replication slippage (refer to Chap. 18 for details).
3. Frameshift mutations.

The markers can be dominant (shows presence and absence) or codominant (can determine homozygous and heterozygous state of diploid organism).

The dominant marker may or may not give locus-specific information, but codominant marker is always locus specific and gives information about the alleles at that locus.

Some properties for ideal DNA marker are highly polymorphic, codominant, frequent, easy, and fast typing. Different types, typing method, and utility of DNA markers are given for understanding of the readers.

6.2.3.1 Single Nucleotide Polymorphism (SNP)

- Each SNP represents a difference in a single nucleotide. As the polymorphism arises from single base substitutions, these are called as single nucleotide polymorphism or SNPs (pronounced snip, plural snips).
- Comparison of genomes of different humans shows gross similarity, however there may be one SNP every 300 nucleotides, and thus the human genome may have roughly ten millions SNPs.
- They are polymorphic. It may be A to G at certain positions or C to T. At some polymorphic sites, a SNP can have three alleles as A, G, or C, or two SNPs may be close to each other.
- They can be easily typed by automated genotyping (pyrosequencing).
- Major SNPs are found in the DNA between genes, where they can act as biological markers, helping scientists locate genes that are associated with disease. Most SNPs have no effect on health or development.
- When SNPs occur within a gene or in a regulatory region near a gene, they may play a more direct role in disease by affecting the gene's function.
- Researchers have found SNPs that may help predict an individual's response to certain drugs, susceptibility to environmental factors

such as toxins, and risk of developing particular diseases.

- Sometimes the SNP can affect a nucleotide that is part of the recognition site of restriction enzymes. This SNP can results in loss or creation of restriction site for the restriction enzyme, known as restriction fragment length polymorphisms (RFLPs) or restriction site polymorphisms (RSPs).

6.2.3.2 Restriction Fragment Length Polymorphism (RFLP)

- Restriction enzymes cut DNA fragments at precise sequences called as restriction site. Any kind of mutation (frameshift, deletion, or addition) may result in creation or loss of restriction sites.
- This results in the variations in the length of the fragments upon digestion, resulting in polymorphism in restriction fragment length, and is known as RFLP.
- A SNP can also give RFLP.
- Association of particular RFLP with a genetic disease can give information about the occurrence of that disease.
- It was hybridization-based marker but now done by PCR. In hybridization based, the DNA is digested with the restriction endonuclease and then its Southern blot is hybridized to a probe spanning the variable sequence. In PCR based, the primers flanking the variable sequence are used to amplify the target sequence and then digesting it with the restriction endonuclease.

6.2.3.3 Microsatellite [Simple Sequence Repeats (SSR)]

- Microsatellites (also called simple sequence repeats) are a class of genetic polymorphism commonly used for mapping and linkage analysis and to trace inheritance patterns.
- Microsatellites are tandemly repeated sequences, where the repeating unit may be 1–10 bp, but typically is one to four nucleotides long (minisatellite, 10–50 repeat units; satellite, 100 or more repeat units). The number of times the unit is repeated in a given microsatellite can be highly variable, a characteristic that makes them useful as genetic markers.
- It usually consists of di- or trinucleotide repeats (ATGATGATG is an example of a trinucleotide repeat (ATG)).
- They can be easily detected by polymerase chain reaction (PCR).
- These markers can be relatively efficient if more than one primer set is combined (multiplexing) in a single reaction. SSRs also have the advantage of displaying great diversity. SSR markers have the advantage of having a large number of alleles for a given locus.
- These polymorphisms are identified by constructing PCR primers for the DNA flanking the microsatellite region. The flanking regions tend to be conserved within the species, although sometimes they may also be conserved in higher taxonomic levels.
- They are also called simple sequence repeats (SSR), sequence-tagged microsatellite sites (STMS), or simple sequence repeat polymorphisms (SSRP).

6.2.3.4 Sequence-Tagged Sites (STS)

- Any DNA sequence that has been mapped to a specific sub-chromosomal location and can be assayed by PCR amplification.
- Single-copy DNA sequences of known map location could serve as markers for genetic and physical mapping of genes along the chromosome. STSs are very useful for making physical maps of human chromosomes. Creating a physical map is much like putting together a large puzzle, where the pieces of the puzzle are pieces of DNA made by cutting up chromosomes.
- STS-based PCR produces a simple and reproducible pattern on agarose or polyacrylamide gel. In most cases STS markers are codominant, i.e., allow heterorozygotes to be distinguished from the two homozygotes.
- The DNA sequence of an STS may contain repetitive elements, sequences that appear elsewhere in the genome, but as long as the sequences at both ends of the site are unique and conserved, researches can uniquely identify this portion of genome using simple tools.

- Thus, in broad sense, STS include such markers as microsatellites (SSRs, STMS).

6.2.3.5 Expressed Sequence Tag (EST)

- Expressed sequence tag is non-polymorphic marker.
- May be a subset of sequence-tagged sites that may be located within DNA sequences known to be transcribed.
- Expressed sequence tags (ESTs) are fragments of mRNA sequences derived through single sequencing reactions performed on randomly selected clones from cDNA libraries
- Thus, an EST is a sequence-tagged site (STS) derived from cDNA.
- These were widely used in cDNA library-based physical mapping.

The genetic maps were created as they would assist in studying the recombination and gene localization and would help in genetic diagnosis. Humans have 23 pairs of chromosomes in their nucleus, of which 22 pairs are autosomes and the 23rd pair is sex chromosome (XX, female; XY, male). These chromosomes vary in their size and shape. The chromosomes have centromere nearly at the center, which divides chromosomes into a long arm (designated q) and a short arm (designated p). The chromosomes can be visualized by special kind of stains, which are able to stain particular and unique portions of each chromosome in the form of stripes, and are known as chromosome banding; each chromosome follows a typical banding pattern. The stain commonly used was Giemsa stain and reverse Giemsa. This banding map of chromosomes or karyotype is called cytogenetic mapping, one of the earliest map (Fig. 6.6). After this the first genetic linkage map (published in 1987) was based on RFLP. There were 393 RFLPs placed on human map. Second-generation map was SSR based. As variation in the tandem repeats of microsatellites arises due to replication slippage resulting in variable number of tandem repeats (VNTRs) (refer to Chap. 18 for details). The microsatellite-based human map had 814 markers. In 1994 integrated genetic map with microsatellite and other markers had marker density close to 1 per megabase. High-density maps were also developed for SNPs. The problem with RFLP maps was that they had large spacing between the markers (>10 cM), and on the other hand, the microsatellite markers were dispersed, very informative, and easy to type with high resolution (1 cM).

6.3 Completion of Human Genome Project

Due to widespread international cooperation and advances in the field of genomics (especially in sequence analysis), as well as major advances in computing technology, a "rough draft" of the genome was finished in 2000. Ongoing sequencing led to the announcement of the essentially complete genome in April 2003, 2 years earlier than planned. The announcement of complete human genome in 2003 was only rough draft for each chromosome with requirement of more details. In May 2006, another milestone was passed on the way to completion of the project, when the sequence of the last chromosome was published.

Researchers have now filled in the gaps and provided far more detail for each chromosome. Most was achieved by comparison of particular DNA sequence across population in the area, which contained anomalies initially, for example, some unstable DNA segments during cloning for subsequent sequencing. The correction of minor errors and mutations cataloging will continue. The scientists have identified some 22,000 "gene loci" (with ~20,000 protein coding genes). Some genes may overlap or code for a number of different products, or are present as pseudogenes, thus estimation of their number becomes more difficult.

Merely 3 % of our entire genome is associated with genes encoding for proteins. The remaining 97 % was previously considered as junk DNA. Now it is known to play an important role in gene regulation and some of it is active and produces RNA molecules playing regulatory roles (ENCODE Consortium).

Undoubtedly, there were enormous benefits of HGP; however the analysis of inherited disorders

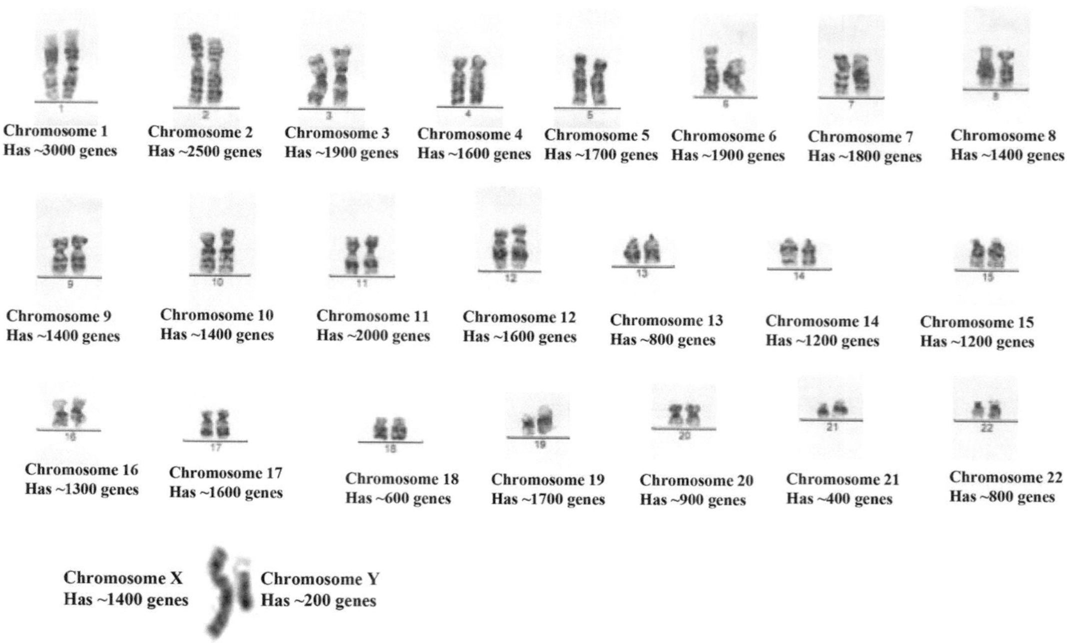

Fig. 6.6 The cytogenetic map of human chromosomes along with approximate number of genes residing on each chromosome

where single/many genes are leading toward the disease would be more interesting and informative in the future.

1. The studies would help in the prenatal disease diagnosis, presymptomatic diagnosis, and association studies for the population, which could be at a risk for a particular disease.
2. The information would be helpful in exploring the functioning and regulation of individual genes.
3. It would improve our understanding of the molecular mechanisms involved in the complex interactions between genetic factors and environmental factors.
4. It will also help to find more drug targets and new therapies for diseases with new and better diagnostic and prognostic markers.
5. It would be probably possible to have individualized drug regimes, designed according to an individual's genetic makeup.
6. One day possibly gene therapy may lead to the possibility of the replacement or repair of defective genes.

The whole of the data generated by the HGP was made freely and rapidly available to all, leading to increased pace of medical discovery. Now more than 1000 disease genes are discovered. The discovery had made possible of finding a gene, suspected of causing an inherited disease. Many genetic tests are available now to diagnose the diseases in humans enabling the people to know and understand their genetic risks or susceptibility for disease. Currently nearly 400 biotechnology-based products resulting from the HGP are in clinical trials.

In this post-genomic era, efforts have been focused on how human genome sequences can specify a person, how the DNA of other organisms is related to us, and how they relate to each other (comparative genomics) and the function, interaction, and expression of genes (functional genomics). The interaction of genes, mutations, and many other factors need to be more covered and unraveled.

Though the human genome sequencing is declared complete, some believe that the genome is yet to be completely sequenced. Each individual except identical twins has unique genome; the complete sequencing involves sequencing multiple variations of each gene and some heterochromatic DNA (about one-fifth of the genome) is unsequenced. Unfortunately due to limitations of existing technology, there are a number of regions of the human genome that can be considered unfinished. (1) The central centromeric regions containing highly repetitive DNA, which are millions of base pairs long, are unsequenced. (2) The terminal ends of chromosomes, the telomeres, are like centromeres with highly repetitive DNA which are not sequenced. (3) The loci with multigene families are very difficult to disentangle with present methodologies and code for important proteins (particularly immune system).

The important challenges of the knowledge are:

The evolution of the genome
The functions of junk DNA
The understanding of the functions of all the genes to form proteins and their regulation
Connection of DNA variation with nonmedical conditions, such as intelligence and personality traits

One major step toward such comprehensive understanding was the development of HapMap in 2005. This would be a listing of common genetic variation, or haplotypes, in the human genome. The advancement of the HapMap (goal is to identify patterns of SNP groups (called haplotypes or "haps")) might be able to tag genetic factors for many common diseases, such as heart disease, diabetes, age-related blindness, obesity, and mental illness.

In an initiative the Cancer Genome Atlas is aimed toward identifying all the genetic abnormalities in all 50 major types of cancer. In another initiative personalized medicine based on each person's genome will lead to a powerful form of preventive and preemptive medicine.

6.4 Comparative Genomics

Comparative genomics is the study of the relationship of genome structure and function across different biological species or strains. It compares the information available to understand the evolutionary process and is promising to yield insights into evolution of species. It exploits both similarities and differences in the proteins, RNA, and regulatory regions of different organisms.

6.5 Functional Genomics

Information obtained by decoding the complete sequence of the genome is indeed the first step toward understanding. However understanding their functional significance, their regulation, and their role in diseases is very complex. These studies are included in "functional genomics [18]."

6.5.1 Genome Annotation

Genome annotation is the process of attaching biological information to sequences. It consists of two main steps:

1. Identifying elements on the genome, a process called gene prediction
2. Attaching biological information to these elements

There are automatic annotation tools which perform this by computer analysis. All the approaches work to complement each other for the better understanding of the genome.

Structural annotation consists of the identification of genomic elements:

- ORFs and their localization
- Gene structure
- Coding regions
- Location of regulatory motifs

Functional annotation consists of attaching biological information to genomic elements:

- Biochemical function
- Biological function
- Involved regulation and interactions
- Expression

These studies may involve experimentation and in silico analysis using a variety of software which have been developed to help scientists.

6.6 Plant Genome Projects

There was great interest to sequence plant genome as they are food providers, they are responsible for the maintenance of carbon and nitrogen cycles, and they have been used in many applications since years. The genome project was initiated in plants when sequencing was done for *Arabidopsis thaliana* (thale cress) in 2000. This was followed by sequencing of *Oryza sativa* (rice) in 2002. Following this many others have been sequenced like *Carica papaya* (papaya), *Zea mays* (maize), *Fragaria vesca* (strawberry), *Solanum lycopersicum* (tomato), *Cajanus cajan* (pigeon pea), and many others. In legumes the genomes of *Glycine max* (soybean), *Medicago truncatula* (barrel medic), and *Lotus japonicus* (bird's-foot trefoil) have been sequenced [3]. Now approximately 55 plant genomes have been published representing 49 different species. Many other genomes are being sequenced and many more are under development. The sequencing techniques helped in ultimate sequencing of many genomes and have enabled rich annotation of the gene networks, and studies of comparative genomics have helped to establish phylogeny and domestication forces, assigning more and more molecular markers, which can help in marker-assisted breeding [15].

Special criteria for choosing the plants for sequencing include (1) size of research community, (2) economic importance of model organisms, (3) small genome size, (4) ploidy (diploid), (5) availability of inbred lines (low heterozygosity), and (6) access to genetic and physical maps, expressed sequence tags (EST)/transcriptome, and other genomic tools. Seventy-three percent (40) of the plant genome publications have been on crop species, and some of these crop species double as model systems, while several were sequenced purely for research such as *Arabidopsis thaliana*, *Arabidopsis lyrata*, *Brachypodium distachyon*, *Physcomitrella patens* (moss), and *Selaginella moellendorffii* (spikemoss). Most (94 %) genomes sequenced to date are angiosperms, of which 36 are dicots and 16 are monocots, while only one gymnosperm (spruce), one bryophyte (moss), and one lycophyta (club moss) have been sequenced.

Since plant genome is 100 times larger than the sequenced avian or mammalian genome with high polyploidy level, thus sequencing the plant genome is quiet a challenging task. Also, the gene content in the plants is complex as is evidenced by the presence of large gene families with abundant pseudogenes. They have high copy chloroplasts and mitochondria, which further complicate assembly of nuclear genome [20].

Due to the complexity, the chances are to assemble the sequencing reads into isolated gene islands among background of high copy repeats; gene sequences may not always be correct due to identical gene families. As in plant genomes, fragmentation is very large and there are chances of misassembly leading to false results [21].

However now it is possible and affordable for the scientific community to sequence and assemble large number of useful or plant genomes of interest and bring the results for final and useful draft assembly. Studying the diversity across populations and comparing non-domesticated with domesticated cultivars would help in the study of their suitability [14].

6.7 Genome Projects for Model Organisms

The technology and tools developed during human genome project helped in characterizing the genomes of other important organisms, which are used in biomedical research. In 2012, GOLD and NCBI added 3736 and 4585 sequenced genomes, respectively. Apart from humans, the genomes of many organisms have been sequenced in recent years. These include mice, rat, fruit fly (*Drosophila melanogaster*), round worm, etc. These along with other representative organisms were considered as model organisms for sequencing projects. Model organisms offer a cost-effective way to follow the inheritance of genes through many generations in a relatively short time. Some of the model genome projects are briefly described. The important ones include the mouse *Mus musculus*, the fruit fly *Drosophila melanogaster*, the worm *Caenorhabditis elegans*, the bacterium *Escherichia coli*, the yeast *Saccharomyces cerevisiae*, the plant *Arabidopsis thaliana*, and many microbes.

Genome projects are important as they help the scientists to perform genetic and biochemical analysis. They also shed light on their growth conditions, and through comparative genomics analysis, their evolution was studied. The projects can highlight:

- Cellular activities
- Disease-causing activities or pathogenesis
- Phylogenetic studies

Bacteria Both pathogenic and nonpathogenic bacteria are there. *Escherichia coli*, a commensal nonpathogenic bacteria in human gut and other vertebrates, is important as it synthesizes vitamin K and B complex. It has offered an understanding different life functions like DNA replication, transcription, translation, etc. Pathogenic bacteria are responsible for a number of diseases. A wide range of bacterial species and strains cause a variety of diseases with different symptoms. A large number of genome projects are finished and launched to obtain more information about pathogenic microorganisms [13].

Archaea They are prokaryotes resembling bacteria but were diverged long back at a very early stage. They are located in extremes of environmental conditions as well as soils and lakes. The interest in the study of archaea is their position as a separate kingdom from other life forms.

Yeast (Unicellular Fungi) They are valuable model organism as the conservation of many key molecules is strongly conserved. They are single-celled eukaryote fungi. Therefore, the study of these might provide important insights of their functions. Among fungi the important organisms for genome project are:

Saccharomyces cerevisiae: Important for baking and simple and easy to grow. Model for cell biology particularly cell cycle control and transcriptional regulation

Saccharomyces pombe: Fission yeast with rapid generation times

Candida albicans: Naturally present in humans but causes infection in patients when their immune system is not active

Protozoa (Unicellular Animals) Primitive animals move with the help of pseudopodia, flagella, and cilia. Many protozoans cause diseases such as:

Entamoeba histolytica: Parasite causing severe gastrointestinal disease

Trypanosomes: Sleeping sickness

Giardia: Severe diarrhea

Plasmodium: Causes malaria

Toxoplasma: Gastrointestinal disease causing damage to internal organs

Other organisms of interest are *Dictyostelium discoideum*, also known as social ameba. Tetrahymena (fresh water ciliate) is the model for cellular and developmental biology.

Multicellular Animals There is vast range of multicellular animals which are being explored. They have been used for understanding developmental biology, cell biology, and gene functions and as disease models.

Caenorhabditis elegans: It is a nematode or roundworm, which may be free living or parasite (causes blindness/elephantiasis and devours crops). *C. elegans* is a model organism for studies on expression, lineage, nervous system, aging, and apoptosis.

Drosophila melanogaster: It is a fruit fly with short life cycle and has been extensively studied over many years for advanced genetic analyses referred to as *Drosophila* genetics.

Fishes: Among fishes zebra fish (*Brachydanio rerio*) is an excellent model of development. Other fishes are pufferfish (*Takifugu rubripes*), used as disease model, and *medaka* (*Oryzias latipes*), used as model of development.

Chick: Among aves it is an important model of development.

Xenopus: It is a model of embryonic development and cell biology. The explored species are *Xenopus laevis* and *Xenopus tropicalis*.

Rat: It is a mammal of choice, which has been extensively utilized for physiological, neurological, pharmacological, and biochemical studies.

Mouse genome sequencing consortium was formed for mouse (*Mus musculus*) genome project. This model had been in debate because of Celera genomics which sequenced its genome, but sequence data was not freely accessible for others. Mouse has the same number of genes as humans. It is very important in comparative genomics analysis where comparison with human genome had been very helpful in defining highly conserved sequences and genome evolution.

The important projects launched for multicellular higher animals were for worms, flies, ascidians (tunicates and sea squirts), fishes, marine animals, frogs, birds, and mammals. Among mammals the important projects were for chimpanzee (close relative of humans), rat, cow, sheep, pig, and dog. The general observation is that we share over 98 % of our genes with chimpanzee.

6.8 Genomics of Pathogens

Hundreds of microbial genome sequences are complete till date. Genetic capacities of the organisms are unraveled by genome sequencing, the complete DNA complement of DNA. The genome information is providing an opportunity to analyze the molecular basis of behavior of microorganism (virulence, commensal (obligate, facultative, and opportunistic)). The novel information are being obtained for determinants of virulence or pathogenecity [13]. The information has strengthened the basic knowledge of three key factors which play an important role in determining the pathogenic potential of the species, namely, (1) bacterial physiology, (2) molecular adaptation to a preferred niche, and (3) genome susceptibility to the uptake of foreign DNA [7, 8].

To address these important issues, (1) the genome capacity of an organism, (2) the potential of genetic capacity, and (3) the susceptibility of this to evolutionary change should be fundamentally questioned.

An important question which requires attention is what creates a pathogen? The answer to this requires the details of the molecular basis of pathogen–host relation with epidemiological, population-level, phylogenetic, ecological, and experimental studies. The genome sequencing has shown that within the taxonomic grouping, there are non-virulent, virulent, and other human pathogens. These pathogens are causative agents of a number of diseases. The information also helps the development of a vaccine or potential drug target.

The selection of index strain is important. The bacterial species may be genetically and phenotypically diverse with pathogenic as well as nonpathogenic strains. Among pathogenic strains, there is significant variety of distinct serotypes or strains causing different diseases, showing selectivity for tissues and host. Post-sequencing of genome, annotation of various intergenic regions, and repetitive sequences are done to predict (1) gene regulation in microorganism, (2) generation of diversity, and (3) effect of selection pressure for evolution.

6.8.1 Properties of Bacterial Genome

Orphan Genes Important general insight to emerge from whole genome sequencing is the number of genes without any known relatives. These are referred to as "orphans" or "ORFans." The striking phenomenon of orphan genes is a big surprise and they are not restricted to species having distantly related genome sequences but are found in strains within a species. Considering pathogenesis the analysis of orphans and taxonomically restricted genes would be important as few may be the reasons for virulence.

Gene Number Another important observation in the relationship between genome size and coding capacity. The case of intracellular pathogen *Mycobacterium leprae* deviates significantly as it is in the process of shedding large number of genes showing reductive genome evolution [4]. In *M. leprae* 49.9 % genome is for functional proteins, 27 % is nonfunctional pseudogenes (acquired mutations), and 23.5 % is purely noncoding. The larger genome of *Mycobacterium tuberculosis* has functional copies of many of these pseudogenes.

Horizontal Gene Transfer As bacteria can acquire foreign DNA, thus horizontally acquired sequences can be found in a single genome in association with mobile elements (evident through departures from average value of G + C content, dinucleotide frequency, and codon composition). The comparative genomic analysis has given tremendous information about genetic transfer between strains within a species. Thus the recipient acquires large number of small sequences with time (mediated by phage or by acquisition of sequences involved in virulence (termed pathogenicity islands (PAIs)).

With this the concept of genetic backbone emerged with a core set of indigenous genes that remain stable over time, and these are being defined by homology comparisons for various evolutionary groups including serogroups, species, etc.

6.8.2 Understanding Genetic Capacity Required To Become a Pathogen

What could define virulence determinant? Complex working of genes together to be pathogenic is important and required information to define virulence factor [24]. It was suggested that the classes of proteins responsible for virulence are:

True virulence factors: present only in pathogens
Genes regulating "true virulence genes" required to colonize the host

The variation is observed and a lot of experimentation is required to answer how many genes are required for virulence [16].

E. coli can become a pathogen by acquiring a single region of DNA, whereas *Salmonella enteria* requires at least 60 genes to be pathogenic. Thus the microbial species living inside host cells have limited opportunities than enteric pathogens residing in complex microbial communities for genetic exchanges.

Steps in the behavior of pathogen require adherence, invasion, multiplication within the host, interference with host defenses, and damage to the host.

Adhesins: Pili/fimbriae; invasins allow bacterial cells to move to their preferred location within the host.
Pabulins: genes which help bacteria scavenge energy and essential nutrients.
Evasins: factors/capsule/surface proteins that bypass host defense mechanism.
Toxins: factors that damage the host.

Genome sequencing provides an overview of all the virulence factors in an organism. Searching for novel virulence factors consists of:

Searching genes homologous to known virulence factors in other species
Identifying direct virulence factors based on their unusual DNA characteristics
Identifying genes found only in pathogens

Homology-based methods for the search of virulence factors are very effective for taxonomic groups but fail where no known copy is found in public database.

M. genitalium genome sequencing identified all copies of the MgPa operon-a key virulence determinants. The analysis of *M. tuberculosis* genome led to the discovery of large PE and PPE gene families with 10 % of total coding capacity of this species, which play a role in virulence and antigenic variation. (The conserved proline–glutamate (PE) and proline–proline–glutamate (PPE) residues at the N termini are present in *Mycobacterium* genome. They are present in higher numbers in their genome and are abundant in the genomes of the pathogenic species).

Simple sequence repeats or microsatellites which are prone for expansion and contraction through the gain or loss via replication slippage are searched. The presence of these sequences in the promoters and coding regions may cause "phase variation" of the gene to which they are associated which gives them an opportunity to rapidly vary phase or undergo rapid switching between distinct phenotypic states and shared by wide range of pathogenic bacteria. They are also called "contingency loci" as they generate random but useful genetic variations. Another method used to analyze novel virulence genes may be subtractive methods to detect unshared genes.

6.8.3 Development of Multiple Drug-Resistant Bacteria

In *Mycobacterium tuberculosis* the development of multiple drug-resistant strains was reported. The patients suffering from MDR strains are also increased where effective treatment of TB becomes very difficult. Tuberculosis was declared as global emergency in 1993 by WHO as it is very difficult to control the disease due to emergence of multidrug-resistant (MDR) strains. Multidrug resistance is defined as resistance to at least rifampin and isoniazid (two important drugs used for treatment). Recently severe form of drug resistance as extensively drug-resistant (XDR) TB has been described. Thus nowadays multidrug therapy is given to control and treat the disease. XDR are defined as having resistance to isoniazid and rifampin, plus any fluoroquinolone and at least one of three injectable second-line drugs (e.g, amikacin, kanamycin, or capreomycin). The XDR TB is of special concern for persons with HIV infection or other conditions that can weaken the immune system.

Possible Reasons

- After discovery of streptomycin in 1943, it was extensively used for treatments in large amounts. Probably due to monotherapy of the drug, resistance to the drug emerged.
- Specific gene mutations were associated with drug resistance.
- Overuse of antibiotics, in diseases like flu, cold, and fever.
- Use in animal feed (as animal growth promoters).
- Antibiotic-resistant genes are present on extrachromosomal molecule, R-plasmid, which are self-transmissible. Antibiotic resistance genes are present as transposable genetic elements.
- Presence of a gene NDM (New Delhi metallo-beta-lactamase) makes bacteria resistant to major groups of antibiotics. (The gene was recognized in Swedish national who acquired the antibiotic-resistant bacteria from India. The gene is widely present in *E. coli* and *Klebsiella pneumoniae* and may be transferred to nonresistant microbes by horizontal gene transfer).

6.8.4 Hepatitis C Virus

Hepatitis C virus (HCV) infection is characterized by evolution to chronicity being the leading cause of end-stage liver disease and hepatocellular carcinoma. It is a global health problem due to the inherent fidelity of viral RNA polymerase. It has RNA genome with 9600 nucleotides with one large open reading frame. The entry of virus in hepatocytes is mediated by its binding to various receptors like CD81, scavenger receptor class B type I (SR-B1), mannose binding lectins, low-density lipoprotein receptor, and glycosaminoglycans. However CD81 and SR-B1 are expressed on other cells as well. Claudin-1 which is a tight junction component highly expressed on liver cells has been recently reported to have key role in entry.

6.8.4.1 Evolution of Hepatitis C Virus

Genotypes

Six clusters of HCV genome are proposed with genotype 1 (predominant), genotype 2 (present in some regions of Africa), genotype 4 (predominant in Egypt), and genotypes 3 and 6 (present in some regions of Asia). They differ from 31 to 33 % among different genotypes (1, 2, 6, 23, 26).

Viral RNA polymerase lacks proofreading activity giving high error rates with every turn of replication (per day turnover rate of 10^{12} virions), thus incorporating many mutations everyday. This high mutation rate leads to heterogeneous viral species which are phylogenetically closely related and are referred as quasispecies.

Genetic Drift

The viral genome is changed continuously due to mutations and is called as genetic drift. As these changes result in the altered protein, thus it is also referred to as antigenic drift. The mutations in the viral genome can be neutral, advantageous, or disadvantageous (deleterious). Deleterious mutations lead to negative selection, while beneficial mutations would undergo positive selection giving viruses the capacity to adapt in the same host or to specific environment. These mutations are major driving force of evolution of HCV. The quasispecies feature of HCV allows it to adjust rapidly in the new environment, where important variants may be present in low number, but according to the environment, the important ones are selected to exist as dominant species.

Effect of Host Immunity

Antibody response in healthy host acts as selection pressure for HCV genome, generating profound sequence diversity with continuous escape from host humoral response. Even $CD8^{+}$T cells against viral proteins lead to its evolution and mutational escape thus prolonging viral persistence in the host.

The virus develops mutations in the regions targeted by $CD4^{+}$ T cells and is present as quasispecies evading detection by cell mediated as well as humoral immunity. The HCV also interferes with innate immunity and interferes with the production of antiviral interferon-α and interferon-β. It can even inhibit the kinase downstream of interferon cascade.

Thus high mutation rate and large number of variants due to its presence as quasispecies are big challenges with important implications for viral persistence, host cell tropism, antiviral drug resistance, and development of HCV vaccine. However HCV protease inhibitors and polymerase inhibitors are being tried for its control. The strategy where many attacks targeted at multiple sites may limit mutational escape and may be effective. The understanding of strategies of virus with better model to study viral cycles is necessary for effective and efficient therapeutic intervention.

6.8.5 Influenza Virus

Influenza viruses that belong to *Orthomyxoviridae* family possess segmented single-stranded negative-sense RNA genome. It has different variants as influenza A, B, and C. Influenza type A contains segmented and eight separate single-stranded RNA molecules encoding for 11 different viral proteins. Influenza B viruses are responsible for many epidemic outbreaks in humans, whereas influenza C viruses have been associated with young children and pigs.

The proteins of influenza are polymerase, structural proteins, and envelope proteins of which its surface protein hemagglutinin (H or HA) and neuraminidase (N or NA) play an important role in the distribution and reproduction of the virus. The hemagglutinin initiates the infection in the host by binding to the host cell surface receptors (sialic acid on host cell surface). After the virus penetrates the cell membrane, its protein neuraminidase (sialidase) cleaves the glycosidic bond formed to the sialic acid residues; thus neuraminidase supports more fresh infections. Hemagglutinin's 16 (H 1–16) different forms and neuraminidase's 9 (N 1–9) different forms have been identified. The names for different subtypes of the virus are derived from the combination of the surface proteins H and N [22].

6.8.5.1 Evolution of Influenza Virus

Influenza A viruses have a wide range of host including mammals, birds, and aquatic birds; these are their natural reservoirs where they cause epidemics of influenza. Influenza B and C viruses are less pathogenic than A and are isolated mainly from humans. A and B subtypes are phylogenetically more related than the C subtype [12].

The Reasons of Their Pathogenesis and High Mutation Rate As influenza virus has very high mutation rate, thus vaccines developed against one strain are not effective against the mutated one. There are some observations for their frequent changes.

High Mutation Rates All A, B, and C have diverged from common ancestors and have formed multiple lineages. The virus has short generation times with its RNA polymerase without proofreading functions due to which nucleotide incorporation is with high rate of errors resulting in high rates of mutations [5].

Selection Pressure of the Host HA protein of influenza is the major target for antibody response in host; thus selection pressure of the host results in ongoing antigenic changes and evasion of immune reactions and hunting for host species which have not countered this virus previously. Thus influenza A can live in many different environments and is adapted for utilizing multiple host species.

Genetic Drift and Antigenic Drift Due to continuous mutations in the genome (genetic drift), the genes encode the antigens with minimal variations (antigenic drift) evading immune responses. Highly pathogenic avian influenza (HPAI) possess multibasic cleavage site (MBCS) which allows HA processing by ubiquitously expressed proteases enabling the systemic spread of the virus and is the cause of high morbidity and mortality in chickens.

Routes of Infection The viral transmission to different hosts can follow different routes (fecal or through breathing or both).

Viral Reassortment The subtypes of low-pathogenicity influenza can switch to highly pathogenic phenotype by acquisition of MBCS during circulation in the poultry. The Hong Kong outbreak of highly pathogenic H5N1 in 1997 with 18 human infections and 6 deaths was due to the appearance of different A1 influenza viruses. This virus was generated after reassortment events occurred between high pathogenic H5N1 and low pathogenic H9N2 causing chicken outbreaks and human cases. The A1V virus has very high mutation rate along with frequent reassortment resulting in diverse viral reservoir.

Genetic reassortment occurs when one cell or host is infected by more than two viral strains. In the process of viral packaging, one or more genetic segments are exchanged between two related viruses. Thus during the viral packaging, mistakes in the combination of nucleotide segments occur as the cellular system is unable to distinguish different strains. The reassortment of complete units of genetic material results in the formation of "reassortants" or "mosaic" viruses (Fig. 6.7). At times, "genetic reassortment" affects the exchange of genome segments encoding the viral surface proteins hemagglutinin and neuraminidase. Thereby, the virus achieves a new antigenic pattern. Thus, the process is called "antigenic shift." Thus influenza virus exists as quasispecies which is continuously evolving.

Year of outbreak	Virus type	Influenza outbreak	Causes
1918/1919	H1N1	Spanish flu	Mutation from bird
1957	H2N2	Asian flu	Reassortment (bird–humans)
1968	H3N2	Hong Kong flu	Reassortment (bird–humans)
1977	H1N1	Russian flu	Reemergence of H1N1 strains
1997	H5N1	Hong Kong avian flu	Reassortment (bird–humans)
2015–2025??	H5N1/ HINI/ H2N2???	?????	Reassortment or mutation?? Bird–human–pig???

This diversification into multiple sublineages has led virus to adapt for different host reservoirs. The continuous adaptation is due to reassortment, wide host range, and selection pressure of the immune system.

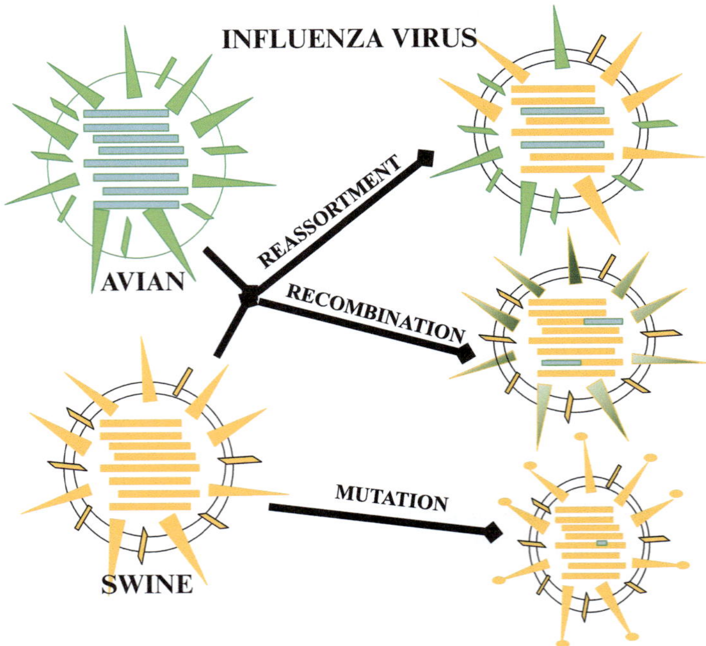

Fig. 6.7 Processes of reassortment, recombination, and mutation which lead to genetic drift and antigenic drift. Due to these the vaccines and immune system are not able to effectively eradicate the virus

> Avian influenza-A subtypes are not able to affect humans and poultry is also not susceptible to human viruses. However pigs play a very important role in the formation of new influenza viruses as they are susceptible to double infections with avian influenza viruses as well as human influenza viruses. Thus new viral variants are produced and transmitted from pigs to humans.

6.8.6 Human Immunodeficiency Virus (HIV)

Human immunodeficiency virus (HIV) belongs to the family of retroviruses and subfamily of lentiviruses. It is the causative agent of acquired immunodeficiency syndrome (AIDS). The human retroviruses belong to two distinct groups: the human T-lymphotropic viruses (HTLV)-1 and HTLV-II and HIV-1 and HIV-2. HIV-1 is the common cause of AIDS having several subisotypes. HIV-2 was first identified in 1986 in West Africa.

HIV virion has an icosahedral structure with numerous external spikes formed by two major envelope proteins gp120 and gp41. On the coat the external protein is gp120 which binds with high affinity to CD4 molecules present on subset of T lymphocytes. After gp120 binds to CD4, its conformational changes facilitate its binding to other coreceptors CCR5 and CXCR4 (seven transmembrane G-protein-coupled receptors). The envelope protein gp120 can also bind with high affinity to DC-SIGN receptor present on dendritic cells.

Another transmembrane envelope protein is gp41 which helps in fusion with host cells. After virus infection in the cells, its reverse transcriptase catalyzes the conversion of RNA into DNA.

6.8.6.1 Evolution of HIV

- Human immunodeficiency virus (HIV) is closely related to simian immunodeficiency viruses (SIVs) infecting primates and feline immunodeficiency virus (FIVs) infecting cats. Surprisingly primates with SIV and wild cats with SIVs live without any harm from these viruses.
- HIV-1 is more closely related to viruses isolated from chimpanzees (*Pan troglodytes trog-*

lodytes is considered a natural reservoir of M and N group) and gorillas.

- There are three major groups of HIV-1:
  - Group M (major): It is responsible for most of the infections in the world.
  - Group O (outlier): A rare form from Cameroon, Gabon, and France.
  - Group N: First identified in a Cameroonian woman with AIDS.

M group has nine subtypes/clades (A, B, C, D, F, G, H, J, and K) and many minor and major circulating recombinant forms (CRFs). When an individual is infected with two or more subtypes, then CRFs are generated by recombining different subtypes to create virus with better selective advantages. The M group of HIV-1 has not only subtypes and CRFs but also sub-subtypes for A (A1 and A2) and F (F1 and F2):

- Thus HIV isolates reveal varying levels of sequence diversity with minor differences in the coding sequence of viral envelope proteins from the same individual to approximately 50 % variations in the isolates from different groups (M, N, and O) of HIV-1.
- Major changes are observed in its hypervariable region by either simple base substitution or insertions or deletions or recombinations or gain or loss of glycosylation sites.
- The reverse transcriptase of HIV is with limited fidelity. Thus the error rate of its RT is very high in each round of replication. This may be responsible for hypermutability of the HIV. The mutation rate of HIV is 65 times higher than influenza virus. Selective pressure of immune system (antibody and cytotoxic T-cell responses) might also contribute to variations. Selection pressure of anti-HIV drugs also led to evolution of viruses which are resistant to the drug.
- Another reason can be extensive viral trafficking leading to more variations and new CRFs. This is due to infection of an individual with viruses of more than one subtype.
- The patients infected with HIV are very prone for infection with *Streptococcus pneumoniae* and *Haemophilus influenzae*, with pneumococcal infection being the earliest serious infection to occur in HIV patients. Other opportunistic pathogens are *Pneumocystis jirovecii*, *M. tuberculosis*, MDR *M. avium*, *Toxoplasma gondii*, *Salmonella*, *Cytomegalovirus*, and many more pathogenic agents.

Monocytes are normal in HIV-infected individuals. They express CD4 molecules and several coreceptors like CCR5, CXCR4, and CCR3; thus they are also target for HIV infection. However HIV has low cytopathicity for cells of monocyte lineage; thus they can serve as reservoirs of HIV infection and resist eradication of HIV by retroviral drugs.

In humans also the mutant CCR5 allele (mutated due to plague epidemic and made its bearers resistant to the disease) also confers resistance to HIV.

6.9 Chapter End Summary

- Human genome project (HGP) was collectively launched by the National Institute of Health (NIH) and the US Department of Energy with a projected time span of 15 years. The multi-institutional collaborative human genome project was completed in 2003, 2 years ahead of its scheduled deadline.
- The project was started with physical mapping and genetic mapping with establishment of central repositories as genome databases for sharing of the information. Celera genomics led by Dr. Venter started parallel sequencing using shotgun approach. The information was continuously uploaded by government-funded project for the ease of all the research laboratories. It also gave rise to the field of bioinformatics.
- The project was helpful to unravel mystery of various diseases. Many disease loci were mapped and tagged with various markers for easy analysis. At the same time, many other

parallel projects were initiated, for example, human proteome, comparative genomics, etc.

- Genome annotation, functional genomics, and proteomic studies aided the analysis of the way genes function and their regulation.
- The genome projects for other organisms were undertaken; some of them serve as model organisms. Mouse map was continuously used for studies of human genes. The microbial genome project threw some light on the evolution of the pathogens and their modes of development of drug resistance. Their continuous evolution is occurring due selection pressure of the host.
- The work on hepatitis C virus showed that the presence of quasispecies and lack of proofreading of its RNA polymerase are major properties leading to its continued evolution. These genome projects have an important role in unraveling the mystery of many life forms.

Multiple Choice Questions

1. The genome projects were possible because of the technique of:
   (a) Chromatography
   (b) X-ray crystallography
   (c) DNA sequencing
   (d) All of the above
2. In humans the genome consists of:
   (a) 22 different DNA molecules
   (b) 25 different DNA molecules
   (c) 24 different DNA molecules
   (d) 23 different DNA molecules
3. Automated DNA sequencers use:
   (a) Chromatographic column
   (b) Spectrophotometric use
   (c) Electrophoretic column
   (d) a and c
   (e) b and c
   (f) All of the above
4. Physical mapping of genome deals with:
   (a) Linkage analysis of the chromosome
   (b) Recombination and crossing over analysis
   (c) Structural analysis of the genome
   (d) All of the above
5. Contiguous clones means:
   (a) Different parts of the chromosomes
   (b) A set of overlapping clones
   (c) A set of cloned DNA in a vector
   (d) None of the above
6. Genome mapping involves:
   (a) Whole genome shotgun approach
   (b) Fluorescence in situ hybridization
   (c) Sequencing of the clones
   (d) All of the above
7. Expressed sequence tag may be used in:
   (a) Restriction library
   (b) Chromosome mapping
   (c) cDNA library
   (d) Genomic DNA library
8. FISH may be utilized to:
   (a) Select the overlap
   (b) Create somatic cell hybrid
   (c) Localize presence of specific DNA
   (d) Construct genetic maps
9. The major benefit of human genome projects is:
   (a) In the study of inherited disorders
   (b) Molecular basis of diseases
   (c) Defining new drug target
   (d) All of the above
10. In human genome project, it was very difficult to sequence centromeres because:
    (a) It is very difficult to clone.
    (b) It has high repetitive DNA.
    (c) Orientation of DNA is very difficult to determine.
    (d) All of the above.
11. Genome project of model organism is important as they:
    (a) Provide easy genetic and biochemical analysis
    (b) Serve as efficient disease model
    (c) Are used to define functions of various genes
    (d) Are helpful in phylogenetic studies
    (e) All of the above
12. Hepatitis C is difficult to eradicate by effective and active immune response because:
    (a) It destroys the antibodies.
    (b) It easily skips targeting by immune system.
    (c) It exists as quasispecies.
    (d) All of the above.

Answers

1. (c); 2. (b); 3. (c); 4. (c); 5. (b); 6. (d); 7. (c); 8. (c); 9. (d); 10. (b); 11. (e); 12. (c)

Review Questions

Q1. What do you understand by genome projects?
Q2. Define genetic and physical mapping of the genome.
Q3. What is gene annotation?
Q4. What are OR Fans? Define their role in pathogenesis.
Q5. What is the role of immune components in hepatitis C virus evolution?
Q6. What do you understand by quasispecies? What advantage does the virus have due to this property?

References

1. Bartosch B, Cosset FL (2006) Cell entry of hepatitis C virus. Virology 348:1–12
2. Bartosch B et al (2003) Cell entry of hepatitis C virus requires a set of co-receptors that include the CD81 tetraspanin and the SR-B1 scavenger receptor. J Biol Chem 278:41624–41630
3. Cannon SB, May GD, Jackson SA (2009) Three sequenced legume genomes and many crop species: rich opportunities for translational genomics. Plant Physiol 151:970–977
4. Cole ST et al (2001) Massive gene decay in the leprosy bacillus. Nature 409:1007–1011
5. Drake JW, Charlesworth B, Charlesworth D, Crow JF (1998) Rates of spontaneous mutation. Genetics 148:1667–1686
6. Evans MJ et al (2007) Claudin-1 is a hepatitis C virus co-receptor required for a late step in entry. Nature 446:801–805
7. Field D, Hood D, Moxon R (1999) Contribution of genomics to bacterial pathogenesis. Curr Opin Genet Dev 9:700–703
8. Field D, Hughes J, Moxon ER (2004) Using the genome to understand pathogenicity. In: Woodford N, Johnson AP (eds) From methods in molecular biology, "Genomics, Proteomics and Clinical Bacteriology". Humana Press, Totowa
9. Groisman EA, Ochman H (1994) How to become a pathogen. Trends Microbiol 2:289–294
10. Gyapay G et al (1994) The 1993–94 Généthon human genetic linkage map. Nat Genet 7:246–339
11. Strachan T, Read A (2011) Human molecular genetics, 4th edn. Garland Science, Taylor and Francis Group, London
12. Jamia Jordens The "rapid" evolution of influenza virus. www.evolution-of-life.com Volkswagen Stiftung
13. Joyce EA et al (2002) Redefining bacterial populations: a post-genomic reformation. Nat Rev Genet 3:462–473
14. Mehrotra S, Goyal V (2013) Evaluation of designer crops for biosafety – a scientist's perspective. Gene 25:241–248
15. Michael TP, Jackson S (2013) The first 50 plant genomes. Plant Genome 6:1–7
16. Moxon R, Tang C (2000) Challenge of investigating biologically relevant functions of virulence factors in bacterial pathogens. Philos Trans R Soc Lond B Biol Sci 355:643–656
17. NIH/CEPH Collaborative Mapping Group (1992) A comprehensive genetic linkage map of the human genome. Science 258:67–86
18. Oliver SG (1996) From DNA sequence to biological function. Nature 379:597–600
19. Olson M et al (1989) A common language for physical mapping of the human genome. Science 29:1434–1435
20. Schatz MC, Witkowski J, McCombie WR (2012) Current challenges in de novo plant genome sequencing and assembly. Genome Biol 13:243–249
21. Schnable PS et al (2009) The B73 maize genome: complexity, diversity, and dynamics. Science 326:1112–1115
22. Sonnberg et al (2013) Natural histrory of highly pathogenic Avian Influenza H5N1. Virus Res 178. doi:10.1016/i.virusres.2013.05.009
23. Timm J, Roggendorf (2007) Sequence diversity of hepatitis C virus: implications for immune control and therapy. World J Gastroenterol 28:4808–4817
24. Wassenaar TM, Gaastra W (2001) Bacterial virulence: can we draw the line? FEMS Microbiol Lett 201:1–7
25. Zhang J et al (2004) CD81 is required for hepatitis C virus glycoprotein-mediated viral infection. J Virol 78:1448–1455
26. Zhang J, Chiodini R, Badr A, Zhang G (2011) The impact of next-generation sequencing on genomics. J Genet Genomics 38:95–109

Some Related websites

http://ghr.nlm.nih.gov/handbook/genomicresearch/snp
http://www.ncbi.nlm.nih.gov/probe/docs/techsts/
web.ornl.gov/sci/techresources/Human_Genome/publicat/jmmbbag.pdf
www.colorado.edu/chemistry/bioinfo/HumanGenomeProject.htm
www.genome.gov

www.ghr.nlm.nih.gov/handbook/hgp
www.knowledgene.com
www.ncbi.nlm.nih.gov
www.ndsu.edu/pubweb/~mcclean/plsc431/students99/boehm.htm
www.nugene.org
www.sanger.ac.uk
www.web.ornl.gov

Some Related Resources

Council for Responsible Genetics
Department of Energy (DOE) Human Genome Project Information
Ensembl Human Genome Browser
ESRC Centre for Genomics in Society (Egenis), University of Exeter
Human Genome News (DOE/NHGRI publication)
Human Genome Organization (HUGO)
HumGen International (ELSI resources)
National Center for Biotechnology Information (NCBI) Human Genome Resources
National Human Genome Research Institute (NHGRI) All About the Human Genome Project (HGP)
Nature Human Genome Collection
NHGRI ELSI Research Program
Nuffield Council on Bioethics
Science Human Genome Special Issue 16 February 2001
Wellcome Trust Sanger Institute Human Genome Project
World Health Organization Genomic Resource Centre

7 Pharmacogenomics and Pharmacogenetics

Abstract

With the information available about human genome and human proteome, it is now well understood that there are a lot of variations between individuals. These minor variations account for many differences like adverse drug reactions, which are responsible for many hospitalizations and casualties. The observed variable effect of drug is due to difference in sensitivity as some people need higher dose and some need lower dose to get similar therapeutic effect, but in some people drug has no therapeutic effects and in some shows strong adverse reactions. Some of these differential effects are due to environmental causes, or the individual's ability to absorb or metabolize a drug may be altered or multiple drug interaction can occur (in people taking multiple drugs). *Pharmacogenetics* is the study of the roles of specific genes in these effects, whereas *pharmacogenomics* is the study of how an individual's genetic makeup affects the body's response to drugs or the personalized medicine deals with the concepts that for a particular disease, the rate of progression of the disease for each person is unique and each person responds in a unique way to drugs. In its broadest sense, personalized medicine includes the detection of disease predisposition, screening and early disease diagnosis, assessment of prognosis, pharmacogenomic measurements of drug efficacy and risk of toxic effects, and the monitoring of the illness until the final disease outcome is known.

7.1 Introduction

With the information available about human genome and human proteome, it is now well understood that there are a lot of variations between individuals. These minor variations account for many differences like adverse drug reactions, which are responsible for many hospitalizations and casualties. The observed variable effect of drug is due to difference in sensitivity as some people need higher dose and some need lower dose to get similar therapeutic effect, but in some people drug has no therapeutic effects and in some it shows strong adverse reactions (Fig. 7.1).

Some of these effects are due to environmental causes, the individual's ability to absorb or

V. Gupta et al., *Basic and Applied Aspects of Biotechnology*, DOI 10.1007/978-981-10-0875-7_7

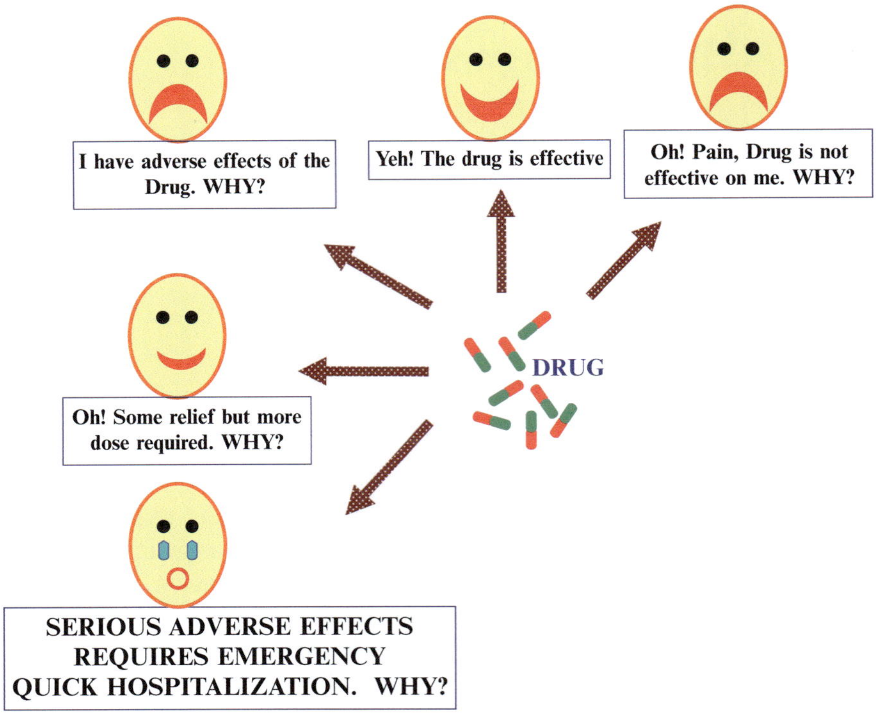

Fig. 7.1 The figure shows the variable effect of the same drug on different individuals suffering from the same disorder

metabolize a drug may be altered, or multiple drug interaction can occur (in people taking multiple drugs). *Pharmacogenetics* is the study of the roles of specific genes in these effects, whereas *pharmacogenomics* is the study of how an individual's genetic makeup affects the body's response to drugs. Or the personalized medicine deals with the concepts that for a particular disease, the rate of progression of the disease for each person is unique and each person responds in a unique way to drugs. In its broadest sense, personalized medicine includes the detection of disease predisposition, screening and early disease diagnosis, assessment of prognosis, pharmacogenomic measurements of drug efficacy and risk of toxic effects, and the monitoring of the illness until the final disease outcome is known.

Pharmacokinetics determines drug delivery to, and removal from, molecular targets or relationship between drug concentration and time. *Pharmacodynamics* is variability in drug actions, despite equivalent drug delivery to effector drug sites (drug concentration and its effects). In other words, pharmacokinetics deals with what is the body's response to the drugs, and pharmacodynamics deals with the response of drug on the body. The genetic variations can affect the pharmacokinetics and pharmacodynamic properties of the drug and variable effects on these responses affect the patient's response to the drug.

Pharmacogenomics or personalized medicine or a recent term given *precision medicine* (Khoury et al. 2012) is helping to device strategies for the coming generation pharmaceutical companies, biotechnology and medical institutes, and academic medical centers [12]. The primary aim is the treatment of the patient with the correct dose of the suitable medicine which is based upon the genes of the individual [8] (Fig. 7.2). With the expanded knowledge of the molecular basis of cancer, it is now known that significant differences in gene expression patterns can guide therapy for a variety of solid tumors and hematologic malignant neoplasms.

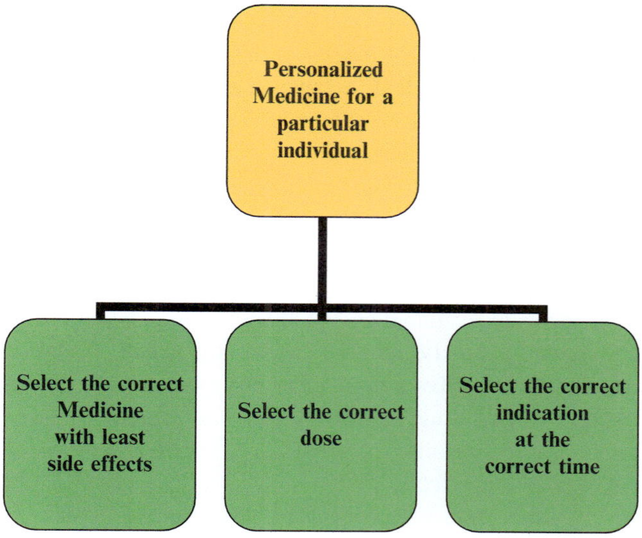

Fig. 7.2 The figure shows advantages of pharmacogenomics studies which leads to complete benefit of the drug to the patients

7.2 Pharmacogenetics

The study of single genetic variant with analysis on responders and nonresponders to the drug or having adverse side effects of the drug (drug toxicity) is the *Pharmacogenetics*. The relationship between genetic defect and abnormal drug response was termed as pharmacogenetics by Vogel in 1959. The response of an individual to a drug differs as it depends upon the genetic and nongenetic factors. These factors may be variations in the target of the drug, genes responsible for the disease, the enzymes that metabolize the drug, and finally clearance of the drug. All these factors may predict the efficacy or toxicity of the drug. Now nearly millions of genetic markers of single nucleotide polymorphism (SNP) type are available for genotyping and phenotyping studies. Their usage can generate clinically useful data, as it can predict the response of an individual because of the presence or absence of a particular genetic variant [26]. In 1999 a consortium to discover human SNPs was formed. Probably high-resolution SNP map would help in the identification of genes for complex diseases such as asthma, diabetes mellitus, atherosclerosis, and psychiatric disorders. SNP technology has been in use in oncology where the efforts for the detection and predisposition for cancer and predicting toxic responses to drugs and selecting the best individual and combination anticancer drugs may be very useful [4].

7.3 Toxicogenomics

Toxicogenomics deals with adverse side effects of the toxic substance/drug termed as adverse drug reactions (ADRs) either due to defect in inherent capacity of the individual to metabolize and detoxify the drug or due to adverse drug–drug interactions. These adverse drug reactions are major cause of death (fifth in the USA) annually and are responsible for number of hospitalizations, which has potential economic burden. Majority of it is because of metabolic activation of the parent drug leading to either immune-mediated toxicity or accumulation of toxic products or reactive products and might affect cardiac function or liver toxicity. Due to this, many drugs are now withdrawn from the market.

7.4 Pharmacodynamics and Pharmacokinetics

Nowadays, usage of a series of bioassays like gene expression profiles and biomarkers help in guiding the dosage, timing, and route of administration for novel and effective therapeutics. These tests are helpful to speed the preclinical and early clinical development of drugs by enhancing the achievement of the ideal therapeutic dose while avoiding dose-related toxic effects, for example, determination of epidermal growth factor receptor (EGFR) in biopsy of skin in patients receiving gefitinib (anti-EGFR small molecule tyrosine kinase inhibitor).

Pharmacokinetics is the analysis of how the drug is made available in the bloodstream, its transportation to the relevant target organ, and metabolization and excretion. Pharmacodynamics is the effect of drug molecule on its molecular target and its signaling or events that determine any therapeutic effects.

The drug processes, which can affect fate of the drug, are:

1. Absorption: Absorption of the drug affects the actual dosage of the drug in the blood.
2. Activation: If the drug is supplied in the form of pro-drug, then it must convert into active drug in the liver by enzymes.
3. Target response: These two together would ultimately affect the dose of the drug and its concentration which is available to act on the target.
4. Catabolism and excretion: The metabolization of the drug in the body would affect its serum life and disposal. Fast metabolizers would readily remove the drug, whereas slow metabolizers would have drug in their system for long durations.

7.5 Pharmacogenomics

It is assumed that variability to drug responsiveness (efficacy, adverse effects, toxicity, maintenance dose) is due to individuals' own genetic makeup (Fig. 7.3) and becomes very serious

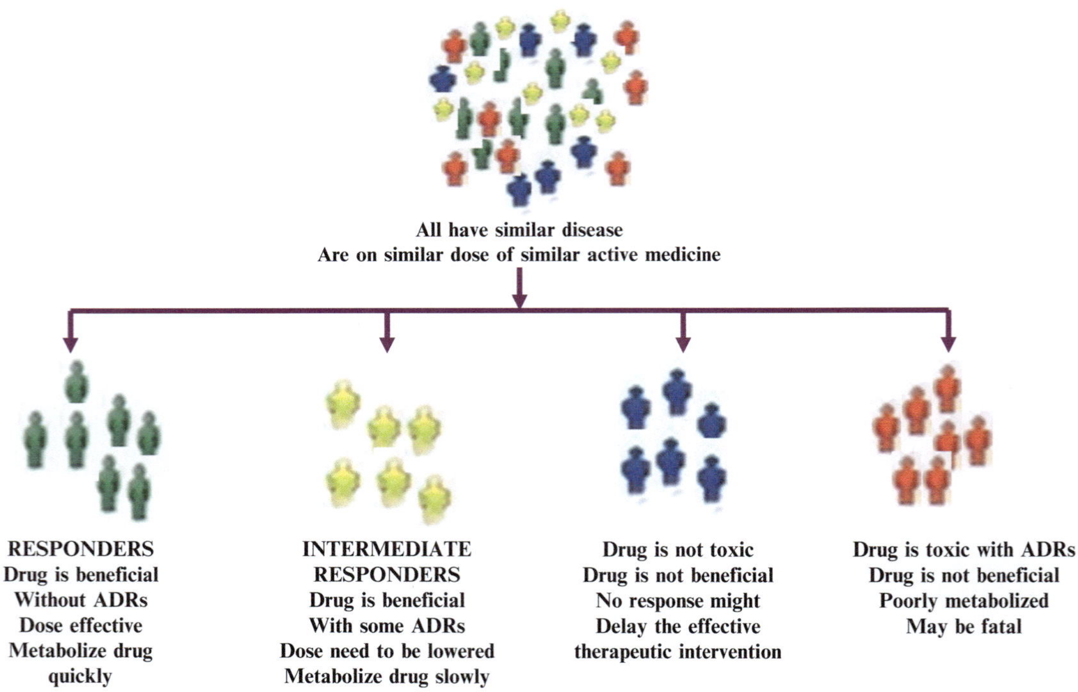

Fig. 7.3 The figure shows that all these individuals are suffering from the same disease; thus, they were given similar drug with the similar dose. Drug response is variable due to genetic or environmental factors; however, individuals in *green* were responders with best effects of drug, and individuals in *yellow* were responsive for the drug, but drug had some adverse side effects in them; thus, they require lower dose; individuals in *blue* had no effect, no ADRs, but due to nonresponsiveness, disease might become problematical and difficult to control; individuals in *red* did not got any benefit from the drug but suffered from serious ADRs requiring hospitalization or sometimes drug might be fatal

where it is responsible for many admissions in the hospital and more than 1 lakh deaths worldwide.

These days, genome-wide association studies (GWAS) are quiet beneficial for survey of the entire genome for association with drug response phenotype [19, 22]. The results for the studies are highly encouraging that once genetic association or association of genes are identified for a particular side effects or adverse effects with a specific drug, the results are immediately helpful for the practitioner. These studies also help to evaluate the dose of the drug, depending upon the metabolization capabilities of the individuals (Fig. 7.4). Another issue pertaining to the need of research in pharmacogenomics is that the effect size for many genetic associations identified to date for pharmacogenomic traits is larger than that of complex diseases. This stronger effect has helped in the identification of pharmacogenomic association with a small size of the samples where association of complex traits requires very large sample size. It is quiet well known that drug response is complex phenotype, which is affected by genetic factors (through regulation of drug-metabolizing enzymes) as well as dosage, diet, age, lifestyle, health condition, environmental status, and socioeconomic condition. The studies have been extremely helpful for some of the drug. The drug administered can be pro-drug, which requires conversion to active component, or the drug which itself mediates effects and afterward it is detoxified and secreted [9].

In some of the drugs which are well characterized with their mechanism of action known, the studies are simple where their potential targets are genotyped, whereas drug with unknown mechanism of action requires complex and many studies to reach to a meaningful conclusion [28].

7.5.1 Metabolism of Drugs

The reaction of metabolism of the drugs is divided into two phases:

Phase I reactions: These involve oxidation, hydroxylation, and hydrolysis reactions producing the biologically active molecule or

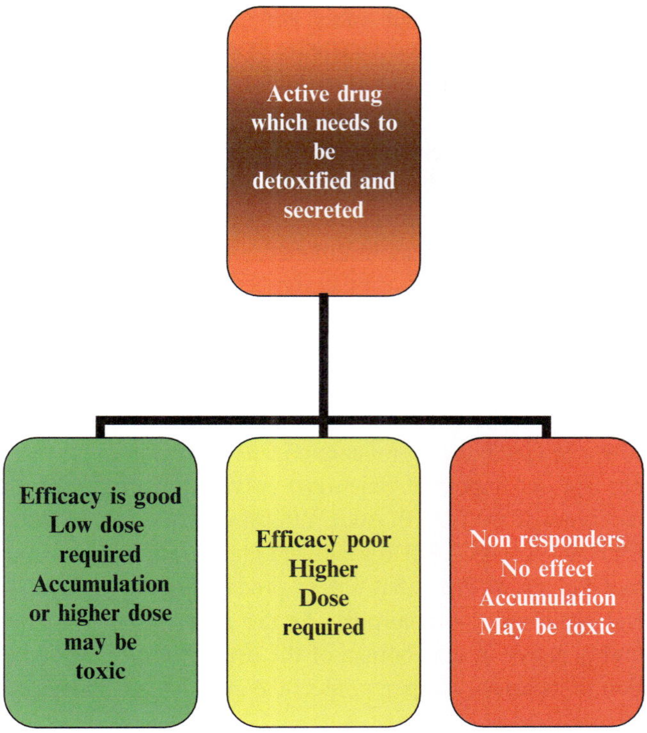

Fig. 7.4 The figure shows the drug which is active. Due to inherent difference, the effectiveness and detoxification vary in different individuals

phase I product can be intermediate in the inactivation and degradation of the drug.

Phase II reactions: These are typically conjugation reactions as acetylation, glucuronidation, or sulfation, and they produce water-soluble compound that is easily excreted [9].

7.5.1.1 Phase I Reactions

Cytochrome P450 (CYP)

The P450 cytochromes are responsible for many phase I reactions involved in the metabolism of the drug. It constitutes a large family of enzymes with iron–sulfur-containing active site. They insert single atom of oxygen atom in many organic compounds resulting in the production of water-soluble hydroxylated products. They have probably evolved to deal with toxic plant metabolites and are mainly present in the liver, but some occur in the small intestine also. Human genome has 57 functional CYP genes, with 1–3 families with 22 different P450 isoforms. Major drugs are lipophilic or are made more water soluble before excretion in the kidneys where CYP 1–3 account for 70–80 % of metabolism and are also involved in other metabolism. They are named as cytochrome P450 family, subfamily, and polypeptide, for example, CYP2D6 (cytochrome P450 family 2, subfamily D, and polypeptide 6). In phase I metabolism, at least ten different P450 enzymes participate. Individual drugs may be substrate for more than one P450 enzymes. Thus, correlating action of one P450 enzyme to a drug is difficult; however, some drugs are metabolized by only one P450 enzyme [11].

The polymorphic xenobiotic-metabolizing CYP enzymes can be divided into two classes: class I with prominent enzymes as CYP1A1, CYP1A2, CYP2E1, and CYP3A4 which are active in metabolizing procarcinogens and drugs but without important functional polymorphism and class II with CYP2A6, CYP2B6, CYP2C9, CYP2C19, and CYP2D6 which are important for metabolism of drugs but not of procarcinogens and are highly polymorphic. As the enzyme is highly active in metabolism of the drugs, variation in its activity largely affects the response of some drugs used to treat hypertension (beta-blockers) and heart diseases and used as psychiatric medicines (tricycle antidepressants).

The xenobiotic-metabolizing P450s are all polymorphic (allelic variants are described on home page www.cypalleles.ki.se). Depending upon different gene variants, they are grouped into four major phenotypes:

1. Poor metabolizers: Defective alleles, lack enzyme activity (splicing defects, gene deletions, amino acid substitutions)
2. Intermediate metabolizers: Heterozygous with decreased enzyme activity
3. Extensive metabolizers: Have two functional alleles
4. Ultrarapid metabolizers: Have more than two active gene copies

Complete absence or very low activity of the enzyme results in the accumulation of certain drugs causing toxicity.

CYP2D6 It is located on chr.22q13 and has nine exons. Poor metabolizers have many kinds of point mutations as nonsense, frameshift, or splice site changes or sometimes complete gene deletions (human genome project reference chromosome 22 was from deletion carrier; thus, the gene was absent in original sequence). Ultramodifiers have increased numbers of CYP2D6 genes. It is involved in metabolism of approximately 25 % of all drugs. It shows enhanced sensitivity to antihypertensive drug debrisoquine and to antiarrhythmic drug sparteins. The patients showed high drug levels in their blood but low urinary levels of the catabolism products due to the low metabolic activity of the enzymes.

Clinical Significance

Morphine Toxicity Variation in its activity significantly affects the fate of codeine, as poor metabolizers are at risk of overdose effects, whereas ultramodifiers may face adverse effects as sedation and impaired breathing. Ultrarapid metabolizers of CYP2D6 results in increased

O-demethylation of codeine to morphine, resulting in very high concentration of morphine with serious morphine toxicity with very small dose of codeine resulting in respiratory depression (Fig. 7.5).

Tramadol Toxicity In this also ultrarapid metabolizer of CYP2D6 generate analgesic opioid receptor agonist O-desmethyltramadol thus better pain control but leads to respiratory depression and nausea. Precautions are required for renal insufficiency.

Antidepressants and P450

Mirtazapine is an antidepressant drug which has dose-dependent sedative effects in sleep disorders. Due to CYP2D6 ultrarapid metabolizers, there are risks on heart rate and blood pressure [9, 27].

Metoclopramide The role of CYP2D6 has been shown in antiemetic drug metoclopramide administration and metabolism of nonsteroidal anti-inflammatory drugs (NSAIDs) by CYP2C9.

CYP2C9 It hydroxylates drugs like NSAIDs (nonsteroidal anti-inflammatory drugs), sulfonylurea urea, inhibitors of angiotensin-converting enzymes, oral hypoglycemic agents, etc. Rare poor metabolizers have an extreme response to hypoglycemic drug tolbutamides (for type II diabetes). Its roles have been described in warfarin.

Warfarin A widely prescribed anticoagulating agent warfarin shows high interindividual variability of dose; thus, if it is not properly monitored, low dose may result in thrombosis and high dose may lead to excessive bleeding. It is a common vitamin K antagonist oral anticoagulant used in prevention of thrombotic disorders by inhibiting vitamin K-dependent clotting pathway. Extra dose might result in cranial hemorrhage, while less dose may result in thrombotic events. GWAS showed that CYP2C9 and VKORC1 may

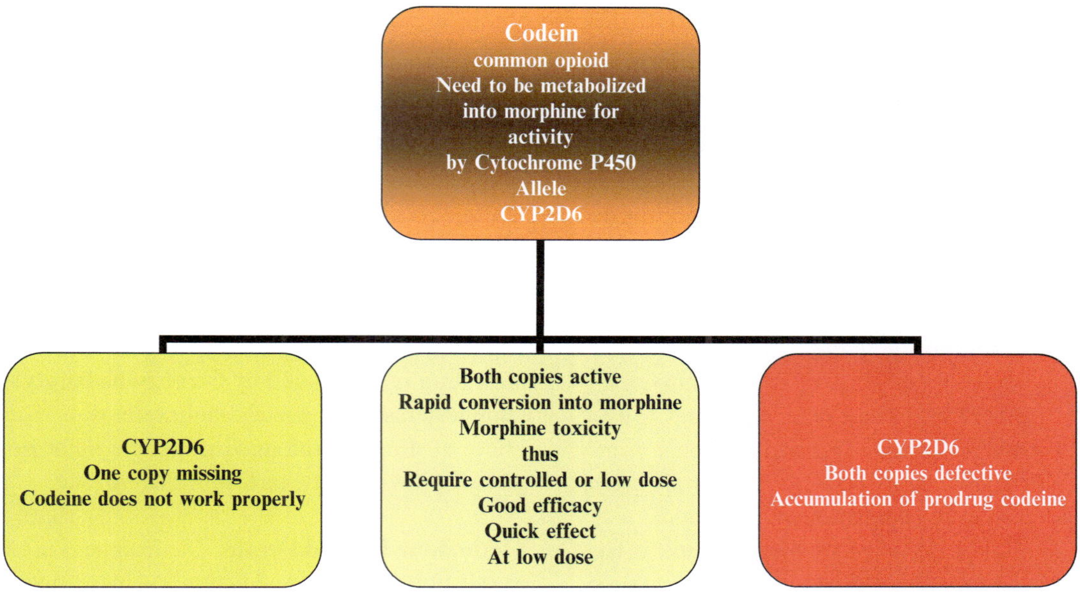

Fig. 7.5 The figure shows that the drug codeine was given which is metabolized to morphine. Due to allelic variation in cytochrome P450 CYP2D6, the response of the individuals differs. The differences are due to genetic components of the individual

be useful in determining optimal dose of warfarin. Genetic-based algorithms efficiently predict dose of warfarin.

CYP2C19 It metabolizes a variety of drugs as mephenytoin, proton pump inhibitors such as omeprazole (treatment of stomach ulcers), proguanil (malarial drug), and certain antidepressants. Poor metabolizers require lower dose, as standard dose of diazepam results in prolonged sedation due to slow demethylation of CYP2C19.

Clopidogrel

Clopidogrel is given to reduce the risk of ischemic stroke and requires transformation into active metabolite by CYP. CYP2C19 which is a reduced function allele faces about one-third reduction of active metabolite of clopidogrel; thus, treated patients are at increased risk for outcome of stroke [17].

Case Study

On May 2006, in Canadian newspaper, a tragic case of morphine poisoning was reported. A 13-day-old baby died, when the mother was prescribed codeine and acetaminophen in her postpartum period. She was actually ultramodifier of CYP2D6 as she carried an extra copy of the gene due to which increased O-demethylation of codeine to morphine occurred. High concentration of morphine was present in the breast milk and in the child's blood [14].

In a similar kind of case, ultramodifier for CYP2D6 faces toxic responses to tramadol treatment. The man had renal insufficiency and developed postoperative respiratory depression with tramadol.

Likewise, in many other studies, it has been found that genes influence either the conversion or the detoxification of drugs in individuals leading to variable response.

(continued)

Thus pharmacogenomics implicated gene screening has been prescribed for few of the drugs before application. In the same line, companies are coming up with solutions like AmpliChip of Roche for diagnostic purpose (Fig. 7.6).

Fig. 7.6 The figure shows the launch of AmpliChip for cytochrome P450 genes, CYP2C19 and CYP2D6

CYP3A4 Approximately 40 % of the drugs are metabolized by P450 CYP3A4 in the liver. However, the activity of this enzyme is variable up to 30-fold due to regulatory effects as the enzyme is inducible.

Suxamethonium It is used in surgery as muscle relaxant, but many patients post usage suffer from prolonged apnea (failure to breathe spontaneously). The normal breathing is resumed after drug is inactivated by the enzyme butyrylcholinesterase (or pseudocholinesterase). The effect of drug is prolonged in people homozygous for low-activity variant of the enzyme.

Methodone Related Deaths Methadone is used as substitute for heroin addiction and in treatment of pain. Fatal poisonings occur at very low concentrations. CYP2B6*6 alleles were associated with slow metabolism.

Human Leukocyte Antigen (HLA)

HLA are group of genes encoding for major histocompatibility complex (MHC) responsible for immune activation. Abacavir is inhibitor of reverse transcriptase and is used in combination with antiretroviral medicines. In 5–8 % individuals receiving its therapy, hypersensitivity reactions develop with fever, rash, fatigue, cough, gastrointestinal symptoms, and shortness of breath (dyspnea) due to a variant of HLA-B, HLA-B*57:01. Routine monitoring of this variant had been recommended by FDA before abacavir treatment.

Statins

The statin drugs, namely, simvastatin, atorvastatin, and rosuvastatin, are main drugs prescribed to lower LDL cholesterol by inhibiting cholesterol synthesis pathway. Statins cause skeletal muscle toxicity, and depending upon severity, it is classified as incipient myopathy, myopathy, or rhabdomyolysis, which is life threatening but rare. In GWAS genetic variants within SLCO1B1 (solute carrier organic anion transporter family member B1) were found to be significantly associated with the risk of developing myopathy. This finding was then replicated in many other studies, the STRENGTH (statin response examined by genetic haplotype markers) showed strong association with simvastatin, modest relationship with atorvastatin, and no significant association with pravastatin.

7.5.1.2 Phase II Reactions

As the reactions of phase II produce excretable water-soluble derivative of the drug involving acetylation, glucuronidation, sulfation, or methylation, the individuals deficient in the enzymes responsible for these reactions are not capable of excreting drugs quickly.

N-Acetylation Humans have two enzymes: NAT1 and NAT2 (2 aryl-N-acetyltransferase, NAT). Both are involved in different spectra of the drug. Frequency variations are present for NAT2. Slow modification leads to prolonged retention of the drug and might lead to bladder cancer, for example, antitubercular drug isoniazid. Isoniazid is present in variable plasma concentration. Slow acetylators are at risk of developing peripheral neuropathy. Several sulfa drugs, antiarrhythmic drug (procainamide), antileprosy drug (dapsone), and antihypertensive drug (hydralazine) show different acetylation rates.

UGT1A1 Glucuronosyl Transferase Some drugs can be excreted after modification to glucuronide conjugates by the action of UDP-glucuronosyltransferases. One of the variant UGT1A1 tackles catabolism of bilirubin (breakdown product of hemes) and anticancer drug irinotecan.

Glutathione S-Transferase (GST) GST is involved in catabolism and detoxification of large groups of xenobiotics and carcinogens. Of the other classes, the gene deletions are common due to unequal crossing over in tandemly repeated gene clusters for two variants GSTM1 and GSTT1. Due to this, their activities can be high, low, or nil. Nil activity is associated with genotoxic effects.

Thiopurine Methyl Transferase (TPMT) It transfers a methyl group from S-adenosyl methionine onto the azathioprine and 6-mercaptopurine (immunosuppressive drug) leading to their inactivation.

7.6 Response of Drug Target

Another aspect of drug metabolism is pharmacodynamics, which is the specific response of a drug target (receptor, enzymes, etc.) to the administered drug.

Beta-Adrenergic Receptors (ADR) ADRB2 gene has two variants P.Arg16Gly and P. Gln27Glu and encodes beta-2 adrenergic receptor. These receptors are targeted for treatment of asthma. Individuals homozygous or heterozygous for Arg16 variants are good responders for albuterol (antiasthmatic drug) than Gly16.

Angiotensin-Converting Enzymes (ACE) ACE (is peptidase) converting angiotensin I into angiotensin II. Angiotensin II is one of the important acute phase proteins which regulates blood pressure. Deletion of Alu sequence in introns shows variable activity when deletion homozygotes have almost double amount of circulating ACE. Thus its inhibitors enalapril and captopril are used for treatment of heart failure.

The other receptors such as serotonin receptor and ryanodine receptor also show variations. Serotonin receptor is an important neurological receptor which serves as target for psychiatric drugs, whereas ryanodine receptor gene encodes a calcium release channel in the muscle sarcoplasmic reticulum responds to halothane or isoflurane (inhalation anesthetics).

7.6.1 Other Applications

Antihypertensive medication: Pharmacogenomic studies are finding applications in usage of antihypertensive medication.

Responsiveness to aspirin was affected due to SNP found in Cox-1 [16].

7.7 Theragnostics and Companion Diagnostics

In the case of breast cancer, trastuzumab (humanized monoclonal antibody) against human epidermal growth factor receptor 2 (HER2) is used [21, 23]. HER2 is expressed on the surface of tumor cells. The FDA approved the simultaneous usage of trastuzumab as diagnostic test (detection of HER2) and approval for its use as therapeutic agent. The simultaneous usage of an agent for therapy and diagnostic test is referred as *theragnostics*.

Companion diagnostics (CDx) first identify specific genetic biomarkers before prescribing specific targeted therapies [1, 24, 25]. Patient profiling is performed which helps to identify the responsive patient population and good efficacy and success of treatment. The first CDx was launched in 1980s and with the success of Herceptin® (trastuzumab) and Gleevec® (imatinib) the field has moved further. Their number is continuously increasing, with 5 diagnostic/drug combinations in 2006 to 63 in 2012. Here some of the FDA-approved companion diagnostics are given, majority of them are for cancer but some are for other disease conditions also [18, 20] (Table 7.1 and text from www.fda.gov).

Because of strong influence of genes on drug susceptibility, the companies are now working on companion diagnostics and therognostics, and many products are also approved (see Table 7.1).

FDA Approval of First Microarray-Based Test: Roche's AmpliChip CYP450 Test

This was the first microarray-based diagnostic test approved in 2005 for detection of genetic variations that can influence drug efficacy and adverse drug reactions. The AmpliChip CYP450 Test has been cleared by the US Food and Drug Administration (FDA) for diagnostic use in the USA. This test, which is powered by Affymetrix microarray technology, analyzes a patient's cytochrome P450 2D6 and 2C19 genotypes from genomic DNA extracted from a blood sample. Test results will allow physicians to consider unique genetic information from patients in selecting medications and doses of medications for a wide variety of common conditions such as cardiac diseases, pain, and cancer. The schematic representation of Roche diagnostics AmpliChip for cytochrome P450 is shown in Fig. 7.6.

Table 7.1 A few companion diagnostic tests approved by FDA. It is an in vitro diagnostic (IVD) device or an imaging tool which provides essential information for the safe and effective use of a corresponding therapeutic product. Its usage with a particular therapeutic product is stipulated in the instructions for use in the labeling of both the diagnostic device and the corresponding therapeutic product, as well as in the labeling of any generic equivalents and biosimilar equivalents of the therapeutic product [2]

Drug trade name (generic name)	Device trade name	Manufacturer	Indications for use (as on FDA website)
Herceptin (trastuzumab); Perjeta (pertuzumab); Kadcyla (ado-trastuzumab emtansine)	HER2 FISH pharmDx Kit	Dako Denmark A/S	Aids assessment of breast and gastric cancer patients before start of Herceptin (trastuzumab) and breast cancer patients for Perjeta and Kadcyla treatment
Erbitux (cetuximab); Vectibix (panitumumab)	The cobas® KRAS mutation test	Roche Molecular Systems, Inc	It is real-time PCR test used for detecting 7 somatic mutations in codons 12 and 13 of the KRAS gene in human colorectal cancer (CRC) tumor tissue. The test is beneficial before treatment with Erbitux® (cetuximab) or with Vectibix® (panitumumab)
Tarceva (erlotinib)	cobas EGFR mutation test	Roche Molecular Systems, Inc	It is a real-time PCR test for the qualitative detection of substitution mutations in exon 19 deletions and exon 21 (L858R) of the epidermal growth factor receptor (EGFR) gene in human non-small cell lung cancer (NSCLC) tumor tissue. The test which is beneficial for treatment with Tarceva® (erlotinib), an EGFR tyrosine kinase inhibitor (TKI), is indicated
Gleevec/Glivec (imatinib mesylate)	Dako c-Kit pharmDx	Dako North America, Inc	It is a qualitative immunohistochemical (IHC) kit system (used on the Dako Autostainer) used for identification of c-kit protein/CD 117 antigen (c-kit protein) expression
Erbitux (cetuximab); Vectibix (panitumumab)	*therascreen* KRAS RGQ PCR Kit P110030	Qiagen Manchester, Ltd	It is a real-time qualitative PCR assay used for the detection of 7 somatic mutations in the human KRAS oncogene, using DNA from colorectal cancer (CRC) tissue before treatment with Erbitux (cetuximab) and Vectibix (panitumumab) based on a KRAS no mutation detected test result
Exjade (deferasirox)	FerriScan K124065	Resonance Health Analysis Services Pty Ltd	The FerriScan R2-MRI system measures liver iron concentration to assist identification and monitoring of non-transfusion-dependent thalassemia patients receiving therapy with deferasirox

Adapted from www.fda.gov medical devices

7.8 Pharmacogenomic Analysis

The analyses of relationship between genes and therapeutic targets are done by two research approaches.

7.8.1 Candidate Gene Approach

This takes into account a limited number of known genes expected as therapeutic target. The polymorphisms are analyzed on candidate genes relevant to the therapeutic phenotype of interest. For example, polymorphism in gene encoding thiopurine S-methyltransferase (TPMT) was identified to decrease TPMT enzyme activity, thus increasing toxicity of 6-mercapto-purine and polymorphism in CYP2C9 and VKORC1 with dose response of warfarin (oral anticoagulant). Warfarin is metabolized via oxidation in the liver by CYP2C9 and mediates its anticoagulating effects by inhibiting VKORC1. However this approach tends to skip unknown genetic variants. The gene expression is also regulated by microRNAs and DNA methylation.

7.8.2 Whole Genome Approach

It is Focused on complete human genome and transcriptome for establishing an association between genes and the drug, for example, NRG3 had SNP which was unknown of having any role in cancer, was found to be associated with platinum agents, and platinum-based therapy was given in cancer patients.

7.9 Pharmacogenomic Developments and Approvals

The aim of personalized medicine is to prescribe the correct dosage and correct drug. However, in the USA, thioridazine (antipsychotic) is marked as contraindicated in patients with low activity of CYP2D6. Antipsychotic aripiprazole and psychostimulant modafinil again indicates CYP2D6 variants. The analysis has very important role in the analysis of responders and nonresponders for initiating medications, thus avoiding adverse reactions with correct drug dose for optimum results [22]. Therefore FDA has approved drug labeling with information on genomic biomarkers and can describe [2]:

- Drug exposure and clinical response variability
- Risk for adverse events
- Genotype-specific dosing
- Mechanisms of drug action
- Polymorphic drug target and disposition genes

Here some of the approved drugs with pharmacogenomic information in their labeling are being listed (from www.fda.gov):

- *CYP2D6*: Amitriptyline (psychiatry); arformoterol (pulmonary); aripiprazole (psychiatry)
- *CYP2C19*: Carisoprodol (rheumatology); citalopram (psychiatry); clopidogrel (cardiology); lansoprazole (gastroenterology)
- *CYP2C9*: Celecoxib (rheumatology); flurbiprofen (rheumatology); CYP3A5-prasugrel (cardiology)
- *UGT1A1*: Arformoterol (pulmonary); indacaterol (pulmonary); irinotecan (oncology)
- *NAT1-2*: Isosorbide and hydralazine (cardiology); rifampin, isoniazid, and pyrazinamide (infectious diseases)
- *CYP2C9; VKORC1; Protein S Deficient*: Warfarin (cardiology or hematology)

7.10 Challenges

- Study design constraint: A comparison of sufferers with and without drug cannot be done as it would be unethical. Thus exposed versus unexposed is very difficult or not possible.
- Pharmacogenomic research is an ongoing research, thus limiting its usefulness.
- Replication of effects with the same drug and resultant phenotype.
- Requirement of same drug as drug–drug interaction is also important; thus, cell lines are in use.

- Genotyping is not possible in seriously ill patients.
- Patients are not too keen on delaying the medication because of genotyping.
- Persons' response to a drug is not affected by single genotype, for example, many P450 enzymes may be involved in metabolism of one drug.

7.11 Chemogenomics

Chemical genomics or chemogenomics is an emerging research field that describes the designing and development of ligands which are target-specific which may be used to study gene and protein functions. This discipline requires the interaction between molecular biologist and synthetic chemist and is referred as chemical biology or chemical genetics [30]. As the knowledge on our genetic makeup is getting advanced, the simultaneous analysis of proteins, genes, and their regulation is getting unraveled, and chemical ligand-based technologies are emerging for understanding of gene and protein function. The landscape of biological research and pharmaceutical industry is changing now with development of target-specific and diversity-based organic synthesis and structure-based chemical ligand. These small chemical compounds can enter the cells and bind to their target proteins creating either loss of function or gain of function phenotype. Nowadays, drug discovery is no more trial-and-error approach but is systematic which is aimed at identifying the genes and proteins as potential therapeutic target. Small, cell-permeable, and target-specific chemical ligands are particularly useful in systematic genomic approaches to study biological questions. On the other hand, genomic sequence information, comparative and structural genomics, when combined with the cutting edge technologies in synthetic chemistry and ligand screening/identification, provide a powerful way to produce target-specific and/or function-specific chemical ligands and drugs.

Chemical genetics may be forward or reverse; in forward genetics, natural product or synthetic compound is screened for its desired effect as inhibition of tumor growth. After its identification, its target is identified, whereas knowledge of the target proteins results in classical reverse genetics. Thus small chemical may be integrated with genomic tools for chemical genomics which when combined with gene expression and protein profiling can produce drug target proteins.

Thus, chemogenomics deals with systematic study of the effects of wide spectrum of small ligands on biological macromolecular targets. As huge data is there, of compounds, targets, and assays and of gene/protein expression levels and binding constants, therefore, information technologies are playing a very important role in planning, analyzing, and predicting chemogenomic data.

7.12 Chemical Kinomics

This is an emerging field which deals with phosphorylation-based complicated cellular signaling networks with the aid of small molecules that can modulate kinase functions. Chemical kinomics is a discipline of chemical genomics that is also referred to as "chemogenomics." As the kinome or kinase family (human genome has 518 (1.7 % of human genes) protein kinase genes) is critically involved in diverse regulation of cellular signal transductions, the improper activation or over- or underexpression of certain protein kinases has been implicated in broad spectrum of human diseases (cancer, inflammatory disease, and autoimmune disorders) [3]. If the kinase is upstream, then this might lead to complete loss of the associated biological function. Approximately 30 kinases have been targeted for drug discovery and development, and over 80 protein kinase inhibitors are known to be undergoing human clinical evaluations [5, 6, 15].

In cancer: small molecule inhibitors against oncogenic protein kinases; chronic myelogenous leukemia (CML) identification and validation of potential target for dasatinib (multi-targeted kinase inhibitor) [7, 13, 20].

7.13 Future Prospects

Now several pharmaceutical companies and genomic institutes are providing FDA the reports of genomic studies with toxicology data. The field itself offers the possibility of performing genetic tests on blood, saliva, or other cells as a source of DNA to minimize serious adverse effects of the drug and improve efficacy. Pharmacogenomics would be very helpful to solve the issues related to drug development, drug selection (by genotyping), and proper dosage determination. Some countries are planning to introduce pharmacogenomics in clinical trial [10]. Overall the requirements in the present scenario are:

- Usage of computational toxicology to predict safety would reduce the number and need for animal testing and human clinical trials.
- Need to take care of genotoxicity and carcinogenicity.
- Requirement of novel technologies so that effects of drug-induced genetic alterations can be monitored.
- Prediction of toxicity by in silico modeling based upon chemical structure, biological attributes, and/or physiochemical properties.

7.14 Chapter End Summary

- We are advancing our knowledge on our genetic makeup. To maximize this, there are lots of techniques which are available for the simultaneous analysis of proteins and genes. Chemical ligand-based technologies are emerging for understanding of gene and functions of the protein.
- The pharmacogenomic studies can help in predicting the response of an individual to a particular drug. Thus the studies would be very helpful in choosing the right patients for treatment with correct drug and optimized dose.
- Cytochrome P450 family of genes largely affect the metabolism and thus response of many of the drugs. Phase I and II reactions occur according to the persons' genetic makeup and affect the individual response to drugs. Companion diagnostics and therognostics are launched for identifying the potential responders before starting the therapy.
- The methods utilized for these studies are simple and waiting to reach clinics to predict the drug dosage and response in clinical practice. Roche has launched the AmpliChip for detection of genes for cytochrome P450 family. Thus in time to come, pharmacogenomics would be a very promising tool for a safe and good future of the mankind. Probably after some time, these technologies would impact prescription of medicines both by clinicians and pharma sector.

Multiple Choice Questions

1. Pharmacogenomics is:
   (a) Study of response of 100 individuals in response to drug
   (b) Study of response of a particular race in response to drug
   (c) Study of response of an individual in response to drug
   (d) All of these
2. Pharmacogenomics is also known as:
   (a) Pharma-related genomics
   (b) Pharmacokinetics
   (c) Pharmacogenetics
   (d) Personalized medicine
3. Pharmacogenetics deals with:
   (a) Study of genome in response to drug
   (b) Study of a gene in response to drug
   (c) Study of changes in genes in response to drug
   (d) Study of mutations in genes in response to drug

4. The field which deals with adverse side effects of the drug is:
   (a) Clinomics
   (b) Toxicogenomics
   (c) Kinomics
   (d) Chemogenomics
5. Genome-wide association studies (GWAS) are beneficial as they:
   (a) Survey the entire genome for association with drug response
   (b) Survey the population for association with drug response
   (c) Survey the responders and nonresponders of a drug by analysis of the genome
   (d) All of these
6. The toxicity to codeine occurred in the individuals because they were:
   (a) Poor metabolizers
   (b) Intermediate metabolizers
   (c) Extensive metabolizers
   (d) Ultrarapid metabolizers
7. Chemical kinomics:
   (a) Deals with complete array of protein kinases
   (b) Deals with molecules affecting kinase functions
   (c) Deals with kinase used in drug discovery
   (d) None of these
8. CYP is highly implicated in pharmacogenomics because:
   (a) Its variation affects genes
   (b) It is involved in drug metabolism
   (c) It is required for activation of the drug
   (d) It is required for excretion of the drug

Answers

1. (c); 2. (d); 3. (b); 4. (b); 5. (d); 6. (d); 7. (a); 8. (b)

Questions

Q1. What is pharmacogenomics?

Q2. What are the clinical applications of pharmacogenomics?

Q3. What is candidate gene approach in pharmacogenomics?

Q4. Define pharmacogenetics, toxicogenomics, and kinomics.

References

1. Agarwal A et al (2015) The current and future state of companion diagnostics. Pharmgenomics Pers Med 8:99–110
2. Analysis by authors utilizing FDA labels at Daily Med. http://dailymed.nlm.nih.gov
3. Bossemeyer D (1995) Protein kinases- structure and function. FEBS Lett 369:57–61
4. Carlson CS, Newman TL, Nickerson DA (2001) SNPing in the human genome. Curr Opin Chem Biol 5:78–85
5. Cohen P (2002) Protein kinases- the major drug targets of the twenty-first century? Nat Rev Drug Discov 1:309–315
6. Fabbro D, Garcia-Echeverria C (2002) Targeting protein kinases in cancer therapy. Curr Opin Drug Discov Dev 5:701–712
7. Fabbro D et al (2002) Protein kinases as targets for anticancer agents: from inhibitors to useful drugs. Pharmacol Ther 93:79–98
8. Ginsburg GS, McCarthy JJ (2001) Personalized medicine: revolutionizing drug discovery and patient care. Trends Biotechnol 19:491–496
9. Strachan T, Read A (2011) Human molecular genetics, 4th edn. Garland Science, Taylor and Francis Group, London
10. Hess P, Cooper D (1999) Impact of pharmacogenomics on the clinical laboratory. Mol Diagn 4:289–298
11. Johansson I, Sundberg MI (2011) Genetic polymorphism and toxicology-with emphasis on cytochrome P450. Toxicol Sci 120:1–13
12. Khoury MJ et al (2012) From precision medicine to precision public health. https://med.stanford.edu
13. Kim JA (2003) Targeted therapies for the treatment of cancer. Am J Surg 186:264–268
14. Koren G et al (2006) Pharmacogenetics of morphine poisoning in a breastfed neonate of a codeine prescribed mother. Lancet 368:704
15. Manning G, Whyte DB, Martinez R, Hunter T, Sudarsanam S (2002) The protein kinase complement of the human genome. Science 298:1912–1934
16. Maree AO et al (2005) Cyclooxygenase-1 haplotype modulates platelet response to aspirin. J Thromb Haemost 3:2340–2345
17. Mega JL et al (2009) Cytochrome p-450 polymorphisms and response to clopidogrel. N Engl J Med 360:354–362

18. Morgan GJ, Pratt G (1998) Modern molecular diagnostics and the management of haematological malignancies. Clin Lab Haematol 20:135–141
19. Ni X, Zhang W, Huang S (2013) Pharmacogenomics discovery and implementation in genome-wide association era. WIREs Syst Biol Med 5:1–9
20. Ozdemir V, Williams-Jones B, Glatt SJ, Tsuang MT, Lohr JB, Reist C (2006) Shifting emphasis from pharmacogenomics to theragnostics. Nat Biotechnol 24:942–946
21. Personalized Medicine Coalition (2011) The case for personalized medicine. Personalized Medicine Coalition, Washington, DC
22. Ritchie MD (2012) The success of pharmacogenomics in moving genetic association studies from bench to bedside: study design and implementation of precision medicine in the post-GWAS era. Hum Genet 131:1615–1626
23. Ross JS, Fletcher JA (1998) The HER-2/neu oncogene in breast cancer: prognostic factor, predictive factor, and target for therapy. Oncologist 3:237–252
24. Ross JS, Ginsburg GS (2003) The integration of molecular diagnostics with therapeutics. Am J Clin Pathol 119:26–36
25. Singh V (2012) Companion diagnostics poised for a breakout. Burrill Rep 2:9, http://www.burrillreport.com/content/BR_0912%20final-v3.pdf
26. Taylor JG, Choi EH, Foster CB et al (2001) Using genetic variation to study human disease. Trends Mol Med 7:507–512
27. Weide JV, Hinrichs JWJ (2006) The influence of cytochrome P450 pharmacogenetics on disposition of common antidepressant and antipsychotic medications. Clin Biochem Rev 27:17–25
28. Xie, Frueh (2005) Pharmacogenomics steps toward personalized medicine. Personal Med 2:325–337
29. Zhang J, Yang PL, Gray NS (2009) Targeting cancer with small molecule kinase inhibitors. Nat Rev Cancer 9:28–39
30. Zheng XFS, Chan TF (2002) Chemical genomics: a systematic approach in biological research and drug discovery. Curr Issue Mol Biol 4:33–43

Some Related Resources

http://www.snp.cshl.org
Pharmacogenetics research network external link NIH
Pharmacogenomic resources
Pharmacogenomics Knowledge BaseExternal Link
Table of pharmacogenomic biomarkers in drug labels external link FDA
www.cypalleles.ki.se
www.genetics.edu.au
www.ncbi.nlm.nih.gov/pubmed

Immunology and Medical Microbiology

8

Abstract

We are continuously exposed to many pathogens through inhalation, ingestion, and touch. The immune system protects us from the majority of these pathogens as flatworms, bacteria, fungi, and viruses. We have also witnessed tremendous progress in the prevention and treatment of infectious diseases; still, they remain a major challenge and are responsible for major cause of death and disability worldwide. The immune system's memory response and vaccination have resulted in complete eradication of many diseases. Our immune system is very adaptive and consists of a variety of cells and molecules, which play an active role in protecting us. It not only protects us from the outside pathogenic agents but also is also capable of recognizing the body's own components. It recognizes them as self and does not induce response against them. It is known as self-/non-self-discrimination. Sometimes due to certain defects or other reasons when the immune system is not able to differentiate self, then it mounts an attack on self-components leading to autoimmunity. The importance of the immune system was recognized by early work of Dr. Edward Jenner and Louis Pasteur; they recognized the abilities of the immune system, and since then the system was gradually being explored and it laid the foundation of immunology.

However, day-by-day microbes are also posing health risks as new strains are continuously being evolved. Many chemotherapeutic agents have been developed to control the spread and infections. However, microbes are also continuously developing the ability of their survival with emergence of new strains and properties. Antibiotic resistance is occurring with all classes of microbes posing a serious clinical problem in managing infections. The diseases like tuberculosis, cholera, and rheumatic fever, which were believed to be eradicated, have ferociously reemerged. The reemergence and new pathogenic agents might be the result of mutations in their genome and changes occurring in the environment. In this chapter, basic concept of the immune system and some of the diseases of the skin, gastrointestinal tract, nervous system, and respiratory system caused by microorganisms are discussed along with sexually transmitted diseases and characterization of pathogens.

V. Gupta et al., *Basic and Applied Aspects of Biotechnology*, DOI 10.1007/978-981-10-0875-7_8

8.1 Introduction

We are continuously exposed to many pathogens through inhalation, ingestion, and touch. The immune system protects us from the majority of these pathogens as flatworms, bacteria, fungi, and viruses. We have also witnessed tremendous progress in the prevention and treatment of infectious diseases; still, they remain a major challenge and are responsible for major cause of death and disability worldwide. The immune system's memory response and vaccination have resulted in complete eradication of many diseases. Our immune system is very adaptive and consists of a variety of cells and molecules, which play an active role in protecting us. It not only protects us from the outside pathogenic agents but also is also capable of recognizing the body's own components. It recognizes them as self and does not induce response against them. It is known as self-/non-self-discrimination. Sometimes due to certain defects or other reasons when the immune system is not able to differentiate self, then it mounts an attack on self-components leading to autoimmunity.

The importance of the immune system was recognized by early work of Dr. Edward Jenner and Louis Pasteur; they recognized the abilities of the immune system, and since then the system was gradually being explored and it laid the foundation of immunology.

However, day-by-day microbes are also posing health risks as new strains are continuously being evolved. Many chemotherapeutic agents have been developed, but on the other hand, microbes are continuously developing the ability of their survival with emergence of new strains and properties. Antibiotic resistance is occurring with all classes of microbes posing a serious clinical problem in managing infections. The diseases like tuberculosis, cholera, and rheumatic fever, which were believed to be eradicated, have ferociously reemerged. The reemergence and new pathogenic agents might be the result of mutations in their genome and changes occurring in the environment. In this chapter, basic concept of the immune system and some of the diseases caused by microorganisms and characterization of pathogens are discussed.

8.2 Introduction to the Immune System

Immune system first recognizes the pathogen and then gives a response against that pathogen. Thus, upon recognition of the pathogenic agent, it triggers the effector response which helps in the elimination of the pathogen.

In our body there are distinct effector responses for pathogens; thus, there are two kinds of immune responses: the innate immune response and adaptive or acquired immune response (Fig. 8.1). Though self-/non-self-recognition is the hallmark of the two responses, the adaptive immune system is much more diverse, is specific, and has memory in contrast to innate immune responses. The innate immune system recognizes and responds nonspecifically against pathogen, while adaptive or acquired immune system mediates specific response and remembers the pathogen after interacting with it [6].

8.3 Immunology and Medical Microbiology

The study of immunology led to the growth of medical microbiology, which deals with identification, and mechanism of action of infectious agents. The disease-causing organisms are called pathogens and their mode of attacking the host and its effects on host is pathogenesis. The major human pathogens are viruses, bacteria, fungi, and parasites which are causative agents of many diseases. The immune system responds differently for these pathogenic agents. Unlike bacteria, fungi, and parasites which are capable of independent growth, the viruses require the host cell for the multiplication.

Whenever there is a state of misfunctioning of any immune components because of genetic defects or because of acquired disease as acquired immunodeficiency syndrome (AIDS) caused by human immunodeficiency virus (HIV-1 and HIV-2), the immune system is not able to respond naturally to infectious agents. These conditions are termed as immunodeficiency where the pathogens that are prevalent in the external

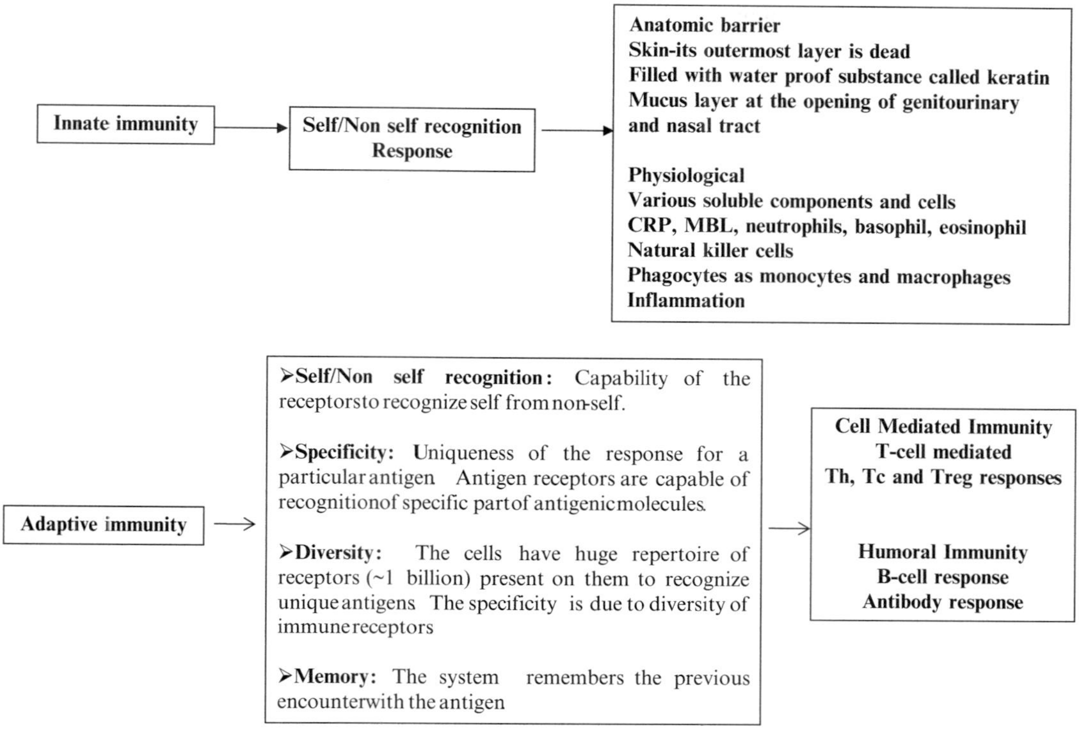

Fig. 8.1 The figure shows the outline of innate and adaptive immune responses

environment but do not cause any harm in normal individual result in severe diseases in these patients, for example, *Candida albicans*. These are known as opportunistic pathogens. Therefore getting infections from these opportunistic agents show the compromised immune system.

Thus, the immune system deals with the majority of pathogenic agents and is well equipped with many cells and molecules to combat them.

8.4 Innate and Adaptive Immune Responses

Innate immunity is the first line of defense system which gives nonspecific response. The immunity is natural (by birth) and quick when the body encounters the pathogen. On the other hand, adaptive immunity is an acquired immunity and takes sometime to mediate the response; the response is highly specific and with memory. When adaptive immune components react for the first time with pathogen, the response is primary; when same pathogen is encountered again by the body, the response is via memory cells with specificity, in less time and heightened as compared to primary response.

Due to memory response of adaptive immune system, the body is given exposure with either weakened or killed or subunit of pathogen. This exposure leads to active clearance of the pathogen by adaptive immunity and induces the formation of memory cells, which prevent subsequent infection with the same pathogen. This is the basis of the field of vaccinology (see Chap. 14); it has led to vigorous decrease in the incidences of the childhood diseases like measles, mumps, and polio and complete eradication of small pox [6].

8.4.1 Innate Immunity

Innate immunity is the first line of defense system of our body which does not give a very specific response. The immunity is natural that is preexisting, is quick, and occurs when the body encounters the pathogens. The innate immune components comprise cells and soluble molecules. The system has physical, chemical, and cellular barriers. The physical barriers include the skin and mucous membrane. The mucous lining includes the mucosal epithelia lining respiratory and genitourinary and gastrointestinal tracts; the mucous which is secreted contains substances like defensins that are capable of killing pathogens or inhibiting their growth.

The skin's outermost layer has dead cells filled with waterproof substance called keratin which does not let viruses to infect and multiply on our body. The skin also has psoriasin, which is capable of lysing *E. coli*. The places from where these pathogens can invade our body have the mucous membrane and have a number of nonspecific mechanisms. Saliva is present in our mouth, tears in eyes, secretion of mucus by epithelial cells and movement of cilia on some of the cells at the lower respiratory tract, and acid and digestive enzymes in the gastric tract. The presence of many nonpathogenic microorganisms as commensals is protective, as they compete with pathogenic agents.

The chemical barriers are various cytokines and soluble proteins. The innate immune components are capable of self-/non-self-discrimination and are able to recognize some specific, conserved, and unique pattern associated with a particular class of microorganisms [6].

These patterns are not present in our body but are typically associated with pathogens and are called pathogen-associated molecular patterns (PAMPs). These PAMPs may be a combination of sugars, certain proteins, lipids, or nucleic acids as peptidoglycan in the bacterial cell wall, flagella of bacteria, lipopolysaccharide (LPS) on gram-negative bacteria, teichoic acids on gram-positive bacteria, chitin, glucans, and zymosan on fungi. These are recognized by some receptors present on our body called as pattern recognition receptors (PRRs) (Fig. 8.2).

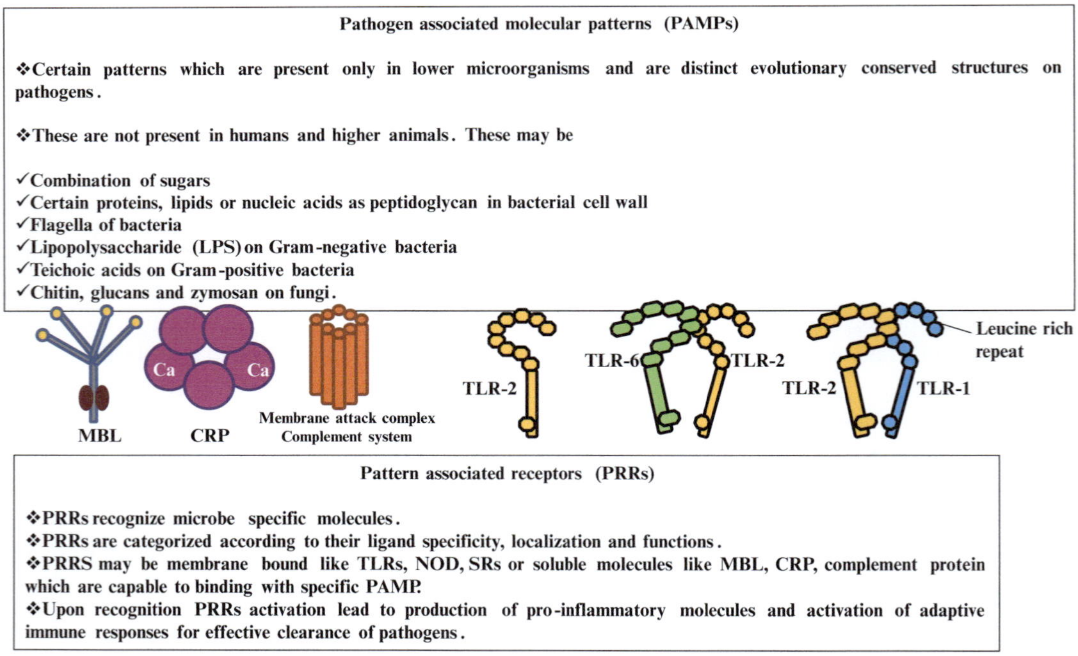

Fig. 8.2 The figure shows various pathogen-associated molecular patterns, which are present exclusively on pathogens. PRRs are present in our body which are capable of recognizing and binding PAMPs. PRRs may be soluble as mannose-binding lectin (MBL), C-reactive protein (CRP), and complement components or may be membrane bound like toll-like receptors (TLRs), NOD, or SRs. Humans have ten kinds of TLRs which are capable of interacting with different PAMPs. TLRs may also form unique binding by interacting with each other as TLR2 and TLR6 and TLR1 and TLR2. TLR have leucine-rich repeats on their extracellular domain

Soluble and membrane bound PRRs

❖**CRP:** Calcium binding, pentraxin family protein, binds pneumococcal protein
❖**Lysozyme:** Cleaves peptidoglycan
❖**MBL:** Binds mannose present on pathogen surface and fix complement cascade.
❖**Complement proteins:** Can directly bind with pathogen associated patterns and fix the complement cascade.
❖**INF:** Antiviral protein secreted by virus infected cell.
❖**Collectins:** Calcium dependent lectins capable of binding oligosaccharide or lipids of microorganisms

✓**Nucleotide binding oligomerization domain (NOD):** NOD1 and NOD2 are present NOD1 binds tripeptide product of peptidoglycan and NOD2 binds muramyldipeptide

✓**Scavenger receptors (SRs):** Present on macrophages and dendritic cells and bind gram negative and positive bacteria and assist their internalization

✓**Toll like receptor (TLR):** Transmembrane receptors rich in leucine rich repeats

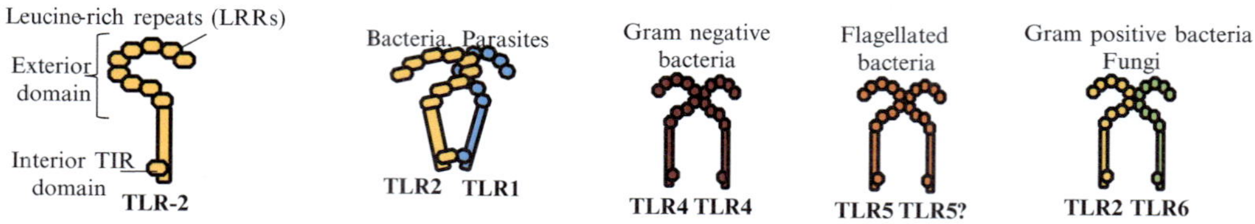

Fig. 8.3 The figure shows the soluble receptor and membrane receptors of innate immunity. *CRP* is C-reactive protein, *MBL* is mannose-binding lectin, *INF* is interferon. NOD, SR, and TLRs belong to the family of membrane receptors. TLR may exist singly or may form homodimer (TLR4) and heterodimer (TLR2 and TLR1)

PRRs may be soluble or membrane bound. Cell surface PRRs are members of the toll-like receptor (TLR) family. Humans have at least ten TLRs which play important roles in innate immune responses (Fig. 8.3).

Toll-Like Receptors

The protein toll was discovered in mutant flies (unable to establish dorsal–ventral axis). Toll is a transmembrane signal receptor protein whose related molecules function in innate immune response and are called as toll-like receptors (TLRs). Toll-like receptors are transmembrane innate immune receptors which are present on cells. These receptors are responsible for binding nonspecifically to many pathogenic patterns. These can also interact among them to form new binding sites for other PAMPs. Major membrane-bound PRR as TLRs have leucine-rich repeat (LRR) motif and IL-1 receptor homology

(continued)

domain. TLRs are divided according to their PAMPs. TLR1, TLR2, TLR4, and TLR6 recognize lipids; TLR3, TLR7, TLR8, and TLR9 recognize nucleic acids.

TLRs recognize PAMPs either directly or via an intermediate PAMP-binding molecule.

TLR1: Is responsible for recognizing triacyl lipopeptide of *Mycobacteria*

TLR2: Recognize peptidoglycans of gram-positive bacteria, GPI-linked proteins of trypanosomes, lipopeptide of *Mycobacteria*, and zymosan of fungi

TLR3: Recognize double-stranded RNA of viruses

TLR4: Recognize lipopolysaccharide of gram-negative bacteria and F-protein of respiratory syncytial virus (RSV)

TLR5: Bind to flagellin of bacteria

TLR6: Bind to diacyl lipopeptide of *Mycobacteria* and zymosan of yeast and fungi

(continued)

TLR7 and TLR8: Bind with single-stranded RNA of viruses

TLR9: Binds with CpG unmethylated dinucleotides of bacterial DNA and herpesvirus

TLR10, TLR11:

Several TLRs also function as dimers of which TLR4 can form homodimers and others can form heterodimers. The dimerization of TLRs affects their specificities also. TLR1/TLR2, TLR3, and TLR9 can bind triacylated lipopeptides, double-stranded RNA (dsRNA), and CpG DNA.

Lectins: Lectins facilitate cell–cell contact. They can bind carbohydrate units. Lectins present on one cell can interact with different carbohydrates which are displayed on the surface of another cell. Their interaction is specific but weak thus many interactions occur. Lectins are divided on the basis of their amino acid sequence and properties. C-type lectins are calcium requiring and bind carbohydrates. For example, selectins are present in three forms: L-form on lymph node vessels, E-form on the endothelium, and P-form on activated platelets.

The variety of soluble factors are shown in Fig. 8.3:

- Lysozyme: It is a hydrolytic enzyme capable of cleaving bacterial cell wall peptidoglycan, present in tears and mucous secretions.
- Interferons: Are group of antiviral proteins.
- Complement: A group of around 20 serum proteins synthesized by the liver which remain in inactive state. They are converted into active form and ultimately form membrane attack complex (ability to damage the membrane of pathogenic organisms) and in the process release proteins which act as anaphylatoxins (enhance inflammation). Complement system can be activated by itself or by innate immune components (MBL) or by adaptive immunity (antibody)

The white blood cells or leukocytes participate in innate and adaptive immunity. Lymphocytes (T lymphocytes and B lymphocytes) are involved in adaptive immune responses, and other leukocytes as neutrophils, eosinophils, basophils, monocytes, dendritic cells, and natural killer cells are important cells of innate immunity (Fig. 8.4). They are again divided into granulocytes and agranulocytes [6]. The cells containing granules have a variety of digestive enzymes in

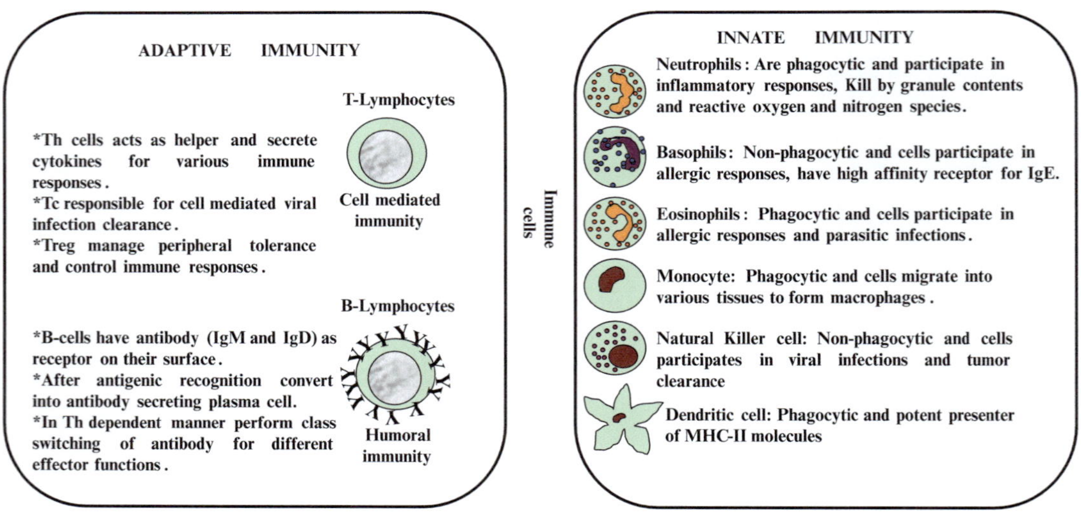

Fig. 8.4 The figure shows various cells involved in innate immunity and adaptive immunity along with their important properties and functions

them for pathogen clearance, while agranulocytes have other modes of action.

Neutrophils: Are called polymorphonuclear cells, take both the acidic and basic stains, are first cells to respond for infections, and are predominant cells involved in inflammation.

Eosinophils: Take up acidic dye eosin. They are cells which respond against parasitic infections. They are phagocytic and also involved in allergic reactions.

Basophils: They take up basic dye methylene blue and have high-affinity receptors of IgE antibody. When they migrate to tissue, they are known as mast cells. They release variety of allergic mediators and are responsible for allergic reactions.

Monocytes and Macrophages: Monocytes are blood phagocytic cells. They have many enzymes which breakdown the pathogenic cells. When monocytes move into tissue, they are known as macrophages. The various macrophages are microglial cells (brain), mesangial cells (kidney), Kupffer cells (liver), osteoclast cells (bone), alveolar macrophages (lungs), and histiocytes (connective tissue).

Natural killer cells: They have the natural capability to kill virus-infected cells. They are mainly responsible for removal of virus-infected and tumor cells.

Some Terms

Antigen: The molecular entities capable of generating immune responses or foreign substances like toxins, pathogens which can bind to the antibodies, or other receptors.

Hapten: Some compounds are antigenic (can bind with antibodies) but are not capable by themselves of inducing a specific immune response, thus are non-immunogenic. Thus, hapten is coupled to an immunogenic carrier, and hapten–carrier conjugate triggers immune responses for hapten antigens, carrier antigens, and new epitopes formed due to hapten–carrier interactions. Dinitrophenol (DNP) is a hapten which cannot elicit antibody response.

(continued)

Immunogen: Antigen which is capable of mounting adaptive immune responses. An antigen may or may not be an immunogen, but an immunogen is always an antigen. Immunogenicity is determined by four properties like foreignness, molecular size, chemical composition, and complexity.

Epitope: Antigenic determinant or epitope is that specific part of antigen which interacts with complementarity-determining region of an antibody. B cell binds surface regions of antigens as they bind directly; however, T cells recognize processed antigen in association with MHC; thus, TCR recognizes hidden or intracellular moieties.

Antibody: Antibodies are effectors of humoral immune responses. It is a Y-shaped glycoprotein present on B cells as part of its receptor and is also secreted to bind to antigenic determinants. It neutralizes the antigen by activating various immune components. There are five classes of immunoglobulins based upon their effector functions: IgM, IgD (both present on B cell as B-cell receptor), IgG, IgE, and IgA.

Toll-like receptors: Toll-like receptors (TLRs) are a class of proteins that play a key role in the innate immune system. They are transmembrane proteins with extracellular domain of leucine-rich repeats (LRRs). Ten human TLRs have been characterized which play an important role in binding to different PAMPs.

Opsonization: When any antigenic molecule is surrounded and bound by antibody from all sides, the binding of antibody triggers phagocytic cells to readily phagocytose the antigen; the coated particles or antibodies are known as opsonins and the process is called opsonization (Fig. 8.5a).

Antibody-dependent cell-mediated cytotoxicity (ADCC): Natural killer (NK) cells express a membrane receptor (CD16)

(continued)

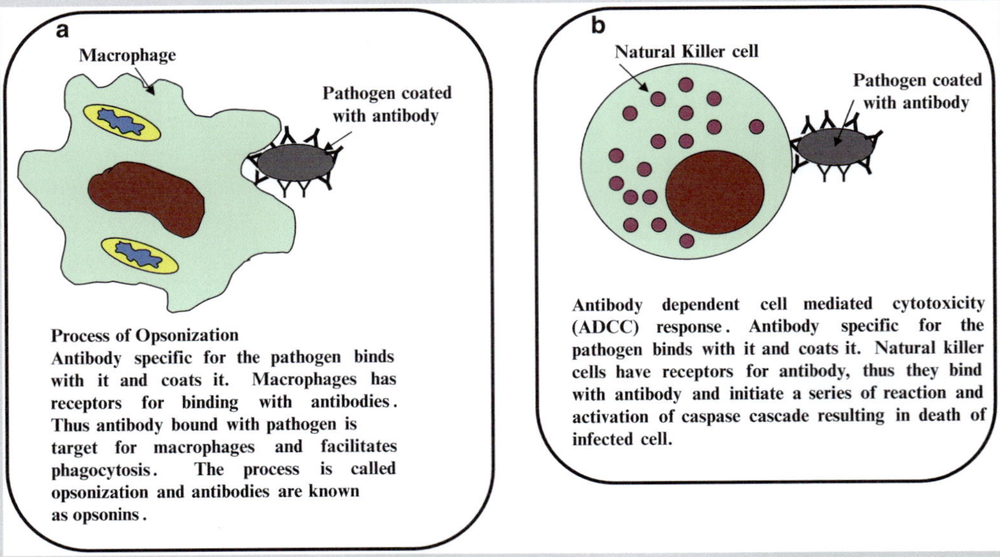

Fig. 8.5 The figure shows two important effector functions of antibody. (**a**) It shows the process of opsonization. When antibody coats a particular pathogen, the process of phagocytosis is facilitated as macrophages have receptor for antibody. It is important for pathogens which resist phagocytosis. The antibody in this is referred as opsonin. (**b**) It shows ADCC response of antibody. Natural killer cells have receptor for constant region of antibody; thus, they can be recruited for elimination of pathogen. The antibody-bound pathogen is rapidly induced for apoptotic cell death like the response mediated by CTL

for a specific portion of the antibody molecule, allowing them to attach to the antibody bound with target. Attachment of NK initiates the cascade of events resulting in the destruction of target cell (Fig. 8.5b). Its granules release perforin and granzyme. Pores formed by perforin assist the entry of granzyme which triggers fragmentation of DNA including viral DNA and cell death.

8.4.2 Adaptive Immune Responses

The adaptive immune response occurs when recognized pathogen challenge occurs. The response can be elicited by many different unique receptors that have variability in antigen-binding regions and conservation in the remaining regions. Thus, the reactions are highly diverse, are highly specific, and are associated with memory (Fig. 8.4). It occurs after 5–7 days of initial antigen exposure. Thus unlike innate immunity which mounts quick response, it takes time. Adaptive immunity is triggered by innate immunity; thus, response is highly interactive with many co-involvements and cooperation to eliminate pathogen effectively. Thus, adaptive immune response can do self-/non-self-discrimination and is highly specific against a variety of pathogens with tremendous diverse memory [6].

Any substance capable of eliciting an adaptive immune response is referred to as an antigen (antibody generator). Adaptive responses can be either cell-mediated or humoral immune responses. Both are mediated by white blood cells called lymphocytes. There are two classes of lymphocytes: the T lymphocytes responsible for cell-mediated immunity and B lymphocytes responsible for humoral immunity whose ultimate effectors are antibodies or immunoglobulins.

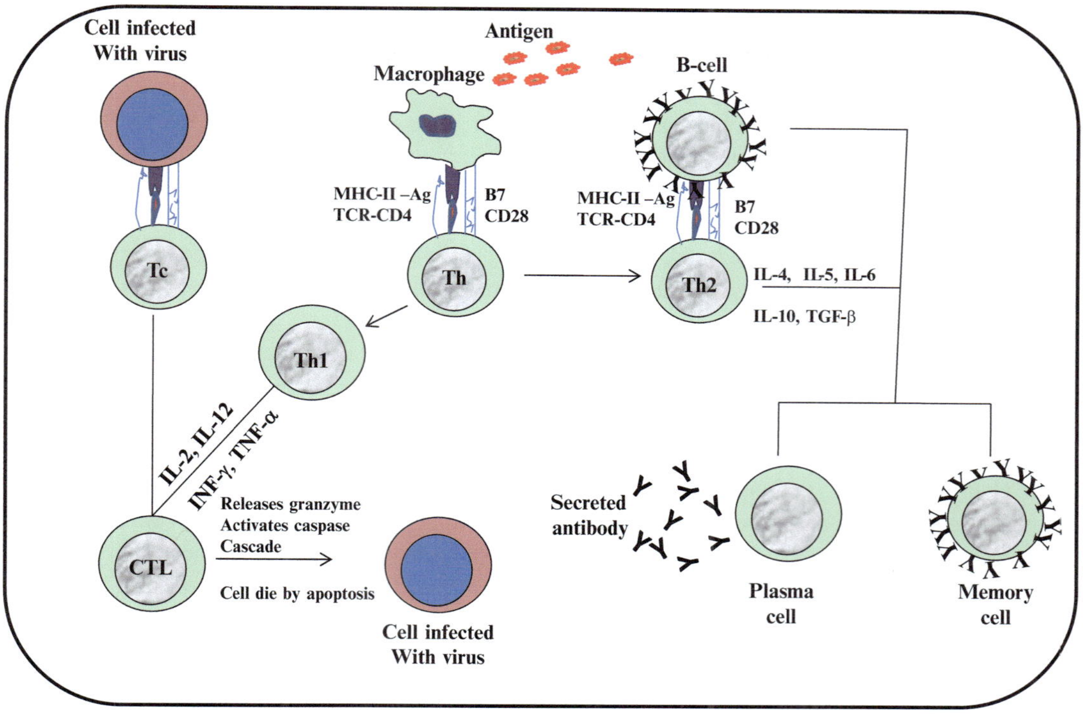

Fig. 8.6 The figure shows that after antigen is encountered by macrophages and B cells, they present it on MHC-II molecule. Antigen-MHC-II is recognized by T-cell receptor (TCR) of Th cells. Th cells produce Th2 population which is helper cell for B-cell response. In the presence of cytokines derived from Th2 and physical contact with T cell, B cell converts into antibody-secreting plasma cell and memory cell and produces all classes of antibody in response to the antigen. Virally infected cell presents endogenous antigen on MHC-I molecule which is recognized by TCR of T-cytotoxic cell. Th cells in the form of Th1 secrete cytokines for CTL. In the presence of Th1-derived cytokines, Tc converts into activated cytotoxic T lymphocytes (CTLs) which activates procaspase cascade and thus kills the virus-infected cell

The humoral branch is mediated by antibodies secreted by activated B cells (called as plasma cells) after they bind any antigenic molecule. The antibodies can be secreted, and binding of antibodies on the viruses and bacteria can either make them very prone for phagocytosis (opsonization), or they can be destroyed by complement-mediated lysis or through antibody-dependent cell-mediated cytotoxicity (ADCC). The B cells after recognition of antigen are activated, and in the presence of cytokines derived from Th cell (Th2 which secretes IL-4, IL-5, IL-6, IL-10, and IL-13) and contact with Th cell, they form antibody-secreting plasma cell, start class switching of antibody, and form memory cells [6].

Cell-mediated response occurs when antigen is presented by cells of the body to T lymphocytes. T cells may be $CD4^+$ T-helper (Th) cells, $CD8^+$ T-cytotoxic (Tc) cells, and CD4 and $CD25^+$ T-regulatory (Treg) cells. Treg plays an important role in the regulation of immune responses. Th and T_C can recognize antigen only when it is presented on another group of polymorphic proteins known as major histocompatibility complex (MHC). MHC-I is responsible for presentation of virus-specific antigens to Tc cells; thus, all the nucleated cells can activate Tc cells which are capable of killing virus-infected cell. MHC-II is specifically presented on antigen-presenting cells (APCs) which activate Th cells which recruit a number of other immune components to clear these pathogens (Fig. 8.6).

Upon recognition of antigen associated with MHC-I, the Tc is activated and forms CTL which releases its granule content rich in perforin and granzyme. This induces the target cell to undergo apoptotic cell death.

8.5 Medical Microbiology

The research is increasingly showing the role of pathogen in several diseases which in earlier days were believed to be noninfectious disease. The usage of nucleic acid amplification technique is able to show the involvement of pathogenic agent in several nervous system disorders which otherwise were considered nonpathogenic (Table 8.1).

The ongoing research might possibly indicate the role of infectious agent in the etiology of rheumatoid arthritis, sarcoidosis or inflammatory bowel disease, and atherosclerosis.

There is emergence of new diseases resulting from changes in the pathogenic agent like changes in coronavirus and onset of severe acute respiratory syndrome (SARS) in 2003 and H5N1 avian influenza virus from poultry farms to humans. These pose a threat as well as challenge to humankind. The understanding of emergence of the disease requires host–pathogen interaction, which is influenced by geography, environment, age, and nutritional status [5].

There is essential requirement of record of the history, behavioral parameters, traveling information, etc. for the effective and early management of the disease. For example, a patient has encountered a pathogenic strain from a geographical area where antibiotic- or drug-resistant strains are predominant. In case a patient has acquired the disease from that region, then the normal medicines would unknowingly delay the intervention in the life cycle of the pathogen resulting in serious condition or death. However, history would provide information about the prevalence of pathogenic strains from that niche for appropriate treatment.

The clinical laboratory or hospital laboratory which undertakes testing of microorganisms should have safety features so that health-care workers remain safe from these infectious agents:

- Microorganisms are studied in clinical hospital laboratories, reference laboratories, and research facilities.
- There should be appropriate posting of signs regarding safety. These are very important to ensure safety.
- Handling of sample with highly contagious microorganism would require high security, appropriate masks, and working in biosafety hoods to avoid splashing or inhaling.
- The laboratory should have well-defined compartments, equipped for each kind of application along with efficient energy-based refrigeration unit for sample and reagent storage. This is important to get accurate results.
- In a hospital microbiology laboratory, the blood cultures are incubated and monitored electronically for bacterial and fungus growth. High-tech instruments are used to aid in the identification of microorganisms. Teamwork is essential in all types of laboratories [5].

The invasion of pathogen can be described in three steps:

1. Colonization: Pathogen presence in or on the host
2. Infection: Attachment and penetration of pathogen inside the host; bypass defense responses of the host immune system
3. Disease: Is caused by action of endotoxin/exotoxin/metabolites secreted by the pathogen in the host

The onset of disease often results in clinical signs and symptoms.

Adhesives of Pathogens The bacterial pathogens often use pili, flagella, or lipopolysaccharide for binding to various host receptors. The virus uses variety of adhesive which can bind to multiple components of the host (Table 8.2).

Table 8.1 Shows the involvement of pathogenic agents in the diseases which were earlier believed to be non infectious diseases

Papillomavirus	Invasive cervical cancer
Helicobacter pylori	Peptic ulcers
Herpesvirus types	Kaposi's sarcoma
Epstein–Barr virus	Hodgkin's disease and certain lymphomas

Table 8.2 Adhesive proteins used by virus for binding to host cell via host receptor

Agent	Adherence factor	Host receptor
HIV	Envelope glycoprotein (gp120)	CD4 and chemokine receptor CXCR4
Measles virus	Hemagglutinin (H) glycoprotein	CD46 and membrane-organizing protein moesin
Herpes simplex virus	Glycoprotein C (gpC)	Heparan sulfate
Influenza virus	Hemagglutinin, neuraminidase	Sialic acid
Epstein–Barr virus	Surface protein	CD21
Adenovirus	Fiber protein	Integrins (vitronectin receptors)

8.6 Gram-Positive and Gram-Negative Infections

Gram staining: This method is based upon the fact that crystal violet is capable of irreversible staining of some bacteria which are referred as gram positive (*Bacillus*, *Staphylococcus*, *Streptococcus*, *Enterococcus*, *Diplococcus*, etc.), and in others the stain can be washed off, that is, they are decolorized and take up counterstain and are referred as gram negative. The stain is taken up due to the cell wall components present in the microorganisms. This test is rapid and gives a good indication about the infectious agent so that before exact identification is done, a choice antibiotic therapy may be started [5].

The culture positivity or negativity confirmation along with antibiotic sensitivity results in better selection of antibiotic so that infection can be efficiently managed. Many modifications of gram staining are now in use to efficiently detect the pathogenic strain.

Tables 8.3 and 8.4 show some gram-positive and gram-negative microorganisms, respectively, along with the consequences of their infections. Their microscopic appearance has been shown for some gram-positive and gram-negative bacteria in Figs. 8.7 and 8.8, respectively.

8.7 Pyrexial Illness

Pyrexia is a physiological response of illness. Pyrexia is also known as fever or hyperthermia and is manifested as elevated temperature. It is due to increase in thermal set point, that is, above the normal body temperature of 97.7–99.5 °F. Fever is a useful defense mechanism, as immune response of the body mediates its response effectively at high temperature. During fever, muscle tone of the body is increased, which presents as chills. Fever is presented by many diseases as tropical or environmental diseases, but one should remember that many common infections like influenza and tuberculosis also occur in the tropics or may be acquired en route to and from. Exotic local febrile patient may also be due to autoimmune or malignant condition. The causes of fever may be bacterial or viral or fungal infections or autoimmune condition or malignancy or folate deficiency:

- *Pathophysiology*: Temperature is ultimately regulated by the hypothalamus; it triggers the fever by inducing pyrogens. The hypothalamus is acted upon by prostaglandin E2, generating systemic response in the body causing heat-creating effect to match a new temperature level.
- *Pyrogen*: Pyrogen is a substance which induces fever; it may be endogenous or exogenous. Exogenous pyrogen may be bacterial lipopolysaccharide (LPS), and endogenous pyrogens are various cytokines like IL-1 (α and β), IL-6, IL-8, TNF-β, macrophage inflammatory protein β, interferon-α, and interferon-β.
- *Diagnosis of pyrexia*: Diagnosis is made by clinical symptoms as increase of temperature, chills, rigor, and increase in muscle tone. For pattern of temperature, diagnosis is confirmed by blood and serological examination (TLC, DLC, ESR, CRP) or by culture and sensitivity test. Pyrexia is managed by treating the cause and antipyretics [5].

Table 8.3 Gram-positive infections and causative organisms

Gram-positive infections	Pathogenic strains	Diseases and symptoms	Morphology
Pneumococcal infections	*Streptococcus pneumoniae*	Otitis media, acute purulent rhinosinusitis, pneumonia, meningitis	Grows in chain
Staphylococcal infections	*Streptococcus aureus*	Toxin- and non-toxin-mediated infections, surgical wound infections, primary bacteremia	Forms grapelike clusters
Streptococcal infections	Group A *Streptococcus pyogenes*	Pharyngitis, post-infectious symptoms of acute rheumatic fever, and glomerulonephritis	Spherical to ovoid, grow in chains
	Group B *Streptococcus agalactiae*	Bacterial sepsis, meningitis in newborns, endometritis, and fever in parturient females	
Enterococci (group D streptococci)	*Enterococcus faecalis, E. faecium*	Urinary tract infections, nosocomial bacteremia, endocarditis	
Corynebacterial infections	*Corynebacterium diphtheriae*	Nasopharyngeal and skin infections, toxin causes systemic toxicity, myocarditis, polyneuropathy	Club-shaped bacillary appearance forming clusters
Listeria	*Listeria monocytogenes*	Food-borne pathogen, causes serious infections in immunocompromised individuals	Rod shaped
Clostridial infections	*Clostridium tetani*	Tetanus: neurologic disorder, increased muscle tone, and spasms due to tetanospasmin (toxin)	Resembles tennis racket
	Clostridium botulinum	Botulism: paralytic disease due to neurotoxins, proceeds with cranial nerve involvement	
	Clostridium perfringens	Gas gangrene (bacteremia) due to active tissue and enterotoxins	

Table 8.4 Gram-negative infections and causative organisms

Gram-negative infections	Pathogenic strains	Diseases and symptoms	Morphology
Meningococcal infections	*Neisseria meningitidis*	Causes life-threatening meningococcal meningitis and fulminant meningococcemia	Gram-negative aerobic capsular diplococci
Gonococcal infections	*Neisseria gonorrhoeae*	Causes sexually transmitted infections as gonorrhea which manifests as cervicitis, urethritis, conjunctivitis	Monococci or diplococci
Haemophilus infections	*Haemophilus influenzae*	Local invasion of mucosal surfaces; otitis media (middle ear through Eustachian tube)	Has variable shape (coccobacillus)
HACEK group infections	*Haemophilus* sp.	Reside in oral cavity and cause local infections of the mouth, sometimes severe systemic infections as bacterial endocarditis	HACEK are fastidious, gram-negative, slow growing, carbon dioxide-requiring organisms
	Actinobacillus actinomycetemcomitans		
	Cardiobacterium hominis		
	Eikenella corrodens		
	Kingella kingae		
Pertussis	*Bordetella pertussis*	Causes violent cough; inspiratory sound at the end of coughing gives the common name as whooping cough for the illness	Aerobic bacilli
Enteric bacilli	*Escherichia coli*	Global pathogens; oropharyngeal colonization may lead to pneumonia	Have extracytoplasmic outer membrane, lipopolysaccharides
	Klebsiella		
	Proteus, Enterobacter		
	Serratia		
	Citrobacter		
Pseudomonas	*Pseudomonas aeruginosa*	Infections in hospital patients	

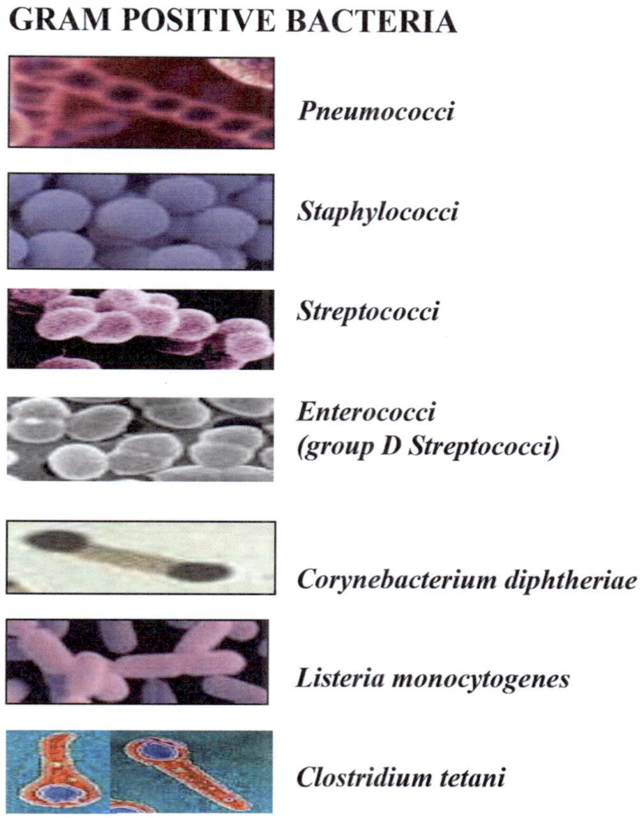

Fig. 8.7 The figure shows the microscopic shape and pattern of various gram-positive bacteria which are causative agent of many human diseases

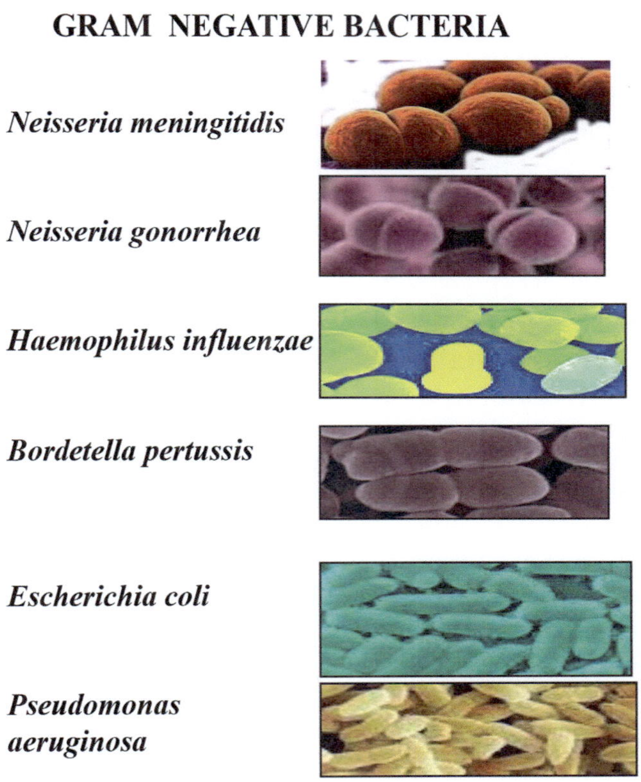

Fig. 8.8 The figure shows the microscopic shape and pattern of various gram-negative bacteria which are causative agent of many human diseases

8.8 Infections of the Gastrointestinal System

The diseases affect the digestive organs because of infectious agents as certain bacteria (*Escherichia coli* (*E. coli*), *Vibrio cholerae*, *Salmonella*, and *Shigella*)), protozoa (*Entamoeba histolytica*), or viruses. Parasites causing gastrointestinal tract infections and infestations are *Entamoeba histolytica*, *Ascaris lumbricoides*, *Taenia saginata*, *Taenia solium*, *Hymenolepis nana*, *Giardia*, *Cryptosporidium*, etc. These infections are often associated with nausea, vomiting, diarrhea, and other gastrointestinal symptoms like dysentery or bleeding diarrhea and may last for 5 or 7 days. The infections may be mild to severe and are either through contaminated food (raw, undercooked, or uncovered food) and/or water (contaminated water).

Gastrointestinal infections present commonly with diarrhea characterized by frequent and watery bowel movement. Virus-induced diarrhea is known as viral gastroenteritis and may be due to *Rotavirus*. Usual GI symptoms are abdominal cramping, followed by diarrhea, fever, loss of appetite, nausea, vomiting, weight loss, dehydration, and mucous or blood in stool. These symptoms typically last for a few days or longer. Viral diarrhea usually goes away without treatment by drinking plenty of fluid to avoid dehydration, while some types of diarrhea require antibiotic treatment to eliminate the causative agents.

8.9 Infections of the Skin and Nail

The skin is the outermost covering of the body which consists of an outermost layer called the epidermis and inner layer the dermis. The epidermis is thin having several layers of epithelial cells of which the stratum corneum being the outermost is composed of dead cells with waterproof protein keratin and creates an effective physical barrier. The inner layer dermis consists of connective tissue and hair follicles, sweat gland, and oil gland duct, containing salt, lysozyme, and sebum (secreted by oil glands) that prevent skin from drying out and also serve as nutrients for certain microorganisms.

Normal microbiota of the skin contains a large number of gram-positive bacteria as *Staphylococcus* and *Micrococcus* which survive antimicrobial properties and his salt concentration of the skin. Microorganisms also reside in the hair follicles and sweat glands. The moistured parts like armpits and legs have higher population of microbes. Other organisms contributing to microbiota are gram-positive pleomorphic rods called diphtheroids like *Propionibacterium acnes* which reside in hair follicles and are anaerobic [5].

Lesions on the skin indicate the microbial disease. They are:

Vesicles: Small, fluid-filled lesions
Bullae: Vesicles larger than 1 cm in diameter
Macules: Flat reddened lesion
Papules: Raised lesions
Pustules: Raised lesions with pus

The important infectious agent may be bacteria or viruses or fungi or parasites. The important bacterial diseases are:

Acne: Common skin disease resulting in inflammatory lesion due to blockage of channels for the passage of sebum to the skin surface. Acne is usually treated with azelaic acid (Azelex), salicylic acid preparation, or retinoids. Inflammatory acne resulting in appearance of pustules and papules is caused by *Propionibacterium acnes*, an anaerobic diphtheroid present on the skin. The treatment for these is phenyl peroxide (available as Benzamycin), antibiotic erythromycin or isotretinoin, or clear light system.

Erysipelas: Appears after *Streptococcus pyogenes* (gram-positive bacteria) infects the dermal layer of the skin. The disease shows eruption of the skin into reddish patches with raised margins. The risk of this may be its progress to local tissue destruction followed by entry into the blood vessel resulting into sepsis. The treatment is β-lactam antibiotics.

Necrotizing fascitis: The invasion of streptococci with excessive tissue destruction results in necrotizing fascitis. It might be caused by exotoxin A, which acts as superantigen. The treatment is penicillin.

Impetigo: Commonly in hospitals, pathogenic strain staphylococci produce enterotoxins, leukocidins, and exfoliative toxins resulting in impetigo of the newborn. Symptoms are thin-walled isolated vesicles that rupture and later crust over. Treatment is hexachlorophene containing skin lotions.

Folliculitis: Caused by *Staphylococcus* and the most common skin microbiota. *S. aureus* enters through the opening in the skin and the hair follicle passage. Infection of hair follicle occurs as pimple on the skin and sty of eyelashes. Serious hair follicle infection is furuncle which is abscess surrounded by inflamed tissue. In case neighboring tissue is invaded, this results in excessive damage (carbuncle) with the symptoms of generalized illness with fever. The invasion of infectious agent to underlying tissue and entry into the bloodstream are serious health risks associated with *Staphylococcus* infection.

Pseudomonas dermatitis and otitis externa result from infection of *Pseudomonas aeruginosa* leading to superficial rash. In otitis externa, infection of the external ear canal occurs. The treatment is fluoroquinolones.

The common viral diseases of the skin are:

Measles: Also known as rubeola in which virus spreads through respiratory route and is highly contagious; the symptoms appear after 10–14 days of virus exposure. The symptoms are reddish macules appearing on the face and spreading to the trunk and extremities. Vaccine of measles, mumps, and rubella (MMR) has almost eliminated the measles.

Rubella (German measles) is caused by *Rubella virus*. The disease has symptoms like measles but is mild and less extensive which usually disappears in 3 days.

Small pox: Caused by *variola virus*. The effective vaccination has almost eradicated the disease. The symptoms are confluent pustules on the skin; the later stage may affect internal organs.

Chicken pox: Results from infection by *varicella zoster* virus. The symptoms are vesicles present on the face, throat, and lower back. The therapy is virostatic drugs. *Shingles* is caused by *herpes zoster*. The symptoms are appearance of vesicles on one side of the waist, face or scalp, or upper chest.

Herpes simplex virus type I infection results in cold sores in which vesicles around the mouth are visible.

Fungal diseases

Ringworm also known as tinea results from infection of *Microsporum*, *Trichophyton*, or *Epidermophyton* species. The symptoms are skin lesions of varied appearance. The treatment is antibiotic griseofulvin or topical applications of miconazole or clotrimazole.

Candidiasis: Results from *Candida albicans* which usually infects the mucous membrane or moist areas of the skin. The topical application of miconazole and clotrimazole is effective.

The other common parasitic infections are scabies (*Sarcoptes scabiei* (mite)). Symptoms are appearance of papules and the treatment is gamma benzene hexachloride or permethrin.

8.10 Infections of the Respiratory System

Diseases of the respiratory system are often manifested by symptoms of dyspnea (shortness of breath), or cough, or chest pain, or abnormality on a chest radiograph. As the symptoms often overlap in various respiratory diseases, the differential diagnosis is required from the history, physical examination, pulmonary function testing, etc., and they may or may not be associated with sputum. The most abundant microflora of the respiratory tract suppresses the growth of the pathogenic bacteria in the upper respiratory tract. There are a variety of symptoms because of infections of various regions of the respiratory system [5].

Pharyngitis: Inflammation of the mucous membranes of the throat or sore throat.
Laryngitis: Infection of larynx.
Tonsillitis: Inflamed tonsils.

> *Sinuses*: Nasal sinuses are cavities in certain cranial bones that open into the nasal cavity lined with a continuous mucous membrane. Infection of the sinus leads to inflamed mucous membrane resulting in heavy nasal discharge, a condition called sinusitis. Blockage of the sinus leaves to internal presence causing pain or sinus headache.

Strep Throat This is caused due to *Streptococcus pyogenes* (group A beta hemolytic bacteria). The symptoms of the disease are inflammation of the mucous membrane along with fever. Lymph nodes located in the neck are enlarged and tender. The disease is diagnosed by indirect agglutination and treated by penicillin.

Diphtheria Caused by *Corynebacterium diphtheriae*. Exotoxin is the cause of symptoms which inhibits translation resulting in tissue and organ damage. The antitoxin therapy and antibiotics are used to cure the disease.

Otitis media This is caused by *Streptococcus pneumoniae*, *Haemophilus influenzae*, *Moraxella catarrhalis*, *Streptococcus pyogenes*, and *Staphylococcus aureus*. In this ear, ache occurs after nose and throat infections. Pus accumulation causes pressure on the eardrum. Treatment is amoxicillin (broad-spectrum penicillin).

Common cold Results from infection of 200 different viruses. Symptoms include sneezing, nasal secretions, and congestion.

Pertussis (whooping cough) Caused by *Bordetella pertussis*. The disease symptoms are because of congestion of the trachea and bronchi due to accumulation of mucous. The bacteria produce toxins and destroy the ciliated cells of the trachea. Vaccination is very effective in reducing the incidences of the disease.

Tuberculosis The causative agent for tuberculosis (TB) is *Mycobacterium tuberculosis*. The bacteria contain lots of lipids which aids in their resistance to drying and antimicrobial agents. *M. tuberculosis* forms lesions known as tubercles. These lesions along with macrophages might calcify and appear clearly on X-ray films. Infection progresses when lesion ruptures releasing bacteria into the lung airways, lymphatics, and cardiovascular system leading to military tuberculosis. In this, the patient suffers from weight loss, cough sometimes with blood, and loss of appetite. Initially effective drug for TB was streptomycin. Currently, the patient is administered with multiple drugs as isoniazid, rifampin, and pyrazinamide. Noncompliance to medication during treatment in part is responsible for emergence of multiple drug-resistant strains which are posing serious problem. The *Bacillus* Calmette–Guerin (BCG; frequent in humans which affects the bones or lymphatics) is being widely used in the world as vaccine.

Pneumonia Caused by *Streptococcus pneumoniae*; apart from this, pneumonia may also be caused by fungi, protozoa, viruses, as well as other bacteria and is known as atypical pneumonia. Pneumonia is named after the respiratory site of the infection:

- Lobar pneumonia: Lobes of lungs are infected.
- Bronchopneumonia: Alveoli of the lungs are infected.
- Pleurisy: Pleural membranes become painfully inflamed.

Disease caused by *S. pneumoniae* is known as pneumococcal pneumonia (formerly *Diplococcus pneumoniae*). The capsules form the basis of serological differentiation of pneumococci into 90 serotypes. The symptoms include high fever, breathing difficulty, and chest pain. Effective drugs are penicillin and fluoroquinolones. Subunit vaccine from purified capsular material of the 23 types of pneumococci has been developed [5].

Respiratory syncytial virus (RSV) The disease is prevalent in infants and infection of the bronchial epidermis occurs during the winter and early spring. The name is derived from syncytium (cell fusion) formation in cell culture. For severe symptoms the antiviral drug ribavirin is administered.

Influenza

The disease is characterized by chills, fever, headache, and muscular aches. Recovery occurs fast and cold-like symptoms appear. Influenza virus has eight separate RNA segments with inner protein and an outer lipid bilayer. Two types of projections embedded in the lipid bilayer are hemagglutinin (H) and neuraminidase (N) spikes. H spikes (about 500 on virion) allow the virus to recognize and attach to cells of the body; this causes agglutination of RBCs. N spikes (about 100 per virus) help the virus to separate from infected cells after intracellular reproduction. Antibodies against H spikes are more important in the resistance of body to the disease. Viral strains are identified by variation in H and N antigens. Different forms of antigens are designated as H1, H2, H3, N1, and N2. The different numbers show substantial alteration in the protein composition of spikes which are referred as antigenic shifts, and they escape from the previous immunity developed in the human body. High mutation rates are characteristic of RNA viruses. The major cause of antigenic shifts is genetic recombination which might occur in infections involving more than one strain and mixing of RNA from strains of animals (swine, horses, and birds) with RNA of human strains. Swines may be infected with both human and fowl influenza, and likely animals involved in shifts therefore are referred as mixing vessels [5].

Minor variation in antigenic makeup is called antigenic drift which might reflect alteration of only single amino acid which is probably due to selective pressure of antibody. The effect of antigenic drift results in low efficacy of vaccine against a particular spike (H2) after 10 years in H2 only. Influenza viruses are also classified according to antigens of their protein coat viz., A, B, and C. A-type cause major pandemics; B viruses are geographically limited and cause mild infections.

These are the reasons that it is not possible to develop an effective vaccine for influenza that can give long-term immunity. The vaccines developed are usually multivalent with 70–90 % protection but duration of protection is not more than 3 years.

Epidemics of the flu are due to a new strain of virus which quickly propagates through population. The cause of death is not influenza virus but often it is secondary bacterial infections (*H. influenza*, *S. aureus*, *S. pneumoniae*).

In 1918–1919 more than 20 million people died in major pandemic of influenza with highest mortality rates in young adults, often dying within a few hours. The infection is restricted to the upper respiratory system but, due to some changes in virulence, results in invasion of the lungs causing viral pneumonia.

The antiviral drugs are amantadine and rimantadine which significantly reduce the symptoms of A-type virus. Inhibitors of neuraminidase-2 zanamivir (Relenza) which can be inhaled and oseltamivir phosphate (Tamiflu) are administered orally.

Fungal diseases: The fungal diseases of the respiratory system are histoplasmosis (inhalation of airborne conidia), coccidioidomycosis (inhalation of airborne arthrospores), and *Pneumocystis* pneumonia (*Pneumocystis jirovecii* causes inflammation and fluid buildup in lungs).

8.11 Infections of the Nervous System

The nervous system consists of the central nervous system (CNS) with the brain and spinal cord and the peripheral nervous system consisting of all the nerves that branch off from the brain and spinal cord. These nerves are the lines which send and receive the signal from the body to the brain (sensory) and from the brain to the body (motor) and thus coordinated the body. The brain and spinal cord are protected by three continuous membranes called meninges (outermost duramater, central arachnoid mater, and innermost pia mater). Between the inner and middle layer is a space called subarachnoid space in which human adults have 100–160 ml of cerebrospinal fluid (CSF).

The blood–brain barrier protects the brain by restricting components of the blood to enter the brain, but some selective substances pass through certain capillaries; lipid soluble molecules can gain entry. Though the CNS has high level of protection, still certain microorganism can gain entry because of trauma, nerve supply, blood, and lymphatics. The inflammation then occurs in the meninges (meningitis) or brain (encephalitis) [5].

Meningitis: Meningitis may be caused by different types of pathogens including viruses, bacteria, and fungi. Its more than 70 % of cases are caused by bacterial species, as gram-positive *Streptococcus pneumoniae* and *Neisseria meningitidis*. They possess capsule that protects them from phagocytosis, and they rapidly multiply in blood and afterward enter the cerebrospinal fluid. The symptoms are fever, headache, and stiffness in the neck followed by nausea and vomiting. The diseases may progress due to the release of endotoxins of the gram-negative pathogens or the release of cell wall components (peptidoglycans and teichoic acids) of gram-positive bacteria. Quick treatment is required with broad-spectrum third-generation cephalosporins.

Tetanus: The disease is caused by an obligatory anaerobic, endosperm-forming, gram-positive *Clostridium tetani*. The symptoms are due to potent neurotoxin, tetanospasmin, which affects relaxation of muscles causing muscle spasms. The jaw muscles are the early target which prevents the mouth from opening known as lockjaw [5].

Spasms in the back muscle cause the head and heel to bow backward (opisthotonos). Death results from spasms of respiratory muscles. Effective vaccination has reduced the incidences. Both attenuated and subunit (tetanus toxoids) vaccines are available:

Botulism: The causative agent is anaerobic, endospore-forming gram-positive *Clostridium botulinum*. Exotoxin produced the most potent of all natural toxins which blocks the release of acetylcholines. The patient undergoes a progressively flaccid paralysis for 1–10 days and may die from respiratory and cardiac failure. Botulism toxin is used for cosmetics as Botox.

Leprosy: Due to infection of *Mycobacterium leprae* in the peripheral nervous system. The organism was isolated by Gerhard A. Hansen and formally known as Hansen's disease. It usually invades cells of the peripheral nervous system. Sulfone drugs as dapsone and other drugs as rifampin and clofazimine are effective [5].

Poliomyelitis: The paralytic form of polio appears in less than 1 % of infected people, whereas majority of the cases are asymptomatic or present with mild symptoms such as headache, sore throat, fever, and nausea. The mode of transmission is through contaminated water and affects the throat (sore throat) and small intestine. The virus enters the lymph nodes and then blood and afterward penetrates the central nervous system. The virus invades and destroys motor nerve cells of the upper spinal cord; sometimes, death can result from respiratory failure. There are three different serotypes of virus. The Salk vaccine uses inactivated polio virus, whereas Sabin is oral polio which is attenuated. The use of vaccination has almost eradicated polio from the population. The use of Sabin/oral polio is not recommended as reversion to virulent state and symptoms in immunocompromised host pose a threat of the disease [5].

Rabies: (Latin, rage or madness) The disease is caused by rabies virus which is transmitted to humans by bite of infected animals (dogs, cats, bats, foxes, etc.). The virus from skeletal muscles and connective tissue moves to CNS where it causes encephalitis. The symptoms are mild and varied and sometimes thought/sight of water can result in fear called hydrophobia. The end stage of disease is due to excessive damage to the nerve cells of the brain and spinal cord. Passive vaccination is preferred as precaution [5].

8.12 Diseases Caused by Prions

Prions are abnormally folded proteins capable of inducing change in the shape of a normal protein. The disease has long incubation times with slow damage without any clinical symptoms. In humans transmissible spongiform encephalitis (spongy/porous degeneration of the brain) known as Creutzfeldt–Jakob disease (CJD) is present. CJD often occurs in families.

8.13 Sexually Transmitted Diseases (STDs) and Congenital Infection

Sexually transmitted diseases (STDs) are prevalent diseases in different parts of the world and are posing major health burden. They not only spread dramatically from one person to another, but they are also passed on to the next generation either as congenital diseases or during child birth [1–3]. Thus, before starting these diseases, one crucial thing would be prevention. Prevention can be achieved by:

- Educating patients who are at risk for STDs.
- Counseling and support for the high-risk patients.
- Detecting asymptomatic or symptomatic patients who are unlikely to seek treatment providing effective diagnosis and treatment of those with STDs.
- Evaluating, treating, and counseling the sexual partners of those with STDs.
- Immunizing patients who are at risk for STDs that are preventable by vaccination.
- Testing for HIV and hepatitis in any patient diagnosed with STD, which is an important preventive measure.
- Counseling about the patient's potential risk for contracting HIV or other STDs should be done before and after testing [7].

Important STDs are syphilis, gonorrhea, HIV, hepatitis B, and chancroid. Others are chlamydial infections, genetic infections with HPV, and genital herpes which can spread very fast. They are more prevalent in individuals with frequent partner change or having multiple partners [1–3].

8.13.1 Syphilis

It is a bacterial disease with *Treponema pallidum* as the causative agent. The disease may be acquired by sexual intercourse or may be transmitted from mother to baby (congenital syphilis) or occasionally by blood transfusion or by nonsexual contacts. Effective prevention and identification of syphilis in pregnant patients requires routine serology to be performed for all pregnant patients at the time of their first prenatal visit. The pathogen causes erosion, ulcer, and vasculitis, and thereafter it may spread into regional lymph nodes. The syphilis can be divided into primary, secondary, latent, and tertiary stages. The lesion, chancre, appears as anogenital ulcer, which can also spread to extra-anogenital sites like the lip, tongue, and tonsils. Secondary syphilis presents itself with rash affecting the palms and sole which later on becomes papules. Tertiary syphilis is manifested with gummatous, cardiovascular, and neurological involvement. Syphilis is diagnosed by identifying treponemes using dark field microscopy or by serological tests. Treatment of choice is intramuscular benzathine penicillin [1–4].

8.13.2 Hepatitis B Virus (HBV)

Sexual transmission is responsible for 30–60 % of new cases of HBV. The risk of perinatal transmission from positive mothers to infants is as high as 85 %. Among these 90 % of infected infants progress to chronic HBV. Immunization is preventive and no other specific treatment for acute HBV exists. Interferon therapy (alpha-2b interferon (Intron A) and lamivudine (Epivir)) is available, but only 40 % is effective in eliminating HBV.

8.13.3 Hepatitis A Virus (HAV)

HAV is transmitted by household and sexual contacts. No treatment is available presently for acute HAV infection except supportive therapy. Passive treatment with antibodies given early can prevent 85 % of new HAV infections. Two vaccines for HAV (Havrix, Vaqta) are given in two-dose series at the interval of 6 months. The first dose gives 99–100 % response, while the second dose provides long-term immunity.

8.13.4 Pelvic Inflammatory Disease (PID)

PID is caused by gonorrhea and chlamydia. PID affects the upper female reproductive tract, developed in 10 % of women. During infection with PID, there are chances of its worsening due to invasion and infection with other infectious agents. PID is presented with mild or nonspecific symptoms and sometimes goes asymptomatic. Missed diagnosis leads to damage to the reproductive tract. Oral antibiotics should be used for its management. Amoxicillin–clavulanic acid (Augmentin) along with doxycycline is effective.

8.13.5 Genital Herpes

Mild or asymptomatic genital herpes is also a common infection in many individuals. The infection can be passed on during delivery; thus cesarean section is recommended for infected females. For the treatment of genital herpes, two new antiviral drugs, famciclovir (Famvir) and valacyclovir (Valtrex), are used in addition to established therapy with acyclovir (Zovirax). The therapy may decrease the severity and duration of a genital herpes outbreak; treatment is effective if started within 24 h of infection.

8.13.6 Gonococcal and Nongonococcal Urethritis

Urethritis: Urethritis can be caused by *Neisseria gonorrhoeae*, *Chlamydia trachomatis*, *Mycoplasma genitalium*, *Ureaplasma urealyticum*, *Trichomonas vaginalis*, herpes simplex virus (HSV), etc. In men it produced urethral discharge or dysuria, and in women it causes internal dysuria (without urinary urgency), pyuria, and absence of *E. coli*. Dysuria which is associated with vulvar herpes or vulvovaginal candidiasis is referred as external as it is painful when comes in contact with urine. Initial evaluation includes urethral discharge, gram staining, and test for *N. gonorrhoeae* and *C. trachomatis*. Quinolines were given for treatment of gonorrhea as quinolines are a simple, safe, and effective treatment for gonococcal infections of any severity. However, because of growing resistance of *Neisseria gonorrhoeae*, the guidelines suggest treatment with cefixime (Suprax) or ceftriaxone (Rocephin). In the absence of gonococci in gram stain, urethritis treatment is directed toward nongonococcal urethritis (NGU). *Mycoplasma genitalium* is a new etiologic agent added to the list of bacteria that cause nongonococcal urethritis (NGU) in addition to *Chlamydia trachomatis*. Azithromycin in a single dose or doxycycline was recommended for the treatment of NGU.

Vulvovaginal candidiasis: Caused by *Candida albicans* with symptoms of vulval itching with or without irritation. It is accompanied by vaginal discharge which is clumped, scanty, adherent plaques with white color. Microscopic evaluation reveals leukocytes, epithelial cells, mycelia, or pseudomycelia in 80 % of positive culture. Topical and oral antifungal like miconazole, clotrimazole or fluconazole are given.

Trichomonal vaginitis: The causative agent is *Trichomonas vaginalis* with symptoms of vulval itching with discharge which is homogeneous, white or yellow, profuse, and purulent and causes erythema of the vaginal and vulvar epithelium. Microscopic examination reveals leukocytes with motile trichomonads in 80–90 % of symptomatic patients. For diagnosis, nucleic acid amplification technology (NAAT) for *T. vaginalis* is used and the treatment is metronidazole.

Bacterial vaginosis: Also formerly known as nonspecific vaginitis caused by *Gardnerella vaginosis*, various anaerobic or non-cultured bacteria, and mycoplasmas. The symptoms include moderate, white or gray, homogeneous, low viscosity, malodorous discharge. Treatment includes metronidazole and clindamycin.

Noncervical human papillomavirus: Major infections of genital human papillomavirus (HPV) infection occur annually with sexual transmission. In most patients with genital HPV, the infection is asymptomatic, subclinical, or unrecognized. The disease may be diagnosed by clinical assessment or biopsy of the lesion. The main goal of therapy for noncervical HPV infection is to treat symptomatic visible lesions by applying podofilox (Condylox) 0.5 % solution or gel. There is no evidence that treatment decreases infectivity or changes the natural course of HPV infections or the risk of development of cervical cancer.

Human immunodeficiency virus (HIV): Is caused by RNA virus of retrovirus family by human immunodeficiency virus type I (HIV-1). Acquired immunodeficiency syndrome (AIDS) is one of the most important sexually transmitted disease. Its epidemic was recognized long back, and despite major advances in prevention and treatment of the disease, the epidemic continues and has devastating effects on human society. The virus critically attacks $CD4^+$ T cells which are active mediators of cell-mediated immunity. The end stage of the disease is severe acquired immunodeficiency with lots of infection occurring in the HIV patients. The typical case of HIV disease then manifests itself as evidenced by certain malignancies and unusual microbial infections that often attack the lungs, intestines, skin, eyes, or the central nervous system.

HIV can be detected in body fluids like blood, semen, and vaginal secretions in the patients. HIV disease spreads mainly by unprotected sexual intercourse, sharing hypodermic needles, blood transfusion, or from mother to newborn.

The prevention, detection, and treatment of STDs is very important. The recommendations of CDC should be integrated in clinical practice to reduce the morbidity and mortality because of STDs.

8.14 Characterization of Pathogens

The practice of clinical microbiology has changed with the advancement in the techniques of molecular diagnostics. The technique of PCR and the machines available are very user-friendly that helps in the detection of pathogen. For detection and characterization of pathogens, many techniques and assays are approved by the US FDA and many more are in the process of development.

In most of the cases, several tests are sought from the clinical laboratory, but occasionally the help of the laboratory is required for exclusion of certain pathogen from the sample, e.g., exclusion of HSV infection in cerebrospinal fluid sample. Help is also required for the best antimicrobial agent for the suppression of growth of pathogen and thus symptoms of the disease.

As new pathogens are evolving and many a times symptoms are confusing, thus the role of clinical microbiologist is very important because the working professional should be aware of the potential pathogens which may be present in the clinical sample.

The molecular diagnostic techniques which are used to detect and characterize the pathogens may be separated into broad categories. The aim of these methods is either to detect a pathogen or exclude the presence of the particular pathogen with an indication about the extent of the presence (titer).

Direct hybridization: These techniques allow for the rapid detection and characterization of bacteria and fungi in the blood sample. In situ hybridization uses either fluorescent probes (fluorescence in situ hybridization (FISH)) or chromogenic probes (chromogenic in situ hybridization (CISH)):

- They may be used for detection of pathogen in blood sample or other clinical samples. It can be used for rapid characterization of bacteria and fungi both in fluids and in histologic section.
- This may be used to differentiate microorganisms of similar morphotypes in positive cul-

ture samples (gram-positive cocci or acid-fast bacilli).

- Before its usage, the preliminary information obtained from traditional techniques is helpful as it helps to make appropriate selection of probes, e.g., differentiating gram-positive or yeast species or presence of *Pseudomonas* or other members of *Enterobacteriaceae* in case of gram-negative culture.
- The FISH assays are currently available in North America to detect and differentiate *S. aureus* from other gram-positive cocci in clusters and *Candida albicans* from other yeasts.
- It can detect common cause of infection in cystic fibrosis and can differentiate *Mycobacteria* or trypanosomes in sleeping sickness. It is also helpful in detecting *Legionella pneumophila* and differentiates filamentous fungi and yeast and yeastlike fungi. It can also help separate the systemic .dimorphic fungi, which appear as yeast (e.g., *Histoplasma capsulatum* and *Blastomyces dermatitidis*) or spherules (*Coccidioides immitis* from *Candida* and *Cryptococcus* species).
- The hybrid capture (Digene) technology has been used for detection of high-risk *HPV subtypes*, *Cytomegalovirus*, *HBV*, *Neisseria gonorrhoeae*, and *Chlamydia trachomatis*.

8.14.1 Nucleic Acid Amplification Technology (NAAT)

Monoplex assays detect a single target and provide present- or absent-type result. For example, PCR performed for specific species may be used to detect presence or absence of that pathogen, e.g., *Legionella pneumophila*.

Used as quantitative methods, it can provide quantitative information (viral loads).

In a multiplex reaction where more targets may be detected, it allows the detection of multiple pathogens simultaneously. Multiplexing in real-time PCR setting is more useful which helps in simultaneous detection and differentiation of respiratory viral pathogens like influenza A, influenza B, and *respiratory syncytial virus* (RSV).

Broad-range nucleic acid amplification is also useful which is used to detect a wide variety of microorganisms. Quantitative data indicated the load of the organism, and positive result indicates the presence of one of the members of that particular group of microorganism.

PCR and NAAT technologies have large diagnostic relevance for detection of microorganisms which are difficult to culture. Samples which do not contain the target sequence (negative) will give minimal fluorescence, for example, detection of *Salmonella typhi* or other broad-range *Salmonella* assays to detect all the members. The assays have been designed to differentiate *HSV types I* and II, BK and JC polyomaviruses, and human herpesvirus 6 types A and B and commonly occurring *Bartonella* species and *M. tuberculosis* from nontuberculous *Mycobacteria*.

Microarray: In this, lots of probes are present on the chip and the sample is labeled and hybridized. The labeled sample is hybridized to the probes and detection of positive signal reveals the presence of that particular target in the sample.

8.15 Chapter End Summary

- The infectious diseases are responsible for major cause of death and disability throughout the world. They are responsible for a number of diseases either involving any local organ (localized or organ specific) or whole body (systemic infections).
- Our immune system is the defense system of our body which protects us from various invading pathogens. The system has two lines of immunity, the innate and adaptive immune responses. Innate immunity is nonspecific immunity and provides first line of defense. This immunity is present since birth.
- Adaptive immunity is specific with specificity, diversity, and memory. Adaptive immunity has T and B lymphocytes as its main components which are responsible for cell-mediated and humoral immune responses, respectively. The main effectors of cell-mediated immune responses are T cells and humoral immune responses are antibodies.

- The immune system protects us from various infections, loss of any component of immune system results in immunodeficiency (due to genetic defect or acquired during lifetime like AIDS), and failure of self-/non-self-recognition results in autoimmunity.
- With the continuous evolution of new diseases and newer causative agents or emergence of mutated pathogens, they are becoming potential challenges for health-care managers. Various diseases of the gastrointestinal tract, skin, nervous system, and respiratory system and sexually transmitted diseases are caused by a number of pathogenic agents.
- Infection of one agent or simultaneous infection of multiple agents is the cause and concern for major infections. With the advanced tools and techniques now, it is possible to characterize and detect these agents for therapeutic intervention.

Multiple Choice Questions

1. Immune system consists of:
   (a) Cell-mediated immunity and humoral immunity
   (b) Innate and adaptive immunity
   (c) Adaptive immunity
   (d) None of these
2. PAMPs are present on:
   (a) Pathogens
   (b) Immune cells
   (c) Body cells of humans
   (d) All of these
3. ADCC response is shown by:
   (a) Neutrophil
   (b) Basophil
   (c) Natural killer cell
   (d) Eosinophil
4. Th1 response is important in:
   (a) Innate immune responses
   (b) CTL-mediated responses
   (c) Humoral immune responses
   (d) All of the above
5. TLR2 is responsible for binding:
   (a) Peptidoglycan
   (b) Double-stranded RNA
   (c) Lipopolysaccharides
   (d) Herpesvirus
6. Body temperature or fever is regulated by:
   (a) Thalamus
   (b) Hypothalamus
   (c) Cerebrum
   (d) Midbrain
7. Normal temperature of the body is:
   (a) 100–101 °F
   (b) 95–97 °F
   (c) 103–104 °F
   (d) 97.7–99.5 °F
8. Pyrexial illness includes all except:
   (a) PUO
   (b) Viral fever
   (c) Bacterial fever
   (d) None of the above
9. Syphilis lesion is known as:
   (a) Chancroid
   (b) Boil
   (c) Abscess
   (d) Laceration
10. Syphilis is caused by:
   (a) *Treponema pallidum*
   (b) *Treponema folliculus*
   (c) *Treponema vulgaris*
   (d) Virus induced
11. Dysentery is caused by all except:
   (a) *Salmonella typhi*
   (b) *Shigella*
   (c) *E. coli*
   (d) *Entamoeba gingivitis*
12. All are skin lesions except:
   (a) Macule
   (b) Papule
   (c) Vesicle
   (d) Glossitis
13. All are sexually transmitted diseases except:
   (a) Syphilis
   (b) HIV
   (c) Malaria
   (d) HSV
14. All of these are symptoms of leprosy excluding:
   (a) Anesthetic patch
   (b) Thickened nerve
   (c) Loss of sweating
   (d) Itching
15. Tuberculosis of lungs is caused by:
   (a) *Mycobacterium leprae*

   (b) *Mycobacterium kansasii*
   (c) *Mycobacterium tuberculosis*
   (d) *Mycobacterium bovis*
16. If the infectious agent is present in the body in latent stage, which technique can detect its presence?
   (a) ELISA
   (b) DNA microarray
   (c) NAAT
   (d) RIA
17. Influenza virus is mutating very fast which leads to emergence of new strains; mutations occur in:
   (a) Viral glycoprotein
   (b) Hemagglutinin
   (c) Reverse transcriptase
   (d) Viral coat
18. MBL is an important component of:
   (a) Adaptive immunity
   (b) Innate immunity
   (c) Pathogen
   (d) Cell-mediated immunity

Answers

1. (b); 2. (a); 3. (c); 4. (b); 5. (a); 6. (b); 7. (d); 8. (d); 9. (a); 10. (a); 11. (d); 12. (d); 13. (c); 14. (d); 15. (c); 16. (c); 17. (b); 18. (b)

Questions

Q1. What is the role of the immune system in the prevention of infection?
Q2. What are soluble factors involved in immune response?
Q3. What is adaptive immune response?
Q4. Give a brief account of PRRs present in the body.
Q5. Write a short note on chancroid.
Q6. What are common cytokines involved in pyrexial illness?
Q7. Why HIV is most commonly associated with tuberculosis?
Q8. Why tubercular infections occur on the upper lobe of lungs?
Q9. Write a note on sexually transmitted diseases.
Q10. Which region is responsible in poliomyelitis?

References

1. Centers for Disease Control and Prevention (1993) Sexually transmitted diseases treatment guidelines. MMWR Morb Mortal Wkly Rep 42:1–102
2. Centers for Disease Control and Prevention (1997) Summary of notifiable diseases, United States. MMWR Morb Mortal Wkly Rep 46:3–87
3. Centers for Disease Control and Prevention (1998) Guidelines for treatment of sexually transmitted diseases. MMWR Morb Mortal Wkly Rep 47:1–111
4. Goh BT (2005) Syphilis in adults. Sex Transm Infect 81:448–452
5. Kasper DL, Braunwald E, Fauci AS, Hauser SL, Longo DL, Jameson JL (eds) (2008) Harrison's principles of internal medicine, 17th edn. Mc Graw-Hill Medical publishers, New York
6. Kindt TJ, Goldsby RA, Osborne BA (2007) KUBY immunology, 6th edn. Freeman and Company, New York
7. Miller KE, Graves JC (2000) Update on the prevention and treatment of sexually transmitted diseases. Am Fam Physician 61:379–386

Some Related Resources

www.lib.ncsu.edu/guides/microbiology/journal.html
www.mhhe.com/biosci/cellmicro/prescott/teachwww.mhtml
www.ncbi.nlm.nih.gov
www.textbookofbacteriology.net/
www.virology.net/garryfavweb22.html
www.waksman-foundation.org/html/microbiology_teachers.html
www.webicina.com/microbiology/

Molecular Diagnostics

9

Abstract

Effective and early management of diseases requires record of the history, behavioral parameters, and travel information. These are helpful for the diagnosis, prevention, and control of the disease. There have been several advancements in the methods for diagnosing infectious diseases. The wide spectrum of tests such as biochemical evaluation, microbiological tools, immunological and molecular biology techniques, etc., is available. Each type of diagnostic technique is strong and reliable in its own sense but poses certain limitations. These limitations may be complemented by using a combination of tests. Older techniques such as microscopy and culturing of organisms from clinical specimens are error-free but are very labor intensive and extremely time consuming. There is a need to develop rapid and sensitive tests that can be used in both high- and low-resource settings. Molecular diagnostics such as Western blot, ELISA, PCR, DNA, and protein microarrays are revolutionizing the clinical practice of infectious diseases. Their effects are significant in acute-care settings where timely and accurate diagnostic tools are critical for patient treatment decisions and outcomes.

9.1 Disease Pathology and Clinical Spectrum

Effective and early management of diseases requires record of the history, behavioral parameters, and travel information. These are helpful for the diagnosis, prevention, and control of the disease. There have been several advancements in the methods for diagnosing infectious diseases. The wide spectrum of tests such as biochemical evaluation, microbiological tools, immunological and molecular biology techniques, etc., is available. Each type of diagnostic technique is strong and reliable in its own sense but poses certain limitations. These limitations may be complemented by using a combination of tests. Older techniques such as microscopy and culturing of organisms from clinical specimens are error-free but are very labor intensive and extremely time consuming. There is a need to develop rapid and sensitive tests that can be used in both high- and low-resource settings. Molecular diagnostics such as Western blot, ELISA, PCR, DNA, and protein microarrays are

V. Gupta et al., *Basic and Applied Aspects of Biotechnology*, DOI 10.1007/978-981-10-0875-7_9

revolutionizing the clinical practice of infectious diseases. Their effects are significant in acute-care settings where timely and accurate diagnostic tools are critical for patient treatment decisions and outcomes.

9.2 Diagnosis of Bacterial, Viral, and Parasitic Diseases

The diagnosis of these agents is done by using many tests either alone or in combination.

9.2.1 Serological Tests

These are serology-based diagnostic tools. They are more sensitive and specific than microscopic tests. There are two categories of these diagnostic tools that are based on antigen-detection assays and antibody-detection assays. These assays include the Western blotting, enzyme-linked immunosorbent assay (ELISA), and all its derived tests such as the Falcon assay screening test-ELISA (FAST-ELISA), dot-ELISA, hemagglutination (HA) test, indirect or direct immunofluorescent antibody (IFA or DFA) tests, complement fixation (CF) test, and immunoblotting and rapid diagnostic tests (RDTs).

9.2.1.1 Western Blot

In a Western blot, the proteins present in a sample are separated according to their molecular weight by gel electrophoresis. A nitrocellulose membrane is placed on the gel, and with the help of electrical current, the proteins are transferred from the gel to the membrane where they adhere. The pattern of protein separation is maintained in the membrane after transfer. The membrane is then probed with specific antibodies (primary antibodies) to determine the presence of the protein. Often a secondary antibody conjugated to biotin or a reporter enzyme is used to enhance the signal and detect the binding of the primary antibody. This procedure is used mainly to determine the presence of an antigen in biological sample with simultaneous determination of the molecular weight of a protein and measure relative amount of a protein present in different samples (Fig. 9.1).

9.2.1.2 Enzyme-Linked Immunosorbent Assay (ELISA)

ELISA is a diagnostic tool that is used in medicine and other industries to detect and quantify specific antigens. The sample with an unknown amount of antigen is immobilized on a solid support, usually a microtiter plate. This is done either nonspecifically by adsorption or specifically by capture by another antibody specific to the same antigen, in a "sandwich" ELISA. After the antigen is immobilized, the detection antibody is added, forming a complex with the antigen. The detection antibody can be covalently linked to an enzyme or can itself be detected by a secondary antibody that is linked to an enzyme. The plate is developed by adding an enzymatic substrate to produce a visible signal which indicates the quantity of the antigen in the sample (Fig. 9.2).

Both Western blot and ELISA are used to detect HIV infection in the blood. They are called indirect tests as they measure the immune system's response to an infectious agent rather than looking for the components of the agent itself. Since ELISA detects HIV antibodies which the body starts to produce between 2 and 12 weeks after becoming infected with HIV, one should wait for at least 3 months to confirm for HIV AIDS. Western blot is the most common method of testing to confirm positive results from ELISA test. It is used more as a confirmatory test as it is difficult to perform and requires high skills. One advantage of Western blot is that it is less likely to give false-positive results as it can effectively distinguish between HIV antibodies and other antibodies.

9.2.1.3 Falcon Assay Screening Test-ELISA (FAST-ELISA)

This test uses synthetic and recombinant peptides to evaluate antibody responses to an antigen. However, this technique is subjected to the same drawbacks as most serology-based tests. Antibodies raised against a peptide from one protein may cross-react with proteins from other species. Moreover, antibodies raised against a

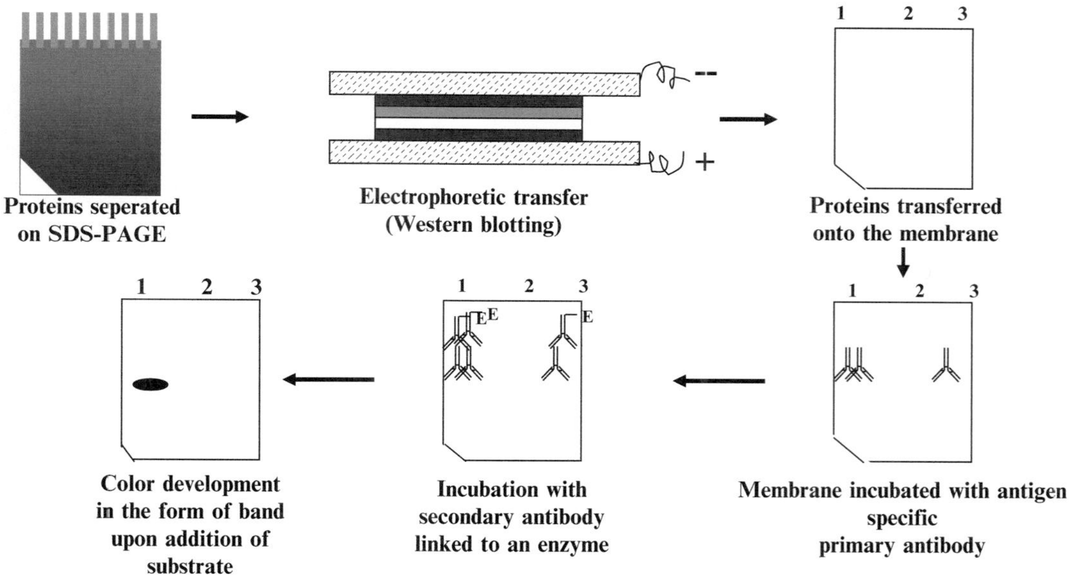

Fig. 9.1 The technique of Western blotting in which proteins separated on SDS-PAGE are transferred from the gel onto a membrane and detected using specific antibodies. Electrophoretic transfer is performed followed by incubation with antigen-specific primary antibody. The blot is washed and subsequently incubated with enzyme-linked secondary antibody. Addition of substrate results in formation of bands, whose intensity is related to the quantity of initial antigen present in the sample. For example, lane 1 and 3 have high and low antigen concentration, where as lane 2 has no detectable antigen

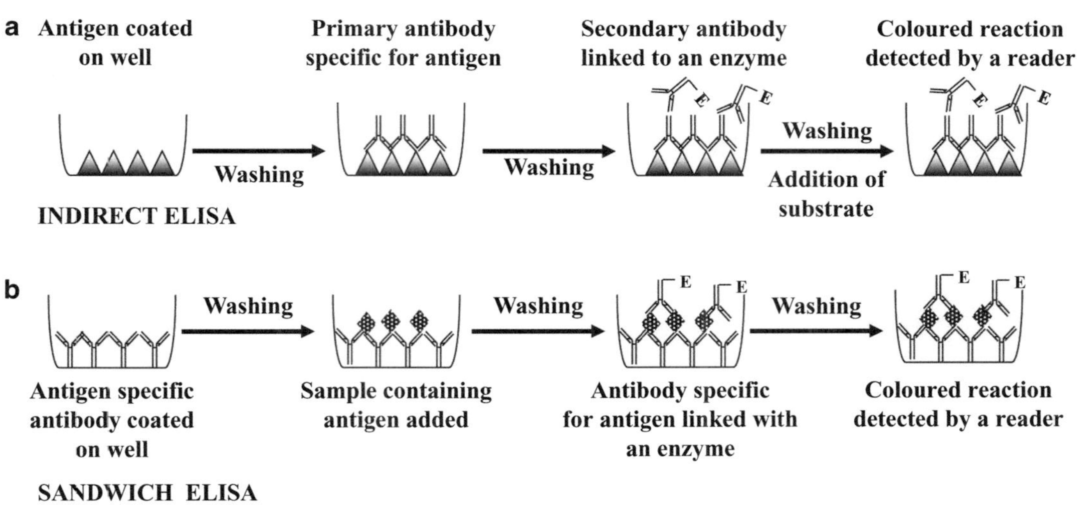

Fig. 9.2 The technique of enzyme-linked immunosorbent assay (ELISA) for the detection of antigen or antibody. (**a**) Indirect ELISA. Here the wells are coated with antigen and incubated with primary or antigen-specific antibody present in the sample. After washing, a secondary antibody linked with an enzyme is added. The binding of the antibody and addition of substrate gives colored product. (**b**) Sandwich ELISA. For authentication of a pathogen, two epitopes are tested by this technique. Antibody specific for a particular epitope of antigen are coated on well followed by the addition of the sample containing antigen. This results in antigen–antibody binding. Binding of another antigen-specific antibody linked with enzyme results in color formation upon addition of the substrate

peptide may react in some assays but not in others as some regions of a peptide may be more immunogenic than others. In the past, the method has been applied to the study of malaria, fasciolosis, schistosomiasis, and taeniasis. Lately it is not used regularly.

9.2.1.4 Dot-ELISA

The main difference between the regular ELISA and the dot-ELISA lies in the surface used to bind the antigen of choice. In the dot-ELISA, the plastic plate is replaced by a nitrocellulose or other paper membrane onto which a small amount of sample volume is applied. The principle is similar to that of immunoblotting. The dotted membrane is incubated first with an antigen-specific antibody followed by an enzyme-conjugated anti-antibody (secondary antibody). The addition of a precipitable, chromogenic substrate causes the formation of a colored dot on the membrane which can be visually read. It is convenient to use, gives rapid results that are fairly easy to interpret, is fast and cost-effective, and hence can be used in the field (e.g., as a dipstick). For all these reasons, the dot-ELISA is extensively used in the detection of human and animal parasitic diseases, including amebiasis, babesiosis, fascioliasis, cutaneous and visceral leishmaniasis, cysticercosis, echinococcosis, malaria, schistosomiasis, toxocariasis, toxoplasmosis, trichinosis, and trypanosomiasis [4].

9.2.1.5 Rapid Antigen-Detection Tests (RDTs)

This test is based on immunochromatographic antigen detection and has been implemented in many diagnostic laboratories as an adjunct to microscopy for the diagnosis of malaria. RDTs consist of capturing soluble proteins by complexing them with capture antibodies embedded on a nitrocellulose strip. A drop of blood sample is applied to the strip and eluted from the nitrocellulose strip by the addition of a few drops of bufer containing a labeled antibody. The antigen–antibody complex can then be visualized directly from the membrane. RDTs are now rapid, stable at temperatures up to 40 °C, easy to use, and cost-effective, thereby providing many advantages over traditional microscopic methods.

9.2.1.6 Luciferase Immunoprecipitation System (LIPS)

This is a modified ELISA-based assay in which serum containing antigen-specific antibodies can be identified by measuring light production. Basically, an antigen of choice is fused to the enzyme reporter Renilla luciferase (Ruc) and expressed as a Ruc-fusion in mammalian cells to allow for mammalian-specific posttranslational modifications. The crude protein extract is then incubated with the test serum and protein A/G beads. During the incubation, the Ruc-antigen fusion becomes immobilized on the A/G beads, which allows the antigen-specific antibody to be quantitated by washing the beads and adding coelenterazine substrate and measuring light production. Some of the advantages of the LIPS technology include its rapidity and accuracy in detecting infected patients. Sensitivity is improved in part by the use of mammalian cells which produce fusion antigens free of contaminating bacterial proteins. In addition, low backgrounds are produced compared to the ELISA.

9.2.1.7 Antibody-Based Diagnosis: Monoclonal Antibodies as Diagnostic Reagents

Monoclonal antibodies (mAb) are derived from identical immune cells that are clones of unique parent cells and can bind to a specific epitope (for further details, refer to Chap. 14). They have been extensively used in biomedical and microbiological research as tools for diagnosis of diseases such as hepatitis, AIDS, influenza, herpes simplex virus infection, chlamydial infection, and treatment of cancer [7, 9]. The monoclonal antibodies being directed against single epitopes are homogeneous and highly specific and can be produced in unlimited quantities. Monoclonal antibodies have tremendous applications in the field of diagnostics, therapeutics, and targeted drug delivery systems, not only for infectious diseases caused by bacteria, viruses, and protozoa but also for cancer and metabolic and hormonal disorders.

In 1975, Kohler and Milstein invented the hybridoma technology. The key idea was to use a line of myeloma cells that had lost their ability to secrete antibodies, fuse these cells with healthy antibody-producing B cells, and select for the successfully fused cells. In hybridoma technology, a myeloma cell rendered drug sensitive through mutation in a growth essential gene, hypoxanthine guanine phosphoribosyl transferase (HGPRT), is chemically fused with immune cells from a host immunized with the antigen of interest, and the resulting cells are grown in medium containing the selective drug. Since the immune cells have a short life span in tissue culture and the myeloma cells are drug sensitive, the only cell that will survive are those myeloma cells which obtained a normal HGPRT gene from the immune cells. Such cells also have a high likelihood of carrying the immune cell's antibody gene resulting in the generation of a hybridoma that can grow continuously in vitro and secrete a single monoclonal antibody (Fig. 9.3).

The diagnosis of any infectious disease often requires the demonstration of the causative organism or presence of a specific antibody. Specific antibody-based tests identify the pathogens associated with the disease. MAbs recognizing unique antigenic determinants on pathogens are developed. This restricted reactivity allows for precise identification of the organism of interest which is the major advantage of MAbs over polyclonal antisera. In case of a pathogen occurring as subtype defined by unique antigenic differences, specific MAbs can be used, whereas conventional antisera needs laborious absorption to remove cross-reactive antibodies. Because of the specificity, homogeneity, and unlimited availability of the MAbs, vast amount of work has been carried out on the production/development

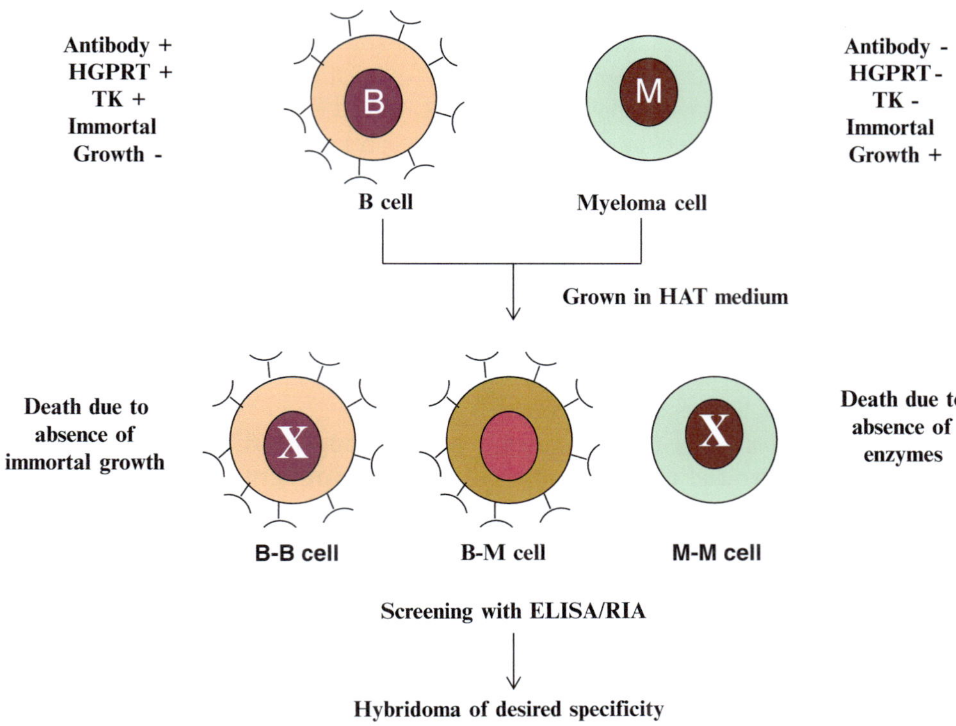

Fig. 9.3 Hybridoma technology. Antibody-secreting B cells that are positive for enzymes (HGPRT and TK) are fused with myeloma cells that have immortal growth. After fusion, the cells are selected on HAT medium. Only B-M fusion cells would survive in presence of aminopterin (HAT) present in the HAT medium. These cells are further screened for desirable specificity of the antibody produced by ELISA or RIA and are used for large-scale production of antibody

of MAbs diagnostic reagent tests against various pathogenic agents.

The immuno-diagnoses of protozoan and parasitic diseases have significantly been improved by MAb technology because the tests involving MAb as diagnostic reagents overcome the limitations of polyclonal antibodies. MAbs were found to be extremely useful in the rapid outbreak of East Coast fever (ECF). MAbs of diagnostic value have also been developed against *Trichomonas vaginalis*, *Leishmania donovani*, *Trypanosoma congolense*, and *Babesia bovis*. Development of monoclonal antibodies for the detection of *Mycoplasma pneumonia* and plum pox virus has been reported.

9.3 Nucleic Acid-Mediated Tests

9.3.1 PCR and Array-Based Techniques in Diagnosis

PCR is the most well-developed molecular technique that has not only been successfully applied for several wide-ranged clinical diagnoses but also has great potential for clinical applications, including specific or broad-spectrum pathogen detection, evaluation of emerging novel infections, surveillance, early detection of biothreat agents, and antimicrobial resistance profiling. PCR-based methods may also be cost-effective relative to traditional testing procedures. Further advancement of technology is needed to improve automation, optimize detection sensitivity and specificity, and expand the capacity to detect multiple targets simultaneously (multiplexing). PCR is the most sensitive and rapid method of detecting pathogens in clinical samples. It is very useful as some of the microorganisms are not easily culturable in vitro or has a very long incubation time. Under these conditions, the diagnostic value of PCR is very important [12].

Traditional PCR procedure includes amplification of specific genes (Fig. 9.4) of the microorganisms and running the product on a gel. The presence of a microbe is confirmed by the presence of a band of appropriate size. Nested, multiplexed, and real-time PCR (RT-PCR) are used for efficiency and quantitation.

Multiplexed PCR allows the detection of multiple sequences in the same reaction tube proving useful in the diagnosis of several infections simultaneously (Fig. 9.5).

RT-PCR system, unlike conventional PCR, allows for the quantification of the original template's concentration through the use of various fluorescent dyes and primers. The concentration is measured through comparison to standard curves. This eliminates the need to visualize the amplicons by gel electrophoresis, thereby greatly reducing the time, risk of contamination, and the introduction of false-positives.

PCR is used to diagnose the presence of several opportunistic pathogens in the cerebrospinal fluid of HIV patients or multiple sclerosis patients [2, 11]. The viral infections that can be determined by this method are *Herpes simplex virus* (type 1 and 2), *Varicella zoster virus*, *Cytomegalovirus*, *Epstein–Barr virus*, and *Japanese encephalitis* virus. Bacterial infection such as *Chlamydia pneumoniae* is also identified. *Mycoplasma* sp. is very difficult to cultivate in laboratory; hence, PCR method is the only reliable method to identify the presence of the samples [8]. DNA probes consisting of cloned ribosomal RNA genes, cDNA to mycoplasmal rRNA, synthetic 16S rRNA oligonucleotide sequences, or cloned mycoplasmal protein genes have been developed and applied as diagnostic tools in a variety of human and animal mycoplasma infections.

9.3.2 Loop-Mediated Isothermal Amplification (LAMP)

Is a unique amplification method with extremely high specificity and sensitivity able to discriminate between a single nucleotide differences. It is characterized by the use of four different primers specifically designed to recognize six distinct regions on a target gene, with amplification only occurring if all primers bind and form a product (Fig. 9.6). The reaction occurs at a constant temperature using strand displacement activity of

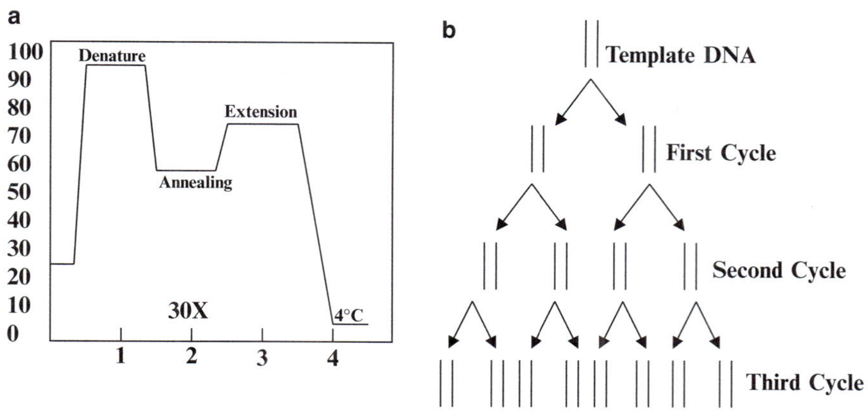

Fig. 9.4 Polymerase chain reaction (PCR). (**a**) shows the PCR cycle where the DNA sequence is amplified using appropriate primers and temperature conditions (denaturation, annealing, and extension). After each cycle, the DNA amplification is shown as in (**b**)

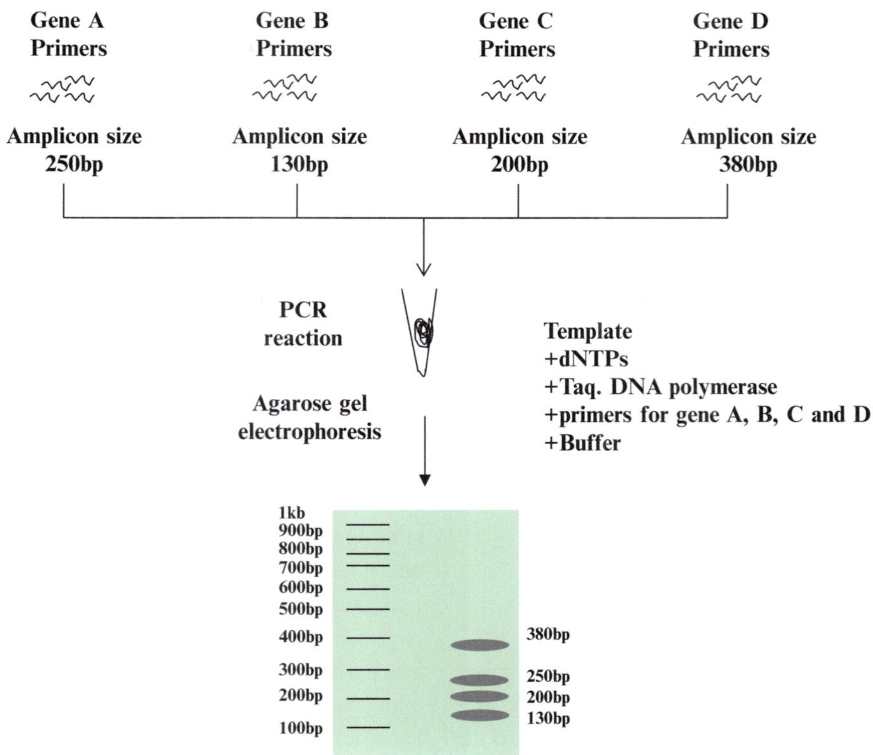

Fig. 9.5 The figure explains multiplex PCR reaction technique. Multiple PCR reactions can be performed in the same tube when the product size of different target amplicons are substantially different from each other and the reaction conditions for all the PCR are similar. The primers for gene A to D, specific for different pathogenic agents, are put together in the same tube and the PCR products are analyzed by agarose gel electrophoresis showing different sized bands

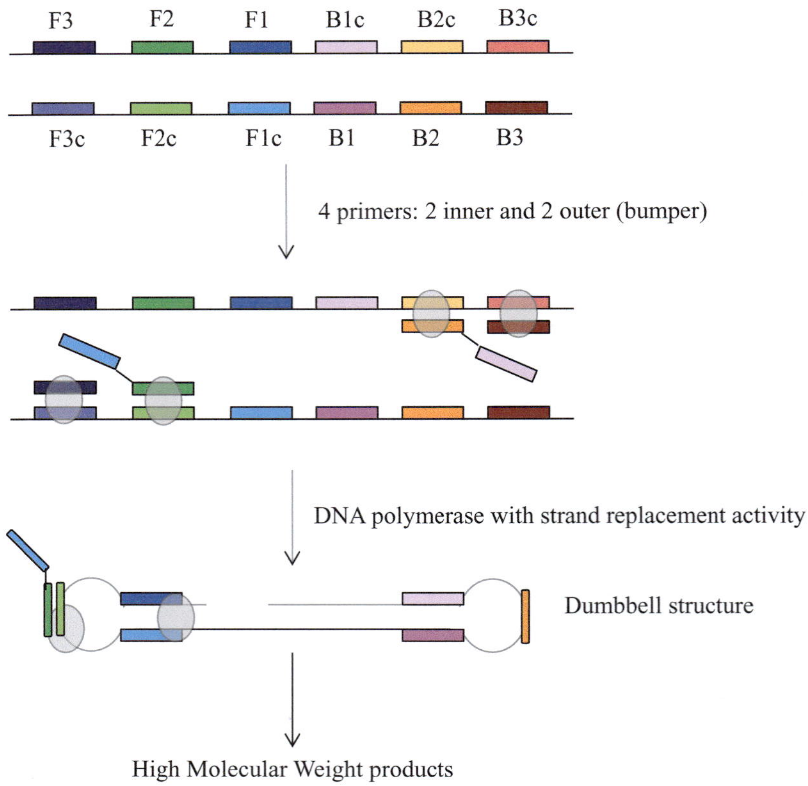

Fig. 9.6 LAMP PCR. In this technique, DNA polymerase with strand displacement property is used to produce single-stranded loop-like templates. 4 primers are used: 2 inner primers and 2 outer primers or bumper primers. The dumbbell-shaped intermediates formed increase the amplification efficiency and the entire reaction takes place at 65° centigrade

DNA polymerase [10]. Amplification and detection takes place in a single step at a constant temperature (65°). It does not require expensive thermo cyclers. The corresponding release of pyrophosphate causes turbidity that is detected visually. Sometimes DNA-intercalating dye is also used. This has been applied for rapid detection of several DNA and RNA viruses such as West Nile and SARS virus. It has also been used for the identification of several parasites.

9.3.3 Luminex xMap Technology

Molecular-based approaches based on nucleic acids offer greater sensitivity and specificity over the existing diagnostic tests. They permit the detection of infections from very low titer samples including those from asymptomatic patients. Luminex technology is a bead-based flow-cytometric assay that allows the detection of various targets simultaneously. The microsphere beads can be covalently bound to antigens, antibodies, or oligonucleotides that will serve as probes in the assay. Up to 100 microspheres are available, each emitting unique fluorescent signals when excited by laser, therefore allowing the identification of different targets. This method has been successfully used for detecting *Cryptosporidium* species. *C. hominis* and *C. parvum* has a single nucleotide difference in the microsatellite-2 region (ML-2) that can be identified only by sequencing which is very time consuming and labor intensive. They can be detected and distinguished by this technology.

However, there are several drawbacks of these methods regarding clinical samples, as PCR is susceptible to inhibitors, contamination, and experimental conditions. The sensitivity and specificity of a PCR assay is dependent on target genes, primer sequences, PCR techniques, DNA extraction procedures, and PCR product detection methods. These might not be optimal in clinical specimens such as blood, urine, sputum, cerebrospinal fluid (CSF), and others. The PCR conditions need to be carefully evaluated and the results confirmed microbiologically. PCR is used for the diagnosis of *HIV-1*, *Hepatitis B and C viruses*, *Human papillomavirus*, *Chlamydia trchomatis*, *Neisseria gonorrhoeae*, *Cytomegalovirus*, *Mycobacterium tuberculosis*, and many others.

9.3.4 Single Nucleotide Polymorphism and Disease Association

Single nucleotide polymorphisms or SNPs (pronounced as snips) are tiny variations in an individual's genetic code. SNPs occur when a single nucleotide (A, T, G, or C) is substituted for another between the members of the same species or between two chromosomes of the same person. When the DNA sequence of a gene differs by only one nucleotide between two individuals, they are called as alleles.

SNP analysis can be done in a single step by using genomic DNA and PCR method (Fig. 9.7). A single SNP analysis can be done by using a specific primer attached to a fluorescence marker, also known as a quenching probe or Q-Probe. When the primer binds with a specific DNA sequence, the fluorescence is quenched due to association with guanine residue. When it dissociates, the fluorescence is acquired. When the primer binds to the wild-type allele, the dissociation occurs at a higher temperature, whereas in a mutant allele, the binding is weak and dissociation takes place at a lower temperature. This change in dissociation curve is analyzed. Two different colors can be used for multiplex analysis.

SNPs occur due to mutation, recombination, and natural selection. SNPs may occur in coding region of genes, in noncoding regions of genes, or in intergenic regions. They are classified into different categories.

9.3.4.1 Synonymous Polymorphism or Silent Mutation

Due to degeneracy of genetic code, the amino acid sequence of the polypeptide might not change. These are known as synonymous polymorphism.

9.3.4.2 Non-synonymous or Replacement Polymorphism

When the changes produce different polypeptides, they are known as non-synonymous or replacement polymorphism. This may result in missense mutation, where a different amino acid is produced, or nonsense mutation, where there is a premature stop codon. Lot of disease mutations are caused by replacement polymorphism.

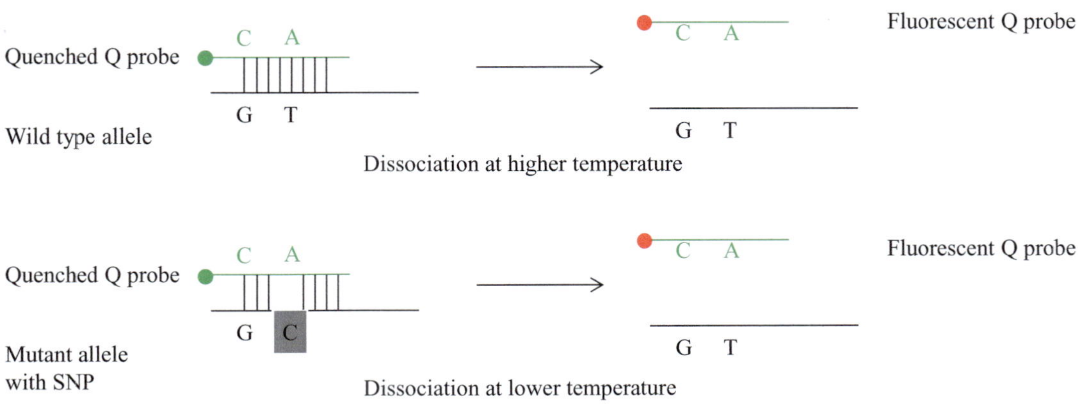

Fig. 9.7 Single nucleotide polymorphism (SNP) analysis. Synthetic oligonucleotide attached to fluorescence marker is incubated with genomic DNA. The fluorescence is quenched due to association to guanine residue. When it dissociates at high temperature, the fluorescence is restored. The dissociation temperature is related to the complementarity of the sequence. The wild type is 100% complementary and hence has a higher dissociation temperature where as the mutant allele with SNP has a low dissociation temperature5 (Source: http://www.aist.go.jp/aist_e/latest_research/2005/20050405/20050405.html)

9.3.4.3 e-SNPs or Expression SNPs

SNPs occurring in noncoding region might affect gene splicing, transcription factor binding, m-RNA degradation, or mutate noncoding RNA.

9.3.4.4 Importance of SNPs

About 99.9 % of DNA sequences are identical between individuals of same species. Out of 0.1 % variation, around 80 % is due to SNPs. Thus they bring about diversity among individuals. This trait is used for DNA fingerprinting in forensic science.

Several diseases are caused by genetic variations in an individual. Genetic factors are responsible for susceptibility and disease progression. SNP profile or haplotype associated with a disease trait may reveal relevant genes associated with a disease state. It provides understanding of many polygenic diseases. In future there are chances that by viewing the SNPs profile of an individual, the physicians might be able to find out the risks associated and plan a personalized medicine. SNPs help in determining the likelihood of a person to develop a particular disease. One of the genes associated with Alzheimer's disease in apolipoprotein E or *ApoE*. It contains two SNPs that result in three possible alleles for this gene: E2, E3, and E4. Each allele differs by one DNA base, and the protein product of each gene differs by one amino acid. Each individual inherits one maternal copy of *ApoE* and one paternal copy of *ApoE*. A person who inherits at least one E4 allele has a greater chance of developing Alzheimer's disease, whereas inheriting the E2 allele reduces the likelihood of developing Alzheimer's. SNPs are not absolute indicators of disease development. *ApoE* is just one gene that has been linked to Alzheimer's. Like most common chronic disorders such as heart disease, diabetes, or cancer, Alzheimer's is a disease that can be caused by variations in several genes. The polygenic nature of these disorders is what makes genetic testing for them so complicated.

9.4 Protein Microarray

Protein microarrays are tools that can be used in both translational as well as basic research. Protein chips can be used for a variety of applications including identification of protein–protein interactions, protein–phospholipid interactions, and substrates for protein kinase. They are used for clinical diagnosis and disease state progression. They can be used to phenotype leukemia cells, identify new protein–protein interactions, screen entire proteomes, and profile hundreds of patient samples. Several arrays are available for specific use. They have been graphically represented in Fig. 9.8. Some of them are discussed here:

9.4.1 Proteomic Arrays

These are high-density arrays and are used to identify novel proteins and protein–protein interactions (Fig. 9.8a). The array library is usually a high-density expression library and the probes are either directly labeled with fluorophores or are tagged with labeled antibodies.

9.4.2 Microspot ELISA and Antibody Arrays

These are used for quantitative profiling of protein expression in clinical samples and cell culture (Fig. 9.8b). These are low-density arrays. Known antibodies are arrayed to capture antibodies from unknown samples. The antigens are either labeled directly or are attached to a secondary antibody. The latter gives a sandwich assay similar to ELISA.

9.4.3 Single-Capture Antibody Arrays

It consists of multiple known antibodies arrayed on a solid surface (Fig. 9.8c). It is used to profile the presence of known antigens from a pool of

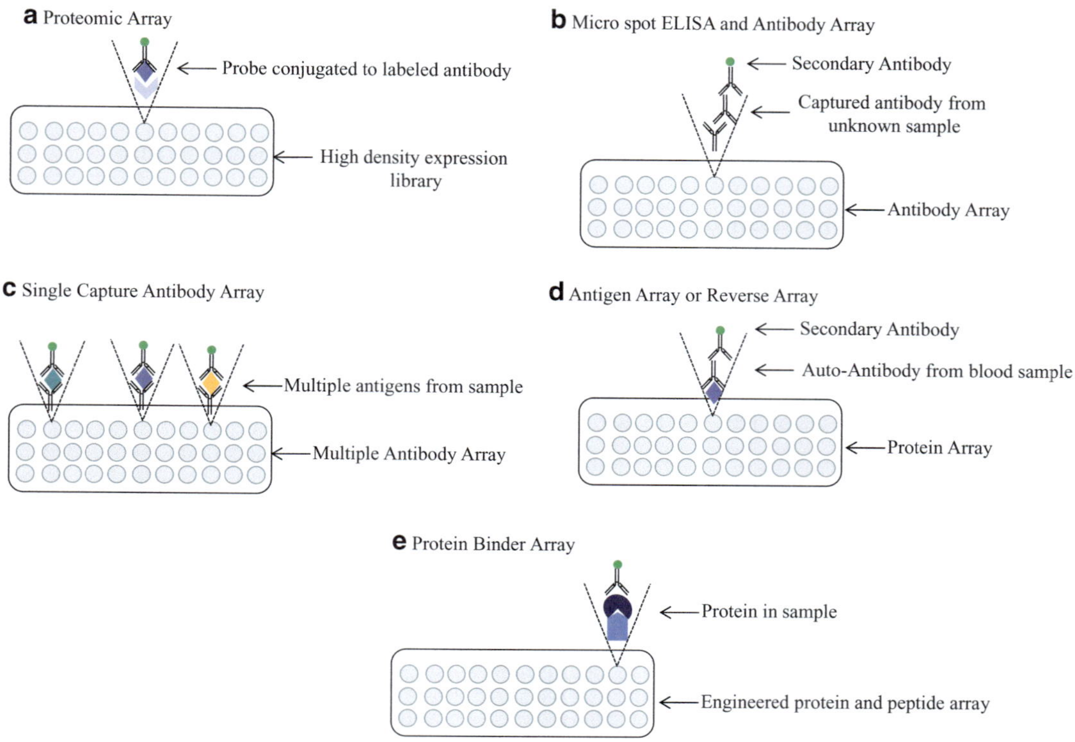

Fig. 9.8 Protein microarray. (**a**) Proteomic array. High-density expression library, probed with samples. This identifies protein–protein interactions. (**b**) Microspot ELISA and antibody arrays. Known antibodies are arrayed to capture antibodies from unknown samples. (**c**) Single-capture antibody arrays. Multiple known antibodies are arrayed on a solid surface, used to profile the presence of known antigens from a pool of samples. (**d**) Antigen arrays or reverse arrays. These are used to detect autoantibodies in samples. These are low-density arrays and are probed with serum or plasma. (**e**) Protein binder assay. Engineered proteins and peptides with various binding motifs are arrayed and are probed with complex samples. Detection with known antibodies helps to identify new binding sites and interactions

samples. Normal and disease samples are used. They are either labeled directly or with haptens.

9.4.4 Antigen Arrays or Reverse Arrays

These are used to detect autoantibodies in clinical and research samples. These are low-density arrays and are probed with serum or plasma (Fig. 9.8d). Reverse arrays are used to probe hundreds of samples to detect the presence of few antibodies. Cell lysates, plasma, and serum are arrayed and are probed with few known antibodies.

9.4.5 Microarray Western

This is an alternative strategy where samples containing several proteins are arrayed on slide and probed with labeled antibodies. Level of number of proteins can be measured simultaneously.

9.4.6 Protein Binder Arrays

This is used to identify novel protein-binding motifs and protein–protein interactions (Fig. 9.8e). Engineered proteins and peptides with various binding motifs are arrayed and are probed with complex samples. Detection with known antibodies helps to identify new binding sites and interactions.

9.5 Isolation, Processing, and Profiling of Proteins and Other Molecules Associated with Disease

Proteomic studies can provide substantial information about clinical state of a disease as they are the final molecular machines of biological processes. They can be used as biomarkers for disease states. Diagnostics use protein and peptide biomarkers from body fluids. All proteomic-based diagnostic efforts seek to identify biomarkers that, alone or in combination, can distinguish between "case" and "control" groups. This can be done in several ways.

9.6 Profiling and Identification of the Protein

9.6.1 Two-Dimensional Gel Electrophoresis

This is a method to identify proteins and peptides in their natural form. Here the proteins are resolved in the first dimension based on pH (a process called isoelectric focusing) and in the second dimension by their molecular weight. This technique is labor intensive.

9.6.2 Mass Spectrometry (MS)

This is an analytical technique where mass-to-charge ratios of particles are measured. It is used to determine the composition of peptides. Proteins from body fluids can be proteolytically cut into small pieces. They are ionized usually to cations by removal of electron. These charged particles are then separated according to their charge and mass. The separated ions are measured and displayed. The resulting spectra can be compared with other peptides in the data base (Fig. 9.9). But in this approach it is difficult to quantitate and study the protein modifications.

9.6.2.1 Matrix-Assisted Laser Desorption Ionization Time-Of-Flight Mass Spectrometry (MALDI-TOF MS)

It is a relatively novel technique in which a coprecipitate of a UV light-absorbing matrix and a biomolecule is irradiated by a nanosecond laser pulse. Most of the laser energy is absorbed by the matrix, which prevents unwanted fragmentation of the biomolecule. The ionized biomolecules are accelerated in an electric field and enter the flight tube. During the flight in this tube, different molecules are separated according to their mass-to-charge ratio and reach the detector at different

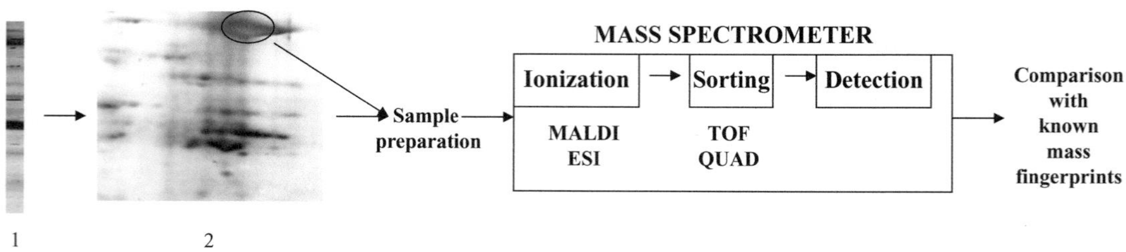

Fig. 9.9 The usage of proteomics approach for diagnostics or profiling. (*1*) First dimensional isoelectric focusing (IEF) gel is used to separate the sample components according to their isoelectric point. (*2*) Second dimensional SDS-PAGE is used which further separates the proteins according to their molecular mass. Sample spots obtained are isolated and prepared for application in mass spectrometer (MS). MS consists of ionization device as MALDI or ESI and mass sorting device as TOF or QUAD and detection is done by a detector. After peptide mass fingerprint is obtained, it is analyzed through comparing the experimentally determined peptide mass fingerprint with known and virtual mass fingerprints using bioinformatics tools

times. In this way each molecule yields a distinct signal. The method is used for detection and characterization of biomolecules, such as proteins, peptides, oligosaccharides, and oligonucleotides, with molecular masses between 400 and 350,000 Da. MALDI-TOF is used for identifying bacterial strains in clinical microbiology laboratories.

The development of automated, high-throughput proteomic technologies such as MS and MALDI-TOF has enabled large numbers of clinical samples to be analyzed simultaneously in a short time. These platforms have made "population-based proteomics" feasible for the first time.

9.7 Nucleic Acid Amplification Technologies (NAAT)

With the use of NAAT, it is now possible to have many copies of target DNA and the technique has advantage of being sensitive, specific, and rapid. It targets the conserve region of the target species. The NAAT test may be planned which would be able to detect single species, strain, or resistance-inducing mutation. Using broad-spectrum probes, the broad categories of the organism may be detected. NAAT has been successfully used in the diagnosis of infective endocarditis as compared to culture technique even when culture reports were negative. In patients with negative sputum smears, the tests based upon NAAT were quiet useful in the clinical diagnosis of tuberculosis.

9.8 Ethics in Molecular Diagnosis

Genetic factors contribute to the risk of several diseases. Genetic testing clarifies the risk of a person suffering from a disease or passing it down to the next generation. However no genetic test can give an absolute answer. It is always percent chance to inherit a disorder. Genetic testing is a rapidly emerging field. There are several disorders that have adult onset and worsen over time. Hence is it advisable for individuals with high risk to undergo genetic testing and counseling. Tests are available for several genetic disorders and cancers. Some of them are sickle cell anemia, Down's syndrome, Huntington's disease, cystic fibrosis, breast cancer, and phenylketonuria. There are three forms of genetic testing. They are diagnostic testing, carrier testing, and predictive testing. Diagnostic testing involves identification of a current disease state. This includes prenatal and newborn genetic testing. Carrier testing includes if an individual carries a particular genetic trait that he or she can pass on to the next generation. Predictive testing determines if a person carries a mutation that can have a late onset of a disorder.

However, genetic testing gives rise to several ethical issues [3]. It might cause potential discrimination regarding social acceptability, job or employment availability, and health insurance coverage. Prenatal testing for genetic disorder may lead to abortion of a fetus. Carriers of genetic mutations ethically should disclose the fact to their life partner or their siblings. But he or she might face social isolation. He or she might not be able to marry and start a family. Similarly if a person is at risk of a late onset of a genetic disorder, the employer might not be willing to hire him or her. The health insurance companies would not want to pay for the medical expenses or might increase the premium [5, 6].

One should also keep in mind that genetic testing cannot give all the answers. For example, it cannot tell about the exact time of onset, penetrance, or person to person variation of a disorder. There are several issues regarding the ethical consideration of genetic testing. Until and unless there are clear laws to protect the individuals, privacy and confidentiality of genetic information should always be protected and individuals wish to be tested or not should be respected [1].

9.9 Chapter End Summary

- Biotechnology has played a very important role in diagnosis and treatment of various bacterial, fungal, viral, and parasitic diseases. It has also helped in identification of early stages of cancer.

- The advancement in molecular techniques has helped in identification of biomarkers that signifies early development and progress of a disease.
- The various tests like serology-based tests and nucleic acid-based tests are diagnostics, but preliminary data from traditional microbiology-based methods are also helpful.
- The serology-based tests may be done either by targeting antigen or antibody and gives an indication about the current state of the disease.
- The tests based upon DNA detection can indicate the presence or absence of the target pathogenic agent but sometimes are unable to predict about the current status of infection.
- The usage and development of monoclonal antibodies have made tremendous advances in biomedicine.
- The usage of nucleic acid-based methods has helped in identifying the presence of certain microorganisms in samples which were earlier believed to be noninfectious.
- The profiling using proteomics tools supplemented with mass spectrometry have helped to characterize many pathogen-specific factors which might be of interest in diagnostics and therapeutics.

Multiple Choice Questions

1. The commonly used method for diagnosis of bacterial infection is:
   (a) Gram staining
   (b) Antibiotic sensitivity
   (c) Biochemical test
   (d) All of the above
2. In indirect ELISA the primary antibody is:
   (a) Enzyme labeled
   (b) Against constant region of antibody
   (c) Fluorescently labeled
   (d) Against antigenic epitope
3. In reverse western technique:
   (a) The sample is affixed on slide.
   (b) The targets are fixed on slide.
   (c) Both the sample and targets are fixed on slide.
   (d) None of them is fixed on slide.
4. PCR is used for detection of:
   (a) Infection with high titer
   (b) Presence of infectious agent
   (c) Antibody
   (d) Antigen
5. Monoclonal antibodies are used in diagnostics because:
   (a) They can be manufactured by hybridoma.
   (b) They can be prepared by plasma cells of the body.
   (c) They are against a specific target.
   (d) None of the above.
6. There is a requirement of engineering of antibody because:
   (a) Engineering makes the antibody more effective.
   (b) Engineering reduces the immune reactions of therapeutic antibody.
   (c) Engineering makes it more specific for its target.
   (d) All of these.
7. The positive PCR reaction specific for a pathogen indicates:
   (a) The pathogen is active.
   (b) The pathogen is not active.
   (c) The presence of the pathogen.
   (d) All of the above.
8. The purpose of a multiplex PCR is:
   (a) Determination of molecular size of the amplicon
   (b) Determination of the activity of the pathogens
   (c) Determination of the presence of a number of pathogenic nucleic acid
   (d) All of the above
9. A specific sequence analyzed by PCR-RFLP in two individuals showed the presence of one band in one and two bands in the other, because of:
   (a) Presence of microsatellite

(b) Presence of satellite
(c) Presence of SNP
(d) Presence of all of them

10. The proteomics studies have become feasible due to:
(a) PCR reaction
(b) RT-PCR
(c) Mass spectrometry
(d) DNA microarray

Answers

1. (a); 2. (d); 3. (b); 4. (b); 5. (c); 6. (b); 7. (c); 8. (c); 9. (c); 10. (c)

Review Questions

Q1. How has the advancement in biotechnological techniques helped in diagnosis of the diseases?
Q2. Discuss a few serological tests for the diagnosis.
Q3. What is the importance of PCR in pathogen detection?
Q4. How are the proteomic assays helpful in aiding diagnostics?
Q5. Discuss DNA microarray technology.
Q6. What is MALDI-TOF?

References

1. Azariah J, Azariah H, Macer DRJ (1997) Bioethics in India, proceedings of the international bioethics workshop in Madras: biomanagement of biogeoresources, University of Madras; Editors:, Copyright Eubios Ethics Institute 1997
2. Cinque P et al (1996) Polymerase chain reaction on cerebrospinal fluid for diagnosis of virus-associated opportunistic diseases of the central nervous system in HIV-infected patients. AIDS 10:951–958
3. Lea DH, Williams J, Donahue MP (2005) Ethical issues in genetic testing. J Midwifery Womens Health 50:234–240
4. Ndao M (2009) Diagnosis of parasitic diseases: old and new approaches. Interdisciplinary perspectives on infectious diseases. Article ID 278246
5. Norrgard K (2008) Diagnostic testing and the ethics of patenting DNA. Nat Educ 1
6. Norrgard K (2008b) Ethics of genetic testing: medical insurance and genetic discrimination. Nat Educ 1(1):82
7. Saleem M, Kamal M (2008) Monoclonal antibodies in clinical diagnosis: a brief review application. Afr J Biotechnol 7:923–925
8. Shmuel R (1994) DNA probes and PCR in diagnosis of mycoplasma infections. Mol Cell Probes 8:497–511
9. Siddiqui MZ (2010) Monoclonal antibodies as diagnostics; an appraisal. Indian J Pharm Sci 72:12–17
10. Tomita N, Mori Y, Kanda H, Notomi T (2008) Loop-mediated isothermal amplification (LAMP) of gene sequences and simple visual detection of products. Nat Protoc 3:877–882
11. Yamamoto Y (2002) PCR in diagnosis of infection: detection of bacteria in cerebrospinal fluids. Clin Diagn Lab Immunol 9:508–514
12. Yang S, Rothman RE (2004) PCR-based diagnostics for infectious diseases: uses, limitations, and future applications in acute-care settings. Lancet Infect Dis 4:337–348

Some Selected Resources

http://en.wikipedia.org/wiki/Monoclonal_antibodies
http://en.wikipedia.org/wiki/Single-nucleotide_polymorphism
http://eubios.info/index.html
http://www.eubios.info/india/BII5.HTM
http://www.differencebetween.com/difference-between-elisa-and-vs-western-blot/
http://www.nature.com/scitable/topicpage/diagnostic-testing-and-the-ethics-of-patenting-709
http://www.nature.com/scitable/topicpage/ethics-of-genetic-testing-medical-insurance-and-651
http://www.nature.com/scitable/topicpage/human-testing-the-eugenics-movement-and-irbs-724
http://www.whatman.com/ProteinMicroarrays.aspx

10 Diagnosis of Specific Diseases

Abstract

There have been great advancements in diagnosis of infectious diseases such as tuberculosis, malaria, and AIDS in the past few decades. Specific techniques have been developed that can efficiently identify the infection at various stages of the disease progression and thus can increase the chances of survival of the patients by appropriate treatment. A number of specifically targeted diagnostic methods and safe therapeutic remedies are being explored for cancer. In this chapter, we have tried to summarize most recent and regularly used methods, important advancements, and the relevant concepts of various tools in context to diagnosis of cancer, tuberculosis, malaria, and AIDS.

10.1 Introduction

There have been great advancements in diagnosis of infectious diseases such as tuberculosis, malaria, and AIDS in the past few decades. Specific techniques have been developed that can efficiently identify the infection at various stages of the disease progression and thus can increase the chances of survival of the patients by appropriate treatment. A number of diagnostic methods and therapeutic remedies are being exploited for cancer that are safer and specifically targeted for its treatment. In this chapter we have tried to include the most recent and regularly used methods to diagnose these diseases with precision.

Immunotherapy is an emerging field that has immense possibility in cancer treatment. We have tried to summarize the important advancements and the relevant concepts of immunological tools in context to cancer diagnosis and treatment.

10.2 Cancer

The uncontrollable division of the cells results in the formation of tumor, and the tumor is called cancer when abnormal cells divide without control and are able to invade other tissues. There are more than 100 different types of cancer. Cancer is a disease caused by many genetic changes that develop over time. Any tissue of the body can develop cancer, but each type has its own unique features despite of the fact that the basic processes that produce cancer are quite similar in all forms of the disease.

Cancer begins when a cell bypasses normal restraints on cell division and begins to follow its

V. Gupta et al., *Basic and Applied Aspects of Biotechnology*, DOI 10.1007/978-981-10-0875-7_10

own agenda for proliferation. All the descendant cells produced by division of abnormal cell also display inappropriate proliferation. A tumor, or mass of cells, produced as a result of proliferation of these abnormal cells may remain within the tissue in which it originated (a condition called in situ cancer) and is called benign tumor. However, one of the central characteristics of a cancer cell is its ability to invade nearby tissue and spread to other parts of the body. Cancers can spread throughout the body by two mechanisms: invasion and metastasis.

Invasion is direct migration and penetration by cancer cells into neighboring tissues. Metastasis is the ability of cancer cells to penetrate into lymphatic and blood vessels, circulate through the bloodstream, and then invade normal tissues elsewhere in the body.

Malignant tumors are capable of spreading by invasion and metastasis, and, by definition, the term "cancer" applies only to malignant tumors. Tumors threaten an individual's life when their growth disrupts the tissues and organs needed for survival.

10.2.1 Mystery of Cancer

People have likely wondered about the cause of cancer for centuries. Its name derives from an observation by Hippocrates more than 2,300 years ago that the long, distended veins that radiate out from some breast tumors look like the limbs of a crab. From that observation came the term karkinoma in Greek and later, cancer in Latin.

10.2.2 Cancer: A Multistep Process

There are many factors which result in the onset of cancer. One of them is oncogenes. In our body many proto-oncogenes are present and they code for proteins involved in molecular pathways that receive and process growth-stimulating signals from other cells in a tissue. Thus the signaling begins with the production of a growth factor responsible for stimulation of cell division. Growth factors move through the spaces between cells and attach to specific receptors located on the surfaces of neighboring cells. Binding triggers a stimulatory signal to proteins in the cytoplasm via signaling intermediates which ultimately activate a set of transcription factors which in turn switch the specific genes on that help move the cell through its cycle.

In multicellular animals, cell division is under strict control of cell cycle checkpoints. The cell cycle takes nearly 20–30 h. It is divided into the preparatory interphase and mitotic or M-phase. Interphase is divided into the G1 phase (in this cell growth, centrosome duplication occurs), S-phase (DNA synthesis), and G2 phase (synthesis of all the factors required for mitosis). In M-phase, mitosis and cytokinesis occur. Transitions between different phases are regulated by different cyclin-dependent kinases (Cdks). Cdks bound to cyclin proteins are active. Different kinds of cyclins are synthesized and degraded at specific points in the cell cycle.

In mammalian cell, the retinoblastoma protein (pRb) and p53 proteins suppress cell division by arresting it in G1 stage. During G1, a regulatory factor E2F controls the synthesis of many proteins required for S-phase. The factor E2F is suppressed at G1 stage when negative regulator pRb is bound to it.

The extracellular signals or mitogens regulate cell division by overcoming intracellular braking system and promote cells through cell cycle. Therefore as G1 progresses, regulatory proteins as cyclinD-Cdk4 and cyclinE-Cdk2 accumulate and phosphorylate pRb. Phosphorylated pRb has very low affinity for E2F; thus E2F becomes free. Free E2F promotes cell to synthesize all factors required for S-phase.

The presence of mitogens increases the rate of cell division by:

- Binds to receptor tyrosine kinase (RTKs) and acts through Ras (small GTPase) and mitogen-activated protein (MAP) kinase.
- Activates transcription factors as Myc, which promotes production of E2F.
- Activation phosphorylates Rb, whose dissociation from E2F promotes cell division.

In order to convert a normal cell into tumor cell, six to seven successive mutations are required. However, occurrence of successive mutations to help develop a tumor and cancer is a rare possibility. Mutated stem cells may be an important target as tumor precursor cells have stem cell-like properties. The genes which are targets for these mutations are oncogenes and/or tumor suppressor genes.

10.2.2.1 Oncogenes

These genes promote normal activity of cell proliferation. Mutations in these make them inappropriately active. A single mutation may make them defective and alter cell behavior.

The mutations in the oncogenes may occur which are capable of preventing all brakes to stop uncontrolled proliferation. Else mutations in tumor suppressor genes may occur which fail to check uncontrolled division and let the cells proliferate indefinitely.

The development of cancer was attributed to oncogenes. These oncogenes were copies of the cellular genes which are present in normal cells called as "proto-oncogenes." These probably accidentally became part of the retroviral genome. Here some of the cellular proto-oncogenes and their viral oncogenes are discussed in Table 10.1.

These proto-oncogenes code for secreted growth factors or cell surface receptors or intracellular signaling components or transcription factors or components of cyclins. Mutations in these result in their becoming oncogenes. They may be activated due to:

- Amplification: Cells may contain multiple copies of normal oncogenes.
- Point mutations: Ras family genes with point mutation in BRAF account for 80 % of malignant melanoma.
- Chromosomal rearrangement: Philadelphia (Ph') chromosome, a small acrocentric chromosome observed in 90 % of patients with chronic myeloid leukemia.
- Translocation: Translocation into a transcriptionally active chromatin region, for example, translocation of MYC close to heavy chain locus of immunoglobulin in Burkitt lymphoma, drives high levels of MYC expression in B cells.

A few human proto-oncogenes that have been altered or mutated in such a way that they promote cell growth in an abnormal or uncontrolled fashion are:

- Proto-oncogene bcl-2 (B-cell lymphoma)
- Proto-oncogene HER2/neu (erbB2) (breast and ovarian cancers)
- Proto-oncogene c-Src (colorectal cancers)
- Proto-oncogene c-Myc (Burkitt lymphoma)
- Tumor suppressor gene BRCA1, BRCA2 (breast and ovarian cancers)
- Tumor suppressor gene p53 (brain tumors; skin, lung, head, and neck cancers)
- Tumor suppressor gene RB (retinoblastoma; bone, bladder, and breast cancers)
- Tumor suppressor gene APC (colorectal cancers)

Table 10.1 Viral oncogenes and cellular protooncogenes for several protein products

Viral oncogenes	Proto-oncogenes	Function	Diseases
v-sis	PDGF B	Platelet-derived growth factor B subunit	Simian sarcoma
v-erbb	EGFR	Epidermal growth factor receptor	Chicken erythroleukemia
v-fms	M-CSFIR	Macrophage colony stimulating factor receptor	McDonough feline sarcoma
v-ras	HRAS	Receptor tyrosine kinase	Harvey rat sarcoma
v-abl	ABL1	Protein tyrosine kinase	Abelson mouse leukemia
v-jun	JUN	AP-1 transcription factor	Avian sarcoma
c-myc	MYC	DNA-binding transcription factor	Avian myelocytomatosis
v-fas	FOS	DNA-binding transcription factor	Mouse osteosarcoma

These are present in retroviruses with normal copies in human genome in the form of protooncogenes

10.2.3 Tumor Suppressor Genes

The cell's ability to attain the capability of ignoring the inhibitory signals for controlled growth and cell division is the cause of cancer. The inhibitory signals normally counterbalance the growth-stimulating pathways. In normal cells, inhibitory messages flow to a cell's nucleus much like stimulatory messages, but when this flow is interrupted, the cell can ignore these normally powerful inhibitory signals.

Apart from controlling proliferation, cells have other systems also that can help them avoid break-free cell division: (1) the DNA repair system, (2) apoptotic cell death, and (3) cell division control.

- DNA repair system operates in every cell of the body, where repair enzyme detects and corrects errors in DNA.
- Apoptotic cell death or system prompting a cell to "commit suicide" is initiated if some essential component is damaged or its control system is deregulated.
- Control on cell cycle ensures that there should be limits on the number of times a cell can divide; thus a cell cannot reproduce endlessly. This system is governed by a counting mechanism that involves the DNA segments at the ends of chromosomes called telomeres. Telomeres shorten every time a chromosome replicates. Once the telomeres are shorter than a certain threshold length, they trigger an internal signal that causes the cell to stop dividing. If the cells continue dividing, the telomeres can be completely lost resulting in damage to DNA adjacent to them. Because DNA ends lacking telomeres are also recognized as inappropriate DNA breaks, the cell repair mechanisms can fuse chromosomes together, a genetic crisis that is inevitably fatal to the cell.

Tumor suppressor genes limit normal cell proliferation. They act to prevent abnormal cell cycle, promote apoptotic cell death when they sense abnormality in the target cell, and maintain genome stability by ensuring accurate replication, repair, and segregation of cell's DNA.

In cancer both the alleles of tumor suppressor genes need to be mutated. The important role of these genes is to maintain normal behavior of the cells. The changes occurring in tumor suppressor genes accounting for tumor may be:

- Silencing of gene by deletion or mutation (loss of heterozygosity): The cancers which are advanced show loss of heterozygosity in nearly one-fourth of the genetic loci specific for cancer, for example, tumor suppressor gene MSH2 may be silenced by mutation; BRCA1 is inactivated in 10–15 % of sporadic breast cancer.
- Point mutations
- Methylation of the promoter: Many genes in our body are under the control of promoters with CpG islands. These promoters are normally unmethylated, but in tumor cells, the methylation of CpG islands of the promoters of certain genes is observed, for example, in tumor suppressor genes MLH1, RASSFIA, HIC1, methylation is the common mode of blocking gene expression.

Progression of cells through the cell cycle is controlled by cyclins and cyclin-dependent kinases. These are regulated at series of checkpoints. Levels of cyclins are important for activation of cyclin-dependent kinases. Certain checkpoints are:

- *G1/S checkpoint*: Controlled by Cdk2/cyclin E. Damage in the DNA prevents progression to further cell cycle stages. Irreparable damage may lead to apoptotic cell death.
- *G2/M checkpoints*: Controlled by Cdk1/cyclin B after activation by the phosphatase Cdc25C.
- *Spindle checkpoints*: Ensures correct separation of chromatids at anaphase. It is mediated by anaphase-promoting complex (APC) or cyclosome. In defective signaling unequal chromatid distribution in two daughter cells destabilizes the genome. APC mutations have been reported in familial and sporadic colon cancer.

Somatic mutations in tumor suppressor genes RB1, TP53, and CDKN2A are the most common changes in tumor cells.

- *pRb*: Retinoblastoma (RB1) gene controls cell cycle. Its gene product pRb is 110kD nuclear protein which binds and inactivates the cellular transcription factor E2F. E2F, when free (phosphorylated pRb is not efficient in binding E2F), stimulates gene expression required for cell cycle progression into S-phase. Viral oncoproteins as adenovirus E1A, SV40-T antigen, and HPV E7 proteins either sequester or degrade pRb.
- *P53*: It is encoded by TP53 and is referred as guardian of the genome. When p53 is nonfunctional or absent, the cells continue to replicate despite DNA damage. Some cellular signals or damaged DNA leads to phosphorylation of p53. Phosphorylated p53 levels are increased with resultant increase in transcription of p53-dependent genes as p21 (an inhibitor of Cdk2), PUMA, BAX, and NOXA, which control apoptosis.
  - *TP53* is one of the major targets of mutation and loss of only this may lead to cancer. P63 and p73 are two relatives of p53. But TP53 loss is critical for onset of cancer. Mutations in TP53 are found in families with dominantly inherited Li-Fraumeni syndrome where affected people suffer from primary tumors; sarcomas; osteosarcomas; tumors of adrenal cortex, breast, and brain; and leukemia.
  - *Mdm2* is a protein which ubiquitylates p53 and targets it for degradation. MDM2 is an oncogene and target of p53. It is amplified in many sarcomas.
- *CDKN2A*: It uses alternative promoters and exons to encode two structurally unrelated proteins.
  - $P16^{INKGA}$ protein: This is the inhibitor of Cdk4/6 and thus maintains pRb in its dephosphorylated and active state and prevents E2F activation.
  - $P14^{ARF}$: This mediates arrest of G1 by destabilizing Mdm2 (which keeps p53 levels low). Loss of p14 ARF leads to high Mdm2 levels and subsequent loss of control on cell cycle.

  In promotion of tumorigenesis, homozygous deletions of CDKN2A gene are common.

10.2.4 Development of Cancer

Cancer does not develop suddenly due to massive alterations in cellular functions resulting from a mutation in one or two genes. But it often develops step-by-step, over time, due to the accumulation of many molecular changes leading to characteristics that eventually produce the malignant state.

The number of cell divisions is high and the time frame involved also may be very long. Time is long but uncertain to accumulate enough mutations to reach a malignant state, and the rates of tumor growth can vary; for example, it can take years for the tumors to be detectable.

10.2.5 The Hallmarks of Cancer

The hallmarks of cancer comprise six biological capabilities acquired during the multistep development of human tumors. The hallmarks constitute an organizing principle for rationalizing the complexities of neoplastic disease:

- Sustaining proliferative signaling
- Evading growth suppressors
- Resisting cell death
- Enabling replicative immortality
- Inducing angiogenesis
- Activating invasion and metastasis

These are due to instability of the genome which generates the genetic diversity that augments their acquisition and inflammation, which gives rise to multiple hallmark functions. Conceptual progress in the last decade has added two emerging hallmarks of potential generality to this list:

- Reprogramming of energy metabolism
- Evading immune destruction

Thus unregulated growth of cells that do not invade other tissues is called benign tumor. Cancer on the other hand is a malignant tumor that is characterized by (1) unregulated cell growth and (2) invasion to other tissues, a process called metastasis. Cancer of epithelial tissues is called carcinoma, non-epithelial tissue (mesenchymal cells) is called sarcoma, and that of leukocytes and lymphocytes are known as leukemia and lymphoma, respectively.

Cancer has strong genetic predisposition and occurs due to mutation in several genes involved in regulation of cell proliferation, survival, DNA repair, angiogenesis, and other mechanisms. Some people are at a greater risk of developing certain types of cancer due to mutation or alteration in the genes that are responsible for developing cancer. These genes are known as susceptibility or risk markers for cancer. The presence of these risk markers indicates that the carrier is likely to develop the disease at some point of his life. With the advancement of technology, there has been tremendous development in the understanding of various kinds of cancer [15]. There is better understanding of tumorigenesis and its progression. Though there is advancement in early detection of cancer, novel therapeutics are required for the prevention of cancer-related death.

In the process of cancer (see Fig. 10.1):

1. DNA damage repair pathways are abnormal.
2. Apoptotic cell death is prevented.
3. Tumors exist under low oxygen tension (hypoxia) and nutrient deprivation.
4. Differentiation is prevented.
5. Cancerous cells have defective cell cycle checkpoints.

10.3 Diagnosis of Cancer

There are a wide number of tests to diagnose cancer. As the mechanism of the disease is getting clearer, new diagnostic, prognostic, and therapeutic tools are gradually being developed. The important tests include the following.

10.3.1 Staging of Cancer

The stage of the cancer means the severity of an individual's cancer based upon the size and the extent (intracompartmental/extracompartmental/nodal) of its spread (metastasis) to other tissues. The staging is very helpful in the choice of treatment and prediction of prognosis. The staging consists of:

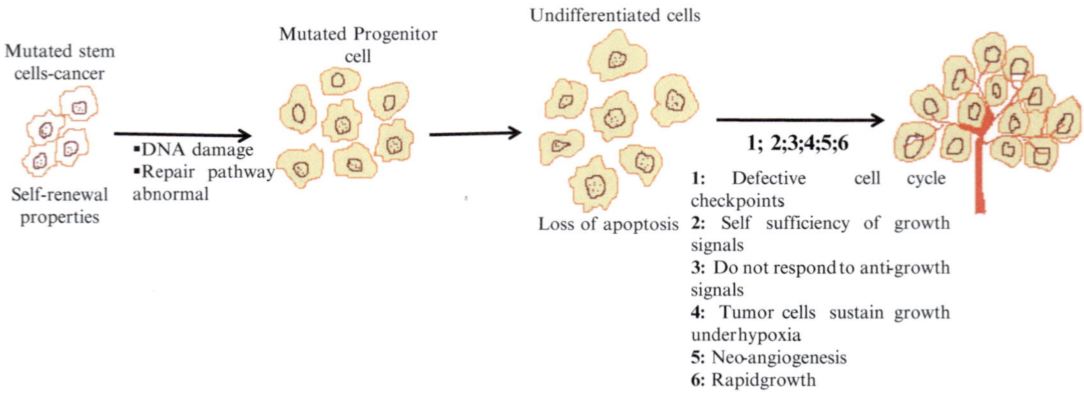

Fig. 10.1 The stem cell cancer. The cells have altered morphology and are self-replicating, and they give rise to mutated progenitor cell as DNA damage repair pathways are abnormal. The cell loses the property of apoptotic cell death control and contact inhibition and does not respond to anti-growth signals or growth factors. Cells divide and grow rapidly under hypoxic condition with angiogenesis

- Determination of the site of the primary tumors and the cell types involved
- Size and its spread to nearby tissue
- Number of primary and metastatic tumors
- Grade of tumor that shows the resemblance of cancerous cells to normal cells

One of the staging systems is the TNM system that has been accepted and is based upon the size and extent (intracompartmental or extracompartmental or nodal) of primary tumor (T), amount of spread to lymph nodes (N), and presence of metastasis (M). Stage O indicates carcinoma in situ; stages I, II, and III indicate a higher number of large-sized tumor with extensive damage; stage IV indicates spread to distant tissues and organs.

Various staging methods are being followed according to the type of cancer. Many other tests like physical examination, imaging, and laboratory tests are utilized to predict stages of cancer [2].

10.3.2 Computed Tomography (CT)

It is also known as computerized tomography or computerized axial tomography (CAT). It is derived from the Greek words *tomos* (to cut or slice or section) and *graphein* (means to write or record). This is an imaging procedure which utilizes special X-ray [1, 21] to create detailed pictures or scans of the internal body parts [12, 31, 32]. The picture in CT procedure shows the organs, bones, and other tissues in thin "slice" of the internal body. Helical CT is a modified version of CT that produces better 3D pictures of areas inside the body and may detect subtle abnormalities [26].

CT is used in many other disorders like circulatory and coronary diseases. In the procedure sometimes contrast agent or dye is used such as iodine or barium (after allergy test) resulting in better images. The scan helps clinicians to detect abnormal growth. CT is used in cancer in many different ways:

- To detect abnormal growth
- To help diagnose the presence of a tumor
- To provide information about the stage of a cancer
- To determine exactly where to perform (i.e., guide) a biopsy procedure
- To guide certain local treatments, such as cryotherapy, radiofrequency ablation, and the implantation of radioactive seeds
- To help plan external-beam radiation therapy or surgery
- To determine whether a cancer is responding to treatment
- To detect recurrence of a tumor

10.3.3 Combined PET/CT

This uses two imaging methods, CT and positron emission tomography (PET). Initially CT is performed to create anatomic pictures of the internal body organs followed by PET to create colored pictures showing functional changes (metabolic function) in the tissue. Combined usage of the two gives complete information about tumor's location, growth, or spread which is helpful in diagnosis, prognosis, and treatment.

Magnetic resonance imaging (MRI) is also very helpful in cancer detection. The technique is as effective as PET/CT scan and is with no risk of radiation exposure. It is helpful in soft tissue tumor delineation and used for brain, spine, muscles, connective tissue, and inside bone.

Ultrasonography is useful in detecting small breast cancers that are not seen in mammogram.

10.3.4 Laboratory Test

Laboratory tests are done on blood, urine, or other tissues for the variations in the levels of specific mediators such as enzymes that are implicated in tumor formation.

10.3.5 Mammograms

This is utilized for the detection of asymptomatic breast cancer. It involves two X-ray pictures or images of each breast helping to detect tumors that cannot be felt. It can detect microcalcifications (tiny deposits of calcium) indicating presence of breast cancer [21, 27].

10.3.6 Pap and HPV Testing

Cervical cancer and several other cancers are caused by the infection of oncogenic human papillomavirus (HPV) [25]. The progression of normal cells toward cancer due to HPV is shown in Fig. 10.2. They are sexually transmitted and may cause anal, vaginal, vulvar, and penile cancers [18]. Sometimes they might cause oropharyngeal cancer. HPV infection can be eliminated by the immune system; however persistent infections may lead to severe pre-cancerous lesions. Untreated lesions, with time might lead to cancer.

Pap smear test is a cytology-based screening for cervical cancers. It detects abnormal cells that might develop into cancer. HPV testing relies upon the presence of DNA or RNA from high-risk HPV types in cervical cells [25].

10.3.7 Pathology Report/Biopsy

The pathology report is a document containing the diagnosis determined by examining cells and tissues under a microscope. The report may indicate the size, shape, and appearance of the specimens. The pathology report requires tissue biopsy [5, 8]. The tissue removed during biopsy is sent for examination and histologic sections are examined for cytological changes (biopsy may be done on cells, tissue, or fluid or entire tumor removed surgically). This report is important in cancer diagnosis and staging.

10.3.8 Tumor Grade

The grading of tumor is based upon how abnormal the tumor cells look under the microscope. It gives an indication of how quickly a tumor is likely to grow and spread. If the cells of the tumor and their organization are close to normal tissue, the tumor is called "well-differentiated." These grow and spread at a slower rate than the tumors that are "undifferentiated" or "poorly differentiated." These have abnormal-looking cells and may lack normal tissue structure. The numerical grade is assigned to most cancers based upon these findings.

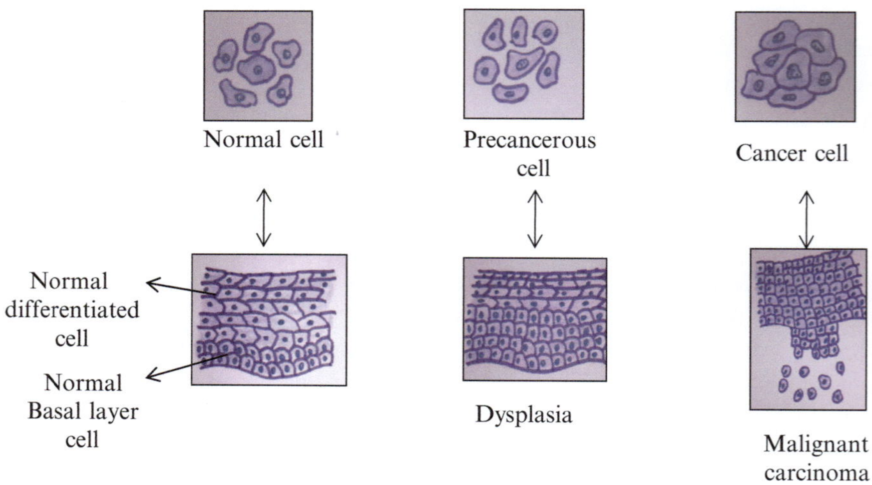

Fig. 10.2 The figure shows the changes in cell morphology and their tissue after infection with human papillomavirus. The tissue is acquired with undifferentiated, metastatic cells

10.3.9 Tumor Markers

Tumor markers are the chemical moieties which are produced by tumor cells or by other cells of the body in response to tumor or cancer [4].

- The presence of tumor marker indicates presence of tumor in the body.
- Each tumor can have different markers. Their levels may alter in more than one type of cancer.
- These markers may be used for the detection, diagnosis, and management of some types of cancer. Their presence in leukemia, breast cancer, melanoma, prostate cancer, and colon cancer might provide immune targets against which the therapeutic agents may be designed and targeted.
- Their level may help the clinicians plan appropriate therapy. The patients' response to therapy can be estimated by using tumor markers. Decrease in marker level suggests positive response to the therapy, whereas no change in marker level suggests nonresponsiveness.
- The abnormal presence of the tumor marker level may suggest cancer but is not enough for diagnosis of cancer. Measurements of tumor markers are usually combined with other tests, such as a biopsy, for accurate diagnosis. In some cases their levels may be helpful to determine the stage of the cancer.
- After treatment, they can be subsequently monitored to check for recurrence of cancer.

> There are clinical practice guidelines from "The American Society of Clinical Oncology" (ASCO), a nonprofit organization, for the tumor markers for breast and colorectal cancer that are helpful to the clinicians. Patient guides are available on the ASCO Web site at http://www.cancer.net/patient/ASCO+Resources/Patient+Guides. Another nonprofit organization "The National Comprehensive Cancer Network®" (NCCN) is an alliance of cancer centers.
>
> (continued)
>
> The National Academy of Clinical Biochemistry (NACB) is a professional organization dedicated to advancing the science and practice of clinical laboratory medicine through research, education, and professional development. The Academy publishes Practice Guidelines and Recommendations for Use of Tumor Markers in the Clinic, which focuses on the appropriate use of tumor markers for specific cancers. More information can be found at http://direct.aacc.org ProductCatalog/Product.aspx

Screening tests are a way of detecting cancer early, before there are any symptoms. A screening test is helpful only if it is highly sensitive and specific. Sensitivity refers to the test's ability to identify people who have the disease. Specificity refers to the test's ability to identify people who do not have the disease. Most tumor markers are not sensitive or specific enough to be used for cancer screening.

Sometimes commonly used tests may not be highly sensitive or specific. For example, prostate-specific antigen (PSA) levels are often used to screen men for prostate cancer, but this is controversial. It is not yet known if early detection using PSA screening actually saves lives. Elevated PSA levels can be caused by prostate cancer or benign conditions, and most men with elevated PSA levels do not have prostate cancer. Moreover benefits of screening PSA for the risk of recurrence or follow-up treatment are not very well defined. However it has certain advantages: (1) It is made only by prostate cells, so a rise in PSA is fairly specific to a prostate problem. (2) The PSA level usually rises even in early cancers, so most prostate cancers can be found at an early stage, when they are most likely to be curable [29].

Another tumor marker, CA 125, is sometimes used to screen women who have an increased risk for ovarian cancer. Scientists are studying whether measurement of CA 125, along with other tests and exams, is useful to find ovarian cancer before symptoms develop. So far, CA 125 measurement is not sensitive or specific enough to be used to screen all women for ovarian cancer. Mostly, CA 125 is used to monitor response to treatment and check for recurrence in women with ovarian cancer.

Cancer researchers are turning to proteomics (the study of protein shape, function, and patterns of expression) in hope of developing better cancer screening and treatment options [15]. Scientists are also evaluating patterns of gene expression for their ability to predict a patient's prognosis.

Unfortunately, none of the tumor markers, including carcinoembryonic antigen (CEA), met the original goal of reliably finding cancer at an early stage. There are few reasons for this:

- These markers are present in almost all control individuals thus their usage for early detection is not conclusive.
- Their levels may be correlated to the presence of cancer, however higher levels alone might not indicate the presence of cancer.
- Some people with cancer never have higher levels of these markers.
- Higher levels are not specific enough as the level of the tumor marker CA 125 can be high in women with gynecologic conditions other than ovarian cancer.

Due to these reasons, these markers are used mainly in patients who have already been diagnosed with cancer to monitor their response to treatment or detect the return of cancer after treatment [6, 9].

Many other tumor markers have been found in recent years and are now under study. Some of these are different from traditional markers, which were proteins found in the blood (Table 10.2).

10.4 Diagnosis of Tuberculosis

Tuberculosis (TB) is an infectious bacterial disease caused by *Mycobacterium tuberculosis* which is aerobic non-spore-forming non-motile single cell bacteria. *Mycobacterium tuberculosis* is the commonest pathogen for pulmonary and extrapulmonary tuberculosis cases, it usually infects lungs, but it can infect other parts of the body such as the kidney, spine, and brain [3]. TB is one of the deadliest diseases; about one-third of the world population is infected by TB and is a leading killer of HIV-infected patients [11]. There are two conditions of TB infection: latent TB infection (LTBI) and TB disease. In case of LTBI, the infected people do not show any symptom of the disease but harbor the bacteria. But they show positive results to tuberculin skin test and TB blood tests. Usually this form of TB is noninfectious. When the immune system becomes weak, latent TB can develop into full-blown TB disease. The typical symptoms of TB are bad cough and dyspnea that lasts for more than 3 weeks, coughing up blood or sputum, pain in chest, weakness, weight loss, loss of appetite, chills, fever, and sweating at night. Treatment of the tuberculosis is challenging from the diagnosis of the disease and then determining susceptibility to drug [24]. Nowadays, emergences of multiple drug-resistant strains are posing problems and high risks associated with disease [28].

TB is suspected in people:

- Who live close to infected individuals
- Who have immunodeficiency such as HIV or any other kind of immunosuppression
- Who have clinical symptoms like unexplained weight loss, loss of appetite, night sweats, fever, and fatigue
- Who have a history of stay in areas where TB is common (most countries in Latin America, the Caribbean, Africa, Asia, Eastern Europe, and Russia)
- Who are using illegal drugs
- Who have lung TB infection that may result in coughing for more than 3 weeks and/or blood in cough (hemoptysis) and pain in chest [34]

Medical history is important as it gives an indication about demographic factors as country of origin, age, ethnic or racial group, occupation, and history of exposure to TB or having HIV or other immunodeficiencies.

Table 10.2 Several markers which are helpful in diagnosis and prognosis of tumor and/or cancer

Tumor marker	Body fluid used	Cancer type	Usage
Alpha-fetoprotein (AFP)	Blood	Hepatocellular carcinoma and germ cell tumors	Diagnose liver cancer
			Follow response to treatment
			Assess stage, prognosis, and response to treatment of germ cell tumors
Beta-2-microglobulin (B2M)	Blood, urine, or cerebrospinal fluid	Multiple myeloma, chronic lymphocytic leukemia, and some lymphomas	Determine prognosis
			Follow response to treatment
Beta-human chorionic gonadotropin (Beta-hCG)	Urine or blood	Choriocarcinoma and testicular cancer	Assess stage, prognosis, and response to treatment
Bladder tumor antigen (BTA)	Urine	Bladder cancer	Not directly used but was used for recurrences
BCR-ABL fusion gene	Blood and/or bone marrow	Chronic myeloid leukemia	Diagnosis and monitor disease status
BRAF mutation V600E	Tumor	Cutaneous melanoma and colorectal cancer	Response to targeted therapies
CA15-3/CA27.29	Blood	Breast cancer	Assess whether treatment is working or disease has recurred
CA19-9	Blood	Pancreatic cancer, gallbladder cancer, bile duct cancer, and gastric cancer	Treatment assessment
CA-125	Blood	Ovarian cancer	Help in diagnosis, assessment of response to treatment, and evaluation of recurrence
Calcitonin	Blood	Medullary thyroid cancer	Diagnosis, assessment whether treatment is working monitoring recurrence
Carcinoembryonic antigen (CEA)	Blood	Colorectal cancer and breast cancer	Check whether colorectal cancer has spread; look for breast cancer recurrence and assess response to treatment determine whether treatment with a targeted therapy is appropriate
CD20	Blood	Non-Hodgkin lymphoma	
Chromogranin A (CgA)	Blood	Neuroendocrine tumors	Help in diagnosis, assessment of treatment response, and evaluation of recurrence help in monitoring for tumor recurrence
Chromosomes 3, 7, 17, and 9p21	Urine	Bladder cancer	
Cytokeratin fragments 21-1	Blood	Lung cancer	Help in monitoring for recurrence help determine treatment and prognosis
EGFR mutation analysis	Tumor	Non-small cell lung cancer	
Estrogen receptor (ER)/progesterone receptor (PR)	Tumor	Breast cancer	Determine whether treatment with hormonal therapy (such as tamoxifen) is appropriate monitor progression and response to treatment
Fibrin/fibrinogen	Urine	Bladder cancer	

(continued)

Table 10.2 (continued)

Tumor marker	Body fluid used	Cancer type	Usage
HE4	Blood	Ovarian cancer	Assess disease progression and monitor for recurrence
HER2/neu	Tumor	Breast cancer, gastric cancer, and esophageal cancer	Determine whether treatment with trastuzumab is appropriate
Immunoglobulins	Blood and urine	Multiple myeloma and Waldenström macroglobulinemia	Help diagnose disease, assess response to treatment, and look for recurrence
KIT	Tumor	Gastrointestinal stromal tumor and mucosal melanoma	Help in diagnosing and determining treatment
KRAS mutation analysis	Tumor	Colorectal cancer and non-small cell lung cancer	Determine whether treatment with a particular type of targeted therapy is appropriate
Lactate dehydrogenase	Blood	Germ cell tumors	Assess stage, prognosis, and response to treatment
Nuclear matrix protein 22	Urine	Bladder cancer	Monitor response to treatment
Prostate-specific antigen (PSA)	Blood	Prostate cancer	Help in diagnosis, assess response to treatment, and look for recurrence
Thyroglobulin	Tumor	Thyroid cancer	Evaluate response to treatment and look for recurrence
Urokinase plasminogen activator (uPA) and plasminogen activator inhibitor (PAI-1)	Tumor	Breast cancer	Determine aggressiveness of cancer and guide treatment
5-Protein signature (Ova1)	Blood	Ovarian cancer	Preoperatively assess pelvic mass for suspected ovarian cancer
Gene signature (Oncotype DX)	Tumor	Breast cancer	Evaluate risk of recurrence
Gene signature (Mammaprint)	Tumor	Breast cancer	Evaluate risk of recurrence

10.4.1 Diagnosis of Latent Infection

Robert Koch in 1891 noticed the skin reaction in tuberculosis patients when they were subcutaneously injected with components of *M. tuberculosis* (old tuberculin (OT)). Tuberculin purified protein derivative (PPD) is being used; however it lacks mycobacterial species specificity. Skin testing with tuberculin-PPD (Tuberculin skin test- TST) is used for detection of latent tuberculosis infection (LTBI). The usage of the "Mantoux tuberculin skin test" (TST) or the TB blood test is common for infection of *M. tuberculosis*. These tests can only tell if there is mycobacterial infection; they cannot identify between LTBI and TB that has progressed to full blown. Physical examinations, microbiological tests, and chest X-Rays can determine whether the person has TB disease.

10.4.2 Mantoux Tuberculin Skin Test

The Mantoux tuberculin skin test (TST) is the standard method for determining whether a person is infected with *Mycobacterium tuberculosis*. It is performed by intradermal injection of 0.1 ml of tuberculin purified protein derivative (PPD) into the inner surface of the forearm. It produces a pale elevation of the skin, or wheal, about 6–10 mm in diameter. The skin reaction is read between 48 and 72 h of the administration. The reaction is measured in millimeters of irritation or swelling (induration) both vertically and horizontally (more important) on the forearm. The test is interpreted according to two factors: measurement in millimeters of the induration and the risk of the person of being infected by TB and progression to the disease if infected. An induration of 5 mm or more is considered positive for HIV-positive individuals, patients on immunosuppressive medicines or with organ transplants, and persons who have been in recent contact with TB patients or whose chest X-ray shows fibrotic changes consistent with initial stages of TB. An induration of 10 mm or more is considered to be positive for immigrants from countries that have high occurrence of TB and use injection drug users, mycobacterial laboratory personnel, people with high-risk clinical conditions, and children, especially less than 4 years old, and adolescents exposed to high-risk adults. Induration of 15 mm or more is considered positive for anyone with or without risk. A false positive may occur for people vaccinated with *Bacillus* Calmette–Guerin (BCG). This vaccine is widely used in countries with high incidence of TB. False-negative results may occur in certain population such as very young children, old people, HIV-positive patients, and people who have been infected very recently. However more tests are required for confirmation.

10.4.3 Chest Radiography

Chest radiograph is used to detect the presence of lesions which can vary in size, shape, density, and cavitation. If TB is suspected in bone or joint then radiographic examination is done for presence of lesions. CT may be done in case the chest radiographic findings are not confirmatory.

10.4.4 Drug Susceptibility Testing

The culture isolates of *M. tuberculosis* should be tested for their sensitivity to isoniazid, rifampin, and ethambutol. In case resistance to one or more of these drugs is present in the culture, expanded testings for susceptibility is required. Susceptibility testing may be done directly with the clinical sample or indirectly with mycobacterial cultures using solid and liquid media. Direct testings are more useful and take about 3 weeks, whereas indirect testings may take nearly 8 weeks.

10.4.5 Microscopy

Microscopically the presence of acid-fast bacilli (AFB) is done on a sputum smear or other specimen indicating the presence of TB. The technique is easy and quick but is not confirmatory as positive results are given by other bacilli as well. Thus culturing might be helpful.

The important consideration of culturing is slow growth requiring incubation period of 4 weeks and drug susceptibility requires further 4 weeks. Other modern rapid methods for isolation include microcolony detection on solid media, septicheck AFB method, microscopic observation of in broth culture (MODS), the BACTEC 460 radiometric system [30], BACTEC MGIT 960 (Becton Dickinson), MB/Bac T system (Organon Teknika), and ESPII culture system (Table 10.3).

Mycobacterium speciation may be carried by phenotypic characterization, biochemical typing, lipid analysis by GC and HPLC.

10.4.6 TB Blood Test (Interferon-Gamma Release Assay-IGRA)

The IGRA test identifies the presence of *Mycobacterium tuberculosis* by measuring immune response to TB bacteria in whole blood.

Table 10.3 Some culture methods and their properties for detection of *Mycobacteria*

S.No.	Culture method	Methodology	Properties
1.	Microcolony detection on solid media	Plates with thin layer of middlebrook 7H11 agar are incubated and examined microscopically [23]	Less sensitive, requires less time, labor intensive with low recovery
2.	Septicheck AFB method	Grown in middlebrook 7H9 broth under enhanced CO_2 (5–8 %) with other nutrients. Contains nonselective media middlebrook 7H11 agar with para-nitro-alpha-acetylamino-b-hydroxy propiophenone (NAP) for *M. tuberculosis* and other part with chocolate agar for contaminants	Requires 3 weeks, simultaneously detects *M. tuberculosis*, non-tuberculous mycobacteria (NTM), and other respiratory pathogens, contaminants Gives better results than other systems
3.	Radiometric BACTEC 460 TB method	Specific for mycobacteria, ^{14}C labeled palmitic acid in 7H12 medium is used. ^{14}C labeled substrate upon metabolization produces $^{14}CO_2$ which is measured by BACTEC system. It is reported in terms of growth index	Faster rate of culture isolation Good correlation in drug susceptibility Results within 8 days Early diagnosis helps in detection in high prevalence areas of TB infection
4.	MGIT 960 mycobacteria detection system	Incubate and monitor 960 mycobacteria growth indicator tube (MGIT) every 60 min for increase in fluorescence and growth detection is based on AFB metabolic O_2 utilization and subsequent intensification of an O_2 quenched fluorescent dye in MGIT	Rapid, accurate, and cost-effective method
5.	MB/BacT system	Based on colorimetric detection of CO_2	Risk of increased contamination, requires longer time for detection of growth
6.	ESP culture system II	Detects pressure changes within head space above broth culture medium in sealed bottle	Should be used with other system

White blood cells of infected persons produce interferon-gamma (INF-γ) in response to antigens produced by *M. tuberculosis*. INF-gamma produced is measured in these assays. An IGRA measures how strong a person's immune system reacts to TB bacteria by testing the person's blood in a laboratory. Two IGRAs are approved by the US Food and Drug Administration (FDA): QuantiFERON®-TB Gold In-Tube test (GFT-GIT) [7] and T-SPOT®TB test. The blood is collected and processed between 8 and 30 h depending on the test and result is obtained within 24 h.

Positive IGRA: This means that the person has been infected with TB bacteria. Additional tests are needed to determine if the person has latent TB infection or TB disease. A health-care worker will then provide treatment as needed.

Negative IGRA: This means that the person's blood did not react to the test and that latent TB infection or TB disease is not likely.

The main advantages are as follows: there is only one visit, results are obtained within 24 h, and the tests are independent of BCG vaccine; however these tests are expensive.

10.4.7 Other Diagnostic Methods

Serodiagnosis methods include capture ELISA for the detection of *Mycobacterium tuberculosis*. With the advancement in the technology, the diagnostics have become quick and automated. The prime objective of the newer methods is precise and timely diagnosis. Today the genotypic and phenotypic methods are available for the early diagnosis of tuberculosis disease. The newer diagnostic technologies include polymerase chain reaction (PCR)-based technologies, fluorescent in situ hybridization, electrochemical detection of DNA, biochips, nanotechnology, and proteomics technology.

10.4.7.1 Serodiagnosis

(a) *Capture/sandwich ELISA*

A quantitative test to detect lipoarabinomannan (LAM) has been developed for the detection of TB in urine specimens (www.icmr.nic.in/ijmr). LAM is a heat-stable, major glycolipid constituent of the cell wall of MTB. This antigen is released from metabolically active or degrading mycobacteria and enters the circulation from where it is filtered in the renal tubules.

(b) *Detection of LAM in sputum*

This test is based on the capture antibody derived from murine source (murine monoclonal antibody against LAM). The rabbit antisera against *M. tuberculosis* are used as source of detector of antibody.

10.4.7.2 Molecular Techniques

(a) *Line probe assays*

Line probe technology is not automated, but is a type of molecular assay that has the appeal of providing detection of specific gene markers without the need for a sophisticated laboratory infrastructure. The technology includes:

1. Extraction of DNA from respiratory specimens or from mycobacteria isolated in culture
2. Amplification of nucleic acid sequences using polymerase chain reaction
3. Hybridization of amplified nucleic acid sequences to a variety of oligonucleotide probes that are immobilized in lines on a solid strip
4. Colorimetric development to mark the nucleic acid probe lines on the immobile strip

The technology has been evaluated for its use in detecting *M. tuberculosis* in respiratory specimens, as well as for detecting drug resistance. Two line probe assays have been developed and evaluated for clinical use.

The first of these assays is the INNO-Lipa Rif. TB (Innogenetics NV, Ghent, Belgium) and the second line probe assay is the GenoType MTBDRplus assay (Hain Lifescience, GmbH, Nehren, Germany) (rapid diagnosis of *M. tuberculosis* infection).

PCR-based line probe assays, such as the Genotype MTBDRplus, are now increasingly used for the rapid detection of drug resistance. Such tests offer sensitive and specific detection of isoniazid and rifampin resistance directly from smear-positive samples or from positive culture samples.

(b) *PCR*

In this, an appropriate sequence specific for *M. tuberculosis* is selected and used with DNA obtained from a patient sample. The PCR allows sequences of DNA presenting only a few copies of mycobacteria to be amplified in vitro such that the amount of amplified DNA can be visualized and identified. With rapid detection methods, the results are available within the day of DNA extraction. The most common target used in the PCR is IS6110, which is specific for the *M. tuberculosis* complex and is present 20 times in the genome, thus offering multiple target amplification.

A fully automated nucleic acid amplification technology (NAAT) is used in the early diagnosis of TB [33], as well as multidrug-resistant TB (MDR-TB) and TB complicated by HIV infection, which are more difficult to diagnose (www.who.int/tb/features). The NAA technique uses chemical, rather than biological, amplification to produce nucleic acid so that within few hours this test distinguishes between *M. tuberculosis* complex and nontuberculous mycobacteria (NTM) in acid-fast bacilli (AFB)-positive specimen [16].

The most significant advancement toward a point-of-care (POC) test for TB has come in the field of nucleic acid amplification with the launch of the GeneXpert MTB/RIF assay [13]. The assay is capable of detecting the *M. tuberculosis* complex while simultaneously detecting rifampin resistance within 2 h (diagnosis and management). Only one automated method for amplifying and detecting nucleic acids has been developed, the Xpert MTB/RIF. This method is designed to be fully automated, self-enclosed system that eliminates the need for most of the laboratory infrastructure needed for nucleic acid amplification for rapid diagnosis of *M. tuberculosis*.

(c) *Multiplex PCR*

Utilizes evaluation of multiple targets. The multiplex PCR is supposed to be less complicated, less time consuming, cost-effective, and superior to the conventional methods. It is also applicable for culture negative samples where strain identification is not possible by conventional approach.

(d) *Loop-mediated isothermal amplification (LAMP)*

LAMP-based assay is another new NAAT which has targeted *gyrB* and *rrs* and more recently the repetitive insertion sequence IS6110 for the detection of the *M. tuberculosis* in clinical sputum specimen [20]. The IS6110-based LAMP assay may be a test with higher sensitivity than assay based on *gyrB* and *rrs*.

Loop-mediated isothermal amplification (Eiken Chemical Company, Tokyo, Japan) is a novel method for amplifying DNA that generates sufficient quantities of nucleic acid for visual detection by use of fluorescent labels. Loop-mediated isothermal amplification is not a commercial assay but rather a molecular method, until there is a commercial assay based on the loop-mediated isothermal amplification method; widespread use of it is unlikely because most resource-limited health-care systems cannot develop, validate, and implement laboratory-developed molecular assays.

(e) *Oligonucleotide microarray*

Microarrays, also known as biochips, have been proposed as new molecular methods for detecting drug resistance in *M. tuberculosis*. This technology allows for the simultaneous detection of many nucleic acid sequences in a sample. The technology allows for detection of nucleic acid sequences of interest, which for *M. tuberculosis* could mean either detection of conserved sequences to identify the presence of the bacterium or detection of other sequences to detect microbial genes that confer drug resistance. Only one commercial assay based on this technology has been evaluated, the TB-Biochip.

(f) *Rapid molecular tests*

The rapid molecular tests have advantages like:

- More rapid than nonmolecular tests
- Potential for high sensitivity and specificity
- Can be manufactured in large quantities
- Decreased cost
- Standardization of field use
- Require less training and infrastructure compared with conventional cultures and susceptibility testing
- Conceptually simple methods
- Easy to manufacture and distribute
- More rapid definitive test results
- Relatively easier to standardize

Disadvantages

- Cost.
- Do not eliminate need for cultures.
- Test limited number of drugs for resistance.
- Require laboratory infrastructure that can accommodate molecular testing.
- Work better with smear-positive than with smear-negative specimens of TB.

10.5 Diagnosis of Malaria

Malaria is the world's most important parasitic infection causing health and developmental challenges in developing and underdeveloped countries. About 3.3 billion people live in areas at risk of malaria transmission in 109 countries. It is the fifth largest cause of death due to infectious diseases worldwide. Malaria is a protozoan disease transmitted by the bite of infected *Anopheles* mosquitoes when they feed on human blood meal. Four species of the genus *Plasmodium* cause nearly all malaria infections in humans (although rare infections involve species normally affecting other primates). These are *P. falciparum*, *P. vivax*, *P. ovale*, and *P. malariae*. Human infection begins when a female *Anopheles* mosquito inoculates plasmodial *sporozoites* from its salivary gland during a blood meal. The early symptoms of malaria are high fever, shaking chills, sweat, muscle pain, and flu-like symptoms. They are not very specific and can be confused with other illness; however infections associated with cerebral malaria are severe, such as severe anemia and acute respiratory distress syndrome. If not treated early, it might be fatal.

The life cycle of malaria parasite involves two hosts: the female *Anopheles* mosquito and human. During blood meal, the mosquito injects sporozoites into humans that infect liver cells and mature into schizonts. These rupture and release merozoites that infect the red blood cells. The ring-shaped trophozoites are formed that mature to schizonts that rupture to release more merozoites in the blood. They undergo asexual reproduction in the erythrocyte. The clinical manifestations of malaria occur during this erythrocytic cycle. Some parasites mature into gametocytes that are ingested by the mosquito and the sexual cycle is completed inside the gut of the mosquito.

Potential threat of malaria epidemics in the status of a low transmission is of major concern because most of the asymptomatic carriers are not treated. Asymptomatic carriers not only act as reservoirs for malaria transmission but also act as risk factors for symptomatic attacks. For elimination of malaria from an endemic area, the identification of asymptomatic infections for treatment is required. Management of malaria requires proper and fast diagnosis which is followed by appropriate treatment. Current diagnosis tool for malaria is mainly based on microscopy or rapid diagnostic tests (RDTs) which are not much sensitive. So we require more advanced technique for the diagnosis purpose (direct blood). The molecular diagnostics would greatly enhance the quality of malaria diagnosis and treatment and would decrease the use of artemisinin-based combination therapy (ACT) which was recommended by the WHO in 2006.

Use of histidine-rich protein 2 (HRP-2)-based rapid diagnostic tests (HRP-2-RDTs) showed a positive response which was further confirmed by microscopy. Ninety percent of these malaria-positive cases are asymptomatic and only half of them were cured with artemisinin-combination therapy (ACT) as confirmed with a follow-up study. Follow-up results were verified by polymerase chain reaction (PCR) if inconsistent findings were observed between RDT and blood films.

Rapid and accurate diagnosis is very important for treatment of affected individuals and prevention of spread of the disease. Some of the prevalent diagnoses are:

10.5.1 Laboratory Diagnosis Method

10.5.1.1 Microscopic Methods

Thick and thin blood smears were prepared using finger-prick blood and were stained with 10 % Giemsa stain for 10 min. A Giemsa-stained blood film provides all the important information required for starting the treatment. A thick blood smear identifies the presence of *Plasmodium* in the blood, whereas a thin blood smear gives fairly accurate information about the species of *Plasmodium* and percentage of patients' red blood cells that are infected by malaria parasite, also known as parasitemia. Though microscopic examination is still considered as a gold standard for diagnosis of malaria, it is relatively expensive as all the clinics around the world, especially in underdeveloped countries, cannot afford to have a microscope. Moreover highly trained microbiologists are required to perform and interpret the test results. So other simpler methods have been developed.

Microtube concentration methods with acridine orange staining: Blood is collected in a specialized tube containing acridine orange, anticoagulant, and a float. After centrifugation, which concentrates the parasitized cells around the float, fluorescence microscopy is performed.

10.5.2 Serology: Indirect Fluorescence Antibody (IFA) Test

Malarial antibody is detected in the blood by IFA test. This is a sensitive method to detect whether a person has been infected by *Plasmodium* even when the parasite count in the blood or parasitemia is too low to be detected by microscopy. Erythrocyte stage *Plasmodium* sp. schizonts are used as antigens. The serum is exposed to the antigen and cognate antibodies, when present, form antigen–antibody complex, which may be probed by fluorescein labeled antihuman antibody. When seen under fluorescent microscope, the parasites give an apple green color showing positive reaction. Species-specific testing is available for *P. falciparum*, *P. vivax*, and *P. malariae*; however *P. ovale* antigens are not always readily available. Since it takes a long time to develop antibodies, it is not recommended for routine diagnosis. This method is used to screen donors for malaria parasite before blood transfusion, to eliminate transfusion-induced malaria as this method is highly sensitive and can detect the presence of *Plasmodium* when the parasitemia is below the detectable level by blood film method. This method is also used to test patients from endemic region with chronic or repeated malaria infection or those whose diagnosis is questionable. There is also some cross-reaction between *Babesia* and *Plasmodium* species.

10.5.3 Rapid Diagnostic Test (RDT)

This detects specific malaria antigens in a person's blood thereby quickly establishing the diagnosis of malaria infection [35]. Malaria RDTs were introduced and more than 60 RDT brands and over 200 different products have been developed. Test strips containing monoclonal antibodies against parasite antigen, histidine-rich protein 2 (PfHRP2), or parasite-specific lactate dehydrogenase (pLDH) are used. These tests can be done in less than 15 min and require very little training. These tests can be very useful in mobile clinics where laboratory facilities are not available. However, RDT may not be able to detect some infections that have very low number of parasites giving rise to false negatives. For determination of prevalence of malaria infection, together with sensitivities and specificities, the predictive values of three diagnostic tests (HRP-2-RDTs, microscopy, and nested PCR) were considered and it was found that only microscopy has poor sensitivity compared to the other tests [19]. But caution should be taken with the positive predictive values (PPV) of the HRP-2-RDTs.

Plasmodium LDH dipstick or card test: A drop of blood is placed on the stick or card, which is then immersed in washing solutions. Monoclonal antibodies capture the parasites and read out as colored bands. One band is genus specific (all malarias) and the other is specific for *P. falciparum*.

10.5.4 Molecular Techniques

Molecular techniques such as PCR to detect *Plasmodium* infections have demonstrated high sensitivity and specificity and have the ability to quantify parasitemia when used in a quantitative real-time PCR format. Therefore, molecular technologies are frequently used in malaria studies and in well-equipped laboratories as the "gold standard" (directly from blood).

(a) *Nested PCR assay based* on the 18S rRNA gene of *Plasmodium* species. Extraction of DNA and PCR assays are carried out blinded to the microscopy results. PCR assays are repeated for all PCR-positive and microscopically negative samples. The lowest limit of detection of *Plasmodium* species is 2 parasites/μL. Nested PCR and real-time PCR are known to have a higher sensitivity and specificity for malaria diagnosis compared with light microscopy.

The analysis of amplicons obtained from conventional PCR formats is elaborate and often requires either a toxic and environmentally hazardous ethidium bromide gel for visualization. An alternative is expensive real-time PCR equipment for analysis. The recent advance of a real-time quantitative PCR technique has proven useful in various applications, including parasite detection, species differentiation, gene expression and regulation, and allelic discrimination, and it is more sensitive than nested PCR.

(b) *Direct blood PCR (db-PCR)* combined with a rapid readout system, nucleic acid lateral flow immunoassay (NALFIA), is used. The direct blood approach circumvents preamplification handling such as DNA extraction. The full blood sample is directly added to the PCR mixture for the amplification of the target DNA of *Plasmodium*. This takes place in less than 1 h. The product can be visualized with NALFIA, which is a rapid immuno-chromatographic test to detect labeled amplicons on a nitrocellulose stick coated with specific antibodies. The amplicons are labeled via specific primers that contain a biotin molecule and a hapten. The detection test is a simple, straightforward, and safe one-step procedure in which the results are visible within 10 min.

The db-PCR is based on a combination of fusion db-PCR buffer and Phire Hot Start II DNA polymerase. The dB-PCR requires two primer pairs, one pair for the amplification of pan-*Plasmodium* and the second pair for amplification of the human housekeeping gene glyceraldehyde 3-phosphate dehydrogenase (GAPDH). The GAPDH gene is used as an amplification control, as well as an internal running control for the NALFIA. A HiFlow 135 nitrocellulose membrane (25 by 5 mm per strip) was used to fabricate the NALFIA sticks for this assay.

The current db-PCR-NALFIA is highly sensitive for the detection of *P. vivax*, as well as *P. falciparum*, but the current format does not differentiate the *Plasmodium* species. The sensitivity of other human *Plasmodium* species—for example, *P. ovale*, *P. knowlesi*, and *P. malaria*—could not be assessed here.

(c) *Loop-mediated isothermal amplification (LAMP)*

PCR-based molecular methods are good for both sensitivity and specificity but too sophisticated and expensive to be applied in most malaria-endemic countries. The recently developed loop-mediated isothermal amplification (LAMP) method is cheaper, simpler, and faster. The LAMP reaction can be conducted under isothermal conditions and is almost as specific and sensitive as the nested PCR method for *Plasmodium*-DNA detection in blood.

The positive LAMP reaction was detected by observing the fluorescence in the reaction tube via the naked eye under a portable UV lamp. In this study, an interval of 10 min was set to determine the threshold time of the LAMP reaction.

Time to obtain test results varied depending on the different methods, i.e., 20 min for HRP-2-RDTs, 60 min for microscopy, 3 h for LAMP, and 8 h for nested PCR. Blood compositions, such as hemoglobin and IgG/IgM, can interfere with the performance of PCR. A rapid extraction of DNA from filter paper may be favorable for LAMP but a conventional method is needed for the nested PCR. LAMP is simpler and faster and can be a potential tool to replace nested PCR. The costs for the LAMP assay are only about a tenth of that for the conventional PCR, the reagents and enzymes are still expensive and may restrict its use in malaria-endemic areas. Both sensitivity and specificity of the LAMP assay were similar to those of nested PCR. With high PPV and NPV, LAMP was the best method for malaria diagnosis. LAMP can be further developed as a point-of-care test.

10.5.5 Drug Resistance Tests

This is done to determine if a particular strain is susceptible to antimalarial drugs. The parasites are grown under increasing concentration of drugs and the concentration at which they do not grow is determined as the end point.

10.6 Diagnosis of Acquired Immunodeficiency Syndrome (AIDS)

HIV/AIDS is secondary immunodeficiency which is caused by retrovirus, that is, human immunodeficiency virus-1. Several methods are available for the diagnosis of HIV [14,17]. They are based upon detection of one or more viral parts or the antibodies that are produced in response to HIV infection [10]. HIV particle is composed of a viral envelope derived from fatty cell membrane of human host cells. There are 72 glycoproteins belonging to gp120 and gp41 family that extend from interior through it (Fig. 10.3). Within the envelope lies the viral core or the nucleocapsid. The outer layer is made out of protein p17 and the inner layer p24. The HIV genome consists of two single-stranded RNA molecules which are associated with several other molecules that are required for HIV integration and propagation in human hosts (Fig. 10.3). The reverse transcriptase that converts RNA to DNA and integrase that helps to integrate its DNA into the host DNA. These proteins and other glycoproteins evoke immune response from the host.

HIV preferentially binds to CD4+ T helper cells with the help of gp120 (Fig. 10.4). It fuses with the cell and converts its single-stranded RNA molecule to double-stranded DNA that integrates with the host genome. It remains latent as a provirus within the host for years. On activation, new virus particles are formed and are released from the host cell by lysis. This causes decrease in T cells and increase in viral load.

There are three stages in infection process. Stage 1 is the acute phase immediately after infection. It is also known as diagnostic window or serological latency. At this stage viral nucleic acid and p24 is detectable in the serum, but antibodies against HIV are not produced. Antibodies are detected about 6–8 weeks postinfection and this phase is known as chronic phase. Detection of host antibodies is known as *seroconversion*. The final stage of HIV infection is the onset of clinical AIDS. It is marked by clinical manifestation such as infection by opportunistic pathogens. The CD4+T cell counts drops to less that 200 per ml. There are different types of diagnostic tests and the goal is to identify HIV as soon as possible. There are two approaches: they either detect the antibodies that are produced in response to HIV or detect the HIV-specific RNA and proteins (p27 antigen).

10.6.1 Tests Detecting HIV-Specific Antibodies

(a) *ELISA test*

Enzyme-linked immunosorbent assay or ELISA can detect the presence of antibodies (IgG) against HIV. The patient's serum is incubated with HIV-specific antigens such as p24, gp41, and/or gp120. Seroconverted HIV-positive patients will have antibodies that will bind to the antigens. The antibodies are detected by either an enzyme-linked anti-human antibody.

(b) *Western blot*

The positive reactions of ELISA tests are confirmed by Western blot analysis. The host antibodies are incubated with viral antigens and the protein complexes thus formed are run on a polyacrylamide gel to separate the proteins. The protein bands are then transferred to nitrocellulose or nylon membrane and probed with radioactive antibodies against specific proteins. They are then identified from autoradiographs.

(c) *Immunofluorescence assay* (*IFA*)

Immunofluorescence assay is an alternative to Western blot. Here the patients' serum is incubated with HIV-infected or HIV-uninfected T cells. Samples that contain antibodies against HIV bind to the infected T cells but not to the uninfected cells. Bound

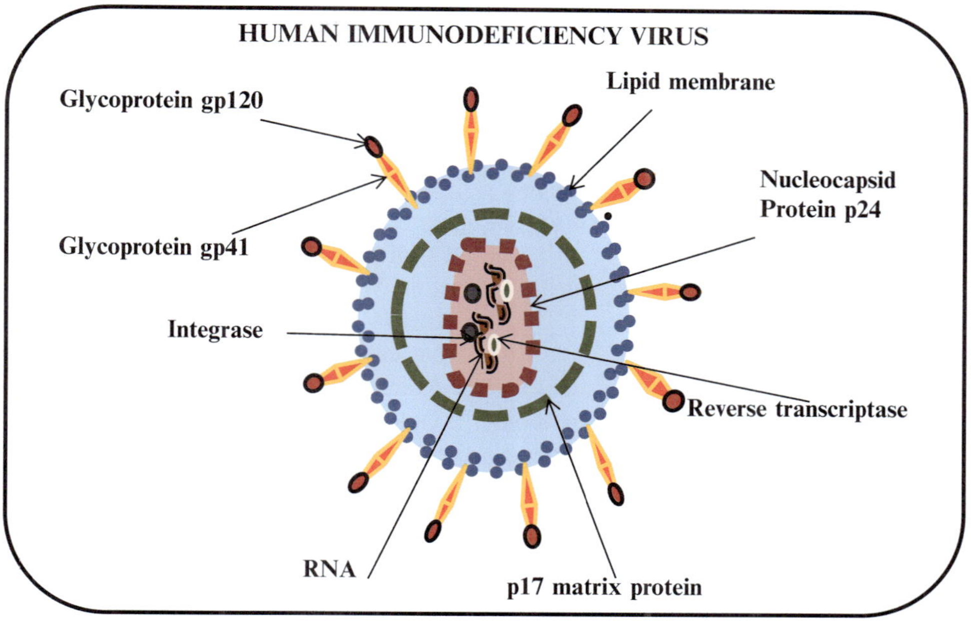

Fig. 10.3 Human immunodeficiency virus-1 (HIV-1). It consists of a host-derived viral envelope. On its lipid envelope it has glycoproteins belonging to gp120 and gp41 family. Nucleocapsid is present inside envelope with outer layer made of protein p17 and the inner layer of p24. Its genome has two single-stranded RNA molecules which are associated with several other molecules as reverse transcriptase and integrase for viral integration and propagation

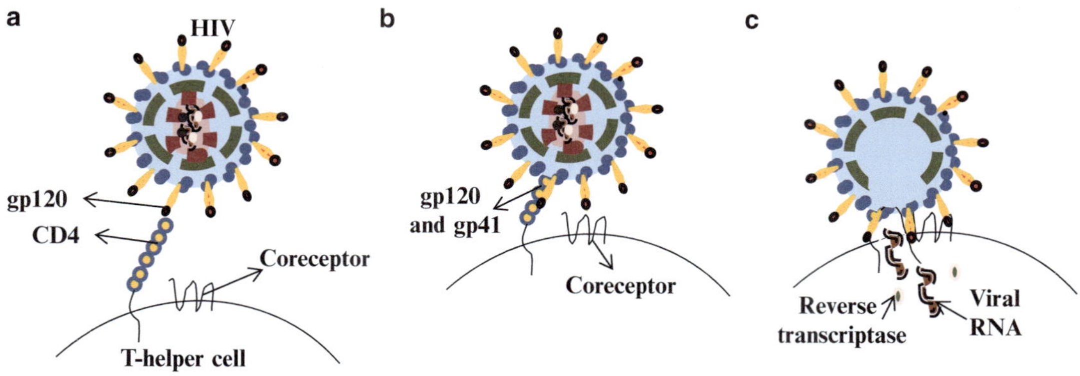

Fig. 10.4 The integration, fusion, and transfer of its RNA into host cell. (**a**) Virus interacts through its glycoprotein gp120 to CD4 receptor. (**b**) Attachment is followed by coreceptor involvement and binding. Binding of gp120 to coreceptor induces changes in gp41 resulting in embedding of virion in host cell membrane and fusion. (**c**) Viral RNA and its associated protein enters inside the cell through fusion pore

antibodies are detected by fluorescently labeled antihuman antibody.

(d) *Detuned assay*

Detuned assay is a dual ELISA that helps to distinguish whether an infection is new or has been there for a while. A recent infection would produce a very low amount of antibody and would require a more sensitive detection criterion or low stringency, whereas long-standing infection would be very easily identified due to the presence of a large quantity of antibody. It is particularly important for public health studies as it gives an idea of spread of the disease within a population.

(e) *Salivary and urine tests*

HIV antibodies can be detected by ELISA and Western blot from both urine samples and saliva. These tests are less invasive but give limited information about the type of infection.

10.6.2 Rapid Test

Several rapid tests have been designed based on the same principles. These are easy to perform and give fairly accurate results within hours. These include agglutination assay, membrane immunoconcentration device, solid phase tests (immuno dots), and immunochromatography. In agglutination test the blood of the patient forms clumps when incubated with hybrid antigen–antibody reagent. In immunoconcentration device, the HIV antibodies are captured and are detected by specific colored reactions. Similarly in solid phase immune Dot assay, the antibodies are captured on a plastic matrix. Immunochromatographic strip test is a one-step method in which the specimen is adsorbed in a pad and immediately reacts with a detection reagent.

ELISA, Western blot, and IFA tests are highly sensitive in patients after seroconversion or after the body has started producing antibody against HIV. But these can give false negative results in cases where the infection has taken place recently and the body has not produced antibodies. For early detection of HIV infection during acute phase of infection, viral load testing is done.

The diagnosis of HIV-1 EIA may be nonspecific at times due to the presence of antibody to class-II antigen, autoantibodies, influenza vaccination, hepatic disease, or acute viral infections [14]. Thus positive EIA requires the usage of one or more confirmatory tests as Western blot for HIV-1. Western blot is most commonly used as confirmatory test as it utilizes the detection of multiple HIV antigens with variable molecular weights. Probing with antibodies results in visualization of antigen–antibody-specific bands:

NEGATIVE WESTERN BLOT + POSITIVE EIA= NO HIV INFECTION

POSITIVE WESTERN BLOT + POSITIVE EIA= HIV INFECTION

Western blot can detect gag, pol, and/or env gene products and positive result is in favor of HIV infection. In 1993, the FDA has established that the Western blot result would be considered positive if two antibodies of the three give positive results for p24, gp41, or gp120.

Refer to Fig. 10.5 for HIV test.

OraQuick Rapid HIV-1 antibody test: Done by blood, plasma, or saliva. Test is highly sensitive and specific. In negative result, HIV infection is not present, and if positive result, confirmation is required by any of the above serological testings.

HIV RNA test and Western blot can also be used for the assessment of patient's response to therapy.

Nucleic acid-based tests: Monitoring the levels of HIV RNA in the plasma of patient is useful as antibody response may be indeterminate or misleading.

Three assays are used for HIV-1 diagnosis

- Reverse transcriptase PCR (RT-PCR) Amplicor
- Branched DNA (bDNA; VERSANT)
- Nucleic acid sequence-based amplification (NASBA-Nuclisens)
- These are important in making the diagnosis of HIV infection; HIV infection confirmation; establishing initial prognosis; determining the need for therapy; and monitoring the effect of therapy.
- PCR based on DNA: For diagnosis of HIV infection, amplifying HIV proviral DNA from PBMCs is useful. This is extremely sensitive as compared to other tests.
- The sensitivity of tests is:
  - RNA-based detection: 40–80 copies RNA/ml of plasma
  - Research laboratory-based RNA assay: 1-few HIV RNA copy/ml
  - PCR-based test: one copy per 10,000–100,000 cells, but specificity may be low.

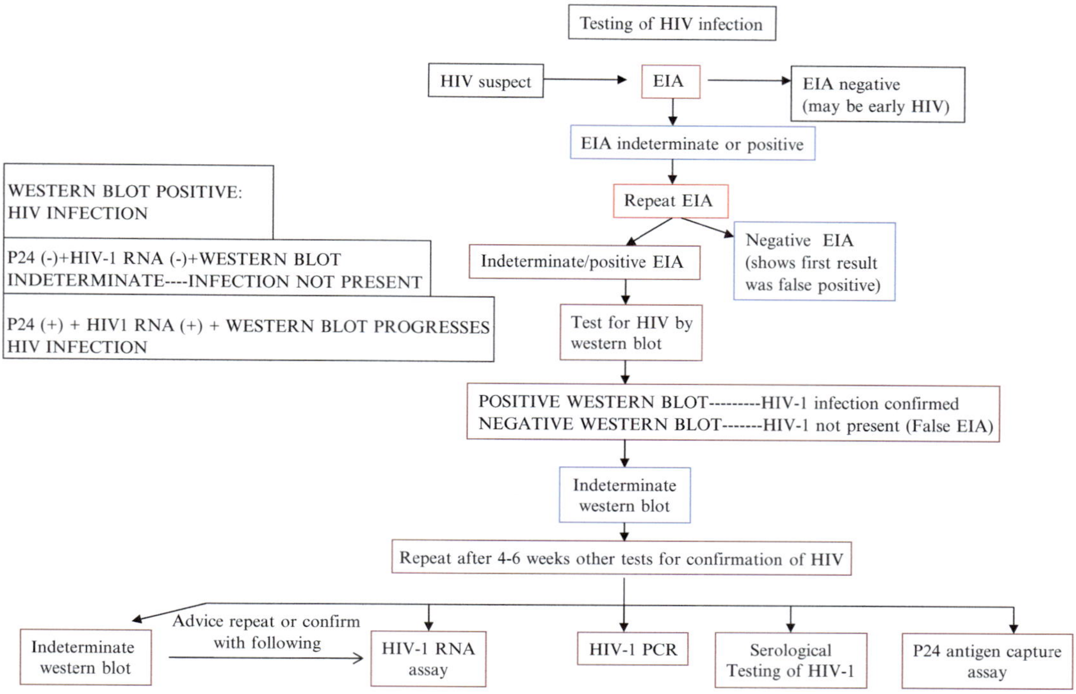

Fig. 10.5 Shows the sensitivity of various tests for diagnosis and confirmation of HIV-1 infection in patients

False positives are reported with each of these techniques. Thus EIA with confirmatory Western blot remains the "gold standard" for HIV diagnosis.

10.6.3 Viral Load Test

This test determines the amount of virus in the blood. This can be done by HIV nucleic acid detection, p24 antigen detection, and peripheral blood mononuclear cell detection.

10.6.3.1 HIV RNA Detection

The presence of viral RNA is detected mainly by three methods: reverse transcription polymerase chain reaction (RT-PCR), branched NA (bDNA), and nucleic acid sequence-based amplification assay.

In RT-PCR assay, the HIV-specific RNAs, extracted from clinical samples, are converted to DNA and are amplified by PCR. They are fused with HIV-specific DNA probe tagged with enzyme. The nucleic acid-enzyme complex is then detected and quantified by a color-producing reaction.

In nucleic acid sequence-based amplification assay, the viral RNA is captured by sequence-specific oligonucleotides tagged with enzyme. In bDNA assay, these oligonucleotides are branched with one branch attached to an enzyme. The enzyme nucleotide complex produces a color that can be detected and quantified. Nucleic acid testing is usually done to determine the viral load in the patient.

10.6.4 Peripheral Blood Mononuclear Cell Culture

Mono nuclear cells are isolated and cultured for several days to isolate HIV. It is very expensive and is only used in infants and patients with very low viral count.

10.7 Chapter End Summary

- Cancer is a life-threatening disease which is affecting large population of the world. Cancer occurs due to uncontrolled proliferation of the cells and their migration to different tissues and organs of the body. The loss of control of cell division may be due to activation of oncogenes or mutations and silencing of tumor suppressor genes. Cancer is a disease caused by many genetic changes that develop over time.

- Uncontrolled proliferation of cells form tumor. If tumor is confined to its site of origin, it is called benign tumor. Migration of tumor to other tissues results in cancer. Early diagnosis of the disease is very important for effective control of the disease.
- The diagnosis of cancer is done as staging or using computed tomography (CT), combined PET/CT, laboratory test, mammograms, Pap and HPV testing, pathology report, and testing for tumor-/cancer-specific marker.
- Tumor markers are used for diagnosis as well as treatment of cancer. These markers comprise mainly the cell surface receptors, signaling molecules, or growth factors which act inappropriately in cancer. However some of them in normal form are present on normal cells. These normal markers are either highly upregulated or mutated on cancerous cells; therefore they cannot be used alone for diagnosis.
- Tuberculosis (TB) is infectious disease caused by *Mycobacterium tuberculosis*. The diagnosis of TB is done by culturing, but since it takes a long time, several fast detection assays have been developed. PCR and recombinant DNA technology tools assist in early diagnosis of TB.
- Malaria is caused by protozoa *Plasmodium* and is transmitted by female *Anopheles* mosquito. The detection of malaria and the infecting species is very important for the control of the disease. For detection of malaria, various traditional and molecular methods have been developed which are sensitive and give accurate results.
- HIV/AIDS is caused by HIV in which T helper cells are destroyed. HIV detection and diagnosis is important for the control of AIDS epidemic. Several antibody-based techniques are utilized for the detection of HIV. But these tests can be done only after 4–12 weeks of infection, when the body starts producing antibody. Initial HIV tests include detection of HIV RNA or nucleic acids. HIV-specific antibodies are detected by specific test like ELISA, Western blotting, IFA, and detuned test. Different rapid diagnostic tests are employed which give result easily and within an hour. HIV RNA detection and p24 antigen is detected in sample for HIV/AIDS diagnosis.

Multiple Choice Questions

1. Which gene is targeted in nested PCR for malaria diagnosis?
   (a) 18S rRNA gene
   (b) 16S rRNA gene
   (c) tRNA gene
   (d) None
2. Mammogram is used for diagnosis of which cancer?
   (a) Blood cancer
   (b) Gastric cancer
   (c) Breast cancer
   (d) Cervical cancer
3. Mantoux tuberculin skin test is a reliable test for the diagnosis of tuberculosis. This test takes how much time to give result?
   (a) 24–48 h
   (b) 48–72 h
   (c) 60–80 h
   (d) >24 h
4. For which cancer Oval-1, CA-125, HE4 are tumor marker?
   (a) Breast cancer
   (b) Ovarian cancer
   (c) Thyroid cancer
   (d) Pancreatic cancer
5. Which molecular test is used for early diagnosis of multidrug-resistant TB?
   (a) Nucleic acid amplification test (NAAT)
   (b) ELISA
   (c) IFA
   (d) Southern blotting

6. Detection of p24 is done by:
   (a) ELISA
   (b) RIA
   (c) IFA
   (d) Both b and c
7. Microscope and rapid diagnostic test technique become insensitive for malaria at which density of parasite?
   (a) 100–150 parasite/μl
   (b) Below 100 parasite/μl
   (c) Above 200 parasite/μl
   (d) None
8. LAM (lipoarabinomannan) present in cell wall of *M. tuberculosis* is detected by which immuno-technique?
   (a) IFA
   (b) RIA
   (c) Northern blotting
   (d) Sandwich ELISA
9. Which gene is targeted for PCR amplification in tuberculosis diagnosis?
   (a) IS6110
   (b) 16S rRNA
   (c) 18S rRNA
   (d) None
10. For malaria diagnosis artemisinin-based combination therapy (ACTs) therapy is recommended by:
    (a) FDA
    (b) ASCO
    (c) WHO
    (d) NACB
11. In the diagnosis of malaria, nucleic acid lateral flow immunoassay (NALFIA) is combined with which technique?
    (a) RT-PCR
    (b) Direct blood PCR (db-PCR)
    (c) Nested PCR
    (d) RIA
12. In HIV infection which cell is depleted?
    (a) CD4 T cell
    (b) B cell
    (c) CD8 T cell
    (d) Natural killer cell
13. Which virus is a major causative agent of many anal, vaginal, vulvar, and penile cancers?
    (a) Human immunodeficiency virus
    (b) Paramyxovirus
    (c) Human papillomavirus (HPV)
    (d) Brome mosaic virus
14. Interferon-gamma release assay-IGRA is used to detect the presence of:
    (a) Malaria
    (b) Tuberculosis
    (c) HIV/AIDS
    (d) Cancer
15. Which stain is used for malaria detection?
    (a) Gram stain
    (b) Giemsa stain
    (c) Acid-fast stain
    (d) None
16. Cancer is characterized by:
    (a) Unregulated cell growth
    (b) Invasion to other tissues
    (c) Undifferentiated mass of cells
    (d) All of the above
17. Cancer cells have:
    (a) Functional DNA repair pathways
    (b) Differentiated cells
    (c) Functional cell cycle check points
    (d) Lack of apoptosis
18. TNM system of cancer staging depends on:
    (a) Time of occurrence of cancer
    (b) Size and extent of primary tumor
    (c) Type of cells involved
    (d) None of the above
19. Tumor markers are:
    (a) Produced by cancerous cells
    (b) Chemical moieties produced in response to cancer
    (c) Used for detection and diagnosis
    (d) All of the above
20. Tuberculosis infection is caused by:
    (a) Mycobacterium
    (b) Candida
    (c) *E. coli*
    (d) *Plasmodium*
21. Latent TB is characterized by:
    (a) Fever
    (b) Blood in cough
    (c) Chest pain
    (d) Positive tuberculin test
22. Most commonly used TB test is:
    (a) Mantoux tuberculin skin test
    (b) Chest X-ray

(c) Microscopy
(d) TB blood test

23. Common target for PCR-based TB diagnosis is:
(a) IS6110
(b) rRNA gene
(c) Housekeeping gene
(d) All of them

24. Malaria is transmitted by:
(a) Culex
(b) Ades
(c) *Anopheles*
(d) All of them

25. During blood meal, the mosquito infects:
(a) Sporozoites
(b) Merozoites
(c) Trophozoites
(d) Schizonts

26. During infection, HIV preferentially binds to:
(a) CD8+ cells
(b) CD4+ helper T cells
(c) Lymphocytes
(d) RBC

27. During acute phase of HIV infection, which of the following can be detected?
(a) HIV-specific antibodies
(b) p24
(c) Viral nucleic acid
(d) b and c

Answers

1. (a); 2. (c); 3. (b); 4. (b); 5. (a); 6. (a); 7. (b); 8. (d); 9. (a); 10. (c); 11. (b); 12. (a); 13. (c); 14. (b); 15. (b); 16. (d); 17. (e); 18. (b); 19. (d); 20. (a); 21. (d); 22. (a); 23. (a); 24. (c); 25. (a); 26. (b); 27. (d).

Review Questions

Q1. How are molecular techniques useful in the diagnosis of diseases?
Q2. What are the imaging techniques used in cancer diagnosis? Explain briefly.
Q3. How do multidrug-resistant strains of tuberculosis develop? Discuss method used for its detection?
Q4. Explain the life cycle of *Plasmodium* with the help of diagram.
Q5. Describe the AIDS virus and different stages in its infection process.
Q6. What are the culture methods used for TB diagnosis?
Q7. Define the following:
(a) Loop-mediated isothermal amplification
(b) Direct blood PCR
(c) Line probe assay
Q8. What are the methods used for detection of viral RNA in HIV diagnosis?
Q9. Describe the TNM system.
Q10. How is malaria detected in the clinical sample (blood)?
Q11. Name some of the tumor markers with examples. What is the relevance of these markers?
Q12. Give details of the tests which are used to detect HIV-specific antibodies.
Q13. Discuss the impairment pathway/process that occurs during cancer.
Q14. How is CT scan useful for cancer detection?
Q15. What are the different screening tests available for the detection of early stages of cancer?
Q16. State how tumor markers can be utilized for cancer diagnosis and what are the drawbacks?
Q17. Describe the molecular techniques utilized for the detection of mycobacterium infection.
Q18. Describe the life cycle of malaria parasite.
Q19. Explain the rapid diagnostic tests for malaria.
Q20. Describe antibody-based diagnosis of AIDS.

References

1. American College of Radiology and Radiological Society of North America (2012) Patient safety: radiation dose in X-Ray and CT exams Radiological Society of North America, Inc. (RSNA)
2. American Joint Committee on Cancer (2010) AJCC cancer staging manual, 7th edn. Springer, New York

3. ATS, CDC, and Infectious Diseases Society of America (2003) Treatment of tuberculosis MMWR 52 (RR-11)
4. Bigbee W, Herberman RB (2003) Tumor markers and immunodiagnosis. In: Bast RC Jr, Kufe DW, Pollock RE (eds) Cancer medicine, 6th edn. BC Decker, Hamilton
5. Borowitz M et al (2004) Pathology and laboratory medicine. In: Abeloff MD, Armitage JO, Niederhuber JE, Kastan MB, McKenna WG (eds) Clinical oncology, 3rd edn. Churchill Livingstone, London
6. Buys SS et al (2011) Effect of screening on ovarian cancer mortality: the prostate, lung, colorectal and ovarian (PLCO) cancer screening randomized controlled trial. JAMA 305:2295–2303
7. Centers for Disease Control and Prevention (2005) Guidelines for the investigation of contacts of persons with infectious tuberculosis and guidelines for using the QuantiFERON®-TB gold test for detecting mycobacterium tuberculosis infection, Coordinating Center for Health Information and Service, Centers for Disease Control and Prevention (CDC), U.S. Department of Health and Human Services, Atlanta, GA. MMWR 54 (No. RR-15)
8. Connolly JL et al (2003) Principles of cancer pathology. In: Bast RC Jr, Kufe DW, Pollock RE et al (eds) Cancer medicine, 6th edn. BC Decker, Hamilton
9. Cramer DW et al (2011) Ovarian cancer biomarker performance in prostate, lung, colorectal, and ovarian cancer screening trial specimens. Cancer Prev Res 4:365–374
10. Fearon M (2005) The laboratory diagnosis of HIV infections. Can J Infect Dis Med Microbiol 16:26–30
11. Global TB/HIV working group. Tuberculosis and HIV. http://www.who.int/en
12. González A et al (2009) Projected cancer risks from computed tomographic scans performed in the United States in 2007. Arch Intern Med 169:2071–2077
13. Gürsoy NC, Yakupoğulları Y, Tekerekoğlu MS, Otlu B (2016) Evaluation of the diagnostic performance of Xpert MTB/RIF test for the detection of Mycobacterium tuberculosis and rifampin resistance in clinical samples. Mikrobiyol Bul 50:196–204
14. Kasper DL, Braunwald E, Fauci AS, Hauser SL, Longo DL, Jameson JL (eds) (2008) Harrison's principles of internal medicine, 17th edn. Mc Graw-Hill Medical publishers, New York
15. Howlader N et al (2013) SEER cancer statistics review, 1975–2010. National Cancer Institute, Bethesda, http://seer.cancer.gov/csr/1975_2010/
16. Isenberg HD, D'Amato RF, Heifets L, Murray PR, Scardamaglia M, Jacobs MC, Alperstein P, Niles A (1991) A collaborative feasibility study of a biphasic system (Roche septi-chek AFB) for rapid detection and isolation of mycobacteria. J Clin Microbiol 29:1719
17. Iweala O (2004) HIV diagnostic tests: an overview. Contraception 70:141–147
18. Katki HA et al (2011) Cervical cancer risk for women undergoing concurrent testing for human papillomavirus and cervical cytology: a population-based study in routine clinical practice. Lancet Oncol 12:663–672
19. Kumudunayana WM et al (2011) Usefulness of polymerase chain reaction to supplement field microscopy in a pre-selected population with a high probability of malaria infections. Am J Trop Med Hyg 85:6–11. doi: 10.4269/ajtmh.2011.10-0337
20. Lee PW et al (2012) Application of loop-mediated isothermal amplification for malaria diagnosis during a follow-up study in Sao Tome. Malar J 11:408
21. Mandelblatt JS et al (2009) Effects of mammography screening under different screening schedules: model estimates of potential benefits and harms. Ann Intern Med 151:738–747
22. Mathews JD et al (2013) Cancer risk in 680 000 people exposed to computed tomography scans in childhood or adolescence: data linkage study of 11 million Australians. Br Med J 346:f2360. doi:10.1136/bmj.f2360
23. Mejia GI, Castrillon L, Trujilo H, Robledo JA (1999) Micro colony detection on 7H11 thin layer culture as an alternative for rapid diagnosis of Mycobacterium tuberculosis infection. Int J Tuberc Lung Dis 3:138–42.
24. Michael L, Wilson MD (2013) Rapid diagnosis of mycobacterium tuberculosis infection and drug susceptibility testing. Arch Pathol Lab Med 137:812–819. doi:10.5858/arpa.2011-0578
25. Moyer VA (2012) On behalf of the U.S. Preventive Services Task Force. Screening for cervical cancer: U.S. Preventive Services Task Force Recommendation Statement. Ann Intern Med. 156:880–91
26. Neumann RD, Bluemke DA (2010) Tracking radiation exposure from diagnostic imaging devices at the NIH. J Am Coll Radiol 7:87–89
27. Pisano ED et al (2005) Diagnostic performance of digital versus film mammography for breast-cancer screening. N Engl J Med 353:1773–1783
28. Ray S et al (2013) Diagnosis and management of miliary tuberculosis: current state and future perspectives. Ther Clin Risk Manag 9:9–26. doi:10.2147/TCRM.S29179
29. Schröder FH et al (2009) Screening and prostate-cancer mortality in a randomized European study. N Engl J Med 360:1320–1328
30. Siddiqi SH, Libonati JP, Middlebrook G (1981) Evaluation of rapid radiometric method for drug susceptibility testing of Mycobacterium tuberculosis. J Clin Microbiol 13:908
31. Smith-Bindman R et al (2009) Radiation dose associated with common computed tomography examinations and the associated lifetime attributable risk of cancer. Arch Intern Med 169:2078–2086
32. U.S. Food and Drug Administration (August 2009) What are the radiation risks from CT? U.S. Department of Health and Human Services, MD 20993

33. Updated Guidelines for the use of nucleic acid amplification tests in the diagnosis of tuberculosis. (2009) MMWR 58(1)
34. World Health Organization (2002) Global tuberculosis control, surveillance planning, financing WHO/CDS/Tuberculosis/295, Geneva
35. Zhao J et al (2012) Adoption of rapid diagnostic test for the diagnosis of malaria, a preliminary analysis of the global find program data, 2005 to 2010. PLoS One 7:e43549. doi:10.137/journal.pone.0043549

Some Related Resources

Global TB/HIV working group. Tuberculosis and HIV.http://www.who.int/en
http://fieldresearch.msf.org/msf/bistream/10144/37314/1/LID%202002%20sept_Malaria%20R&D%20agenda.pdf
http://seer.cancer.gov/csr/1975_2010/
http://www.cancer.gov/cancertopics/factsheet/Detection/Pap-test
http://www.cancer.gov/cancertopics/factsheet/Detection/pathology-reports
http://www.cancer.gov/cancertopics/factsheet/Detection/PSA
http://www.cancer.gov/cancertopics/wyntk/cancer
http://www.cancer.org/TreatmentTreatmentsandSideEffects/TreatmentTypes/Immunotherapy/index
http://www.cdc.gov/malaria/diagnosis_treatment/diagnosis.html
http://www.cdc.gov/tb/topic/testing/default.htm
http://www.georgiahealth.edu/itss/edtoolbox/7390/Cancer1/HIV%20testing.pdf
http://www.mayoclinic.com/health/tuberculosis/DS00372/DSECTION=tests-and-diagnosisdiagnosis

11 Molecular Therapeutics

Abstract

With the advancement in genomic and proteomic studies, there is an increase in the understanding of disease processes, their causes, and mediators involved in it. The pharmaceutical companies are targeting the biomolecules for treatment of the diseases (such as proteins, enzymes, ligands, receptors, and genes) that are involved either during the onset of the disease or in its progression. These biological drugs are very specific and have fewer side effects as compared to their chemical counterparts. The biotechnology-derived products (1) are safe, (2) are targeted, (3) have high efficacy, (4) are with better diagnostic and therapeutic potential, and (5) show good response. Hence, biomolecules are increasingly being used as drugs. In this chapter, the concept and uses of immunomodulators, immunosuppressors, interferons, protein therapeutics and monoclonal antibodies, their uses, and therapeutic potential are discussed.

11.1 Introduction

With the advancement in genomic and proteomic studies, there is an increase in the understanding of disease processes, their causes, and mediators. The pharmaceutical companies are targeting the biomolecules (such as proteins, enzymes, ligands, receptors, and genes) that are involved either during the onset of the disease or in its progression for treatment of the diseases. These biological drugs are very specific and have fewer side effects as compared to their chemical counterparts. The biotechnology-derived products (1) are safe, (2) are targeted, (3) have high efficacy, (4) are with better diagnostic and therapeutic potential, and (5) show good response. Hence, biomolecules are increasingly being used as drugs. In this chapter, we have tried to summarize biomolecules used for therapeutic purposes.

11.2 Immunostimulants

These are biomolecules or cytokines that are responsible for the activation of the immune system. They increase the ability of the immune system to fight infectious agents and cancer. These can act on different components of the immune system, such as adaptive or innate immunity, and mediate powerful responses. The immunostimulants may be specific for a particular antigen or might nonspecifically boost the entire immune

V. Gupta et al., *Basic and Applied Aspects of Biotechnology*, DOI 10.1007/978-981-10-0875-7_11

system. The purpose of these immunostimulants is promotion of immune functions like phagocytosis, complement activation, or activation of other pathways [9]. These find applications in cancer therapy and as adjuvants in vaccines ensuring activation of adaptive immunity. Some of the important immunostimulants are:

1. Adjuvants: These compounds are administered with vaccines to increase their immunogenicity and result in the activation of adaptive immune response. Even non-immunogenic hapten is capable of activating adaptive immunity in the presence of an adjuvant (for details, refer to Chap. 14).
2. Levamisole (Ergamisol): This is used in atopic diseases and helps to restore function of T and B lymphocytes, macrophages, and other leukocytes.
3. Thalidomide (Thalomid): This is used in urinary bladder tumor and papillary tumors.
4. Interferon (INF): They constitute a group of important natural molecules that are produced in response to viral infection. They are being used as therapeutics against cancer. (Refer to Interferon as therapeutic agents Sect. 11.4.)
5. Interleukin-2 (IL-2) (aldesleukin; Proleukin): They are natural cytokines produced by Th cells. They enhance proliferation and activation of T lymphocytes and macrophages.
6. Usage of antibody: *In passive immunity*, they are used as preformed antibody when there is a risk of infection. They act like normal antibody and promote clearance of the pathogen. *For therapeutics*, monoclonal antibodies (mAbs) are being used against many diseases especially cancer and autoimmunity (see Sect. 11.6). They promote the clearance of cells to which they bind, thus they are capable of specifically targeting tumors.

11.3 Immunosuppressors

The agents or drugs that are used to suppress or reduce the immune response are known as immunosuppressors. Immunosuppressants may prevent the recognition or the response of the immune system. They are used in organ transplantation (where they help to prevent rejection of foreign (allograft) graft or organ) and in the case of autoimmunity (where immune system targets own body components). However these drugs have potential drawbacks as they have harmful side effects such as fever, pain, difficulty in urination, nausea, vomiting, loss of appetite, and uncontrolled bleeding due to injury or infection. Their usage makes the individual susceptible to other infections and cancer. The physician prescribing these must be aware of the patient's health conditions such as pregnancy, allergies, kidney, or liver disease or any ongoing bacterial or viral infection. These drugs are contraindicated under some conditions, thus should be taken under strict medical supervision. Some of the immunosuppressors in clinical usage are:

- Antimetabolites such as azathioprine (Imuran) and mycophenolate mofetil (MMF). Azathioprine is a purine antimetabolite; however, their usage posttransplantation has potential risk of developing malignancy.
- Macrolides {cyclosporine (Neoral), tacrolimus, sirolimus} inhibit production of IL-2. Cyclosporin is a high molecular weight undecapeptide which is a calcineurin inhibitor and is eliminated by hepatic metabolism. It blocks activation of T lymphocytes. Sirolimus (Rapamune), a macrolide antibiotic, inhibits G1 to S phase cell division; it is also being explored for its usage as an immunosuppressive drug. Tacrolimus is active against Th cells and inhibits production of IL-2. It may cause nephro- and neurotoxicity.
- Corticosteroids (prednisolone) inhibit production of IL-1 and IL-6 by macrophages and can inhibit all stages of T-cell activation. It is routinely used as an immunosuppressor. However, it has negative side effects such as hyperlipidemia, osteoporosis, infections, Cushing disease, and so many others.
- Antibody: Polyclonal antilymphocytic antibody is used for lysis of the immune cells. Some of the examples are horse antithymocyte globulin (Atgam), rabbit antithymocyte globulin (thymoglobulin), muromonab-CD3

with trade name OKT3 which blocks T-cell function, anti-CD25 antibody (chimeric murine–human basiliximab with trade name Simulect) and daclizumab with trade name Zenapax which is a humanized antibody against IL-2 receptor alpha chain (CD25), and anti-CD20 rituximab with trade name Rituxan and Mabthera, against CD20 antigen present on B cells.

11.4 Interferons

Interferons (INF) are a group of proteins that are produced when a cell is infected by a virus. However the protection or resistance to viral infection by INF is short term and does not last forever. They are produced and secreted by cells of immune system (leukocytes, NK cells, fibroblast and epithelial cells). Three classes of INFs have been identified (INF-α, INF-β, INF-γ). There are two types of interferons: type I interferon includes INF-α and INF-β and type II interferon includes INF-γ. INFs modulate the response of the immune system against viruses, bacteria, and cancer. They do not kill virus-infected cells, but boost the immune system response (immunomodulators) for clearing the infection. They reduce the growth of cancer cells by inducing cell differentiation, inhibiting proliferation, inhibiting angiogenesis, and exerting immunoregulatory functions. They have profound effects on the immune system as they increase the expression of class I and class II MHC molecules, augment the activity of NK cells, and make the cell a potential target for T-cytotoxic cell attack.

The interferon acts by:

- Induction of 2'-5′-oligo (A) synthetases. This requires double-stranded RNA for its activation. Activated synthetase in the presence of double-stranded RNA polymerizes oligo (A) and thereby activates RNAse-L. Activation of RNAse-L degrades all single-stranded RNA (rRNA, mRNA, vRNA).
- Induction of serine–threonine kinase (PKR) which is activated by double-stranded RNA and phosphorylates and negatively regulates the translational initiation factor eIF-α. This causes shutting off of protein synthesis in infected cells.
- Induction of Mx proteins which belong to the family of GTPases. They are important in inhibiting the replication of influenza virus and vesicular stomatitis virus (VSV).
- All these effects are directed against the infected cell causing viral and cellular dysfunction (Fig. 11.1). Table 11.1 shows the mechanisms used by the virus to evade the immune system.

Interferons α and β are produced by all cells, whereas γ is produced by NK and T cells. Side effects of interferon include flu-like symptoms after each injection, such as fever, chills, headache, muscle aches, and pain. The symptoms vary from mild to severe. Tissue damage at the site of injection may occur. Depression and suicide have been reported; however, it is not clear if that is due to the therapy or the disease.

11.5 Proteins as Therapeutic Agent

In humans, nearly 25,000–35,000 genes encode for more than 1 lakh functional proteins. Thus, the proteome is highly variable. The proteins are further modified after their production by post-translational modifications such as cleavage, phosphorylation, acylation, and glycosylation. Inappropriate folding, modification, or mutation causes the protein to become nonfunctional and leads to disease. Proteins are used as drugs and several protein therapeutics have been recommended by the FDA for clinical usage. More are in the process of development [1]. These are the various advantages of using protein therapeutics:

1. Being a biomolecule, it is highly specific and is naturally present; thus, it does not interfere with other pathways.
2. Their role cannot be mimicked by any chemical compound.
3. As they are natural, their presence does not lead to adverse side effects on the host body.

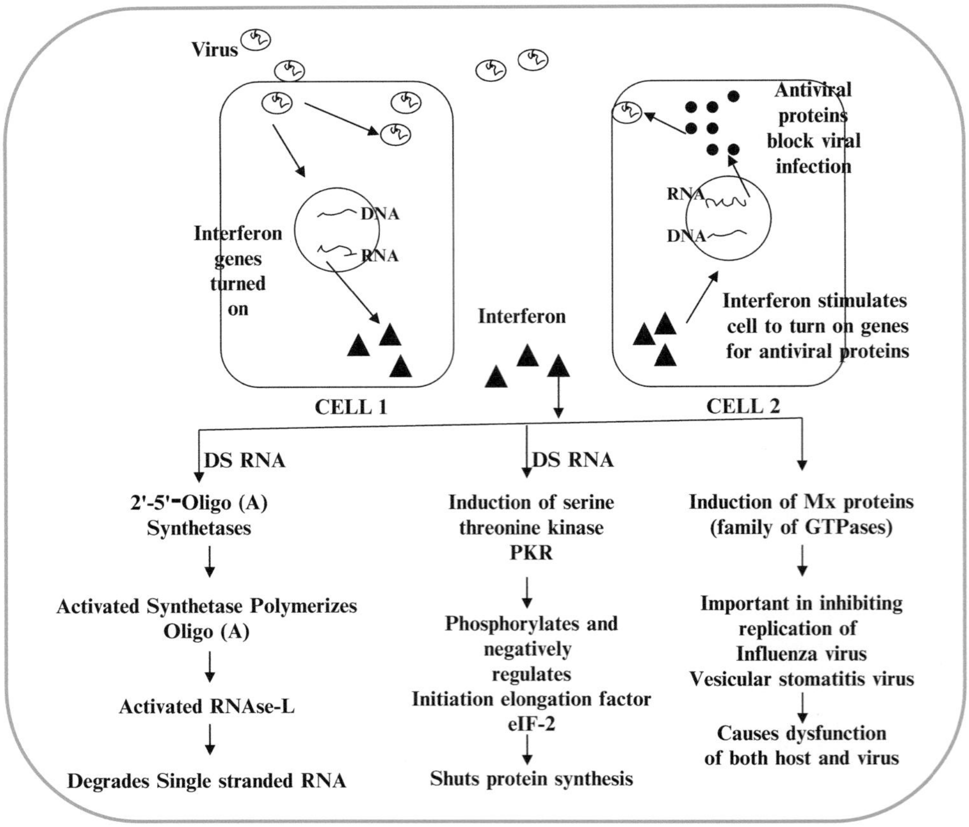

Fig. 11.1 Shows the events which occur after virus infects a live cell. Cell 1 is virus infected and produces interferon. This interferon is released and acts on nearby cells. The double-stranded RNA (dsRNA) and interferon stimulate RNAse-L; it degrades single-stranded RNA, stimulates PKR, inhibits eIF-2, and shuts down protein synthesis. For some of the virus, activation of Mx proteins leads to the inhibition. In whole sequence of events, Cell 1 is killed by virus, whereas cell 2 is protected

4. They are well tolerated and do not lead to any immune response in the recipient.
5. Protein therapy for nonfunctional or mutated gene has its own limitations and is currently not available for many of the diseases.
6. The time taken for approval of protein therapeutics is faster than other drugs.
7. Since the manufacturers can obtain the right protection and patent on protein therapeutics, there is financial motivation for developing protein drugs [6].

Protein therapy is effective, safe, and life saving. Earlier the proteins were purified from their native sources but that had potential risk of transmission of unknown pathogens. Nowadays, various systems for protein production have evolved, such as bacteria, fungi, insect cell, mammalian cell, human cell, and transgenic animals [17]. The potential advantages, drawbacks, and risks are detailed in Chap. 4. The trend of production of proteins started with the production and usage of insulin for the treatment of diabetes. The recombinant insulin was the first protein therapeutic which was approved by the US FDA in 1982 for the treatment of diabetes mellitus.

The protein therapeutics may be divided into four groups on the basis of their function and application. Groups I and II are approved as protein therapeutics, whereas III and IV are approved as vaccines and diagnostic agents [6].

Table 11.1 Evasion of immune response by virus

Virus type	Effects on immune components for infecting the host
Adenoviruses	Prevent TNF-induced lysis and block HLA class I expression. It evades the action of T-cytotoxic cells by inhibition of MHC I expression
Herpes simplex virus (HSV) and cytomegalovirus (CMV)	Blocks MHC I expression and thus evades the action of T-cytotoxic cells
Epstein–Barr virus (EBV)	Has IL-10 homologue that inhibits natural killer (NK) cell and T-cell response
Vaccinia virus	Has soluble receptor for INF-α and binding proteins for INF-γ, IL-1, IL-18, and TNF
	It affects host's innate and adaptive immune responses
	It also encodes caspase inhibitor that inhibits the ability of $CD8^+$ cytotoxic cells to kill virus-infected cells
Pox virus and herpesvirus	Chemokine-binding proteins inhibit cellular inflammatory responses

11.5.1 Group I

The proteins playing the function of enzymes or regulators are in represented in group I. They are further divided into Ia, Ib, and Ic.

Ia: These are used in metabolic or endocrine defects arising due to genetic mutation, aberrant protein expression, or defect in posttranslational modifications (see Table 11.2Ia).

Ib: The therapeutics in this group augment or increase the existing pathways of hematological, endocrine, or immune response. Table 11.2Ib, shows the group which includes the therapeutic factors like interferon or growth factors required for augmentation in existing activity.

Ic: The use of naturally occurring proteins to modify the pathophysiology of human diseases. This requires a thorough understanding of protein function in humans and other animals. Protein therapeutics [6] with novel functions are shown in Table 11.2Ic

Some other therapeutics with novel functions (Ic) include:

- Papain (Accuzyme; Panafil) is obtained from papaya fruit and utilized in debridement of necrotic tissue
- Both L-asparaginase (Elspar) and PEG asparaginase (Oncaspar) are nonrecombinant and are used in acute lymphocytic leukemia.
- Lepirudin (Refludan) is used in heparin-induced thrombocytopenia and is obtained from salivary gland of medicinal leech *Hirudo medicinalis*.
- Streptokinase (Streptase) is used in transmural myocardial infarction and is produced by group C β-hemolytic streptococci.

11.5.2 Group II

They are proteins having targeted effects and are subdivided into two groups, IIa and IIb. This class includes antibodies.

IIa: These products interfere with a molecule in an organism. The binding might result in blocking the function of the normal target, and thus either the target is destroyed or a function post binding is stimulated. Table 11.3IIa shows the compounds present in group IIa as therapeutic agents used in various disorders.

IIb: They deliver proteins or compounds to the targeted site. Thus, they are referred as target-specific therapeutics. The delivery of the therapeutic agent to its target poses the greatest challenge. The specialized transport of biomolecules under normal conditions is performed by various proteins. The therapeutics used for this purpose are:

- Denileukin diftitox (Ontak).
- Ibritumomab tiuxetan (Zevalin) is a mAb against CD20 and linked to radioactive yttrium (Y-90) which targets CD20-positive non-Hodgkin's lymphoma cells.

Table 11.2Ia Shows the important therapeutics where protein function is abnormal or deficient

Indication	Therapeutic agent	Trade name	Disease	Function
Deficiency of hormones	Insulin	Humulin, Novolin	Diabetes mellitus	Maintains blood glucose
	Insulin human inhalation	Exubera[b]	Diabetes mellitus	Insulin for inhalation with faster action
	Growth hormone (GH)	Genotropin, Humatrope, Norditropin	Deficiency disorders of growth hormone	Regulates many hormones (somatomedins) for growth, development, and maintenance
	Somatotropin			
	Mecasermin	Increlex	Deficiency of GH or severe insulin-like growth factor 1 (IGF-1) deficiency	Induces chondrocyte growth, mitogenesis, and organ growth
Thrombosis disorders	Factor VIII	Bioclate, Helixate, Kogenate, ReFacto	Blood-clotting disorder (hemophilia A)	Blood-clotting factor
	Factor IX	BeneFix	Blood-clotting disorder (hemophilia B)	Blood-clotting factor
Deficiency of important enzymes	β-Glucocerebrosidase	Cerezyme, Ceredase[a, b]	Gaucher disease	Hydrolyzes glucocerebroside to glucose and ceramide, recognized by endocytic receptors on macrophages, and accumulates lipids in Gaucher disease (anemia, thrombocytopenia, hepatomegaly)
	Alglucosidase-α	Myozyme	Pompe disease (glycogen storage disease type II)	Hydrolyzes α-1,4- and α-1,6-glycosidic linkages of lysosomal glycogen
	Adenosine deaminase (pegademase bovine, PEG-ADA)	Adagen[a]	SCID (severe combined immunodeficiency) due to ADA deficiency	Prevents accumulation of adenosine by metabolizing it
	α -1-Proteinase inhibitor	Aralast[a], Prolastin[a]	Congenital α-1 antitrypsin deficiency	Purified from pooled plasma, it prevents elastase-mediated destruction of pulmonary tissue
	Pancreatic enzymes (lipase, amylase, protease)	Acro-Lase[a], Cotazym[a], Creon[a], Donnazyme[a], Pancrease[a], Viokase[a], Zymase[a]	In cystic fibrosis, chronic pancreatitis, pancreatic insufficiency, or duct obstruction, gas, bloating	Purified from hogs and pigs; digests food

[a]Produced without usage of recombinant DNA technology
[b]The drug is being withdrawn

Table 11.2Ib The table shows important therapeutic agents which are used to augment the responses of the body

Indication	Therapeutic agent	Trade name	Disease	Function
Hematopoiesis	Erythropoietin (EPO), Epoetin-α	Epogen, Procrit	Anemia	Stimulates production of erythrocytes
	Darbepoetin-α	Aranesp	Anemia	Stimulates production of erythrocytes; it is modified EPO with longer half-life
	Filgrastim (granulocyte colony-stimulating factor, G-CSF)	Neupogen	Neutropenia (due to AIDS, chemotherapy)	
	Pegfilgrastim (Peg-G-CSF)	Neulasta	Neutropenia	Stimulates production, differentiation, and migration of neutrophils
				Stimulates neutrophil production
	Sargramostim (granulocyte-macrophage colony-stimulating factor; GM-CSF)	Leukine	Leukopenia (due to AIDS)	Stimulates proliferation and differentiation of neutrophils, eosinophil, and blood monocytes
Fertility	Human follicle-stimulating hormone (FSH)	Gonal-F, Follistim	In in vitro fertilization (IVF)	Helps to enhance ovulation
	Human chorionic gonadotropin (HCG)	Ovidrel	In in vitro fertilization (IVF)	Stimulates the rupture of ovarian follicle and ovulation
	Lutropin-α	Luveris	Infertility due to deficiency of luteinizing hormone (LH)	r-LH; support FSH-induced follicular development
Immunomodulators	Interferon (IFN)	Infergen	Chronic HCV infection	Immunoregulator
	Type I α-IFN alfacon		Hairy cell leukemia, chronic myelogenous leukemia, Kaposi's sarcoma, chronic HCV infection	Immunoregulator
	IFN-α2a	Roferon-A		
	Peginterferon-α2a	Pegasys	Chronic HCV	PEG-conjugated for increased half life
	IFN-α2b	Intron A	HBV, HCV, melanoma, Kaposi's sarcoma, follicular lymphoma, hairy cell leukemia, condylomata acuminate	Immunoregulator
	PegIFN-α2b	Peg-Intron	HCV	PEG-conjugated for increased half life
	Interferon-αn3 (IFN-αn3)	Alferon N[a]	Condylomata acuminate (genital warts by human papillomavirus)	Nonrecombinant mechanism unknown
	Interferon-β1a (rIFN-β)	Avonex, Rebif	Multiple sclerosis	Antiviral and immunomodulator
	Interferon-β1b (rIFN-β)	Betaseron	Multiple sclerosis	Antiviral and immunomodulator
	Interferon-γ1a (rIFN -γ)	Actimmune	Chronic granulomatous disease	Increased inflammatory and antimicrobial response
	Aldesleukin (interleukin-2 (IL-2)), epidermal thymocyte-activating factor (ETAF)	Proleukin	Metastatic renal cell carcinoma	Stimulates –T, -B, and –NK cells

(continued)

Table 11.2Ib (continued)

Indication	Therapeutic agent	Trade name	Disease	Function
Growth factors and thrombosis	Alteplase (tissue plasminogen activator, tPA)	Activase	Acute myocardial infarction, ischemic stroke, pulmonary embolism	Promotes fibrinolysis by binding fibrin and converting plasminogen to plasmin
	Reteplase (deletion mutant of tPA)	Retavase	Acute myocardial infarction	Deletion of domain binding to inhibitor
	Tenecteplase	TNKase	Acute myocardial infarction	tPA with greater specificity by replacement of amino acids
	Factor VIIa	NovoSeven	Hemorrhage with hemophilia A and B	Initiates coagulation cascade
	Recombinant human bone morphogenetic protein 7 (rhBMP-7)	Osteogenic protein 1	Tibial fracture nonunion, lumbar spine fusion	Mechanism unknown
	Palifermin (keratinocyte growth factor, KGF)	Kepivance	Severe oral mucositis	Stimulates keratinocyte growth in skin
	Becaplermin (platelet-derived growth factor, PDGF)	Regranex	Used in diabetic ulcers	Promotes wound healing by enhancing granulation tissue formation

[a]Produced without usage of recombinant DNA technology

Table 11.2Ic This table shows the protein therapeutics with novel functions

Indication	Therapeutic agent	Trade name	Clinical usage	Function
Cosmetic	Botulinum toxin type A	Botox	Dystonia, cosmetic uses	Prevents acetylcholine release
	Botulinum toxin type B	Myobloc	Dystonia, cosmetic uses	Prevents acetylcholine release
Skin ailment	Collagenase	Collagenase Santyl	Chronic dermal ulcers and severely burned areas	Digests collagen in necrotic base of wounds
Drug supplement	Hyaluronidase (bovine, ovine)	Amphadase (bovine), Hydase (bovine)	Used for enhanced absorption and dispersion of injected drugs	Catalyzes hydrolysis of hyaluronic acid for faster drug absorption
	Hyaluronidase (recombinant human)	Hylenex	Increases the absorption and dispersion of injected drugs	Catalyzes hydrolysis of hyaluronic acid for faster drug absorption

Table 11.3IIa This table shows the therapeutic agents used in various disorders

Indication	Clinical agent/(Trade name)	Usage
Cancer therapy	Trastuzumab (Herceptin)	Breast cancer; humanized mAb IgG1 that targets HER2/neu cell surface receptor highly expressed on cancerous breast cells
	Rituximab (Rituxan)	Used singly or in combination for follicular CD20 B-cell non-Hodgkin's lymphoma; also used in rheumatoid arthritis in combination with methotrexate; it is chimeric human/mouse IgG1 antibody binding CD20 present on B cells
	Alemtuzumab (Campath)	Used in B-cell chronic lymphocytic leukemia; this is humanized mAb directed against CD52 present on T and B cells
	Panitumumab (Vectibix)	Used in metastatic colorectal cancer, this is human mAb that binds EGFR
	Cetuximab (Erbitux)	Used in colorectal cancer and head and neck cancer; the mAb is humanized binding EGFR
	Bevacizumab (Avastin)	Used in colorectal cancer and non-small cell lung cancer; the mAb is humanized binding all forms of VEGFA
Antibodies or immunoregulators	Abatacept (Orencia)	Rheumatoid arthritis (RA); fusion protein of extracellular domain of CTLA4 with Fc portion of human IgG1 (inhibits T-cell activation by binding CD80 and CD86 preventing interaction with CD28)
	Anakinra (Antril, Kineret)	Used in RA; it is a recombinant IL-1 receptor antagonist
	Adalimumab (Humira)	Used in RA, Crohn's disease, ankylosing spondylitis, psoriatic arthritis (PsA). It binds TNF-α and prevents its binding to its receptors
	Etanercept (Enbrel)	Used in RA, juvenile RA, PsA, and plaque psoriasis; it is a dimeric fusion protein between soluble TNF receptor and Fc portion of IgG1
	Infliximab (Remicade)	Used in RA, Crohn's disease, ankylosing spondylitis, psoriatic arthritis (PsA), and plaque psoriasis. Chimeric mAb that binds and neutralizes TNF-α and prevents ongoing inflammation
	Alefacept (Amevive)	Used in chronic plaque psoriasis and is a dimeric fusion protein that binds CD2 of lymphocytes and prevents interaction with LFA3
	Efalizumab (Raptiva)	Used in plaque psoriasis and is a humanized mAb against CD11a
	Natalizumab (Tysabri)	Used in relapsing multiple sclerosis and binds integrins preventing their interaction with VCAMs
	Eculizumab (Soliris)	Humanized mAb used in paroxysmal nocturnal hemoglobinuria and targets complement protein C5
Other important indications	Pegvisomant (Somavert)	Used in acromegaly and is PEG-conjugated recombinant human growth hormone which blocks GH receptors
	Abciximab (ReoPro)	Used in prevention of cardiac ischemia in unstable angina and is a Fab fragment of chimeric human/mouse mAb 7E3 that inhibits platelet aggregation
	Crotalidae polyvalent immune Fab (ovine) (CroFab)	Antivenom for Western diamondback, Eastern diamondback, Mojave rattlesnakes, and water moccasins. Consists of a mixture of Fab of IgG that binds and neutralizes venom toxins of ten clinically important North American Crotalidae snakes
	Ranibizumab (Lucentis)	Used in neovascular age-related macular degeneration and is a humanized mAb binding to VEGFA

(continued)

Table 11.3IIa (continued)

Indication	Clinical agent/(Trade name)	Usage
HIV infection	Enfuvirtide (Fuzeon)	Advanced HIV infection in adults and children >6 years of age; it is a 36 amino acid peptide that inhibits HIV entry into host cells by binding the envelope protein gp120/gp41
Prevention of transplant rejection	Muromonab-CD3 (Orthoclone OKT3)	Used in acute renal graft rejection or steroid-resistant cardiac or hepatic allograft rejection; it is a murine mAb that blocks the T cell by binding to its coreceptor CD3
	Daclizumab (Zenapax)	Used to prevent rejection of allograft in renal transplantation; it is a humanized IgG1 mAb that binds CD25 and prevents IL-2-mediated activation
	Basiliximab (Simulect)	Used to prevent rejection of allograft in renal transplantation; it is chimeric human/mouse IgG1 that binds CD25 and prevents IL-2-mediated activation
	Antithymocyte globulin (thymoglobulin)	Used in acute kidney transplant rejection and aplastic anemia; it selectively depletes T cells

- Gemtuzumab ozogamicin (Mylotarg) which has a small-molecule chemotherapeutic agent calicheamicin (antibiotic which can destroy DNA of cancer cells) linked with mAb against CD33. This is delivered to CD33-positive cells of acute myeloid leukemia, thus promoting their clearance.
- Tositumomab and ^{131}I-tositumomab (Bexxar and Bexxar I-131).

11.5.3 Group III

They are the vaccines, subdivided into three groups, IIIa, IIIb, and IIIc. They are being exploited for their usage for broad protection against infectious agents, and some of the FDA approved can protect against multiple infectious agents and includes synthetic, recombinant, and purified protein components. These represent subunit vaccines which are nonpathogenic, are safe, and can induce immunity.

IIIa: These protect against the deleterious foreign agent. The usage of hepatitis B surface antigen (HBsAg), which is a noninfectious component of hepatitis B virus (HBV) as a vaccine, protects the host from natural HBV infection:

- *Hepatitis B* surface antigen (HBsAg) vaccine marketed as Engerix and Recombivax HB. The vaccine is prepared by recombinant DNA technology using the gene encoding for surface antigen of the virus [7, 8].
- *Human papillomavirus* (HPV) vaccine marked as Gardasil. The vaccine is prepared by recombinant DNA technology using genes encoding for capsid proteins for four strains.
- OspA marketed as LYMErix which is a noninfectious lipoprotein on *Borrelia burgdorferi* outer surface gives protection against Lyme disease (however, the licensed vaccine was withdrawn from the market even after the success of phase III clinical trials).

IIIb: These are utilized for treating conditions of autoimmunity like erythroblastosis fetalis.

Rhophylac, which is an anti-rhesus (Rh) IgG, neutralizes Rh antigens that might activate antibody response against them in Rh-negative individuals. This is used in preventing obstetric complication in an Rh-negative mother during pregnancy.

IIIc: They are used in the therapy of cancer. These vaccines are promising as they can target specific cancers.

11.5.4 Group IV

These include the proteins in diagnostics. They are not used to treat a disease but are very useful in the diagnosis, both in vivo and in vitro.

Injecting purified protein derivative (PPD) can tell whether an individual has been exposed to *Mycobacterium tuberculosis*. If the immune response is activated, then it shows that the subject has been exposed to mycobacterium. Some examples are enlisted below (Table 11.4).

The protein therapeutic products are finding their usage in day-to-day clinical practice. The advancement in their manufacturing and engineering gives better therapeutic outcome [18]. They are safer, are effective, have improved bioavailability, and have lower immunogenicity.

11.6 Monoclonal Antibodies

An antibody specific for a particular epitope of the antigen is called a monoclonal antibody (mAb). It is used in several clinical applications. As the production of monoclonal antibodies requires priming of host with pathogens, therefore it is difficult to obtain human-derived antibody-secreting B cells. Monoclonal antibodies are usually obtained from non-human sources (mouse/horse). Antibodies are glycoproteins produced by the B lymphocytes of the immune system in response to specific components of foreign

Table 11.4 The list of some proteins which are useful in *in vivo* or *in vitro* diagnosis of diseases

In vivo diagnostics	Usage	Trade names
Exposure to *Mycobacterium*	Noninfectious recombinant purified protein derivative (PPD) from *Mycobacterium* in injected	DPPD
Defective secretion of growth hormone (GH)	Recombinant fragment of growth hormone–releasing hormone (GHRH) to stimulate release of GH from pituitary	Geref
Serum thyroglobulin in thyroid cancer	Use of thyroid-stimulating hormone (TSH) stimulates thyroid epithelial cells to take up iodine and secrete thyroglobulin, triiodothyronine, and thyroxine	Thyrogen
Prostate cancer diagnosis	Capromab pendetide, indium-111-labeled anti-PSA antibody which detects intracellular PSA	ProstaScint
Tumor of neuroendocrine and lymphoma	Indium-111-labeled octreotide	Octreoscan
Colon and ovarian cancer detection	Satumomab pendetide, indium-111-labeled mAb against tumor-associated glycoprotein (TAG-72)	OncoScint
Colon and breast cancer	Arcitumomab which is a technetium-labeled anti-carcinoembryonic antigen (CEA) antibody	CEA-Scan
Lung cancer (small cell) detection	Nofetumomab being technetium-labeled antibody for small cell lung cancer	Verluma
Acute venous thrombosis	Apcitide which is atechnetium-labeled synthetic peptide and binds receptors on activated platelets	AcuTect
In vitro diagnostics	**Usage**	**Trade names**
HIV infection	HIV antigens which detect presence of specific antibody by ELISA or Western blot	Enzyme immunoassay, OraQuick, Uni-Gold

Hybridoma Technology

The technology for the production of monoclonal antibodies was introduced by Georges Kohler and Cesar Milstein in 1975 and is known as "hybridoma technology." The key steps for using hybridoma technology for production of monoclonal antibody are discussed here (see Fig. 11.2).

- For obtaining B cells which can recognize and secrete the desirable antibodies, we require exposure of the immune system to the antigen. The exposure would trigger adaptive immune response resulting in the production of B cell specific for antigen.
- Antigen of choice is purified and injected in the mouse in two doses, (1) initial priming of the animal and (2) booster dose after 10 days. As the immune system of the animal recognizes the antigen as foreign, thus it mounts an attack by generation of T- and B-cell responses.
- B cells which are antibody secreting are separated, and as these cells have finite life span, they are allowed to fuse with myeloma cells resulting in the formation of hybridoma with immortal growth properties.
- In the next step, this hybridoma needs selection for B-M cell fusion (B-B and M-M fusion need to be discarded). Fusion

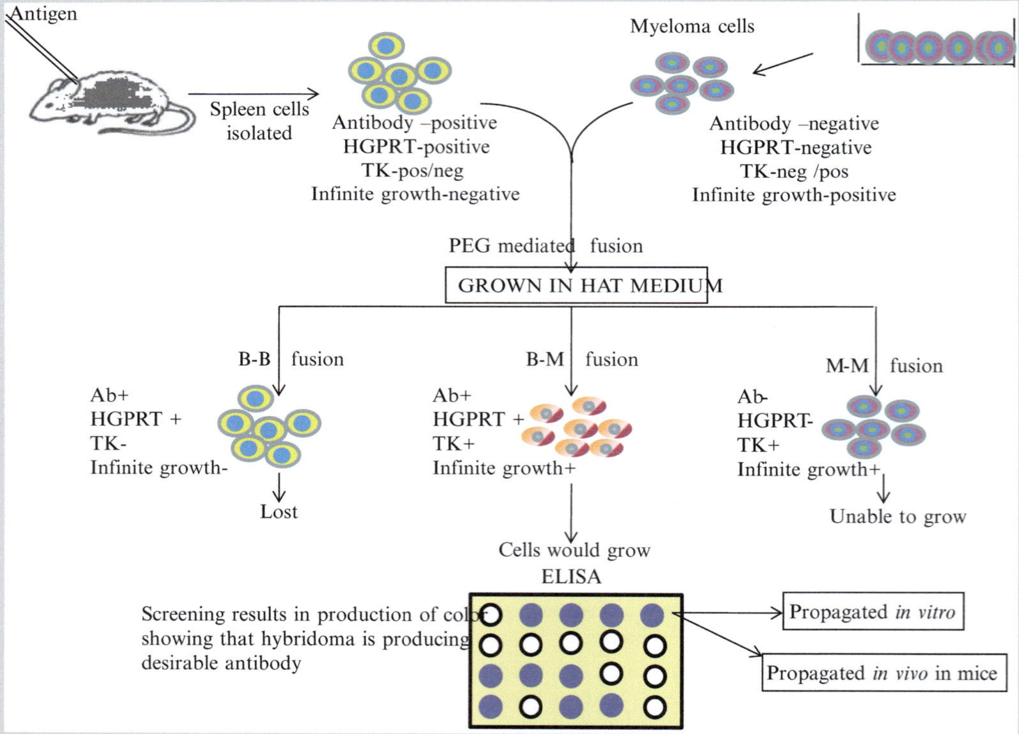

Fig. 11.2 The figure shows the hybridoma technology for production of monoclonal antibodies. The mice are immunized with desirable antigen and spleen cells are collected containing primed B cells. These B cells are antibody secreting, hypoxanthine–guanine phosphoribosyltransferase (HGPRT) positive, thymidine kinase positive or negative, and immortal growth negative. B cells are allowed to fuse with myeloma cells which are HGPRT and antibody negative. The fusion product is grown on HAT (hypoxanthine, aminopterin, and thymine) medium where aminopterin inhibits de novo DNA biosynthesis pathway. Only B-M fusion is able to grow in HAT medium which is further screened by ELISA for desirable antibody-producing hybridoma and propagated for antibody production

(continued)

is done by putting these cells in a medium called as HAT (hypoxanthine–aminopterin–thymidine).

- Here, aminopterin present in the medium is a blocker of de novo biosynthetic pathway of DNA biosynthesis. Thus, this pathway is nonfunctional, but an enzyme known as hypoxanthine–guanine phosphoribosyltransferase (HGPRT) when present in cells allows the synthesis of nucleotides from the precursor hypoxanthine present in the medium.
- Another enzyme thymidine kinase (TK) can utilize thymine present in the medium.
- Thus, B cells are HGPRT+, TK+−, antibody+, and immortal growth −, whereas myeloma cells are HGPRT−, TK∓, antibody−, and immortal growth+.
- The incubation in HAT results in killing of B-B fusion as they do not have immortal growth properties and sometimes are negative for TK too. M-M fusion is not able to grow due to absence of HGPRT.
- B-M fusion has all the properties as HGPRT+, TK+, antibody+, and immortal growth+; thus, they are able to survive in HAT medium. However, hybridoma produced initially is tetraploid but loses extra chromosomes in subsequent divisions.
- Screening is done for the selection of the appropriate hybridoma; secreting antibody of choice is selected by adding these cells, preferably single cell, to the microtiter plate; and using ELISA or RIA, they are screened for appropriate antibody or culturing each hybridoma and selection by ELISA, RIA, or Western blotting.
- Once the appropriate hybridoma which is secreting desirable antibody is identified, it can be cultured indefinitely for harvesting the desirable antibody.

proteins or antigens known as antigenic determinants or epitopes.

As the binding of antibody is highly specific to its epitopes, thus antibodies can be efficiently utilized for:

1. Detection of a particular antigen (ELISA, RIA, Western blotting)
2. Targeting antigen for antibody-mediated clearance
3. Using antibodies for passive immunization where preformed antibodies can bind with the antigens and facilitate their clearance

Non-human mAb works well with in vitro diagnostics, but when introduced to the patient, they are recognized as foreign (due to non-human origin) and evoke immune response resulting in clearance of the therapeutic antibody.

Identification of epitope or antigenic determinants of specific causative organism or a specific antibody (generated in the host) against the antigenic epitope is required for diagnostics. The mAbs are highly specific and can determine the presence of a specific antigen. They are also used for purification of various active therapeutic agents. The purity may be more than 99 % by usage of affinity column packed with specific mAb. Their various applications include:

Phage Display Library for Monoclonal Antibody Production

As the virus has only two components to run its machinery, that is, nucleic acid (DNA/RNA) and proteinaceous coat, thus any gene when fused with its coat gene is expressed on its surface. The genes which encode variable region of antibodies (derived from B cells of immunized mice or human patient who had the infection) are fused to coat protein genes of bacteriophage by inserting foreign DNA phage genome (capsid proteins from pIII, pVI, pVII, pVIII, and pIX are used to display proteins).

- The bacteriophage is allowed to infect bacteria and it induces the production of its protein in the bacterial cell. Along with its coat protein, it also produces the product of cloned antibody gene and displays it on its surface (Fig. 11.3).
- This results in generation of antibody variable region on phage.
- In this the libraries are prepared from non-immunized naïve human antibody fragments. The genes are amplified, assembled, combined randomly, and cloned to produce unique combinatorial library of variable region of antibody or ScFv.
- The collection of all the representative recombinant bacteriophage, where each one of them may display a different antigen-binding domain on its surface, is known as a phage display library.
- These can be then screened for appropriate target antigens (possibly can bind unlimited antigenic determinants).

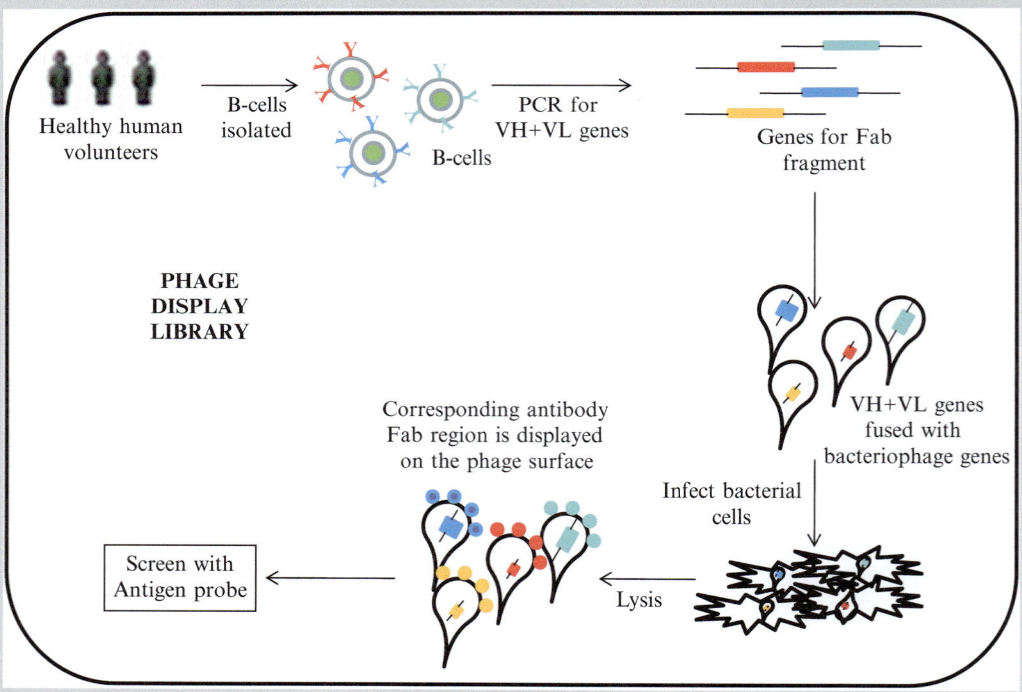

Fig. 11.3 The figure shows phage display library. In this B cells are isolated from healthy human volunteers, and PCR amplification for variable region of light chain and heavy chain is done resulting in different combinations of variable regions of heavy and light chains. These are randomly cloned in bacteriophage virus giving unique combinations of Fab region (with variable light chain and heavy chain) of antibodies. After phage infects bacteria, these antibody Fab regions are displayed on its coat. Screening with desirable antigen results in selection of phage with appropriate antibody derived from humans. This can then be fused with constant region human genes for complete antibody or may be used as single chain fragment variable (ScFv) for therapeutic purposes

(continued)

- Then, target phages are recovered and used to infect fresh bacteria. Each phage isolated in this way will produce a monoclonal antigen-binding particle analogous to a monoclonal antibody.
- Then, appropriate selection genes unique for an antigenic epitope are recovered and can be used for the production of complete antibody.
- Then, these reconstructed genes may be introduced into suitable host, and antibodies may be obtained.

Limitations of phage display libraries are:

1. Complete antigen-specific monoclonal antibody repertoire genes are not obtained in a given phage library.
2. The variable or Fab region of antibody obtained by phage library may not be true representation of in vivo natural immunoglobulins.
3. The technology is complicated and time requiring.

11.6.1 Diagnosis of Pathogen

The diagnosis has significantly improved for many pathogens after advancement in mAb production and commercialization. They are highly useful in detection of East Coast fever (ECF), *Trichomonas vaginalis*, *Leishmania donovani*, *Trypanosoma congolense*, and *Babesia bovis*. They specifically target the pathogen-associated antigens. The routine techniques for diagnosis are enzyme immunoassay, radioimmunoassay, and so on.

11.6.2 Viral Diseases

The studies are being done for the presence of specific viral antigen or antibody against virus. For example, the antibody against HIV or hepatitis C virus may be detected by using mAb against these infections in the host (refer to group IV therapeutics).

11.7 Catalytic Antibodies or Abzymes

Apart from numerous functions of antibodies such as neutralization, agglutination, complement fixation, and activation, they also possess enzymatic activity [10]. Long back it was recognized that any chemical entity or newly synthesized compounds are able to generate an antibody against itself with high degree of specificity. Antibodies with enzymatic properties have been described in human autoimmune disorders, such as autoimmune thyroiditis, systemic lupus erythematosus (SLE), scleroderma, rheumatoid arthritis (RA), multiple sclerosis (MS), and acquired hemophilia (AH). Antibodies isolated from these conditions were able to specifically hydrolyze thyroglobulin, DNA, RNA, myelin basic protein (MBP), and factor VIII (VIII) or factor IX (IX), respectively. Monoclonal antibodies have been discovered to act like enzymes, and these are called catalytic antibodies or abzymes (ab: antibody, zymes: enzymes). These abzymes speed up the chemical reaction like catalysts.

Antigen–antibody binding is similar to enzyme–substrate complex and both involve weak, noncovalent interactions. Antibodies like enzymes are capable of stabilizing the transition state of the bound antigen, thus reducing the activation energy for chemical modification. To authenticate this, the hapten carrier complex resembling the transition state of the ester was injected in mice. Then, antibody-secreting B cells were isolated and hybridoma was generated for synthesizing monoclonal anti-hapten antibody. The antibody thus obtained, when incubated with ester substrate, mediated the hydrolysis by 1,000-fold acceleration. The reaction was highly specific. Thus these antibodies were referred as catalytic antibody [10] or abzymes. Goal in antibody research is finding an antibody which:

- Can cut peptide bonds at specific amino acid residues
- Has the ability to dissolve blood clots
- Cleave viral glycol proteins at specific sites so that they can block viral infectivity

11.8 Antibodies as In Vitro and In Vivo Probes

As the antibody is highly specific for its antigenic epitope, it has tremendous applications in the field of medicine. Because of its specificity, it is routinely in use in various diagnostic techniques like ELISA, RIA, FACS, and immunofluorescence. Usage of these helps us to determine presence or absence of pathogen (see details in Chap. 9). Using them under in vivo conditions results in analysis of antigen to which they are specific. In in vivo condition, they have dual role, both diagnostic and therapeutic. If specific tumor marker is known, then antibody against that marker after suitable labeling would be able to detect the presence of that marker along with giving a clue on metastasis if any [14].

11.9 Cancer Detection and Therapy

For the detection of cancer, it is very important to identify the tumor- or cancer-specific marker (refer to Chap. 10). This marker would help in detection of the cancer, and it can be targeted by the antibody for its selective removal. The usage of labeled antibody has tremendous potential to detect as well as target the cancerous cells. But diagnosis of cancer is recommended by using many techniques; the presence of the marker should not be treated as presence of cancer; rather markers are more helpful in predicting the response of drugs on the cancer.

11.9.1 Cancer Therapeutics

The markers associated with cancer are being explored and identified. The antibody against these tumor markers is helpful in detecting their presence and specifically targeting the cells for clearance.

However, studies have shown that antibodies produced by the hybridoma technology lead to treatment-associated anaphylaxis when injected in humans. Unfortunately, obtaining human antibody is very difficult and suffers from many technical problems; therefore, the efforts are directed toward engineering antibody with the help of recombinant DNA technology. Phage display and other techniques are being utilized to generate human antibodies [3]. Scientists have reduced the immunogenicity of these antibodies by systematically replacing amino acid sequences on the mouse mAbs with the human counterpart forming humanized antibodies. Antibodies can also be a toxin or radioisotope conjugated for effective clearance of the target cell. Various antibodies engineered to solve these problems are shown in Fig. 11.4. Several steps are followed to humanize a mouse-derived antibody:

- Its modeled structure is compared to that of human immunoglobulin.
- Closest match is identified.
- Genetic engineering approaches are used to place the specificity genes (complementarity determining regions, CDRs) from the mouse-derived hybridoma to corresponding regions of matched human immunoglobulin cDNA.
- As the CDR regions are small, therefore the antibody thus produced has specificity to the epitope like the original mouse antibody.
- The effector region of immunoglobulin is derived from human genes so it is less likely to be recognized as foreign in humans.
- Affinity is tested by its ability to compete with parental antibody.

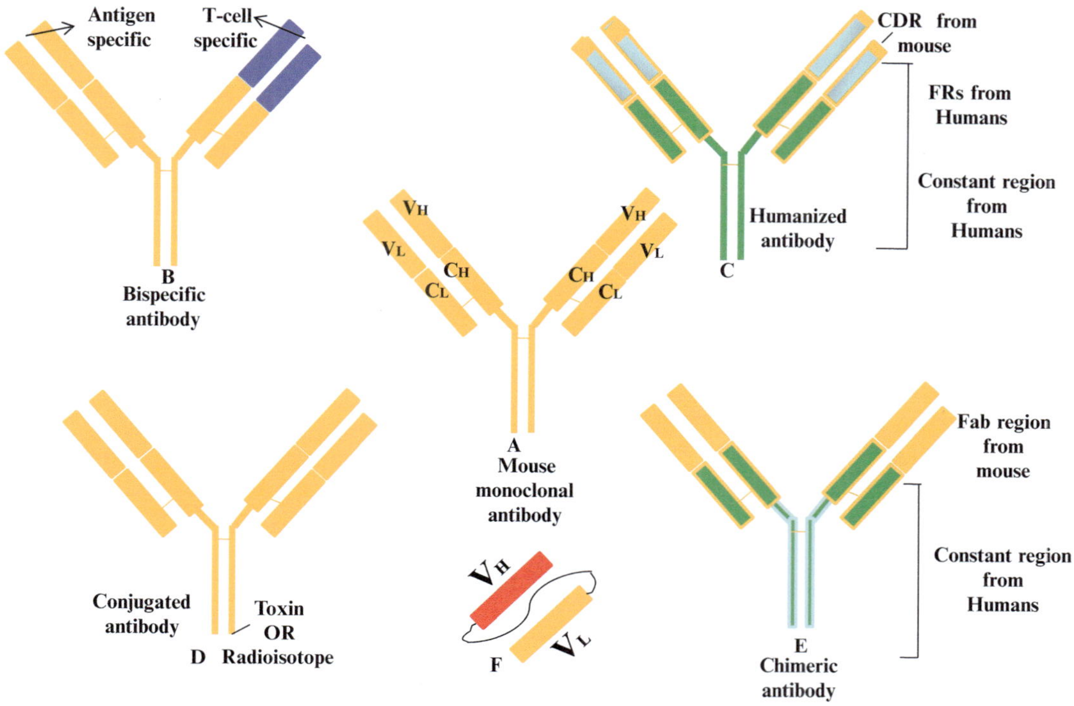

Fig. 11.4 The figure shows the antibody (**a**) with all heavy chains and light chains derived from mouse; (**b**) the bispecific antibody, one Fab, is antigen specific and the second is T cell specific; (**c**) the humanized antibody, the complementarity determining region (CDR), is from mouse and the rest is all derived from human; (**d**) the antibody conjugated to toxin or radionucleotide for efficient clearance of the antigen; (**e**) the chimeric antibody where variable region is derived from mouse and constant region from humans; (**f**) single chain fragment variable

The successful mAb should have the following properties: (a) high affinity to the target, (b) able to promote clearance of the target, (c) low immunogenicity, and (d) high yield and cost-effective. Thus, their production requires lots of standardizations and optimizations for generating effective antibodies.

Now it is also possible to design and prepare the genes encoding for immunoglobulin molecule where variable region from one species and constant region from other species are fused. The technology has paved a new way for development of unique antibodies with desired specificities. The antibody engineering aims to (a) reduce the antibody to its minimal functional size, (b) have bispecific antibody, and (c) develop humanized antibody (see Fig. 11.4).

Genetic engineering approaches have enabled development of single chain fragment variables (ScFvs) which are monovalent and highly reduced in size as compared to antibody. Fv is smaller than Fab as it does not have the first constant region with disulfide bonds that link the heavy and light chain together. ScFvs of different specificities can be linked together resulting in the production of bispecific antibodies which are capable of binding two different receptors on single or different cells. In this way, the antibody is designed to enhance the activation of various components of the immune system to potentiate the effects of the antibody. Toxin molecule can be fused with ScFv for augmenting the clearance of target cells through antibodies.

Since the process of generating humanized antibody is a highly cumbersome, complicated, and multistep process, scientists are trying to develop a mouse with large human immunoglobulin loci into mouse germline.

XenoMouse^{R} (Abgenix Inc., Fremont, CA)

HuMab Mouse[R] (GenPharm, Medarex, San Jose, CA) carries both human V_H and V_L repertoire.

The antibodies developed from these are fully human. If lambda V_L chain is introduced in these, then one day a full human repertoire of antibodies would be generated from mice without any issue of immune responses or allergic reaction. Scientists have demonstrated that it may be possible to produce an antibody without fusing them with myeloma cells using transgenic mice H-2kb-tsA58 (bacteriophage immunized) and ImmortoMouse (Charles River Laboratories, Wilmington, MA).

11.9.2 Cancer–Tumor Immunology

Cancer is one of the leading causes of death these days. It is caused by uncontrolled division, growth, and migration of these cells to different sites of body cells. Great progress has been made in the field of cancer immunology.

11.9.3 Tumor Antigens

Though the tumor cells are essentially body cells with unregulated growth and should be tolerated by the immune system as they are "self" cells, often they express antigens that are usually not present in that particular cell type or environment or in the developmental stages. Antigen glycolsphingolipid GD2, a disialoganglioside, is usually expressed in the outer membranes of neuronal cells that are not exposed to the immune cells due to the blood–brain barrier. GD2 is expressed on tumor cells such as neuroblastomas, medulloblastomas, astrocytomas, melanomas, small cell lung cancer, osteosarcomas, and soft tissue sarcomas [15].

Other kinds of tumor cells exhibit cell surface receptors that are not present in healthy cells. These receptors activate signal transduction pathways that cause unregulated cell growth. ErbB2 is a receptor that is constitutively expressed in breast cancer cells.

These can be targets for immunotherapy [4].

11.9.4 Immune Response to Tumors and Tumor Evasion of the Immune System

The immune system uses natural killer (NK) cells and cytotoxic T cells to identify and destroy abnormal cells. However, some tumor cells evade the immune system and form cancer. Tumor cells often have reduced MHC class II molecules on their surfaces and evade T cells. Some tumor cells produce substances to inhibit immune response such as cytokine TGF-β that suppresses macrophages and lymphocytes. Immunological tolerance might develop against some tumors.

11.9.5 Cancer Immunotherapy

Immunotherapy is a treatment that stimulates patients' own immune system to recognize, attack, and destroy malignant cells. A long time back in the late 1800s, William Coley, a New York surgeon, first noticed that getting infection after a surgery helped some cancer patients. He started treating cancer patients with certain kinds of bacteria which were known as Coley toxins. However, this therapy was overshadowed by other forms of cancer treatments such as radiation therapy and chemotherapy. Immunotherapy has developed and is being used to treat certain forms of cancer. However, it is associated with many side effects as capillary leakage to low level autoimmunity [16]. It works best for smaller and early stages of cancer [5].

This can be done either by immunizing the patient with cancer vaccine so that the immune cells are trained to recognize the cancerous cells and specifically destroy them or by directly administering antibodies against these cells. The alternative approach is cell-based immunotherapy. Here, natural killer (NK) cells, cytotoxic T lymphocytes (CTL), lymphokine-activated killer (LAK) cells, and dendritic cells (DCs) either are activated in vivo by administering certain cytokines such as interleukins or are isolated, enriched, and transfused to patients to combat cancer.

11.9.6 Antibody Therapy

This is a passive immunotherapy where monoclonal antibodies made in the laboratory are injected into the patients directly instead of depending on the patient's immune system to fight the disease. It is the most widely used form of cancer therapy. They bind to the antigens expressed by the cancer cells and help in recruiting the other components of the immune system (see Fig. 11.5).

Two types of monoclonal antibodies are available, naked monoclonal antibodies and conjugated monoclonal antibodies. Naked mAbs are those without any drugs or radioactive material attached to them, whereas conjugated monoclonal antibodies are those joined to a chemotherapy drug, radioactive particle, or a toxin.

(a) Naked monoclonal antibodies:

These are the most commonly used antibodies and work by attaching to specific antigens. They have different mechanisms of action. Depending on the mechanism of action, they can be grouped as markers for destruction, activation blockers, and other mechanisms.

1. *Marker for destruction*

   Some naked mAbs attach to cancer cells and act as a marker for the body's immune system to destroy them (Table 11.5).

2. *Activation blockers*

   These naked mAbs do not interact with the body's immune system. They block the specific receptors or antigens that help the cancer cells to grow. These are also referred as targeted therapies (Table 11.6).

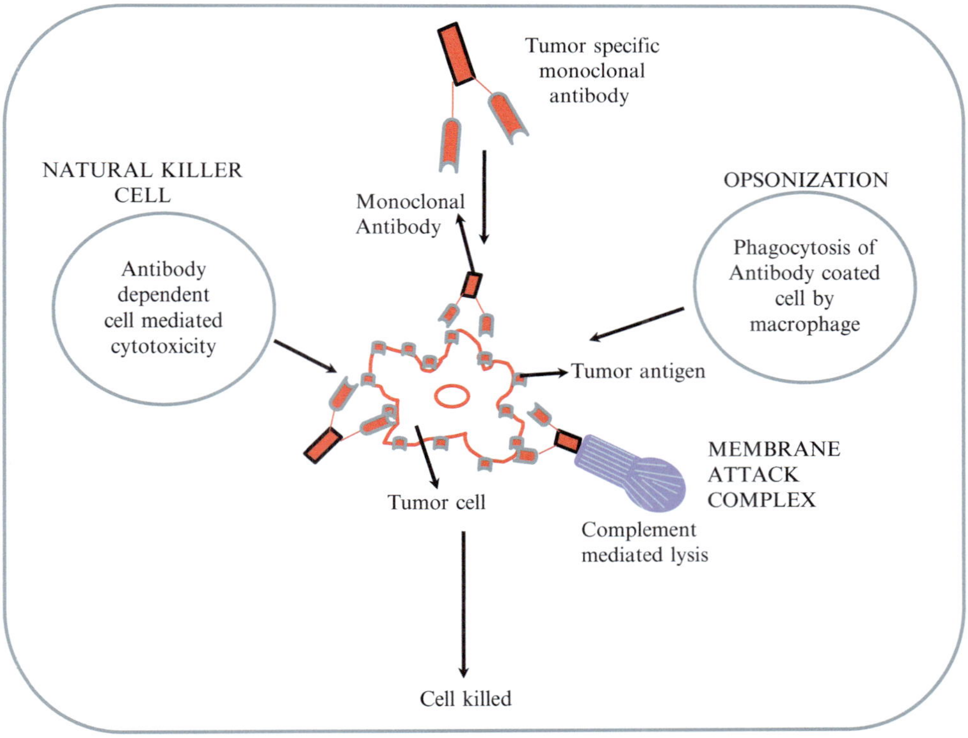

Fig. 11.5 The figure shows the binding of monoclonal antibody with the specific tumor antigen. After binding, (1) coating of antibody on tumor cell allows macrophage to phagocytose the antibody-coated cell by the process of opsonization, (2) antibody bound to tumor cell interacts with NK cells for ADCC response, and (3) antibody fixes complement system which kills the cell by membrane attack complex, resulting in death of tumor cell

Table 11.5 Monoclonal antibodies directed against cellular markers upregulated on cancer cells for promoting destruction by host immune system

mAb name Trade name®	Cancer type	Target	Mechanism of action
Rituximab	B-cell non-Hodgkin's lymphoma	CD20 antigen on B cell	Labels cells so that the immune system can attack them
Rituxan®	Chronic lymphocytic leukemia (CLL)	CD20	
Ofatumumab	Chronic lymphocytic leukemia	CD52 antigen in B cells and T cells	
Arzerra®	B-cell chronic lymphocytic leukemia		
Alemtuzumab			
Campath®			

Table 11.6 Monoclonal antibodies and their targets for treatment of cancer by checking growth of cancer cells

mAb name Trade name®	Cancer type	Target	Mechanism of action
Trastuzumab Herceptin®	Breast and stomach cancer	HER2/neu protein	HER2/neu protein is present in a large amount on tumor cells. When activated, this protein helps the tumor to grow. The mAb blocks the activation
Cetuximab Erbitux®	Colorectal cancer, head and neck cancer	EGFR	Blocks the activation of EGFR
Panitumumab Vectibix®	Colorectal cancer	EGFR	Blocks activation
Bevacizumab Avastin®	Colorectal, lung, breast, and kidney cancer and glioblastomas	VEGF	VEGF is made by tumor cells to attract new blood vessels to feed them. This mAb attaches to VEGF and blocks the signaling

Table 11.7 Modes of cancer treatment by blocking activity or inhibiting immune regulatory mechanisms

mAb name Trade name®	Cancer type	Target	Mechanism of action
Denosumab Xgeva®	Bone cancer	Rank ligand	Rank ligand is made by cancer cells when they attack bone. This mAb binds to rank ligand and stops bone destruction
Ipilimumab Yervoy®		CTLA4, antigen found on regulatory T cells and cytotoxic T cells	It lowers regulatory T cells and increases immune response. It binds to cytotoxic T cells and activates them

3. *Other mechanisms*: Table 11.7 shows other mechanisms for cancer treatment.

(b) *Conjugate monoclonal antibodies*

These are attached to drugs (chemo-labeled antibodies), toxins (immunotoxins), or radioactive substances (radiolabeled antibodies). The mAbs deliver the substances to the target cells. They are also known as tagged, labeled, or loaded antibodies (see Fig. 11.6a, b).

1. *Radiolabeled antibodies* (Table 11.8)

Apart from treating cancer, these are also used for detection and screening.

2. *Immunotoxins*

These mAbs are attached to bacterial toxins such as diphtheria toxin (DT) and pseudomonal enterotoxin (PE400) or to plant toxins such as ricin A or saporin. Clinical trials are going on (Fig. 11.6a).

(c) *Other immunotherapies*

Instead of using mAbs, toxins are linked to growth factors as several cancer cells have a large number of growth factor receptors. The growth factor-toxin binds to these receptors and kills cells (Table 11.9).

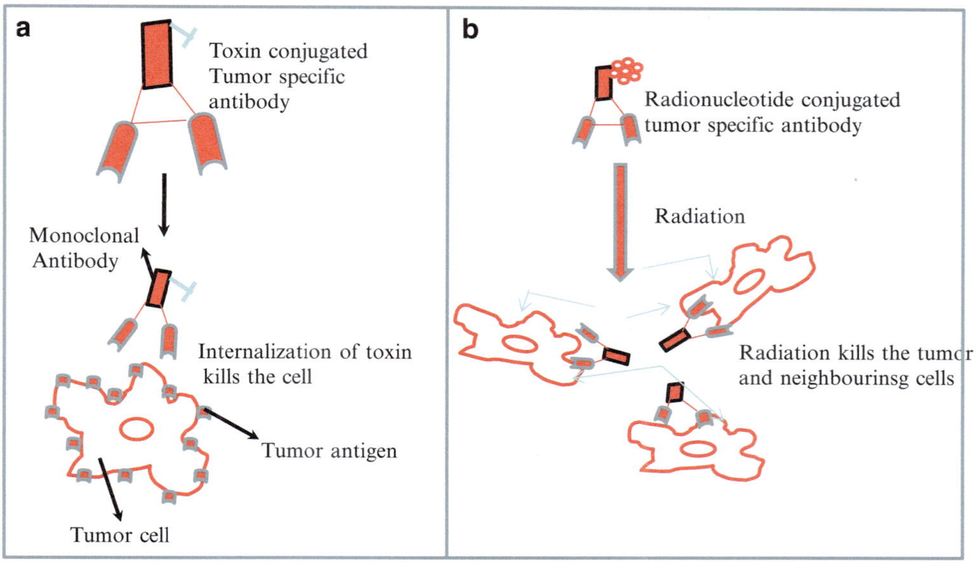

Fig. 11.6 (**a**) The figure shows toxin-conjugated antibody. Toxin is internalized by the cell and it mediates its killing. (**b**) Radionucleotide-conjugated antibody. Binding of antibody allows the radioisotope to kill the tumor-specific cell

Table 11.8 Radio-conjugated monoclonal antibodies for destruction of cancerous cells

mAb name Trade name®	Cancer type	Target	Mechanism of action
Ibritumomab tiuxetan Zevalin®	B-cell non-Hodgkin's lymphoma	Cancerous B cells	Radiation therapy
Tositumomab Bexxar®	Non-Hodgkin's lymphoma	B cells	Radiation therapy

Table 11.9 Toxin linked growth factor for control of cancer cell growth in the body

Growth factor-toxin name Trade name®	Cancer type	Target	Mechanism of action
Denileukin diftitox Ontak®	Skin lymphoma Mycosis fungoides		IL-2 attached to diphtheria toxin

11.10 Cancer Vaccines

Cancer vaccines are a biological response modifier. They are either preventive (prophylactic) or therapeutic vaccines [11–14]. They may work like traditional vaccines by stimulating the immune system's ability to fight infections, or they can help the immune system restore its function to fight infections and diseases. The FDA has also approved a vaccine for HBV infection leading to liver cancer. Vaccines against hepatitis B virus (HBV) may lower risk of getting liver cancer. Vaccines against human papillomavirus (HPV) help prevent cervical, vaginal, vulvar, and anal cancer. These vaccines do not target cancer cells but prevent infections that might cause cancer.

True cancer vaccines have cancer cells, cell parts, or pure antigens that increase immune response against cancer cells that are already present in the body. Table 11.10 Shows cancer vaccines.

(a) *Tumor cell vaccines*

These are made from actual cancer cells that have been removed during surgery. They

Table 11.10 Vaccines for cancer which are helpful in activation of host immune system for control or preventing onset of cancer

Cancer therapy/vaccine name Trade name	Cancer type	Target	Mechanism of action
Sipuleucel-T Provenge Cancer treatment vaccine	Prostate cancer	Prostate cancer cells	WBC from patient removed and exposed to prostate cancer cells. These cells are given back to the patients. Boost the immune system cells
Cervarix® VLPs made from HPV types XVI and XVIII	70 % of cases of cervical cancer and anal, vaginal, vulvar, penile, and oropharyngeal cancers	Protects against cervical cancer	
Gardasil® human papillomavirus quadrivalent VLPs (types VI, XI, XVII, and XVIII), recombinant	Cervical cancer 70 % of cases of cervical cancer 70 % of cases of vaginal cancer 50 % of cases of vulvar cancer	Protects 80 % of anal cancer cases and 90 % of genital warts cases	Protects 11- or 12-year-old children from two types of HPV

are irradiated and treated with chemicals or new genes to make them be recognized as foreign. These cells are then injected back to the patients. They can be "autologous," where they are derived from the same person, or "allogenic," where the cells come from others.

(b) *Antigen vaccines*

Antigen vaccines boost immune response by using only one antigen rather than the whole tumor cell containing several antigens. They are specific for certain types of cancer.

(c) *Dendritic cell vaccines*

Dendritic cells are special antigen-presenting cells (APCs) that help the immune system recognize cancer cells. Dendritic cell vaccines are made by removing the patient's dendritic cells and exposing them to the cancer cells. Then they are put back into the patient's body. These dendritic cells that have cancer antigens on their surface are better able to help the immune system to recognize and destroy the cancer cells.

(d) *DNA vaccines*

Tumor cells or antigens injected as vaccine causes desired immune response but gradually become less effective over time. DNA vaccines are cells that carry the gene for a specific antigen and can continuously produce antigens over the time. These are some of the several approaches the scientists use to combat cancer using immunotherapy. Several of them are already in the market and there are several more in clinical trial.

11.11 Advancements in Therapy of AIDS (Human Immunodeficiency Virus)

For the control of symptoms and progression of HIV, highly active antiretroviral therapy (HAART) has shown potential for control of HIV infection before the late stages. HAART is an anti-HIV cocktail with a combination of three or more drugs. These are antiretroviral medications with protease inhibitors. The treatment is very effective (1) for slowing HIV virus replication and (2) retarding the HIV spread in the body; thus, the therapy aims to reduce HIV load of the virus to undetectable levels in the blood and facilitate recovery from the disease.

HAART regime requires (1) daily intake of drugs, (2) adherence to drugs for viral control, and

(3) continuous access to treatment. New advanced therapies based on stem cells have shown promising response for limiting HIV infection.

However new therapeutic modalities were highly in demand, as the present therapies are associated with chronic inflammation and immune dysfunction. Sometimes, cryptic viral replication persists in dispersed lymphoid organs during treatment.

Non-nucleoside Reverse Transcriptase Inhibitors (NNRTI): They are delavirdine (Rescriptor, DLV), efavirenz (Sustiva, EFV), and nevirapine (Viramune, NVP) and can block the spread of HIV by preventing it from infecting new cells. They may be prescribed in combination with other antiretroviral therapeutics.

Inhibitors of fusion: These drugs act by blocking the viral fusion inside the cell preventing its replication, for example, enfuvirtide (Fuzeon or T-20).

Some combinations of non-nucleoside reverse transcriptase inhibitors are being used for the treatment of HIV infection. Protease inhibitors are also listed which are helpful in the treatment of HIV infection. The general side effects with all therapeutics include fever, fatigue, body aches, diarrhea, nausea, vomiting, and weakness. Skin rashes, dizziness, headache, loss of appetite, mild stomach cramps or pain, and trouble sleeping also occur with many treatments. Table 11.11 shows the therapeutics used for control of HIV.

HIV-resistant cell: Various approaches are being explored for targeting different aspects of HIV replication like (1) targeting cellular genes necessary for viral replication; (2) RNA interference (RNAi), using ribozyme to reduce CCR5 RNA levels; (3) targeting HIV gene expression (critical genes required for HIV infectivity as Tat and Rev may be targeted using ribozyme (small catalytic RNA molecules); and (4) introduction of genes that interfere with HIV replication like host restriction factors and fusion inhibitors. All viral transcripts including tat, rev, gag, pol, nef, vif, env, vpr, as well as LTR are susceptible to RNA interference. In human system the replication of HIV is restricted to cells which have CD4 and either CCR5 or CXCR4.

The novel targets are being explored for AIDS therapy owing to the fact that major host of HIV is human (recent report shows new world monkeys may also be explored), with some who are resistant to HIV infection. Thus HIV entry pathway components are being explored as possible targets.

CD4: It is one of the important components of T-helper cells. As all HIV strains are capable of binding CD4, therefore therapies involving soluble forms of CD4 showed good inhibition under in vitro conditions but not in in vivo conditions. HIV-1 was resistant to soluble CD4 due to strong and stably attached gp120 and lower affinity for CD4.

IL-16: Under in vitro conditions, IL-16 inhibits HIV promoter activity. In vivo, IL-16 levels are increased followed by HIV infection but decline during late stages of the disease. IL-16 may be toxic and proinflammatory under in vivo conditions. No clinical trials have been reported as yet using IL-16.

Sulfated Sugars: Sulfated sugars like heparin and dextran sulfate may block HIV infection in vitro by interacting with gp120 (heparin) or preventing gp120 binding with CD4 (dextran sulfate). They can also block other retroviruses using different receptors. Some studies have reported benefits of intraperitoneal dextran-2-sulfate administration by reduction in viremia. However, in vivo they do not interfere with HIV replication and rather might be acting via macrophages.

gp41: It may be a potential target as leucine zipper-derived peptides, and membrane proximal α-helix of gp41 could inhibit infection in vitro. To inhibit HIV entry, a gp41-derived protein, C46, was developed (structurally similar to fusion inhibitor, enfuvirtide) which was able to effect entry of HIV.

Coreceptors: The involvement of coreceptors in HIV infection has provided new therapeutic targets to explore. CCR5 coreceptor is required for viral entry, thus homozygous CCR5 (carrying deletion of 32 bp of the CCR5 gene) is largely

Table 11.11 A few therapeutics used for control of HIV

Drugs	Trade name	Mode of action	Side effects
Nucleoside reverse transcriptase inhibitors (NRTI)			
Abacavir	Ziagen, ABC	Nucleoside analog reverse	Loss of appetite and respiratory symptoms
Didanosine	Videx, dideoxyinosine, ddI	Transcriptase inhibitors; inhibit reverse transcriptase of HIV virus, preventing its proliferation	Pancreatitis; peripheral neuropathy
Emtricitabine	Emtriva, FTC		Burning, tingling, or pain in the hands, arms, feet, or legs
Lamivudine	Epivir, 3TC		Burning, tingling, or pain in the hands, arms, feet, or legs
Stavudine	Zerit, d4T		Peripheral neuropathy include a sharp, burning pain sensation in the hands or legs
Tenofovir	Viread, TDF		Liver or kidney failure and pancreas disease
Zalcitabine	Hivid, ddC		Oral ulcers and peripheral neuropathy
Zidovudine	Retrovir, ZDV, or AZT		Neutropenia, with risk of infection of the lungs, kidneys, blood, and skin
Combinations of NRTIs			
Zidovudine and lamivudine	Combivir	Combinations are more effective to suppress HIV replication	Fever, fatigue, body aches, diarrhea, nausea, vomiting, weakness, burning, tingling, or pain in the hands, arms, feet, or legs, loss of appetite, and respiratory symptoms
Zidovudine, lamivudine, and abacavir	Trizivir		
Abacavir and lamivudine	Epzicom		
Tenofovir and lamivudine	Truvada		
Protease inhibitors (PI)			
Amprenavir	Agenerase, APV	They inhibit viral replication at the later stage of viral life cycle	Numbness around the mouth and abdominal pain
Atazanavir	Reyataz, ATV		Yellowing of the eyes or skin, change in heart rhythm, diabetes, and high blood sugar, diarrhea, infection, nausea, and blood in the urine
Fosamprenavir	Lexiva, FOS		Rash, nausea, and diarrhea
Indinavir	Crixivan, IDV		Kidney stones, high sugar, and fat levels in the blood, and onset or worsening of diabetes
Lopinavir	Kaletra, LPV/r		Abnormal stools or bowel movements, liver problems
Ritonavir	Norvir, RIT		Constipation, indigestion, flatulence, pancreas disease, worsening of diabetes
Saquinavir	Fortovase, Invirase, SQV		Stomach and intestinal problems, sleep disturbance including insomnia, anxiety, sex drive disorder

resistant to HIV infection; therefore, CCR5 may be a good target [2].

11.12 Chapter End Summary

- Technology advancements have helped in the development of not only drug targets but also potential therapeutic targets. To achieve desirable results, there are certain chemical mediators which are utilized.
- Immunostimulants are compounds which increase the efficacy of the immune system, for example, cytokines or adjuvants (are used in vaccine and are responsible for increasing immunogenicity). They can also be used in passive immunity. When the pathogen is present in the system and needs quick elimination, then preformed antibodies may be given which help in the elimination of pathogens by opsonization or ADCC or complement-mediated lysis.
- Immunosuppressors are used to suppress immune responses and are used in inflammation and transplantation. However, the use of immunosuppressants may lead to increased incidences of infections.
- Interferons are antiviral compounds which are secreted from our cells in response to viral infection. Interferon therapy is used in many clinical conditions like cancer, multiple sclerosis, viral infections, etc.
- The various protein therapeutics are used, of which the majority of them are produced by recombinant DNA technology. The therapeutics are grouped in various categories depending upon their clinical requirement.
  Group I includes the protein therapeutics which are essentially replacement molecules. When there is deficiency or genetic defect of enzyme or regulators, they are supplied from outside for normal functioning of the body, for example, insulin in diabetes patients.
  Group II therapeutics includes the proteins having targeted effects as monoclonal antibody for therapy of cancer and other disorders.
  Group III therapeutics includes vaccines which are useful in prevention of infection by generating memory response of immune system.
  Group IV includes proteins in diagnostics.
- The antibodies have gained attention because of their unique property of specificity. Because of this property, they are used for detection as well as treatment. They have several nonclinical applications also as in affinity purification, cell detection, or in vitro diagnostics. They are the major components in group II therapeutics. By genetic engineering approaches, these may be made chimeric, humanized, bispecific, or toxin/radioactive molecule conjugated antibody for increased efficacy.
- Cancer immunotherapy is used to treat malignancies. They are based on antibodies and cancer vaccines.

Multiple Choice Questions

1. Biotechnology-derived therapeutic products are:
   (a) Less immunogenic
   (b) Effective
   (c) Safe
   (d) All of the above
2. Immunostimulants:
   (a) Stimulate the immune response
   (b) Stimulate the effects of immune cells
   (c) Stimulate complement cascade
   (d) All of the above
3. Immunosuppressors are used in:
   (a) Cancer therapy
   (b) Bacterial infection
   (c) Autoimmunity
   (d) None of these
4. An example of an immunostimulant is:
   (a) Antibody
   (b) Hapten
   (c) Adjuvant
   (d) None of these
5. In organ transplantation, which therapy is helpful?
   (a) Immunostimulants
   (b) Interferons
   (c) Immunosuppressors
   (d) All of the above
6. Corticosteroids are used for suppression of inflammation because they:

   (a) Inhibit B-cell function
   (b) Inhibit IL-1 and T-cell function
   (c) Inhibit the activity of neutrophils
   (d) All of the above
7. Interferons are secreted from:
   (a) All cells of the body
   (b) Lymphoid cells
   (c) T cells
   (d) B cells
8. Multiferon is:
   (a) Natural type I interferon
   (b) Type II interferon
   (c) Interferon-gamma
   (d) All of the above
9. The antibody used for treatment of breast cancer is:
   (a) Adalimumab
   (b) Rituxan
   (c) Trastuzumab
   (d) All of the above
10. The important therapeutic usage of antibody is in:
   (a) Passive immunization
   (b) Cancer therapy
   (c) Cancer diagnostics
   (d) All of the above

Answers

1. (d); 2. (d); 3. (c); 4. (c); 5. (c); 6. (b); 7. (a); 8. (a); 9. (c); 10. (d)

Review Questions

Q1. What are immunosuppressors?
Q2. What is the function of interferon?
Q3. What is the mechanism of action of interferon?
Q4. What are the clinical values of monoclonal antibodies?
Q5. What is replacement protein therapy?
Q6. Explain cancer immunotherapy with reference to antibody-mediated therapy.

References

1. Brannigan JA, Wilkinson AJ (2002) Protein engineering 20 years on. Nat Rev Mol Cell Biol 3:964–970
2. Clapham PR, McKnight A (2002) Cell surface receptors, virus entry and tropism of primate lentiviruses. J Gen Virol 83:1809–1829
3. Hammers CM, Stanley JR (2013) Antibody phage display: technique and applications. J Invest Dermatol 134:e17. doi:10.1038/jid.2013.521
4. Kirkwood JM et al (2012) Immunotherapy of cancer in 2012. CA Cancer J Clin 62:309–335
5. Klebanoff CA, Gattinoni L, Restifo NP (2006) CD8+ T-cell memory in tumor immunology and immunotherapy. Immunol Rev 211:214–224
6. Leader B, Baca QJ, Golan DE (2008) Protein therapeutics: a summary and pharmacological classification. Nat Rev 7:20–39
7. Lollini P, Cavallo F, Nanni P, Forni G (2006) Vaccines for tumour prevention. Nat Rev Cancer 6:204–216
8. Mullins DW, Sheasley SL, Ream RM, Bullock TNJ, Fu YX, Engelhard VH (2003) Route of immunization with peptide-pulsed dendritic cells controls the distribution of memory and effector T cells in lymphoid tissues and determines the pattern of regional tumor control. J Exp Med 3(198):1023–1034
9. Murphy K, Travers P, Walport M (eds) (2008) Janeway's immunobiology, 7th edn, Garland Science. Taylor & Francis Group, LLC, New York
10. Nevinsky GA, Bunneva VN (2010) Natural catalytic antibodies ins norm, autoimmune, viral, and bacterial diseases. Scientific World Journal 10:1203–1233
11. Pazdur MP, Jones JL (2007) Vaccines: an innovative approach to treating cancer. J Infus Nurs 30:173–178
12. Reichert JM (2003) Trends in development and approval times for new therapeutics in the United States. Nat Rev Drug Discov 2:695–702
13. Ribas A, Butterfield LH, Glaspy JA, Economou JS (2003) Current developments in cancer vaccines and cellular immunotherapy. J Clin Oncol 15:2415–2432
14. Sharma P, Wagner K, Wolchok JD, Allison JP (2011) Novel cancer immunotherapy agents with survival benefit: recent successes and next steps. Nat Rev Cancer 11:805–812
15. Wang R, Rosenberg SA (1999) Human tumor antigens for cancer vaccine development. Immunol Rev 170:85–100
16. Weber JS et al (2015) Toxicities of immunotherapy for the practitioner. J Clin Oncol 33:2092–2099
17. Wurm F, Bernard A (1999) Large scale transient expression in mammalian cells for recombinant protein production. Curr Opin Biotechnol 10:156–159
18. Zoller MJ (1991) New molecular biology methods for protein engineering. Curr Opin Biotechnol 2:526–531

Some Related Resources

http://www.cancer.gov
http://www.cancer.org/Treatment/TreatmentsandSideEffects/TreatmentTypes/Immunotherapy/index
http://www.georgiahealth.edu/itss/edtoolbox/7390/Cancer1/HIV%20testing.pdf
http://www.mct.aacrjournals.org
http://www.ucsfhealth.org/
www.cancermoleculartherapeutics.org
www.cancermoleculartherapeutics.org/save_date.html
www.massgeneral.org/research/resourcelab.aspx
www.ncbi.nlm.nih.gov/entrez

Rational Drug Designing

12

Abstract

Almost all the pharmaceutical companies across the world dedicate a significant portion of their revenues into research and development (R&D). This primarily goes into new drug design, testing, and finally the approvals of the drugs for human consumption. The drug may be a small molecule, which can interact with biomolecules (targets) and may activate or inhibit the functions of the molecules resulting in the improvement in the symptoms of the disease. The designing of the drug may involve the molecules, which may bind the target biomolecules like enzyme–ligand or enzyme–inhibitor complexes. The target molecule may be a signaling/biochemical intermediate having a key role in the pathogenesis of the disease or a crucial molecule resulting in infection or survival of the pathogenic agent. Targets of the drug may be proteins or nucleic acids (DNA/RNA) having well-defined secondary structures like bacterial ribosomes or HIV genome portions, or in cases of human protein malfunctioning, G-protein-coupled receptors (GPCRs) may also prove effective. Even other molecules, which can affect the functionality of ion channels, proteases, kinases, and nuclear hormone receptors, may also be used. The target-based designing involves designing and developing drug, which is capable of modulating the function of these. Designing drug against pathogenic proteins is aimed to find critical molecule, which is unique to the pathogen whose suppression or modulation is done in a way to completely inhibit its life cycle for effective elimination. However, it is very difficult to identify the target diseases as cancer because the molecules involved in it may be mutated cellular proteins responsible for performing critical cell functions. Targeting these proteins results in loss of function, which can be difficult to recover, and another challenge is drug-induced disruptions of oncogenic complexes. However, the introduction of drugs has resulted in dramatic improvement in the quality of patient's life. The discoveries have provided safe and effective drug targets and have been responsible for decreasing human morbidity and mortality. A typical drug discovery and development cycle takes years from the time of conceptualization. It is also a very expensive process. Computer-aided drug designing (CADD) is one of the approaches

V. Gupta et al., *Basic and Applied Aspects of Biotechnology*, DOI 10.1007/978-981-10-0875-7_12

that holds potential for optimizing the whole process and can save billions of dollars for the pharmaceutical companies. It also holds potential for India to generate millions of trained professionals who can assist in the drug design process.

12.1 Introduction

Almost all the pharmaceutical companies across the world dedicate a significant portion of their revenues into research and development (R&D). This primarily goes into new drug design, testing, and finally the approvals of the drugs for human consumption. The drug may be a small molecule, which can interact with biomolecules (targets) and may activate or inhibit the functions of the molecules resulting in the improvement in the symptoms of the disease. The designing of the drug may involve the molecules, which may bind the target biomolecules like enzyme–ligand or enzyme–inhibitor complexes. The target molecule may be a signaling/biochemical intermediate having key role in the pathogenesis of the disease or a crucial molecule resulting in infection or survival of the pathogenic agent. Targets of the drug may be proteins or nucleic acids (DNA/RNA) having well-defined secondary structures like bacterial ribosomes or HIV genome portions, or in cases of human protein malfunctioning, G-protein-coupled receptors (GPCRs) may also prove effective. Even other molecules, which can affect the functionality of ion channels, proteases, kinases, and nuclear hormone receptors, may also be used. The target-based designing involves designing and developing drug, which is capable of modulating the function of these. Designing drug against pathogenic proteins targets to find critical molecule, which is unique to the pathogen. Usage of this should result in suppression or modulation of pathogen to inhibit its life cycle for its effective elimination. However, it is very difficult to identify the targets of cancer, as they may be mutated cellular proteins responsible for performing critical cell functions. Targeting these proteins results in loss of function, which can be difficult to recover, and another challenge is drug-induced disruptions of oncogenic complexes. However, the introduction of drugs has resulted in dramatic improvement in the quality of life. The discoveries have provided safe and effective drug targets and have been responsible for decreasing human morbidity and mortality. A typical drug discovery and development cycle takes years from the time of conceptualization. It is also a very expensive process. Computer-aided drug design is one of the approaches that holds potential for optimizing the whole process and can save billions of dollars for the pharmaceutical companies. It also holds potential for India to generate millions of trained professionals who can assist in the drug design process.

12.2 Modes of Drug Discovery

Traditional approaches to drug discovery rely on systematic synthesis and screening of a large number of compounds to identify the ones that will be effective in treatment:

- The traditional approach is based on identifying the structure of the target.
- Once the structure of the target is established, the next stage entails identification of the lead that will bind the target and act as an inhibitor.
- The inhibitor acts by interacting with the active site through one of the two—protein–protein or protein–nucleic acid—interactions leading to the propagation of signaling events and/or alteration of metabolic processes.
- For example, the metabolic changes that can be induced are by inhibiting enzymes, block ion channels, and favor or oppose receptors. The modulation of the biological functions is achieved by one of the following:

(a) Inhibiting the function with small molecules whose competitive binding affinity is greater than their natural ligands that bind to the active sites
(b) Activating biomolecules that are functionally deregulated in diseases
(c) Inhibiting the bimolecular interactions by small molecules

While the number of 3D structures has increased manifolds in the last few years, and the computer-aided methodologies have increased the pace of drug discoveries, the developments are still inadequate as the 3D structures of many important targets are still unknown [9, 22]. Some of the biomolecules have more than one structure bound to different molecules, making the task of target identification more challenging. Rational drug design as a stream targets to address these shortcomings of existing structure-guided drug design [15]. The new approach combines applied chemical and biological streams to streamline drug discovery, design, development, and optimization. Given the advancements in methodologies, computer-aided applications can be applied to further optimize and speed up the process.

The computational methodology optimizes the drug design process by understanding the specific molecular recognition events of the target macromolecule with molecular structure of the treatment approaches, thereby leading to design of improved leads for the target. Requirements of conversion of drug lead into a drug are determined by its effectiveness and its tolerance by the human body with minimal or no side effects.

12.3 Lead Compound Identification and Optimization

Computer-aided drug design approaches have widely been employed in lead identification and lead optimization stages of drug development. The outline of drug development has been shown in Fig. 12.1. Over and above the benefits of reduced time and cost, this method can also be used to identify/design new inhibitors or for optimization of absorption, distribution, metabolism, toxicity, or excretion profile of identified molecules.

The starting point for drug design and development is the lead compound. The lead compound is analyzed for its advantages and disadvantages, its being a potential drug, its biological activity, and the presence of side effects, but it acts as a starting point for the development of a better agent. The biological activity is desirable in the lead compound; thus, bioassays are used for the analysis of its biological activity.

Drug design can be categorized into two streams: structure-based drug design and ligand-based drug design.

Structure-based drug design (SBDD) is the approach where the structural information of the drug target is exploited for the development of its inhibitor. Identifying the structure of the receptor (cell membrane receptors and transmembrane receptors are specialized integral membrane proteins that take part in communication between the cell and the outside world) is the first step in the SBDD process [1]. The next stage in SBDD is identification of active site of drug targets which, when characterized from a structural point of view, will provide information on its binding features (Fig. 12.2). The information of the active site composition and the orientation of various amino acids at the binding site are deployed to design ligands specific to that particular target.

Ligand-based drug design (LBDD) is used in the absence of the receptor 3D information, and it relies on knowledge of molecules or ligands that bind to the biological target of interest.

12.4 Structure-Based Drug Design

The structure-based drug design (SBDD) involves the steps of cloning, purification, sequence, and structure determination of the target protein (Fig. 12.2). The structural determination may be done by X-ray crystallography, NMR or homology modeling, or threading or ab initio method. SBDD leverages applications of

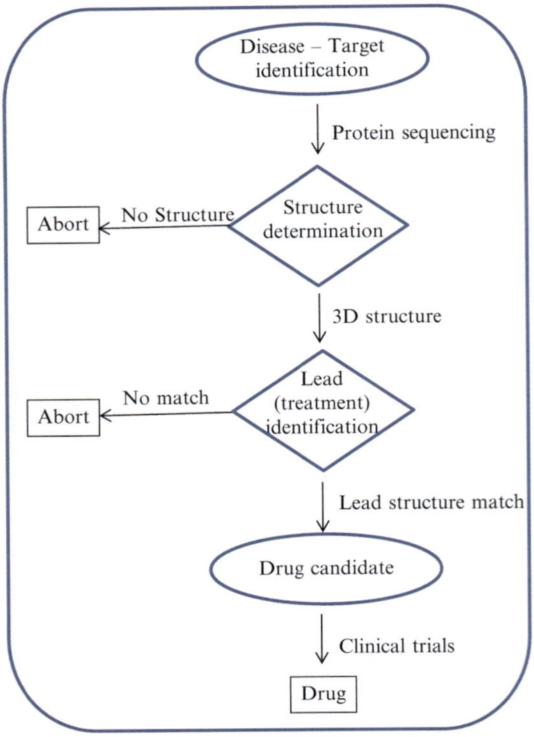

Fig. 12.1 The outline of drug development steps

computing speed, ligand scoring functions, and the combinatorial chemistry to identify binding structure that can be utilized for developing a drug [11].

12.4.1 Cloning a Drug Target

The process of SBDD starts with the cloning of the target for the drug. The objective of the exercise is to determine the possible structure of the target and possible inhibitor binding sites [2, 4, 6]. Commonly used techniques used for the identification of receptor structure are X-ray crystallography or NMR.

12.4.2 X-Ray Crystallography

This technique requires purified and properly crystallized material suitable for study. The crystallization in itself is very challenging and requires a lot of hard work, expertise, and specialized techniques.

When X-ray fall on crystal, its diffraction pattern is recorded from many directions. Superimposition of these images is helpful in predicting structure of the crystal. X-ray crystallography allows the determination of the three-dimensional structures of protein molecules (Fig. 12.3.):

- Structure-based drug design (SBDD) uses this information to find out how small molecules (ligands) interact with a protein's surface. It may help in the understanding of the features of known drugs that bind to certain proteins. It also helps in designing new molecules that fit the ligand-binding site of a protein in competition with the natural ligand. It may involve thinking of a strategy of using the inner shape of a lock to build a key that selectively fits just that lock.
- By design and optimization, an inhibitor molecule can be made to complement important cavities of a protein in terms of its shape and chemical properties.

The crystallography is done in solid phase; therefore, three-dimensional structure obtained may vary from its natural phase. Nonetheless, the technology has been and would be pioneer in determining the structure of the effective drug.

Some of the benefits of X-ray crystallography are:

- This technique is capable of providing very high-resolution structures which are needed to determine precise atomic-level descriptions of ligand-binding sites.
- Once crystallization techniques have been worked out for one complex, they should be fairly similar with subsequent complexes.

Hence X-ray crystallography is better suited for studying small samples of molecules that have been screened by a previous method. This technique, however, is not suited for library screenings.

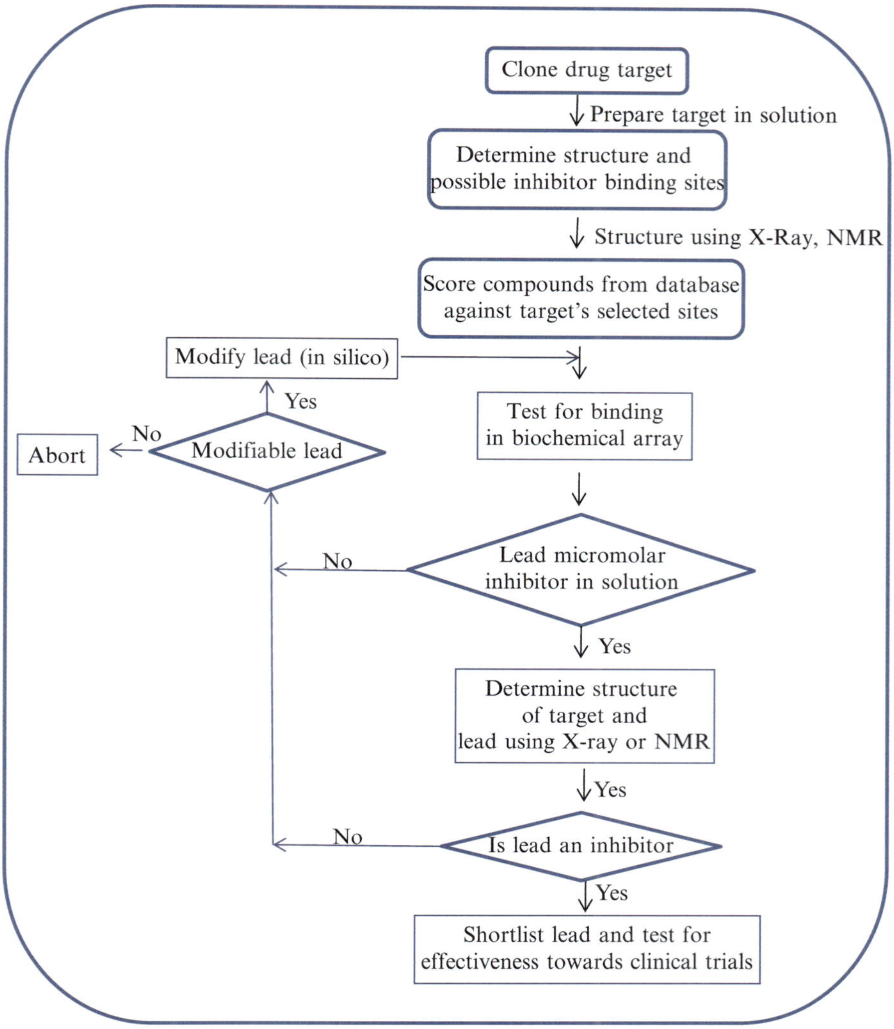

Fig. 12.2 Schematic representation of structure-based drug design

12.4.3 Nuclear Magnetic Resonance (NMR) Spectroscopy

The requirement of pure and good crystal and variation in three-dimensional structure from its natural condition are limitations of X-ray crystallography. NMR is the technique of choice, which uses soft radiation (radio waves) and examines molecules in the liquid phase. NMR is one of the important techniques, which is used to determine the structure of organic compound in solution. The technique works well with the elements having unpaired spin of proton. As the unpaired proton spins in the nuclei, it generates a magnetic field having a magnetic moment. In the presence of external magnetic field, the spin of nuclei may be either aligned with the external field (lower energy) or opposed to the external field (higher energy). As the field strength is increased, the resonance occurs (having a particular external field strength and frequency of radio wave) giving a signal which is read as NMR spectra.

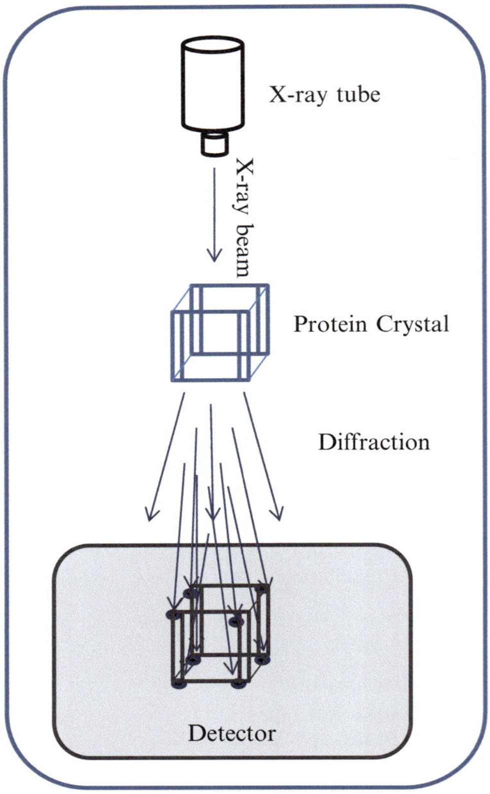

Fig. 12.3 The basic working of X-ray crystallographic instrument. The X-rays fall on the ordered arrangement of molecules called as crystal and are diffracted in different directions. They are then detected on the detector. Crystal is rotated and analyzed from all the directions, and images are then superimposed to reveal the actual arrangement of crystal molecules

The advantages of using NMR are that it also helps in determining complexes and the nature of the solution in which the drug is kept. It is suited for rapid screening of large number of molecules as potential ligands using SAR by NMR [24]. It can also be very quick at mapping residues that are altered following ligand binding.

The data obtained is not as precise as X-ray crystallography. For this reason, it is limited to smaller macromolecules, more or less kDa.

In the absence of structural information, there are three approaches for the prediction of the tertiary structure of proteins. (1) homology modeling (comparative modeling), (2) threading, and (3) ab initio method.

12.4.4 Comparative or Homology Modeling

Targets without 3D structure require homology modeling. If no experimentally determined structure is available, a homology model can also be explored. A homology model works well, if an empirically determined 3D structure is available for a sufficiently similar protein (30 % or better sequence identity). This is helpful in defining 3D coordinates of proteins with the help of software [29]. It detects and aligns sequences with template structures and represents similarity of the residues at topologically corresponding points in the reference protein [10]. In the condition where the structural information is available, the modifications may be done for desired properties. It can use software that arranges the backbone of the sequence identically to this template. However, this technique is moderately accurate for the positions of alpha carbons in the 3D structure, in regions where the sequence identity is high. It is inaccurate for the details of side chain positions and for inserted loops with no matching sequence in the solved structure (Fig. 12.4.).

A Homology Modeling Routine Needs Three Items of Input

- The sequence of the protein with unknown 3D structure, the "target sequence."
- A 3D template is chosen by virtue of having the highest sequence identity with the target sequence. The 3D structure of the template must be determined by reliable empirical methods such as crystallography or NMR and is typically a published atomic coordinate "PDB" file from the Protein Data Bank (PDB).
- An alignment between the target sequence and the template sequence.
- All other methods for judging protein structures, such as stereochemical soundness (bond lengths, bond angles, planarity, and packaging) and residues in the most favored regions of the Ramachandran plot, apply to analyzing a homology model as well as to experimentally derived models.

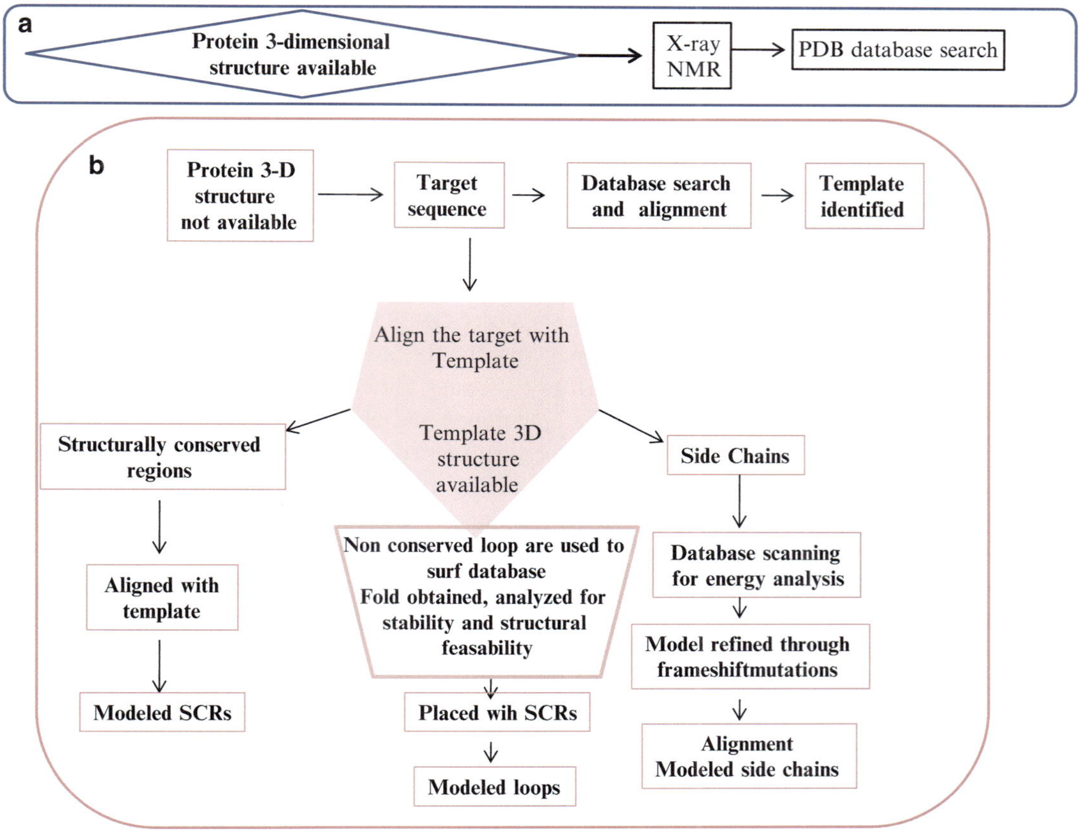

Fig. 12.4 (**a**) Modeling of the target when its three-dimensional structure is available. (**b**) shows the outline of homology modeling when three-dimensional structure of the target is not available. *SCR* structurally conserved regions, *3D* three-dimensional structure, *X-ray* X-ray crystallography, *NMR* nuclear magnetic resonance

It aligns structurally conserved regions (SCRs) using the fact that evolutionary-related proteins share a similar structure. It is based on conservation of highly similar structural conformation than the amino acid sequence of the protein and minor alterations in sequence normally result in little variation in the 3D structure. Approximately, 80,000 experimental protein structures are present in the Protein Data Bank (http://www.rcsb.org/pdb).

12.4.4.1 Template Fold Recognition

In this, the target sequence is compared and matched with the sequence of known structures from PDB database. The sequence thus obtained is the template sequence (template sequence is obtained from database which shows similarity to the target sequence). Using computer programs, the sequences are matched for the structurally conserved region (SCR) of the target and template. Following which, SCRs are modeled by computer algorithms.

12.4.4.2 Building of a Model

Model building by rigid-body assembly involves the division and evaluation into conserved core regions or SCR, non-conserved loops connecting the conserved regions, and side chains present in the backbone. The non-conserved or the loop regions are then used by the software to surf database for all possible folds. The folds are also analyzed for stability and structural feasibility [7]. When matched, the loops are also placed on the computer-generated models of SCR. Then, for the side chains present in the SCR, the software scans the database with analysis of energies and then places it on the modeled structure. The model is refined through a series of amino acid residue substitutions, insertions, and deletions for tuning alignment, modeling loops, and side chains [14]. They are further verified by computer programs using Ramachandran plot for possible conformations [13].

12.4.4.3 Loop Modeling

Non-conserved regions between SCRs are loops where insertion and deletion often occur. Loops determine the functionality as they contribute to active and binding sites. The loop modeling requires accuracy and consistency with the protein structure. Loop predictions are important for the construction of protein backbone, with conformational length and space and status of side chains [12, 16].

- Database method of loop prediction measures the separation and orientation of the backbone.
- Construction method involves extensive database search.
- Scaling-relaxation method relies on sampling of full segment along with its end-to-end distance measurement.
- Molecular mechanics/molecular dynamics verify the conformation by Ramachandran plot.

12.4.4.4 Modeling of Side Chain

Side chains of proteins may exist in a small number of rotamers (low-energy conformation) [26].

12.4.4.5 Validation of Model

Models are validated for proper protein stereochemistry, such as symmetry checks and geometry checks (chirality, bond lengths, bond angles, torsion angle models, salvation) by first category validation models as PROCHECK and WHATIF. The validation of fitness of sequence to structure along with assigning of score for each residue in its current conformation by second category models as VERIFY3D and PROSAII.

Homology modeling may be done by MODELLER [23], RAMP, SwissModel, PrISM, and COMPOSER. The homology model can predict the atomic details of the target proteins.

Using the structural information obtained through the above techniques, the structure is then prepared for drug design programs by first adding hydrogen atoms usually absent in crystal structures determined with data at a resolution lower than 1.0 A. The protonation and tautomeric states of residues as well as the state of histidine residues should be assigned. Small molecules such as ions and water molecules can be included during the lead generation phase in cases where they play structural roles that are crucial for the confirmation of the target; otherwise, they are usually removed to allow any potential lead to occupy their positions.

12.4.5 Threading/Fold Recognition

Threading/fold recognition: Used for sequence identity of approximately 30 %. In this, the software checks for available folds in the Protein Data Bank (PDB), it checks if the sequence can adopt any known fold. It can only help to predict the fold of the protein [28].

12.4.6 Ab Initio Method

Ab initio predictions start with the assumption considering the free energy minimum for the native structure of a protein. Thus it searches for conformations having minimal energy for each amino acid sequence. Thus, in this, from the primary structure, the software analyzes the probability of one amino acid being in particular conformation, according to which it generates secondary structure and finally the three-dimensional structure of the protein.

12.5 Ligand-Based Drug Design

In this, the model of biological target is created based upon the knowledge of the ligand that what can bind to it or can interact with it. In the condition where the structural information is available, the modifications may be done for desired properties [5]. Targets without 3D structure require homology modeling.

12.6 Drug Targets

Drug targets are usually proteins that are bound during the drug design process. However, there are also the cases where the drug is designed against RNA targets with well-defined secondary structure for, e.g., the bacterial ribosome, the receptors for the G-proteins, or the genomes.

The choice of the target is usually dependent on few factors like:

1. Drug targets should have a unique function in the pathogen. It should only be present in the pathogen and should be inhibited by a small molecule.
2. The target should be a part of a crucial cycle in the cell, and its elimination/blocking should lead to the desired effect required for curing a disease.
3. The target should be unique, i.e., no other pathway should be able to supplement the function of the target and overcome the presence of the inhibitor/facilitator.
4. The target molecule should be able to be inhibited by binding a small molecule. The target generally has a well-defined binding pocket.

12.6.1 Identification of Target Site

Target sites are pocket or protuberance with a variety of potential hydrogen bond donors and acceptors, hydrophobic characteristics, and sizes of molecular surfaces. The ligand-binding site can be the active site, as in an enzyme, an assembly site with another macromolecule, or a communication site necessary in the mechanism of the molecule. RNA secondary structural elements can provide excellent target sites since they are species specific, bind ligands, and can be specific for a disease state. Target sites for protein–protein interactions can be difficult to locate as these surfaces are often flat, large, and hydrophobic. Techniques like co-crystallization can be used for these situations.

12.7 Identification of Lead for Drug Designing

Once the structure and the target site are identified, there are two major methodologies for developing a lead—computer-aided methods and the experimental methods. Experimental methods are high-throughput screening with combinatorial chemistry, in which thousands of compounds are tested for biochemical effects.

12.7.1 Computer-Aided Drug Designing (CADD)

For computer-aided methods, there are two major challenges that need to be overcome:

- High degree of computational accuracy required to predict significant changes in binding affinity. Clearly, one of the first challenges for calculating reliable protein–ligand binding affinities is a high-quality molecular model.

- The number of degrees of freedom in a large drug molecule can be numerous to provide a significant hurdle for sampling all of the possible energetically accessible conformations. Even a modest-sized protein has many times more rotatable bonds that results in a nonlinear explosion of combinations. The most common computational approach for addressing the sampling problem is molecular dynamics and Monte Carlo simulations.

Having handled these challenges, the computer-aided methods are divided into three major categories:

Inspection – a method in which the known molecules that bind the site, such as substrates or cofactors in the case of enzymes or peptides in the case of protein–protein or protein–nucleic acid interactions, are modified to become inhibitors based on maximizing complementary interactions in the target site.

Virtual screening – a method involving database of available small molecules which are docked into the region of interest in silico and scored based on predicted interactions with the site.

De novo generation – a method involving small fragments of molecules, such as benzene rings, carbonyl groups, and amino groups, that are positioned in the site, scored, and linked in silico. The final compounds, created in silico from the linked fragments, are synthesized in the laboratory.

12.8 Docking Method

The need for a rapid search for small molecules that may bind to targets of biological interest is of crucial importance in the drug discovery process [8]. One way of achieving this is the in silico or *virtual screening* of large compound collections to identify a subset of compounds that contains relatively many hits against the target, compared to a random selection from the collection [17]. The compounds that are virtually screened can stem from corporate or commercial compound collections or from virtual compound libraries. If a three-dimensional (3D) structure or model of the target is available, a commonly used technique is structure-based virtual screening (SBVS). Here, a so-called docking program is used to place computer-generated representations of a small molecule into a target structure (or in a user-defined part thereof, e.g., the active site of an enzyme) in a variety of positions, conformations, and orientations [25]. Each such docking mode is called a "pose." In order to identify the energetically most favorable pose (also referred to as "pose prediction"), each pose is evaluated ("scored") based on its complementarity to the target in terms of shape and properties such as electrostatics. A good score for a given molecule indicates that it is potentially a good binder. This process is repeated for all molecules in the collection, which are subsequently rank-ordered by their scores (i.e., their predicted affinities). This rank-ordered list is then used to select for purchase, synthesis, or biological investigation only those compounds that are predicted to be most active. Assuming that both the poses and the associated affinity scores have been predicted with reasonable accuracy, this selection will contain a relatively large proportion of active molecules, i.e., it will be "enriched" with actives compared to a random selection.

Despite the technical challenges in reliably predicting the binding mode of a molecule and its binding affinity relative to other compounds, in many cases, docking campaigns have yielded significant hit rate improvements compared to random screening.

12.8.1 Importance of Understanding Solution and Flexible Ligand

There is a lot of emphasis on the crucial effects of including protein and ligand flexibility in the docking and scoring process. Most proteins and most ligands are quite flexible in solution and may experience a full ensemble of possible conformations. As a result, lead generated from a single, rigid structure may have differing results in solutions than in silico. In order to account for the landscape of protein and ligand conformations, several drug design algorithms incorporate protein and/or ligand flexibility. However, modeling molecular flexibility especially for the target macromolecule drastically increases the computation time required.

Many programs that allow protein flexibility incorporate information from multiple protein structures. Ensembles of structures can be experimentally determined such as NMR ensembles or multiple crystal structures [20].

Solvent effect: solvent plays an important role in ligand bindings in several ways. It can be done in the following ways:

Order water molecules seen in the structure can be incorporated into the designed ligand, effectively increasing ligand binding by increasing the entropy of the system. Order water molecules can be treated as bound ligands, and contacts with them can be maximized.

- Accurate determination of the relative free energies of all of the molecular species in solution and the corresponding microscopic binding-free energies for all of the molecular species binding with the protein is very important to study the binding of a protein with multiple molecular species of a ligand.
- Fully polarizable continuum model (FPCM) method allows accurate determination of the solvent effects in the first-principles quantum mechanism (QM) calculations on molecules in solution.
- The combined use of the FPCM-based QM calculations and other computational modeling and simulations enables us to accurately account for a protein binding with multiple molecular species of a ligand in solution.
- Based on the computational modeling of the detailed protein–ligand interactions, possible new drugs may be designed rationally either as small-molecule ligands of the protein or engineered proteins that bind/metabolize the ligand. The computational drug design has successfully led to discovery and development of promising drugs [18, 21].
- After discovery of possible drug, they are modified to enhance binding. Upon evaluation of their potency, they are examined for reactivity, bioavailability, toxicity, drug resistance, metabolism, and immune reactivity. After validation and safe preclinical trials, the lead can be processed for clinical trials.

12.8.2 Algorithms Underlying Various Docking Programs

Many excellent drug design software methods available are capable of either virtual screening or de novo generation. A large number of docking programs and search algorithms have been published. One criterion for classifying the underlying algorithms is the way the ligands are treated during docking. In some of these algorithms, the ligand is built up incrementally, starting from a docked "base fragment." Programs that follow this approach include Hammerhead, DOCK, and FlexX. In other programs, such as AutoDock, Genetic Optimization for Ligand Docking (GOLD), ICM-Dock, and QXP, the ligand is treated in its entirety.

In addition to ligand flexibility, it is desirable to keep at least part of the receptor flexible in order to allow for conformational changes that are necessary to accommodate the ligand, a phenomenon referred to as "induced fit." Because it is computationally expensive, few docking program allow protein flexibility [3]. Exceptions are the latest versions of AutoDock, FlexE, QXP, Affinity, and the latest version of ICM-Dock. The way flexibility is handled differs from program to program. For example, FlexE uses multiple receptor conformations, Affinity allows any selection of atoms to be mobile with a user-defined tethered buffer region between the fixed and mobile regions, and QXP allows user-defined parts of the protein, for example, selected side chains or a particular loop, to move. Another criterion to classify docking programs would be according to the search strategy employed. Roughly speaking, one could distinguish between programs trying to maximize shape complemen-

Table 12.1 Overview of docking algorithms

Program	Algorithm
AutoDock	Lamarckian GA
DOCK	Shape matching (sphere images)
DOCK (NWU version)	Shape matching (sphere images)
FlexX	Incremental constructions
FRED	Shape matching (Gaussian functions)
Glide	Description matching/MC
GOLD	Genetic algorithm
Hammerhead	Incremental construction
ICM	MC minimization
LigandFit	Shape matching
QXP	MC minimization, tree searching, and pruning
SLIDE	Descriptor matching
Surflex-Dock	Surface-based molecular similarity

tarity—often based on geometric criteria—and programs incorporating an energy-driven or stochastic algorithm. Well-known representatives of the former group are DOCK, FlexX, and FRED. Among the latter group, programs such as AutoDock, ICM-Dock, QXP, and GOLD can be found [19].

A quick overview of the docking algorithms is provided in Table 12.1.

Though majority of the software come with built-in scoring, however they are designed to work across a large set of target proteins, but they need not necessarily be the best function for a particular target. For this reason, one can optimize a scoring function, e.g., by experimental binding affinities/inhibition constants against a set of compounds; this information can be used to fine-tune the scoring function to that target.

One can also modify the best affinity prediction through rescoring. In rescoring, the poses generated by the docking program are scored and then one of more alternative scoring functions are applied to those poses, thereby improving the affinity function of the target.

Despite its active usage in docking, it has its own deficiencies in terms of precisely identifying the protein binding site. It may be used to determine the most probable docking pose for a ligand in a protein binding site where a crystal structure is unavailable [22]. Here the goal is just to find the correct orientation and conformation of the ligand in the protein. It is important to consider docking-based virtual screening more as a filter than any kind of ordered list. Docking does filter out compounds that fit the active site poorly, but it does not differentiate well between weak and potent binders.

12.9 De Novo Generation

De novo is a Latin expression meaning "from the beginning." Active site of drug targets when characterized from a structural point of view will shed light on its binding features. The information of active site composition and the orientation of various amino acids at the binding site can be used to design ligands specific to that particular target. Computational tools that can analyze protein active site and suggest potential compounds are extensively used for de novo design methods. Many promising approaches with the goal of ligand design are reported. Some of the prominent methods for the computer-aided ligand design are:

1. Fragment location methods: To determine desirable locations of atoms or small fragments within the active site.
2. Site point connection methods: To determine locations ("site points") and then place fragments within the active site so that those locations are occupied by suitable atoms.
3. Fragment connection methods: Fragments are positioned, and "linkers" or "scaffolds" are used to connect those fragments and hold them in a desirable orientation.
4. Sequential buildup methods: Construct a ligand atom by atom or fragment by fragment.

5. Whole-molecule methods: Compounds are placed into an active site in various conformations, assessing shape and/or electrostatic complementarity.
6. Random connection methods: A special class of techniques combining some of the features of fragment connection and sequential buildup methods, along with bond disconnection strategies and ways to introduce randomness.

12.10 Drug Lead Evaluation

Once a small molecule is identified as potentially binding to the target molecule, it is evaluated before proceeding further. It is important to consider that the ranking assigned by the scoring function is not always indicative of a true binding constant, since the model of the target–ligand interaction is inherently an approximation. Both the solvent effect and the effects of target and ligand flexibility are usually imprecisely described. Usually, several molecules which scored well during the docking run are evaluated in further tests since even the top-scoring molecule could fail in in vitro assays.

- Leads are first evaluated visually with computer graphics and can often be optimized at this step for increased affinity.
- Leads are also evaluated for their likelihood to be orally bioavailable: less than five hydrogen bond donors and less than ten hydrogen bond acceptors, a molecular weight less than 500, and a calculated log of the partition coefficient less than 5.
- There is also a phenomenon to be tested that the number of rotatable bonds should be less than ten in order to increase the potential for oral bioavailability.

Other factors like chemical and metabolic stability and the ease of synthesis can also factor into decision to proceed with a particular candidate lead.

Drugs developed should also be evaluated for safety problems, adverse toxic side effects, cardiac toxicity, and development of drug resistance. It also implies that a relatively high docking score or binding affinity does not necessarily mean that a substance is going to be a potent inhibitor for that target or it would be free from undesirable side effects. These criteria do provide valuable information for drug design. The methods like gene expression profiling/protein profiling and advanced computational tools can be used to gain insights to overcome adverse side effects of drugs.

Lastly, promising leads reenter the structural determination process to find the exact binding mode and to evaluate any further optimization that becomes evident.

In CADD, a computer is an essential tool apart from other physical techniques of crystallization and spectroscopy, and the process is very complex requiring scientists from many disciplines working together.

12.11 Drug Discovery

The drug may be produced either through chemical synthesis or may be based on biopolymer.

Sometimes drugs can be accidentally discovered but in practice they are developed with organized effort to unravel ways to treat the diseases. The process also helps in the improvement of existing drug with more safety and efficacy. However the advantage of the designing would be not to affect any other molecule leading to minimal or no side effects, which may be achieved by the screening, where large number of synthetic chemical compounds or naturally derived products are used for desirable effects. The screening process requires an appropriate procedure and was used as the method of choice before, but it suffers from many problems like:

1. Screening a large number of compounds; however, in case the selection of appropriate pH/concentration is somehow not maintained, the effective compound can remain in inactive state and might skip the screening.
2. Requirement of cell lines or animal model for efficacy testing may not reflect the natural human host.
3. It requires too much time.

With the advancement in tools and technology and the understanding of genetics and biochemistry of the disease process, it has become somewhat easier to interrupt or check the disease by inhibition of the mediators. It has helped in defining potential drug targets leading to designing of drug, which can specifically interact, with the target molecule. In this, the knowledge of the normal pathway and the pathway operative in disease state with the understanding of exact step being altered is required along with three-dimensional structure of the molecules involved in the process [27]. This knowledge is used for defining specific targets as effector molecules, enzymes, receptors, or nucleic acids for designing of drug.

The drug should possess certain properties:

1. Efficacy (less amount should be sufficient)
2. Safety (it should be with low toxicity or minimal side effects)
3. Persistence in the body (should stay in the system for prolonged exposure)
4. Route of administration and cost-effectiveness

12.12 Chapter End Summary

- Traditional drug design relies on systematic synthesis and screening of large number of compounds to identify the ones that will be effective in treatment. This process can be very long and time consuming.
- Computer-aided drug design can help identify the lead and can significantly reduce the time for drug design process.
- The process of drug design entails identification of 3D structure of the target site. The lead is identified based on its efficacy to bind the target site. Based on the scoring of the binding, the leads are identified and further evaluated for drug candidates.
- Structured-based drug design relies on structural information of the target for identification of the lead, whereas ligand-based (de novo) drug design relies on knowledge of molecules or ligand that binds the target.
- Some of the prominent methods for identifying 3D structure are X-ray crystallography and NMR. In the absence of a 3D structure, a homology model is applied, if an empirically determined 3D structure is available for a sufficiently similar protein (50 % or better sequence identity).
- Once the target site is identified, the lead identification is done using inspection, virtual screening, or de novo design methods.
- The lead with the highest score needs not necessarily be the candidate for drug development as it is also based on a lot of other factors like efficacy, safety and reaction to the body, and cost.

Multiple Choice Questions

1. Metabolic changes in the body can be induced by:
   (a) Increasing the energy function of the enzymes
   (b) Identifying the structure of the enzyme and then changing their Ph
   (c) Inhibiting enzymes and blocking ion channels
   (d) None of the above
2. Which of the following is NOT a de novo drug design method?
   (a) Fragment location methods
   (b) Site point connection method
   (c) Fragment connection methods
   (d) Linker dislocation methods
3. Which of the following is a FALSE statement?
   (a) Target sites are pocket or protuberance with a variety of potential hydrogen bond donors and acceptors, hydrophobic characteristics, and sizes of molecular surfaces.
   (b) Homology model works when a 3D structure can be easily identified.
   (c) Virtual screening method involves database of available small molecules which are docked into the region of interest in

silico and scored based on predicted interactions with the site.
(d) Docking has its own deficiencies in terms of precisely identifying the protein binding site.

4. Which of the following explains X-ray crystallography?
   (a) When the crystals are tested for their wavelengths, then they match with the target.
   (b) When crystals are passed through X-rays, the patterns of the targets and leads matches.
   (c) When X-rays is applied on the crystal, its structure changes resulting in identification of the right compound.
   (d) When X-rays fall on crystal, its diffraction pattern is recorded from many directions. Superimposition of these images is helpful in predicting structure of the crystal.
5. Which of the following is NOT a characteristic of the drug target?
   (a) The targets should have a unique function in the pathogen. It should only be present in the pathogen and should be inhibited by a small molecule.
   (b) The target should be a part of a crucial cycle in the cell, and its elimination/blocking should lead to the desired effect required for curing a disease.
   (c) The target should be unique, i.e., no other pathway should be able to supplement the function of the target and overcome the presence of the inhibitor/facilitator.
   (d) The target should be able to change the metabolism of the pathogen by blocking the pathway from multiple routes.
6. Drug designing is important because:
   (i) It can give a more effective treatment.
   (ii) It helps to target a specific biological mediator.
   (iii) It is very cost-effective.
   (iv) It gives safer compounds as compared to general inhibitors.
   (a) Both (i) and (ii)
   (b) Both (i) and (iii)
   (c) Both (iii) and (iv)
   (d) (i), (ii), and (iv)
7. A lead compound is:
   (a) Lead (Pb)-containing drug
   (b) The target biomolecule
   (c) The enzyme complex which is used for suppression of the pathway
   (d) Initial compound on which designing and development of drug is based
8. Lead compound search requires:
   (a) A preexisting drug
   (b) Nanotechnology for delivery
   (c) Structure–activity relations
   (d) A bioassay
9. Term used for drugs having similar structure to a known drug:
   (a) Me-too drug
   (b) Analog drug
   (c) Derivative drug
   (d) None of the above
10. Nowadays, novel compounds (dolostatins) for many ailments as cancer are obtained from:
   (a) Medicinal plant
   (b) Marine sources
   (c) Transgenic animals
   (d) None of these
11. A detection system patented by Pharmacia Biosensor "BIAcore" used to detect target–ligand binding is based upon:
   (a) X-ray crystallography
   (b) Nuclear magnetic resonance
   (c) Surface plasmon resonance
   (d) Fluorescence-activated cell sorter
12. Drug obtained from natural sources is:
   (a) Cephalosporin
   (b) Artemisinin
   (c) Ceftriaxone
   (d) Isoniazid

Answers

1. (c); 2. (d); 3. (b); 4. (d); 5. (d); 6. (d); 7. (d); 8. (d); 9. (a); 10. (b); 11. (c); 12. (b)

Review Questions

Q1. Explain the process of lead identification for drug design process.
Q2. What is the difference between structure-based drug design and ligand-based drug design?

Q3. Explain the process of X-ray crystallography for identification of 3D structure.
Q4. Explain homology modeling for structure identification.
Q5. What are the various methods for de novo generation?
Q6. Explain various methods of computer-aided drug design.

References

1. Anderson AC (2003) The process of structure-based drug design. Chem Biol 10:787–797
2. Baker D, Sali A (2001) Protein structure prediction and structural genomics. Science 294:93–96
3. Carlson HA, Mccammon JA (2000) Accommodating protein flexibility in computational drug design. Mol Pharmacol 57:213–218
4. Chandonia JM, Brenner SE (2005) Implications of structural genomics target selection strategies: Pfam5000, whole genome, and random approaches. Proteins 58:166–179
5. Merz KM, Ringe D, Reynolds CH (eds) (2010) Drug design structure and ligand based approaches. Cambridge University Press, Cambridge
6. Funkhouser T (2005) Structure based drug design. (https://www.cs.princeton.edu)
7. Gromiha MM, Selvaraj S (eds) (2002) Recent research developments in protein folding, stability and design. Research Signpost, Trivandrum
8. Hilllisch A, Pineda LF, Hilgenfeld R (2004) Drug Discov Today 9:659–669
9. Joseph-McCarthy D (1999) Computational approaches to structure-based ligand design. Pharmacol Ther 84:179–191
10. Krieger E, Nabuurs SB, Vriend G (2003) Homology modeling. In: Bourne PE, Weissig H (eds) Structural bioinformatics. Wiley-Liss, Hoboken
11. Kroemer RT (2007) Structure-based drug design: docking and scoring. Curr Protein Pept Sci 8:4
12. Lee J, Lee D, Park H, Coutsias EA, Seok C (2010) Protein loop modeling by using fragment assembly and analytical loop closure. Proteins 78:3428–3436
13. Lesk AM (2003) Introduction to bioinformatics. Oxford University Press, New Delhi
14. Lesk AM, Chothia CH (1986) The response of protein structures to amino-acid sequence changes. Philos Trans R Soc Lond Ser B 317:345–356
15. Mandal S, Moudgil M, Mandal S (2009) Rational drug design. Eur J Pharmacol 625:90–100
16. Michalsky E, Goede A, Preissner R (2003) Loops in proteins (LIP)-a comprehensive loop database for homology modelling. Protein Eng 16:979–985
17. Nabuurs SB, Wagener M, de Vlieg J (2007) A flexible approach to induced fit docking. J Med Chem 50:6507–6518
18. Peitsch MC (1996) ProMod and Swiss-Model: internet-based tools for automated comparative protein modelling. Biochem Soc Trans 24:274–279
19. Peitsch MC (1997) Large scale protein modelling and model repository. Proc Int Conf Intell Syst Mol Biol 5:234–236
20. Petros AM et al (2006) Discovery of a potent inhibitor of the antiapoptotic protein Bcl-XL from NMR and parallel synthesis. J Med Chem 49:656–663
21. Sali A, Blundell TL (1993) Comparative protein modelling by satisfaction of spatial restraints. Mol Biol 234:779–815
22. Samudrala R (2002) Lessons from blind protein structure prediction experiments In: Gromiha MM, Selvaraj S (eds) (Recent research developments in protein folding, stability and design. Research Signpost, Trivandrum
23. Sanchez R, Sali A (1997) Evaluation of comparative protein structure modeling by MODELLER-3. Proteins 1:50–58
24. Shuker SB et al (1996) Discovering high affinity ligands for proteins: SAR by NMR. Science 274: 1531–1534
25. Sutcliffe MJ, Haneef I, Carney D, Blundell TL (1987a) Knowledge based modeling of homologous proteins, part I: three-dimensional frameworks derived from the simultaneous superposition of multiple structures. Protein Eng 5:377–384
26. Sutcliffe MJ, Hayes FR, Blundell TL (1987b) Knowledge based modeling of homologous proteins. Part II: rules for the conformations of substituted side chains. Protein Eng 1:385–392
27. Wei R (2005) Successes and failures in structure-based drug design. http://spider.iwr.uni-heidelberg.de/
28. Westbrook J, Feng Z, Chen L, Yang H, Berman HM (2003) The Protein Data Bank and structural genomics. Nucleic Acids Res 31:489–491
29. Yang AS, Honig B (1999) Sequence to structure alignment in comparative modeling using PrISM. Proteins 37:66–72

Some Selected Resources

genome.wellcome.ac.uk
http://scc.acad.bg/ncsa/articles/library
http://www.bicpu.edu.in/seminar/05/pdf/sel.pdf
molpath.ucsd.edu/PDF
www.ncbi.nlm.nih.gov/pubmed
www.udel.edu/chem/bahnson
http://scc.acad.bg/ncsa/articles/library

Drug Targeting and Delivery 13

Abstract

Drug delivery is defined as mechanisms to introduce pharmaceutical compounds to human in order to achieve therapeutic effects. We have come a long way since chewing medicinal plants and inhaling soot from medicinal substance were the only form of drug delivery. These approaches lacked consistency and uniformity of drug delivery. Since then there has been a continuous effort to discover and improve drug delivery routes and drug delivery systems. Conventional drug delivery system includes drug delivery via oral route as solutions, suspensions, emulsions, and tablets. Some are delivered systemically via injections and intravenous application. Medications are applied topically as lotions and gels. Nasal route is used for drug delivery to lungs by inhalers and nebulizers. Apart from antibiotics, vaccines, and chemical compounds, modern medicine includes recombinant DNA, insulin, interferon, interleukin, erythropoietin, tissue plasminogen activator, and other peptides and macromolecules as drugs that require efficient drug delivery systems. Traditional drug delivery systems suffer from various limitations such as low bioavailability, intolerance, toxic side effects, reduced plasma half-life, higher concentration, and low efficacy. The hydrophilic drugs have difficulty in passing through the cell membrane. Systemically delivered drugs reach all the organs irrespective of the affected organ. This causes toxic side effects on the healthy cells. The drugs tend to degrade fast in the plasma so higher doses of drug are required and hence it becomes toxic with reduced efficacy and are expensive. The biological barriers exclude the drug from reaching the affected cells and tissues. Efficient drug targeting can improve drug delivery efficacy, reduce side effects, and lower treatment cost. Hence, much effort is given on the development of novel carriers that would meet the requirement of drug delivery systems. The main areas of research are to increase bioavailability of the drugs, increase plasma half-life, and target to specific organs or cells. This would result in lowering the dose, which would also lower drug-induced toxicity, protect bystander cells and organs from adverse side effects, and reduce medical expenses. In this chapter, we will discuss the biological barriers, advances in drug delivery systems, drug targeting, and their application in diseases.

V. Gupta et al., *Basic and Applied Aspects of Biotechnology*, DOI 10.1007/978-981-10-0875-7_13

13.1 Introduction

Drug delivery is defined as mechanisms to introduce pharmaceutical compounds to human in order to achieve therapeutic effects. We have come a long way since chewing medicinal plants and inhaling soot from medicinal substance were the only form of drug delivery. These approaches lacked consistency and uniformity of drug delivery. Since then there has been a continuous effort to discover and improve drug delivery routes and drug delivery systems. Conventional drug delivery system includes drug delivery via oral route as solutions, suspensions, emulsions, and tablets. Some are delivered systemically via injections and intravenous application. Medications are applied topically as lotions and gels. Nasal route is used for drug delivery to lungs by inhalers and nebulizers. Apart from antibiotics, vaccines, and chemical compounds, modern medicine includes recombinant DNA, insulin, interferon, interleukin, erythropoietin, tissue plasminogen activator, and other peptides and macromolecules as drugs that require efficient drug delivery systems. Traditional drug delivery systems suffer from various limitations such as low bioavailability, intolerance, toxic side effects, reduced plasma half-life, higher concentration, and low efficacy. The hydrophilic drugs have difficulty in passing through the cell membrane. Systemically delivered drugs reach all the organs irrespective of the affected organ. This causes toxic side effects on the healthy cells. The drugs tend to degrade fast in the plasma so higher doses of drug are required and hence it becomes toxic with reduced efficacy and are expensive. The biological barriers exclude the drug from reaching the affected cells and tissues. Efficient drug targeting can improve drug delivery efficacy, reduce side effects, and lower treatment cost. Hence, much effort is given on the development of novel carriers that would meet the requirement of drug delivery systems. The main areas of research are to increase bioavailability of the drugs, increase plasma half-life, and target to specific organs or cells. This would result in lowering the dose, which would also lower drug-induced toxicity, protect bystander cells and organs from adverse side effects, and reduce medical expenses. In this chapter, we will discuss the biological barriers, advances in drug delivery systems, drug targeting, and their application in diseases [26].

13.2 Biological Barriers to Drug Delivery

The biological barriers such as the skin, mucosal membrane, lung surfactant, and blood–brain barrier (BBB) are designed to restrict the entry of foreign molecules and allow selective molecules to enter. A major challenge in the drug delivery field is to enhance transport of therapeutics across these barriers. Tissue-specific transporters are being used to carry larger molecules across biological barriers. The barriers are temporarily perturbed by the use of molecules that weakens the barrier, e.g., by targeting tight junctions. An important aspect is to target drugs to pathogenic multidrug-resistant microorganisms that often employ multiple strategies to evade drugs and host immune system. Some of the barriers that prevent drug delivery to the desired cells and organs are discussed below.

13.2.1 Skin Barriers

The major function of the skin is to maintain a barrier between the external environment and the internal milieu. It prevents excessive water and electrolyte loss and provides first line of defense against physical, chemical, and microbial assault [9]. The unique structure of the skin helps it to perform these specialized functions. It is a multilayered organ consisting of the dermis, epidermis, and subcutaneous fat tissue. The epidermis is the outermost layer of the skin and comprises of four distinctive layers consisting of sequential differentiation stages of the keratinocytes, the major cell type in the epidermis. The layers include the superficial stratum corneum (SC), stratum granulosum (SG), stratum spinosum (SS), and the innermost undifferentiated basal layer or stratum basale (SB) [40]. The SG, SS, and SB are viable regions, whereas the SC consists of nonviable final differentiation product. It is composed of flattened, anucleate corneocytes, surrounded by multiple planar lamellae sheets, enriched in ceramides, cholesterol, and free fatty acids. The structure of the SC is often referred to as a "bricks in mortar" structure, in which the corneocytes are the bricks and the lipids surrounding them are the mortar. The localization of these highly hydrophobic lipids within the extracellular domains of the SC inhibits the water loss. This lipid matrix acts as the main barrier for diffusion of substances through the skin blocking the entry of toxic compounds and allergens. Additionally antimicrobial peptides are delivered to the SC intercellular domains via secretion of lamellar body contents [12]. The dead cells provide protection against pathogen colonization as well as UV light invasion. They work as permeability barrier and antimicrobial barrier. Any kind of damage to the skin due to injury, burn, or chronic psoriasis increases the chances of infection immensely.

13.2.2 Mucus and Surfactants

Mucus is an aqueous heterogeneous mixture of glycoprotein that covers most of the application sites of drugs. It is secreted by specialized goblet cells of columnar epithelium. The major constituents of mucus are water and high molecular weight glycoproteins called mucin. Mucin contains heavily O-glycosylated serine/threonine-rich tandem repeat domains that form network. Other components include immunoglobulins, lipids, DNA, electrolytes, inorganic salts, enzymes, and mucopolysaccharides. It covers most of the epithelial surfaces and protects underlying tissue against water loss, infiltration of pathogens, foreign molecules, and extracellular environment. At the same time, it allows diffusion of specific molecules inside the cell. Properties and function of mucus layer vary at the different sites. In gastrointestinal mucosa it helps with smooth passage of food. In airway passage it maintains hydrated layer for gaseous exchange. Mucin can be of two kinds, secreted mucin and bound mucins. Protective function of mucus depends on the structure and thickness of the layer. Gastrointestinal mucosa has a thick mucus layer. It protects against the acidic environment. Mucus has a high turnover and washes out the adhered foreign molecules. It is usually the first barrier that drugs must overcome in order to be absorbed and gain access to the circulatory system and their target. It affects the absorption of drugs that are delivered via oral, pulmonary, vaginal, and nasal routes. Solubility and lipophilicity of the drug determine the absorption rate across the mucus layer and biological membrane.

Drug needs to permeate the mucus layer to reach the underlying epithelial layer. For a while mucoadhesive drugs were used that would adhere to the mucus layer for a long period of time and provide more opportunity for the drug to diffuse in. However, since the turnover rate of mucus layer is very high, the chances of the drugs to be washed away are also higher. So now the approach is to use mucopenetration that would deposit the drug to the epithelial layer. The micro-particulate carriers are coated with substances that can easily penetrate the mucus layer such as PEG. Mucolytic agents such as proteases are also used for penetration of the drugs.

13.2.2.1 Lung Surfactant and Alveolar Air Blood Barrier

Pulmonary route provides an attractive noninvasive drug delivery route. It is an appealing route for drug administration due to large surface for drug dispersion, a low content of drug-metabolizing enzymes, and a high vascularization for systemic drug delivery. However the presence of cellular and noncellular barriers is a challenge for appropriate drug delivery.

The noncellular barriers include respiratory mucus and alveolar fluids, while the presence of immune cells restricts the nanoparticle delivery by phagocytosis and clearance:

(a) Respiratory mucus

The respiratory mucus is a defense mechanism against inhaled dusts, toxins, allergens, and microbes. It lines the respiratory epithelium from the nose to the terminal bronchioles.

The mucus layer lies on the tips of the cilia which are surrounded by a periciliary fluid layer. Inhaled materials are captured in this blanket of mucus and continuously transported by the cilia to the esophagus. This process is known as mucociliary clearance mechanism and the physical properties of respiratory mucus determine the efficiency of clearance. It is mainly composed of a three-dimensional network of cross-linked mucin chains which gives the mucus viscoelastic properties. These are secreted by goblet cells and submucosal glands.

(b) Alveolar fluid

The alveolar epithelium is covered with a thin continuous layer of pulmonary surfactant which comprises phospholipids and specific surfactant-associated proteins. These are synthesized and secreted by type II alveolar cells. The major function of pulmonary surfactant is to reduce surface tension at the air–water interface of the terminal airways, thereby reducing the tendency of alveoli to collapse. This is also known as bronchoalveolar lavage fluid (BALF) and is composed of 90 % lipids and 10 % proteins. They reduce the surface tension during inhalation and expiration and protect against pathogens. The inhaled pathogens are opsonized by immunoglobulins and subsequently eliminated via alveolar macrophages. The alveolar fluid represents a critical immunological barrier to nanoparticle (NP)-mediated drug delivery. It stops the diffusion of NPs by sterically obstruction or by binding to them. The non-cross-linked molecules bind to the surface of the NPs and cause aggregation further impeding their capacity to move through the meshes of the biopolymer network in mucus. Additionally the mucus blanket is continuously removed via mucociliary transport or coughing. Therefore, the NPs should be able to cross the mucus layer before they are cleared from the respiratory tract [37].

13.2.2.2 Gastrointestinal Mucosa

The human gut is lined with multilayered mucosal barrier that defends against bacterial infestation and toxin interaction [23]. This also impedes transfer of drugs across the mucosal membrane. The gastrointestinal mucosa consists of three layers. The mucus layer surrounds the lumen and is the contact point of microbes and toxins. Underneath the mucus layer lays the glycocalyx which is a filamentous layer of branched carbohydrates. It is followed by the epithelial cell layer held together by tight junctions. The mucus layer varies in its properties and structure at different sites. It is discontinuous in the small intestine but forms thick continuous layer in the colon with upper loose layer and lower adherent layer. It protects from bacterial infestation. The mucus contains mainly MUC2 mucin which is a large and complex glycoprotein. It is made up of a protein core containing tandem repeats that are rich in proline, serine, and threonine. The protein core is highly O-glycosylated. MUC2 dimerizes via C-terminal cysteine–cysteine disulfide bonds. This assembly forms mucus gel that prevents pathogens from accessing the epithelium, acting as lubricant and a densely packed, sievelike barrier. The glycocalyx is a uniform filamentous layer made up of glycoproteins and glycolipids that protrude from the apical surface of the epithelial layer. It forms a network covering the epithelial layer. It forms a physical barrier for enteric bacteria. It has a fast turnover and clearance and helps in the removal of attached pathogens. The epithelial layer of the gut mucosa is held together by desmosomes, adherens junctions, and tight junctions. Tight junctions are made up by interaction between proteins, occludin, claudin, tricellulin, and junctional adhesion molecules. They help in maintaining the integrity of the surface. Apart from the structure, the epithelial cells produce antimicrobial peptides called defensins that have microbial activity. Though they are very effective against microbial infestation, they form a barrier for drug transfer and delivery [8, 38].

13.2.3 Blood–Brain Barrier (BBB)

Blood–brain barrier protects the brain from toxic compounds by acting as a diffusion barrier from the blood to brain. The molecular components and the transport systems present in BBB hinder the entry of harmful toxins present in the blood as well as provide an effective efflux system to protect the neural tissue. This however makes drug delivery to the brain very challenging. Effective treatment for brain diseases such as multiple sclerosis, encephalitis, neurological disorders, stroke, and tumor is difficult to achieve. The BBB is a complex cellular network of brain endothelial cells, basal membrane, pericytes, and astrocytes. Brain endothelial cells also known as brain microvascular endothelial cells (BMECs) form tight junctions and hinder the movement of molecules across the membrane. They have efflux transporters that are capable of effluxing small lipophilic molecules that are able to diffuse into the cells. The basal membrane restricts the movement of solutes and consists of laminin, type IV collagen, and fibronectin. The pericytes are embedded in the basal membrane. These are contractile cells and regulate BBB specific gene expression in BMEC. They inhibit immune cells from damaging BBB and also reduce vascular permeability. The tight junction of BBB consists of three integral membrane proteins: occludin, claudin, and junction adhesion molecules. Cytoplasmic accessory proteins are also present. Together they impede molecular movement across the barrier. There are large number transporters that help in efflux of molecules. They are of two types: ATP-binding cassette (ABC) transporters and solute carrier (SLC) transporters. ABC transporters are ATP-driven efflux pumps. They are mainly localized in the luminal sides of brain capillaries and impede brain uptake of lipophilic molecules, potentially toxic metabolites, xenobiotics, and drugs. These are P-glycoprotein (P-gp), breast cancer resistance protein (BCRP), and multidrug resistance proteins (MRPs). SLC transporters on the other hand require electrochemical or concentration gradients of solute. The major SLC transporters include proton-coupled oligopeptide transporters, monocarboxylate transporters, organic anion polypeptide transporters, organic ion (anion and cation) transporters, and nucleoside transporters. They hinder drug delivery to the tumors.

13.2.4 Microbial Biofilm

Biofilms are community of microorganisms attached to a surface and form barrier to drug delivery. Bacterial biofilm is formed when single or multiple species of bacteria grow together in a form of a community encased by extracellular exopolysaccharide matrix. They exhibit unique properties not seen in their single cell, planktonic counterpart. They show increased resistance to antibacterial agents [22]. Biofilm has become a major problem in medical industries as they can form on medical implants such as pacemaker, catheter, lenses, and artificial hips [10]. It is very difficult to treat the infection due to their unique physiological and phenotypic properties. Multiple mechanisms are involved that make these surface-attached communities (are) more resistant to bactericides.

The bacterial community in a biofilm is embedded in an exopolysaccharide matrix or glycocalyx. This prevents the access of antibacterial agents to cells. Studies have shown that commonly used disinfectant such as chlorine did not reach greater than 20 % of the bulk media's concentration within a mixed *Klebsiella pneumoniae* and *Pseudomonas aeruginosa* biofilm. The thickness of the biofilm also determines the rate of diffusion of the drug. Increase in cell density and barrier formation are the major causes for physical exclusion of the drugs.

The bacterial community in a biofilm is slow growing as they do not have access to enough nutrients. The slow growth rate of the cells in the community makes them resistant to antibiotics as most of them work on fast-growing cells [32]. This also creates heterogeneity within the population due to gradient of nutrition availability, waste product production, respiratory activities, and signaling factors produced among the cells within the same community. Acridine orange staining method can identify regions of biofilms that contain rapidly or slowly growing cells based on their relative RNA–DNA content as fast-growing cells contain more RNA. Similarly protein synthesis and respiratory activity also differ at different areas of the biofilm. This difference in phenotype within the community makes them differentially sensitive to the drugs.

Physiological changes such as initiation of stress response, multidrug resistance (MDR), quorum sensing, and change in outer membrane protein (OMP) are the other reasons for persistent infection.

Due to slow growth and limited access to nutrient, general stress response is initiated within the biofilm community. Stress response creates physiological changes within the cells and protects them from environmental changes such as heat shock, cold shock, and chemicals. Alternative σ-factor RpoS is the central regulator of stress response and known to be expressed in the high cell density and stationary phase cultures. They produce osmoprotectant trehalose and catalase that protects from hydrogen peroxide. *E. coli* lacking RpoS do not form proper biofilm. Increased resistance of biofilm community has been indicated in quorum sensing, and increase in MDR pumps and change in the outer membrane protein have been implicated. Mutant in quorum-sensing system in *P. aeruginosa* was unable to form a biofilm and excessive sensitive to SDS has been reported. Eliminating microbes that live as a community in a biofilm has become a major challenge for drug industries [20].

13.2.5 Drug Efflux Pumps

Acquisition of multidrug resistance (MDR) by cancerous cells and microbes has become a serious problem in health-care industry. MDR refers to resistance to lethal doses of multiple structurally diverse drugs and chemical compounds. Hence it has become increasingly difficult to treat bacterial infections and cancer. Though there are several mechanisms by which the cells become resistant to a particular drug such as mutation of the target, decreased permeability, and drug metabolism, the major cause for MDR is drug extrusion by active transporters. These transporters utilize energy either by the hydrolysis of ATP or proton gradient to efflux the chemical compounds.

The multidrug transporters have broad specificity. The efflux pumps [41] present in microbes and mammalian cells can be classified to five families: the resistance-nodulation-division (RND) family, the major facilitator superfamily (MFS), the ATP (adenosine triphosphate)-binding cassette (ABC) superfamily, the small multidrug resistance (SMR) family [17], and the multidrug and toxic compound extrusion (MATE) family. The majority of the transporters belong either to MFS or ABC superfamily:

(a) Resistance-nodulation-division superfamily: RND proteins are present both in prokaryotes and eukaryotes. However the best characterized protein is AcrB from *E. coli* that increases resistance to several antibiotics. *E. coli* and other gram-negative bacteria have an outer membrane and a cytoplasmic membrane separated by a periplasmic space. Hence there is a need to expel drugs from the cell as well as the periplasmic space. The functional AcrB protein exists as a trimer with each unit having 12 transmembrane domains and a periplasmic domain. Two helper proteins TolC and AcrA are required to expel drug. TolC forms the pore across the periplasmic space and outer membrane and AcrA have a role in membrane fusion. Together this tripartite transporter transports drugs through the periplasm and outer membrane using proton gradient.
(b) Major facilitator superfamily (MFS): MFS transporters comprise one of the largest families of active transporters. They are energized by electrochemical proton gradient. A well-studied example of this family is EmrD from *E. coli*. It has 12 transmembrane α-helices organized as two bundles of six helices each. They form a hydrophobic cavity in the lipid bilayer. The internal alpha helices are involved in drug recognition and translocation. Proton movement induces a conformational change that facilitates drug release. The transporter then returns to its native state and binds drug.
(c) ATP-binding cassette (ABC) superfamily: ABC transporters utilize ATP binding and

hydrolysis to transport the substrate across the membrane. It is universally present in all cells and the best studied example is mammalian P-glycoprotein that is overexpressed in drug-resistant cancer cells. ABC transporters are also implicated in antibiotic resistance in bacteria, fungi, and parasitic protozoa. They are responsible for herbicide resistance in plants. The minimal functional unit of all ABC transporters consists of four domains. There are two cytoplasmic nucleotide-binding domains (NBDs) and two transmembrane domains (TMDs). The NBD binds and hydrolyzes ATP. The TMDs consist of multiple membrane-spanning α-helices and form the pathway through which substrates cross the membrane. These four domains can be fused into multi-domain polypeptides in a variety of ways [35]. Bacterial multidrug transporters commonly form homodimers of molecules comprising one NBD and one TMD, whereas mammalian P-glycoprotein has all four domains fused into a single polypeptide. Sav1866 from *Staphylococcus aureus* is an example of ABC transporter.

(d) Small multidrug resistance (SMR) superfamily: SMR proteins are a small family of transporters and are restricted to prokaryotes. They are smallest drug transporters with only four α-helices and no extra membrane domain. However they function as dimers. In *E. coli*, SMR protein EmrE confers resistance to a wide variety of hydrophobic cationic molecules, including antibiotics. It is a dimer and utilizes proton gradient energy to pump out drugs.

(e) Multidrug and toxic compound extrusion (MATE) family: The MATE transporters belong to a larger multidrug–oligosaccharidyl-lipid–polysaccharide (MOP) flippase superfamily. These transporters are widespread in bacteria, higher animals, and plants. It was first identified in *Vibrio parahaemolyticus* as Na-cationic antiport, named as NorM. It has 12 transmembrane domains and pumps out cationic dyes, fluoroquinolones, and aminoglycosides into the periplasmic space. Many of these transporters use Na or H or both as energy source. Another example is EmmdR, from *Enterobacter cloaca*.

The transporters were evolved to protect the cells from environmental toxins and expel the harmful metabolites. However multidrug transporters have become a problem for drug delivery. Different mechanisms are being developed to combat MDR [30]. The drugs are being modified so that they are not recognized by the transporters or induce changes in their affinity. Another approach is to use inhibitors against the transporters. The two pumps commonly involved in cancer therapy resistance, P-glycoprotein (PGP) and MDR-associated protein (MRP), are targeted by antisense oligos to reduce their expression [46]. Chemical compounds Incel (biricodar dicitrate, VX-710) and VX-853 that block PGP and MRP are used as combination drugs.

13.3 Drug Delivery System (DDS)

Drug delivery systems are mechanisms to deliver drugs to their targets consistently and uniformly. Conventional drugs are delivered orally or systematically and are plagued with a lot of problems. These are partly overcome by lipid- or polymer-based drug delivery systems that have improved pharmacological and therapeutic properties of the drug. Poor solubility of a drug decreases bioavailability. Encasing drugs in lipid micelle or liposome increases solubility by providing appropriate hydrophobic and hydrophilic environment. Regulated drug release can reduce or eliminate tissue damage by accidental extravasation of cytotoxic drugs. It protects drugs from premature degradation and reduces dosage in contrast to free drugs. It can alter pharmacokinetics of the drug and reduce clearance by the kidney. Free drugs have widespread distribution and poor bioavailability. Thus drug delivery systems have the advantages of delivering encapsulated drugs with better bioavailability. They demonstrate higher stability, controlled release, and long persistence in the system. Additionally they can be efficiently delivered at target tissue reducing the dosage and toxic side effects.

13.3.1 Nanoparticle-Mediated Delivery

Nowadays nanotechnology-based systems are used for improved efficacy and efficient drug delivery [25, 27]. The common nanosystems used are hydrogel, cyclodextrins, liquid crystalline phase, and nanoparticulate pharmaceutical drug delivery systems (NDDSs). NDDSs include liposomes, polymeric nanoparticles, polymeric micelles, metal nanoparticles (silica, gold, silver, and other metal nanoparticles), carbon nanotubes, solid lipid nanoparticles, niosomes, and dendrimers (as classified by Torchilin 2014) [45]. Drug targeting can be improved by ligand-mediated drug delivery.

The classification of polymeric systems is based upon their linear polymer chain in solution formed by cross-linking. Several drug delivery systems have evolved [44]. They are as follows.

13.3.1.1 Hydrogels

These are hydrophilic in nature and are formed by network of polymeric chains promoting the release of the drug through the spaces which are created in the network during dissolution of the polymeric matrix (Fig. 13.1). They have high water absorption capability and are easy to handle. Drug release from hydrogels depends upon the initial concentration of the drug, its solubility, and its interaction with the polymer [31]. Some examples of hydrogels are:

- Poly(vinyl alcohol) tetrahydroxyborate (PVA-THB) hydrogels demonstrate controlled release, bioadhesion, and low toxicity and hence have therapeutic potential for topically treating acute and chronic wounds [28].
- Chitosan-based hydrogels can release drug under different environmental stimuli [4].
- Cyclodextrin with linear polymer with PEG and transferrin has been approved for commercial use in melanoma therapy [36].
- Injectable hydrogels are explored for cancer therapy. Poly(ε -caprolactone-co-lactide)–poly(ethylene glycol)–poly(ε -caprolactone-co-lactide) (PCLA–PEG–PCLA) block copolymer used for incorporating paclitaxel has shown to be effective in suppressing tumor [33].

13.3.1.2 Cyclodextrins

Cyclodextrins (CDs) belong to the family of natural cyclic oligosaccharides with α -(1-4)-linked glucopyranose subunit bonds produced from starch via enzymatic conversion using cyclodextrin glycosyl transferases (CGTases) [21]. They can form drug complexes and are biocompatible, with less toxicity and high flexibility in the delivery profile (Fig. 13.1). Studies demonstrate their efficiency in drug delivery in melanoma [39, 47].

13.3.1.3 Liquid Crystalline Phases

These nanostructured systems exist as liquid crystals. They have the advantages of controlled release of drug as well as protection of active

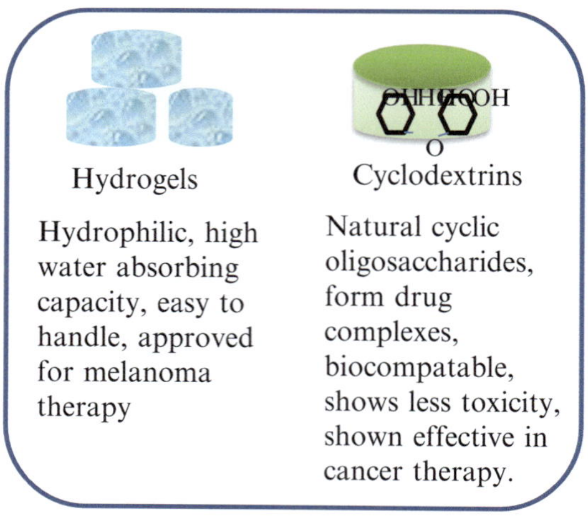

Fig. 13.1 The general schematic appearance and features of hydrogels and cyclodextrins

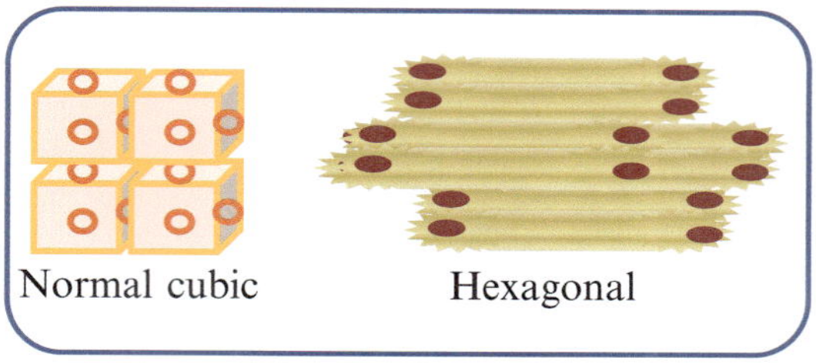

Fig. 13.2 Normal cubic and hexagonal appearances of liquid crystalline phase. Their properties include controlled drug release, impart protection from thermal degradation, and photo-bleaching. Variation in surfactant concentration leads to lamellar, cubic, or hexagonal forms. Used in cancer treatment

ingredients from thermal degradation or photo bleaching (Fig. 13.2). They can compartmentalize drugs in their inner phase droplets having different physicochemical properties and can induce alterations in the biological properties of the incorporated substances. They may be classified as lysotropic and thermotropic. Lysotropic is formed through adding solvents, whereas thermotropic is temperature dependent.

Due to changes in surfactant concentration, different liquid crystalline forms such as lamellar, hexagonal (hexasomes), and cubic (cubosomes) can be generated. Identification and classification of liquid crystalline materials may be done by polarized light microscopy. Liquid crystals have been evaluated for release of gallic acid in cancer treatments.

13.3.1.4 Liposomes

Liposomes have liquid core surrounded by phospholipid bilayer. Liposomes can be formed by sphingomyelins or lecithins and synthetic lipids as dimyristoyl, distearoyl, dipalmitoyl, and dioleoyl. They are 10 nm to several-micrometer spherical nanostructured molecules with defined shape and size (Fig. 13.3). Liposomes are formed by an external phase having double phospholipid membranes and an aqueous internal phase with the capability to encapsulate both hydrophobic and hydrophilic compounds. Their advantages are biocompatibility, low toxicity, non-immunogenicity, and improved drug efficacy. They fuse to cell membrane and deliver the drug. They are used as carriers of antifungal, antibacterial, and cancer drugs:

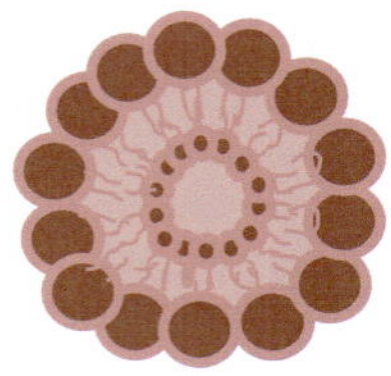

Fig. 13.3 Structure of liposomes. They have liquid core surrounded by phospholipid layer. They are biocompatable, nontoxic, and non-immunogenic and used for packaging of drugs such as amphotericin B, ampicillin, polymyxin B, ciprofloxacin, anticancerous (daunorubicin and doxorubicin), insulin, and DNA. Used for drug delivery in fungal and bacterial infections, cancer cells, and diabetes

- Liposome is used to deliver anticancer drug doxorubicin. Doxorubicin can induce myocardial toxicity. However when delivered in a liposomal preparation, the off-target toxicity is significantly reduced.
- Liposomes consisting of phosphatidylcholine, phosphatidylethanolamine, oleic acid, and cholesteryl hemisuccinate were developed by encapsulating a purified recombinant T4 endonuclease V [48]. When delivered to the skin, they tend to improve DNA repair.

13.3.1.5 Other Nanoparticles (NP)

Nanoparticles are delivery vehicles to transport therapeutics safely to their appropriate destination and release them gradually. NPs comprise of particles ranging from 10 to 100 nm in diameter. They are coated with biodegradable polymers that allow slow delivery of the drug from the core via diffusion or erosion. Drug release rate depends on biodegradation of the polymeric coat, diffusion of the drug, and its solubility and adsorption.

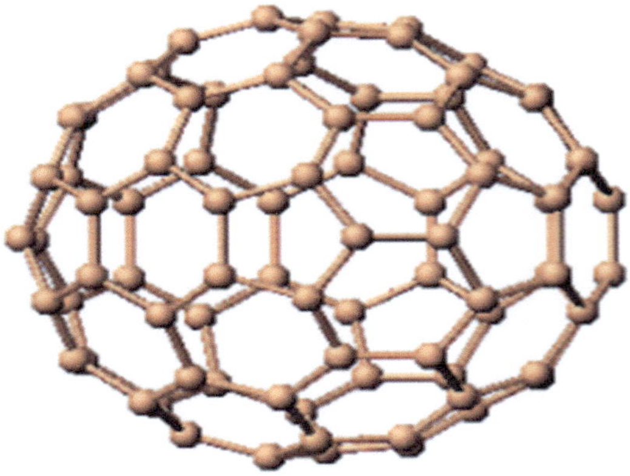

Fig. 13.4 Fullerenes are carbon-containing structures resembling graphite. They have low immunogenicity and high solubility and used for drugs as amino acid derivatives C60, fulleropyrrolidines, fullerene C60, metallofullerol, and carboxyfullerenes. Used in HIV research as protease inhibitor, replication inhibitor, and reverse transcriptase inhibitor and also used for neurological diseases and cancer

Fig. 13.5 The structure of solid lipid nanoparticles. They are made of physiological lipids as triglycerides, waxes, and steroids which exist as solid phase at room temperature. They are stable, non-immunogenic, and biodegradable and used for drugs as isoniazid, rifampicin, clotrimazole, pyrazinamide, tobramycin, miconazole nitrate, econazole nitrate, doxorubicin, tamoxifen, and insulin. Used to deliver inhalable antibiotics in tuberculosis, antibiotics for gram-positive and gram-negative bacteria, cancer, inflammatory diseases, and diabetes

There are currently several types of nanoparticles that are used for therapeutics, diagnostics, and imaging. They vary in size, shape, chemical composition, and surface properties. They are broadly divided into the following groups:

(a) Fullerenes: These are hollow spherical, ellipsoidal, or tubular structures that are composed of carbon molecules (Fig. 13.4). They are commonly known as buckyballs or buckytubes and have structures similar to graphite. They have diverse biological uses such as in gene delivery system as they have low immunogenicity and high solubility. The carbon nanotube can absorb light of near-infrared wavelength. This property is used for imaging, biosensing, and therapeutic applications. These can be photosensitized to produce heat to kill tumor cells and microbes. Various derivatives of fullerenes have been used as in HIV research as protease inhibitor, replication inhibitor, and reverse transcriptase inhibitor. They have been used for the study of free radicals and oxidative stress and neurological diseases such as Parkinson's, leukemia, and bone cancer.

(b) Solid lipid nanoparticles: They are made out of physiological lipids including fatty acids, triglycerides, waxes, steroids, and partial glycerides. They are solid at room temperature. They are small in size, have large surface area, and can interact with lipid bilayer and hence form an attractive model for drug delivery. Several surfactants are used to stabilize the lipid dispersion. The main advantage is they do not require organic solvents and they are stable, non-immunogenic, and biodegradable. They have been used as controlled delivery system for several diseases. They have been used to deliver inhalable antibiotics in tuberculosis; antibiotics against gram-positive and gram-negative bacteria and mycoplasma; antifungal agents; insulin delivery; breast cancer, colorectal cancer, and carcinoma drugs; inflammation; and immunity (Fig. 13.5).

(c) Nanostructured lipid carriers: Solid and liquid lipids are blended to form nanostructured lipid carriers [43]; however, they are in solid form at body temperature. They also form lipid-drug conjugate nanoparticles. The drugs are released by diffusion as well as degradation of the lipid particles. They are ideal for triggered drug release. They have a large number of applications in water-soluble drug administration. They are also used in food, agriculture, and cosmetic industries.

(d) Nanoshells: These are spherical nanoparticles containing dielectric core covered with thin metal shells such as gold, silver, and silicon. They are extensively used for bio-imaging for their optical properties and also for therapeutics (Fig. 13.6). They have resistance to chemical and thermal degradation. They can be easily conjugated to biomolecules and act as carriers. Gold-based nanoparticles have been used for treatment of rheumatoid arthritis. Gold interacts with near-infrared radiation based on surface plasmon scattering or surface plasmon resonance. This property is used for its detection at a very low concentration. Gold NPs emit heat when excited by minimum near-infrared radiation. This property is used as therapeutics to kill cancer cells.

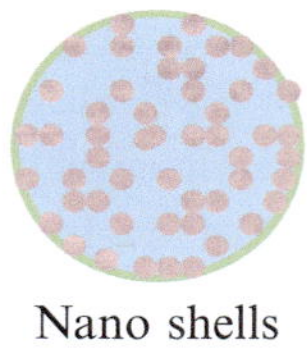

Fig. 13.6 Nanoshells which are spherical nanoparticles covered with metal shells. Drugs packaged may be minoxidil, flurbiprofen, fluticasone propionate. Used in water-soluble drug administration and release of anti-inflammatory drug, cosmetic, biochemical purposes, bioimaging, and therapeutics in rheumatoid arthritis

(e) Quantum dots (QD): These are semiconductor nanocrystals and core–shell nanocrystals. They contain interphase between different semiconductors. They have unique optical properties and are used for imaging and detection. They serve as structural scaffolds where hydrophobic drugs can be embedded. Hydrophilic therapeutics such as small interfering RNA, antisense oligonucleotide, and antibody peptides can be attached and delivered (Fig. 13.7a).

(f) Supra-magnetic nanoparticles: These are made from iron oxides and are attracted to magnetic field. They are coated with antibodies and act as carriers of drug to specific cells (Fig. 13.7b). Their magnetic properties are used to home them to a specific location. They get heated in the presence of an externally applied AC magnetic field. These characteristics make them attractive for many applications, ranging from various separation techniques and contrast-enhancing agents for MRI to drug delivery systems,

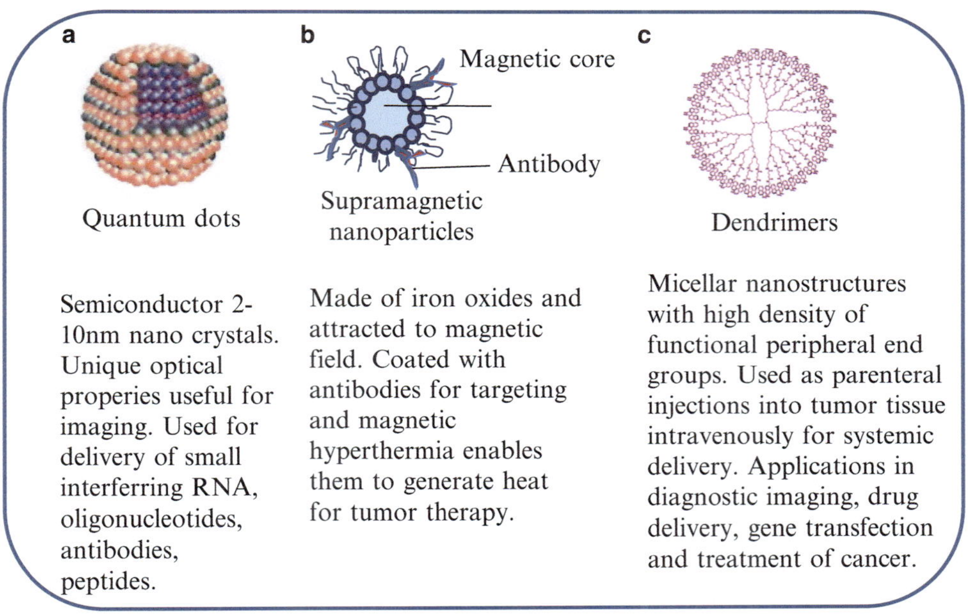

Fig. 13.7 The schematic diagram and features of (**a**) quantum dots, (**b**) supra-magnetic nanoparticles, and (**c**) dendrimers

magnetic hyperthermia (local heat source in the case of tumor therapy), and magnetically assisted transfection of cells.

(g) Dendrimers: These are unimolecular, monodisperse, micellar nanostructures. They are about 20 nm in size, with a well-defined, regularly branched symmetrical structure and a high density of functional end groups at their periphery. Different polymers used are polyamidoamine (PAMAM), melamine, poly l-glutamic acid (PG), polyethyleneimine (PEI), polypropyleneimine (PPI), and polyethylene glycol (PEG), and chitin. The structure of dendrimers consists of three distinct architectural regions as a focal moiety or a core, layers of branched repeat units emerging from the core, and functional end groups on the outer layer of repeat units. They are known to be robust, covalently fixed, three-dimensional structures possessing both a solvent-filled interior core (nanoscale container) and a homogeneous, mathematically defined, exterior surface functionality. Dendrimeric vectors are most commonly used as parenteral injections, either directly into the tumor tissue or intravenously for systemic delivery. Dendrimers may be used in two major modalities for targeting vectors for diagnostic imaging, drug delivery, gene transfection and detection, and therapeutic treatment of cancer and other diseases (Fig. 13.7c).

13.3.2 Cells as Drug Delivery Vehicle

Stem cells, red blood cell (RBC), and immune cells such as dendritic cells (DC), monocytes, macrophages, neutrophils, and lymphocytes are highly mobile and have intrinsic homing property to the site of injury or tumor. These cells have been exploited as drug carriers in cell-mediated drug delivery system. They offer several advantages such as targeted drug transport, increase of drug half-life, controlled release, and diminished immunogenicity and cytotoxicity. They can migrate across impermeable barriers and release their drug cargo at sites of infection or tissue injury.

(a) RBC: RBCs have several advantages as drug vehicle such as low immunogenicity, long half-life and is biodegradable. Hence they are used as carrier erythrocytes or carrier RBCs. They have prolonged life span and can protect encapsulated drug from degradation. Loading of RBC is comparatively simple [42]. The drug-loaded NPs can be electroporated; however this might rupture the cell membrane. Other approach is to deliver by osmosis. Here the drugs are mixed in hypotonic solution and they enter the RBC by osmosis. Membrane active drugs can be attached to the surface by co-incubation. The RBC can be loaded via bio-bridge such as biotin and avidin reaction. The most attractive property of RBC as a vehicle is its long life span and semipermeable membrane that allows diffusion of the drugs gradually over a period of time. Several antiparasitics, antiretroviral agents, antibiotics, and cardiovascular drugs have been loaded into the RBCs to avoid their rapid clearance and achieve the sustained release and improved biodistribution. RBC turnover takes place in the liver and spleen, a property used to target drugs to those organs. RBC is used for immunotherapy as antigen delivery system. It is degraded by macrophages and DCs and presented to T cells. HIV regulatory protein, TAT, was coated on the RBC membrane. RBC–TAT effectively delivered antigen to DC and induced specific CD4+ and CD8+ T-cell responses in vitro.

(b) Stem cell: Mesenchymal stem cells (MSC) and neuronal stem cells have homing property that is used for cell-based drug delivery. They have the ability to migrate toward tumor microenvironment and have been used to transfer drugs such as TNF-related apoptosis inducing ligand (TRAIL), interferon-β (IFN-β), and IL12/18, to inhibit the growth of the tumor. They are used to load NPs and transfer them to the site.

(c) Immune cells: Macrophages, neutrophils, and lymphocytes are recruited to the tumor site or site of injury. These cells are used as

Trojan horse to deliver therapeutics. Since macrophages capture foreign bodies, it is easy to load them with drugs. RNA-loaded liposomes, magnetic NPs, gold nanoshells, imaging agent-loaded NPs, quantum dot, and drug-loaded NPs are easily phagocytosed by macrophages and can be transferred to tumor tissue. They can migrate through the blood–brain barrier (BBB) to reach brain tumors

However there are certain limitations to cell-based delivery systems such as low drug-loading capacity, lysosomal degradation of therapeutics, and premature release and safety that need to be addressed while developing cell-based drug delivery system for human. Factors affecting cell-based drug delivery are:

(a) Drug loading: The major problem in cell-based drug delivery is that these cells can efficiently degenerate and clear the drug. To avoid premature clearance of the drug, they are incorporated into polymeric nano-carriers such as liposomes, nanoparticles, micelle, nanofibers, and nanotubes. These protect the drug from degradation and help in regulated release. The nano-carrier has a core and a shell. The core or the central part carries the drug that is protected by the polymer shell. The shell coating defines properties such as cell internalization and loading and stability and circulation time. Charged nanoparticles are rapidly taken up by the cells with the help of membrane receptors such as mannose receptors, Fc receptors, and complement receptors. Adsorption and internalization of positively charged nanoparticles are much higher. Positively charged nanoparticles with antiretroviral drugs, indinavir (IDV), ritonavir (RTV), and efavirenz (EFV), are used to load immune cells. Nanozymes is a family of nanoparticles that comprise of redox enzyme, catalase, and variant synthetic polymers. The loading capacity of nanozymes comprising of positively charged mono-polymers such as polyethyleneimine (PEI) and poly-L-lysine (PL) is greater. Electrostatically neutral and hydrophilic polyethylene glycol (PEG) is the most common shell-forming polymer currently used. However it is not suitable for cell-mediated delivery as it has limited cellular uptake [2] and is resistant to phagocytosis. Modification of PEG shell with streptavidin facilitates targeting to biotinylated T-cell markers such as anti-CD3 or peptide/major histocompatibility (MHC) complexes. Particle size and shape also determine cellular uptake. Usually smaller carriers are picked up by the cells easily; however, it limits the amount of drug taken in.

(b) Drug preservation: The stability of the drug inside the cell is a major concern. The nanoparticles can end up in lysosomes and are degraded. The cationic nanoparticles are better than the anionic ones as they are more resistant to phagosomal acidification and are resistant to lysosomal destruction. Loading of nanozymes containing positively charged block copolymers such as PEI-PEG and PL-PEG protects the enzyme in macrophages. A positively charged block copolymer prevents phagosome-lysosomal clearance functions and, as a result, enzyme degradation. Drug degradation can be avoided by surface attachment of the drug to the cell. Streptavidin-coated nanoparticles attach to the biotinylated plasma membrane. Red blood cells (RBCs) with attached drugs are also used as vehicle for carrying drugs. Glycoprotein A covalently conjugated to the surface of the RBCs may provide extended half-life, controlled volume of distribution, and multivalent therapeutic interactions. However, general limitations of the "back pack" approach include decreased loading of cell carriers, impeded drug release at the disease site, as well as increased immunogenicity and toxicity.

(c) Drug release: Controlled release of drugs modulates the rate of appearance, dose, and duration of drug exposure at the sites. Drug release from cell carrier can be modulated by utilizing the cellular response to various conditions. Targeting the cell carriers to the site and ensuring prolonged stay cause effective drug unloading. The immune cells liberate the drugs by exocytosis. Increased intracellular Ca^{2+} ion and hypothermia triggers can be used for drug release.

(d) Applications of cell-based drug delivery: The pathophysiology of a disease can be utilized for homing drug-loaded cell carriers to the disease site. It is especially crucial in the case of central nervous system disorders when cells need to penetrate the blood–brain barrier (BBB) to mediate therapeutic effect [5]. In neurological diseases, such as Alzheimer's disease (AD), Parkinson's disease (PD), prion disease, meningitis, encephalitis, and HIV-associated neurocognitive disorders (HAND), inflammation occurs due to extensive recruitment of mononuclear phagocytes. These cells efficiently cross the BBB due to their margination and extravasation properties causing barrier breakdown as a consequence of brain inflammation. These cells can be loaded with a required drug and administered intravenously to reach the brain. Similarly migration of inflammatory-response cells toward injury sites can be used to deliver drugs to infarcted myocardium, spinal cord injury, and cerebral ischemia.

Neuronal stem cells (NSC) are suggested drug delivery vehicle to CNS. They are highly migratory and migrate to areas of brain pathology including ischemic and neoplastic brain lesions that are commonly present in AD, PD, brain cancer, stroke, and multiple sclerosis.

Hypoxia is widespread in malignant human tumors due to their poorly organized vasculature. The cytokines released by tumor cells in response to hypoxia and other physiological stresses usually attract macrophages and monocytes. Hence monocytes facilitate anticancer drug delivery to these cells, avoiding indiscriminate drug distribution and decreasing severe toxicity.

Drug incorporated magnetic nanoparticles loaded into the cell carriers can be used for targeted delivery by application of local magnetic fields.

13.3.3 Extracellular Vesicles (EV)

These are small vesicles that are secreted by all types of eukaryotic cells. They deliver lipids, proteins, and RNA and alter the response of the receiving cells. These are used as DDS as they can be targeted to specific cell types. Depending on the cellular origin, they can be classified to shedding vesicles or ectosomes and exosomes. The ectosomes are directly derived from the plasma membrane, while exosomes are derived from endosome and secreted by vesicular bodies.

(a) RBC-derived micro vesicles: During the process of RBC aging, nano-sized vesicles are produced containing denatured hemoglobin, membrane proteins, and toxins that are produced as oxidative waste. These are phagocytosed by macrophages. These are used as vehicles to carry drugs to different tissues. The drug is loaded by osmosis or co-incubation. The main advantage is the small size and the ability to be endocytosed. The circulation time and cell phagocytosis efficiency of drug-loaded RBC MVs are greater.

(b) Tumor cell-derived microparticles: These are used to deliver chemotherapeutic drugs to the tissue. The tumor cell line is grown in the presence of the drugs and apoptosis is induced. The microparticles generated are loaded with the drug. Methotrexate is successfully delivered via this method.

(c) Mesenchymal stem cell-derived micro vesicles: MSC-based micro vesicles can be used as vehicle. They are hypo-immunogenic and tumor specific. However missed cells have greater risk of adverse effects and genetic mutation. Therefore, scientists are trying to remove the genetic material from it.

(d) Exosomes: Exosomes are naturally occurring membrane vesicles with intracellular origin. They are rich in cholesterol, sphingomyelin, and ceramide and have protein markers of tetraspanins (CD63, CD9) and TSG101. They mediate immune response and provide recipient cells with new functional properties. Since they are natural vehicles with targeting properties, they are excellent as vehicles to transport drugs. Exosomes are released by immune cells such as DC and macrophages and contain antigenic materials and peptide MHC complex for antigen presentation. This property is used for cancer immune therapy and the tumor peptides are pulsed into it. They are stable in circulation and can reach the cancer tissue due to the enhanced permeation retention (EPR) effect. They can be actively transported to selected tissue.

(e) Vascular endothelial cells: Human umbilical vascular endothelial cells (HUVEC) are cultured with the NPs or quantum dots. These are internalized by the cells and extracellular vesicles are generated. In this method we can select the parental cell type with certain surface properties to target the desired tissue.

(f) Bacterial outer membrane vesicle (OMV): These are derived from gram-negative bacteria and are similar in structure to EV. They are 20–250 nm in diameter and have lipid bilayer, periplasm, and membrane protein. OMVs were usually pathogenic due to the presence of toxins and lipopolysaccharides (LPS) on the surface. However, the OMVs isolated from attenuated bacterial strains do not show pathogenicity or toxicity and are used as the vaccine or drug carriers as they have various endogenous antigens and are natural adjuvants.

13.4 Drug Targeting

Conventional drug delivery has broad distribution. This causes suboptimal dosing to the affected organs and cells and toxicity to the bystander cells. Recent development has focused on targeted delivery of drugs to the affected tissue [1]. In this aspect nanoparticles (NP) loaded with drugs are considerably useful [16]. This is achieved by coating the surface of the NP with broad range of molecules that can be homed to specific site (Fig. 13.8). Targeting approaches of NP are of two types, passive and active targeting.

13.4.1 Passive Targeting

Passive targeting causes normal bio-distribution of NP with in body without any aid. This method uses natural physiological processes for drug delivery. When delivered intravenously, NPs are opsonized and are engulfed by macrophages. The macrophages have the inherent property to accumulate to the site of infection or injury. This property is utilized to deliver drugs in atherosclerosis and other inflammatory diseases. The NPs have a natural tendency to accumulate in the liver or spleen for clearance. This is used to treat hepatic disorders such as leishmaniasis. However this clearance mechanism could be deterrent to cancer treatment. The NPs are coated with PEG. This reduces protein adsorption on the surface and prevents opsonization. This increases bioavailability and lowers clearance of the drug. Passive targeting is favored in tumors as they have leaky vasculature and the particles can easily pass through it. The tumor vessels have weak endothelial cell junction and poor lymphatic drainage that promotes accumulation of NP. This process is known as enhanced permeation retention (EPR) effect. The NP can be actively transported to tumor cells rather than normal cells.

13.4.2 Active Targeting

Active targeting is achieved by coating the NPs with ligands that facilitate homing, binding, and internalization of the drug to the target. Different moieties have been used as targeting agents that are expressed in target cells. These are vitamins, carbohydrates, aptamers, peptides [3], proteins, and antibodies. However antibody-mediated targeting has become the major focus of research especially in cancer therapeutics.

In cancer immunotherapies activate the components of immune system to destroy cancerous tissue. These therapeutic agents are specific monoclonal antibody. The antibody binds the specific target and facilitates its clearance by immune components or it can deliver toxic molecules causing death of tumor cells. After the antibody

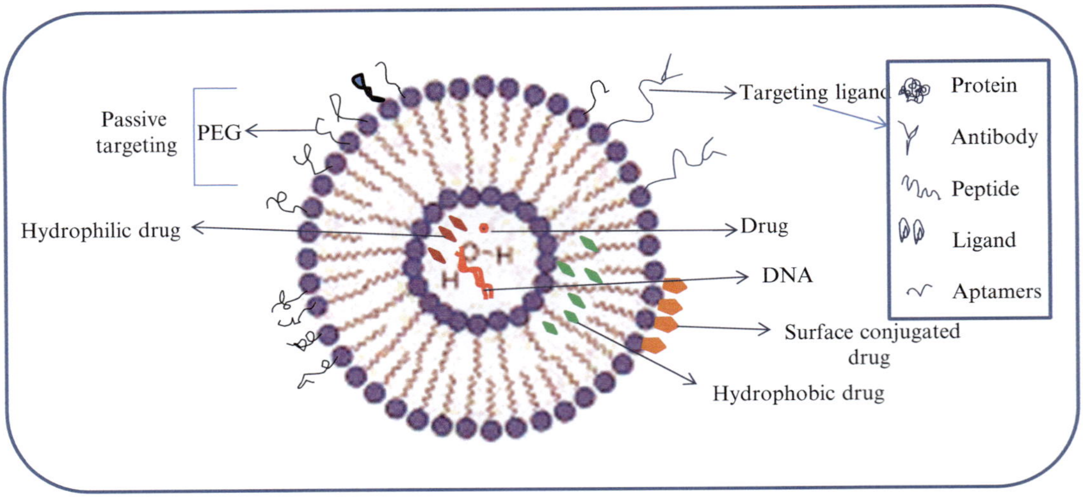

Fig. 13.8 The packaging and targeting of drug using liposomes as carriers. The presence of PEG ensures prolonged persistence and increased bioavailability of drugs. Inside the core hydrophilic drugs can be packaged. Hydrophobic drug can be packaged between the bilayer. Efficient targeting can be done by coating the nanosome with homing molecules as antibodies, proteins, peptides, ligand or aptamers, or other molecules

binds to its target, it triggers the immune destruction of cancer cells, and toxin-conjugated antibody can deliver the toxin to the cell mediating its killing. This does not affect the nontarget cells. Antibodies can also be conjugated to delivery vehicles as nanoparticles for efficient destruction of the target.

13.4.2.1 Antibody-Conjugated Nanoparticles

In this approach, antibodies are conjugated to drug-loaded NPs (Fig. 13.8). Antibodies possess exceptional specificity toward target antigens that have been exploited to transfer drug-loaded NPs to cancerous cells. They are used for delivery of cytokines, enzymes, and toxins for chemotherapy. They are also used for imaging and diagnostic uses when attached to fluorophores. Some of the antibody-NP conjugates used in cancer therapy are discussed below:

(a) EGFR/HER1: EGF receptor is a member of human epidermal receptor family that binds to EGF or TGF-α. The signaling promotes proliferation, migration, and angiogenesis. This receptor is overexpressed in many solid tumors. Hence antibody strategies are used for EGF receptor activation. Cetuximab is used in colorectal and head–neck carcinomas and panitumumab in colorectal cancer. Antibody-conjugated liposomes and NPs have been developed for diagnostic purpose when conjugated with gold particles [11].

(b) Her2/neu: Human EGF receptor 2 is a membrane tyrosine kinase receptor. It is linked to signal transduction of cell growth and differentiation. It is a target in breast cancer. Trastuzumab is used for HER2-positive cancer. HER2 is also the target for drug delivery. Trastuzumab-conjugated NP is internalized fast and induces cytotoxicity.

(c) Vascular endothelial growth factor receptors: VEGFR are tyrosine kinase receptors and when activated promote angiogenesis and vasculogenesis. They are upregulated in angiogenic tumors. Anti-VEGF quantum dots are used for early diagnosis. Supraparamagnetic iron oxide NPs conjugated with anti-VEGF receptor antibody are used as diagnostics. Gold-conjugated NPs have been used for treatment of B-cell chronic lympholytic leukemia. It is also used for radiotherapy where I^{131} is targeted. This improves localization.

(d) Prostate-specific membrane antigen: PSMA is nonsecreted type II transmembrane glycoprotein that is expressed in high level in prostate cancer. Anti-PSMA radioactive conjugate is used for diagnosis and drug delivery to prostate cancer.
(e) Transferrin receptor (Tfr): It is a membrane carrier protein that regulates intracellular iron levels. It is overexpressed in proliferating cells. The ligand is transferrin. Anti-transferrin antibody conjugate is used as drug delivery target. Tfr is expressed in blood–brain barrier.
(f) Dendritic cell receptors: Dendritic cells (DC) are important antigen-presenting cells. They internalize, process, and present antigens to T cells. The main receptors are C-type lectins, integrin, and FcR. Antigen-loaded particles and liposomes targeted to DC are utilized to increase T-cell-mediated immune response.

13.4.2.2 Aptamers

Aptamers are short, synthetic oligonucleotides that bind specifically to molecular targets such as protein, nucleic acids, cells, and tissues with affinities and specificities that are comparable to antibodies [29]. The term aptamer is derived from the Latin word "apta" that means "to fit." Aptamers specifically fit to the target and are evolving as a new family of therapeutic [19]. The aptamers are screened by using a selection procedure known as systematic evolution of ligands by exponential enrichment (SELEX) that is similar in principle to phage display. In this method a random sequence library of ssDNA or ssRNA that spans 20–100 nucleotides in length is constructed. The randomization of nucleic acid sequences provides a diversity of 4n, with n corresponding to the number of randomized bases. Usually each library has 1×10^{13} to 1×10^{14} members. Each random sequence region is flanked by constant sequences which are used for capture or PCR priming. The initial diverse pool of aptamers is then exposed to a target molecule, with the expectation that a portion of the aptamers would specifically bind to it. Nonbinding aptamers are washed away, while candidate aptamers with high target-binding affinity are enriched at each selection round by PCR amplification (DNA aptamers) or RT-PCR followed by in vitro transcription (RNA aptamers). The enriched pool of aptamers is then exposed to the target again, and the process is repeated. After multiple rounds of target selection and enrichment, aptamer pools show increase binding affinity and converge to one or more consensus sequences. Finally, individual aptamer clones can be generated and tested for target-binding affinity and specificity [29].

High-affinity aptamers that target protein families including cytokines, proteases, kinases, cell surface receptors, and cell adhesion molecules have been identified. Clinically important targets such as von Willebrand factor (vWF), platelet-derived growth factor (PDGF), E-selectin, vascular endothelial growth factor (VEGF), nuclear factor kB (NFkB), tenascin-C, and prostate-specific membrane antigen (PSMA) have been identified.

Aptamers can be used as targeted drugs similar to monoclonal antibodies [18]. They have several advantages over monoclonal antibodies. They are produced chemically and are not prone to contamination by bacteria or virus. They are easily scalable. They are non-immunogenic. They are small in size and have easy entry to cells. They are stable and can be conjugated to fluorescent dye and functional groups.

However, aptamers are prone to degradation by nucleases present in the serum. The phosphodiester backbone is vulnerable to serum ribonucleases at pyrimidine residues and the 5′- and 3′-termini are susceptible to exonucleases. To overcome exonuclease degradation, aptamers can be chemically synthesized and capped with modified or inverted nucleotides to prevent terminal degradation. Modified oligonucleotides can also be incorporated within the aptamer, either during or after selection, for enhanced endonuclease stability. Some modified nucleotide triphosphates, particularly 2′-O-modified pyrimidines, can be efficiently incorporated into aptamer transcripts by T7 RNA polymerases. Common chemical modifications included during selection are 2′-amino pyrimidines and 2′-fluoro-pyrimidines. Locked nucleic acids (LNAs) can be utilized to help stabilize aptamer structures. Polyethylene glycol (PEG) incorporation prolongs circulation time resulting in a more favorable pharmacokinetic profile.

13.4.2.3 Hormones

Hormones are used as therapies to slow or stop the growth and progression of hormone-sensitive tumors. Some tumors are dependent on hormones for their growth and proliferation. The hormone therapy acts by inhibiting the hormone production in the body or interfering with its action. This therapy is approved for breast cancer and prostate cancer. Breast cancers are affected by female hormones such as estrogen and progesterone. The cancerous cells may be categorized as estrogen receptor positive (ER+) or progesterone receptor positive (PR+) or double positive. Thus hormone treatment aims to stop these hormones going inside the affected cells. The hormone therapies used are tamoxifen, aromatase inhibitors, and luteinizing hormone (LH) blockers. Likewise prostrate cancer is dependent upon the male hormone testosterone for its growth. The therapy is intended to reduce or stop the production of testosterone in the body which can either slow down or stop the growth of the cancer, like luteinizing hormone (LH) blockers, antiandrogens, and gonadotropin-releasing hormone (GnRH) blockers.

13.4.2.4 Signal Transduction Blockers

These interfere with the activities of molecules participating in signaling reactions, thus blocking the cell response to a signal. However the cancerous cells bypass the signals which can control their growth and continue to divide uncontrollably. The tumor cells have increased division capability, loss of contact inhibition, metastasis, and angiogenesis. Some of these processes involve the activity of kinases. These have emerged as key regulators of many aspects of neoplasia; thus they may be attractive targets for anticancer therapy [24]. This has led to the development of inhibitors of protein kinases for controlling cancer. These inhibitors interfere with the inappropriate signaling. Signal transduction targets for the cancer treatment include the receptor and non-receptor tyrosine kinases (TKs) and receptor inhibitor for the growth factors as epidermal growth factor receptor (EGFR) inhibitors, the Ras/Raf/MAPK pathway, and the phosphatidylinositol-3 kinase (PI3K)/Akt/PTEN pathway.

13.4.2.5 Gene Expression Modulators

They are capable of modifying the protein function which controls gene expression like antisense RNA (see targeting of viral infections in Sect. 13.5.3).

13.4.2.6 Apoptosis Inducers

By the process of apoptosis, the body regulates the number of cells by removing unwanted, abnormal, and aged cells. Cancer cells however are not under the regulation of programmed cell death or apoptosis. The inducers of apoptosis are an interesting target that can induce death of cancer cells. Tumor necrosis factor-related apoptosis-inducing ligand (TRAIL) may be a promising candidate for cancer therapeutics. This can specifically and selectively kill cancer cells without affecting other normal cells.

13.4.2.7 Inhibitors of Angiogenesis

Angiogenesis (growth of new blood vessels) is very important for growth of tumor as they require oxygen and nutrients. Any therapeutic molecule which can target the angiogenesis process and interfere with it may block the growth of tumor. The inhibitors that interfere with angiogenesis may be antagonists of growth factors such as vascular endothelial growth factor (VEGF), which promotes angiogenesis.

13.5 Application in Diseases

13.5.1 Drug Delivery to the Brain

Delivery of drug in the brain is a big challenge; some strategies are being devised for efficient delivery of drug in the brain. Tumors of the brain may be classified in two groups: primary brain tumor, where the cancer initiates in the brain, and secondary brain tumor, where the cancer initiates in other parts of the body and metastasizes to the brain. Primary brain tumor can arise from brain cells, meninges, glands, and nerves. Glioma, the common type of tumor, arises from glial tissues. Depending on the cell types involved, they can be astrocytoma, oligodendroglioma, and ependymomas. In all the cases, the blood–brain barrier restricts drug entry and impedes success in chemotherapy. Improved drug delivery systems are implemented to benefit patients:

Intra-arterial delivery: Intra-arterial delivery of drugs such as monoclonal antibodies is increased to the tumor region by transient disruption of BBB.

Convection-enhanced diffusion (CED): Forceful delivery of fluid into the brain CED technique increases drug delivery to the tumor region. It is used in transcranial brain drug delivery.

Microdialysis: Passive diffusion of a drug across the BBB by microdialysis is an efficient method of intra-tumoral drug delivery.

Receptor-mediated endocytosis and exocytosis: Monoclonal antibodies targeted to receptors facilitate the entry of the therapeutic compounds across the BBB of brain tumors.

BBB disruption: This can be done by several methods and is useful to increase drug delivery to the tumor. This can be achieved by osmotic disruption, the use of chemicals, and MRI-guided focused ultrasound BBB disruption technique.

Modulation of transporters: The P-glycoproteins (P-gp) of the ABC drug efflux transporters, breast cancer resistance protein (BCRP), and multidrug resistance proteins (MRPs) are present in brain tumors that cause efflux of drugs [7]. Several inhibitors have been discovered that are co-administered with the chemotherapy agents. They modulate the expression of the transporters and effective drug delivery to the tumor niche.

Trans-nasal drug delivery route: Olfactory and trigeminal nerves of the nasal mucosa are anatomically connected to the central nervous system. The drugs administered through this path reach the cerebrospinal fluid (CSF), spinal cord, and brain parenchyma without any surgery.

13.5.2 Targeted Therapies for Cancer

In targeted therapies for cancer, the drugs or molecules used are capable of blocking the growth and spread of cancerous tissue by either binding with specific targets or molecules which are involved in growth, development, and spread of cancer or promoting their immune clearance. Targeted therapies are also referred to as "precision medicines" or "molecularly targeted drugs or therapies."

The differences between targeted therapeutics and general therapeutics are tabulated in Table 13.1.

FDA has approved many therapies to treat specific types of cancer. After identification of the target, the therapeutic agent or lead drug is used; this may interfere with target's ability to promote cancer cell growth and survival:

- Therapies are aimed toward a target that is specifically present or associated with cancer cells but not with normal cells. Targets are more abundant in cancer cell as compared to normal cells. For example, high levels of human epidermal growth factor receptor 2 (HER2) are expressed on certain breast cancer and stomach cancer. Therapy for breast cancer is directed against HER2 receptor, for example, trastuzumab (Herceptin®) is approved for treatment of certain breast and stomach cancers overexpressing this receptor.
- The therapy can also target proteins which are altered or mutated in cancer cells and cause progression of the disease. For example, cell growth signaling protein (BRAF) is present in a mutated form (BRAF V600E) in many melanomas [13]. The approved therapeutic against mutated target is vemurafenib (Zelboraf®). This product is approved to treat patients with metastatic or inoperable melanoma with mutated BRAF protein.
- Chromosomal abnormalities (may be fusion gene with parts of two different chromosomal segments) may also be targeted where fusion protein may drive cancer progression. For example, BCR–ABL fusion protein (made from pieces of two genes that are joined

Table 13.1 Differences between targeted and conventional therapies for effective control of the diseases

	Targeted therapies	Conventional therapies
1	Act on specific molecular targets	Act on general cells or in cancer therapy act on rapidly proliferating cells whether normal or cancerous
	Associated with the disease	
2	The drugs usually interact with the target	Drugs usually kill the cells with nonspecific side affects
3	Drugs usually block growth and division of target cells (cytostatic)	Drugs usually kill the target cells (cytotoxic)
4	Requires identification of potential target having key role in the pathogenesis of the disease	Intended for eliminating the function but have serious side effects also
4	None of the other cells are affected by drug except the target-bearing cells	Have nonspecific side effects on nontargets also
5	As targeted thus low dose is effective for therapy	Non-targeted thus higher dose required to achieve therapeutic goal
6	More effective	Less effective

together and promotes growth in few types of leukemic cells) is targeted by imatinib mesylate (Gleevec®).

The therapeutic agent may be monoclonal antibodies or small molecules where they can act externally and internally from within the cell, respectively. For the development of drugs for specific target, refer to Chap. 12. Monoclonal antibodies are engineered to make them humanized by replacing maximum mouse portion with corresponding portion of human antibodies. This is important as it can prevent anti-isotypic responses in the host body destroying the antibodies.

Before starting with targeted therapy, it is very important to determine whether the particular target is present in the patient or not. If the target is present, the therapy can be started and is restricted to those patients where mutation in specific gene codes for the target molecule.

The FDA-approved targeted therapies for treatments of cancer are shown in the Table 13.2. Some of the therapies are approved for more than one type of cancer.

13.5.2.1 Limitations of Targeted Cancer Therapies

There are certain limitations of targeted therapies:

- Cancer cells may become resistant to the therapy. Resistance may be due to (1) change in the target by mutation, (2) improper interaction of the therapeutic agent with its target, and (3) tumor that bypasses the requirement of the target. Thus combination therapies are more effective.
- Sometimes it is difficult to develop drugs against the targets, for example, Ras (a signaling protein) is mutated in many cancers, but its inhibitors could not be designed.
- In some cases targets are present in normal cells as well but abundantly present in cancer cells; blocking these have an effect on other cells as well.

13.5.3 Targeted Therapies for Viral Infections

Targeted delivery is very helpful to eliminate viral infections. Since virus multiplies within the host cells, targeting therapeutics to control them is seemingly difficult. Genetic medicines aiming to stop viral multiplication hold hope for future treatment. These include antisense DNA, RNA, aptamers, and ribozymes which may target viral genome in a sequence-specific manner and inhibit its replication. Some of the drugs are enlisted below:

- Vitravene (fomivirsen) is an antisense drug and is the first in the category that has been approved by US FDA for treatment of cytomegalovirus (CMV) retinitis. CMV is an opportunistic infectious agent which infects under immunosuppressive state and in AIDS patients. Infection of the eye can lead to degeneration of the retina and blindness. Formivirsen is the phosphorothioate oligonucleotide which inhibits CMV replication.

Table 13.2 FDA approved therapies along with their targets for treatment of cancer

Cancers	Targets	Therapeutic agent
Adenocarcinoma (stomach gastrointestinal junction)	mAB for human epidermal growth factor receptor-2 (HER2) fully human mAb for VEGFR 2	Trastuzumab (Herceptin®)
		Ramucirumab (Cyramza®)
Basal cell carcinoma	Hedgehog signaling pathway targeting agent	Vismodegib (Erivedge™)
		Sonidegib (Odomzo®)
Brain cancer	Humanized mAb blocking angiogenesis by inhibiting VEGF-A	Bevacizumab (Avastin®)
	Immunosuppressor, mTOR inhibitor	Everolimus (Afinitor®)
Breast cancer	Immunosuppressor, mTOR inhibitor	Everolimus (Afinitor®)
	Antagonist of estrogen receptor	Tamoxifen, toremifene (Fareston®)
	mAb against HER2 receptor	Trastuzumab (Herceptin®)
	Antagonist of estrogen receptor	Fulvestrant (Faslodex®)
	Inhibiting the synthesis of estrogen	Anastrozole (Arimidex®)
	mAb for HER2-positive breast cancer	Pertuzumab (Perjeta™)
	Antibody-drug conjugate (antibody linked to cytotoxic agent)	Ado-trastuzumab emtansine (Kadcyla™)
	Inhibits cyclin-dependent kinases CDK4 and CDK6	Palbociclib (Ibrance®)
Cervical cancer	Humanized mAb blocking angiogenesis by inhibiting VEGF-A	Bevacizumab (Avastin®)
Colorectal cancer	Chimeric mAb against EGFR	Cetuximab (Erbitux®)
	Fully human mAb against EGFR	Panitumumab (Vectibix®)
	Humanized mAb blocking angiogenesis by inhibiting VEGF-A	Bevacizumab [6] (Avastin®)
	Recombinant fusion protein inhibiting VEGF	Ziv-aflibercept (Zaltrap®)
	Multi-kinase inhibitor	Regorafenib (Stivarga®)
	Fully human mAB for VEGFR 2	Ramucirumab (Cyramza®)
Dermatofibrosarcoma protuberans	Tyrosine kinase inhibitor	Imatinib mesylate (Gleevec®)
Endocrine/neuroendocrine tumors	Somatostatin analog inhibits release of its target hormones including GH, TSH	Lanreotide acetate (Somatuline® Depot)
Head and neck cancer	Chimeric antibody against EGFR	Cetuximab (Erbitux®)
Gastrointestinal stromal tumor	Tyrosine kinase inhibitor	Imatinib mesylate (Gleevec®)
	Small-molecule inhibitor of receptor tyrosine kinase	Sunitinib (Sutent®)
	Multi-kinase inhibitor	Regorafenib (Stivarga®)
Giant cell tumor of the bone	Fully human mAb against RANK ligand	Denosumab (Xgeva®)
Kaposi sarcoma	Form of Vit-A	Alitretinoin (Panretin®)
Kidney cancer	Humanized mAb blocking angiogenesis by inhibiting VEGF-A	Bevacizumab (Avastin®)
	Kinase inhibitor	Sorafenib (Nexavar®)
	Small-molecule inhibitor of receptor tyrosine kinase	Sunitinib (Sutent®)
	Multi-targeted receptor tyrosine kinase inhibitor	Pazopanib (Votrient®)
Leukemia	Carboxylic acid form of Vit-A	Tretinoin (Vesanoid®)
	Tyrosine kinase inhibitor	Imatinib mesylate (Gleevec®)
	Chimeric mAb against CD20 of B cells	Rituximab (Rituxan®)
	Humanized mAb against CD52 of mature lymphocytes	Alemtuzumab (Campath®)
	Fully human mAb against CD20 of B cells	Ofatumumab (Arzerra®)

(continued)

Table 13.2 (continued)

Cancers	Targets	Therapeutic agent
Liver cancer	Kinase inhibitor	Sorafenib (Nexavar®)
Lung cancer	RTK of EGFR and erbB2	Afatinib dimaleate (Gilotrif®)
	Humanized mAb blocking angiogenesis by inhibiting VEGF-A	Bevacizumab (Avastin®)
	EGFR inhibitor	Gefitinib (Iressa®)
	Immunomodulator	Nivolumab (Opdivo®)
Lymphoma	Radioisotope labeled mAb against CD20 of B cells	Ibritumomab tiuxetan (Zevalin®)
	Inhibitor of Bruton's tyrosine kinase of B cells	Ibrutinib (Imbruvica™)
	Chimeric mAb against CD20 of B cells	Rituximab (Rituxan®)
Melanoma	Inhibitor of B-Raf (regulates cell growth)	Dabrafenib (Tafinlar®)
	Targets programmed cell death receptor	Pembrolizumab (Keytruda®)
Multiple myeloma	Proteasome inhibitor	Bortezomib (Velcade®)
	Proteasome inhibitor	Carfilzomib (Kyprolis®)
Ovarian (epithelial/fallopian tube/primary peritoneal cancers)	Humanized mAb blocking angiogenesis by inhibiting VEGF-A inhibits poly ADP-ribose polymerase (PARP), an enzyme involved in DNA repair	Bevacizumab (Avastin®) Olaparib (Lynparza™)
Pancreatic cancer	Tyrosine kinase inhibitor	Erlotinib (Tarceva®)
Prostate cancer	Microtubule inhibitor	Cabazitaxel (Jevtana®)
	Isotope of radium with half-life of 11.4 days and is similar to calcium	Radium 223 chloride (Xofigo®)
Thyroid cancer	Tyrosine kinase inhibitor	Cabozantinib (Cometriq™)
	Kinase inhibitor of VEGFR	Vandetanib (Caprelsa®)

- A second-generation antisense oligonucleotide, mipomersen (Kynamro; Isis Pharmaceuticals/Genzyme), was approved by US Food and Drug Administration (FDA). This inhibits apolipoprotein B100 and was approved for homozygous familial hypercholesterolemia (HoFH) (a rare genetic disorder that leads to excessive levels of low-density lipoprotein (LDL) cholesterol).
- Herpes simplex virus (HSV) infections are very common and affect nearly 90 % of the world population. HSV-1 causes skin lesions and involves other organs like the eyes (ocular herpes) or central nervous system (herpes encephalitis, herpes meningitis) which may be serious. The therapies based on RNA interference may provide a hope for treatment of the disease.

13.6 Side Effects of Targeted Therapies

Several side effects are associated with targeted therapies:

- Diarrhea and liver problems are common with occurrence of hepatitis and elevated liver enzymes.
- Skin problems such as acneiform rash, dry skin, nail changes, and hair depigmentation are seen in patients.
- Hampered blood clotting and wound healing.
- High blood pressure.
- Gastrointestinal perforation.
- Immunosuppression and impaired sperm production.

However, certain side effects of some targeted therapies are being reported such as patients developing acneiform rash (skin eruptions resembling acne) while being treated with inhibitors of signal transduction, erlotinib (Tarceva®) or gefitinib (Iressa®) (both target EGFR). Patients who develop rash tend to respond better than those without rash. Side effects may vary in adults and children [15, 34].

13.7 Chapter End Summary

- Drug delivery is defined as mechanisms to introduce pharmaceutical compounds to human in order to achieve therapeutic effects.
- Conventional drug delivery system includes drug delivery via oral, nasal, dermal, and intravenous route. However they have various limitations such as low bioavailability, intolerance, toxic side effects, reduced plasma half-life, higher concentration, and low efficacy.
- A major challenge in the drug delivery field is to enhance transport of therapeutics across biological barriers such as the skin, mucosal membrane, lung surfactant, and blood–brain barrier (BBB) that are designed to restrict the entry of foreign molecules. Multidrug-resistant pumps present in tumor cells and microbes have further compounded the problem of drug transportation to the targets. The microbe communities in biofilm restrict drug entry and are difficult to treat.
- Drug delivery systems (DDS) are mechanisms to deliver drugs to their targets consistently and uniformly. The important features of DDS are to increase plasma half-life of the drugs, reduce immunogenicity, increase bioavailability, and reduce side effects.
- Several DDS have been developed depending on the requirements. Some of them are nanoparticles, cyclodextrins, liquid crystalline phases, and liposomes. The drugs are either encased within the core or coated on the surface of the delivery vehicles. They act as vehicles to transport drugs to appropriate target cells and tissues.
- Cells are used as vehicles for drug delivery such as RBC, stem cells, and immune cells. They are less immunogenic and have long half-life. Extracellular vesicles derived from RBCs, tumor cells, mesenchymal stem cells, exosomes, vascular endothelial cells, and bacterial outer membrane vesicles are other attractive drug delivery systems.
- Targeted drug delivery aims at delivering drugs only to the affected cells and tissue. This reduces the toxicity to bystander cells. There are several approaches to achieve targeted delivery. Different moieties have been used as targeting agents that are expressed in target cells. These are vitamins, carbohydrates, aptamers, peptides, proteins, and antibodies. Drug-loaded nanoparticles attached to specific antibodies are targeted to cells expressing the antigens or receptors.
- Many targeted delivery approaches have been approved for cancer therapy and viral infections. They hold promises for next-generation treatment.

Multiple Choice Questions

1. Efficient drug delivery is required for:
   (a) Insulin
   (b) Paracetamol
   (c) Inhalers
   (d) All of the above
2. Drug delivery is important as conventional approaches result in:
   (a) Low bioavailability
   (b) No side effects
   (c) High efficacy
   (d) None of these
3. Which statement about the biological barriers is true?
   (a) They help in active transport of the drug.
   (b) They allow only selective molecules to pass through them.

   (c) They enhance the effect of the drug on the target.
   (d) They improve efficacy of the drug.
4. In the layers of skin, “bricks in mortar” refers to:
   (a) Stratum corneum
   (b) Stratum granulosum
   (c) Stratum spinosum
   (d) Stratum basale
5. Brain microvascular endothelial cells:
   (a) Form tight junctions
   (b) Have efflux transporters
   (c) Restrict molecule entry
   (d) All of these
6. Microbial biofilm is:
   (a) Deposition of microbes on the lens
   (b) Matrix protecting bacterial colony
   (c) Increased resistance to antibacterial agent
   (d) All of these
7. Plane polarized light can be used for identification of:
   (a) Dendrimers
   (b) Liposomes
   (c) Liquid crystalline phase
   (d) Hydrogels
8. Using cells as drug delivery vehicles has advantages of:
   (a) Biodegradability and low immunogenicity.
   (b) Improve efficacy of drug.
   (c) They release the drug slowly.
   (d) All of the above.
9. Trastuzumab targets:
   (a) Estrogen receptor
   (b) Epidermal growth factor receptor
   (c) Vascular endothelial growth factor receptor
   (d) All of these
10. Aptamers are:
   (a) Ligands
   (b) Oligonucleotides
   (c) Hormones
   (d) All of these

Answers

1. (a); 2. (a); 3. (b); 4. (a); 5. (d); 6. (d); 7. (c); 8. (a); 9. (b); 10. (b)

Review Questions

Q1. Why is drug delivery important?
Q2. What are biological barriers to drug delivery?
Q3. What are microbial biofilms?
Q4. What are nanoparticle-based delivery systems?
Q5. What are aptamers?
Q6. What do you understand by targeted delivery?

References

1. Allen TM, Cullis PR (2004) Drug delivery systems: entering the mainstream. Science 303:1818–1822
2. Batrakova E, Gendelman HE, Kabanov AV (2011) Cell-mediated drugs delivery. Expert Opin Drug Deliv 8:415–433
3. Bechara C, Sagan S (2013) Cell-penetrating peptides: 20 years later, where do we stand? FEBS Lett 587:1693–1702
4. Bhattarai N, Gunn J, Zhang M (2010) Chitosan-based hydrogels for controlled, localized drug delivery. Adv Drug Deliv Rev 62:83–99
5. Bhowmik A, Khan R, Ghosh MK (2015) Blood brain barrier: a challenge for effectual therapy of brain tumors. BioMed Res Int. doi:10.1155/2015/320941
6. Cai J, Ma H, Huang F et al (2013) Correlation of bevacizumab-induced hypertension and outcomes of metastatic colorectal cancer patients treated with bevacizumab: a systematic review and meta-analysis. World J Surg Oncol 11:306
7. Cancer multidrug resistance (2000) Nature biotechnology vol 18 Supplement (Reprinted from Nat. Biotechnol (1999)17:94–95)
8. Cone RA (2009) Barrier properties of mucus. Adv Drug Deliv Rev 61:75–85
9. Elias PM (2008) Skin barrier function. Curr Allergy Asthma Rep 8(4):299–305
10. Encyclopedia of Medical Devices and Instrumentation (2006) Second edition. In Webster JG (ed) Copyright Wiley, India
11. Fay F, Scott C (2011) Antibody-targeted nanoparticles for cancer therapy. Immunotherapy 3:381–384
12. Feingold KR, Elias PM (2014) Role of lipids in the formation and maintenance of the cutaneous permeability barrier. Biochim Biophys Acta 1841:280–294
13. Flaherty KT, Infante JR, Daud A et al (2012) Combined BRAF and MEK inhibition in melanoma with BRAF V600 mutations. N Engl J Med 367:1694–1703
14. Gao Y et al (2014) Nanotechnology-based intelligent drug design for cancer metastasis treatment. Biotechnol Adv 32:761–777
15. Gore L, DeGregori J, Porter CC (2013) Targeting developmental pathways in children with cancer: what price success? Lancet Oncol 4:e70–e78

16. Guterres SS, Alves MP, Pohlmann AR (2007) Polymeric nanoparticles, nanospheres and nanocapsules, for cutaneous applications. Drug Target Insights 2:147–157
17. Higgins CF (2007) Multiple molecular mechanisms for multidrug resistance transporters. Nature 446:749–757
18. Jayasena SD (1999) Aptamers: an emerging class of molecules that rival antibodies in diagnostics. Clin Chem 45:1628–1650
19. Keefe AD, Pai S, Ellington A (2010) Aptamers as therapeutics. Nat Rev Drug Discov 9:537–550
20. Kim L (2001) Riddle of biofilm resistance. Antimicrob Agents Chemother. doi:10.1128/AAC.45.4.999-1007
21. Laza-Knoerr AL, Gref R, Couvreur P (2010) Cyclodextrins for drug delivery. J Drug Target 18:645–656
22. Mah TFC, Toole GAO (2001) Mechanisms of biofilm resistance to antimicrobial agents. Trends Microbiol 9:34–39
23. Merga Y, Campbell BJ, Rhodes JM (2014) Mucosal barrier, bacteria and inflammatory bowel disease: possibilities for therapy. Dig Dis 32:475–483
24. Mikalsen T, Gerits N, Moens U (2006) Inhibitors of signal transduction protein kinases as targets for cancer therapy. Biotechnol Annu Rev 12:153–223
25. Mohanraj V, Chen Y (2006) Nanoparticles—a review. Trop J Pharm Res 5:561–573
26. Molema G (2001) Drug targeting: basic concepts and novel advances. In: Molema G, Meijer DKF (eds) Drug targeting: organ-specific strategies, vol 12. Wiley-VCH Verlag GmbH, Weinheim. doi:10.1002/352760006X.ch1
27. Mudshinge SR, Deore AB, Patil S, Bhalgat CM (2011) Nanoparticles: emerging carriers for drug delivery. Saudi Pharm J 19:129–141
28. Murphy DJ, Sankalia MG, Loughlin RG, Donnelly RF, Jenkins MG, Mccarron PA (2012) Physical characterisation and component release of poly(vinyl alcohol)-tetrahydroxyborate hydrogels and their applicability as potential topical drug delivery systems. Int J Pharm 423:326–334
29. Ni X et al (2011) Nucleic acid aptamers: clinical applications and promising new horizons. Curr Med Chem 18:4206–4214
30. Pagès JM, Amaral L (2009) Mechanisms of drug efflux and strategies to combat them: challenging the efflux pump of Gram-negative bacteria. Biochim Biophys Acta 1794:826–833
31. Patel MP, Churchman ST, Cruchley AT, Braden M, Williams DM (2013) Delivery of macromolecules across oral mucosa from polymeric hydrogels is enhanced by electrophoresis (iontophoresis). Dent Mater 29:e299–e307
32. Penesyan A, Gillings M, Paulsen IT (2015) Antibiotic discovery: combatting bacterial resistance in cells and in biofilm communities. Molecules 20:5286–5298
33. Peppas NA, Bures P, Leobandung W, Ichikawa H (2000) Hydrogels in pharmaceutical formulations. Eur J Pharm Biopharm 50:27–46
34. Petrelli F, Borgonovo K, Cabiddu M, Lonati V, Barni S (2012) Relationship between skin rash and outcome in non-small-cell lung cancer patients treated with anti-EGFR tyrosine kinase inhibitors: a literature-based meta-analysis of 24 trials. Lung Cancer 78:8–15
35. Ramsey JD, Flynn NH (2015) Cell-penetrating peptides transport therapeutics into cells. Pharmacol Ther. doi:10.1016/j.pharmthera.2015.07.003
36. Rigon RB et al (2015) Nanotechnology-based drug delivery systems for melanoma antitumoral therapy: a review. BioMed Res Int article ID 841817. doi: 10.1155/2015/841817.
37. Sanders N, Rudolph C, Braeckmans K, De Smedt SC, Demeester J (2009) Extracellular barriers in respiratory gene therapy. Adv Drug Deliv Rev 61:115–127
38. Sigurdssona HH, Kirchb J, Lehrb CM (2013) Mucus as a barrier to lipophilic drugs. Int J Pharm 453:56–64
39. Singh M, Sharma R, Banerjee UC (2002) Biotechnological applications of cyclodextrins. Biotechnol Adv 20:341–359
40. Smeden JV, Janssens M, Gooris GS, Bouwstra JA (2014) The important role of stratum corneum lipids for the cutaneous barrier function. Biochim Biophys Acta 1841:295–313
41. Sun J, Deng Z, Yan A (2014) Bacterial multidrug efflux pumps: mechanisms, physiology and pharmacological exploitations. Biochem Biophys Res Commun 453:254–267
42. Tan S, Wu T, Zhang D, Zhang Z (2015) Cell or cell membrane-based drug delivery systems. Theranostics 5:863–881
43. Thomas S, Shanks R, Sarathchandran C (2013) In: Andrew W (ed) Nanostructured polymer blends, 1st edn. Elsevier, Oxford
44. Tiwari G et al (2012) Drug delivery systems: an updated review. Int J Pharm Investig 2:2–11
45. Torchilin VP (2014) Multifunctional, stimuli-sensitive nanoparticulate systems for drug delivery. Nat Rev Drug Discov 13:813–827
46. Tse MT (2013) Regulatory watch: antisense approval provides boost to the field. Nat Rev Drug Discov 12 doi:10.1038/nrd3963
47. Valle EMMD (2004) Cyclodextrins and their uses: a review. Process Biochem 39:1033–1046
48. Wolf JP et al (2000) Topical treatment with "liposomes containing T4 endonuclease V protects human skin in vivo from ultraviolet-induced upregulation of interleukin-10 and tumor necrosis factor- α . J Invest Dermatol 114:149–156

Related Sources

http://www.atdbio.com/content/13/Oligonucleotides-as-drugs
http://www.fda.gov/Drugs
www.cancerresearchuk.org

14 Vaccine

Abstract

Humans have tried to fight diseases by the usage of vaccines. Vaccination is a powerful and cost-effective weapon for the prevention of diseases. Because of worldwide vaccination programs, the humans have not only successfully used vaccines to get rid of many childhood diseases but also some of the diseases like smallpox are completely eradicated. Any preparation of virus, bacteria, or their subunits, which is able to impart protection in the individual in which they are administered by activating the immune system, is known as vaccine. As immunization drive starts from childhood therefore it has substantially lowered the morbidity and mortality of children from many infectious diseases.

Vaccination is a mode of active immunization in which attenuated live microorganism or killed microorganism or its specific component is administered in humans. Following exposure to modified pathogenic agent (not able to harm and cause diseases in humans), the immune system is activated. Immune systems' innate and adaptive branches effectively neutralize the pathogenic agents and its T and B cells owing to their memory response remember their first encounter with the pathogen. Upon subsequent encounter in the due course of time with the same agent, the immune systems' adaptive branch (T- and B-cell response) effectively and quickly mounts a heightened response and facilitates antigen clearance from the body.

However, despite of tremendous developments in science and technology, some diseases (AIDS, malaria, multiple drug-resistant TB, leishmaniasis) are still big challenges for which we do not have effective vaccines. Vaccines need to be developed for these diseases, which are cause of millions of death worldwide. The chapter is designed to give the insights about the properties of vaccines, adjuvants, attenuated, inactivated, subunit, and nucleic acids as vaccines. This also covers edible vaccines, vaccination for cancer, reverse vaccinology, and generation of vaccines.

V. Gupta et al., *Basic and Applied Aspects of Biotechnology*, DOI 10.1007/978-981-10-0875-7_14

14.1 Introduction

Humans have tried to fight diseases by the usage of vaccines. Vaccination is a powerful and cost-effective weapon for the prevention of diseases. Because of worldwide vaccination programs, the humans have successfully used vaccines to get rid of many childhood diseases. Some of the diseases like smallpox are completely eradicated. Any preparation of virus, bacteria, or their subunits, which is able to impart protection in the individual in which they are administered by activating the immune system, is known as vaccine. As immunization drive starts from childhood therefore it has substantially lowered the morbidity and mortality of children from many infectious diseases.

Vaccination is a mode of active immunization in which attenuated live microorganism or killed microorganism or its specific component is administered in humans. Following exposure to modified pathogenic agent (not able to harm and cause diseases in humans), the immune system is activated. Immune systems' innate and adaptive branches effectively neutralize the pathogenic agents and its T and B cells owing to their memory response remember their first encounter with the pathogen. Upon subsequent encounter in the due course of time with the same agent, the immune systems' adaptive branch (T- and B-cell response) effectively and quickly mounts a heightened response and facilitates antigen clearance from the body.

However, despite of tremendous developments in science and technology, some diseases (AIDS, malaria, multiple drug-resistant TB, leishmaniasis) are still big challenges for which we do not have effective vaccines. Vaccines need to be developed for these diseases, which are cause of millions of death worldwide.

14.2 Vaccine Technology: Role and Properties of Adjuvants

Vaccine designing is multidisciplinary approach and requires scientific understanding of the disease. The future lies in developing the vaccines for the diseases for which there is no cure and improving the present vaccines to make them more safe, efficient, and cost-effective. The discovery and release of a successful vaccine is costly and time consuming, and there are high risks associated with their development [20] as:

1. After tough scrutiny, compounds are screened for their potential as vaccines.
2. Promising vaccines can have unacceptable side effects.
3. The major formulations fail in preclinical or phase I trial.
4. Attenuated/live vaccines pose threats because of reversion or disease when administered in immunocompromised host.
5. Developing cost-effective and safe vaccine which can be efficiently delivered to population at risk.
6. It should be approved for human use.

With an increasing competition in the industry and new pathogenic strains, a rational and systematic approach is necessary for the development of vaccine, which is stable, safe, and efficacious. Therefore, it becomes essential to list and analyze all factors that can affect the outcome of the responses:

1. Thorough understanding of immunological responses
2. Characterization of antigen
3. Selection of appropriate adjuvant
4. Suitable trials
5. Analysis of product contact material
6. Monitoring stability, safety, and efficacy

1. *Thorough understanding of the immunological responses:*

   It requires an understanding of viral/bacterial pathogens with subsequent development of hypothesis about the characteristic of a successful vaccine. The success depends upon identification of the most relevant mechanism of the immune protection and a thorough understanding of pathogenesis of the disease. Although this information is highly essential, many a times the work starts without this knowledge because for many infectious diseases the critical immunological protection is unknown (HIV, malaria). Although a few components are strongly implicated in protection, precise requirement for protective immunity and their role is not fully resolved which is the subject for intense investigation. The immune system mechanism of prevention and clearance of infections is different from establishing protection after subsequent exposure than for vaccination designing needs vaccine capable of efficient priming and clearance and memory in the host as a defense strategy.

2. *Identification and characterization of antigen:*

   The next step of designing is the identification of potential antigen for desirable specificity. The use of subunit vaccines as purified proteins/surface polysaccharides/synthetic peptides is continuously being tried. The usage requires detailed knowledge of antigen and technology to produce antigen of interest. It becomes essential to list and analyze all factors that can affect the efficacy of antigen.

   The antigen is characterized biophysically for optimum temperature and pH, which can maintain it in an appropriate biological state. The stability is crucial for the shelf life of the vaccine; therefore, stability excipients need to be investigated and incorporated into vaccine formulation.

3. *Antigen interaction with adjuvant:*

   The effectiveness of vaccine depends upon its potential immunogenicity where it can evoke adaptive immune responses in the recipients. Unlike attenuated vaccines, in the subunit vaccine, the success depends upon the selection of an appropriate adjuvant for evoking desired antigen-specific immune response. Selection of adjuvant is critical step as it interacts with the immune system through various mechanisms and strongly influences the development of antigen-specific T-helper cell, T-cytotoxic cell, or antibody-mediated responses. However, for evoking desired antigen response, aluminum gels or aluminum salts (as aluminum hydroxide, aluminum phosphate, and aluminum potassium sulfate) have been in use since the 1930s, while monophosphoryl lipid A was used since 2009 in Cervarix (vaccine in the USA). Other adjuvants as toll-like receptor agonists (LPS mimics, flagellin, CpG motifs), saponins (QS-21), and cytokines (type-I interferons and IL-12) are also being tried for their efficacy and usage.

Adjuvants (Latin, adjuvare: to help) are the substances which when mixed with antigens enhance the immunogenicity of antigens [11]. The mode of action of adjuvants is not very clear, but they might act by prolonging exposure of antigen in the recipient, inducing costimulatory signals, and increasing local inflammation and stimulation of nonspecific proliferation of lymphocytes. The most commonly utilized adjuvants are aluminum salt and squalene-based oil-in-water emulsions. Other adjuvants as toll-like receptor agonists, saponins, and cytokines are in development.

Aluminum-containing adjuvant is widely used and has an excellent safety profile. They play an important role in prolonged persistence and uptake of antigen, recruitment of antigen-presenting cells (APCs), and stimulating differentiation of APCs and T cells.

Emulsions: Dispersion of two immiscible liquid phases as water-in-oil and oil-in-water were utilized as adjuvants. Water-in-paraffin was first evaluated by Freund in 1940. Antigen in water-in-paraffin oil emulsion with *Mycobacterium tuberculosis* was Freund's complete adjuvant (FCA) and without mycobacterium was incomplete Freund's adjuvant (IFA). It enhanced immunogenicity; however, water-in-oil emulsion adjuvants exhibited high reactiveness and their use was discontinued.

(continued)

In recent time, squalene-based oil-in-water emulsions are the main focus. Squalene is biodegradable (precursor of cholesterol) and its primary source is shark liver oil. They act by stimulating the release of chemokines that attracts monocytes and granulocytes, promotes monocyte maturation into dendritic cells, and enhanced antigen uptake with subsequent activation of T- and B-cell responses. Squalene-based emulsion are MF5 (Novartis Vaccines and Diagnostics), AS03 (GlaxoSmithKline), and AF03 (Sanofi Pasteur).

4. *Suitable trials*: Before the drug is used for the purpose of humankind, it undergoes series of steps to ensure the safety and efficacy. After in vitro studies, nowadays, drugs are tested in cell lines. Cell lines provide an alternative to use animals for the drug in question. After satisfactory results are obtained in cell lines' experiments, the drug is tested on animal models before clinical trial. After successful animal and clinical trial, the drug is approved fit to use in human beings.
5. *Analysis of product contact material*: It is important to evaluate impact of the steps and material on formulations. Impact of filter membranes and product contact material should be evaluated. The best conditions and material (which is contamination-free) should be chosen for appropriate folding and efficacy of the antigen for the release of safe and reliable vaccine.
6. *Monitoring stability, safety, and efficacy:* Vaccines are used to immunize human beings (infants, children, and adults); therefore, safety aspects are extremely important. Microbial contamination is an important threat; therefore, aseptic processing and sterile filtration of vaccine formulation are required. Aluminum-containing adjuvant cannot be sterilized due to particle size (greater than 0.2 mm), so they must be sterilized prior to formulation and should be handled aseptically.

   It is also important not to lose the antigen preparation in the process of sterilization at the same time removing all microbial contamination, so choice of filter membrane is done to ensure efficacy and safety.

14.3 Prophylaxis

Prophylaxis: It is the mode, which is adopted by people to protect themselves from the diseases. Prophylaxis deals with preventive measures to control the outbreak or epidemic of infectious agents. There are various modes like sanitation, vaccination, and many others to control the disease.

Immunization drive and vaccines: Immunization refers to induction of immune response either by active or passive mode, while vaccination refers to administration of vaccine for the purpose of immunization. Immunization is the program promoted by world health organization (WHO) as health initiative having major impact on world's health. Immunization has been integrated into routine health-care system in countries resulting in better control of disease-related morbidity and mortality in newborns and children and has been instrumental in saving millions of lives worldwide.

The usage of immunization drive has seen extraordinary success; however, the constituent of vaccines (mercury compound thimerosal, formerly used as preservative) has been widely condemned in some countries as the cause of neurodevelopmental disorders, autism and attention deficit hyperactivity disorder, diabetes, and a few allergic and autoimmune diseases. However, due to all these concerns, the

(continued)

vaccines' safety and effectiveness are under constant study. As vaccines are administered in children, their safety aspects are of prime importance. There were several concerns with the usage of vaccines: (1) too many vaccines administered early in life; (2) safety concerns with measles, mumps, and rubella (MMR) combination vaccine; and (3) the presence of thimerosal as preservative. Since 1990, thimerosal has been removed from all the vaccines routinely used in childhood vaccination program with flu being exception. For other concerns like development of autism and MMR vaccines, the thorough research has been conducted and studies have found vaccines as safe and effective modes to prevent infectious diseases. The studies do not show any link between autism and MMR vaccine, thimerosal, or administration of multiple vaccines in one time or fevers. However, the safety aspects are continuously being monitored. For details of safety assessment of vaccines, refer to http://www2.aap.org/immunization/families/faq/vaccinestudies.pdf.

Resistance to vaccination is growing; however, risks posed to population need to be balanced with benefits and creating more safe vaccine for public use. In the absence of vaccination, the only measure to prevent the diseases can be eradication of causative agent. Immunity to infections can only develop either by infection itself or by active immunization. The benefits of immunization are complete eradication of smallpox and rare or no occurrence of childhood diseases like tetanus, diphtheria, and measles mumps.

14.3.1 Passive and Active Vaccination

Passive immunity: Passive immunity is achieved when the body does not participate in the clearance and neutralization of pathogenic agents. In this, preformed antibodies are transferred to the recipients, and because of these, the pathogen clearance is promoted without activation of immune system. Passive immunity to diphtheria, tetanus, polio, rubella, streptococci, and mumps occurs due to transfer of maternal antibodies present in colostrum and milk to the infant [9]. It can also be achieved by injecting recipients with preformed antibodies (Fig. 14.1).

The passive dose is required in (1) congenital immune system defects, (2) exposure to a disease which might worsen rapidly, and (3) time which is too short to allow active involvement of immune system.

Passive immunization is given to individuals who counter rabies, tetanus, botulism, diphtheria, or hepatitis and is also used to protect from poisonous snake and insect bite. Caution is to be taken when injecting foreign (horse or mouse) antibody in humans as they cause strong anti-isotypic antibody response that might lead to serious complications. Passive immunization gives short-term protection; therefore, active immunity is essential for long-term memory response.

Active immunization: Active immunization refers to the activation of the body's immune system resulting in the memory responses. Active immunization can occur if an individual encounters a particular pathogenic agent by the use of vaccine (attenuated, inactivated, or subunit vaccines) (Fig. 14.2).

Apart from natural infection and attenuated vaccines, other modes require multiple boosters. Successful active immunization means:

- Activation of adaptive immune response (T- and B-cell response) for generation of protective memory.
- Subsequent exposure of the same pathogen leads to heightened immune response because of memory with short-time period.
- Memory response results in elimination of pathogen (Fig. 14.3) on subsequent encounter.
- Prevention of the disease for which the recipients are actively immunized.

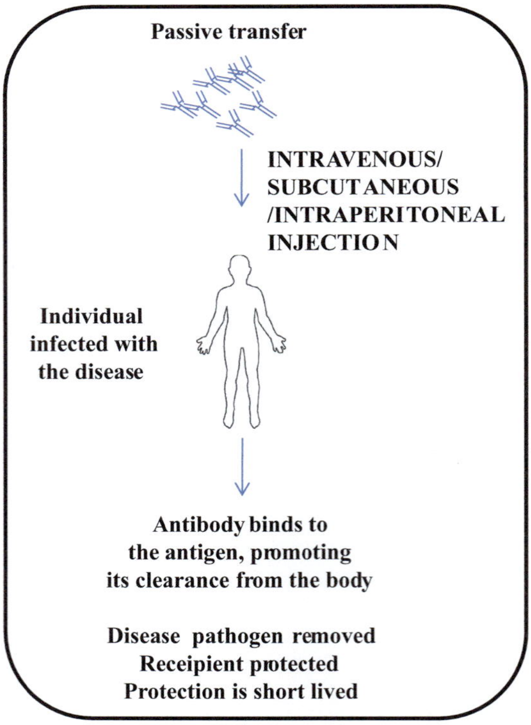

Fig. 14.1 The passive transfer of antibody for the therapeutic purpose. In this the body does not participate actively for removal of the infectious agent, but removal of pathogen is due to transfer of antibody

The immunization program supported by government has resulted in dramatic decrease or total eradication of the childhood diseases, which were otherwise a major cause of early death worldwide.

Any invader in the body is recognized by our vigilant immune system components. In immune response, the antigen initially is recognized by innate immune components and B cells (humoral immunity). The natural killer (NK) cells, neutrophils, and macrophages are all activated. A few specialized phagocytic cells phagocytose and process the antigen and present it to T cells for their activation. Upon activation, the T-cytotoxic (Tc) cells mediate response along with NK cells with the assistance of T-helper (Th) cells which induce suicidal cascade in target cell leading to its killing. B cells upon activation and with assistance of Th cells release heavy amounts of antibody that can trigger phagocytosis (opsonization) by macrophages, cell death (antibody-dependent cell-mediated cytotoxicity, ADCC), or complement-mediated lysis for efficient clearance of pathogen [1, 8]. B and T cells encounter the antigen and some of them remember the encounter with a particular antigen resulting in production of memory cells. These memory cells participate and mount a heightened immune response upon repeat exposure to the same pathogen. Immunization aims to induce production of sufficient memory cells that can remember and eliminate pathogens [1, 8].

14.3.2 Routes of Vaccine Administration

Vaccines can be administered by injections or oral intake. However, now newer delivery methods have evolved like via inhalation route or patch application. The vaccine can be administered by inhalation, for example, inhaled vaccine (aerosol) for Ebola virus [5] has been successful and effective in monkeys and inhaled vaccines for influenza has made available in the form of nasal spray. A patch can also deliver the vaccine where the patch may contain a matrix of very fine and tiny needles,

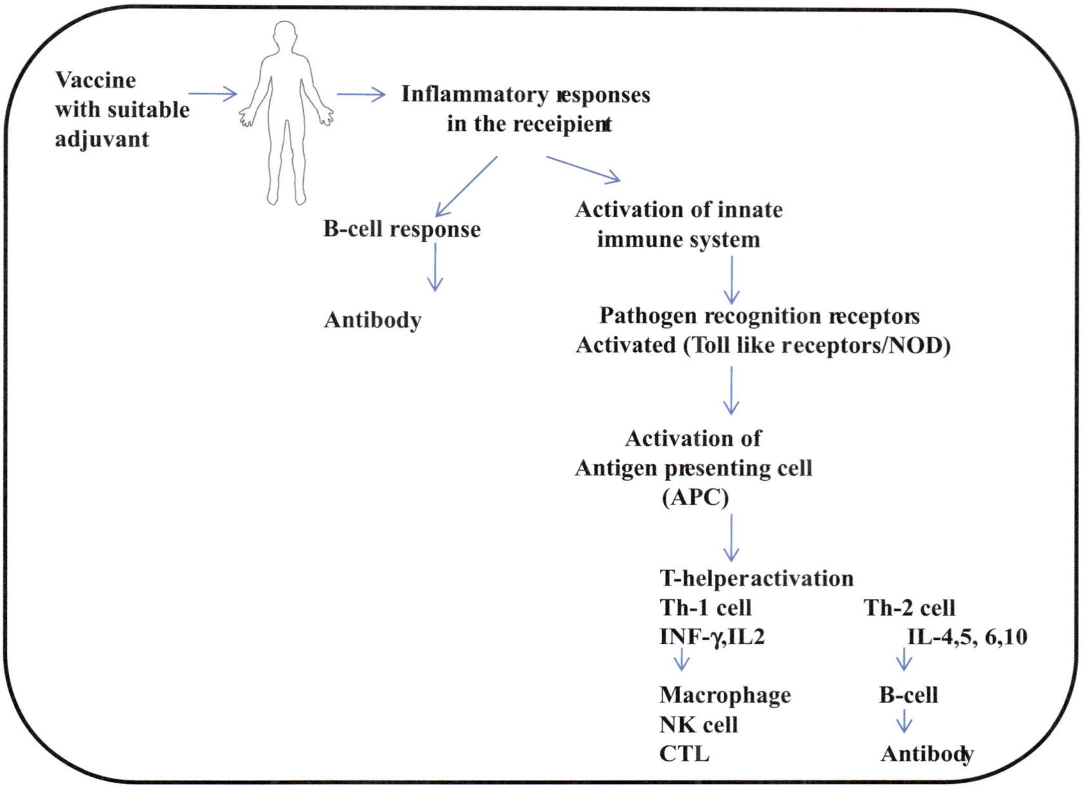

Fig. 14.2 The active immunization involves activation of innate immunity and adaptive immunity. Adjuvants help in increasing immunogenicity and inflammation. Complete activation of adaptive immunity with the help of innate immune components activate T and B cells

which can be helpful in delivery without syringe. The patch method may be useful in areas where medically trained professionals are not available.

Clinical Trials

It is a set of protocols that are conducted for approval of the drug for usage on human beings. The tests are conducted to analyze the safety and efficacy of the drug only after preliminary satisfactory information has been obtained for nonclinical safety. In the protocol, treatment is assigned by the investigator on research subject/volunteer and their outcomes are measured.

Clinical trials are designed for assessment of safety and effectiveness of:

- Discovered new drug
- Dosage of drug

(continued)

- Exploring more and/or different targets for existing drug
- Testing combination for effective management of disease
- Analysis that drug is more effective than preexisting drugs

Phase I, Safety: Experimental drug is evaluated for its safety, physiologically acceptable dosage, and its side effects on small group of human volunteers (20–80).

Phase II: Evaluation of testing protocol on a larger group of people (more than 100) for further assessment of its safety.

Phase III: Drug is tested on large groups of people (1,000–3,000) for evaluation of its effectiveness, monitoring side effects,

(continued)

comparing it to commonly used treatments, and collecting information that will allow the experimental drug or treatment to be used safely.

Phase IV: Additional information on drug's risk, benefits, and optimal use for "post-approval" studies.

Clinical trials are conducted in phases. The trials at each phase have a different purpose and help scientists answer different questions.

14.4 Attenuated Vaccine

Attenuation is achieved by growing the pathogenic agents in abnormal culture conditions and selecting the mutants surviving under abnormal condition. These mutants are maintained in those conditions for very long time period where their normal growth properties are altered such that their virulence/pathogenicity is reduced but they are live. Attenuated vaccines are live virus/bacterial vaccines, which are capable of transient growth in the host [8].

Vaccination Initial Studies

Inducing immunity to protect from the diseases was earlier attempted by Chinese and Turks. However, later on in 1798, English physician Edward Jenner observed that milkmaids who had contracted a disease called cowpox were immune to smallpox. Cowpox is mild than smallpox which results in disfiguring scars and is fatal. Jenner hypothesized that if fluid from a cowpox pustule is introduced into normal people, it may protect them from subsequent infection with smallpox. Then to test his assumption he deliberately inoculated an 8-year-old boy with fluid from cowpox pustule and later on infected the child with

(continued)

smallpox. As predicted, the child was protected from smallpox. The technique, which could impart protection from smallpox by inoculation with cowpox, spread quickly.

However, it took long time to apply this to other diseases. Louis Pasteur was working on bacteria causing fowl cholera. He had succeeded in growing it in culture and showed that chickens injected with bacterial culture developed cholera. Then he went for summer vacation and after he was back, he injected the chickens with his old culture of bacteria. The chickens became ill but they recovered. He then grew fresh culture with intention of injecting it in fresh fowl, but due to nonavailability of fowl, he used the previously injected chickens. The results were surprising as chickens were completely protected from the disease. Then he assumed that aging has weakened the virulence of the pathogen, and these attenuated strains when administered are capable of imparting protection for the disease. These attenuated strains were called as vaccine (from the Latin word vacca, meaning "cow") [8].

The advantages of attenuated vaccines are:

- Single dose is effective, and boosters are not essentially required because immune system is exposed to the live organism for prolonged period.
- Efficiently activates adaptive immune responses.

Disadvantages:

- Storage conditions are to be strictly maintained.
- Reversion is one of an important risk.
- Vaccines can cause serious complications in immunocompromised patients (Table 14.1).

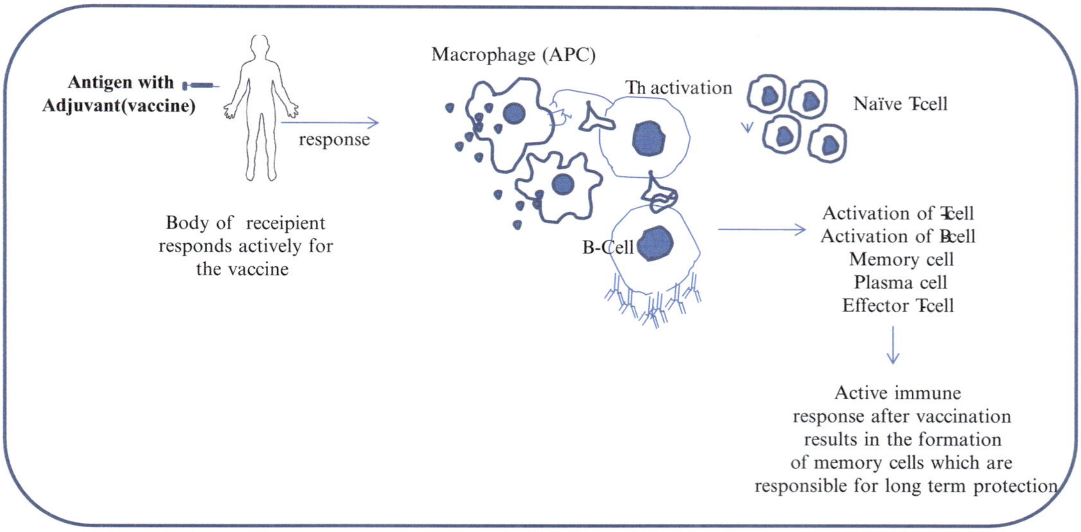

Fig. 14.3 The subunit vaccine where active vaccination involves administration of vaccine along with participants for clearance of antigen with production of memory cells

14.5 Inactivated/Killed Vaccine

Due to problem of reversion of pathogenic agent in attenuated vaccines, the live virus/bacteria were inactivated by heat or chemical treatment. These inactivated vaccines have microorganisms incapable of replication but is able to activate immune system for memory response. Controlled inactivation is done for production of vaccine where microorganism is properly inactivated (that is killed), but at the same time, care is taken to maintain the surface structure of epitope (for specificity and memory).

Heat inactivation is not preferred because it causes denaturation and thus loss of native structure of the epitopes. Chemical inactivation using formaldehyde or various alkylating agents has been successfully used. In contrast to attenuated vaccine where one dose is sufficient for long-term immunity, inactivated killed vaccines require boosters with appropriate adjuvants for activating immune system of the host. The humoral immune activation is the predominant immune system branch, which responds for inactivated/killed vaccines.

The complications of inactivated vaccines were observed when Salk vaccine for polio was prepared and formaldehyde failed to kill the entire virus in two lots resulting in outbreak of paralytic polio in a high percentage of recipients.

14.6 Subunit Vaccines

Subunit vaccines consist of purified macromolecules derived from pathogens. The various subunit vaccines are given with appropriate adjuvants to activate humoral and cell-mediated immunity (Fig. 14.4).

Inactivated exotoxins: Several strains of bacteria cause disease by production of exotoxins like diphtheria, tetanus, and cholera toxins. These toxins produce disease symptoms. The exotoxins are prepared and purified and then they are inactivated with formaldehyde to produce toxoid. These toxoids when used as vaccine generate antitoxoid antibodies, which can neutralize the natural toxins when encountered by pathogenic agents and protect the vaccine.

Table 14.1 Some of the attenuated vaccines along with their properties

Attenuated vaccine/trade name	Microorganism	Attenuated by	Disease/precautions/approval	Advantages	Disadvantages
Bacillus Calmette–Guérin (BCG) (BCG vaccine)	*Mycobacterium bovis*	Bacillus of Calmette and Guérin (BCG) strain of *Mycobacterium bovis* are grown in ox bile medium for attenuation	Tuberculosis, used experimentally in 1921	Capable of replication so elicits strong immune response Impart lifelong immunity	Reversion to mutant form Storage is crucial Sometimes severe complications are observed in immunocompromised hosts
Typhoid (Vivotif®)	*Salmonella typhi* (Ty21a)	Grown under controlled conditions in medium of yeast extract, an acid digest of casein, dextrose, and galactose	Typhoid fever	Boosters are not essentially required	
Measles, mumps, and rubella (German measles) (MMR II) Measles, mumps, rubella, and varicella for chicken pox (ProQuad)	Measles virus, mumps virus (paramyxoviruses), and rubella virus (togavirus) Measles virus, mumps virus (paramyxoviruses), and rubella virus (togavirus) and varicella	Live virus attenuated by multiple passages through chick embryo cell culture	Usage avoid during or before pregnancy		
Polio virus	Sabin polio (3 strains of polio)	Grown in monkey kidney epithelial cells	Approved in 1961 and trivalent oral vaccine approved in 1963		
Rotavirus RATARIX Rota Teq (pentavalent)	Rotavirus-induced gastroenteritis due to (G1 and non-G1 types)	Grown in cell culture approved in 2008	Some health impact needs to be studied		
Varicella zoster (Varivax)	Varicella causing chicken pox	Attenuated by passage in human cells. FDA approved in 1995	Some health impact needs to be studied		
Yellow fever (YF-Vax)	Yellow fever virus	Grown in chick embryos	Marked reduction in worldwide cases		
Smallpox (ACAM2000)	Causative agent are two strains of virus, Variola major and Variola minor. Vaccine is prepared by pox type virus Vaccinia virus	Used experimentally by E. Jenner in 1796	Disease eradicated in the late 1970s		

Recombinant DNA technology is playing an important role where exotoxin gene is cloned and expressed and toxin is obtained in sufficient quantity. In the absence of vaccination against a few dangerous microbial agents, these antitoxoids are given which are usually produced in horses or other animals. These antitoxoids neutralize the natural toxins and protect the individual.

14.6.1 Capsular Polysaccharides

Some pathogenic bacteria possess hydrophilic polysaccharide capsule, which has antiphagocytic properties due to which it skips from defense response of the host, and cause diseases. These polysaccharide moieties when used as vaccine develop antibodies against them. These antibodies when encounters the live bacteria with capsule results in opsonization and subsequent clearance from the body. *Streptococcus pneumoniae* consists of many antigenically different capsular polysaccharides. Administration of these purified potent polysaccharides along with suitable adjuvant leads to formation of opsonizing antibodies. Vaccine for *Neisseria meningitidis* (bacterial meningitis) consists of capsular polysaccharide.

14.6.2 Viral Glycoproteins

Virus coat containing glycoproteins (envelope protein) have been tested for their potential as vaccines. Glycoprotein D from herpes simplex virus type 2 vaccine prevented herpes in many recipients. However, envelope protein from HIV-1 has been tested, but successful results were not obtained. The antigenic genes (from viral, bacterial, and protozoan pathogens) are expressed in bacterial, yeast, or mammalian cells and are used for vaccine production. The recombinant antigen vaccine in use is the surface antigen of hepatitis B virus (HBsAg) .

14.7 Synthetic Peptides as Vaccine

The construction of an effective synthetic vaccine requires the identification of potent antigenic epitope which when targeted may induce protective immunity against pathogenic microorganism. It is observed that large epitopes, which can form tertiary conformation, are efficient, for example, two proteins of VPI coat protein of foot and mouth disease virus (FMDV) and synthetic peptide vaccine for malaria (three short peptides representing sequences from three blood stage antigens were used). Care is taken to use a mixture of conformational epitopes to produce a successful antitoxin vaccine. However, if synthetic peptides capable of targeting humoral and cell-mediated immune responses may be produced by chemical synthesis, it can revolutionize the field of vaccinology.

- Subunit vaccines can be very effective vaccine, but identifying the most active and effective epitope is needed. Once the antigenic epitope has been identified, then they can be used as vaccine.
- A disadvantage with subunit peptide vaccine is the immunogenicity of peptides as it is very difficult to activate both cell-mediated and humoral immune responses.
- A disadvantage of polysaccharide subunit vaccines is their inability to activate Th cells. They activate B cells in a thymus-independent type 2 (TI-2) mode with little class switching, no affinity, and little development of memory cells.

14.8 Conjugate Vaccines

Conjugate vaccines were proposed where polysaccharide antigen can be conjugated to some sorts of protein carrier and this complex would be effective in activation of Th cells. The vaccine for *Haemophilus influenzae* type b (Hib) consists of

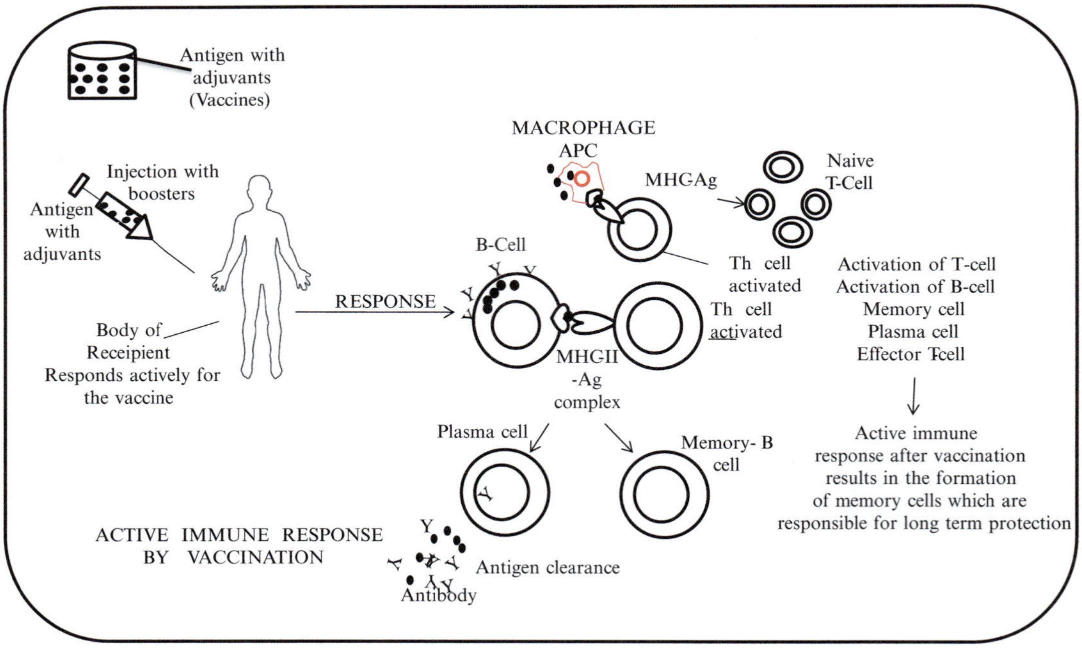

Fig. 14.4 The active vaccination involves administration of vaccine along with suitable adjuvants. In this the body's immune system participates for clearance of antigen with production of memory cells

type b capsular polysaccharide covalently linked to a protein carrier, tetanus toxoid. The conjugate of polysaccharide–protein is more immunogenic than any of them alone as it is efficiently able to activate both the humoral and cell-mediated immunity and enables class switching and affinity maturation.

Promising results were obtained by using polysaccharide component of fungi (*Aspergillus fumigatus* and *Candida albicans*).

14.9 DNA Vaccines

The present strategy is exploring the utilization of plasmid DNA encoding antigenic proteins as potential vaccine. Muscle cells take up the DNA and the antigenic proteins are expressed in the cells activating both cell-mediated and humoral immune responses. The advantages are:

1. Protein is expressed in its native conformation.
2. Long-term exposure of immune system to the antigen.
3. Efficient activation of cell-mediated and humoral immune systems leading to memory response.
4. There is no requirement of specific storage condition like refrigeration of plasmid DNA, and it reduces the cost and is easy to handle.
5. The genes expressing cytokines can be ligated to the vector, and the cytokine produced at the site may activate the desirable branch of immune system.
6. Vaccine is safe, without any infectious agent.
7. Easier to produce in large quantity.

The disadvantage of DNA vaccine is that they can be used for only protein antigens, and for the nonprotein antigens (polysaccharide) they cannot be used.

14.10 Edible Vaccines

The edible vaccines may be vaccines derived from plants or animals by using the tools of recombinant DNA technology (Fig. 14.5).

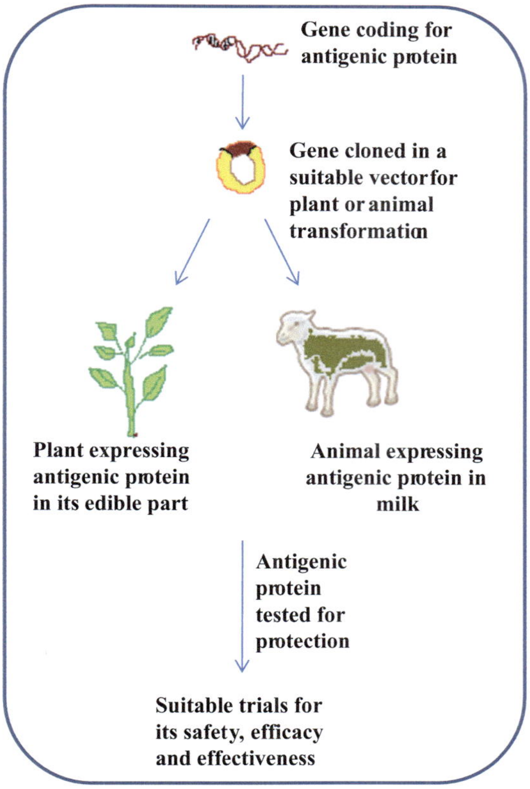

Fig. 14.5 The production and usage of edible vaccine produced either in plant or in animal source. The vaccine is evaluated for effectiveness and upon clinical trials may be recommended for human usage

For plant-derived vaccine, the genes encoding immunogenic proteins of an infectious agent can be transferred into either the nuclear genome of a plant system or the chloroplastic system, such that the plant is capable of producing the desired immunogenic protein subunit vaccines [10, 18]. The approach uses freeze-dried plant cells for bioencapsulation of vaccine antigens that protect them in the stomach from acids and enzymes but are released to the immune system in the gut when plant cell walls are digested by bacteria that colonize the gut. These vaccines have the following advantages over other vaccines:

- Plants provide an inexpensive source of edible vaccines.
- They can be given in natural condition.
- Do not require fermentors for large-scale production.
- Do not require processing, purification, storage, and packaging.
- Do not require storage and sterile delivery.
- Cloning in chloroplastic system results in high production.
- They stimulate mucosal immune system, the first line of defense effectively.
- It increases the value of plants as novel sources of medicinal drugs (biopharming).
- Plant pathogens normally do not infect humans; thus, unlike mammalian pathogens, they are safe.

Potato and tomato plants have synthesized antigens from *Norwalk virus*, enterotoxigenic *E. coli*, *Vibrio cholerae*, and *hepatitis B virus*. A recently completed human study has shown that a recombinant bacterial antigen, subunit B of heat labile enterotoxin, produced in a potato and eaten resulted in production of both serum antibodies (IgG and IgA) and mucosal antibodies (sIgA) to the antigen. Many vaccines in plants are in the process of development, and vaccine antigens for different bacterial, viral, or protozoan pathogens have been expressed in chloroplasts.

- In mice, chloroplast-derived anthrax conferred 100 % protection.
- Oral vaccine of F1-V antigens (*Yersinia pestis*) without adjuvant conferred greater protection (88 %).
- Oral immunization of malarial vaccine antigens fused to the cholera antigen (CTB-AMA1/CTB-Msp1) conferred prolonged immunity (50 % life span) with antigen-specific titers of IgA and IgG1.
- Vaccine against cholera toxin B subunit, hepatitis B surface antigen, *E. coli* heat labile enterotoxin, and *Norwalk virus* capsid protein have been developed.
- Almost all tested vaccines showed immunologic responses in animals and tested humans.
- Vaccine for respiratory syncytial virus (RSV) has been developed.
- In the chloroplasts, the presence of chaperones or enzymes results in assembly with suitable posttranslational modifications.
- The chloroplast expression system has been successfully used to produce vaccine antigens

against cholera, anthrax, plague, *Canine parvovirus*, CSFV, EBV, FMD, HIV antigens p24, HEV, HPV, *Rotavirus*, amebiasis, and malaria.

The technique has applications but low yield is major limitation. Enough yield of recombinant vaccine is not obtained in plant tissue, and plant-specific glycans might alter the properties of recombinant proteins. However, no transgenic plant-based vaccine has moved beyond phase I clinical trial, thus highlighting the need to explore new technologies.

14.11 Vaccines for Cancer

The vaccines for cancer are preventive which are used to trigger memory B-cell formation in the body of host [13]. They are targeted for hepatitis B virus (HBV) and human papillomavirus (HPV) where the immune response activation and memory may impart protection for these agents [14]. Activation of memory response results in antibody production via humoral immune response and apoptotic death of target cell mediated by killer T cells (Tc) (cell-mediated response). Cancer vaccines are biological response modifiers which may be preventive (tend to prevent development of cancer) or therapeutic (tend to treat an existing cancer by stimulating host immune responses). Two types of cancer-preventive vaccines are available, and one cancer treatment vaccine has recently become available.

The US FDA has approved Gardasil® (for HPV types 6, 11, 16, 18) and Cervarix® (for HPV types 16, 18), which aim to impart protection against infection due to human papillomavirus (HPV), which are causative agents of 70 % cases of cervical cancers. They are also responsible for vaginal, vulvar, anal, penile, and other cancers. The vaccine can be administered in children above 9 years till 26 years [4, 14, 16].

- Gardasil® (manufactured by Merck & Company) protects against HPV types 6, 11, 16, and 18 where 6 and 11 are responsible for causing many cases (90 %) of genital warts in males and females. The vaccine has noninfectious virus-like particles (VLPs) and is called quadrivalent as it targets four HPV types [4].
- Cervarix® (manufactured by GlaxoSmithKline) protects against types 16 and 18 and thus is a bivalent vaccine. It consists of VLPs made with proteins from HPV types 16 and 18.
- A cancer-preventive vaccine that protects against hepatitis B virus (HBV) has also been approved by FDA in 1981, as chronic HBV infection may lead to liver cancer [19].
- Treatment vaccines for cancer are therapeutic which help in treatment of cancer either to delay its progression or stop their growth, inducing shrinkage of tumor or complete elimination [13]. As immune system does not recognize cancer cells as foreign and antigenic, thus, it does not mount an attack on them. As cancer cells carry normal self-antigens, thus, they skip immune responses. Sometimes cancer-associated antigens are induced by cancer cells, but they rapidly lose these due to mutations, and even if immune system recognizes them, they tend to suppress and skip attacks by Tc and NK cells. Thus cancer treatment vaccine should be capable of stimulating specific immune response against target cancer cells or strengthened immune responses, which can bypass the barriers imposed by cancer cells. The treatment vaccines are designed to perform these tasks.
- FDA approved cancer treatment vaccine sipuleucel-T (Provenge®, manufactured by Dendreon) in 2010 for metastatic prostate cancer. The vaccine is designed to stimulate immune response against prostatic acid phosphatase (PAP), which is present on most prostate cancer cells.

Virus-Like Particles (VLPs)

Virus-like particles (VLPs) are self-assembled structure from viral antigens. These VLPs resemble the natural virus but lack the viral genome. Thus, when they are present in the body of the host, the body generates immune response against VLPs. Thus, when natural virus encounters the body, the immune system mounts heightened response and facilitates its clearance. They are safe, stable, and immunogenic. For their production, nowadays, plants have been widely explored due to their ability for large-scale low-cost production with eukaryotic processing machinery for correct posttranslational modification and folding with minimal or no risk of transmission of animal/human pathogens and without any requirement of fermentation reactions. They are widely used as vaccinating agents for protection against viral antigens.

Many other vaccines are being analyzed in clinical trials for protection or active cancer prevention. The DNA- and RNA-based therapeutic vaccines are also being exploited for cancer antigens for effective elimination by immune system [17].

Cancer vaccines are also administered with suitable adjuvants. As adjuvants, the attenuated bacillus Calmette–Guérin (BCG) or bacterial Detox B may be used for potent immune response. Other substances as keyhole limpet hemocyanin (KLH) (sea animal-derived protein), emulsified oil as montanide ISA–51, or cytokines may also be used with vaccine. Common adjuvant cytokines which are being tried in cancer vaccines include interleukin-2 (IL2, aldesleukin), interferon alpha (INF-α) [2, 3, 6], and GM-CSF (sargramostim).

Unlike other vaccines, cancer vaccines are associated with side effects as inflammation at the injection site with pain, redness, swelling, itchiness, warming of the skin, and occasionally a rash may occur. Flu-like symptoms, fever, chills, weakness, nausea, vomiting, and muscle aches, may also occur.

14.12 Generations of Vaccines

The generations of vaccines grossly consider (1) microorganism usage, (2) production of important part derived from it, and (3) usage of technology for engineering and DNA-based vaccines.

First-generation vaccines required cultivation of pathogenic organisms. The pathogen is then weakened by attenuation (attenuated vaccines) or inactivated by heat or formaldehyde (inactivated vaccines) before administering into the humans. These vaccines have been in use since long time. They have been very effective in the control and eradication of many diseases. As there was a risk of reversion in attenuated vaccines and inactivated vaccines also required precautions and boosters, thus, scientists explored more safe vaccines alternative by the use of recombinant DNA technology leading to second-generation vaccines.

Second-generation vaccines utilize only specific part of the antigen thus referred as subunit vaccines. These vaccines comprise of inactivated exotoxins (toxoids) or bacterial capsular polysaccharides or viral glycoproteins or proteins conjugated with carriers.

The third-generation vaccines are naked DNA or DNA vaccines, which can be administered directly or may be packaged in a recombinant virus or bacteria [12]. DNA-based vaccines have been approved for West Nile virus in horses, hematopoietic virus in Salmon, growth hormone–releasing hormone in swine and food animals, and melanoma for dogs. The animal studies showed good response which led to clinical trials in humans also for HIV-1, cancer, human papillomavirus (HPV), and so on, but they were poorly immunogenic with very low or nonexistent antibody titers though they were well tolerated and safe. They are still at experimental stage.

14.13 Reverse Vaccinology or Genome-Based Vaccine Development

The conventional methods of vaccines like attenuated, inactivated, and subunit vaccines have been very successful. The vaccination program has successfully eradicated smallpox and resulted in the virtual disappearance of diseases like diphtheria, tetanus, poliomyelitis, pertussis, measles, mumps, rubella, and invasive *Haemophilus influenzae B*. This has increased the life quality and expectancy. However, there are still many challenges like (1) difficulty in cultivating some microorganisms, (2) finding a suitable animal model for disease as HIV-1, (3) recognizing suitable vaccine candidate for some diseases, and (iv) failure to purify the specific antigen.

14.13.1 Reverse Vaccinology

With the advancement in sequencing of genome and availability of proteome data, it was possible to identify the suitable vaccine candidates without conventional vaccine research. The technique of reverse vaccinology deals with analyzing the potential surface-exposed proteins by computational methods from the total proteins, which may be encoded by microorganism. It starts from genome rather than microorganisms [15]. This technology overcomes the problems associated with non-cultivable microorganisms and to antigens that are not expressed in in vitro conditions.

In the case of *Meningococcus B*, the approach of reverse vaccinology through computational analysis allowed the prediction of proteins that could be surface-exposed or homologous to known factors associated with virulence and pathogenesis. This leads to selection of 570 potential vaccine candidates, of which successful cloning and expression could be done for 350. These proteins were purified and used for mice immunization, and out of these 28 proteins showed positive results when analyzed by ELISA. Further confirmation of these as suitable vaccine candidates, their presence and conservation were tested in a panel of 31 strains of *N. meningitides*. Five of these showed strong conservation, and in less than 2 years, reverse vaccinology achieved its goal. That is identifying surface-exposed proteins in *N. meningitidis B*, which were able to induce protection and cross-reactivity among distantly related strains and serotypes, suitable to be used in a universal vaccine against this microorganism. The success was extended to other microorganisms, such as *Streptococcus pneumoniae*, *Porphyromonas gingivalis*, *Chlamydia pneumoniae*, *Bacillus anthracis*, *Streptococcus agalactiae*, *Streptococcus pyogenes*, ExPEC, and many others, to find universal vaccine.

One of the limitations of the reverse vaccinology approach is the inability in identifying nonproteic antigens such as polysaccharides or glycolipids, which have been components of many successful vaccines.

14.14 Chapter End Summary

- Vaccine is preparation of the pathogenic agent, which can give protection against that agent by evoking protective responses in the body. These responses are activation of adaptive (humoral and cell-mediated) responses which generates memory cells.
- The vaccines are often supplemented with adjuvants, which when mixed with the pathogenic agent makes the agent more immunogenic. They efficiently activate T- and B-cell responses.
- The development of vaccines not only requires understanding of the pathogen but also the development of immune responses in the host. After appropriate selection, animal trials, safety report, and preliminary approval for clinical trials, the studies are done on humans.
- Passive vaccination involves administration of preformed antibacterial or antiviral agents where immune system is not actively activated. Active vaccination involves activation of host immune system to give effector and memory response.

- The attenuated vaccines are prepared by growing the pathogenic agent under abnormal culture conditions for prolonged period. They are live vaccines, in which the organism is capable of growing inside the host but its pathogenicity is highly reduced. They are effective in single dose but may revert to virulent form.
- In inactivated vaccines, the microorganism is killed by either heat or chemical. Chemical inactivation is preferred, as it does not alter the structure of the epitope. They are safe, stable, and easy to transport but are required to be administered as injections.
- The subunit vaccines are in use, which are purified macromolecules of pathogen. They are given with appropriate adjuvants. Subunit vaccines include toxoids (inactivated exotoxins), bacterial polysaccharides, and viral glycoproteins. These may be produced by recombinant DNA technology and their administration is safe.
- Generation of vaccines is according to their part used as vaccinating agents. First-generation vaccines have complete pathogen as attenuated and inactivated. Second-generation vaccines are subunit vaccines and third-generation vaccines are nucleic acid-based vaccines. Reverse vaccinology is done for microbial agents which are difficult to culture.

Multiple Choice Questions

1. Vaccines have helped humans from:
   (a) Protection from many infectious agents
   (b) Development of new diseases
   (c) Removal of all pathogens
   (d) All of the above
2. This might be one of the disadvantages of inactivated vaccine:
   (a) High immune reactions
   (b) Risk of reversion to a pathogenic form
   (c) Requirement of boosters
   (d) All of the above
3. Development of successful vaccine requires:
   (a) Characterization of the pathogen
   (b) Knowledge about the immune response
   (c) Development of trials and approvals
   (d) All of the above
4. An adjuvant is used for making the agent:
   (a) Antigenic
   (b) Tolerogenic
   (c) Allerogenic
   (d) Immunogenic
5. The commonly used adjuvant is:
   (a) Freund's complete adjuvant
   (b) Alum
   (c) Incomplete Freund's adjuvant
   (d) None of the above
6. In passive mode of immunization:
   (a) Antibodies are given
   (b) Antibacterial agent is given
   (c) Drug is given
   (d) All of the above
7. Attenuated vaccines are:
   (a) Live
   (b) Killed
   (c) Subunit
   (d) Pathogenic
8. Inactivated vaccines are:
   (a) Live
   (b) Killed
   (c) Subunit
   (d) Pathogenic
9. Vaccination is given for evoking:
   (a) Macrophage response in the body
   (b) Production of more antibody
   (c) Production of memory T and B cells
   (d) None of the above
10. Clinical trials are done:
    (a) After development of vaccine
    (b) After animal trials and approval
    (c) After approval
    (d) None of the above
11. The subunit vaccine is:
    (a) Protein subunit of bacteria
    (b) Surface glycopolysaccharide of pathogen
    (c) Toxoid
    (d) All of the above

12. The technology of reverse vaccinology is useful in:
    (a) Pathogenic microorganisms
    (b) Noncultivable organism
    (c) Nonpathogenic organisms
    (d) All of these

Answers

1. (a); 2. (c); 3. (d); 4. (d); 5. (b); 6. (d); 7. (a); 8. (b); 9. (c); 10. (b); 11. (d); 12. (b)

Review Questions

Q1. What do you understand by prophylaxis?
Q2. What is active and passive immunization?
Q3. What are vaccines?
Q4. Define attenuated and inactivated vaccines.
Q5. What are subunit vaccines?
Q6. Why subunit vaccines are considered more advantageous as compared to other vaccines?
Q7. Define edible vaccines.
Q8. What is reverse vaccinology?

References

1. Abbas AK, Lichtman AH (2009) Basic immunology, 3rd edn. Saunders Elsevier, Philadelphia
2. Alm, Gunner V (2003) Role of natural interferon-alpha producing cells (plasmacytoid dendritic cells) in autoimmunity. Autoimmunity 36:463–472
3. Decatris, Marios (2002) Potential of interferon-alfa in solid tumours. Biodrugs 16:261–268
4. Doorbar J (2006) Molecular biology of human papillomavirus infection and cervical cancer. Clin Sci 110:525–541
5. Fleur NS (2015) Inhaled ebola vaccine stops virus in monkeys, Study finds. The New York Times (July 13)
6. Goodsell, David S (2001) The molecular perspective: interferon. Oncologist 6:374–375
7. Ilchmann K et al (n.d.) Vaccine development & the BWC. http://hsp.sussex.ac.uk/sandreviews/
8. Kindt TJ, Goldsby RA, Osborne B (2007) KUBY immunology, 6th edn. Freeman and Company, New York
9. Kit Y-Y, Semenov DV, Nevinsky GA (1996) Phosphorylation of different human milk proteins by human catalytic secretory immunoglobulin A. Biochem Mol Biol Int 39:521–527
10. Korban SS, Krasnyanski SF, Buetow DE (2002) Foods as production and delivery vehicles for human vaccines. J Am Coll Nutr 21:212S–217S
11. Krensky AM, Flavio V, Bennett WM (n.d.) Chapter 52. Immunosuppressants, tolerogens, and immunostimulants. In: Brunton LL, Lazo JS, Parker KL (eds) Goodman & Gilman's the pharmacological basis of therapeutics, 11th edn: http://www.accessmedicine.com/content.aspx?aID=951722
12. Liu M (2011) DNA vaccines: an historical perspective and view to the future. Immunol Rev 239:62–84
13. Lollini PL, Cavallo F, Nanni P, Forni G (2006) Vaccines for tumour prevention. Nat Rev Cancer 6:204–216
14. Lowy DR, Schiller JT (2006) Prophylactic human papillomavirus vaccines. J Clin Invest 116:1167–1173
15. Moriel DG, Scarselli M, Serino L, Mora M, Rappuoli R, Masignani V (2008) Genome-based vaccine development a short cut for the future. Hum Vaccin 4:184–188
16. Parkin DM (2006) The global health burden of infection-associated cancers in the year 2002. Int J Cancer 118:3030–3044
17. Scheibner V (1993) Vaccinations 100 years of orthodox research. New Altantean Press, Santa Fe
18. Semiromi AD, Samson N, Daniell H (2009) The green vaccine: a global strategy to combat infectious and autoimmune diseases. Hum Vaccine 5:488–493
19. U.S. Centers for Disease Control and Prevention (2005) A comprehensive immunization strategy to eliminate transmission of hepatitis B virus infection in the United States: recommendations of the Advisory Committee on Immunization Practices (ACIP) Part 1: immunization of infants, children, and adolescents. Morb Mortal Wkly Rep 54(No. RR–16):1–31
20. World Health Organisation (2006) Development of new vaccines. Fact Sheet no 289, Revised December 2006

Some Selected Resources

immunostimulants.blogspot.com
www.cdc.gov/vaccines/events/niiw/ed-resources/partner-resources.html
www.cdc.gov/vaccines/hcp/patient-ed/conversations/index.html
www.dshs.state.tx.us › Immunization Branch
www.hrsa.gov/vaccinecompensation/
www.immunizationinfo.org/parents/evaluating-information-web
www.immunize.org/vaccine-safety-resources.pdf
www.vaccines.gov › More info › Guide to vaccines websites

15 Embryo Transfer Technology

Abstract

In vitro fertilization (IVF) is formation of embryo in culture condition. IVF tools have dramatically improved in the years. IVF was initially indicated for women with tubal factor infertility, but in the present scenario, it is the treatment of choice for all causes of infertility where the conventional or conservative methods fail. The chapter covers various factors responsible for male and female infertility, technique of in vitro fertilization, preimplantation genetic diagnosis and factors such as ovarian hyperstimulation syndrome, and ethical issues associated with the field.

15.1 Introduction

In vitro fertilization (IVF) is formation of embryo in culture condition. IVF tools have dramatically improved in the years. IVF was initially indicated for women with tubal factor infertility, but in the present scenario, it is the treatment of choice for all causes of infertility where the conventional or conservative methods fail.

15.2 Infertility

Infertility is the inability to conceive after 12 months of unprotected intercourse. In a study prediction based upon fecundability is the probability of achieving pregnancy in one menstrual cycle (20–25 % in healthy young couple). With the fecundability of 0.25, 98 % of couples should be able to conceive within 13 months. In the modern society, the infertility rates are stable but the number of couples without children has increased. Infertility may be due to reduced conception rates or the need for medical help; however, male factor is responsible in 25 %, female factor in 58 %, and without any reason in 17 %.

Initial evaluation of the possible cause of infertility is very important. Initial evaluation requires discussion about the timing and frequency of intercourse, prevalence of risk factors (smoking, alcohol, caffeine, obesity), and finally the investigations (semen analysis in male, confirmation of ovulation in the female, tubal patency in female). Psychological aspects should be assessed, as stress management is essential as it influences the associated processes; counseling may be done if required [2].

V. Gupta et al., *Basic and Applied Aspects of Biotechnology*, DOI 10.1007/978-981-10-0875-7_15

15.2.1 Male Infertility

Male factor infertility nowadays can be treated. The reasons and their treatment options are available.

In normal condition, there is pulse of luteinizing hormone (LH) in every 1–3 h. Elevated LH suggests primary defect at testicular level, with problems in sperm motility and transport. Low LH suggests defects at hypothalamic–pituitary level. Low testosterone leads to male factor infertility. High follicle-stimulating hormone (FSH) indicates damage to the seminiferous tubules.

Patients with primary testicular defects and problems associated with sperm motility are suitable candidates for IVF. Initial treatment is dependent upon sperm concentration and motility. If the sperm count is $15–20 \times 10^6$/ml with normal motility, it is mild male factor infertility; if counts are $10–15 \times 10^6$/ml with 20–40 % motility, then it is moderate male factor infertility; if counts are less than 10×10^6/ml, then it is a severe defect. In moderate male factor defect, intrauterine insemination or simultaneous treatment of the female partner with clomiphene or gonadotropins might help; these may require IVF with or without intracytoplasmic sperm injection (ICSI). In severe condition, IVF with ICSI or the use of donor sperm should be done.

15.2.2 Female Infertility

In females the infertility may be due to any of these reasons:

- Abnormalities in menstrual function due to (1) ovulatory dysfunction (if FSH level at the third day of menstrual cycle is less than 10 IU/ml, then it indicates inadequate ovarian oocyte reserve) and (2) abnormalities of the uterus or outflow tract
- Low FSH, LH and estradiol, and prolactin due to hypothalamus or pituitary involvement
- Polycystic ovarian syndrome (PCOS) (in this hyperandrogenism and irregular menstrual cycles occur)
- Ovarian (low estradiol and increased FSH)
- Uterine tract abnormality

In females, it should be ascertained that the amenorrhea (no menstrual cycle) is primary or is occurring after normal puberty and menarche.

15.3 Indications for IVF

When conservative treatment fails, it indicates the requirement of IVF.

IVF or ICSI is indicated in significant male factor defect or tubal disease; IVF using donor oocytes is indicated in patients with premature ovarian failure or woman with advanced age.

Success rates depend upon severity and age of the woman. In female of less than 40 years, 18–24 % success is expected per cycle; however, in female of more than 40 years of age with very low oocyte and fertilization ability, the success is moderately decreased. For IVF to be successful, the age of the woman is also very important where successful IVF decreases each year after the age of 40.

15.4 Intracytoplasmic Sperm Injection (ICSI)

- Poor quality semen: ICSI worked well when semen specimens are of extremely poor quality. In this, the sperms are immobilized in polyvinylpyrrolidone (PVP), or their tail is crushed and then a single spermatozoon is aspirated into a microneedle. This sperm is injected directly into the ooplasm of the oocyte, which has been stripped of its surrounding cumulus mass. This is followed by fertilization the next day, which is evident by the presence of the male and female pronuclei (Fig. 15.1).
- Obstructive azoospermia: In the congenital defect where the vas deferens is absent or due to vasectomy or arrest of sperm maturation due to nonobstructive azoospermia, ICSI is helpful. In these conditions, sperms are obtained by microsurgical or percutaneous aspiration from the epididymis or the testes but yields are extremely low.
- Presence of antisperm antibody: The presence of antisperm antibodies results in degradation

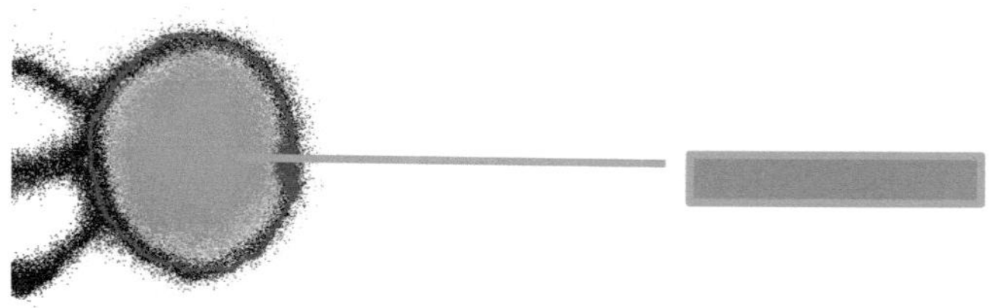

Fig. 15.1 Intracytoplasmic sperm injection (ICSI). In this process oocyte is held and a single sperm is injected into each oocyte after retrieval of oocyte. After the fertilization process is over, the fertilized egg or embryo is put in the mother's womb

of sperms; thus, ICSI is quiet helpful. ICSI is also used when there has been poor fertilization in a previous IVF cycle despite a normal semen analysis. It also improves fertilization rates with oocytes matured in vitro or cryopreserved.

- Rescue ICSI of unfertilized eggs following conventional IVF may be done; however, its indications for its use should be justified.

15.4.1 Concerns About ICSI

- No natural selection thus genetically abnormal sperm may participate in IVF.
- Meiotic spindle apparatus may be disrupted resulting in aneuploid embryo.
- Sperms used from chromosomal defects leading to male infertility because of Y chromosome are passed on to all the offspring.
- Reports suggest that ICSI might induce de novo chromosomal anomalies, usually involving the sex chromosomes, but the phenotypes are normal.
- Mental development is also normal in children conceived through ICSI.
- Polygenic background appearing at birth as congenital anomalies.
- Assisted reproductive technology might also affect epigenetic characteristics of the male gamete and the female gamete or might have an impact on early embryogenesis.
- It might be also associated with an increased risk for genomic imprinting abnormalities.

The technique came into being when in 1978 the first "test-tube baby" was born. The successful birth of the child was the outcome of the collaboration between Oldham gynecologist Patrick Steptoe and two Cambridge doctors, Robert Edwards and Barry Bavister. They developed a technique by which an egg taken from a woman's ovary could be fertilized in a test tube and then returned to a womb to grow. Following this at 11:57 p.m., July 25, 1978, Louise Brown, the first test-tube baby, was born at a small local hospital in Oldham in the industrial north of England.

However, the future of IVF was in dark till fertility specialists gathered to discuss about it. Afterward in 1990 the "Human Fertilization and Embryology Act" came.

After that enormous number of people have benefited from the process, though many bioethics and biosafety issues surround it.

> In 1978 Steptoe and Edwards [9] reported the first birth resulting from IVF. Louise Brown entered the world in 1978. As the first baby born via in-vitro fertilization (IVF), she opened a whole new arena of hope to infertile couples, in addition to aggravating a hotbed of moral and ethical debate. (CNN 1999) [1]

15.5 Technique of IVF

IVF was initially performed with the single dominant ovarian follicle; however, it occurred to be inefficient with poor pregnancy rates. Afterward the protocols using "superovulation" with parenteral gonadotropins were adopted for the induction of multiple follicles. Multiple oocytes gave an opportunity to select best to transfer. Other embryos left might be cryopreserved for future embryonic transfer, eliminating the need for another in vitro fertilization [3].

15.5.1 Superovulation

Superovulation (for multiple follicle maturation) may be achieved by performing three basic steps:

- Prevention of spontaneous luteinizing hormone (LH) by suppression of the pituitary (requires 10–14 days), which otherwise leads to ovulation before the oocyte can be retrieved. This is achieved by subcutaneous injection of leuprolide acetate (analog of gonadotropin-releasing hormone) with dose of 0.5 mg once a day (OD) starting around cycle day 21.
- After pituitary suppression, administration of recombinant follicle-stimulating hormone (FSH) once a day with follicular monitoring through transvaginal ultrasonography and serum estradiol levels.
- After the lead follicle achieves a mature size of 18–20 mm, human chorionic gonadotropin is given subcutaneously for endogenous LH surge required for final oocyte maturation.

However, there are conflicting data regarding the benefits of older urinary gonadotropin products (FSH and LH) and the recombinant products.

15.5.2 Retrieval of Oocyte and IVF

- Oocytes are retrieved by transvaginal ultrasound-guided needle aspiration of follicular fluid after 34–38 h of delivery of human chorionic gonadotropin under sedation. The fluid is examined for the presence of oocytes with its cumulus mass of granulose cells. (Oocyte retrieved may be 10–20.)
- In the next step, oocytes are placed in the culture medium (based on human fallopian tube fluid) and incubated at 37 °C.
- After this 1–2 lakhs sperms are added in a small drop of media to the oocytes or by direct injection of a single sperm using intracytoplasmic sperm injection (ICSI).
- After 12–20 h fertilization is observed with male and female pronucleus in 65 % of oocytes (lower fertilization suggests defects in one or both the gametes).
- After 3 days of retrieval and fertilization, embryos are transferred into the uterus via a small flexible transcervical catheter (Fig. 15.2).

The implantation of embryo into the endometrium is critical and is the limiting stage for successful IVF. The transfer at blastocyst stage, that is, 5 days after oocyte retrieval, might result in successful implantation. The implantation may be done in the biological mother or surrogate mother (Fig. 15.3.). Best embryos are known to divide in culture condition [5] while arrested division is observed in embryos, which are of poor quality.

Unfortunately avoiding multiple pregnancies is problematic; thus, the American Society of Reproductive Medicine guidelines recommended the implantation of one to two embryos in women under 35 years of age, three embryos in women 35–37 years, four in women 38–40 years, and five in women over 40 years, although fewer embryos may be transferred. The ultimate goal is to transfer a single embryo. Exogenous progesterone is given for endometrial receptivity and is continued until gestational week 8–10.

A series of subcutaneous injections of human chorionic gonadotropin can be used to increase endogenous progesterone production, but this is associated with an increased risk of ovarian hyperstimulation syndrome.

15.5.3 Drawbacks of IVF

IVF is (1) expensive and (2) inconvenient and (3) has high percentage of multiple pregnancies, (iv) neurological sequelae [6], and (v) multiple birth defects [6]. However, the incidences match the natural process of child birth [4]. In the trials, IVF is associated with multiple gestation, twins in 31 % participants, triplets in 6 %, and higher multiples induced in 0.2 %.

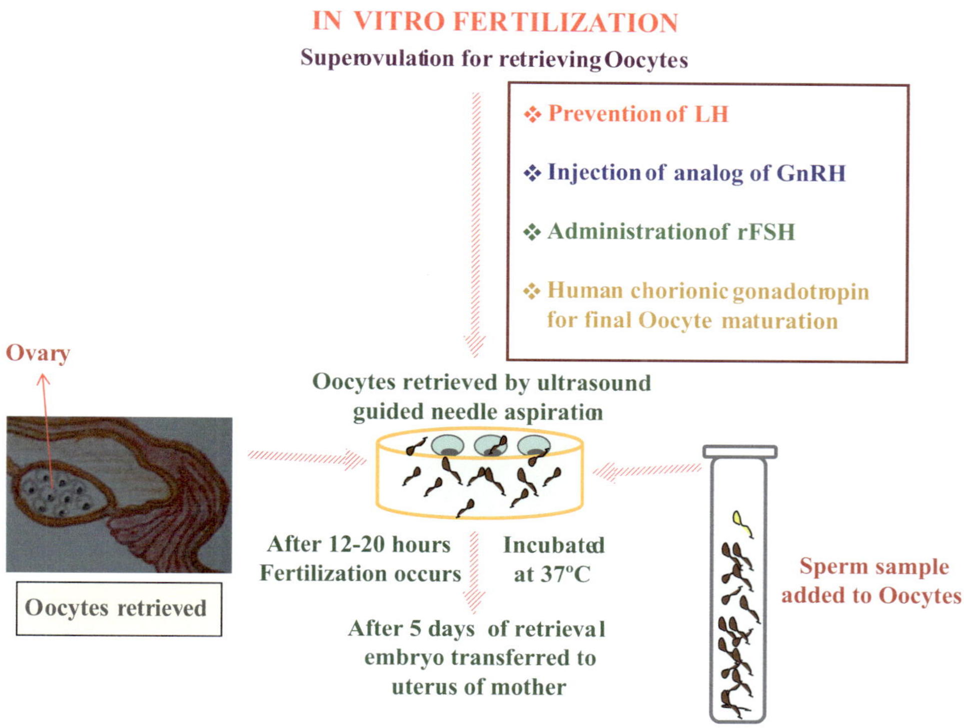

Fig. 15.2 The process of in vitro fertilization. Initially superovulation is performed with hormonal therapy as shown. After that mature oocytes are retrieved by ultrasound-guided needle aspiration. They are then maintained on culture medium which is based upon human fallopian tube fluid. They are incubated for 37 °C. Simultaneously semen with sperm is collected from the male partner or from the sperm bank. One to 2 lakhs sperms are added to oocytes. After 12–20 h, after fertilization is observed or after 3–5 days of retrieval, the embryo is transferred to the uterus by a small needle

15.5.4 Risk of Ovarian Hyperstimulation Syndrome

The highest medical risk to women undergoing IVF is life-threatening ovarian hyperstimulation syndrome. Mild ovarian hyperstimulation syndrome leads to lower abdominal discomfort about 1 week after oocyte retrieval; it is common due to the presence of multiple functional cysts. These cysts may resolve spontaneously within 2 weeks or may last for several weeks with pregnancy. These are produced due to vascular permeability leading to the loss of intravascular fluid into the third space resulting in electrolyte imbalances, prerenal oliguria and renal failure, tense ascites with respiratory compromise, and, rarely, adult respiratory distress syndrome and hypercoagulability with thromboembolism. Ovarian hyperstimulation syndrome is self-limited, and treatment is supportive until spontaneous resolution occurs, usually by 8–10 weeks of gestation.

15.6 Developments in Assisted Reproductive Technology

Many techniques involving micromanipulation of sperm and embryos are in use for successful IVF with minimal risks; these include intracytoplasmic sperm injection (ICSI), assisted embryo hatching, preimplantation genetic diagnosis (PGD), in vitro maturation (IVM), and oocyte and ovarian tissue cryopreservation. However, technical expertise is required to continue the process with minimal time lag between steps [8].

15.6.1 Assisted Hatching

Embryo hatches at the stage of blastocyst by dissolving the zona pellucida for implantation in the endometrium. Zona pellucida, which is an acellular matrix of glycoproteins, carbohydrates, and zona-specific proteins surrounding the oocyte

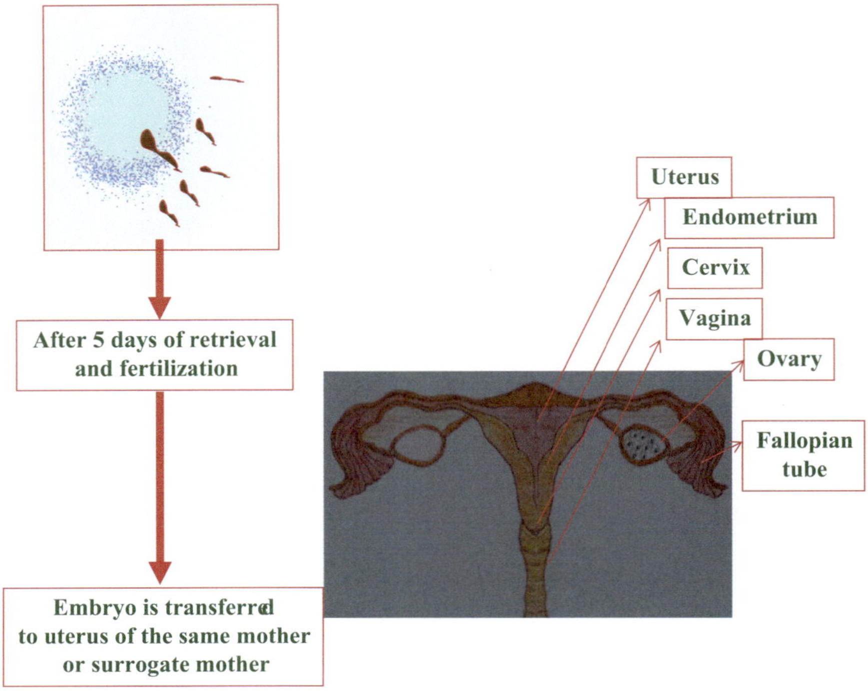

Fig. 15.3 The process of fusion of sperm and ovum. After 3–5 days of retrieval of ovum, the embryo is transferred to the uterus of the mother or surrogate mother. However, implantation into the endometrium is a critical step in the process of in vitro fertilization

and early embryo, plays an important role in sperm binding and fertilization, prevents polyspermic fertilization, and aids compaction of the blastomeres. Several techniques like partial zona dissection, thinning an area with acid Tyrode's solution, noncontact infrared diode laser, or piezoelectric micromanipulator have been developed to assist hatching and improve IVF pregnancy rates. However with assisted hatching, no significant improvement was observed limiting its use in routine practice.

15.6.2 Preimplantation Genetic Diagnosis

In this analysis, embryos are tested for the presence of genetic anomalies at 3–5 days after fertilization, and only unaffected embryos are transferred to the maternal uterus. The technique offers good prospects to couples at risk for conventional prenatal diagnosis [7].

Prenatal diagnosis may be recommended:

- To couples with history of congenital diseases
- Aged couples (>35 years older female)
- IVF along with ICSI

15.6.3 In Vitro Maturation (IVM) of Oocytes

IVF, which involves hormones to stimulate oocytes to mature, is associated with risks of multiple pregnancy, ovarian hyperstimulation, inconvenience, and monitoring. With IVM, the maturation process occurs in vitro. Therefore, using oocytes in the ovary, which are arrested in meiotic prophase I until ovulation and require supplementation of endogenous or exogenous FSH, stimulates the follicles to reach maturation. The process of "maturing" allows fertilization of the oocytes to occur. The final maturation process is under the influence of LH and results in the completion of meiosis I and formation of the first polar body. Meiosis II then proceeds to metaphase II, the stage at which the oocytes are typically retrieved for conventional IVF.

The results would improve if numerous immature oocytes are aspirated from unstimulated ovaries followed by IVM, which would reduce the cost by avoiding gonadotropins. The indications for IVM are:

- Patients with polycystic ovary syndrome (are at higher risk for ovarian hyperstimulation)
- Poor responders or patients wishing to avoid gonadotropins (risk of side effects of gonadotropins induced breast cancer)
- Oocyte donors
- Cancer patients who wish to preserve their fertility (cancer therapy might make them sterile)

After maturation, a single sperm is injected by ICSI. The pregnancy rates are comparable to conventional IVF.

15.6.4 Cryopreservation of Oocyte

It would be an achievement if unfertilized oocytes may be frozen in natural conditions without developing any anomalies. The advantages of oocytes freezing would be many:

- Female in their 20s may cryopreserve their own oocytes, and after their career goals are achieved, they can achieve motherhood.
- Females treated for cancer who are facing the risk due to damage of ovarian failure due to radiation or chemotherapy.

The oocytes are cryopreserved by vitrification process, exposed to cryoprotectants, plunged into liquid nitrogen, and transplanted back into the host (autotransplantation) for cycles of IVF.

Limitations: With cryopreserved oocytes, pregnancy is rarely achieved; conditions for freezing, maintenance, and thawing should be excellent; There is risk of development of cancer.

Surrogacy: Is the arrangement where a woman agrees to become pregnant and gives birth to the child. After the child is born, it is handed over to its parents or the person concerned. The surrogate mother may be the nonbiological mother of the child or some other woman's womb may be used for someone else's baby. Medically surrogacy is an option for the woman who cannot bear the embryo for implantation, gestation, and birth.

Artificial insemination: When the sperms of the husband or donor (in case of male infertility) are placed into the female's uterus or cervix using artificial means, it is called artificial insemination.

Gamete intrafallopian transfer: This technique is used when the woman who wants to have the child of her own has at least one normal fallopian tube. Eggs are placed in this tube along with a man's sperm to fertilize there.

Ectopic pregnancy: When implantation of embryo occurs outside the womb, it is referred as ectopic pregnancy. Though IVF is done very carefully, embryo takes some time to attach to the womb and thus may get implanted outside the womb (to fallopian tubes).

15.7 Ethical Issues in IVF

1. The whole process of IVF is unnatural: The methodology bypasses the natural method. The new life is created in the laboratory.

   Reasoning: It aims to overcome the problems associated with normal child birth as every woman wants to have child of her own.
2. There are many risks associated with IVF like multiple pregnancy and genetic defects for the newborn.

   Reasoning: In natural process also the mutations occur and abnormalities are there, but IVF once completely established probably would be able to bypass these limitations.
3. For IVF, single females may also approach thus, what about the marital relationship, and it also involves process of masturbation in males.

   Reasoning: For the couples who cannot have a child of their own, IVF is an option to experience parenthood. Masturbation is part of the process whose aim is novel for collection of sperms.

4. For childless couples, adoption is a better solution.

   Reasoning: If the couples will not be able to give the love and affection to the orphan child, then it would become even more problematic; in that case, IVF is a practical solution, and the care of orphans would be open to the families who can love, support, and accept the child.

5. The technique is expensive, it is not affordable for many, embryo is exposed to unnatural environment, and unused embryos are destroyed.

   Reasoning: The case of infertility should be treated as disease.

15.8 Future Prospects

It is a very useful technique. The limitations are gradually being overcome day by day. The technique with IVM, ICSI, and PGD may allow for higher and error-free pregnancies. The disadvantages because of routine IVF may be resolved like ovarian hyperstimulation syndrome. Successful cryopreservation and retrieval would be a boon as in any condition oocyte may be preserved and used afterward.

15.9 Chapter End Summary

- IVF has become the treatment option for all causes of infertility.
- If the conventional methods fail, then it offers an opportunity to childless couples to have child of their own.
- In this technique, the females are given hormonal therapy for hyperovulation. Then oocytes are retrieved from the females through aspiration. They are maintained in culture medium and then sperms (1–2 lakhs) are added. After fertilization, at 5 days upon retrieval, the embryo is placed in the body of either biological mother or surrogate mother.
- The technique is expensive and leads to high rate of multiple pregnancies.
- There is risk of ovarian hyperstimulation.
- However, the advances in the technology are being developed to take care of the present risks involved with IVF and make it a safe technique of the future.

Multiple Choice Questions

1. Male infertility is due to:
   (a) Testicular defects
   (b) High sperm count
   (c) High sperm motility
   (d) All of these
2. Female infertility is due to:
   (a) Problems associated with the hypothalamus and pituitary
   (b) High follicle-stimulating hormone
   (c) High estrogen, progesterone, and amenorrhea
   (d) All of these
3. Sperm donor is helpful in cases of:
   (a) Azoospermia
   (b) Presence of anti-semen antibody
   (c) Decreased sperm motility
   (d) None of these
4. A risk with ICSI might lead to:
   (a) Inferior embryo
   (b) Aneuploid embryo
   (c) Defects in the Y chromosome are passed on without selection
   (d) All of these
5. Superovulation means:
   (a) Generation of superior ovum
   (b) Generation of multiple follicles
   (c) Generation of ovum without meiosis
   (d) All of the above
6. Ovarian hyperstimulation syndrome is due to:
   (a) Female infertility

(b) In vitro maturation
(c) Treatment of gonadotropins
(d) All of these

7. In vitro maturation is the technique of choice in:
   (a) Polycystic ovarian syndrome
   (b) Cancer patients
   (c) In skipping gonadotropin therapy
   (d) All of the above
8. Cryopreservation of oocytes might be helpful in:
   (a) Storage improves quality of oocytes, which may be used afterward
   (b) Patients suffering from disease can donate their oocyte for later use
   (c) Cryopreservation results in generation of designer babies
   (d) All of the above

Answers:
1. (a); 2. (a); 3. (a); 4. (d); 5. (b); 6. (c); 7. (d); 8. (b)

Review Questions

Q1. What do you understand by infertility?
Q2. What is IVF?
Q3. What is ovarian hyperstimulation syndrome?
Q4. What are assisted reproductive technologies?
Q5. How would cryopreservation be helpful in the IVF?

References

1. CNN (1999) Financing infertility by Nelson R Copyright 1999 WebMD edition.cnn.com/HEALTH/women/9905/19/financing.infertility/
2. Fertility: assessment and treatment for people with fertility problems. NICE clinical guideline CG156 – issued: Feb 2013
3. Goldberg JM, Falcone T, Attaran M (2007) In vitro fertilization update. Cleve Clin J Med 74:329–338
4. Hart R, Norman RJ (2013) The longer-term health outcomes for children born as a result of IVF treatment. Part II-mental health and development outcomes. Hum Reprod Update 19:244–250
5. Kirkegaard K, Agerholm IE, Ingerslev HJ (2012) Time-lapse monitoring as a tool for clinical embryo assessment. Hum Reprod 27:1277–1285
6. Lambert RD (2002) Safety issues in assisted reproduction technology. The children of assisted reproduction confront the responsible conduct of assisted reproductive technologies. Hum Reprod 17:3011–3015
7. Mastenbroek S, Twisk M, Van Der Veen F, Repping S (2011) Preimplantation genetic screening: a systematic review and meta-analysis of RCTs. Hum Reprod Update 17:454–466
8. Nargund G (2009) Natural/mild assisted reproductive technologies: reducing cost and increasing safety. Womens Health 5:359–360
9. Steptoe PC, Edwards RG (1978) Birth after the reimplantation of a human embryo. Lancet 2:366

Some Selected Resources

www.ivf.com/ivffaq.html
www.mayoclinic.com/health/in-vitro-fertilization/MY01648
www.nlm.nih.gov/medlineplus/ency/article/007279.htm
www.webmd.com/infertility-and-reproduction/.../in-vitro-fertilization

16 Stem Cell Biology and Its Clinical Application

Abstract

Stem cells are cells which regardless of their source have the properties of division and renewing for long periods through mitotic cell division, are dedifferentiated (unspecialized) with unique capacity to produce unaltered daughter cells (self-renewal), and generate specialized cell types (potency) under certain physiologic or experimental conditions. Stem cells are the "building blocks" for every type of cell in the body, capable of maturing into any tissue type including the pancreas, blood, or neuronal cells, i.e., they are master cells from which all specialized cells and tissues in our body are derived under certain physiologic or experimental conditions. These unspecialized cells have two important characteristics that distinguish them from other cells in the body. First, they can replenish their numbers for long periods through cell division. Second, after receiving certain chemical signals, they can differentiate or transform into specialized cells with specific functions, such as a heart cell or nerve cell. Stem cells play an important role of repair and replace of damaged cells in our body. Identification of stem cells requires the separation and purification of cells, usually based on a combination of specific cell surface markers. Stem cell therapy has the potential to dramatically change the treatment of human diseases. With the discovery of the processes like induced pluripotency and hope of using somatic cells for therapy, the world is waiting to treat the disorders for which no cure exists.

16.1 Introduction

Stem cells are cells which regardless of their source have the properties of division and renewal for long periods through mitotic cell division, are dedifferentiated (unspecialized) with unique capacity to produce unaltered daughter cells (self-renewal), and generate specialized cell types (potency) under certain physiologic or experimental conditions (Fig. 16.1). Stem cells are the "building blocks" for every type of cell in the body, capable of maturing into any tissue type including the pancreas, blood, or neuronal cells, i.e., they are master cells from which all specialized cells and tissues in our body are derived under certain physiologic or experimental

V. Gupta et al., *Basic and Applied Aspects of Biotechnology*, DOI 10.1007/978-981-10-0875-7_16

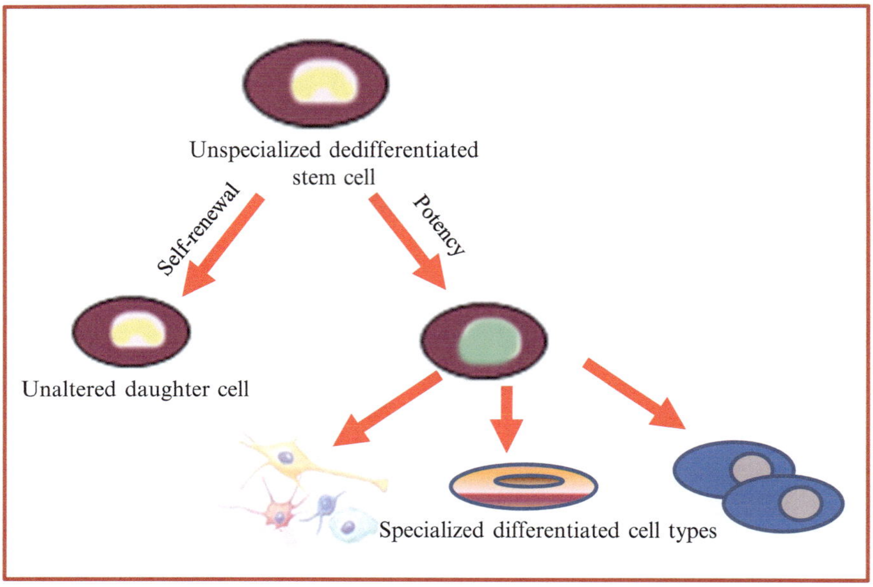

Fig. 16.1 The basic properties of stem cells which are self-renewal and potency. Due to self-renewal capability, the stem cells maintain its number and generate specialized cell types by property of potency

conditions. These unspecialized cells have two important characteristics that distinguish them from other cells in the body. First, they can replenish their numbers for long periods through cell division. Second, after receiving certain chemical signals, they can differentiate or transform into specialized cells with specific functions, such as a heart cell or nerve cell. Stem cells play an important role of repair and replace of damaged cells in our body [9]. Identification of stem cells requires the separation and purification of cells, usually based on a combination of specific cell surface markers. However, the lack of specific cell surface markers for other types of stem cells has made it difficult to isolate them in large quantity.

16.2 Stem Cell Classification

Totipotent stem cells (derived from early embryos) can form an entire organism autonomously. A fertilized egg (zygote) and cells produced in the first few divisions of the fertilized egg are totipotent stem cells (Fig. 16.2).

Pluripotent stem cells (PSC) are descendants of the totipotent stem cells of the embryo (undifferentiated inner cell mass of blastocyst) and can form almost all the cell lineages (endoderm, mesoderm, and ectoderm), including germ cells (Fig. 16.3).

Multipotent stem cells are descendents of pluripotent stem cells (derived from the fetal tissue, cord blood, cord tissue matrix, and adult tissue). They are antecedents of specialized cells in particular tissues which can produce only cells of a closely related family of cells but cannot form all the cell lineages (e.g., hematopoietic stem cells present in the bone marrow, differentiate into red blood cells, white blood cells, platelets) of the body. Their differentiation ability is limited but they are promising tools for cell-based therapies.

Oligopotent stem cells can form more than one cell lineage but are more restricted than multipotent cells. Oligopotent cells are sometimes called progenitor cells or precursor cells; however, these terms are often more strictly used to define partially differentiated or lineage-committed cells (as myeloid progenitor cells) that can divide into different cell types but lack self-renewing capacity (Fig. 16.3).

Unipotent cells or *monopotent cells* can produce only one cell type, but have the property of

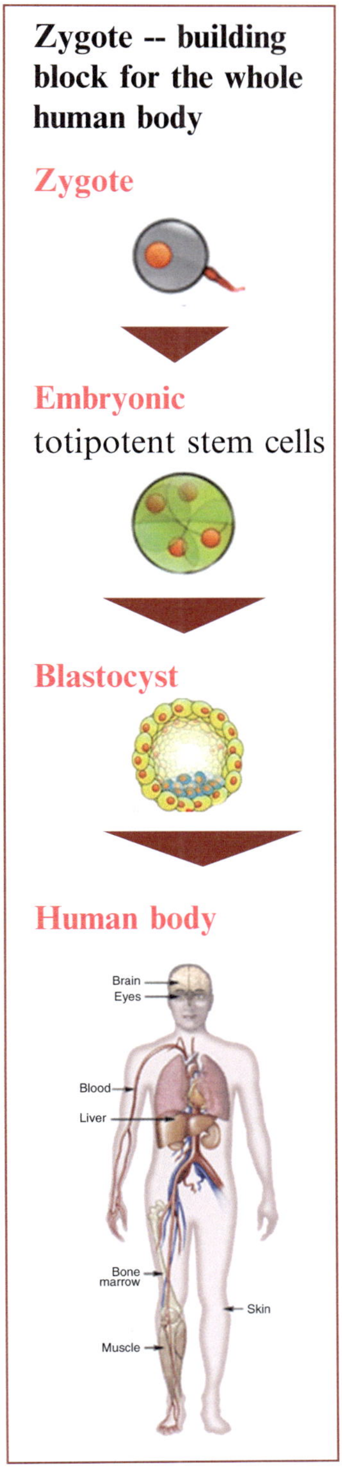

Fig. 16.2 The capability of zygote to create a complete body. The zygote is the only totipotent cell in animals

self-renewal, for example, spermatogonial stem cells, but maintain the same cell type (Fig. 16.4).

The types of mammalian stem cells are embryonic stem cells, adult stem cells, and induced pluripotent stem cells. These have the properties of self-renewal and are able to differentiate into specialized cells upon induction.

16.2.1 Embryonic Stem Cells

Embryonic stem cells (ESCs) are totipotent cells present in blastocysts (4–5-day-old embryo and consisting of 50–150 cells) and are cultures of cells derived from the epiblast tissue of the inner cell mass (ICM) or earlier morula stage and require specific signals for correct differentiation into all of the specialized embryonic tissues. During development, ES cells can give rise to all the cell types of the three primary germ layers: ectoderm, endoderm, and mesoderm. In other words, they can develop into each of the more than 200 cell types of the adult body when given sufficient and necessary stimulation for a specific cell type (refer to Fig. 16.2).

The human ESC is characterized by the presence of several transcription factors and cell surface proteins. The transcription factors *Oct-4*, *Nanog*, and *SOX2* form the core regulatory network that ensures the suppression of genes that lead to the differentiation and the maintenance of pluripotency. These stem cells hold great promise for treating many diseases like diabetes, Parkinson's disease, Alzheimer's disease, spinal cord injury, heart failure, and bone marrow failure and understanding early human development.

Identification of Embryonic Stem Cell

Identification of embryonic stem cells is important. The laboratories which maintain stem cells use these tests:

- Embryonic stem cells can be maintained undifferentiated in culture conditions by subculturing for many months.

(continued)

- They express unique transcription factors such as *Oct-4*, *Nanog*, and *SOX2*, which maintain them in undifferentiated conditions with self-renewal property. Detection of these (especially *Oct-4*) help to establish identity of the cells.
- They should be checked for revival after freezing and thawing.
- Their chromosomes are examined under the microscope for damage or change of chromosome number.
- Test for pluripotency: When embryonic stem cells are injected in an immunosuppressed mouse, they form a benign tumor called teratoma. Teratoma has various differentiated or partially differentiated cells. This shows that embryonic stem cells can differentiate into multiple cell types, i.e., they are pluripotent.

16.2.2 Adult Stem Cell

Multipotent adult stem cells are present in adult tissues including that of brain, skin, and skeletal muscle stem cells. In adult organisms, stem cells and progenitor cells act as a repair system for the body, replenishing specialized cells, but also maintain the normal turnover of regenerative organs, such as the blood, skin, or intestinal tissues (Fig. 16.3). The term adult stem cell refers to any cell, which is present in different organs in a developed organism that has two properties: the ability to divide and create another cell like itself and also divide and create a cell more differentiated than itself. A great deal of adult stem cell research has focused on clarifying their capacity to divide or self-renew indefinitely and their differentiation potential. Adult stem cells are somatic cells that are being explored to treat leukemia and related bone/blood cancers through bone marrow transplants. Adult stem cells like mesenchymal autologous bone marrow-derived stem cells (BMCs) [2], human myoblasts, and peripheral blood-derived stem and progenitor cells are used for treatment of end-stage heart failure. Adult stem cells are also used in the bone and cartilage repair and wound healing.

Multipotent stem cells are also present in amniotic fluid. Amniotic stem cells are multipotent and non-tumorigenic and can differentiate into many lineages.

16.2.3 Induced Pluripotency

They are reprogrammed cells given pluripotent capabilities. Using genetic reprogramming with transcription factors, pluripotent stem cells equivalent to embryonic stem cells have been derived from human adult skin tissue. In the 2006 study, the usage of four transcription factors such as *Oct-4*, *Sox2*, *Klf4*, and *c-Myc* was capable of reprogramming mouse fibroblast cells to embryonic cell-like state in several weeks of culture, where the cells were induced to express genes for maintenance of ESCs [14]. However, the usage of *c-Myc* has the risk of tumor growth in some cases limiting the usage of induced pluripotent stem cells (iPSC) in therapy, and the factor being dispensable could be easily displaced. Induced pluripotent stem cells were highly similar to ESCs by either these four genes or a combination of *Oct-4*, *Sox2*, *Nanog*, and *Lin28*. Human keratinocytes respond better than human fibroblast cells for iPSC production [19, 20].

Nuclear Reprogramming Certain conditions, which are responsible for changes in the expression of nuclear genes affecting the differentiated status of completely differentiated cell, are referred as nuclear reprogramming. Nuclear reprogramming results in generation of an unspecialized dedifferentiated pluripotent or progenitor cell. The cell completely loses its differentiation status or can be induced to switch to convert into cells of other lineages (transdifferentiation). Many transcription factors and appropriate conditions are required for reprogramming of differentiated somatic cells.

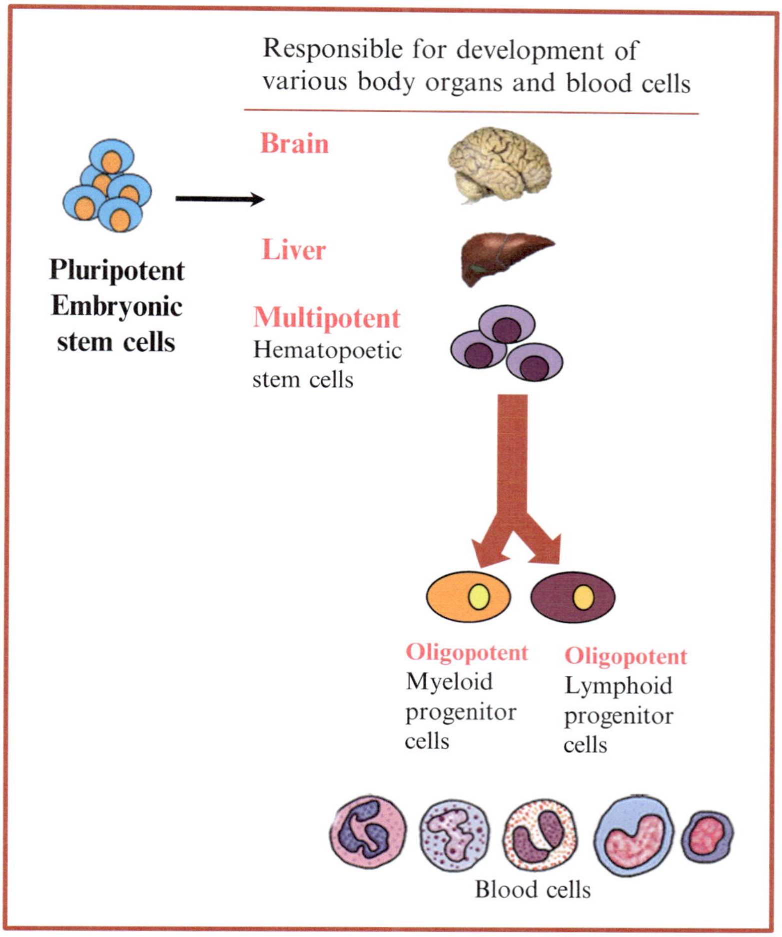

Fig. 16.3 The pluripotent embryonic stem cells which can give rise to the germinal layers, and subsequently they can form various body organs and blood cells

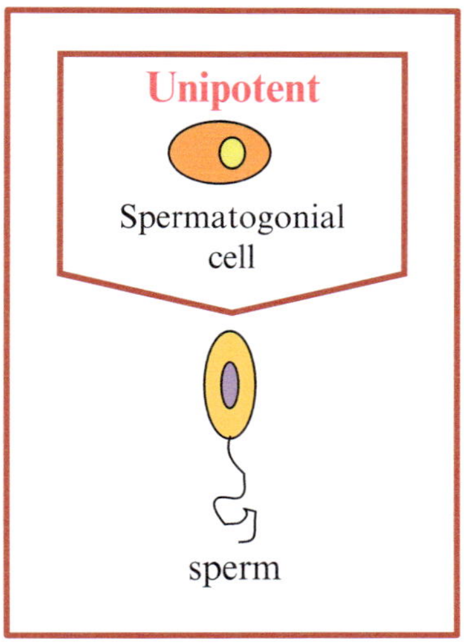

Fig. 16.4 The unipotent stem cell, which can give rise to only one kind of cell. Spermatogonial cell gives rise to the sperm

Though induced pluripotent stem cells (iPSCs) are derived from adult tissue [24], they demonstrate important properties of pluripotent stem cells and are very close to ESCs but are with reduced efficiency. iPSCs show the expression of stem cell markers, telomerase activity, transcriptional and epigenetic patterns resembling ESCs, and reactivation of pluripotency genes and can contribute to most of the different tissues when injected into mouse embryos at an early stage of development. iPSCs can differentiate into other cell types including their ability to contribute to the germline like ESCs [11, 25] but with the reduced efficiency and greater variability when differentiating into neural cells.

In adult stem cell therapy, bone transplant is used. Medical researchers may derive stem cell to treat cancer, Parkinson's disease, spinal cord injuries, ischemic encephalopathy, cerebral palsy, spinal cord injury, cardiomyopathy, amyotrophic lateral sclerosis, multiple sclerosis, and multiple damage. The concerns for using iPSCs are the patient safety issues and reduced efficiency requiring more research. The recent development of iPS cells has been called a bypass of the legal controversy of using ESCs [7].

16.3 Stem Cell Plasticity

Because of their combined abilities of unlimited expansion and pluripotency, embryonic stem cells remain a theoretically potential source for regenerative medicine and tissue replacement after injury or disease. It is believed that once a cell is differentiated, their phenotypes are stable. Most adult stem cells are lineage restricted (multipotent) and are generally referred to by their tissue origin (mesenchymal stem cell, adipose-derived stem cell, endothelial stem cell). However, there are a number of reports showing that tissue stem cells, which are thought to be lineage-committed multipotent cells, possess the capacity to differentiate into cell types outside their lineage restrictions known as *transdifferentiation* (Fig. 16.5). Hematopoietic stem cells (HSCs) can be converted into neurons as well as germ cells. This possibility may provide a means to use tissue stem cells derived directly from a patient for therapeutic purposes, eliminating the use of embryonic stem cells. Nevertheless, strict criteria and rigorous validation are required to establish tissue stem cell plasticity.

16.4 Stem Cell Division and Differentiation

The specific factors and conditions that allow stem cells to remain unspecialized is an important area of research. Understanding the signals in a mature organism causes a stem cell population to proliferate and remain unspecialized until the cells are needed for repair of a specific tissue. Scientists are just beginning to understand the signals inside and outside cells that trigger stem cell differentiation. The internal signals are controlled by a cell's genes, which are interspersed across long strands of DNA, and carry coded instructions for all the structures and functions of a cell. The external signals for cell differentiation include chemicals secreted by other cells, physical contact with neighboring cells, and certain molecules in the microenvironment. Still questions about stem cell differentiation remain as:

1. What are the internal and external signals for cell differentiation and are they similar for all kinds of stem cells?
2. What are the specific sets of signals that promote differentiation into specific cell types?
3. Unlike adult stem cells, embryonic, or pluripotent, stem cells are not restricted to any particular tissue or organ and are capable of producing all cell types. By studying how these cells develop into mature cells, such as those that make up our bone, blood, and skin, one can learn how those cells function and what goes wrong when they are diseased.

Nevertheless there are certain tests that measure the cell's fundamental properties (see identification of embryonic stem cells).

To ensure self-renewal, stem cells undergo two types of cell division. Symmetric division, which gives rise to two identical daughter cells both endowed with stem cell properties. Asymmetric division, on the other hand, produces only one stem cell and a progenitor cell with limited self-renewal potential. Progenitors can go through several rounds of cell division before terminally differentiating into a mature cell.

It is possible that the molecular distinction between symmetric and asymmetric divisions lies in differential segregation of cell membrane proteins (such as receptors) between the daughter cells. Stem cells remain undifferentiated due to environmental cues in their particular niche. Stem cells differentiate when they leave that niche or no longer receive those signals, for example, studies in *Drosophila* germarium have identified the signals *dpp* and *adherens* junctions that prevent germarium stem cells from differentiating.

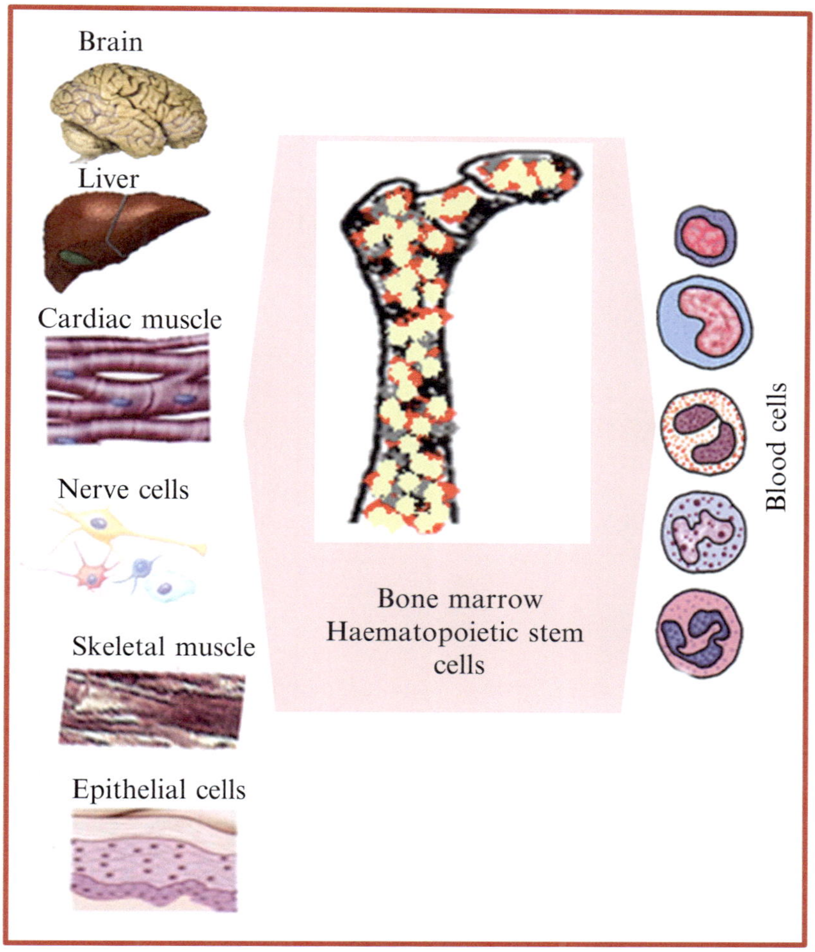

Fig. 16.5 Transdifferentiation of hematopoietic stem cells into various body parts and cells

Case Study

Tremendous success is obtained in using human embryonic stem cells (ESCs) to treat laboratory animals. The research is struggling hard to find its way out of the laboratory to the patients. A success story of ESCs was curing macular degeneration and other causes of blindness, repairing heart damage, treating type I diabetes and treating strokes in mice, restoring blood flow to "ischemic" limbs in rats, and restoring a normal rhythm to damaged hearts in pigs. However, due to ethical issues, there is restriction on ESCs and at many places government agencies are not funding research with ESCs.

Geron Corp. was expected to start implanting neural cells derived from human ESCs into patients with spinal cord injuries. The trial was put on hold, but fortunately FDA notification enables Geron to move forward with the world's first clinical trial of a human embryonic stem cell (hESC)-based therapy in man. FDA had granted clearance of the company's investigational new drug (IND) application for

(continued)

GRNOPC1 in patients with acute spinal cord injury.

The phase I multicenter trial was designed to establish the safety of GRNOPC1 in patients with "complete" American Spinal Injury Association (ASIA) Impairment Scale grade A subacute thoracic spinal cord injuries. Unfortunately the company decided to pull off entirely from stem cell business. Geron's withdrawal leaves Advanced Cell Technology as the only company now conducting a clinical trial involving human embryonic stem cells (*The New York Times*).

Age-related macular degeneration has no treatments available and has two predominant forms, wet and dry. Dry AMD accounts for almost 90 % of all cases and includes a breakdown or thinning of the layer of retinal pigment epithelial (RPE) cells in the patient's macula, the region at the center of the retina responsible for high acuity vision [16]. Over time, the progressive loss of RPE cells and accompanying loss of photoreceptors can cause severe vision loss and even blindness. Advanced Cell Technology (ACT), Inc. is being allowed, and the US Food and Drug Administration (FDA) has cleared the company's investigational new drug (IND) application to treat dry age-related macular degeneration (AMD) using RPE cells derived from human embryonic stem cells (hESCs). ACT is now permitted to initiate a phase I/II multicenter clinical trial to treat patients with dry AMD, the most common form of macular degeneration in the world.

In 2014, Ocata Therapeutics (formerly Advanced Cell Technology, Inc.) announced positive outcomes of usage of human embryonic stem cells (hESCs) from its early-stage clinical trial on 18 patients for the treatment of dry age-related macular degeneration and Stargardt disease. Ten out of 18 subjects showed vision improvements (with some participants reporting dramatic improvements) and halted the progression of the disease in 17 of the 18 subjects.

16.5 Therapeutic Roles of Stem Cells

In particular, embryonic cell lines, autologous embryonic stem cells generated through therapeutic cloning, and highly plastic adult stem cells from the umbilical cord blood [10] or bone marrow [12] are promising candidates for stem cell therapy [17]. Stem cell research is the promising area in which researchers can investigate the possibility of cell-based therapies to treat disease, which is often referred as regenerative/reparative medicine where they can potentially lead to treatments for diabetes, cancers, heart diseases, blood diseases, Parkinson's disease, multiple sclerosis, and Alzheimer's disease [8]. Stem cells have tremendous potential to improve human health. They can be developed as cell-based therapies and can be targeted in tumors to provide new cancer therapies. They can be used to provide improved understanding of disease mechanisms. Research on stem cells promises to lead to innovative cell transplantation therapies. It can also have a major impact on understanding and improving the regenerative capacity of the human body, to achieve knowledge about how an organism develops from a single cell and how healthy cells replace damaged cells in adult organisms.

However after trials are approved using embryonic stem cells, they being totipotent cells, require specific signals for correct differentiation. Differentiating ES cells into usable cells while avoiding transplant rejection is just a few of the hurdles that embryonic stem cell researchers still face.

16.6 Tissue Development and Disease

A number of experiments over the last several years have shown the plasticity of stem cells, which include blood cells becoming neurons [13], liver cells that can be made to produce insulin, and hematopoietic stem cells that can develop into heart muscle. Therefore, exploring the possibility of using adult stem cells for cell-based therapies has become a very active area of investigation by researchers.

The study of cancer stem cells could lead to insights into cancer cell biology. Recent studies indicate that cancers are continually replenished by a small population of cancer stem cells that are capable of self-renewal. Researchers are working to identify and isolate the rare cancer and leukemia stem cells from the majority of human cancers and leukemia's and discover the cancer genes in these cancer stem cells. By studying the genes involved in self-renewal of adult stem cells, it may be possible to identify new molecular targets for drug and immune therapies, thereby developing new therapeutic approaches to killing cancer stem cells with the goal of moving these findings into clinical trials.

16.7 Stem Cell Replacement

Stem cells offer the possibility of a renewable source of cell replacement for all organs. Therapeutic cell replacement may be achieved by:

1. Injection of stem cells directly into the damaged organ or into circulation (bone grafts, skin grafts)
2. In vitro differentiation of stem cells followed by transplantation into a damaged organ, for example, pancreatic islet cells for diabetes whereas cardiomyocytes to treat ischemic heart disease
3. Stimulation of endogenous stem cells to facilitate repair

16.8 Regenerative Therapies

- Different stem cells include embryonic stem cells (ESCs), umbilical cord blood stem cells, organ-specific somatic stem cells, and somatic stem cells, which can generate cell types specific for the required target rather than the donor organ.
- ESCs tend to develop chromosomal abnormalities and can form teratomas if they are not committed to the desired cell types before transplantation. Somatic cell nuclear transfer may be an alternative method for creating ES cell lines that are genetically identical to the patient.
- With increased understanding of the ability of these tissue or organ-specific stem cells to replenish or repair damaged or congenitally abnormal tissues or organs, the tissue-specific stem cells may one day be used to replenish cells damaged by Parkinson's disease, Alzheimer's disease, multiple sclerosis, or diabetes.
- Another interesting goal of stem cell research is to use embryonic stem cell for better understanding and treatment of diseases, which are lineage restricted. This might give us a clue regarding their development into mature cells of our body along with their differentiation, function, and their possible role in diseases. Their basic understanding might offer therapeutic solutions to the existing medical problems.
- Adult stem cells may be used in veterinary medicine to treat tendon and ligament injuries in horses. The use of adult stem cells in research and therapy is not controversial as embryonic stem cells, because the production of adult stem cells does not require the destruction of an embryo. Additionally, in some instances, adult stem cells can be obtained from the same patient (an autograft), thereby minimizing the risk of rejection.
- Mesenchymal stem cells (MSCs) are multipotent mesenchymal stromal cells, which are self-renewing. They can be found in almost all postnatal organs and tissues but can be isolated from the bone marrow and also from adipose tissue, umbilical cord blood, compact bone, and other tissues. The main functional characteristics of MSCs are their immunomodulatory ability, capacity for self-renewal, and differentiation into tissues of mesodermal origin. Previous studies have shown that MSCs are able to differentiate into several cell types, including cardiomyocytes, vascular endothelial cells, neurons [3], hepatocytes, epithelial cells, and adipocytes, making them a potentially important source for the treatment of debilitating human diseases. Such

multipotent differentiation characteristics coupled to their capacity for self-renewal and capability for the regulation of immune responses [1, 23], described MSCs as potentially new therapeutic agents for treatment of the complications of diabetes mellitus (DM).

- Pluripotent stem cells have already been used experimentally to treat mice with diabetes. A set of growth factors are discovered that induced pluripotent stem cells to develop into insulin-producing cells normally found in the pancreas. When they implanted these cells into the diabetic mice, the implanted cells produced insulin in a biologically normal way and treated the diabetes. This work is in investigational stage, but has tremendous scope for providing cure for diabetics.
- In bone marrow transplants, blood-forming stem cells regenerate the blood of transplant recipients. The use was made of purified stem cells rather than whole bone marrow taken from the patient before chemotherapy. This avoided re-injecting patients with their own cancer cells. Isolating adult stem cells from a variety of tissues in addition to the blood and brain stem cells could also help in other areas of cancer treatment. Thus damage from radiation can be reduced by supply of stem cells.
- Pluripotent stem cells, like adult brain stem cells, might also replace nerves damaged in spinal cord injuries or cells lost in neurodegenerative diseases. Parkinson's disease (PD) is a very common neurodegenerative disorder that affects more than 2 % of the population over 65 years of age and is caused by a progressive degeneration and loss of dopamine-producing neurons, which leads to tremor, rigidity, and hypokinesia (abnormally decreased mobility). Studies have shown that mouse embryonic stem cells were directed to differentiate into dopaminergic neurons due to introduction of Nurr 1 gene. These dopaminergic neurons when transplanted into the murine Parkinson's disease model reinnervated the murine brain with dopamine release, improving the motor function. The successful generation of an unlimited supply of dopamine neurons could make neurotransplantation widely available for Parkinson's patients at some point in the future.
- Studying mouse pluripotent stem cells carrying disease-causing mutations has already greatly enhanced scientific and medical knowledge of how genetic diseases develop. This knowledge may be applied to further this understanding by studying human pluripotent stem cell lines carrying mutations found in such genetic disorders such as cancer, Parkinson's disease, Alzheimer's disease, Lou Gehrig's disease, adult and juvenile diabetes, autoimmune diseases, allergic disorders, and early onset heart and cardiovascular disease. By studying how specific genetic mutations cause a cell to become diseased and how the proteins made by the mutated genes fail to function properly, researchers hope to generate drugs or therapies that make up for the genetic defects behind many diseases.

16.9 Disease-Specific Stem Cell Approach

Diabetes mellitus- Islet cells and pancreas transplantation for type I diabetes is cell-based approach. Successful therapy would depend on developing the source of cells that can be amplified and have the ability to synthesize, store, and release insulin when it is required primarily in response to changes in the glucose level. However on one hand, the proliferative capacity must be tightly regulated to avoid excessive expansion of β-cells, and on the other hand, the rate of apoptosis must be controlled to prevent loss of cells [18, 21].

Cautions of Stem Cell Therapy

"Hemacord" is the only FDA-approved stem cell product. Hemacord (a cord blood-derived product) is manufactured by the New York Blood Center, which is used in patients with disorders affecting the body's blood-forming system. Biologics license application (BLA) for cord blood hematopoietic progenitor cells, manufactured by Clinimmune Labs, University of Colorado Cord Blood Bank, has also been granted for cord blood product.

However patients need to be beware of the unscrupulous providers of stem cell treatments that are illegal and potentially harmful. FDA is concerned about the patients hoping the cure of their diseases of which cure is not available may leave them vulnerable to these stem cell providers and cautions the patients to consider the approval of stem cell treatment.

Considering even autologous stem cell treatment in the USA, patient should enquire:

- Physician for FDA approval for therapy.
- Is he (patient) a part of an FDA-regulated clinical study?
- In autologous cells, there are safety risks, including risks introduced when the cells are manipulated after removal.
- Incorrect placement of stem cells in the body can result in tumors.

In other countries also, safety, effectiveness, and regulatory approval of the stem cell product and therapy should be sought (www.fda.gov/).

16.9.1 Nervous System

Human embryonic stem cells can be induced to generate neural stem cells to give rise to neurons, oligodendroglia, and astrocytes. Transdifferentiation of the bone marrow and adipose stem cells into neural stem cells has been reported. Neurologic disorders have been targeted for stem cell therapies including spinal cord injury, amyotrophic lateral sclerosis [15], stroke, traumatic brain injury, and Parkinson's disease. Both embryonic stem cells and bone marrow-derived stem cells are able to facilitate remyelination after experimental spinal cord injury.

16.9.2 Liver

Transplantation of hepatocyte demonstrates that it can potentially substitute for organ transplant.

16.9.3 Heart Disease

The heart has the ability to achieve low levels of cardiomyocyte regeneration. The regeneration is accomplished by cardiac stem cells resident in the heart. The isolation, characterization, and amplification of these ex vivo might provide an ideal source of stem cells for therapeutic use.

However, in a number of experiments over the last several years regarding the *plasticity* and *transdifferentiation* of stem cells, researchers aim to develop new medical strategies capable of extending the capacity for growth and healing present in embryos into later stages of life [4]. Medical researchers believe that stem cell therapy has the potential to dramatically change the treatment of human disease. A number of adult stem cell therapies already exist; in the future, medical researchers anticipate being able to use technologies derived from stem cell research to treat a wider variety of diseases [22] (Table 16.1).

Table 16.1 The potential uses of stem cell therapy in various regenerative therapies and various disease states

Pathological condition	Stem cell therapy approach
Lymphomas and malignancies	Autologous stem cell transplantation currently going on
Ischemic heart disease and cardiomyocyte regeneration [5]	Cell replacement of cardiomyocytes
Diabetes mellitus	Islet cells
Nervous system	Differentiation of embryonic cells or somatic
Spinal cord injury	Cells into neural cells
Amyotrophic lateral sclerosis [6]	
Stroke	
Traumatic brain injury	
Batten disease	
Parkinson's disease	
Liver	Hepatocyte transplantation
Skin	Skin transplantation
Eye	In the future stem cell therapy might offer cure
Cartilage	
Bone	
Kidney	
Lung	
Endometrium	
Vascular endothelium	
Smooth muscle	
Striated muscle	
Tooth implantation	

Success Story

Stem cell therapies offer the potential to treat diseases or conditions for which few treatments exist. After approval for phase I trial, the usage of stem cells has slowed the progression of amyotrophic lateral sclerosis, often called ALS or Lou Gehrig's disease, a devastating condition with a 2–5-year survival rate, in a small group of patients.

Lou Gehrig's disease, ALS, is a fatal neurodegenerative disease with no known cure. It causes the deterioration of specific nerve cells in the brain and spinal cord called motor neurons, which control muscle movement. As the illness progresses, patients lose their ability to walk, talk, and breathe.

Their use extended the lives of patients with amyotrophic lateral sclerosis (ALS) and significantly improved the quality of their lives. Human fetal neural stem cells, cultivated by Neuralstem, have substantially slowed muscle degeneration in six patients with ALS.

The phase I trial (18 patients), conducted at Emory University, involved the administration of human fetal spinal cord stem cells to the lower (lumbar) spinal region of 12 ALS patients, the upper cervical region of three ALS patients, and both regions of three patients. The trial's progress led to the beginning of phase II trial.

The success of the trial surprised everyone with the most impressive responder, Atlanta resident Ted Harada, who could not walk without a cane. One month after receiving one million cells, 500,000 on either side of his lower spine, he had abandoned the cane and participated in a 2.5-mile walkathon, actions considered unprecedented in medical circles.

Now, a new approval from the US Food and Drug Administration paves the way for University of Michigan to become the second site in the trial after the first phase of the trial has taken place at Emory University. The FDA approval of a phase II trial was announced by Neuralstem, the company whose product the trial is testing. The phase II trial would evaluate the efficacy and safety of the stem cell injections with higher doses of cells.

In another success story, the Tisch MS Research Center of New York announced that it has received investigational new drug (IND) approval from the Food and Drug Administration (FDA) to commence a phase I trial using autologous, mesenchymal stem cell-derived neural progenitor

(continued)

cells (MSC-NPs) in the treatment of multiple sclerosis (MS). MS is a chronic human autoimmune disease of the central nervous system that leads to myelin damage and neurodegeneration and affects approximately 2.1 million people worldwide.

16.10 Controversy Surrounding Stem Cell Research

There is a widespread controversy over human embryonic stem cell research because of destruction of embryo for creation and usage of stem cells. Therefore human embryonic stem cell research is controversial as starting a stem cell line requires the destruction of a human embryo. Recent researches demonstrating the production of induced pluripotent stem cells (iPSCs) from adult stem cell lines using a single-cell biopsy may allow production of stem cells without embryonic destruction. It is not the entire field of stem cell research, but the specific field of human embryonic stem cell research that is at the center of an ethical debate.

Opponents of the research argue that embryonic stem cell technologies can fundamentally devalue human life. Those in the pro-life movement argue that a human embryo is a human life and is therefore entitled to protection.

Contrarily, supporters of embryonic stem cell research argue that such research should be pursued because the resultant treatments could have significant medical potential. The excess embryos from fertility clinics can be obtained with consent and used for the research. The ensuing debate prompted authorities around the world to limit laboratories of using embryonic stem cells. Pluripotent stem cells unlike ESCs have unstable genetic complement, probably due to intrinsic and extrinsic factors leading to genetic mutation, less effective surveillance mechanisms, or combination of both.

Thus, alternative cell source replacement may be embryonic stem cells, induced pluripotent stem cells, and mesenchymal stem cells; however, differentiated cells are functionally immature with very low efficiency, for example, usage of cells for replacement of β-cells of pancreas shows poor glucose-stimulated insulin secretion compared with native β-cells.

The stem cell research requires translation of preclinical results into therapies with prime consideration of safety and efficacy. The studies also need to be focused on side effects, tumor-forming potential, and their differentiation. Nevertheless, this is the fascinating field holding many opportunities to produce results in clinical subjects and provide hopes in areas where traditional medicine does not offer therapeutic opportunities. The promise of stem cell therapies is an exciting one, but significant technical hurdles remain that will only be overcome through years of intensive research.

Thwarting a Stem Cell Scheme

In 2011, three men in the USA were convicted on charges of criminal activity related to unapproved (FDA) manufacturing, selling, and using of stem cells. Of the three, one of the accused was a licensed midwife who was operating a maternity care clinic in Texas and obtained umbilical cord blood from birth mothers and telling them that it was for research purposes but instead sold the cord blood to an Arizona-based laboratory who then sent the blood to a university consultant at South Carolina.

The Arizona laboratory owner was convicted for unlawful introduction of stem cells into other patients by performing unapproved stem cell procedures on people suffering from multiple sclerosis, cancer, and other conditions. The university consultant, an assistant professor, was manufacturing stem cell products using university facilities. The three made more than $1.5 million for patients for promised treatments.

Thus, one needs to be aware of these kinds of frauds as they give false hopes to patients and their relatives suffering from incurable diseases.

FDA's OCI worked the case with the Federal Bureau of Investigations and the Internal Revenue Service's Criminal Investigations Division.

16.11 Chapter End Summary

- The self-renewable cells found in organism that can divide and differentiate into diverse and specialized cell types are known as stem cells.
- There are two broad categories of stem cells—embryonic and adult somatic stem cells. These cells can be grown under in vitro condition and can be transformed into specialized cell types.
- Based on potency, the stem cells may be (1) totipotent (can differentiate into embryonic and extraembryonic cell types), (2) pluripotent (can differentiate into nearly all cell types), (3) multipotent (can differentiate into number of closely related family of cells), (iv) oligopotent (can differentiate into a few cells), and (v) unipotent (can produce only one cell type).
- Stem cells can undergo two types of cell division: (1) symmetric division giving rise to two identical daughter cells both with stem cell properties and (2) asymmetric division producing only one stem cell and a progenitor cell with limited self-renewal potential.
- The external growth factors lead to reprogramming of cells with appropriate signals activating specific transcription factors.
- Stem cell therapy has the potential to dramatically change the treatment of human diseases. Adult stem cell treatments have been used for many years to successfully treat leukemia.
- Early applications of adult stem cells have focused on intravenous delivery of blood progenitors known as hematopoietic stem cells (HSCs) and on mesenchymal stem cells (MSCs).
- For both cell lines, direct injection or placement of cells into a site in need of repair may be the preferred method of treatment, as vascular delivery suffers from a "pulmonary first-pass effect" where intravenous injected cells are sequestered in the lungs.
- The use of adult stem cells in research and therapy is not considered as controversial as the use of embryonic stem cells, because the production of adult stem cells does not require the destruction of an embryo.
- The stem cell therapy holds tremendous potential in the cure of many diseases; however, even in autologous transplantations, the cells that contain their new host's nuclear DNA could still be rejected by the individual's immune system due to mitochondrial DNA. Tissues made from a person's stem cells could therefore be rejected, because mitochondrial genomes tend to accumulate mutations.

Multiple Choice Questions

1. The controversy surrounding human stem cell research is not the ability to end the disease, but rather:
   (a) Some feel that the researchers are playing God.
   (b) There are no clear standards in guiding the research.
   (c) The use of frozen embryos to conduct the research.
   (d) The use of sources other than embryos to conduct the research.
2. What makes stem cells desirable for therapeutic use?
   (a) They develop into different organs
   (b) They develop into nerve cells and brain
   (c) They are found in reproductive tissue
   (d) They develop into many different tissue types
3. Stem cells which can differentiate into any cell lineages:
   (a) Progenitor cells
   (b) Multipotent cells
   (c) Pluripotent cells
   (d) Oligopotent cells
4. If the genome of each and every cell is similar, then why cannot cells other than stem cells be used for differentiation into various tissues?
   (a) Stem cells contain sex chromosomes
   (b) Stem cells contain different genome than other cells

   (c) As development proceeds genetic makeup of cells changes
   (d) As cells develop some of the genes can be turned off permanently
5. Stem cells have the property of:
   (a) Dedifferentiation and self-renewal
   (b) Differentiation and self-renewal
   (c) Nonself-renewal and dedifferentiation
   (d) None of these
6. Stem cells can offer cure for:
   (a) Genetic diseases
   (b) Amputated body part
   (c) Metabolic diseases
   (d) Repair injuries
7. Which of the following are potential therapeutic uses of adult stem cells?
   (a) Replace neurons after an accident
   (b) Regenerate cells of hematopoietic system
   (c) Repair damage to heart muscles
   (d) All of the above
8. Which of the following is a source for embryonic stem cells?
   (a) Sperms
   (b) Liver tissue
   (c) Inner cell mass of blastocyst
   (d) Oocyte
9. A progenitor cell can differentiate into:
   (a) All tissue types
   (b) Only a type of lineage
   (c) Tissues of closely related family
   (d) All the cell lineages
10. A totipotent cell can form:
   (a) An entire organism
   (b) Only cells of closely related family
   (c) Only hematopoietic cells
   (d) Only sperms and ovum
11. What is the role of adult stem cells in the body?
   (a) They provide the source of cells for treatment of diseases
   (b) They act as repair system for the body
   (c) They control the working of organ where they are present
   (d) All of the above
12. What determines the conversion of the one type into other types of cells?
   (a) Growth factors and signals present in the surrounding
   (b) Only transcription factor activation
   (c) Binding to the other cell types leading to differentiation
   (d) None of the above
13. Transdifferentiation is:
   (a) Ability of cell to develop into mature tissue
   (b) Ability of cell to develop into totipotent cell
   (c) Ability of cells to develop into gametes
   (d) Ability of cells to differentiate into cells of other lineages
14. Which of these can be called stem cells?
   (a) Umbilical cord blood
   (b) Early embryos
   (c) Sperm and ovum
   (d) Adult differentiated tissues
15. Embryonic stem cells can be obtained from:
   (a) Enucleated cells
   (b) All cells
   (c) Sperm and ovum
   (d) Inner cell mass
16. Which of the following is a disadvantage of adult stem cells over embryonic stem cells?
   (a) They grow very fast
   (b) Easy to obtain
   (c) They are very small therefore difficult to manipulate
   (d) Difficult to locate
17. Somatic cell nuclear transplantation uses material from:
   (a) Umbilical cord blood
   (b) Adult cell
   (c) Blood cells
   (d) None of the above

Answers

1. (d); 2. (d); 3. (c); 4. (d); 5. (a); 6. (d); 7. (b); 8. (c); 9. (c); 10. (a); 11. (b); 12. (a); 13. (d); 14. (a); 15. (d); 16. (d); 17. (b)

Review Questions

Q1. What are stem cells?
Q2. What are therapeutic roles of stem cells?
Q3. Why stem cell research is into too much debate and controversy?

Q4. What are the properties of embryonic stem cells of which they are much sought for therapeutic uses?
Q5. What are mesenchymal stem cells?
Q6. Define transdifferentiation and plasticity of stem cells.
Q7. What are ethical and social issues surrounding stem cell research?
Q8. What kind of diseases can be potential targets of stem cell therapy?

References

1. Androutsellis-Theotokis A, Leker RR, Soldner F et al (2006) Notch signaling regulates stem cell numbers in vitro and in vivo. Nature 442:823–826
2. Bone Marrow Transplantation and Peripheral Blood Stem Cell Transplantation (2010) In: National Cancer Institute Fact Sheet web site. National Institutes of Health, U.S. Department of Health and Human Services, Bethesda. Cited 24 Aug 2010
3. Cristofanilli M, Harris VK, Zigelbaum A, Goossens AM, Lu A, Rosenthal H, Sadiq SA (2011) Mesenchymal stem cells enhance the engraftment and myelinating ability of allogeneic oligodendrocyte progenitors in dysmyelinated mice. Stem Cells Dev 20:2065–2076
4. Eisenberg LM, Eisenberg CA (2003) Stem cell plasticity, cell fusion, and transdifferentiation. Birth Defects Res C Embryol Today 69:209–218
5. Ghodsizad et al (2013) Clinical application of adult stem cells for therapy for cardiac disease. Cardiovasc Ther 31:323–334
6. Glass JD, Boulis NM, Johe K, Rutkove SB, Federici T, Polak M, Kelly C, Feldman EL (2012) Lumbar intraspinal injection of neural stem cells in patients with amyotrophic lateral sclerosis: results of a phase I trial in 12 patients. Stem Cells 30:1144–1151
7. Goldthwaite CA (2011) The promise of induced pluripotent stem cells (iPSCs). U.S. Department of Health & Human Services. stemcells.nih.gov/info/Regenerative_Medicine
8. Gurtner GC, Callaghan MJ, Longaker MT (2007) Progress and potential for regenerative medicine. Annu Rev Med 58:299–312
9. Harrison's principles of internal medicine, (2008) 17th edition, Eds: Kasper DL, Braunwald E, Fauci AS, Hauser SL, Longo DL, Jameson JL. Mc Graw-Hill Medical publishers, United States of America
10. Kang KS, Kim SW, Oh YH et al (2005) A 37-year-old spinal cord-injured female patient, transplanted of multipotent stem cells from human UC blood, with improved sensory perception and mobility, both functionally and morphologically: a case study. Cytotherapy 7:368–373
11. Kang L, Wang J, Zhang Y, Kou Z, Gao S (2009) iPS cells can support full-term development of tetraploid blastocystcomplemented embryos. Cell Stem Cell 5:135–138
12. Karanes C, Nelson GO et al (2008) Twenty years of unrelated donor hematopoietic cell transplantation for adult recipients facilitated by the National Marrow Donor Program. Biol Blood Marrow Transplant 14:8–15
13. Lindvall O, Kokaia Z (2006) Stem cells for the treatment of neurological disorders. Nature 441:1094–1096
14. Maherali N, Sridharan R, Xie W et al (2007) Directly reprogrammed fibroblasts show global epigenetic remodeling and widespread tissue contribution. Cell Stem Cell 1:55–70
15. Riley J, Federici T, Polak M, Kelly C, Glass J, Raore B, Taub J, Kesner V, Feldman EL, Boulis NM (2012) Intraspinal stem cell transplantation in amyotrophic lateral sclerosis: a phase I safety trial, technical note, and lumbar safety outcomes. Neurosurgery 71:405–416
16. Schwartz et al (2014) Human embryonic stem cell-derived retinal pigment epithelium in patients with age-related macular degeneration and Stargardt's macular dystrophy: follow-up of two open-label phase 1/2 studies. Lancet 7:509–516
17. Singec I, Jandial R, Crain A, Nikkhah G, Snyder EY (2007) The leading edge of stem cell therapeutics. Annu Rev Med 58:313–328
18. Stepniewski et al (2015) Induced pluripotent stem cells as a model for diabetes investigation. Sci Rep 26:8579
19. Takahashi K, Yamanaka S (2006) Induction of pluripotent stem cells from mouse embryonic and adult fibroblast cultures by defined factors. Cell 126:663–676
20. Takshashi et al (2007) Induction of pluripotent stem cells from adult human fibroblasts by defined factors. Cell 131:861–872
21. Toyoda et al (2015) Cell aggregation optimizes the differentiation of human ESCs and iPSCs into pancreatic bud-like progenitor cells. Stem Cell Res 14:185–197
22. Weissman IL (2000) Stem cells: units of development, units of regeneration, and units in evolution. Cell 100(1):157–168
23. Yong Z (2012) Stem cell educator therapy and induction of immune balance. Curr Diabetes Rep 12:517–523
24. Yu J, Vodyanik MA, Smuga-Otto K et al (2007) Induced pluripotent stem cell lines derived from human somatic cells. Science 318:1917–1920
25. Zhao XY, Li W, Lv Z et al (2009) iPS cells produce viable mice through tetraploid complementation. Nature 461:86–90

Some Selected Resources

clinicaltrials.gov/show/NCT01469832
http://www.advancedcell.com

http://www.biosciencetechnology.com/
http://www.fda.gov/
http://www.meih.org/DepartmentSubCat.asp?subcatid=186
http://www.ul.edu.lb/lu/default.aspx
Clive Cookson "Generous Staffing And Permissive Laws Aid Asia's Largest Stem Cell Effort", "Scientific American" 27 June 2005
Stem cell therapy center, Lebanon
www.ncbi.nlm.nih.gov/pubmed

Gene Therapy

17

Abstract
Genes are regions on chromosomes which carry specific information for the production of proteins. Advancement in science, especially in medical research, has led to the identification and isolation of genes responsible for human diseases. A person carrying the defect in a specific gene might develop disease, which may be inheritable. In gene therapy, nucleic acid is the therapeutic agent. In this, the defective genes responsible for the disease are genetically modified in patients in order to achieve a therapeutic goal and return the individual to good health. The therapy is achieved by using a viral vector or gene delivery vehicles. The steps involved in the therapy in which the gene of interest enters the target cell via viral vector and begins expression are referred to as transduction. Gene therapy can be used to treat diseases caused by mutations in the patient's own DNA (inherited disorders, cancers) as well as infectious diseases and is particularly valuable in cases where no conventional treatment exists or where that treatment is inherently risky.

17.1 Introduction

Genes are regions on chromosomes which carry specific information for the production of proteins. Advancement in science, especially in medical research, has led to the identification and isolation of genes responsible for human diseases. A person carrying the defect in a specific gene might develop disease, which may be inheritable. In gene therapy nucleic acid is the therapeutic agent (Fig. 17.1). In this the defective genes responsible for the disease are genetically modified in patients in order to achieve a therapeutic goal and return the individual to good health. The therapy is achieved by using a viral vector or gene delivery vehicles. The steps involved in the therapy in which the gene of interest enters the target cell via viral vector and begins expression are referred to as transduction.

Gene therapy can be used to treat diseases caused by mutations in the patient's own DNA (inherited disorders, cancers), as well as infectious diseases, and is particularly valuable in cases where no conventional treatment exists or where that treatment is inherently risky. Identification of genes responsible for a genetic disease is important: identification of genes that caused the disease can be done by locating the genetic loci in the human genome and is carried out by linkage analysis or positional cloning.

V. Gupta et al., *Basic and Applied Aspects of Biotechnology*, DOI 10.1007/978-981-10-0875-7_17

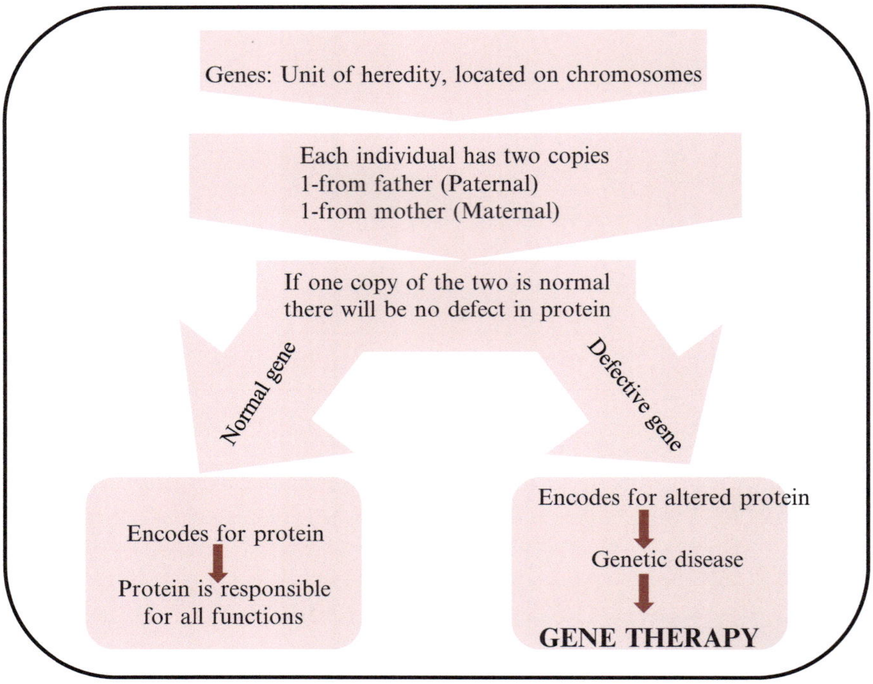

Fig. 17.1 The function of normal gene and effect of defective gene, which causes disease. The individuals with defective gene are the candidate for gene therapy

After localization of the gene, genome sequence is determined in order to identify the gene:

(a) Gene identification for an understanding of biochemical basis of diseases
(b) Identification of the defect leading to the disease phenotype

Gene therapy offers best therapeutic modalities in molecular medicine by correcting cellular dysfunction as in cystic fibrosis, combined immunodeficiency syndromes, hemophilia, and muscular dystrophies. The therapy has shown successful results in the severe combined immunodeficiency disease. The technique is one of the most powerful and has the potential to cure many diseases for which there is no cure or treatments [16]. Over 1,800 trials for gene therapy have been reported with some ongoing, completed, or approved worldwide.

There are two basic approaches to gene therapy:

- Germline therapy
- Somatic cell therapy

17.2 Germline Therapy

In the germline therapy, the modification is done in germ cells that are sperms or eggs, producing a permanent and genetically transferable modification. The functional gene is introduced into either of them, which then integrate in the selected germ cell, and subsequently the therapeutic gene is present in each cell of the body. The fertilized egg may be used to insert the therapeutic gene and reimplanted into the mother. If successful, the functional gene is present and expressed in all cells of the resulting individual.

It is usually carried out by microinjection of DNA into the isolated egg cell. Theoretically it could be used to treat any inherited disease. As the changes are made into germ cells or fertilized egg, therefore the changes are heritable, that is, they are passed on to future generation. The technique of making permanent changes might prove effective in the genetic diseases, but due to ethical and technical reasons, their use has been prohibited in humans.

17.3 Somatic Cell Therapy

The therapy involves modification of somatic cells. The therapy can be done either in vivo or ex vivo. Somatic cell gene therapy can be achieved by ex vivo transfer, where therapeutic genes are transferred into the somatic cells .of a patient involving manipulation of ordinary cells, usually ones that can be removed from the organism, transfected, and then placed back in the body. In a few gene therapy clinical trials, cells from the patient's blood or bone marrow are removed and grown in the laboratory with concomitant exposure to virus with the functional gene. The virus infects the cell with insertion of the gene of interest into the genome. These cells after perpetuation in the laboratory may be returned back to the patients. As the modification is done outside the body, the technique is referred as ex vivo. In contrast in vivo refers to the modifications occurring in the cells inside the body. In this the gene of interest is transferred with suitable vector inside the body of the patient (Fig. 17.2).

The current therapy or trials are for somatic cell therapy. The technique is promising for inherited blood diseases (like hemophilia, thalassemia) with genes being introduced into stem cells from the bone marrow, which give rise to all the specialized cell types in the blood. The strategy is to prepare a bone extract containing several billion cells, transfect these with a retrovirus-based vector, and then reimplant the cells. Subsequent replication and differentiation of transfectants leads to the added gene being present in all the mature blood cells. As the somatic cells are being modified, therefore the effects of the therapeutic gene would not be heritable.

Modification of somatic cells may be done for:

- Genetic disease therapy that often involves supplying a functional copy of defective gene to treat conditions like loss of function
- Supplying a gene that can achieve a similar biologic effect using an alternate pathway
- Repair of the mutant gene to correct mutation in situ
- Supplying an antisense oligonucleotide to inhibit function of the gene, for example, inhibition of gene expression of pathogen by targeting its essential functions
- Targeting specific cells for removal
- Downregulating a harmful response through siRNA

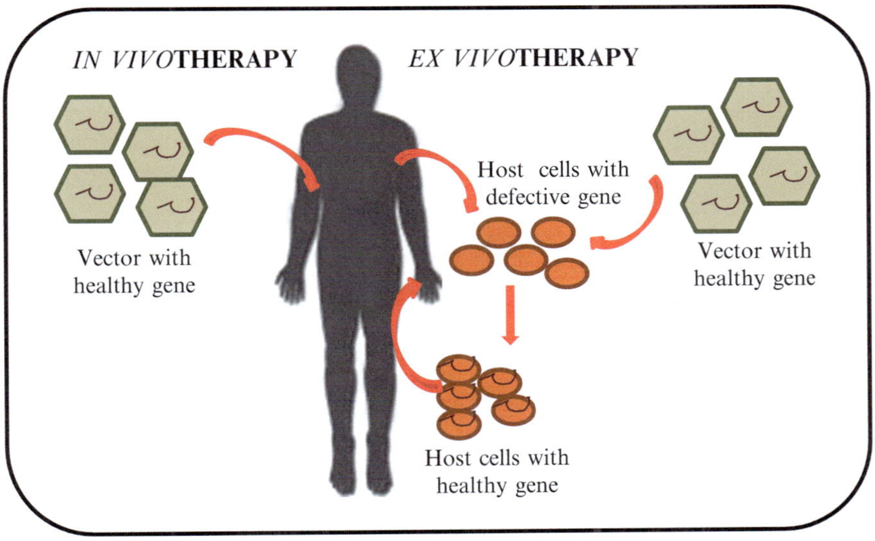

Fig. 17.2 The in vivo and ex vivo approaches for gene therapy

Long-term gene expression is achieved either by transducing stem cells with an integrating vector. At times integration of viral vectors into the host genome may have long-term gene expression. Therapeutic gene can be integrated into the stem cell using viral vectors thus ensuring the presence of gene in all progeny cells. However, this can have potential safety concerns, as virus integration is random.

17.4 Intracellular Barriers to Gene Delivery

Extensive research has shown that cellular factors might contribute to the uptake and motility of incoming molecules. The cytoskeleton provides an intracellular highway for molecular traffic from one subcellular location to another.

The transfer of healthy gene in the nucleus requires cell entry, intracellular motility, and nuclear delivery of gene therapy vectors. Vector design should be critically considered for efficient intracellular vector trafficking (events that mediate the movement of a gene delivery vector from the cell surface to the nucleus). Nonviral vectors are safe as compared to viral vectors because they lack specific immune response; however, short duration and limited efficacy are major hurdles [3]. Organically modified silica or silicate (Ormosil) has been used as DNA vectors and can perform targeted delivery of DNA. Cytosolic release of heterologous DNA is a requirement for nuclear translocation, but the major problem encountered in DNA–polycation complex by endocytosis is that large fraction is targeted to the lysosomal compartment resulting in its degradation. This allows only a small fraction of internalized plasmid DNA, but diffusional and metabolic barriers further decrease the number of intact plasmid.

Nonviral methods can transfer nucleic acids into cells known as transfection. Certain nonviral vectors undergo efficient cell binding and internalization but poor gene expression due to the intracellular barriers (inefficient internalization, endosomal escape, cytosolic trafficking, and nuclear delivery), thus less efficient gene transfer.

Viruses have evolved mechanisms to overcome these barriers; therefore, studies on viral trafficking can benefit the development of both viral and nonviral gene transfer vectors. There are various alternative routes of cell entry as evidenced by the reports on the plant and bacterial toxins. These may help in the delivery of vectors (viral or nonviral).

17.5 Virus-Mediated Gene Transfer

In ~70 % of clinical trials, viral vectors are in use, as delivery of the gene using viral vectors is very efficient. Some viruses have broad tropism; they can infect many different kinds of cells. Some are capable of binding to specific receptors present on few cells, thus having narrow tropism, the example being the herpesvirus, which specifically infects cells of the central nervous system.

The usage of virus for transfer of DNA into cells of humans or other animal cells is known as transduction. Some viruses are capable of fusing with the host cell membrane to release their genome into host cells, while others can undergo receptor-mediated endocytosis. The virus along with therapeutic DNA either can integrate into the genome for stable transgene expression or can remain in extrachromosomal positions in the cells (transient expression). These viruses have variable insertional capacities and they require the genes responsible for synthesis of coat proteins and directing its packaging into the capsid (see Table 17.1).

The virus vector offers extremely high transfection frequency, enabling a large proportion of the stem cells in a bone marrow extract to receive the new gene. However their usage is limited as they are immunogenic and elicit intense inflammation of endothelial cells. On the other hand, the nonviral delivery of vehicles, such as plasmids and antisense oligonucleotides, has been associated with a lower transfection efficiency and transient expression of the gene product because of various reasons.

Table 17.1 The properties of various viral vectors for use as vehicles for gene transfer

	Viral vectors	Properties
1.	Retrovirus	Random but stable insertion in dividing cells with long-term expression
		Insertional mutagenesis might occur
2.	Lentivirus	Stable insertion; therefore, persistent gene transfer
		Induce oncogenesis
3.	Adenovirus	High packaging capacity in slow or nondividing cells with short-term expression
		Genome is bigger
		Strong immune response may cause cardiotoxicity or brain damage
4.	Adeno-associated virus (AAV)	Insertion at a very specific site in slow or nondividing cells with short-term expression
		Looked on as having major potential in gene therapy
		Limited packaging capacity
5.	Simian virus 40 (SV40)	Infects several mammalian species
		Undergoes lytic and lysogeny cycle
		Large packaging constraints

Viral vectors
Adenovirus
Double-stranded DNA virus having cloning capacity of 7.5 kb
Causative agent of eye, intestinal, and respiratory infections
Upon infection, their DNA remains transiently in the host
High efficiency of transduction and can transduce dividing and nondividing cells
High level of gene expression
The genetic material doesn't integrate in the host genome
Immunogenicity is their biggest concern and repeated administration is required due to their transient nature
Engineered serotype 5 is a common vector (E1a and E1b regions deleted)
Adeno-associated virus (AAV)
Belongs to the family of parvovirus
Single-stranded DNA virus with cloning capacity of nearly 4.5 kb
The wild-type virus is capable of inserting foreign DNA at specific site (chr-19)
Recombinant AAV with therapeutic genes fuses its ends with the help of inverted terminal repeats (ITR) and is maintained in episomal form
Recombinant AAV is expressed for a long time in host cells and possesses high level of gene expression
AAV can infect dividing and quiescent or nondividing cells; therefore, the possibilities are being explored for gene transfer in the cells of the nervous system or brain
Clinical trials are ongoing with recombinant AAV for muscular dystrophies, hemophilia B, alpha-1 antitrypsin deficiency, and lipoprotein lipase deficiency
Herpes simplex virus
Has double-stranded DNA as genetic material and is a human neurotropic virus
It is examined for gene transfer in the nervous system
It can live life long as nonintegrated extrachromosomal elements in the sensory ganglia with cloning capacity of more than 30 kb
The wild-type HSV-1 virus is able to infect neurons. Infected neurons are not rejected by the immune system
It is being explored for its potential for nervous system disorder
Vaccinia virus
Ankara (MVA)-attenuated vaccinia virus used safely as smallpox virus
Infect dividing and nondividing cells with cloning capacity of 25 kb
Non-integrating with transient level expression
Used to transfer suicide genes to kill tumor cells in cancer gene therapy
Retrovirus
RNA viruses having reverse transcriptase
Integrating viruses with 7–8-kb cloning capacity
These have limited natural hosts
First vectors used for gene therapy
Virus attaches to the dividing cell and efficiently delivers its genes with stable transfer (integrates in the host genome) and gives long-term gene expression
The studies are going on where viral genome may be manipulated for exchange of disease-causing genes with therapeutic genes and then transferred into suitable host
Q vectors or self-inactivating (SIN) retrovirus is engineered to minimize the risks

Viral vectors
As they can infect only dividing cells, thus they are used in cancer treatment
Lentivirus
RNA viruses
Integrating viruses with 8 kb cloning capacity
High efficiency of gene transfer
Infect dividing and nondividing cells and provide long-term gene expression
From HIV virus, unnecessary genes are removed for generating safe packaging lines
Self-inactivating lentivirus offers more safety

As pathogenic agents, viruses can attach to the cells and deliver their genes. The studies are going on where viral genome may be manipulated for exchange of disease-causing genes with therapeutic genes and transfection of these in the host. Scientists have shown that intravenously injected recombinant adenovirus vectors encoding a beta-galactosidase reporter gene were successfully delivered to normal rat myocardium using microbubbles and transthoracic 1.3 MHz diagnostic ultrasound, at a mechanical index of 1.5, delivered at a burst of three frames of ultrasound every four to six cardiac cycles. However, transfection was not observed if the adenovirus was administered in the same dose without microbubbles or if the adenovirus was administered with microbubbles but in the absence of ultrasound. Importantly, using the same model, the authors confirmed that plasmid transgene expression can be directed to the heart, with an even higher specificity than viral vectors, and that this expression can be regulated by repeated treatments.

17.6 Nonviral Vectors

The DNA/gene delivery methods, which do not use viruses, use natural or synthetic compounds. These have minimal toxicity and immunogenicity, can be easily produced, and if required can be repeatedly given in the patients:

- Nonviral vectors like liposomes or nanoparticles are safe but less efficient as compared to viral vectors.
- Their usage for gene delivery result in poor transfer rates and low level of transgene expression.
- Delivery of gene is very problematic in nondividing cells.
- They elicit minimal immune reactions.
- Their delivery in the cell can be done by microinjection or lipid-mediated gene transfer.

Generally, nonviral methods were associated with having less efficiency with short-lived expression; however, the usage of some physical methods has reached the efficiency and expression duration with good clinical effects. The efficiency of nonviral vectors is limited by anatomical barrier like epithelial, endothelial cell linings and the extracellular matrix surrounding the cells. These prevent direct access of the vector to the target cells. Another problem is the presence of blood- and tissue-specific phagocytes, which are responsible for clearance of DNA-loaded particles, further reducing their efficiency. However, researches are going on for developing nonviral chemical vectors. Physical methods for successful in vivo gene transfer have been achieved, but still there is a lot to be done to achieve the transfection frequencies comparable to viral vector (Table 17.2). Several viral and nonviral vectors are compared for efficient delivery of the gene across these barriers to target cells in Table 17.2.

17.7 Overview of Inherited and Acquired Diseases for Gene Therapy

Majority of gene therapy protocols are approved for cancer. Apart from cancer, the therapy can be used for vascular diseases, severe combined immunodeficiency (SCID), muscular dystrophy, alpha-1 antitrypsin deficiency, hemophilia B, Leber's congenital amaurosis, and factor VIII/IX deficiency.

Somatic cell therapy also has potential in the treatment of lung diseases such as cystic fibrosis. Transgenic animals are generated by incorporation of genes into eggs or early embryos.

Gene transfer can be carried out in cultured cells, which are then reintroduced into the patient,

Table 17.2 The comparison of viral and nonviral vectors

	Viral vectors	Nonviral vectors
Vectors	Retroviruses, lentivirus, adenovirus, adeno-associated virus (AAV), herpes simplex virus-1 (HSV-1), simian virus 40 (SV40), *Alphaviruses*	Transposons, liposomes, Ormosil (organically modified silica or silicate), naked DNA, site-specific integrase
Genome	DNA/RNA	No genome
Immunogenicity	Mild to severe	No immune response
Advantages	Persistent gene transfer	Persistent gene expression
	Highly effective	Transfects many cell types
	Wide host cell range	
Disadvantages	Mutagenesis, oncogenesis, toxicity	Blocked by intracellular barriers
		Early stage of development

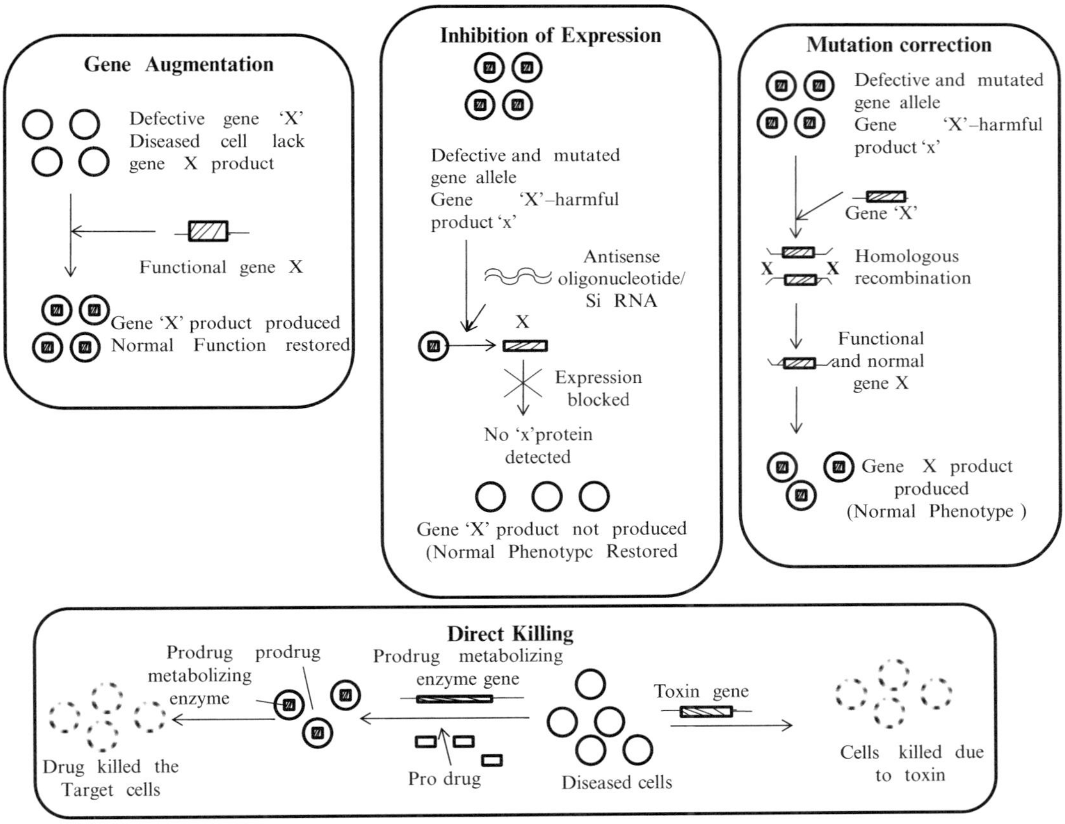

Fig. 17.3 Various modes of gene action useful for cancer therapy

or DNA can be transferred to the patients in vivo, directly or using viral vectors.

The following strategies are used (Fig. 17.3):

- Gene augmentation therapy (GAT), where DNA is added to the genome with the aim of replacing a missing gene product
- Gene targeting to correct mutant alleles
- Gene inhibition therapy, using techniques such as antisense RNA expression
- The targeted ablation of specific cells [30]

Therapeutic gene transfer effectively generates transgenic human cell clones, and, for this reason, only somatic cells can be used as targets. The prospects of germline transmission in humans raise serious concerns. As an alternative to permanent gene transfer, transient gene therapy can be achieved using oligonucleotides, which can disrupt gene expression at many levels but do not permanently change the genetic material of the cell.

17.8 Attempts at Human Gene Therapy

The gene therapy has seen the times of rejoice and setbacks. Many laboratories are working hard to bring gene therapy readily available for the cure of the patients [22]. More than 1,800 trials have been approved with approximately 3 % in phase III clinical trials. Of these trials almost two-third are for cancer. However, many of the trials are terminated, some are active but not recruiting participants, some are completed, or ongoing. Other conditions as monogenic disorders, cardiovascular diseases, and infectious diseases account for a less number of ongoing trials. Achieving therapeutic goal in monogenic diseases has been an attractive gene therapy goal. In these, the disease occurs due to deficiency of a single gene product:

- The first gene therapy trial was done in 1990 for adenosine deaminase-severe combined immunodeficiency (ADA-SCID). Wonderful results were obtained in gene therapy trial for SCID using γ-retroviral vectors. Adenosine deaminase-severe combined immunodeficiency (ADA-SCID) is inherited as an autosomal recessive disorder. It can be treated by the use of hematopoietic stem cells (HSCs) as the target for transduction. The treated children were responding well and no complications were reported at 4 years follow-up. Monogenic adenosine deaminase (ADA) is inherited as an autosomal recessive gene and its deficiency may be lethal [2]. In the absence of the enzyme, adenosine is not metabolized resulting in high accumulation of 2′-deoxyadenosine. This compound is toxic to immune cells. The therapy for this uses modified the retroviral vector, where T cells are modified under in vitro condition. The cells, which receive functional genes upon selection, are reintroduced into the patient. However, the modified T cells can only last for as long as 6 months.
- However, in 1999, the treatment for ornithine transcarbamylase (OTC) deficiency was done using adenovirus vectors by intrahepatic injection resulted in death of an 18-year-old participant due to intense immune reactions and multiple organ failure (see case study) [29].
- In 2000 therapeutic effect from gene transfer were observed for X-linked severe combined immunodeficiency (SCID) resulting from mutations in the gene IL2RG (encodes γc subunit of cytokine receptor) required for normal development of T and NK cells. There was complete reconstitution of the immune system with development of functional T-cell and B-cell responses, and remarkable gains in growth occurred in some. This gene therapy trial was performed at two places, one in France and the other in the UK in 20 boys. However, the approach resulted in major setback when 5 of the 20 treated children developed leukemia, with one death due to insertional mutagenesis and activation of cellular oncogenes. However, in the long-term follow-up of nine boys (French trial) who must be teenagers now, four had leukemia, of which one died. The other three were successfully treated with chemotherapy and were among the seven children who exhibited long-term immune reconstitution. Gene-corrected T cells lasted for more than 10 years. These children with functional immune systems have responded to vaccination and live normal lives.
- In 2006 chronic granulomatous disease (CGD) treatment was started in two adult patients using γ-retroviral gene therapy. CGD is recessive immunodeficiency condition that affects the phagocytic function due to defective NADPH phagosome oxidase of the immune phagocytic cells [26, 34]. Despite initial success, both the patients showed silencing of transgene along with myelodysplasia due to

insertional activation of cellular oncogenes followed by death of one patient 27 months after therapy from sepsis [34].

- In 2006 therapy for hemophilia B was started using AAV2 vectors to hepatocytes for factor IX but immune response destroyed the transduced cells [24].
- Year 2009 witnessed successful gene therapy for ADA deficiency [2] and another report of successful gene therapy for central nervous system disorder and X-linked adrenoleukodystrophy using lentivirus vector.
- Another report of positive outcomes of gene therapy was described using retinal injections of recombinant AAV for treating Leber's congenital amaurosis (a form of childhood blindness). The therapy resulted in vision gain till 1 year.

Case Study

Great caution is warranted as gene therapy is pursued. It is very difficult to predict the response of the body after high doses of viral vectors [12]. Unfortunately, Jesse Gelsinger, an 18-year-old participant in Philadelphia in a clinical trial for ornithine transcarbamylase (OTC) deficiency in September 1999, became unexpectedly ill. The participant died after administration of adenovirus (in the liver) with OTC gene which resulted in intense inflammatory response against the viral vector. This publicized case led clinical trials using adenoviral vectors on hold. That was the dark period for gene therapy with lots of criticism and skepticism. All the trials after that are under strict regulation of the National Institute of Health and Food and Drug Administration (FDA).

However in less than 1 year after Gelsinger case, report of successful gene therapy trial in two children suffering from a severe combined immunodeficiency (SCID-XI) by Cavazzana-Calvo (2000) [5]

(continued)

using Moloney murine leukemia virus gave hope.

In another trial of 2002, two of the ten children treated for severe combined immunodeficiency (SCID-X1) developed leukemia like clonal lymphocyte proliferation. These major setbacks in the trials illustrate that the therapy needs safety regulation and intense animal experimentation before clinical trials. However, trials in the UK and Australia for SCID-X1 with different protocols were allowed. Nevertheless gene therapy was highly successful in treatment of cancer and gave encouraging results.

Animal studies using recombinant AAV infused in a hemophilic dog model into the skeletal muscle or liver resulted in long-term (>5 years) expression of factor VIII or factor IX. However, AAV vector with factor IX in skeletal muscles was safe in patients suffering from hemophilia, but the levels of factor IX in circulation were low (>1 %) with short periods. AAV vector expressing factor IX resulted in >5 % circulating levels but only for 6–10 weeks resuming the baseline levels thereafter [24].

The early phase trials are under way for age-related macular degeneration resulting from neovascularization by inhibitors such as angiostatin or inhibiting the expression of vascular endothelial growth factor (VEGF) by RNAi-mediated knockdown.

The success and setbacks have been occurring with gene therapy trials for SCID [11, 15, 18], chronic granulomatous disease, and many others. The field is still in its infancy but has promising potential to revolutionize the future of medicine.

Another important monogenic disorder, cystic fibrosis (CF), is lethal. In this, normal cell membrane functioning is disabled as the gene product, "transmembrane conductance regulator" (CFTR) acting as chloride channel, function is lost. These chloride channels are present in epithelia of the airways, the alimentary canal, etc. The virus (ade-

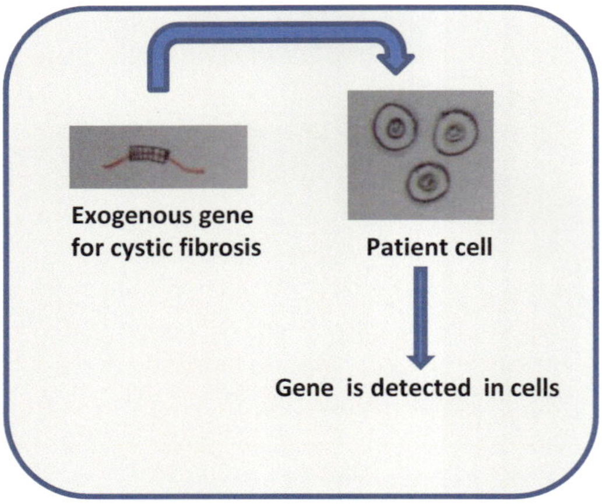

Fig. 17.4 The effect of healthy gene in the monogenic disorder as cystic fibrosis where gene is detected in the transfected cells

novirus) which infects the respiratory tract are modified and transferred to the host (Fig. 17.4). The genetic material of adenovirus does not incorporate into the host genome; rather, it is maintained in episomal form for only few weeks. As adenovirus-mediated gene transfer is transient, the use of adenovirus as a vector for treatment of CF would require repeated administration.

Duchenne muscular dystrophy (DMD) is a monogenic disorder in which progressive wasting of skeletal muscles and later on cardiac muscles occurs. Restoration of dystrophin gene is beneficial for DMD patients. However the gene is large and efficient delivery to skeletal and cardiac muscles is a big challenge.

Other diseases where gene therapy is likely to give results are Wiskott–Aldrich syndrome, chronic granulomatous disease, sickle cell anemia, and thalassemia.

Clinical trials using recombinant AAV vectors (which have the advantage of maintaining itself in episomal form, minimizing risks associated with insertional mutagenesis) are ongoing for several diseases. Muscular dystrophies, α-1 antitrypsin deficiency, lipoprotein lipase deficiency, and hemophilia B are several disease models. Hemophilia is the promising disease model as active clotting factors can be synthesized in a variety of tissue types, but human trials for proper and long-term expression need to come. The trials were successful in dog model.

The potential of gene therapy for treatment of angina or myocardial ischemia to achieve therapeutic angiogenesis is also considered. Major transgene used has been vascular endothelial growth factor (VEGF) because of its specificity for endothelial cells. Long-term effects of transgene are not required as after vascularization local expression is favored. Studies have suggested beneficial effects but larger study population is required for definite conclusion.

17.8.1 Gene Therapy for Cancer

More than 600 gene therapy protocols have been proposed of which >80 % are for cancer than cardiovascular diseases and a limited number for monogenic disorders.

Cancer is caused by the somatic mutation of cellular genes. These genes include oncogenes, tumor suppressor genes, and DNA repair genes. The generation of cancer is currently thought to be a multistep process of genetic alterations that vary according to the type and stage of cancer. Since cancer is a genetic disease, gene therapy could be applied in treatment of tumors. The most advanced applications of gene transfer technology in medicine are in gene therapy for cancer. Gene therapy is less harmful than conventional cancer therapy because of their targeted delivery. Cancer trials are basic safety trials. For therapy of cancer, various options are being explored (Fig. 17.5) and some of them are shown in Fig. 17.5.

Researchers are studying several methods for treating cancer with gene therapy. Some approaches target cancer cells, to destroy them or prevent their growth. Others target healthy cells to enhance their ability to fight cancer. In some cases, researchers remove cells from the patient, treat the cells with the vector in the laboratory, and return the cells to the patient. In others, the vector is given directly to the patient. Some gene therapy approaches being studied are described below:

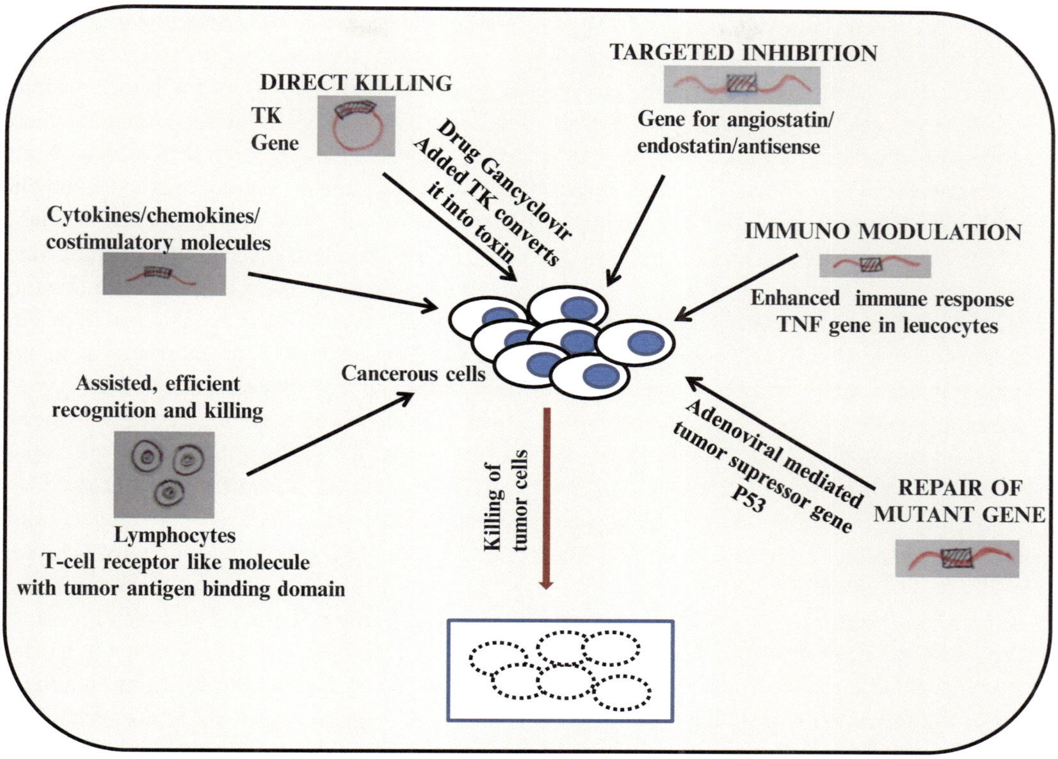

Fig. 17.5 The basic mechanisms of effects of insertion of foreign gene in the host for treatment of cancer. The gene can mediate assisted recognition and killing, direct killing, targeted inhibition, immunomodulation, or repair of gene defect in cancer patients

- Replacing an altered tumor suppressor gene that produces a nonfunctional protein (or no protein) with a normal version of the gene [23]. Because tumor suppressor genes (as TP53) play a role in preventing cancer, restoring the normal function of these genes may inhibit cancer growth or promote cancer regression. Another local approach uses adenoviral-mediated expression of the tumor suppressor P^{53}. The strategy has shown complete and partial recovery in squamous cell carcinoma of the head and neck, esophageal cancer, and lung cancer.
- Introducing genetic material to block the expression of an oncogene whose product promotes tumor growth. Short RNA or DNA molecules with sequences complementary to the gene's messenger RNA (mRNA) can be packaged into vectors or given to cells directly. These short molecules, called oligonucleotides, can bind to the target mRNA, preventing its translation into protein or even causing its degradation.
- Improving a patient's immune response to cancer. In one approach, gene therapy is used to introduce cytokine-producing genes into cancer cells to stimulate the immune response to the tumor.
- Inserting genes into cancer cells to make them more sensitive to chemotherapy, radiation therapy, or other treatments.
- Inserting genes into healthy blood-forming stem cells to make them more resistant to the side effects of cancer treatments, such as high doses of anticancer drugs.
- Introducing "suicide genes" into a patient's cancer cells. A suicide gene is a gene whose product is able to activate a "pro-drug" (an inactive form of a toxic drug), causing the toxic drug to be produced only in cancer cells in patients given the pro-drug. Normal cells, which do not express the suicide genes, are

not affected by the pro-drug. Local delivery of a pro-drug or a suicide gene increases sensitivity of tumor cells to cytotoxic drugs. The strategy was injection of an adenoviral vector expressing thymidine kinase (TK) gene into the tumor (Fig. 17.5). Cells, which take up and express the TK gene, are killed after administration of ganciclovir, which is phosphorylated to a toxic nucleoside by TK.
- Inserting genes to prevent cancer cells from developing new blood vessels (angiogenesis). Gene transfer strategy has been developed for tumor angiogenesis. These include expression of angiogenesis inhibitors as angiostatin and endostatin, or by use of siRNA for targeted inhibition of gene expression or by assisted killing by immune system cells or by direct killing of disease cell.

The proposed gene therapy clinical trials, or protocols, must be approved by at least two review boards at the researchers' institution before they can be conducted. Gene therapy protocols must also be approved by the FDA, which regulates all gene therapy products [35]. In addition, gene therapy trials that are funded by the National Institutes of Health must be registered with the NIH Recombinant DNA Advisory Committee.

17.8.2 Gene Therapy and the Central Nervous System

Gene therapy in the affected regions of the brain might resume the normal functioning by synthesis of the gene product. The several nervous system diseases may be potential diseases where gene therapy holds promise.

17.8.3 Gene Therapy and Orthopedics

Gene therapy can ensure efficient protein delivery by release of a desired product at therapeutic levels. The gene may be inserted by using in vivo or ex vivo approach. The gene expression required at the target site of the bone, cartilage, ligament, and tendon is of appropriate and specific cytokines and growth factors, which can accelerate healing. Several growth factors, including insulin-like growth factors 1 and 2 (IGF-1 and IGF-2), transforming growth factor-β (TGF-β), platelet-derived growth factor (PDGF), acidic and basic fibroblast growth factors (α-FGF and β-FGF), and the bone morphogenetic proteins (BMPs), have been shown to contribute to bone development and regeneration process. These factors induce the migration, proliferation, and differentiation of osteoprogenitor cells. Gene therapy is also being used to treat ligament injury and rheumatoid arthritis.

17.8.3.1 Gene Therapy and HIV

HIV-1 has become one of the most important viral pathogen and causative agent of AIDS. Control of its rapid spread and treatment are very big challenges. Some of the approaches may be:

- Making cells resistant to HIV infection or inhibition of HIV replication.
- Using antisense RNA by the use of retroviruses for tat and rev mRNAs or to pol and env mRNAs.
- Retroviruses encoding ribozyme (RRz2) directed against regulatory region.
- Decoy RNAs where retroviruses encoding RRE sequence sequester all rev protein and prevent HIV replication.
- T-cell modification so that T-cytotoxic cell can express chimeric T-cell receptor for cytotoxic response to HIV-infected cells. ScFvs are also being explored for control of HIV infection.

Other Diseases

Areas of interest for gene therapy include therapies for HIV and neurodegenerative disorders like Parkinson's and Alzheimer's disease. Some of the approved gene therapy trials are mentioned in Table 17.3.

Table 17.3 This table lists the success of gene therapy for a few diseases along with the corporate launchers (some are approved; others are in trials or under development)

	Research institute	Target diseases
1.	SiBiono GeneTech	(Approved by the Chinese FDA)
		Gendicine—for head and neck cancer (Ad-p53 gene therapy)
2.	Celladon	(FDA granted breakthrough status)
		MYDICAR is intended to reduce the risk of heart failure in patients with a deficiency of the enzyme SERCA2a
3.	Introgen Therapeutics Inc.'s Advexin	(Approved in some countries)
		INGN241 and Advexin are in late stage development (Ad5CMV-p53 gene) A replication-defective adenoviral CMV vector that encodes a wild-type p53 gene
		Therapy is for recurrent, inoperable squamous cell carcinoma of the head and neck (phase III trial)
		Is also indicated at non-small cell lung cancer, breast cancer, esophageal cancer, and oral cancer
4.	Acuity Pharmaceuticals	(Under investigational new drug (IND) application)
		Cand5; wet age-related macular degeneration (siRNA-based therapy)
5.	Isis Pharmaceuticals, Inc.	Antisense oligonucleotides (AONs) (approved AON)
		Vitravene® (fomivirsen)
		Eye treatment due to cytomegalovirus retinitis in patients [25] with acquired immunodeficiency syndrome (AIDS)
		(Brand name not available in the USA; generic medicine may be available)

17.9 Gene Doping

Gene doping is a gene therapy for restoring muscle function lost because of disease or age to improve performance. The nontherapeutic use of cells, genes, and genetic elements or the modulation of gene expression, having the capacity to improve athletic performance, is defined as gene doping by the World Anti-Doping Agency (WADA).

Gene doping could involve the recreational use of gene therapies intended to treat muscle-wasting disorders. Many of these chemicals may be indistinguishable from their natural counterparts. In such cases, nothing unusual would enter the bloodstream so officials would detect nothing in a blood or urine test. For example, gene doping could be used to provide athletes a source of erythropoietin (EPO), a hormone that promotes the formation of red blood cells that is already widely abused in sports. Another candidate gene is insulin-like growth factor 1 (IGF-1) which partly controls the building and repair of muscles by stimulating the proliferation of satellite cells.

German scientists have developed a blood test that can reliably detect gene doping even after 56 days: "For the first time, a direct method is now available that uses conventional blood samples to detect doping via gene transfer." Gene doping may be used to improve sport performance, but is not limited to humans may be used and in the sport of horseracing.

Case Studies

The gene therapy trial was done in 1989 by Rosenberg et al. where they used a retrovirus to introduce the gene coding for resistance to neomycin into human tumor-infiltrating lymphocytes before infusing them into five patients with advanced melanoma. Since then more than 1500 clinical trials are underway for over 100 genes.

The first adverse event of a gene therapy treatment was the death of 18-year-old

(continued)

Jesse Gelsinger who was suffering from ornithine transcarbamylase (OTC) deficiency. The death occurred because of totally unexpected and devastating inflammatory reaction to the adenovirus-based vector. All the trials by the US Food and Drug Administration (FDA) were kept on hold [35]. In its final judgment in 2005, the US Department of Justice held the University of Pennsylvania (Trial Institute) responsible and ordered them to pay $517,000 settlement.

In 2000 France reported the successful treatment of a rare form of X-linked severe combined immunodeficiency (SCID-X1) resulting in early block in T- and natural killer (NK) lymphocyte differentiation. Unfortunately at the end of 2002, two of the ten children developed leukemia like clonal lymphocyte proliferation, which was related to retrovirus vector, halting the trial.

The UK trial using the same approach but different protocol to treat SCID-X1 was allowed and gave successful results. A similar trial in Australia for SCID-X1 also continued. The French trial was started again, but sadly one of the children developed a proliferative condition involving a different oncogene.

17.10 Recent Developments in Gene Therapy

1. Use of liposomes as vehicles for delivery of genes into the brain for Parkinson's disease. This would be a great achievement as it is difficult for other vectors to bypass the blood–brain barrier.
2. In the case that site-specific integration of a specific gene may be achieved in the host, then random integration and its serious consequences may be avoided.
3. Nonviral delivery modes are being tried to efficiently deliver specific genes for monogenic disorders [32].

Other strategies include RNA interference and therapeutic ribozymes.

17.10.1 RNA Interference Technology

The gene therapy approach which is attracting attention is RNA interference (RNAi) approach which may be used to block the expression of a chosen gene where gene function is not desirable or is harmful. Some clinical conditions are caused by gain of function or dominant negative effects. The strategy can work in the treatment of certain cancers, for example, silencing of oncogenes or autoimmune diseases. This can also be used in infectious disease where the therapy can be aimed for the silencing of infectious agent gene. Approximately, >11 trials of RNAi for various conditions like age-related macular degeneration, asthma, Huntington's disease, and hepatitis C are in process. However, there are challenges to this technology: (1) RNA thus introduced is very prone for its degradation by nucleases, (2) very little or no intact RNA reaches its target site, (3) and antisense oligonucleotides (AONs) can be applied directly to diseased tissue [21]. For example, Vitravene® (fomivirsen) in 1998 became the first FDA-approved therapeutic AON for eyes for treatment of cytomegalovirus retinitis in patients with acquired immunodeficiency syndrome (AIDS). However, delivery to the target cell is a constraint, and their conjugation with peptides for targeting and delivery can make them potentially immunogenic.

17.10.2 Therapeutic Ribozyme

Therapeutic ribozymes or RNA enzymes which can cleave RNA molecules can have therapeutic opportunities. Genetically modified ribozymes are being created where their natural RNA cleaving domain is fused with an antisense RNA sequence of specific target gene. The modified ribozyme thus can bind to the target gene RNA and cleave it. Their effectiveness in clinical practice is a matter of concern; however, the scientists are trying to use them as effective therapeutic molecules with minimal side effects.

17.10.3 Antisense Oligonucleotides

Antisense oligonucleotides (AONs) can hybridize to a sense target sequence, promoting its cleavage by RNase-H and specific gene expression knockdown. It can be used as therapeutic to knockdown genes involved in cancer, inflammatory diseases, and viral infections [17, 36].

Vitravene is a registered antisense oligonucleotide used to treat cytomegalovirus-induced retinitis. Other AONs for cancer and inflammatory diseases are in phase II and III clinical trials.

The therapies using antisense-mediated exon skipping [1] are promising therapeutic interventions for Duchenne muscular dystrophy (DMD). In this, AONs are used to restore cryptic splicing, to change levels of alternatively spliced genes. In DMD an exon is skipped in order to restore a disrupted reading frame. This results in the generation of internally deleted, but largely functional, dystrophin proteins and would convert a severe DMD into a milder Becker muscular dystrophy phenotype. This has become the most promising therapeutic tool for DMD, and a successful first-in-man trial has recently been completed [8].

The approach offers therapeutic opportunities for diseases where mutations often induce cryptic splice sites such as the Hutchinson–Gilford progeria syndrome [31].

17.10.4 Zinc Finger Nuclease

Zinc finger nucleases (ZFNs) may be engineered [28] which utilize the modular Cys2His2 zinc finger DNA-binding moiety linked to the catalytic subunit of the type II restriction enzyme FokI. By appropriate arrangement of zinc finger modules, virtually any sequence of DNA can be targeted for nucleolytic cleavage [4, 19, 20, 27, 28, 33]. The strategy was targeting of two ZFN monomers so that they can bind adjacent sites on complementary DNA strands flanking the target sequence, allowing the dimerization of the nuclease domain FokI, which is required to cleave double-stranded DNA. The technology may be used to engineer nuclear genomes in order to add, correct, or disrupt genes and is currently available commercially and via various protocols for in-house manufacture [17].

This approach can be used to correct a pathogenic mutation in IL2RG (X-linked SCID locus) with high efficiency in cultured human cells. They can also be used for $CD4^{+}$ T-helper cells which are resistant to HIV infection.

Approved Gene Therapy Product

The drug, called Gendicine, was first approved for clinical use in gene therapy by the State Food and Drug Administration (SFDA) of China to treat head and neck squamous cell carcinoma (HNSCC) in January 2004. Its launch is also surrounded by debates about regulatory concerns over its clinical safety and efficacy. Gendicine is the true hybrid of biotech age and is a recombinant adenovirus with some genetic pieces removed and others added.

Adenoviruses are a class of viruses that elicit respiratory, intestinal, and eye infections in humans. One of the main added components of Gendicine is the p53 gene—a so-called "tumor suppressor" gene or what others have termed the "guardian of the genome" because it is known to express an oncoprotein that kills tumor cells. For years, p53 and its mutations have been studied for their relevance to the cause and prevention of cancer. In addition to the original adenovirus and the p53 components, Gendicine also contains gene segments from a *Rous sarcoma virus* and bovine growth hormone.

17.11 Risks and Problems Involved in Gene Therapy

(a) Gene transfer mediated by viral vectors may target many cell types, and random insertion might lead to cancer or other complications in the host.

(b) Immune response—as the defense response is always ready to tackle the intruder, thus it reduces effectiveness of the therapy, and subsequent repetition of therapy elicits intense inflammatory reactions which may be fatal to the host.
(c) Overexpression of the inserted gene producing too much of the proteins may also trigger inflammation or an immune response.
(d) Chance of transmission of the virus from the patient to other individuals or into the environment.
(e) Long-lasting therapy is not achieved by gene therapy. For long-lasting effects, the transferred DNA must remain functional and stable, but because of this problem, the patient will have to undergo multiple rounds of gene therapy.
(f) The vectors used as delivery vehicles may also cause toxicity or might prove to be oncogenic.
(g) The therapy cannot address the disorders due to effects on multiple genes.
(h) Chance of inducing a tumor (insertional mutagenesis)—if the DNA is integrated in the wrong place in the genome, for example, in a tumor suppressor gene, it could induce a tumor. This has occurred in clinical trials for X-linked severe combined immunodeficiency (X-SCID) patients, in which hematopoietic stem cells were transduced with a corrective transgene using a retrovirus, and this led to the development of T-cell leukemia in 3 of 20 patients.

17.12 Potential of Gene Therapy

The new therapies like gene therapy [7] and stem cell therapy are in the course of development and are likely to become increasingly important with the current concerns of safety of gene transfer. So far highest gene therapy trials are done for cancer. The second targeted disease is cardiovascular disease followed by inherited monogenic disease. Other diseases where gene therapy holds promise is infectious diseases, neurological diseases, ocular pathologies, and inflammatory diseases.

The Center for Biologics Evaluation and Research (CBER) regulates cellular therapy products, human gene therapy products, and certain devices related to cell and gene therapy. None of the human gene therapy products have been approved by CBER. More than 320 gene therapy clinical trials are going on which are regulated by the Office of Cellular, Tissue, and Gene Therapies (OCTGT) in the Center for Biologics Evaluation and Research of the US Food and Drug Administration (FDA) (Table 17.4).

Some of the gene therapy products, which are approved and in trials for various diseases [9, 10, 13, 14], are being discussed in Table 17.5.

The setbacks in the clinical trials for cystic fibrosis and severe combined immunodeficiency (SCID) perceived as ideal gene therapy targets [6] have caused termination of many research projects and retraction of participants. The drug, called Gendicine, was used to treat head and neck squamous cell carcinoma (HNSCC).

Table 17.4 Some important gene therapy trials (some are approved, some ongoing, some completed, or some terminated) for the various clinical conditions [13, 14]

S. No.	Disease conditions	Diseases/deficiency
1.	Monogenic disorders	Alpha-1 antitrypsin
		Becker's muscular dystrophy
		Chronic granulomatous disease
		Cystic fibrosis
		Duchenne muscular dystrophy
		Familial amyotrophic lateral sclerosis
		Gaucher disease
		Hemophilia A and B
		Leukocyte adherence deficiency
		Ornithine transcarbamylase deficiency
		Severe combined immunodeficiency
		Wiskott–Aldrich syndrome

(continued)

Table 17.4 (continued)

S. No.	Disease conditions	Diseases/deficiency	
2.	Infectious diseases	Adenoviral infection	
		AIDS/HIV	
		Cytomegalovirus infection	
		Epstein–Barr virus	
		Hepatitis B	
		Hepatitis C	
		Influenza	
		Japanese encephalitis	
		Tetanus	
3.	Cardiovascular diseases	Angina	
		Myocardial ischemia	
		Peripheral vascular disease	
		Pulmonary hypertension	
		Vascular complications (diabetes)	
4.	Neurological diseases	Alzheimer's disease	
		Carpal tunnel syndrome	
		Epilepsy	
		Multiple sclerosis	
		Myasthenia gravis	
		Parkinson's disease	
		Peripheral neuropathy	
5.	Ocular diseases	Diabetic macular edema	
		Glaucoma	
		Macular degeneration (age related)	
		Retinitis pigmentosa	
		Superficial corneal opacity	
6.	Cancer	Adenocarcinoma of lings	Lymphoma
		Astrocytoma	Multiple myeloma
		Breast	Nasopharyngeal carcinoma
		Cervix	Neuroblastoma
		Colon	Ovary
		Colorectal	Pancreas cancer
		Germ cell	Post-hepatitis liver cancer
		Glioblastoma	Prostate
		Glioma	Renal
		Leukemia	Sarcoma
		Liver metastasis	Skin melanoma
7.	Other diseases	Chronic renal disease	
		Inflammatory bowel disease	
		Rheumatoid arthritis	

Table 17.5 Gene therapy products (some are approved by various government regulatory offices while most are in clinical trials)

	Research institute	Target diseases
1.	SiBiono GeneTech	(Approved)
		Gendicine—for head and neck cancer
2.	Boehringer Ingelheim Pharmaceuticals, Inc. (Ridgefield, CT)	(Approved)
		Gilotrif (afatinib)—for late-stage (metastatic) non-small cell lung cancer (NSCLC)
3.	Merck	(Approved)
		Gardasil anticancer vaccine (against HPV types VI, XI, XVI, XVIII)
4.	Canji Inc. (Schering-Plough Corporation)	Ovarian cancer, hepatocellular cancer, and colorectal cancer
5.	Oxford Biomedica plc	Metastatic and colorectal cancer, also working on breast cancer and pancreatic cancer
6.	Cell Genesys's unique technology	GVAX vaccine targeting five cancers but more advanced for prostate cancer
7.	Introgen Therapeutics Inc.'s Advexin	(Approved in some countries)
		INGN241 and Advexin are in late stage development
		Therapy is for recurrent, inoperable squamous cell carcinoma of the head and neck (phase III trial)
		Is also indicated at non-small cell lung cancer, breast cancer, esophageal cancer, and oral cancer
8.	AnGes MG, Inc. in Japan	Cardiovascular system (peripheral and coronary artery disease, hemodialysis access, peripheral vascular disease)
	Centelion SA (Sanofi-Aventis Group)	
	US-based Valentis, Inc.	
9.	US-based Avigen, Inc.	Parkinson's disease
10.	Targeted Genetics, Inc.	Stopped research on cystic fibrosis continues with treatment for arthritis
11.	GenVec	HIV vaccine
12.	Acuity Pharmaceuticals	Wet age-related macular degeneration (siRNA-based therapy)
13.	Celladon	Congestive heart failure
14.	Ceregene	Central nervous system

(continued)

Table 17.5 (continued)

17.13 Chapter End Summary

- Gene therapy is a treatment that involves introducing DNA/genetic material into a person's cells to fight or prevent disease.
- Gene therapy is under development for a number of diseases, such as severe combined immunodeficiencies, hemophilia, Parkinson's disease, cancer, and even HIV, through a number of different approaches; a gene can be delivered to a cell using a carrier known as a "vector."
- The most common types of vectors used in gene therapy are viruses. The viruses have high efficiency of gene transfer at high levels with transient or stable transgene insertion.
- The viruses used in gene therapy are altered to make them safe, although safety risks still exist with gene therapy as chromosomal integration is random thus they can accidently activate oncogenes.
- Several nonviral vectors are under development and are with lower risks; however, efficiencies of transfection are low with only transiently weak expression.
- The therapy can be somatic or germline with the latter being heritable. Only somatic cell therapy is allowed because of ethical concerns.

- The gene therapy has several approaches like replacing the defective copy or silencing of existing gene.
- The major ongoing trials for gene therapy are for cancer with very few of them addressing monogenic and infectious diseases. Almost all the cells of the human body are potential target for gene therapy.

Multiple Choice Questions

1. The process used in human cells in which normal genes are inserted to correct disorders is:
   (a) Recombinant subunit vaccine
   (b) Stem cell therapy
   (c) Gene therapy
   (d) Attenuated live viral vaccine
2. The commonly used vehicles used to carry healthy gene into host cells are:
   (a) Fungi
   (b) Bacteria
   (c) Viruses
   (d) Human cells
3. Which of the following techniques involve the use of fatty bubble as a transport vehicle?
   (a) *Agrobacterium*
   (b) Gene gun technology
   (c) Liposome
   (d) Microinjection
4. An adenovirus-carrying nasal spray with a functional human CFTR gene that is used to treat cystic fibrosis is an example of which type of gene therapy?
   (a) Ex vivo
   (b) In vivo
   (c) In situ
   (d) In vitro
5. Cell type which would not be a direct target for gene therapy is:
   (a) Muscle
   (b) Cardiac
   (c) Red blood
   (d) Nerve cell
6. Viral vectors most commonly used in clinical gene therapy trials for nervous system are:
   (a) Adeno-associated viruses
   (b) Herpesvirus
   (c) Vaccinia virus
   (d) Adenoviruses
7. In general, which one of these is not a risk factor in gene therapy using adenoviruses?
   (a) Toxicity due to administered dose
   (b) Insertional mutagenesis
   (c) Overexpression triggering immune response
   (d) None of the above
8. Which disease can be an effective target for gene therapy?
   (a) Osteoarthritis
   (b) Duchenne muscular dystrophy
   (c) Diabetes type II
   (d) Hypersensitive reactions
9. In a successful gene therapy, the healthy gene inserted into a target cell must:
   (a) Switch off constitutive gene expression in the cell
   (b) Form integral part of mRNA molecule
   (c) Kill the defective copy of gene
   (d) Be able to make correct amount of protein
10. Majority of gene therapy trials target:
   (a) Neurodegenerative disorders
   (b) Severe combined immunodeficiency
   (c) Cancer
   (d) Acquired immunodeficiency
11. When gene therapy is done in germline cells, it is:
   (a) Not heritable
   (b) Heritable
   (c) Unrelated to heritability
   (d) Seldomly heritable
12. First gene therapy trial was done in:
   (a) 2000
   (b) 1990
   (c) 1988
   (d) 1993
13. Germline therapy can correct a genetic defect in:
   (a) Patient only
   (b) Patient and his spouse

(c) Parent of the patient
(d) Patient and all of his or her descendants

14. Gene therapy in myoblast could be used to correct:
(a) Cystic fibrosis
(b) Parkinson's disease
(c) Muscular dystrophy
(d) Severe combined immunodeficiency

15. What is the main objective of a phase I clinical gene therapy trial?
(a) Assessment of the safety of the gene therapy product
(b) Evaluation of the optimal doses of the gene therapy product
(c) Provision of scientific proof of successful treatment
(d) Provision of preclinical safety data for clinical studies

Answers

1. (c); 2. (c); 3. (c); 4. (b); 5. (c); 6. (b); 7. (b); 8. (b); 9. (d); 10. (c); 11. (b); 12. (b); 13. (d); 14. (c); 15. (a)

Review Questions

Q1. What is gene therapy?
Q2. What is the role of gene therapy in the treatment of cancer?
Q3. How many different diseases can be potential target for gene therapy?
Q4. Which vectors can be used for delivery of genes to the cells?
Q5. Which are the potential risks associated with gene therapy?
Q6. Which was the first disease approved for treatment with gene therapy?
Q7. What are the ethical and social issues with the gene therapy?

References

1. Aartsma-Rus A, Ommen GJV (2007) Antisense-mediated exon skipping: a versatile tool with therapeutic and research applications. RNA 13:1609–1624
2. Aiuti A et al (2009) Gene therapy for immunodeficiency due to adenosine deaminase deficiency. N Engl J Med 360:447–458
3. Al-Dosari MS, Gao X (2009) Nonviral gene delivery: principle, limitations, and recent progress. AAPS J 11:671–681
4. Carroll D, Morton JJ, Beumer KJ, Segal DJ (2006) Design, construction and *in vitro* testing of zinc finger nucleases. Nat Protoc 1:1329–1341
5. Cavazzana-Calvo et al (2000) Gene therapy of human severe combined immunodeficiency (SCID)-X1 disease. Science 28:669–672
6. Cavazzana-Calvo M, Fischer A (2007) Gene therapy for severe combined immunodeficiency: are we there yet? J Clin Invest 117:1456–1465
7. Cotrim AP, Baum BJ (2008) Gene therapy: some history, applications, problems and prospects. Toxicol Pathol 36:97–103
8. Deutekom et al (2007) Local dystrophin restoration with antisense oligonucleotide PRO051. N Engl J Med 357:2677–2686
9. Edelstein ML, Abedi MR, Wixon J, Edelstein RM (2004) Gene therapy clinical trials worldwide 1989–2004 – an overview. J Gene Med 6:597–602
10. Edelstein ML, Abedi MR, Wixon J (2007) Gene therapy clinical trials worldwide to 2007-an update. J Gene Med 9:833–842
11. Fischer A et al (2004) LM02 and gene therapy for severe combined immunodeficiency. N Engl J Med 350:2526–2527
12. Gene Therapy Advisory Committee 12th Annual Report (pages 10 and 13–15). Available: http://www.advisorybodies.doh.gov.uk/genetics/gtac/GTAC12thannualReport.pdf
13. Gene therapy clinical trials worldwide (2006) J Gene Med, New Jersey, Wiley www.abedia.com/wiley/indications.php
14. Ginn SL et al (2013) Gene therapy clinical trials worldwide to 2012-an update. J Gene Med 15:65–77
15. Hacein-Bey-Abina S, von Kalle C, Schmidt M et al (2003) A serious adverse event after successful gene therapy for X-linked severe combined immunodeficiency. N Engl J Med 348:255–256
16. Kasper DL, Braunwald E, Fauci AS, Hauser SL, Longo DL, Jameson JL (eds) (2008) Harrison's principles of internal medicine, 17th edn. Mc Graw-Hill Medical publishers, New York
17. Hausen P, Stein H (1970) Ribonuclease H—an enzyme degrading enzymes with zinc finger DNA-recognition domains. Eur J Biochem 14:278–283
18. Herzog RW (2010) Gene therapy for SCID-X1: round 2. Mol Ther 18:1891
19. Kim YG, Cha J, Chandrasegaran S (1996) Hybrid restriction enzymes: zinc finger fusions to Fok I cleavage domain. Proc Natl Acad Sci U S A 93:1156–1160
20. Kim S, Lee MJ, Kim H, Kang M, Kim JS (2011) Preassembled zinc-finger arrays for rapid construction of ZFNs. Nat Methods 8:7

21. Kurreck J (2003) Antisense technologies. Improvement through novel chemical modifications. Eur J Biochem 270:1628–1644
22. Leiden JM (1995) Gene therapy—promise, pitfalls, and prognosis. N Engl J Med 28:871–873
23. Lin E, Nemunaitis J (2004) Oncolytic viral therapies. Cancer Gene Ther 11:643
24. Manno et al (2006) Successful transduction of liver in hemophilia by AAV-Factor IX and limitations imposed by the host immune response. Nat Med 12:342–347
25. Marwick C (1998) First "antisense" drug will treat CMV retinitis. JAMA 280:871
26. Otto et al (2006) Correction of X-linked chronic granulomatous disease by gene therapy, augmented by insertional activation of MDS1-EVI1, PRDM16 or SETBP1. Nat Med 12:401–409
27. Papworth M, Kolasinska P, Minczuk M (2006) Designer zinc-finger proteins and their applications. Gene 366:27–38
28. Pearson H (2008) Protein engineering: the fate of fingers. Nature 455:160–164
29. Raper SE, Chirmule N, Lee FS et al (2003) Fatal systemic inflammatory response syndrome in a ornithine transcarbamylase deficient patient following adenoviral gene transfer. Mol Genet Metab 80:148–158
30. Sadelain M et al (2003) Targeting tumours with genetically enhanced T lymphocytes. Nat Rev Cancer 3:35
31. Scaffidi P, Misteli T (2005) Reversal of the cellular phenotype in the premature aging disease Hutchinson–Gilford progeria syndrome. Nat Med 11:440–445
32. Shah PB, Losordo DW (2005) Non-viral vectors for gene therapy. Clinical trials in cardiovascular disease. Adv Genet 54:339–361
33. Smith J, Bibikova M, Whitby FG, Reddy AR, Chandrasegaran S, Carroll D (2000) Requirements for double-strand cleavage by chimeric restriction enzymes with zinc finger DNA-recognition domains. Nucleic Acids Res 28:3361–3369
34. Stein et al (2010) Genomic instability and myelodysplasia with monosomy 7 consequent to EVI1 activation after gene therapy for chronic granulomatous disease. Nat Med 16:198–204
35. Takefman D, Bryan W (2012) The state of gene therapies: the FDA perspective. Mol Ther 20:877–878
36. Zamecnik PC, Stephenson ML (1978) Inhibition of Rous sarcoma virus replication and cell transformation by a specific oligodeoxynucleotide. Proc Natl Acad Sci 75:280–284

Some Selected Resources

history.nih.gov/exhibits/genetics/sect4.htm

http://www.fda.gov/biologicsbloodvaccines/cellulargenetherapyproducts

http://www.fda.gov/biologicsbloodvaccines/cellulargenetherapyproducts

National Cancer Institute Factsheet

Some text directly from National Cancer Institute Factsheet as source

Successful gene therapy trial in Parkinson's Disease

www.cfgenetherapy.org.uk/genetherapy/successes.html

www.ncbi.nlm.nig.gov/pubmed

Forensic Medicine

18

Abstract

Forensic science is a link between science and law. As the criminals became smarter, the methods of identifying criminals like photographs and fingerprint did not worked. The forensic science came as a rescue for crime investigating agencies and police, for seizing the criminals and culprits and resolving disputes related to paternity and crime. Years back, it was realized that small portions from the skin can be used for identification of an individual by unique signatures known as DNA fingerprint that laid the foundation for this interesting and highly technical branch of forensics. The important functions of forensic laboratories are to undertake diverse laboratory examination of physical evidence material and therefore to provide reliable and irrefutable circumstantial evidence providing or strengthening the links in the chain of evidences to detect the guilty or clear the innocent and responsibility of providing a meaningful report in a meaningful time frame. It helps in the investigation of crime through scientific guidance by using appropriate scientific aid within and beyond the laboratory. This chapter focuses on identification of various body fluids, technique of DNA fingerprint, DNA profiling, and DNA methylamine-based methods for individual identification.

18.1 Introduction

Forensic science is a link between science and law. As the criminals became smarter, the methods of identifying criminals like photographs and fingerprint did not worked. The forensic science came as a rescue for crime investigating agencies and police, for seizing the criminals and culprits and resolving disputes related to paternity and crime. Years back, it was realized that small portions from the skin could be used for identification of an individual by unique signatures known as DNA fingerprints that laid the foundation for this interesting and highly technical branch of forensics. The important functions of forensic laboratories are to undertake diverse laboratory examination of physical evidence material and therefore to provide reliable and irrefutable circumstantial evidence providing or strengthening the links in the chain of evidences to detect the guilty or clear the innocent and responsibility of providing a meaningful report in a meaningful time frame. It helps

V. Gupta et al., *Basic and Applied Aspects of Biotechnology*, DOI 10.1007/978-981-10-0875-7_18

in the investigation of crime through scientific guidance by using appropriate scientific aid within and beyond the laboratory [13].

18.2 Collection of Specimen

Collection of specimen or biological fluid is very crucial and requires extensive care. This sample can provide useful information about the scene of the crime [1]. Crime investigators should follow certain precautions (listed on FBI website for protection of particular information). The sources of materials from the crime scene may be dirty laundry, a cigarette butt, used mug or glass, hair, fresh or dried bloodstain, dried semen or saliva, and teeth [4,13]. These all need to be handled with precautions as all the living things have DNA; therefore, attention is required while collecting and preserving the evidence:

- Wearing of disposable gloves with frequent changes.
- Use of disposable instruments.
- Talking, sneezing, coughing, or touching anything that might contain the investigator's own DNA should be avoided.
- Careful handling to avoid contamination.

The identification of stain present at the scene of crime is important. Various tests are used to identify the type of stain. The nature of accident or crime may be predicted from the pattern of bloodstains.

Blood consists of cells and plasma with a pH of 7.4. It contains three types of cells: (1) RBCs, (2) WBCs, and (3) platelets. The tests of forensic importance employ RBCs which are circular biconcave and nonnucleated in man and mammals (except camel in which they are oval and nonnucleated). In pisces, amphibians, reptiles, and aves, they are oval and nucleated. The preliminary tests are used initially so that identification can be made; later on these are followed by confirmatory tests. There are number of tests utilized by forensic experts apart from circumstantial evidence.

18.2.1 Physical Examination

It is very important to do the examination of clothing, pattern of spot (size, shape, and direction), character of stain, and age and condition of stain. It is also important to evaluate the nature of stain, the species of stain (human or other animals), and properties of the blood.

Chemical properties of blood: The various properties of blood like blood group antigens, blood proteins, and enzymes are determined by the genetic makeup of the individual, which are inherited from parents. Since different individuals have different alleles, therefore, their tests are used routinely for stain grouping.

18.2.2 Blood Groups

1. ABO system: the codominant blood group antigens present in humans (A, B, AB, and O).
2. MN system: present as alleles (M, N, and MN).
3. Rh system: Rh factor is named as antiserum in rabbit against red cells of rhesus monkey.

Dilute or tampered blood spots require expensive chemicals and other tests. The various tests are mentioned herewith which are utilized for detection of blood spot.

Heme (hemoglobin) possesses peroxidase-like activity leading to breakdown of hydrogen peroxide with oxidizing agent production. This oxidizing agent then reacts with a variety of agents (benzidine, orthotolidine, leucomalachite green, leucocrystal violet, or phenolphthalein) with color production. The tests are mentioned in Table 18.1.

18.2.3 Spectroscopic Analysis

This is done with fresh blood, two absorption peaks of oxyhemoglobin between D and E lines of oxyhemoglobin. Addition of reducing agent results in one broadband of red hemoglobin

Table 18.1 The table shows various tests employed for determination of hemoglobin of blood

Benzidine test	Benzidine in glacial acetic acid in the presence of hydrogen peroxide results in greenish blue color. Might give false positives (plant peroxidases)
Phenolphthalein test (Kastle–Meyer test)	Phenolphthalein is reduced by zinc dust in strongly alkaline medium. Reduced phenolphthalein in the presence of hydrogen peroxide and by the action of catalase gives pink or purple color. It is highly sensitive (detection as dilute as 1 in 10,000)
	Is nondestructive for the samples
Takayama's hemochromogen crystal test	Microscopic test—acicular-shaped salmon pink hemochromogen crystals are formed by Takayama's reagent (sodium hydroxide, pyridine, and glucose)
Hemin crystal/Teichmann's test	It is of less practical value in which hemoglobin is converted into hemin, which in the presence of halogens forms brown rhombic crystals
Luminol (alternate light source)	Chemiluminescence of luminol is enhanced by iron. Even washed bloodstains are detected; however, duration of illumination is less (30 s). It reacts with copper or bleaching agents.
Bluestar® Forensic	Can detect bloodstains by chemiluminescence and gives sensitive and stable results. Does not damage DNA
OneStep ABAcard® HemaTrace test strip	Immunochromatographic test based upon monoclonal antihuman hemoglobin antibody

between D and E. Carboxyhemoglobin is like oxyhemoglobin but unaffected by reducing agent.

Methemoglobin Due to poisoning by nitrites, phenacetin, and sulfanol, it shows bands, one in red, two similar to oxyhemoglobin, and faint fourth one in green region.

18.2.4 Electrophoretic Separation

Electrophoretic field-based grouping of blood protein is also done:

1. Haptoglobin (Hp): Alpha-2-glycoprotein, which occurs in homozygous or heterozygous combination.
2. Globulin Gc: Codominant with Gc1-1, Gc1-2, and Gc2-2.
3. Gm and Inv immunoglobulins: Located on immunoglobulins.
4. Hemoglobin types: Fetal, adult, or abnormal.

18.2.5 Semen

The thick, yellowish white secretion with seminal odor is semen. The secretions of seminal vesicles and prostrate have high choline and lecithin (seminal vesicle) and acid phosphatase and spermine (prostrate) with testis-derived spermatozoa and epithelial cells. The detection of semen can be done by the following tests as in Table 18.2.

18.2.6 Saliva

Saliva is secreted by the salivary gland of the mouth and is detected by alpha-amylase activity, which hydrolyzes polysaccharides into small sugar molecules by starch iodine test. Due to variability in the presence of amylase (little/no/good concentration), the tests may be supplemented with other tests.

Alpha-amylase can also be detected by Phadebas® test, which has amylopectin–procion red. However, two different forms of alpha-

Table 18.2 The table shows tests for detection of semen [3, 10, 12]

UV light-based wood lamp (320–400 nm)	Semen detection. Specificity is low and might give false positives
Bluemaxx TM BM500	Highly sensitive to semen
Polilight®	Detect several body fluids including semen
Lumatec® Superlite 400	Semen and saliva both can be detected
Acid phosphatase test	AP is secreted by the prostate gland. 500–1,000× higher activity than other body fluids
Florence test	Microscopic in which Florence reagent (potassium iodide and iodine) results in dark brown crystals resembling hemin
Barberio's and creatine phosphokinase test	Used for presence of spermine and enzyme activity in spermatozoa
Detection of prostate-specific antigen (PSA)	PSA can be detected by OneStep ABAcard® PSA which uses monoclonal antihuman PSA
	Antibody

amylase which are indistinguishable in their enzyme activity are found in the human body, AMY1 (saliva, breast milk, and perspiration) and AMY2 (pancreas, semen, and vaginal secretion).

18.2.7 Other Body Fluids and Components

Urine: Urine fluoresces in the presence of UV light and ammonia smells due to bacterial degradation.

Milk: It is detected in the colostrum on undergarments (e.g., brassiere or blouse) and is associated with crimes of pregnancy, abortion, or concealment of birth.

Hair: Hair is examined for texture, cuticle, medulla, pigment, etc.

18.3 DNA Fingerprinting

The crucial stage of forensic medicine was being taken care by the advance tools of biotechnology like DNA profiling and recently emerging RNA profiling and DNA methylation. The world is grateful to the DNA analysis, which has helped to solve high-profile crime. The tools enabled investigators to do DNA analysis from hairs, bloodstains, semen, or other biological materials from the scene of the crime. These tools are now commonly referred as DNA profiling (formerly a different technique of DNA fingerprinting was in use). Before going into details of the techniques, first we should understand the basics, which lead to the unique profile of DNA.

DNA fingerprinting: Sir Alec Jeffreys of the University of Leicester in the mid-1980s introduced the method for analyzing DNA for the identification of individuals called DNA fingerprinting. The DNA fingerprint is important, as, except identical twins, no two individuals can have the same fingerprint. The genome is a constant feature in terms of location of genes with their average separation. However, differences, which lead to polymorphism, may be because of:

- Slippage of replication
- VNTRs
- SNPs
- RFLP

18.3.1 Slippage of Replication

The replication slippage may arise from mispairing of template strand or a newly synthesized strand during DNA replication (Fig. 18.1). The short tandem repeats are considered very prone for slippage. The figure shows normal replication bubble in the region. The slippage can occur due to the region of non-pairing with one or more repeats of the newly synthesized strand (backward slippage) or the parental template strand (forward slippage). These cause insertion and deletion of the resulting sequence, respectively [5]. However the error may be corrected by repair enzyme. The slippage of replication or polymerase slippage may result in variable number of tandem repeats (VNTRs) (Fig. 18.1).

This can also occur because of unequal crossing over, where the nonallelic homologous recombination occurs predominantly in

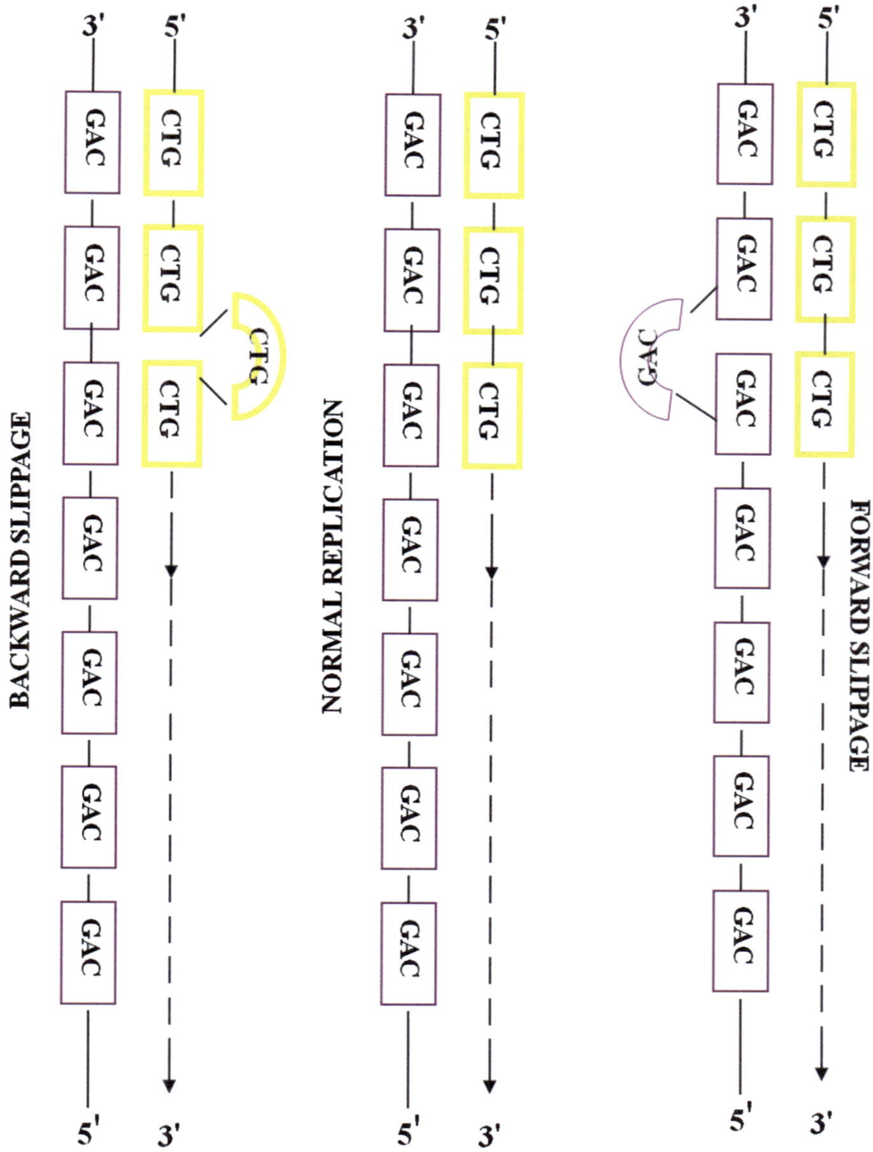

Fig. 18.1 Slippage of replication due to slipped-strand mispairing of the complementary copy of the DNA at short tandem repeats (STRs). Backward slippage and forward slippage lead to mispairing of single or more repeat motifs in the newly synthesized and parental strand, respectively, leading to insertion in backward and deletion in forward slippage

regions where there are tandem repeats of moderate- to large-sized sequence with high homology between the repeats between non-sister chromatids. These can result in tandemly duplicated locus as compared to original singly copy locus.

18.3.2 Satellite DNA

The occurrence of repeated noncoding human DNA is in blocks of tandem repeat motifs. These motifs may be simple (1–10NT) or complex (10–100NT). Three subclasses of repeats are present:

(1) satellite, (2) minisatellite, and (3) microsatellite DNA.

Satellite DNA Very large array of tandemly repeated DNA with simple or complex sequence, which makes most of the heterochromatic regions and occurs notably close to centromeres. Three human satellite DNAs, satellite I, II, and III, have been separated by buoyant density gradient centrifugation. These satellites may be 5–171 bp spanning 100 kb to several Mb size.

Minisatellite, DNA It has motifs of moderate size, which are dispersed over nuclear genome. Hypervariable minisatellite DNA sequences of short tandem repeats are organized in over 1,000 arrays (0.1–20 kb long) and are highly polymorphic. They may have variable repeat units in size with common core sequence (GGGCAGGAXG where X is any nucleotide) and are found near telomeres but can also occur at other chromosomal locations. Hypervariable minisatellites are reported to be the hot spot of homologous recombination. They have a major role in DNA fingerprinting where DNA probe with common core sequence can hybridize with multiple minisatellite DNA loci on all chromosomes, which result in individual specific unique pattern. Separate families of minisatellite sequence are present at telomeric DNA where 3–20 kb tandem hexanucleotide units are present which are added by telomerase.

Microsatellite Commonly referred as simple sequence repeats (SSRs) which have repeat motifs of less than 10 bp. They are present throughout the genome for over 60 Mb (2 % of the genome). Arrays may have dinucleotide, trinucleotide, tetranucleotide, or more nucleotide motifs, but dinucleotide repeat of CA/TG is common (0.5 % of the genome) and others are rare. The polymorphism in these arises from replication slippage.

18.3.3 Single Nucleotide Polymorphism

The single base changes occurring in two or more individuals are referred as single nucleotide polymorphism (SNP/snips). These mutations may be transition or transversion. Many SNPs occur in the intronic region, but sometimes they may lead to serious disorders where single base change occurring in the gene may result in incorporation of wrong amino acid and therefore nonfunctional protein, for example, sickle cell anemia, thalessemia etc. SNPs occur throughout the entire genome with a frequency of once every 1,000 bases. The presence of SNP may also result in creation or loss of restriction enzyme site. Due to loss or creation of the restriction site, polymorphism may be observed.

18.3.4 Restriction Fragment Length Polymorphism (RFLP)

RFLP is a codominant DNA marker. RFLP analysis involves the separation of DNA followed by restriction digestion with the desired enzyme. The digested DNA is electrophoresed and transferred onto the membrane (Southern blotting). The DNA is hybridized with radiolabeled probes against the target sequences, the size may vary because of loss or creation of restriction site.

The DNA fingerprinting invented by Jeffreys et al. in 1985 [6, 7] involves RFLP with radiolabeled probes against hypervariable dispersed repetitive sequences. This repeated sequence is dispersed throughout the genome, that is, they vary in their genomic position and can be located at different positions in the genomes of different people. The repeat initially utilized had the sequence GGGCAGGAXG (X is any nucleotide) obtained from myoglobin gene (Fig. 18.2).

The sequence is present in many minisatellite spreads across the genome. Hybridization of the Southern blot with radiolabeled probe against target sequence reveals a series of bands, each

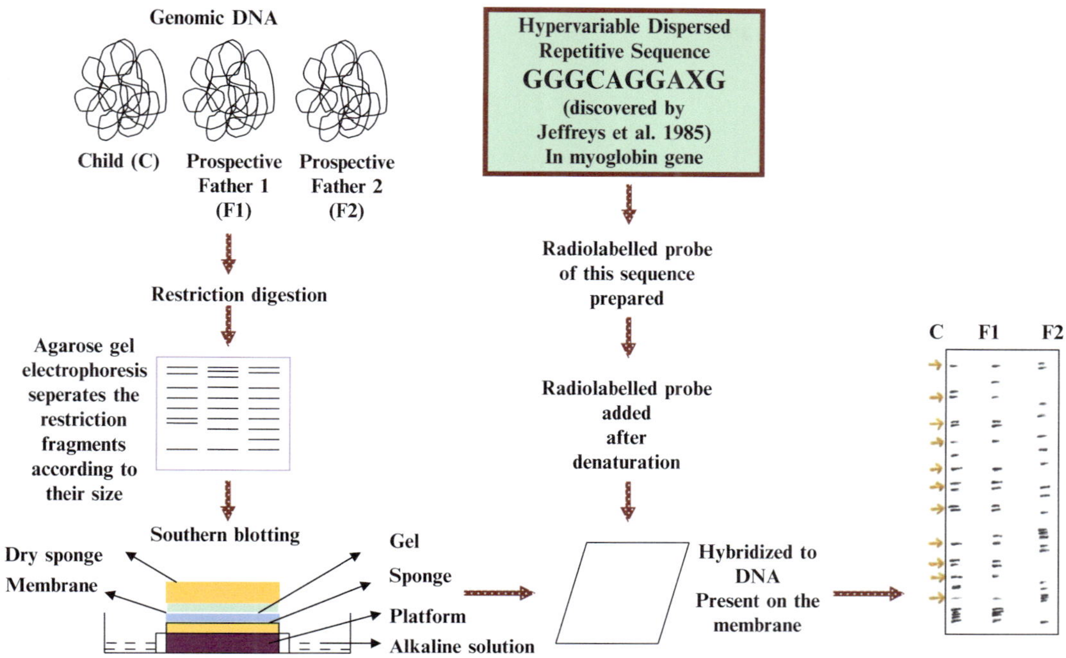

Fig. 18.2 The figure shows the technique of DNA fingerprinting. Genomic DNA was isolated from the child, prospective fathers 1 and 2. After restriction digestion, Southern blotting was performed. Radiolabeled probe was prepared for the minisatellite as shown. The membrane was hybridized with the probe prepared. Upon developing and obtaining DNA fingerprint, it is observed that prospective father F1 is actually the father of the child

band representing a restriction fragment that contains the repeat. The polymorphism between two samples may result from:

- Variation in the occurrence of restriction site
- Variable number of tandem repeats

The methodology revolutionized the medical identification but suffers from three limitations for wide usage in routine crime investigation:

- The technique requires large amount of DNA (several micrograms) for Southern blotting. The minute amount of sample obtained from the scene of crime may not be sufficient for DNA fingerprinting.
- It is not possible to ascertain that which pairs of bands represent alleles as the bands are matched by position and intensity. The gel is divided into many bins; the matching and scoring for bands are done for each bin.
- Although the sites of repeat sequence are hypervariable but with a limit, there is a small chance that two unrelated individuals can have the near similar fingerprint, which can lead to wrong acquittal.

18.4 DNA Profiling

The powerful tool of DNA profiling avoids all these limitations of the DNA fingerprinting. The DNA profiling uses microsatellite with tri- or tetranucleotide repeat polymorphism, which can be typed by multiplex PCR (Fig. 18.3). But other variant as minisatellites or SNPs may also be used.

18.4.1 Multiplex Polymerase Chain Reaction (PCR)

The technique of interest can be multiplex PCR [5]. The amplification is done with multiple pairs of primers with one for each target sequence. This technique may be of importance

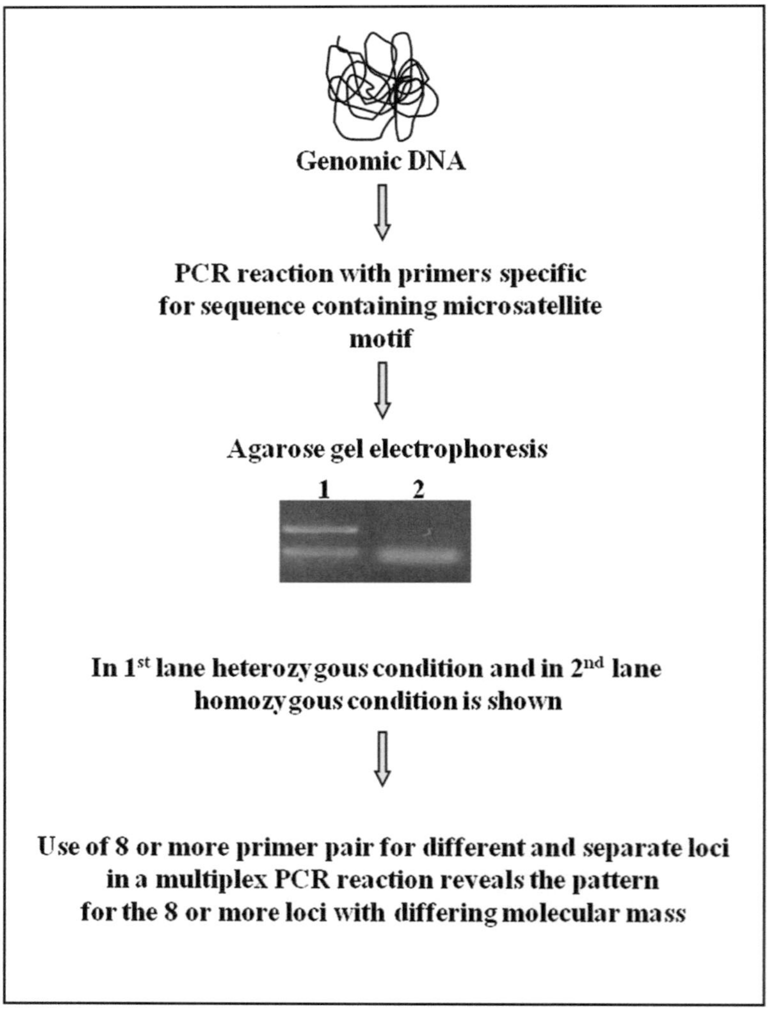

Fig. 18.3 The figure shows the technique of DNA profiling. In this DNA is targeted for the presence of microsatellite motif containing sequence using specific primers. A combination of eight or more primer pairs results in specific amplification and thus absolute monitoring. It is highly impossible that two different individuals would contain same repeats at 8–10 sites which are analyzed

in working on multiple loci for giving definitive results.

Variable number of tandem repeats (VNTR): The human genome has huge intronic region, which has unique sequences and replicative sequences.

The microsatellite or short tandem repeat (STR) is a short sequence of 1–10 nucleotides in length that is repeated several times in tandem array of which dinucleotide repeat CA is most common in human genome with 5–20 motifs.

The advantage is that even a single cell left at the scene of crime can be typed. The variation in the number of repeats may be due to errors in replication as described above, and many versions of STR may exist with differing number of repeats in different DNA sequences. The typing is allele specific and is done using the PCR primers that bind to DNA sequence containing the repeat. The product is analyzed on agarose gel electrophoresis.

The band shows the test allele or alleles present in the DNA sample with two copies of each STR (paternal and maternal) and a single copy that is a homozygous condition. The combined genotyping at 10–15 unlinked highly polymorphic loci may give definitive result. In the population, the profiling is done using a panel of nine STRs which can be typed in a single multiplex PCR using a series of primer pairs. Each primer pair leads to amplification of STR which has different size and therefore different position on agarose gel (18.3). The tool is helpful in solving paternity disputes and assists in crime-related investigations.

Mystery of Missing Romanovs For about 300 years, the Romanovs ruled the Russia. Following Bolshevik revolution in 1917, the last ruling Tsar, Nicolas II, his wife Tsarina Alexandra, and their five children, Olga, Tatiana, Maria, Anastasia, and the crown prince Alexei, were held along with four loyal members of the royal family (their family doctor, valet, maid, and cook). In July 1918 the royal family and their staff were executed.

In the late 1970s, a mass grave containing the remains of the dead royal family (with three children) and their loyal members was found. DNA extraction from bones (bodies recovered were little more than the collection of bones) followed by STR typing by only two primers provided sufficient data for male and female parents and of the children. The remains were also matched from living relatives of the Romanovs through mitochondrial DNA (maternal inheritance). The studies showed that the remains were those of Tsar Nicolas II, Tsarina Alexandra, and three of their daughters.

The two children missing from the mass grave, Alexei and one of his sisters, Anastasia, became a mystery for many years. In between several women claimed to be a Romanov princess. In 2007 the archeologists discovered a few bone fragments about 70 m away from the first grave. Thorough investigation revealed that the bones belonged to two people, one female and the other male. Similar royal jewelry was worn by the female. They were then tested for mitochondrial DNA and autosomal STR, and the results showed that the second grave of one female and one male child was that of missing Romanov children of Tsar Nicolas II and Tsarina Alexandra. This was the most publicized case solved by DNA fingerprinting, and the advances in the field were capable of identifying the entire Romanov family.

(continued)

Woman in the Suitcase In November 2002 in North Yorkshire, an almost naked and decomposed body of a young Asian woman, in a suitcase, was discovered. The forensic analysis confirmed that the body was from Southeast Asia of a 21-year-old woman Hyo Jung Jin, who was studying French at the University of Lyon. However, after a trip to London, she did not return and her brother put her details on a Korean missing person's website. Later her details were noticed by South Korean police officer who contacted the authorities. Her records of fingerprint were obtained from the Korean embassy, and the police was able to link Hyo Jung Jin's death to that of In Hea Song (another Korean student whose almost naked body was found in a cupboard). Her identity was confirmed by matching her DNA with her parents. Money was withdrawn from both women's bank card. Both women were associated with the same house with same landlord Kyo Soo Kim. The tape used for both murders was tracked to four branches of Tate Gallery in the UK. One roll of tape was found in Kyo Soo Kim's home along with other evidences of blood. In 2003, the judge found Kyo Soo Kim guilty of the two murders.

DNA profiling may also have certain issues, as there are rare chances of fortuitous match but:

- Close relatives may give good match with the profiles.
- In case the sample obtained from the crime scene is not able to give a full profile due to small quantity or minor degradation resulting in failed amplification of some genes, the quantity of DNA is below threshold level.
- PCR may amplify the DNA in these conditions, but the profile is a partial representation of the original sample.
- In case the sample obtained from crime scene has a mixture of several DNA samples, then

identification of individual profile is extremely difficult from a mixed sample.

The markers in the Combined DNA Index System (CODIS) (in the USA) and Second-Generation Multiplex Plus (SGM+) (in Europe) or amelogenin as sex markers are chosen as they may only help to identify a person. These help to ascertain ethnic groups as allelic frequency differs between ethnic groups.

There still are many issues such as courtroom issues, independence of genotypes, and ethical and political issues associated with the field.

CODIS and SGM+ Markers

STRs are used to reveal polymorphism in DNA profiling. In the USA, 13 marker sets are selected referred to as Combined DNA Index System (CODIS) markers which are present on 13 different autosomes along with amelogenin (AMEL) on X chromosome (AMELX) and AMELY used to ascertain individuals and their sex. In the UK the Second-Generation Multiplex Plus (SGM+) markers are used. These consist of ten core autosomal STR and AMELX and AMELY. Both have eight core loci from different targets and AMEL as common loci.

18.5 Future Prospects

Years have seen many methods and techniques for the identification of samples available from crime scene. The newer methods with RNA profiling or DNA methylation detection have been proposed which are specific to forensically relevant body fluids [5, 9].

18.5.1 RNA Profiling

Even tools based upon RNA technology are coming up with wide application. The RNA analysis can give an estimation of the delay in postmortem, age of bloodstain and wound, cause of death, and source of body fluids [2, 5]. Scientists have identified the messenger RNA specific for blood and saliva and showed stable expression pattern after almost 6 months of storage and reliable expression in 16-year-old bloodstains [8, 11]. The advantage of identification of body fluid is simultaneous extraction of DNA and RNA from a particular stain. A multiplex reaction can detect several body fluids with data on multiple gene expression. Quantitative reverse transcriptase PCR (qRT-PCR) might be appropriate for detection of relative gene expression levels along with end point PCR for detection of tissue-specific transcript. In multiplex RT-PCR reaction, the scientists have used body fluid-specific genes, i.e., for blood, β-spectrin (SPTB), porphobilinogen deaminase (PBGD), d-aminolevulinate synthase 2 (ALAS-2), or hemoglobin A (HBA); for saliva, statherin (STATH) and histatin 3 (HTN3); for semen, protamine 1 (PRM1), protamine 2 (PRM2), or kallikrein (KLK); for vaginal secretion, human beta-defensin1 (HBD-1) and mucin 4 (MUC4); and for menstrual blood, matrix metalloproteinase 7 and 10 (MMP7 and MMP10). Housekeeping gene GAPDH was used for normalization of the expression of body fluids. However, several mRNA markers proposed for vaginal secretions gave false positives; thus, 16S–23S rRNA intergenic spacer region (ISR) of *Lactobacillus crispatus* and *Lactobacillus gasseri* predominant in the vagina of women might be a potential tool for identification of vaginal fluids.

Though mRNA usage was successful, inadequate environmental conditions such as humidity and temperature may alter mRNA stability. Thus, microRNAs (miRNAs) which are noncoding RNA of 18–22 nucleotides in length and regulate gene expression at posttranscriptional level are known to have tissue-specific expression pattern. They have the advantage of being small thus less prone for degradation. The scientists have demonstrated that miRNAs have the ability to identify body fluids with as little as 50 pg of total RNA. However complete validation and appropriate models are required for miRNA quantification in forensic application.

18.5.2 DNA Methylation

DNA methylation-based profiling methods are able to give body fluid identification due to tissue-specific methylation patterns. Methylation occurs at fifth position of the pyrimidine ring of cytosine in CpG dinucleotides and affects chromatin structure, thus inhibiting gene expression. Methylation patterns are different due to chromosome segments called tissue-specific differentially methylated regions (tDMRs); their detection would allow identification of tissue or cell types of DNA samples. The genomic loci have been identified which are differently methylated by using methylation-sensitive restriction enzyme PCR (MSRE-PCR) with methylation-sensitive restriction enzyme digestion of DNA followed by multiplex PCR of specific loci with fluorescently labeled primers and subsequent detection. Wasserstrom et al. [14] developed a kit DNA source identifier (DSI)-semen TM which is based upon detection of semen-specific DNA methylation patterns in five genomic loci using MSRE-PCR. Age-related methylation changes in semen-specific tDMRs can predict body fluid from young or elderly man as they are susceptible to change by aging. However, apart from semen application, DNA methylation studies would require more markers and careful monitoring for future application.

18.6 Chapter End Summary

- Forensic medicine deals with the examination of the biological material present at the scene of crime. Though the field has wide applications, the technique is capable of precisely identifying biological material.
- DNA fingerprinting which was based upon satellite DNA revolutionized the world. The technique was based upon RFLP and probing the regions by using the minisatellite sequence probe. As the analysis involved random sequences, it required very high amounts of DNA and was time and labor intensive and thus has become obsolete.
- Nowadays even more powerful DNA profiling is used. It used several primers for STS markers in a multiplex reaction. It is easier and PCR-based method thus is advantageous over DNA fingerprinting. It can be performed by single cell left at the crime scene.
- Even single cell can be typed. Recently use is made of miRNA and DNA methylation which can give an indication about the source of body fluid.
- However, the forensic analysis which was based upon DNA profiling was excellent in making personal identification and with development of markers related to biological fluids made the overall forensic evaluation very powerful.
- In the times to come, forensic science would greatly improve because of advances in genetics, epigenetics, and molecular biology.

Multiple Choice Questions

1. Minisatellite DNA used by Jeffery had the sequence:
   (a) CCCGGCGGXTC
   (b) GGGCAGGAXG
   (c) TTCCGGGXTCG
   (d) CGCCTGCCAXC
2. Microsatellites are:
   (a) Repeat motifs of ten nucleotides
   (b) Repeat motifs of di- or trinucleotides
   (c) Repeat motifs of 100 bp
   (d) Repeat motifs of 1,000 bp
3. Multiplex PCR is:
   (a) PCR reaction specific for single loci
   (b) PCR reaction with multiple sets of random primers
   (c) PCR reaction with multiple and specific primers
   (d) None of these
4. DNA fingerprinting refers to:
   (a) DNA analysis by using satellite DNA
   (b) DNA analysis using RNA probe
   (c) DNA analysis using protein probe
   (d) All of these
5. The mRNA analysis might be useful for identification of:
   (a) An individual at the crime scene
   (b) A population in the city
   (c) Identification of body fluids
   (d) All of the above

6. Which of these can give approximation of the age on the sample?
   (a) DNA fingerprinting
   (b) DNA footprinting
   (c) DNA methylation
   (d) All of these
7. Kastle–Meyer test is based upon:
   (a) Reaction of benzidine in glacial acetic acid
   (b) Reaction of phenolphthalein in alkaline medium
   (c) Immunochromatographic test
   (d) Chemiluminescence of luminol
8. For identification of vaginal samples, which method might be useful?
   (a) DNA analysis of human
   (b) RNA analysis of human
   (c) Bacterial DNA analysis
   (d) None of these
9. MicroRNA (miRNA) has:
   (a) Tissue-specific expression
   (b) Small in size
   (c) Less prone for degradation
   (d) All of these
10. VNTRs might arise due to:
   (a) Slippage of replication
   (b) Duplication of microsatellite
   (c) Deletion of microsatellite
   (d) All of these

Answers

1. (b); 2. (b); 3. (c); 4. (a); 5. (c); 6. (c); 7. (b); 8. (c); 9. (d); 10. (d)

Review Questions

Q1. What is the importance of forensics?
Q2. What tests can be used for identification of blood?
Q3. What is DNA fingerprinting?
Q4. What do you understand by slippage of replication?
Q5. What is DNA profiling?

References

1. An JH, Shin KJ, Yang WI, Lee HY (2012) Body fluid identification in forensics. BMB Reports 545–553
2. Bauer M (2007) RNA in forensic science. Forensic Sci Int Gen 1:69–74
3. Fielder A, Rehdorf J, Hilbers F, Johrdan L, Stribl C, Benecke M (2008) Detection of semen (human and boar) and saliva on fabrics by a very high powered UV-/VIS- light source. Open Forensic Sci J 1:12–15
4. Greenfield A, Sloan MA (2003) Identification of biological fluids and stains. In: James SH, Nordby JJ (eds) Forensic science: an introduction to scientific and investigative techniques. CRC Press, Boca Raton, pp 203–220
5. Human Molecular Genetics (2011) Forth edition by Strachan T and Read AP, Garland Science, London and New York
6. Jeffreys AJ, Wilson V, Thein LS (1985) Individuals-specific fingerprints of human DNA. Nature 314:67–73
7. Jeffreys AJ, MacLeod A, Tamaki K, Neil DL, Monckton DG (1991) Minisatellite repeat coding as a digital approach to DNA typing. Nature 354:204–209
8. Juusola J, Ballantyne J (2007) mRNA profiling for body fluid identification by multiplex quantitative RT-PCR. J Forensic Sci 52:1252–1262
9. Kayser M, De Knijff P (2011) Improving human forensics through advances in genetics, genomics and molecular biology. Nat Rev Genet 12:179–192
10. Nelson DG, Santucci KA (2002) An alternate light source to detect semen. Acad Emerg Med 9:1045–1048
11. Nussbaumer C, Gharehbaghi-Schnell E, Korschineck I (2006) Messenger RNA profiling: a novel method for body fluid identification by real-time PCR. Forensic Sci Int 157:181–186
12. Vandenberg N, Van Oorschot RA (2006) The use of Polilight in the detection of seminal fluid, saliva, and bloodstains and comparison with conventional chemical-based screening tests. J Forensic Sci 51:361–370
13. Virkler K, Lednev IK (2009) Analysis of body fluids for forensic purposes: from laboratory testing to non-destructive rapid confirmatory identification at a crime scene. Forensic Sci Int 188:1–17
14. Wasserstrom A, Frumkin D, Davidson A, Shpitzen M, Herman Y, Gafny R (2012) Demonstration of DSI-semen-A novel DNA methylation-based forensic semen identification assay. Forensic Sci Int Genet (in press)

Some Selected Resources

http://dx.doi.org/10.1016/j.fsigen.2012.08.009
journals.lww.com/amjforensicmedicine
Legal and Forensic Medicine – Springer
www.forensicmed.co.uk
www.thefreedictionary.com/forensic+medicine

Environmental Biotechnology 19

Abstract

In this chapter, the reader would learn about the scope and advantages of environmental biotechnology and its role in establishing sustainable environment practices. Eco-friendly that is pollution-free and sustainable environmental conditions are what everyone wants. However, due to extreme negligence and industrialization, human beings have exploited the environment badly. With the help of biotechnological tools, the issues like the monitoring, assessment, and control of the pollution by bioremediation may be implemented. The sustainable environmental practices would require (1) reduction in the production and consumption of material inappropriate for the environment, (2) creating the environment safe for not only present but for future generations, and (3) leading life which is healthy and in accordance with nature.

Now as legislative is established for environmental protection, there has been coordination and support of the number of countries for setting up standards for industry and compliance to these standards. Thus in general, environmental biotechnology deals with the role of biotechnology in maintaining suitable environment, where the environment is clean with clean water and pollution-free air with restoration of the natural resources.

19.1 Introduction

In this chapter the reader would learn about the scope and advantages of environmental biotechnology and its role in establishing sustainable environmental practices. Everyone wants to stay in conditions which are pollution-free. Only sustainable environmental practices can help reduce the damage to the environment. But due to extreme negligence and industrialization, human beings have exploited the environment badly. With the help of biotechnological tools, the issues like the monitoring, assessment, and control of the pollution may be implemented. The sustainable environmental practices would require (1) reduction in the production and consumption of material inappropriate for the environment, (2) creating the environment safe for not only present but for future generations, and (3) leading life which is healthy and in accordance with nature.

Now as legislative is established for environmental protection, there has been coordination

V. Gupta et al., *Basic and Applied Aspects of Biotechnology*, DOI 10.1007/978-981-10-0875-7_19

and support of the number of countries for setting up standards for industry and compliance to these standards. Thus in general environmental biotechnology deals with the role of biotechnology in maintaining suitable environment, where environment is clean with clean water and pollution-free air with restoration of the natural resources [7, 11].

19.2 Pollution

Pollution is the condition whereby the presence of unwanted and harmful substances referred as pollutants pose risk to the health of not only human beings but also other animals and plants living in those conditions. The effects of pollutants can be harmful and toxic. These affect directly or indirectly through plants (polluted water used for irrigation purposes) and then via animals would enter the food chain.

Xenobiotics The compounds present in air, water, soil, or sediments which are foreign to the earth's biosphere are called xenobiotic compounds. These compounds may be polycyclic aromatic hydrocarbons, azo compounds, nitroaromatic compounds, aromatic hydrocarbons, halogenated aliphatics or synthetic polymers, and many others.

The pollutants present may be two types: (1) biodegradable and (2) nonbiodegradable.

Biodegradable pollutants: These pollutants can be broken down and converted into harmless and simpler substances by the activity of microorganisms. These contaminants may be degraded or transformed by microorganisms (bacteria (aerobic or anaerobic) or fungi) which have enzymes for their degradation or transformation. The substances include discharges (urine, fecal matter) from animals, plant residues (agricultural remains, paper, and wood), and dead remains of plants and animals.

Nonbiodegradable: The pollutants which cannot be broken down into simpler and harmless substances are referred as nonbiodegradable. These include heavy metals (lead, arsenic, mercury, etc.), plastics, dichlorodiphenyltrichloroethane (DDT), other insecticides, and pesticides.

The major damage suffered by the environment is due to anthropogenic activity, urbanization, and industrialization resulting in increased pollutants which not only are deleterious for other life forms but also for human beings. These conditions are responsible for many health issues.

19.2.1 Greenhouse Effect

The activities of humans like industrialization, burning of fossil fuel like coal and oil, urbanization over the past several decades have resulted in warming of our earth. The effects are due to raised levels of gases known as greenhouse gases. One of the important greenhouse gases is carbon dioxide (CO_2) along with methane (CH_4) and nitrous oxide (N_2O). The levels of CO_2 have increased from 280 parts per million to 379 parts per million in nearly 150 years. They affect the environment by (1) trapping the sun's energy in the lower atmosphere leading to increase in the temperature; (2) these gases block heat from escaping (Fig. 19.1). There are certain long-lived gases that are present in the atmosphere but are not responsive to the changes in temperature and are described as "forcing climate change." [17] However, gases such as water vapor, which responds physically or chemically to temperature changes, are described as "feedbacks." Gases contributing to greenhouse effect are:

(a) Water vapor: It is an important greenhouse gas, but it acts as feedback to the climate because increased water vapor in the atmosphere results in increased clouds and precipitation.

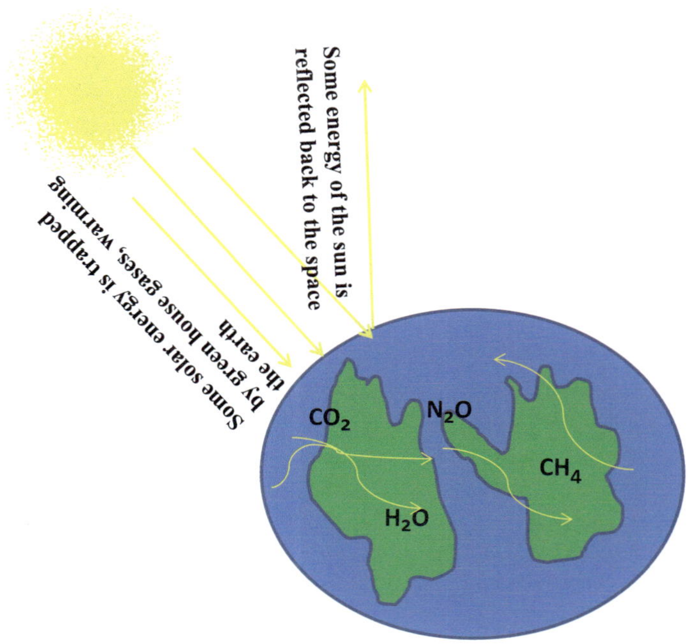

Fig. 19.1 Greenhouse effect where the greenhouse gases present on earth atmosphere trap the energy of the sun resulting in warming of the earth

(b) Carbon dioxide: A minor but important component which is a by-product of respiration and is released in volcanic eruptions, deforestation, and burning of fossil fuels. It is one of the most important "forcings" of climate change.

(c) Methane: It is a hydrocarbon produced by human activity (production and transportation of coal, decomposition of wastes in landfills, manure management) and natural source. It is a very active greenhouse gas but is relatively less abundant in the atmosphere.

(d) Nitrous oxide: A powerful greenhouse gas that is produced by agricultural activity as soil cultivation practices, industrial activities, combustion of fossil fuel, biomass burning, and many others.

Chlorofluorocarbons (CFCs): They were important widely used refrigeration compounds. They were responsible for destruction of ozone layer which protects the earth from harmful UV rays of the sun. Thus, they now are under strict regulation.

19.2.2 Water Pollution

Water is one of the important indispensable factors for life. Water is available to the man by rain and the flowing rivers and safe drinking water is required by all life forms. The water is said to be safe when it is free of any pathogenic virus or bacteria and free from chemicals or industrial wastes or dyes. Nature has provided us with pure water. This natural pure water has essential dissolved gases (carbon dioxide, oxygen, nitrogen) and minerals (calcium, magnesium, sodium). But when the discharges from various fabricated units are discharged into the water, the water becomes contaminated.

The contamination is due to:

1. Sewage discharge (may have pathogens apart from organic matter)
2. Residual insecticides and pesticides (due to soil erosion)
3. Industrial waste (may have heavy metals, dyes, nonbiodegradable pollutants)

These all when discharged into the water make it heavily polluted and unfit for drinking purposes. When water becomes polluted, the total suspended solids (TSS), total dissolved solids (TDS), and biological oxygen demand (BOD) increase. Let us have a brief discussion about the properties of the wastewater. The color, odor, and solids in wastewater are due to various kinds of wastes disposed in it as domestic and/or industrial discharges or due to natural decay process going on and soil erosion. The wastewater can have many chemical components such as organic (carbohydrate, fats, phenolics), inorganic (heavy metals, nitrogen, methane, oxygen, hydrogen sulfides), and biological (plants, animal remains, microorganisms). Some of the terms are elaborated for readers:

Sewage: Waste matter from domestic or industrial organization or agricultural activities, individually or combined with organic, inorganic, and toxic compounds, carried in sewers or drains for conversion into harmless products or dumping purposes.

Industrial waste is the waste produced by industrial activity which includes any material that is rendered useless during a manufacturing process such as that of factories, mills, and mining operations. It has existed since the start of the Industrial Revolution.

Biochemical oxygen demand (BOD) is the amount of dissolved oxygen needed by aerobic biological organisms in a body of water to break down organic material present in a given water sample at certain temperature over a specific time period. As the pollution levels increase, the BOD of the water increases.

Total dissolved solids (TDS) are a measure of the combined content of inorganic and organic substances contained in a liquid in molecular, ionized, or microgranular (colloidal sol) suspended form.

Dissolved solids refer to any minerals, salts, metals, cations, or anions dissolved in water. Total dissolved solids (TDS) comprise inorganic salts (principally calcium, magnesium, potassium, sodium, bicarbonates, chlorides, and sulfates) and some small amounts of organic matter that are dissolved in water.

Chemical oxygen demand (COD): The capacity of water contaminants to consume oxygen during the process of decomposition of the pollutants especially the organic matter and conversion into inorganic wastes as CO_2, H_2O, and NH_4 is COD. The COD test procedure is based on the chemical decomposition of organic and inorganic contaminants, dissolved or suspended in water. The result of a chemical oxygen demand test indicates the amount of water-dissolved oxygen (expressed as parts per million or milligrams per liter of water) consumed by the contaminants, during 2 h of decomposition from a solution of boiling potassium dichromate. The higher the chemical oxygen demand, the higher the amount of pollution in the test sample. For the contaminants that can be oxidized biologically, the biological oxygen demand (BOD) method is used.

Industrial pollutants: Various industries add pollutants to the environment particularly water. Polluted water is responsible for many critical clinical conditions and is responsible for many diseases.

Agricultural pollutants: These are major contaminants of soil and water. For increasing the yield, the farmers use chemical fertilizers which either stay in the soil or are carried to water bodies. These increase the load of various salts and result in salt imbalance and disturbance to the nearby ecosystem. Apart from fertilizers, insecticides and pesticides are used for protecting crops from infestation. These are chemical compounds used intentionally in the agricultural fields and several other places (homes, parks, schools) to destroy the fungi, rodents, and insects/pests. They are deliberately used to prevent damage to the crops by pests and thereby when soil erodes, they are carried toward the water bodies. These are highly poisonous compounds and used as insecticides, fungicides, herbicides, fumigants, or rodenticides. The problem with their usage is that they are not specific for a particular range of organisms; rather, they are deleterious for all forms of life including man.

Polychlorinated hydrocarbons (PCHs): DDT; PCBs; lindane; aldrin; organophosphates, parathion; carbonates, sodium thiocarbamates.

PCHs are very effective and able to control pests and insects; however, they are persistent threat for the environment. They move from environment/medium into the body of living organism especially in fat cells (*bioconcentrated*) and movement along the food chain with high concentration in topmost organism as humans (*biomagnifications*). They pose risk as they are (1) toxic compounds and (2) difficult to remove and (3) they themselves and their degradation products are toxic.

Case Study

One of the most popular and known pesticides was DDT (dichlorodiphenyltrichloroethane). DDT was the first chemical pesticide which found very wide acceptance to solve pest problem across the world. The insecticidal action of DDT was discovered by Swiss chemist Paul Hermann Muller in 1939. After that it was largely used as agricultural insecticide and during World War II to control malaria and typhus fever. Muller was awarded the Nobel Prize in medicine for discovery of DDT. DDT was used all over the world for killing pests and insects.

However due to its wide usage and impact on the environment, the American biologist Rachel Carson in 1962 published the book *Silent Spring*. The book targeted the environmental impact of DDT, nondiscriminate DDT spraying in the USA, and questioned the logic of releasing large amounts of potentially dangerous chemicals into the environment without a sufficient understanding of their effects on ecology or human health. The book claimed that DDT and other pesticides had been shown to cause cancer and that their agricultural use was a threat to wildlife, particularly birds. Its publication was a seminal event initiated for the protection of the environment and resulted in huge outcry

(continued)

from public. It led to the constitution of Environmental Protection Agency (EPA) to review the effects of DDT. Justification of its environmental hazards, human health hazards, and effects on animals led to a ban on the usage of DDT in agriculture in 1972.

The DDT and its breakdown product DDE have been linked with cancers, miscarriage and low birth weights, male infertility, nervous system damage, and liver damage. It acts by opening sodium ion channels in neurons leading to their spontaneous firing and death followed by spasms.

The properties which led to outcry for ban on DDT were:

1. It was a persistent organic pollutant.
2. It was adsorbed on sediments.
3. Its breakdown products DDE and DDD are toxic and persistent.
4. DDT and its metabolites are readily taken up by living beings and these are neither metabolized nor break in the body.
5. It is highly lipophilic and very resistant to metabolism.
6. It has half-life of 6–10 years in the human body.
7. It is toxic to all animals and interferes with endocrine processes.

Due to public outcry and reports under Stockholm convention, a worldwide ban on agricultural usage of DDT was formalized. However, the use of DDT in disease vector control was practiced but remained controversial. It was very effective in reducing deaths due to malaria.

The incidences of waterborne infections can increase with the high contamination by infectious organisms including viruses, bacteria, protozoa, and parasites. These organisms are transmitted due to fecal waste prevalently present in the areas with poor sanitary conditions. (1)

These organisms can cause various diseases such as hepatitis A, hepatitis E, polio, diarrhea, typhoid, dysentery, cholera, amebiasis, and giardiasis. (2) High fluoride concentration in water results in mottling of dental enamel. (3) The mineral and heavy metal abundance in water accumulates at unusual sites and is the cause of many diseases. (4) Stagnant water provides breeding opportunities to harmful insects which are the major cause of outbreak of many diseases like yellow fever, dengue, malaria, and filarial disease. (5) Pesticides (DDT, heptachlor, dieldrin), polychlorinated biphenyls (PCBs), and dioxins act as endocrine disrupters which not only affect normal functioning of the body but also affect sexual development and fertility [10].

19.2.2.1 Other Types of Pollutants

Polychlorinated biphenyls (PCBs): These are used in insulators, transformers and capacitors, and paint additives. They are highly stable and have been reported to be responsible for liver damage and calcium metabolism. They affect the reproductive system of seals. They have been banned in 1979.

Acid rains: The acid rain is due to the presence of the pollutants sulfur dioxide and nitrogen oxides, which when combined with water form sulfuric acid and nitric acid. They can cover big area. It is due to increased pollution from burning (fossil fuels and agricultural wastes), forest fires, and agricultural waste incineration leading to brown cloud of ash, acids, and other particles present in air. It not only affects the rainfall pattern but also affects life forms (algae, amphibian, fishes and other lower animals, microorganisms). It removes soil nutrients making them unavailable to the plants and dissolves toxic metals. Acid rain also results in acidification of the lakes which is detrimental for the growth of animals and plants living in them.

The increasing acid rains results in nutrient pollution, which are responsible for the growth of algae. This alga thus grows vigorously utilizing major oxygen needed by other life forms and the process is known as eutrophication. This occurs when untreated wastes are discharged into water bodies. Serious eutrophications were observed in shallow estuaries and bays. They also produce toxic chemicals which are detrimental to other marine animals, birds, and mammals. Many algal blooms can easily be seen from space because their skeletons are highly reflective, making the water appears milky. The blooms of these kinds of algae (harmful algal blooms (HABs)) are called red tides or brown tides, for example, high algal blooms in the gulf coast of Florida due to *Gymnodinium breve*, which produces a toxin called brevetoxin. In 1996 one of these HABs killed 149 manatees:

Heavy metals:

The heavy metals such as cadmium, lead, mercury, arsenic, and chromium are environmental threats. Cadmium is used in batteries and poses threat to human health (kidney damage, bone effects and fractures) at a very low concentration. Mercury exposure to the population is primarily via fish food. Mercury does not pose a serious risk to general people; however, high fish-consuming population may be at risk. It can cause neurological damage and risk to developing fetus in pregnant women. Lead is a serious air pollutant whose emission occurs from petrol. Exposure of lead causes neurotoxic problems in children. Exposure to arsenic is responsible for increased risk of skin cancer and skin lesion such as hyperkeratosis and pigmentation; exposure by inhalation can cause lung cancer. Selenium poisoning is also a serious problem. Chromium is used in industrial processes, which causes severe health effects in humans.

Case Study

Selenium poisoning was reported from California in the 1970s. Due to high concentration of selenium and poor drainage, the crops were dying. After appropriate drainage facility, the selenium was carried to evaporation pool. In one instance, this was done at Kesterson reservoir which was

(continued)

close to its national wildlife. However after those evaporation pools were created, they noticed that large numbers of bird's eggs failed to hatch, with some showing complete breeding failure and death of many adult birds. Selenium toxicity resulted in extreme problems with wildlife and crops.

In 1989, Kern National Wildlife Refuge in the Tulare basin reported high rates of deformity. In California, Salton Sea suffered buildup of nutrients and salts. In 1999 it resulted in death of eight million fishes in a single day.

Cadmium, arsenic, boron, uranium, and pesticides also get concentrated and may be contributing to the problem.

19.2.3 Air Pollution

It is another important health problem worldwide. It occurs due to increased industrialization, increased automobile exhausts, deforestation, and greenhouse gases. Pollutants in air can cause chronic medical conditions such as allergies, asthma, COPD, and thus high pressure on pulmonary and cardiac system. As air pollution is responsible for adverse effects on health, thus protective measures are an essential part, apart from monitoring, assessment, and control measures by the government. Individual actions can be (1) more indoor activity, (2) maintaining clean indoor air (use of filters and cleaners), (3) avoiding outdoor activity in an environment with high air pollution (extreme traffic time, industrial operating timings), and (4) use of respirators [16].

19.2.4 Noise Pollution

Noise pollution is due to extreme noise and is responsible for disturbing behavioral pattern of animals. It can result in ear damage and persistent irritation in animals.

19.2.5 Electromagnetic Pollution

With the advent of technology, there has been continuous increase in the usage of many devices which have electromagnetic fields, and thus we are continuously exposed to electromagnetic radiations. These radiations which are present due to mobiles and wireless telecommunication services, radio communications, and smartphones are quiet strong.

Some scientific reports have linked these radiations as the cause of certain types of cancer and the effect of these on pineal gland and its hormone melatonin which might lead to sleep disorders, depression, and cancers. However exact effect and radiation influence needs further research and scientific verification to come to some meaningful conclusions.

Due to recent surge in all kinds of pollution including industrial, agricultural, and automobile exhaust, the world is witnessing death of many living organisms, reduced reproductive capacities, human health hazards, and increased imbalances in the nature (melting of glaciers, uneven rainfall, and increased sea levels).

Case Study

The climate of the earth is changing with an alarming rate. The major changes are due to greenhouse gases as carbon dioxide. These are increase in temperature, melting of glaciers, rise in sea level, and altered rainfall patterns. Due to these changes, the plants and animals are getting affected:

- Their geographical distribution is getting affected.
- Early blossoming of trees (cherry trees).
- Shoreline species of some invertebrates (limpets, snails, and sea stars) are shifting their range northward in California. Species unable to change their range due to lack of food or resting places are adversely affected.

(continued)

- Many species of amphibians, reptiles, and birds are affected.
- A coalition of US government agencies said in the 2000 report that global warming could end cold winters in the Northeast and wipe out the alpine meadows of the Rockies, as well as Florida's coral reefs.

19.3 Environmental Monitoring

As there has been extensive damage to the environment, thus it is very important to monitor and assess the quality of the changing environmental conditions. The monitoring is essential so that environmental impact assessment can be done for its effects on various life forms as well as effect or influence of human activity on the environment, for appropriate protection of the environment. There are different methods for monitoring of different components of the environment.

The monitoring can be done by physicochemical and biological methods. Contaminant monitoring involves the regular and periodic evaluation of various chemicals in water, soil, sediment, and air over a fixed duration of time.

19.3.1 Water Quality Monitoring

The monitoring of water quality should have clear objectives. The monitoring is different for various target substances or living organism present in water in question. It is a challenging task as monitoring programs are useful for only a set of parameters. Different instruments and programs are required for different measurements.

These continuously require revision due to additions like acid rain, halogenated hydrocarbons, greenhouse gases, synthetic hormone analogs, and many more into the water bodies.

Grossly monitoring can be of several aspects:

- Biological: Review of natural plants and animals living in a particular environment (would be variable for each study). Monitoring of salmonoid fish (brown trout) or salmon, whose decline is an early indication of pollution. Monitoring of pathogens would be important.
- Radiological: Assessment and control of exposure to ionizing radiation or radioactive substances.
- Microbiological: Most commonly monitored organisms are pathogens such as bacteria, virus, fungus, and other microbial agents.

19.3.2 Air Quality Monitoring

Monitor fuel gases (oxygen, carbon monoxide, carbon dioxide) and compliance with air emission standard lay down by the United States Environmental Protection Agency's Acid Rain Program or other federal or state-permitted standards:

- Monitoring of the environment is a big requirement because only after the assessment, the processes can be initiated which can lead to eco-friendly practices for sustainable environment.
- Monitoring, assessment, and treatment of polluted water, air, and solid waste can be addressed by biotechnological tools.
- Many virulent microorganisms are present in polluted water which are responsible for various clinical diseases.
- Monitoring of these is very important and removal is essential. *Cryptosporidium parvum* is present as contaminant in drinking water and is capable of causing diseases at very low level.
- Detection of waterborne pathogen sometimes requires pre-enrichment, to reach to detectable limits. Detection limit of waterborne pathogens can be 10–1,000 organism/ml (MERIFLUOR *Cryptosporidium–Giardia* test from Meridian Biosciences).

19.3.3 Biomarker/Bioindicator

- Biomarker is any biological molecules which when analyzed can give indication about the presence of specific microbial protein or nucleic acid, enzyme or specific protein, or histo-cytopathological behavioral.

- Therefore determination of contaminant level along with assessment of various parameters of biological system can be useful in reflecting changes in environmental conditions.
- Some organisms or communities can be analyzed for their measurable biological function and/or chemical composition in changing environmental conditions.
- The analysis suffers from constraint like availability of such sensitive biological material, which has, broad applicability is reliable, easy to use, and good for quality control.

19.3.4 Biosensors

- The biosensors are devices which can qualitatively and quantitatively detect the presence of variety of molecules (proteins, pollutants, toxins) in the sample.
- Biosensor is a broad term that refers to any system that detects the presence of a substrate by the use of biological component which then gives the signal that can be quantified.
- It contains immobilized biological material which has biological affinity to the target molecules (like enzymes, antibody, hormone, or nucleic acid).
- After detecting and interaction with affinity target molecule, it produces measurable signal.
- The affinity molecule is associated with microchip device used for estimation of the target molecule present in the environment.
- It is in turn attached to the amplifier which is capable of detecting the signal.
- Important biosensors used in environmental monitoring are:

Some Important Biosensors

- Enzyme biosensor: In this enzymatic activity can predict catalytic transformation of pollutant or modification in enzyme activity due to the presence of a particular pollutant or complete suppression of enzymatic activity by pollutant.
- Immunoassay biosensors: In this antibody specific to pollutant (pesticide/toxin) is used to detect presence of low quantities of that pollutant. It can also detect the presence of low concentration of triazines, malathion, or carbamates.
- BOD biosensor: BOD sensor used the yeast *Trichosporon cutaneum* and detects level of organic pollutants. Phenol oxidase can detect phenol; acetylcholine esterase can detect organophosphorous compound in water.
- Phytochelatins are responsible for chelation of heavy metals (such as Hg^{2+}, Cd^{2+}, Pb^{2+}, Cu^{2+}, Zn^{2+}).
- Chloroplast D1 protein: Photosynthetic activity can be used for herbicides.
- Whole-cell biosensors are based on chlorophyll fluorescence or enzymes (phosphatase and esterase) inhibition.
- Plants are also used as biological indicators as sensitive and resistant white clover (*Trifolium repens*) and *Centaurea jacea* (brown knapweed) as model species.
- Invertebrate species like insects and crustaceans can also be used for biomonitoring.

The detection of the signal can be electrochemical, fluorescence or optical, or optical electronic or acoustic (Fig. 19.2).

19.3.5 Nanoparticle-Based Detection

- High sensitivity fluorescent dye doped nanoparticles can enhance the signal by 10^5–10^6 times thus allowing detection at low concentration. Nanoparticle biotech instrument can be deployed at the environmental site.
- Nanoparticles (1–100 nm) in diameter have unique properties and widely used in biomedical, electronic, environmental, paramedical, cosmetic, and energy. They are capable of detecting biotic (pathogens) and abiotic components (toxins, metal ions, and organic pollutants).

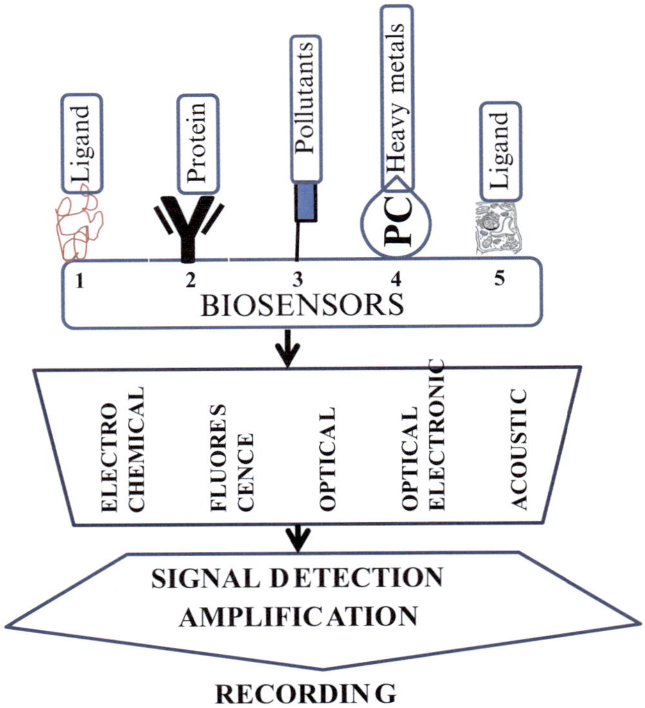

Fig. 19.2 Biosensors used for detection of environmental contaminants. The detection is done by either electrochemical, fluorescence or optical, or optical electronic or acoustic. The biosensors can use any of the following probes: *1* enzyme, *2* antibody, *3* BOD, *4* phytochelatins, *5* cell

- Nanoparticles are incorporated into nanosensors. They provide rapid, high-throughput detecting ability on a portable device. They are considered potential sensing material due to physical confinement of electrons at nanoscale.
- Surface-modified nanocolloids such as gold nanoparticles (GNPs), magnetic nanoparticles (MNPs), quantum dots (QDs), and carbon nanotubes show target-specific binding properties.

19.3.6 Continuous Emissions Monitoring System (CEMS): Used for Air Quality Monitoring

- Small sample of fuel gas is extracted by pump via sample probe.
- Dilution-extractive probe dilutes the sample with clean and dry air (100:1) as the sample can be sticky.
- Diluted sample moves through a sample line (umbilical) to a manifold.
- Gas analyzers extract the sample and measure concentration by using either infrared, UV absorption, chemiluminescence, fluorescence, or β-ray absorption.
- The gas exits out and data acquisition and handling system (DAHS) receives signal output from each analyzer in order to collect and record emissions.
- Internal quality assurance check is achieved by introduction of certified concentration of gas to the sample probe.

19.3.7 Particulate Matter Sampler

- Measures mass concentration or chemical composition of particulates in the ambient air.
- Known volume of air is drawn through the filter. Filter is weighed on an analytical balance before and after sampling.
- Difference in weight is divided by volume of air pulled through the filter which gives the mass concentration of the particulate.

19.3.8 Portable Emission Measurement System (PEMS)

- Lightweight portable instrument and is useful for assessment of air
- Used for various mobile standard such as CO_2, NO_x, particulate matter, carbon monoxide, and hydrocarbons
- Helps reduce the load of the vehicular pollutants upon appropriate testings and allowed limits

19.4 Biotechnology and Environment

- Biotechnological tools can provide powerful monitoring of ecosystem.
- By bioremediation toxic pollutants may be converted into non/less toxic compounds.
- Provide sustainable environment with renewable sources of energy.

19.5 Bioremediation

Biological means living organism and *remediation* to give solution; thus, collectively the term *bioremediation* means the usage of living organism (microorganism (microbial remediation or simply bioremediation) or plants (phytoremediation)) to give solutions to the problems of pollution of water, soil, or oil spills. In this the toxic pollutants by the action of enzymes present in various microorganisms can be converted into less toxic or simpler nontoxic compounds. The US Environmental Protection Agency (USEPA) has defined bioremediation as "managed or spontaneous practice in which microbiological processes are used to degrade or transform contaminants to less toxic or non toxic forms, thereby remediating or eliminating environmental contamination" (USEPA 1994; Tally 2005). The process of biotechnology involves removal, separation, destruction/degradation, containment, or immobilization. The first three are aimed for reduction or removal of pollutants and the last one controls migration of the contaminant. The destruction/degradation is done by microorganisms, whereas immobilization may be achieved by rendering contaminant insoluble or its chelation.

19.5.1 Need for Bioremediation

- With multiple human activities, there is increased pollution of air (with high CO_2, NO_x, SO_2, greenhouse gases, and particulate matter), pollution of water (industrial, domestic, and agricultural wastes, dyes, nutrients, and biological contaminants), and pollution of soil (disposal of hazardous waste, spraying of pesticides and insecticides or fertilizers).
- Many of these contaminants can be removed, degraded, or transformed into nontoxic substances by the activity of microbes, plants, and animals under suitable conditions by the process of bioremediation for the betterment of the environment.
- As many of the pollutants are toxic and their accumulation in the environment can be deleterious for all life forms, thus bioremediation can eliminate the risks and hazards of these chemical wastes.

19.5.2 Microorganisms Involved in Bioremediation

- The usage of bacteria, yeasts, fungi, algae, and protozoa for environmental cleaning is bioremediation, whereas the use of plants for transformation or containment is referred as phytoremediation.
- These microorganisms have the ability to degrade most hazardous chemicals. They can live as free organisms or communities (consortia).
- Bacteria and fungi are capable of degrading complex molecules making the waste free of many dangerous contaminants, whereas algae and plants can absorb many metals and minerals from the environment.
- These organisms are active in activated sludge processes, lagoons, ponds, wet lands, anaerobic wastewater treatment, bioleaching, phytoremediation, land farming, or slurry

reactions, and the process may be aerobic or anaerobic.

- Aerobic bacteria like *Alcaligenes*, *Pseudomonas*, *Rhodococcus*, *Mycobacterium*, and *Sphingomonas* require appropriate oxygen and degrade pesticides, hydrocarbons such as alkanes, and polyaromatic compounds using these as their carbon and energy source. The anaerobic bacteria do not require oxygen and are mostly used for pretreatment of water, polychlorinated biphenyls, or trichloroethylenes.
- Important bacterial species active in bioremediation are *Streptococcus*, *Bacillus subtilis*, *Vibrio cholerae*, *Spirillum volutans*, *Caulobacter*, *Gallionella*, *Rhodomicrobium*, *Sphaeratilus natans*, and *Streptococcus*. Actinomycetes involved are *Micromonospora*, *Streptomyces*, and *Nocardia*. Blue-green algae (*Cyanobacteria*) *Anabaena* and others such as *Volvox* and red, green, and brown algae are important organisms. Fungal species have the ability to degrade extremely diverse and toxic pollutants. Fungal varieties involved are white rot fungus, *Phanerochaete chrysosporium*, *water molds*, *Neurospora crassa*, *Saccharomyces cerevisiae*, *Agaricus*, and *Penicillium*. Methylotrophs utilize methane as their carbon and energy source.

19.5.3 Factors Affecting Bioremediation

The factors which affect the process are (1) the pollution and pollutant and (2) the remediating agent and the conditions for its growth:

I. The foremost factor that affects the process is the nature and type of contaminant and its physical state (solid, liquid, gas), concentration, and aggregation.

   Contaminant concentrations are important as they influence microbial activity, as high concentrations can be detrimental to the bacteria and low concentrations can prevent the process of degradation.

   Contaminant bioavailability indicates the availability of the contaminants whether they are freely available or sequestered by other molecules or diffused. The bioavailability affects microbial reactions.

   Site characteristics include conditions like:

   - *pH*: It affects the ultimate bioavailability of nutrients. pH is a very important parameter which governs the growth of microorganism.
   - *Redox potential*: The concentration of electron acceptors and oxygen affects the ultimate reactions.
   - *Temperature*: It is one of the important factors which affect microbial growth.
   - *Nutrients*: Nutrients are important growth determinants; thus, optimal concentration should be present for microbial activity.

   The microorganisms use the contaminants by aerobic or anaerobic process as their energy source or contaminants are co-metabolized with an energy source. The process involves redox reactions for production of energy like an energy source (electron donor), an electron acceptor (oxygen, manganese, or carbon dioxide), and nutrients. Their redox potentials provide an indication of the relative dominance of the electron acceptor classes which along with nutrients are very critical components.

   Bioremediation may be conducted in situ or ex situ. In situ treatment refers to soil or groundwater treatment without any removal or transportation [18]. This has benefits of being economical but controlling or manipulating the contaminant environment becomes difficult. Ex situ involves transportation of the contaminated media to a treatment area.

   Thus for good results of any bioremediation program, conceptual site model (CSM) to evaluate the potential for applying bioremediation is developed. In this the nature and extent of contaminants, characteristic of site, conditions of contaminants, and biodegradative potential are established. This helps in establishing the requirement of bioaugmentation or biostimulation or natural conduct of the process.

II. The process is largely affected by the availability of natural or engineered microbes with optimum enzyme activity for degradation or transformation of the pollutant.

For maximum output, the availability of air (oxygen) for aerobic microbes, temperature, moisture content, nutrient levels, and pH play a very important role. The optimum temperature range is 20–30 °C and a pH of 5.9–9.0 (depending upon microbial species).

19.5.4 Process of Bioremediation

The process of bioremediation involves

$$\text{Organic matter} + O_2 \xrightarrow{\text{Cells, nutrients}} CO_2 + H_2O + \text{new cells}$$

Biotransformation: The metabolic modification of the molecular structure of the compounds resulting in either the loss or change of their original properties with minimal or no loss of molecular complexity is biotransformation.

Biotransformation affects the solubility, mobility, or toxicity of the compound. Co-metabolism: In this the microorganisms which are growing on one compound can transform other chemicals which are not their energy or carbon source. The process can result in only minor modification or incomplete or complete degradation. In this process, the primary substrate usually induces production of enzyme which can fortuitously alter the structure of other compound. However the process does not have any benefits for the microorganism.

The products' bioconversion may also be degraded by other microorganisms. All these processes have tremendous potential for ecological improvement; however, with their increasing usage, some of them have become persistent or are metabolic dead-end products which when entering food chain result in biomagnification.

19.5.5 Types of Bioremediation

Bioremediation may be of three types which are used to decontaminate the environment:

Biodegradation: In this the organic compounds are broken into simpler compounds and finally to water, carbon dioxide or oxides, or salts of elements.

Mineralization: When complete breakdown of organic compounds into inorganic components occurs, it is referred as mineralization. Although complete or ultimate biodegradation and complete mineralization are used interchangeably, however biodegradation involves the formation of inorganic compounds along with the biomass.

Biostimulation: In the process of biostimulation, the bacteria are motivated for initiation of the process by addition of nutrients and other necessary substances. The process enhances the microbial growth rates; thus, they act readily and efficiently on the contaminants.

Bioaugmentation: The process is used at special site for removal of contaminants. However controlling the growth of the microorganism for removal of contaminant is difficult.

Microbial activity can be enhanced by either providing optimum conditions like air, organic substances, and other necessary conditions (biostimulation) or by addition of appropriate microorganisms (bioaugmentation). Biostimulation is efficient when the microbial species are already present in the contaminants, but other conditions are not very favorable, and bioaugmentation is efficient when contaminant-degrading bacteria are not naturally present or present at very low levels at the site. These processes are effective to treat soil and groundwater.

Intrinsic Bioremediation: The process of intrinsic bioremediation takes place in soil and water because these two places are always full of contaminants and toxins. This process is also called as natural attenuation. It also means the use of the microorganisms to remove the harmful substances from soil and water. Especially those sites are treated with this method, which are underground, for example, underground petroleum tanks. It is difficult to know if there is a leakage in the petroleum pipes. Contaminants and toxins find their way to enter in these sites and create harmful effects on the petrol. Therefore, only microorganisms can destroy the toxins and clean the tanks. Great care should be taken if some leakage occurs in the petroleum tanks or pipes because it may damage the human health.

Depending upon the natural ability of microorganisms and conditions, the process of bioremediation can be:

- *Complete*: Results in detoxification of pollutants to carbon dioxide, water, and harmless inorganic salts. *Incomplete bioremediation* results in detoxification of the initial product. The final product can be less toxic but sometimes more toxic than the original pollutant. For example, the incomplete degradation of tri- and tetrachloroethylene can lead to production of more toxic and carcinogenic compound vinyl chloride.
- *Spontaneous or natural biodegradation* occurs normally; however, in the case of non-feasibility of natural biodegradation due to lack of nutrients, oxygen, or suitable bacteria, supplementation of these may be required for better and effective process to occur and may be called as *engineered in situ bioremediation*.
- *Aerobic or anaerobic bioremediation*: Bioremediation may be aerobic (requiring oxygen as electron acceptor and carbon substrate as electron donor) or anaerobic (reaction which occurs in the absence of oxygen). The anaerobic reactions may be fermentation, reductive dechlorination, or methanogenesis. In these reactions sulfate or nitrate or carbon dioxide or organic compound or oxidized materials act as electron acceptors in place of oxygen.
- *Cometabolic bioremediation* is the process where microbes degrade the contaminants by the side reaction, and they do not require the contaminant as carbon or energy source (EPA 2006).

19.5.6 In Situ Bioremediation

In in situ remediation, the reactions can be performed in their original place of occurrence without requirement of transportation, such as soils and groundwater. As it does not require transportation, thus, it is more cost-effective, convenient, and cleaner with minimal alteration and exposure of contaminated material; however, it is time consuming and is more effective in loose soil.

Intrinsic bioremediation: In this natural microbial population is utilized with supplementation of nutrients and oxygen for their optimal activity.

Engineered bioremediation: When conditions are not favorable, the site requires introduction of suitable microorganisms as well as nutrients for their growth.

The in situ process can be achieved by:

Bioventing: In this the oxygen is pumped in the form of air. The air is passed through the wells. Wells are injected into the contaminated soil. Microorganisms grow and air along with nutrients and phosphorous is passed through wells for optimal microbial growth.

Injection of hydrogen peroxide: It is done in contaminated water by pumping using series of pipes to circulate shallow contaminated areas of soil.

19.5.7 Ex Situ Bioremediation

It requires excavation and replacement of contaminated soil or groundwater for bioremediation process to occur. Ex situ bioremediation is helpful for phenols, cresols, hydrocarbons, aromatic hydrocarbons, and semi-volatile compounds. The ex situ can be done in solid phase or slurry phase. Slurry-phase remediation involves the mixing of contaminated material (soil) with water and other components in the bioreactor tank. It is mixed continuously for optimal activity of microorganisms with additions of oxygen and nutrients.

After completion of the process and the separation of soil and water, and after appropriate testing, it is replaced in the environment.

Solid-phase system may use waste material from leaves, animal remains and agricultural waste, and domestic or sewage sludge.

Land farming: Simplest, cheapest, and most basic form of bioremediation. In this the contaminated soil is spread over a pad (having system for collection of toxic residual liquids) in layers of 0.3 m thickness in large areas. The soil is turned periodically for mixing of air allowing microorganisms present in the soil to break down the contaminants.

Turned window: Excavated soil is piled up in 1.5–2 m height and aerated by regular turning of machines. Nutrients are added to the windows

Soil biopiles: In this piles of soil are placed in a pile up to 3 m in height. The aeration is provided by either a vacuum pump or air blower systems. The vapors can be collected and treated by activated carbon. As their construction can be tall, thus, they require less space and temperature can be controlled. It is the method of choice in the decontamination of odorous or volatile compounds. Temperature plays an important role in determining rate of microbial degradation. Though they are effective for large range of contaminants and give good flexibility, they are not effective for heavy metals or chlorinated hydrocarbons as trichloroethylene (TCE). They are also difficult for non-permeable soil as they are aerobic.

Composting: In this the waste is mixed in straw, hay, or corn cobs for optimal water and air. In static pile, piles of aerated compost are maintained using blowers or mechanically agitated in the vessel. Compost in the treatment tank is turned periodically for aeration. In window composting, compost in long piles is mixed by a tractor.

Compost

Normal tropical soils are highly deficient in all necessary nutrients of the plants. Large amount of these nutrients is available in domestic wastes and agricultural by-products. The formation of compost utilizes various environmental wastes as wastewater treatment solids, wood preservatives, chlorinated hydrocarbons, pesticides, and heavy metals, and microorganisms convert these into less toxic products which are normally not available to the plants growing on it (Fig. 19.3).

The benefits of compost are:

- Improves water dynamics of soil (water infiltration, percolation, and water holding)
- Reduces irrigation needs
- Reduces leaching
- Reduced fertilizer requirements
- Reduced soil loss from erosion due to improved soil structure

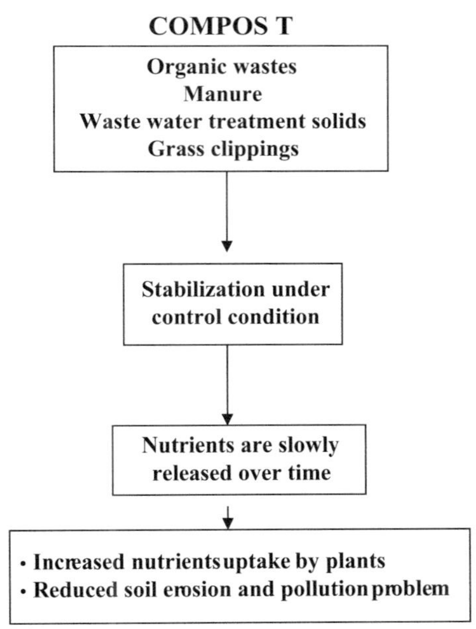

Fig. 19.3 For formation of compost, the wastes are stabilized under control condition. The stabilization may take up nitrogen and ties it in compost's organic matter. Microorganisms in compost break contaminants such as chlorinated hydrocarbon, wood preservatives, solvents, pesticides, petroleum products, etc. into less toxic or harmless products. The heavy metals such as lead are bound to compost in a way that it is not available to the plant

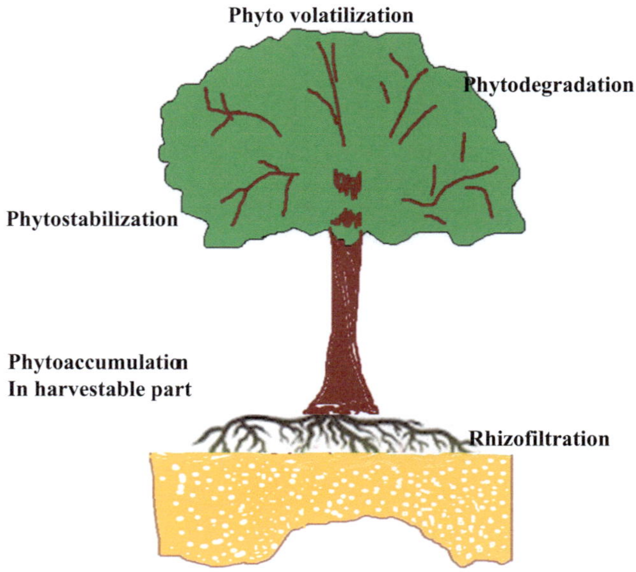

Fig. 19.4 The process of phytoremediation. The plant can use any of these modes individually or collectively to decontaminate the soil

As nutrient composition of each compost can vary, thus analysis of its contents is very important.

Vermicomposting

- Vermicomposting utilizes certain species of earthworms. Earthworms can enhance the process of waste conversion and produce better end product.
- Earthworms are active at 10–32 °C, the temperature inside the moist organic material.
- The process is advantageous as the material passes through the earthworm gut, resulting in significant transformation.
- The earthworm castings (worm manure) are rich in microbial activity and plant growth regulators.
- Vermicompost provides all nutrients in readily available form and increases nutrient uptake by the plants, improving plant growth and yield.
- There are 3,600 types of earthworms and they are either burrowing or non-burrowing.

Burrowing

- Live deep in the soil.
- Are pale and 20–30 cm long.
- Life span of 15 years.
- They eat 90 % soil and 10 % organic wastes.
- Come only at night.
- For example, *Pertima elongate* and *Pertima asiatica*.

Non-burrowing

- Live in the upper layer of soil.
- Are red or purple and 10–15 cm long.
- Life span of 28 months.
- They eat 10 % soil and 90 % organic wastes.
- Conversion is faster.
- For example, *Eisenia fetida* and *Eudrilus eugeniae*.

19.5.8 Phytoremediation

All the metals are present in the earth. Many of them such as Cu, Fe, Mg, Mn, Ni, Zn, etc. are essential for plant cells. However higher concentration of these metals and low concentration of some other metals considered as heavy metals such as Ag, Al, As, Cd, Cr, Cs, Hg, Pb, Sr, and U exhibit phytotoxicity and are therefore considered as pollutants [3, 12, 13]. Due to industrialization, mining and smelting, sewage sludge treatment, warfare, and military training, waste disposal sites and indiscriminate agricultural fertilizer use have caused significant addition of previous toxic metals to soils (Fig. 19.4).

- *Phytoextraction/phytoaccumulation*: It is the uptake and accumulation of metals from soil into the plants harvestable part. The plants can accumulate heavy metals into the roots and aboveground shoots or leaves from a widespread area. The capacity of plants to tolerate potential consequences caused by the extracted/accumulated metals decides the effectiveness and success of phytoremediation system. Nowadays, hydroponic systems with plants having very high uptake of the contaminants (hyperaccumulators) by roots (rhizofiltration) with poor translocation to the shoots are being explored for heavy metals and radionuclides from water [20, 21]. For example, *Brassica juncea* and *B. nigra* have high metal-accumulating ability, *B. carinata A. Br* can accumulate lead, and *Thlaspi caerulescens* can accumulate Zn and Cd.
- *Phytotransformation or phytodegradation*: In this organic contaminants from soil, sediments, or water are taken up and transformed to a more stable, less toxic form by the action of enzymes. The process involves the breakdown by metabolic enzymatic processes. In this the contaminants are degraded into simpler compounds which can be integrated in plant tissues. Phytotransformation can be employed for remediation of contamination of chlorinated solvents (as trichloroethylene), ammunition wastes, and herbicides, petrochemical sites and storage areas, fuel spills, landfill leachates, and agricultural chemicals.
- *Phytostabilization*: In this the pollutants are taken up by the plant and are adsorbed or bound into the plant structure that forms a stable mass of plant rendering them immobile. The mobility of heavy metals in soil is reduced by minimizing soil erosion, reducing contaminant solubility or bioavailability by accumulation or adsorption by plant roots. Alkalizing agents, organic matter, and phosphates can decrease solubility of metals preventing leaching into deep water table. It is useful at sites with shallow contaminants.
- *Phytodegradation or rhizodegradation*: In this the pollutants are degraded by the presence of proteins and enzymes of plant and their symbiont as bacteria or fungi.
- *Rhizofiltration*: Removal and uptake of contaminant by plant root. Effective in natural wetlands and estuary area, surface water and groundwater, industrial and residential effluents, downwashes from power lines, storm waters, acid mine drainage, agricultural runoffs, diluted sludges, and radionuclide-contaminated solutions. Plants can effectively remove toxic metals such as Cu^{2+}, Cd^{2+}, Cr^{6+}, Ni^{2+}, Pb^{2+}, and Zn^{2+} from aqueous solutions.
- *Phytovolatilization*: In this after uptake of the contaminant from soil and groundwater, the plant evaporates or transpires it, for example, selenium, mercury, and volatile hydrocarbons, although it may not lead to complete remediation. Phytovolatilization reduces groundwater contamination.
- *Vegetative cap*: In capping the cover is placed over contaminated material such as landfill wastes or contaminated soil. Caps prevent people and wildlife from coming in contact with contaminant. In this rainwater is evaporated/transpirated by plants to prevent leaching contaminants from disposal sites.

19.5.9 Mode of Phyto-tolerance

Phyto-tolerance includes processes such as exclusion, compartmentalization, complexation, and the synthesis of metal-binding proteins and/or metal ion chelation which are defense strategies evidenced in plants under metal stress.

Plants have been reported to avoid the damaging effects of metal toxicity, by binding of heavy metals to cell wall and immobilization, exclusion of the plasma membrane, expression of more general stress response mechanisms such as stress proteins (heat shock proteins), and metal chelation and compartmentalization [5]. In particular, chelation is the most widespread intracellular mechanism for the maintenance of low concentrations and detoxification of free metals in plant cytoplasm. The chelation of metals can be done by non-thiol compounds such as histidine, nicotianamine, and organic acids or by thiol compounds such as glutathione (GSH) [1], phytochelatins (PCs), and metallothioneins (MTs) [8]. The thiol compounds are with sulfhydryl (-SH) group which can bind a variety of metals [2].

Glutathione (GSH) has a wide distribution in plant cell compartment (in the cytosol, endoplasmic reticulum, and vacuole). It maintains cellular redox homeostasis and antioxidant defense [1, 2]. It is also involved in chelation and detoxification of free metals. It is the precursor for the synthesis of phytochelatins (PCs)

Phytochelatins [PCs, γ-glutamyl (Glu)-cysteinyl (Cys)] are important nonprotein, metal-binding (and metal detoxifying), S-rich, thiolate peptides. They are nonribosomal peptides and directly synthesized enzymatically in response to varied metals from GSH by phytochelatin synthase (PCS).

Metallothioneins (MTs) are –SH compounds with cysteine-rich small and low molecular weight (4–8 kDa) gene-encoded polypeptides. Based on sequence similarities and phylogenetic relationships, plant MTs are divided into four subfamilies, type 1–4. Each type displays different spatial and temporal expressions. They have two binding domains α and β, which consist of Cys-clusters. Covalent binding of metal atoms involves sulfhydryl cysteine residues. The N-terminal part of the peptide is designated as β-domain and has three binding sites for divalent ions, and the C-terminal part (the α-domain) has the ability to bind four divalent metal ions. They are widely distributed in both prokaryotic and eukaryotic organisms.

19.5.10 Uses of Bioremediation

The process is very useful for remediation of various contaminants present in various sources (Fig. 19.5).

19.5.10.1 Wastewater Treatment

It is grouped as (1) sewage (domestic wastewater mixed with commercial and industrial water), (2) commercial and industrial wastewater, and (3) agricultural wastewater.

Municipal wastewater was treated by activated sludge process. Municipal sewage treatment plants and filters for treatment of contaminated gases were into being for long. They were very effective. Effective treatment of water by microbial action is dependent upon source of it.

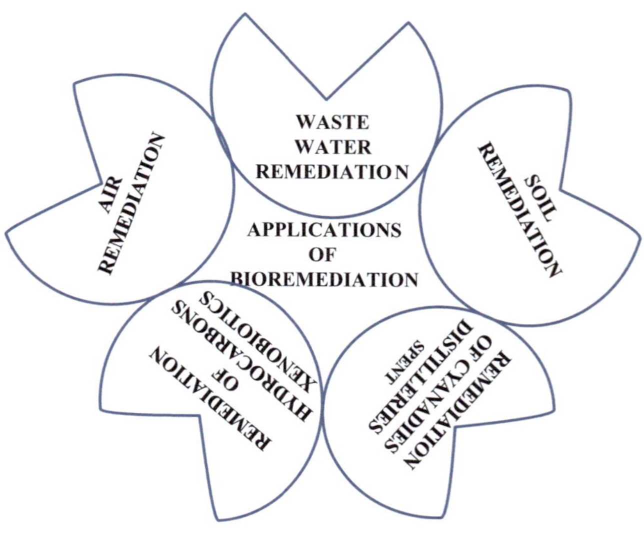

Fig. 19.5 Potential applications of bioremediation process for remediation of various contaminants present in different parts of the nature

As water receives all kinds of contaminants, thus, its quality parameters are highly diverse. The treatment should be adequate to pollution loading, and selection of appropriate microbial consortia is important along with treatment scheme for removal of non-settleable colloidal solids, organic wastes, heavy metals, and chlorinated compounds [9].

The contaminated water may itself be detrimental for microbial growth, thus requiring pretreatment before microbial remediation.

The treatment involves collection of wastewater in treatment plant. It utilizes large amounts of wastewater in a continuous system. Initially physical method as sedimentation (settling of solids by gravity) is done after which "clarified" effluent is removed. Sometimes aeration and filtration are done. Sometimes water is with undesirable waste and then the wastes are held and mixed with other wastewater. This is termed as equalization. After this chemical treatment, other processes such as chlorination (adding chlorine, a strong oxidizing chemical), neutralization, and coagulation are performed.

Biological treatment involves using aerobic or anaerobic or their combination for reduction of biodegradable organic content, reduction/removal of recalcitrant organics, removal of heavy metals, and removal and inactivation of pathogenic microorganisms.

Aerobic biotreatment: Utilizes municipal and industrial wastewater where microbial cells utilize organic material in the presence of oxygen. The treatment can occur in suspension (activated sludge) or anchorage dependent (fixed film).

Anaerobic treatment: Usually done as pretreatment process which is done for minimizing the oxygen demand and excessive formation of sludge. High loads of wastewater treated by anaerobic methods produce low quantities of biological excess sludge with high treatment efficiency, low capital costs, methane production, and minimal nutrient requirement.

The methods are suitable from discharges from distilleries, breweries, paper mills, or petrochemical plants. It may be done by sulfate reduction (for removal and recovery of heavy metals and sulfate denitrification for removal of nitrates) and bioremediation for toxic pollutants.

Upflow anaerobic sludge blanket treats sulfur-rich wastewater, whereas purple non-sulfur bacteria generate large amount of useful biomass with little carbon dioxide. This biomass may be utilized for agricultural purposes and feed for fish and animals.

Activated sludge: Done in aeration tank that allows suspended growth of bacterial biomass. They have better working efficiency in pure oxygen instead of air and can operate at higher biomass concentration. Domestic wastewater is treated by aerobic sludge process as they have proteins, carbohydrates, fats and oils, and urea along with trace levels of pesticides. It can be done in single tank or as combined process with continuous feed or discontinuous feed as fed batch reactors.

Fixed film: Trickling filters support attached growth of biomass. Biofilm reactors are applied in variants as trickle filter, rotating disk reactor, and airlift reactor. They can be done as fluidized bed reactors, fixed bed reactors, or trickling filter.

Advanced biotreatment: Utilized for water reuse purposes. Membrane technology is a recommended one as it combines biological and physical processes. It is applied for organic wastewater treatment. Usually wastewater treatment will involve collecting the wastewater in a central, segregated location (the Wastewater Treatment Plant) and subjecting the wastewater to various treatment processes. Most often, since large volumes of wastewater are involved, treatment processes are carried out on continuously flowing wastewaters (continuous flow or "open" systems) rather than as "batch" or a series of periodic treatment processes in which treatment is carried out on parcels or "batches" of wastewaters. While most wastewater treatment processes are continuous flow, certain operations, such as vacuum filtration, involving as it does, storage of sludge, the addition of chemicals, and filtration and removal or disposal of the treated sludge, are routinely handled as periodic batch operations.

Membrane bioreactor (MBR) can be aerobic or anaerobic and is applied for removal of dissolved organic substances, broken down by microorganisms. MBR is capable of complete removal of suspended solids with removal of particle-bound micropollutant and pathogenic bacteria and virus removal by size exclusion. It also stabilizes sludge and provides conditions for selective growth of specific microorganisms for degradation of hazardous substances. MBR is cost-effective, reliable, and user-friendly technology with good output.

Molecular techniques: Involves removal of xenobiotics and use of molecular probes for detection of pathogens and parasites. In this, selection of appropriate microorganisms or engineered strains may improve the process.

Metal removal: Metals like iron, copper, cadmium, nickel, and uranium can be complexed by bacteria as *B. licheniformis* and *Zooglea ramigera*. The bacteria use various mechanisms like adsorption to cell surface, complexation and solubilization of metals, volatilization, precipitation, or intracellular accumulation. Some microorganisms involved in metal removal are *Aeromonas* and *Flavobacterium* (they transform selenium to volatile alkylselenides); *E. coli*, *B. cereus*, and *Aspergillus niger* (cadmium accumulation) [19]; and *Enterobacter cloacae* (reduces chromium).

19.5.10.2 Soil Bioremediation

Utilized for soil contaminant degradation which can be performed either in situ or ex situ, in situ bioremediation can be done by bioaugmentation, bioventing, and biosparging, whereas ex situ can be done by land farming, composting, biopiles, windrows, or using bioreactor [9]. For optimum results conditions required would be such that they promote maximum microbial growth and activity requiring efficient control of moisture, temperature, oxygen, and pH and availability of nutrients. The processes can be anaerobic or aerobic.

Composting: Biological treatment of organic compounds under aerobic conditions can be done by composting. Composting results in transformation of problematic waste into useful product "compost" which can be used as soil conditioner, fertilizer, biofilter, or fuel. Oil remediation is done by soil composting that involves controlled decomposition of contaminants by bacteria and fungi into humus-like product. Composting can lead to waste stabilization, volume and mass reduction, drying, elimination of phytotoxic substances, and better sanitation.

Phytoremediation is presently commonly used to remove metals from contaminated soil [5].

Anaerobic treatment accelerates natural decomposition under optimum conditions. The waste is generally organic and can be treated separately or together in co-fermentation of organic wastes with agricultural and/or industrial waste. The process consumes less energy and produces low excess sludge.

Mechanical-biological treatment: In this the waste material undergoes a series of mechanical and biological steps to reduce the volume and load and stabilization.

19.5.10.3 Bioremediation of Gases

In waste gas treatment, odors and volatile organic compounds (VOC) need low-cost processes. Biofilters are important biological systems with high efficiency and are cheap. Biofilters and biotrickling filters were developed as reliable and cost-effective technology for treatment of polluted airstreams. The biodegradation of pollutants by microorganisms leads to products which are harmless, and as diverse consortia of microorganisms are in use, thus they help in the removal of complex pollutants. These can also be utilized for control of odor in aerobic treatment processes.

19.5.10.4 Biodegradation of Hydrocarbons

All petroleum hydrocarbons can be oxidized to water and carbon dioxide. However, the rate is variable depending upon the nature and physical and chemical properties of the contaminants. Hydrocarbons can be processed by both aerobic and anaerobic reactions. Many fungal strains such as *Pichia*, *Rhodosporidium*, *Candida parapsilosis*, and *Rhodosporidium toruloides* can act on hexadecane and kerosene (naphthalene); algae such as *Cyanobacteria* and *Microcystis aeruginosa* can also act on benzene, toluene, and naphthalene.

19.5.10.5 Bioremediation of Xenobiotics

Xenobiotics (xenos: foreign, strange) are non-natural chemically synthesized compounds which are foreign on the earth. Xenobiotic degradation is brought about by fungi and bacteria. They use these compounds as their source of carbon, energy, nitrogen, or sulfur. However certain xenobiotic compounds are either resistant to microbial degradation or result in production of more toxic compounds as compared to their original substrate. The microbial reduction of tetrachloroethylene and trichloroethylene results in the production of vinyl chloride (a carcinogenic compound).

19.5.10.6 Bioremediation of Chlorophenols

These are aromatic ring structures with chlorine atom and hydroxyl group at the benzene ring and include monochlorophenols, polychlorophenols, or chloronitrophenols. These are extremely toxic due to their carcinogenic and mutagenic properties which pose persistent risk to the environment. Their major source of contamination is industrial wastes, pesticides, and/or herbicides which can enter water bodies and pose threat. Due to their carcinogenicity, they are in hit list of priority pollutants. The biodegradation of some of these is summated [4] in Table 19.1.

Bacterial degradation of MCP and poly-CPs:

- Several pathways have been implicated in the degradation of these.
- The degradation results in the formation of chlorocatechols or chlorohydroquinones.
- CAPs and CMPs are highly toxic and need appropriate bioremediation measures.

19.5.10.7 Biodegradation of Some Other Contaminants

Radionuclides: Uranium or thorium affects the environment and has long half-life; therefore, their removal or biosorptive accumulation would be an important and desirable effect for environmental cleanup. Their biosorption is done using bacteria, fungi, yeasts, algae, and natural materials; however, mechanism is not fully understood.

Cyanide: Industrial wastes with cyanide need appropriate attention and treatment. Cyanide is removed by biological oxidation or hydrolytic reactions or reductive reactions with high degradation ability. Fungi and bacteria perform bioremediation and use cyanide as source of nitrogen and carbon. In biological treatment of cyanide, bacteria convert free and metal complex cyanides to bicarbonate and ammonia.

Distillery spent: Bioremediation processes are also helpful in distillery spent. It involves aerobic and anaerobic treatment. Remediation of gasoline, ethers, benzene, etc. occurs by aerobic degradation.

Textile azodyes are degraded by anaerobic treatment (white rot fungi) and aerobically by bacterial consortia [14].

19.5.11 Advantages of Bioremediation

- Reduces adverse effects of pollutants on the environment.
- Valuable minerals may be reused.
- Often useful products can be generated.
- Other advantages include stabilization of the waste, reduced volume in the waste material, destruction of pathogens in the waste material, and production of biogas.

Table 19.1 The degradation of chlorophenols

Type of chlorophenols	Precursor	Microorganism	Enzymes	Product
Poly-CPs	2,4-DCP	Breakdown product of herbicide	2,4-Dichlorophenoxyacetate–a--ketoglutarate dioxygenase	2,4-DCP
Dichlorophenols (DCPs)		2,4-Dichlorophenoxyaceticacid (2,4D)	2,4-DCP-hydroxylase	↓
Trichlorophenols (TCPs)		2,4-DCP is source of carbon and energy for *Rhodococcus opacus 1G*, *Pseudomonas* sp. (*DP-4* and *NCIB9340*), *Rhodococcus erythropolis*	3,5-Dichlorocatechol dioxygenase	3,5-Dichlorocatechol
Tetrachlorophenols (TeCPs)		Calcitrants due to two or more chlorine atoms		↓
Pentachlorophenol (PCP)				2,4-Dichloromuconic acid
				↓
				↓
				↓
				3-Oxodipic acid
Chloro-nitrophenols	2C4NP	2C4NP is carbon and energy source for	2C4NP-dehalogenase	4-Nitrophenol
2-Chloro-4-nitrophenol (2C4NP)		*Burkholderia* sp. *SJ98*		↓
4-Chloro-2-nitrophenol (4C2NP)		*Burkholderia* sp. *RKJ800*		↓
4-Chloro-3-nitrophenol (4C3NP)		*Arthrobacter nitrophenolicus* SJCon		↓
2-Chloro-5-nitrophenol (2C5NP)				↓
2-Chloro-3-nitrophenol (2C3NP)				↓
				Maleylacetate
Chloro-aminophenols	4C2AP	Used in manufacture of dyes	Deaminase	4-Chlorophenol
4-Chloro-4-aminophenol (2C4AP)		*Burkholderia* sp. *RKJ800*	4CC-1,2-dioxygenase	↓
				↓
				Cis,cis-chloromuconic acid

Case Study

Using bioremediation for cleaning of the environment is eco-friendly and the best remedy. The engineering of microorganisms results in the production of genetically modified organisms referred as "superbugs" (superbugs are also used for multiple drug-resistant bacteria) and has opened new ways to clean the atmosphere.

In 1970 Dr. Ananda Mohan Chakrabarty's group engineered and created a new bacterial strain by insertion of plasmid capable of utilizing toxic organic chemicals (camphor, hexane, octane, xylene, toluene, and naphthalene). This was referred as "superbug" and was patented in 1980 for cleaning oil spill. Dr. Ananda explored and discovered a method for genetic cross-linking that was capable of placing all four plasmid genes in place and produced a new, stable, bacteria species (now called *Pseudomonas putida*). This was capable of consuming oil faster than the initial strains of four oil-eating microbes. This was granted the first US patent for genetically modified organism (two utility patents were granted to pure bacterial cultures, patented by Louis Pasteur).

The modified bacteria of Dr. Chakrabarty were granted a patent in the UK, before the US patent. The patent was initially denied by the Patent Office; however, the United States Court of Customs and Patent Appeals overturned the decision in Chakrabarty's favor. His landmark research has opened new avenues on genetic modification experiments.

19.5.12 Limitations of Bioremediation

Bioremediation uses microbial agents to reduce, degrade, or transform the pollutants on the contaminated sites. It is highly helpful in cleaning of contaminated sites and safe disposal of contaminants. However for the process to be successful:

- There is requirement of pre-evaluation of contaminated site for selection of efficient and right microbes with other conditions like nutrients, pH, and temperature.
- The eventual uses of chemical pollutants, xenobiotics, which pose resistance to their degradation, are posing difficulty in bioremediation process. It is very difficult to degrade heavy metals, radionuclides, and some chlorinated compounds [6].
- Sometimes microbial degradation of contaminants may produce toxic metabolites; sometimes these are more toxic than original metabolite.
- More research and funding are required for bioremediation work so that the process can be tailored for specific polluted sites as each site has specific requirement.
- Bioremediation techniques do not result in the production of high-value products; thus, capital investment in this research area is very low. Commercial activity in research and development in bioremediation is very low.
- The usage of genetically modified organisms is also under regulatory bodies. The free launch of GMOs at the remediation site might also pose risk for environment.
- There is another limitation of having trained human person to handle the bioremediation work at specified sites. Thus, the bioremediation program requires integration of various fields such as biotechnology, microbiology, geology, engineering, hydrogeology, soil science, and project management.

Still there are tremendous potential and opportunities for bioremediation techniques.

19.6 Integrated Pest Management and Biopesticides

Pests: Pests are small insects that damage or interfere with our crop plants, affecting their yield and impact human and animal health. They may transmit diseases. Pest may be weed, bird, rodent, insect, tick, mite, nematode, bacteria, virus, or fungus. They cause harm to the crop by eating them and causing diseases in them.

Integrated pest management: The process used to solve pest problems with minimal impact on life forms and the environment is referred as "integrated pest management or IPM." IPM is used to manage all pests' problems, in urban, agricultural, wildland, or natural areas (technical definition of IPM). Pest management includes: tackle pest infestations, growing crop with resistance to pests, inclusion of factors that affect pest growth, and ability to survive.

IPM strategies: (1) field monitoring; (2) pest identification, identify the pests which are present; (3) monitoring and assessment of pest numbers and damage, their population and damage assessment; (4) providing guidelines for timely action; (5) preventing pest problems and best management route; and (6) record keeping.

In case pest requires control, IPM's role is to execute the most effective strategy at appropriate time for best outcome. It uses any measure alone or in combination. Its tools are biological, cultural, physical/mechanical, and chemical management. The IPM uses the following strategy to monitor pests:

Biological control: The usage of biological tools, which can remove or eliminate pests. These may be pest pathogens, predators, parasites, or competitors. Many organisms have natural enemies, which can be exploited for their removal.

Cultural control: These include practices which reduce the establishment, reproduction, dispersal, and thereby survival of pests. The modified irrigation practices may help in control of pests.

Mechanical and physical controls: They are capable of killing pest directly. They can be used to make the environment highly unsuitable for pest, for example, traps for rodents, steam sterilization of soil, mulches for weed removal, etc.

Chemical control: Pesticides are used in chemical control. However, safe pesticides (or the one with least harm) are used when it is urgent to use them or they can be used in minor quantity with other control measures for best results. Pesticides are used in a way to minimize their harmful effects on environment and other life forms.

Therefore, IPM practices focus on pest management with combination of methods and use of absolutely nontoxic methods. In case nontoxic does not work, it uses pesticides with least toxicity. Pesticides with minimal toxicity are noncarcinogenic and include boric acid, desiccant dusts (diatomaceous earth and silica gel), and nonvolatile insect and rodent baits.

19.6.1 Biopesticides

Pesticides were largely used to check infestation of pest and insect on agricultural crops. Increased use of pesticides has led to development of resistance in the pests; resistance development leads to higher sprays of pesticide creating economic burden on the farmer and extra load of pesticides on the environment affecting pollution of soil, water and air, and health problems.

Biopesticides are biological solutions for pest control. They offer new generation safe products for sustainable agricultural practices. They are alternative of chemical pesticides. Judiciously using them minimizes pest resistance, eliminates adverse effects on environment, and is friendly for other life forms.

Biopesticides are:

- Natural materials like plants, microorganisms (bacteria, fungi, and virus), and some minerals. Studies of soil microbiology and ecology had led to the identification of many different microorganisms that act as antagonists or hyperparasites of pathogens and insect pests.

- Their usage can largely affect pest infestations negatively.
- They affect environment positively.
- Result in high yields of the crops.
- Do not leave any product or residue, which are harmful to the plant.
- Do not affect nontarget organisms.
- Are cost-effective alternatives.

Plant biopesticides: These are plant extracts used as biopesticides (see Table 19.2). Nicotine was used in the seventeenth century to control plum beetles.

Microbial biopesticides: They may include a pathogen or parasite that infects the target. They may also act as inducers of plant resistance or competitors of pests. They include bacteria, fungi, viruses, and protozoa. They kill pests by either their toxins or causing diseases in them.

Biopesticides and bioinsecticides are nontoxic, eco-friendly, and cost-effective solution for pest management. Biopesticide is a widely applicable term, which includes all biocontrol agents. Bioinsecticide is used for living organisms and the formulations from them for insect control.

Bioinsecticides are composed of bacterial strains, crystal protein, and inert fillers. The mutant *B. thuringiensis* producing crystal protein were successfully tested on insects affecting cotton, cabbage, maize, sunflower, and pigeon pea. Spores of *Bacillus thuringiensis* (Bt) are widely used as biocontrol agents. Bt was isolated by Japanese biologist Shigetane Ishiwata from a diseased silkworm. It was then rediscovered after 10 years by Ernst Berliner in Thuringen, Germany. It is the most widely used biopesticide. In 1938, first commercial Bt-derived product Sporeine was launched in France.

Cry genes from *Bacillus thuringiensis* have been successfully used to solve pest infestations. Biological control of pests has emerged as an important and safe alternative to pest management.

Examples of a few microbial biopesticides:

- *Epidinocarsis lopezi* (parasitoid wasp) used to control cassava mealy bug, brown plant hopper in paddy rice, and stem borers in sugarcane.
- *B. thuringiensis* control diamond back moth (*Plutella xylostella*) causing damage to vegetables, and stem borer of maize.
- Fungus *Trichoderma* and *Gliocladium* for quick wilt disease.
- Fungus *Trichoderma* spp. is effective in controlling plant diseases by stimulation of plant host defenses.
- *Nuclear polyhedrosis virus* (NPV) acts as pathogen and causes disease in bollworm

Table 19.2 List of some plant derived biopesticides

Biopesticide	Component/part producing	Botanical name	Target
Plant derived	Azadirachtin from leaves, seeds	*Azadirachta indica*	Targets some 200 insects, inhibits feeding, insect growth regulator, inhibits metamorphosis
	Plant extract	*Dysphania ambrosioides*	Breaks exoskeleton of pests adversely affecting its respiratory system; targets aphids, whiteflies, mites, leafhoppers
	Caffeic acid, alkaloids from seeds, leaves, bark, roots	*Albizia lebbeck*	Acts as biopesticide
	Chalcones and aurones from flowers	*Butea monosperma*	Termiticidal
	Saponins from seeds	*Madhuca indica*	Repellent and biopesticide
	Annonine from seeds and leaves	*Annona squamosa*	Biopesticide

infestating cotton. Sandoz is marketing a product based on it as Elcar.
- *Non-occluded baculovirus* is a biopesticide.
- *Trichogramma* is a minute wasp that is insect egg parasites.
- *Bacillus thuringiensis var. israelensis* (Bti) is extremely lethal to Diptera insects.
- *Agrobacterium radiobacter* strain K84 is naturally present in soils and plant root. It is a biopesticide used to control crown gall in greenhouse and nursery environment.

Biochemical biopesticides: Have variety of mechanisms. They may interfere with growth, feeding, development, or reproduction of a pest or pathogen. Others may be repellants or barrier for host–pest interactions.

Nearly 100 active ingredients of biopesticides are registered with the US EPA Biopesticides division. The total biopesticide trade in 2009 was 1.6 billion US dollars. It is expected to reach 3.5 billion dollars in the coming years. India being an agricultural country has vast potential for biopesticides as its annual utilization of pesticides is more than 1 lac tons.

A few limitations of biopesticides are:

- Due to their high specificity, it is very important to identify the pest.
- Pest clearance may require multiple biopesticides.
- The variability in results is obtained as they are affected by biotic and abiotic factors.

19.7 Role of Biotechnology in Innovative Products

The environment is being heavily contaminated with hazardous and toxic substances reaching very high concentrations posing health risks.

In this the bioremediation can offer an eco-friendly solution for decontamination of the environment.

The industrial effluent is a mix of many toxic metals and substances, making detoxification extremely difficult; thus, microbial cultures are required which can sustain these extreme contamination conditions along with protecting themselves from their toxicity for efficient decontamination.

Multiple heavy metal mix can interact resulting in synergism, antagonism, or additive effect (in terms of toxicity) making detoxification extremely difficult. This requires specialized microbial cultures. Presently the focus is on using fungi for bioremediation of industrial effluent. The fungi not only have a strong potential for use in non-sterile open environment, but its mycelial growth provides it a competitive advantage over its single-cell counterparts (such as bacteria and yeasts). Their high surface-to-cell ratio maximizes both mechanical and enzymatic contact with the substrate. The degradative enzymes are extracellular, enabling fungi to tolerate higher levels of toxic chemicals.

One of the fungal species *Aspergillus lentulus* is capable of simultaneous removal of various hazardous metals such as Cr, Cu, and Pb bearing high potential for industrial applications. It is also tolerant to the thermal, alkali, and halo characteristics of wastewater; thus, it has efficacy for removal of textile dyes making it industrially more useful [15].

Another concern is usage of wastewater contaminated with heavy metals and pesticides for irrigation purposes, leading to accumulation and toxicity in plants and then to animals and humans. Thus it requires:

- Development of bioremediation technique which can remove heavy metal and pesticide from wastewater.
- Selection of appropriate microorganism for application.
- The microorganism should be able to resist extremes of pH, temperature, and salinity.
- Microorganism with multiple metal tolerance along with capacity to degrade or transform xenobiotics would be very helpful.
- This would require very efficient sampling and standard remedy for best output.

19.8 Chapter End Summary

- The field of environmental biotechnology came into being after man realized that he has already caused major damage to the nature and environment.
- Due to urbanization and industrialization, many pollutants, toxic substances, heavy metals, plastics, radionuclide, and many others were released in the environment which are harmful for all life forms. Thus, this area of biotechnology aims to detect, monitor, and treat the pollutant abundance on the earth for sustainable environment practices, to check the manufacturing and release of substance which can cause irreversible damage to the environment as chlorofluorocarbons, DDT, and so on.
- Everything is getting severely affected due to rising pollutants levels. Not only water, air, and soil are among the most badly affected. The rise in greenhouse gases, persistent xenobiotics, and plastics are increasing the load and posing risk for all.
- Their detection and monitoring becomes essentially important to assess the quality and keep preventive measures in place to prevent further degradation.
- Biosensors, biomarkers, and nanosensors are sensitive measures to detect the presence and quantify the pollutant level.
- Bioremediation being a very important process can be performed by microbes, algae, and plants for either degradation or transformation or accumulation of the contaminant for its effective clearance.
- Various modes of bioremediation along with phytoremediation, factors affecting the process and performing life forms, are important for effective reactions to occur.
- By the process of bioremediation, wastewater, soil, air, oil spills, xenobiotics are treated to maintain the environment clean.
- The process has several advantages, limitations, and shortcomings for effective maintenance of the environment.
- Integrated pest management is very important for safe and effective control of pest with minimal adverse effects on the environment. Since years, plant extracts have been used as effective pest-controlling agents. Many microbial and plant-derived compounds are used to control pest. Their usage results in removal of pest, protecting the environment from pesticide hazards and sustainable agricultural and environmental conditions.

Multiple Choice Questions

1. Environmental biotechnology deals with:
   (a) Assessment of environment
   (b) Monitoring of environment
   (c) Bioremediation
   (d) All of the above
2. Which of the following statement about xenobiotics is most appropriate?
   (a) Useful nonnatural compounds
   (b) Are highly toxic
   (c) Are resistant for microbial action
   (d) None of the above
3. Biodegradable pollutants are:
   (a) Which are boiling sensitive
   (b) Which can be bioremediated
   (c) Which can be stored
   (d) All of the above
4. Greenhouse gases are:
   (a) CO_2, CO, O_2 and H_2O
   (b) CO, NO_2, H_2O
   (c) CO_2, CH_4, N_2O and H_2O
   (d) CO_2, H_2S, N_2 and H_2O
5. Feedback gas in greenhouse effect is:
   (a) Carbon dioxide
   (b) Water vapor
   (c) Hydrogen sulfide
   (d) Methane
6. Increased BOD of a pond is indicative of:
   (a) Better water quality
   (b) Highly polluted water
   (c) Fresh water
   (d) None of the above

7. Bioconcentration means:
   (a) Increased concentration of compound in question in nature
   (b) Increased loss of compound from animal body
   (c) Increased deposition of compound in animal body
   (d) All of the above
8. The book *Silent Spring* revealed:
   (a) The elaboration of spring season
   (b) The impact of sewage on spring
   (c) The impact of DDT on environment
   (d) None of the above
9. Eutrophication results in:
   (a) Depleting nutrient and oxygen in water bodies
   (b) Vigorous growth of algae
   (c) Can cause red tides
   (d) All of the above
10. An example of biosensor is:
    i. Usage of animal
    ii. Usage of heavy metals
    iii. Usage of plant
    iv. Usage of phytochelatins
       (a) All are correct
       (b) i, ii, and iii are correct
       (c) i, iii, and iv are correct
       (d) ii, iii, and iv are correct
11. The process of bioremediation involves:
    (a) Transformation
    (b) Degradation
    (c) Mineralization
    (d) All of the above
12. In situ remediation involves:
    (a) Performing reaction in bioreactor
    (b) Performing reaction in treatment plant
    (c) Adding microbes at the contamination site
    (d) None of the above
13. Phytoremediation is very efficient at:
    (a) Decontaminating oil spills
    (b) Decontamination of heavy metals
    (c) Wastewater treatment
    (d) Decontamination of DDT
14. Activated sludge involves:
    (a) Aerobic reactions
    (b) Anaerobic reactions
    (c) Filtration
    (d) None of the above
15. Dr. Ananda Chakrabarty engineered strain of which of the following that was useful in clearing oil spills?
    (a) *Escherichia coli*
    (b) *Saccharomyces cerevisiae*
    (c) *Pseudomonas putida*
    (d) All of the above

Answers

1. (d); 2. (a); 3. (b); 4. (c); 5. (b); 6. (b); 7. (c); 8. (c); 9. (d); 10. (c); 11. (d); 12. (c); 13. (b); 14. (a); 15. (c)

Review Questions

Q1. What do you understand by environmental biotechnology?
Q2. Define pollution?
Q3. What is the role of biomarkers and biosensors in environmental monitoring?
Q4. Why is environmental monitoring important?
Q5. What is bioremediation?
Q6. Give a brief account of phytoremediation.
Q7. What are xenobiotic compounds?
Q8. What is in situ and ex situ bioremediation?
Q9. Write environmental hazards of DDT and any two air pollutants.
Q10. Define the terms: biomagnification, bioabsorption, and biotransformation.
Q11. What are biopesticides?
Q12. Describe integrated pest management.

References

1. Anjum NA, Umar S, Chan MT (2010) Ascorbate-glutathione pathway and stress tolerance in plants. Springer, Dordrecht 323–336.
2. Anjum et al (2015) Jacks of metal/metalloid chelation trade in plants-an overview. Front Plant Sci 6
3. Anjum NA, Singh HP, Khan MIR et al (2015b) Too much is bad – an appraisal of phytotoxicity of elevated plant-beneficial heavy metal ions. Environ Sci Pollut Res 22:3361–3382
4. Arora PK, Bae H (2014) Bacterial degradation of chlorophenols and their derivatives. Microb Cell Factories 13:31

5. Blaylock M, Ensley B, Salt D, Kumar N, Dushenkov N, Raskin I (1995) Phytoremediation: a novel strategy for the removal of toxic metals from the environment using plants. Biotechnology 13:468–474
6. Boopathy R (2000) Factors limiting bioremediation technologies. Bioresour Technol 74:63–67
7. Caplan JA (1993) The worldwide bioremediation industry: prospects for profit. TIB Tech 11:320–323
8. Cobbett C, Goldsbrough P (2002) Phytochelatins and metallothioneins: roles in heavy metal detoxification and homeostasis. Annu Rev Plant Biol 53:159–182
9. Cunningham SD, Shann JR, Crowley DE, Anderson TA (1997) Phytoremediation of contaminated soil and water. In: Kruger EL, Anderson TA, Coats JR (eds) Phytoremediation of soil and water contaminants, vol 664, ACS symposium series. American Chemical Society, Washington, DC
10. Fisher A, Walker M, Pam P (2010) DDT and DDE: sources of exposure and how to avoid them (PDF). Retrieved 2 Dec
11. Gavrilescu M (2010) Environmental biotechnology: achievements, opportunities and challanges. Dynamic biochemistry, process biotechnology and molecular biology. Global Science Books Miki cho, Ikenobe 3011-2, Japan 4: 1–36
12. Hassan Z, Aarts MGM (2011) Opportunities and feasibilities for biotechnological improvement of Zn, Cd or Ni tolerance and accumulation in plants. Environ Exp Bot 72:53–63
13. Jarup L (2003) Hazards of heavy metal contamination. Br Med Bull 68:167–182
14. Kaushik P, Malik A (2009) Fungal dye decolourization: recent advances and future potential. Environ Int 35:127–141
15. Kaushik P, Malik A (2010) Alkali, thermo and halo tolerant fungal isolate for the removal of textile dyes. Colloids Surf B: Biointerfaces 81:321–328
16. Laumbach et al (2015) What can individuals do to reduce personal health risks from air pollution? J Thorac Dis 7:96–107
17. Mike L (2009) Solar change and climate: an update in the light of the current exceptional solar minimum. Proc R Soc 466: 303–329
18. Miller R (1996) Phytoremediation, technology overview report, ground-water remediation technologies analysis center, Ground-Water Remediation Technologies Analysis Center William Pitt Way Pittsburgh vol 3
19. Mishra A, Malik A (2012) Simultaneous bioaccumulation of multiple metals from electroplating effluent using Aspergillus lentulus. Water Res 46:4991–4998
20. Mishra A, Malik A (2013) Recent advances in microbial metal bioaccumulation. Crit Rev Environ Sci Technol 43:162–1222
21. Padmavathiamma PK, Li LY (2007) Phytoremediation technology: hyper-accumulation metals in plants. Water Air Soil Pollut 184:105–126

Some Related Resources

CDC – NIOSH Pocket Guide to Chemical Hazards – DDT
DDT General Fact Sheet – National Pesticide Information Center
DDT Technical Fact Sheet – National Pesticide Information Center
U.S. Environmental Protection Agency, 2000
U.S. Environmental Protection Agency, 2004
U.S. Environmental Protection Agency, 2006
IPCC Forth Assessment Report, 2007
http://advancinggreenchemistry.org
http://ei.cornell.edu/biodeg/bioremed/
http://www.agrilife.in/biopesticides.htm
http://www.beyondpesticides.org/
http://www.bioremediationstudy tripod.com
http://www.climate.nasa.gov/causes
http://www.fao.org/agriculture/crops
http://www.ipm.ucdavis.edu/
http://www.vertasefli.co.uk
Pesticide residues in food 2000: DDT
Scorecard: The Pollution Information Site – DDT
World Health Organization, The WHO Recommended Classification of Pesticides by Hazard, 2005.

Plant Biotechnology and Agriculture

20

Abstract

Agricultural biotechnology is the term used in crop and livestock improvement through biotechnology tools. Biotechnology encompasses a number of tools and elements of conventional breeding techniques, bioinformatics, microbiology, molecular genetics, biochemistry, plant physiology, and molecular biology. The biotechnological tools that are important for agricultural biotechnology include conventional plant breeding, tissue culture and micropropagation, molecular breeding or marker-assisted selection, and genetic engineering and GM crops. In this chapter, readers would learn about the role of biotechnology in crop improvement and the major applications of the field.

Biotechnology has given a new dimension to scientific innovations, offering efficient and cost-effective means to produce a diverse array of novel, value-added products and tools. The present and future focus is on continuing improvement of agronomic traits such as yield and abiotic stress resistance in addition to the biotic stress tolerance of the present generation, crop plants as biomass feedstocks for biofuels and "biosynthetics," value-added output traits such as improved nutrition and food functionality, and plants as production factories for therapeutics and industrial products. From a consumer perspective, the focus on value-added traits, especially improved nutrition, is of greatest interest. Both traditional plant breeding and biotechnology-based techniques are needed to produce plants with the desired quality traits. Continuing improvements in molecular and genomic technologies are contributing to the acceleration of product development.

With almost 870 million people estimated to suffer from chronic hunger worldwide, undernourishment represents a major problem that severely affects people in developing countries. In addition to undernourishment, micronutrient deficiency alone can be a cause of serious illness and death. Large portions of the world population rely on a single, starch-rich crop as their primary energy source, and these staple crops are generally not rich sources of micronutrients. As a result, physical and mental health problems related to micronutrient deficiencies are estimated to affect around

V. Gupta et al., *Basic and Applied Aspects of Biotechnology*, DOI 10.1007/978-981-10-0875-7_20

two billion people worldwide. The situation is expected to get worse in parallel with the expanding world population. Improving the nutritional quality of staple crops seems to be an effective and straightforward solution to the problem. Conventional breeding has long been employed for this purpose, but success has been limited to the existing diversity in the gene pool. However, biotechnology enables addition or improvement of any nutrient, even those that are scarce or totally absent in a crop species. In addition, biotechnology introduces speed to the biofortification process compared to conventional breeding. Genetic engineering was successfully employed to improve a wide variety of nutritional traits over the last decade.

20.1 Introduction

All the living beings are continuously evolving according to changing environmental conditions. But these evolutionary changes are not detectable unless major alterations in the form of new traits are observed. After his own evolution, man gradually learned to grow, breed, and domesticate plants and animals to suit his needs. Subsequently, after many years of growing and using crops, the field of agriculture came into being. Then, many years later, after potential for biotechnological tools and techniques was realized, agricultural biotechnology was established. Thus, agricultural biotechnology is the term used in crop and livestock improvement through biotechnology tools. Biotechnology encompasses a number of tools and elements of conventional breeding techniques, bioinformatics, microbiology, molecular genetics, biochemistry, plant physiology, and molecular biology. The biotechnological tools that are important for agricultural biotechnology include conventional plant breeding, tissue culture and micropropagation, molecular breeding or marker-assisted selection, and genetic engineering and GM crops. In this chapter, readers would learn about the role of biotechnology in crop improvement and the major applications of the biotechnology for the improvement of agronomic traits [23, 60].

Biotechnology has given a new dimension to such innovation, offering efficient and cost-effective means to produce a diverse array of novel, value-added products and tools. The present and future focus is on continuing improvement of agronomic traits such as yield and abiotic stress resistance in addition to the biotic stress tolerance of the present generation. Engineering of crop plants may be done for:

1. Biomass feedstocks for biofuels and "biosynthetics"
2. Value-added output traits such as improved nutrition and food functionality
3. Production of therapeutics and industrial products

From a consumer perspective, the focus is on value-added traits, especially improved nutrition. Both traditional plant breeding and biotechnology-based techniques are needed to produce plants with the desired quality traits. Continuing improvements in molecular and genomic technologies are contributing to the acceleration of product development.

With almost 870 million people estimated to suffer from chronic hunger worldwide, undernourishment represents a major problem that severely affects people in developing countries [7]. In addition to undernourishment, micronutrient deficiency alone can be a cause of serious illness and death. Large portions of the world population rely on a single, starch-rich crop as their primary energy source, and these staple crops are generally not a rich source of micronutrients. As a result, physical and mental health problems related to micronutrient deficiencies are estimated to affect around two billion people worldwide. The situation is expected to get worse in parallel with the expanding world population. Improving the nutritional quality of staple crops seems to be an effective and straightforward solution to the problem. Conventional breeding has long been employed for this purpose but success has been limited to the existing diversity in the gene pool. However, biotechnology enables addition or improvement of any nutrient, even those that are scarce or totally absent in a crop species. In addition, biotechnology introduces speed to the biofortification process compared to conventional breeding. Genetic engineering was successfully employed to improve a wide variety of nutritional traits over the last decade.

"Functional food" components are of increasing interest in the prevention and/or treatment of at least four of the leading causes of death in the USA: cancer, diabetes, cardiovascular disease, and hypertension. "Functional foods" have been defined as any modified food or food ingredient that may provide a health benefit beyond the traditional nutrients it contains [20]. The term "nutraceutical" is defined as "any substance that may be considered a food or part of a food and provides health benefits, including the prevention and treatment of disease." Some of the nutrient-related correlation links are:

- Dietary fat and fiber to the prevention of colon cancer
- Folate to the prevention of neural tube defects
- Calcium to the prevention of osteoporosis
- Psyllium to the lowering of blood lipid levels
- Antioxidant to the scavenging of reactive oxidant species and protection against oxidative damage of cells that may lead to chronic disease

On the functionality side, there is a mirror component from the perspective of the genetic makeup of the individual doing the consuming. This field of personal response to nutrients is further divided into two thematic subsets with subtle differences. "Nutrigenomics" is the prospective analysis of differences among nutrients in the regulation of gene expression, while "nutrigenetics" is the analysis of genetic variations among individuals with respect to the interaction between diet and disease. These spheres of enquiry are designed to provide nutritional recommendations for personalized or individualized nutrition [22].

20.2 Conventional Plant Breeding

Many years ago, man started domestication of crops and animals. In the process man started selecting plant or animal varieties with better traits. After prolonged selections, the varieties were chosen which were with higher yield of seeds and fruits, best suited in a particular environment, and pest resistant, with wide adaptability. These were referred as cultivated varieties or "cultivars." Man domesticated them, grew them, and stored their seeds for next season. The developments in agriculture led to easy and settled life for human beings. Agricultural practices have undergone tremendous changes, many of which have made food and fiber production more efficient and safer.

Before the principles of genetics were established, the farmers had started selective breeding of crops. However, after Gregor Mendel's discovery of heritability of traits, segregation of characters and independent transfer of characters led to the understanding that plants and animals acquire traits from parents, which created the potential for people to selectively breed crops and livestock. That is how plant breeding came into being, and man started selective mating and cross-pollination of the plants and animals for better traits and improved varieties. These discoveries revolutionized agriculture by launching the development of selective crossbreeding with a comprehensive understanding of the underlying mechanisms of inheritance that dramatically increased the productivity and quality of the plants we grow for food, feed, and fiber.

20.2.1 Selective Crossbreeding

In the traditional breeding approach, the new varieties are developed either by selecting plants

Fig. 20.1 Plant breeding: selective cross between high-yielding, dwarf, and pest-sensitive variety and a variety with pest resistance, which is tall, and with low yield. The resultant offspring is tall, high-yielding, and resistant variety

with desirable characteristics or by combining qualities from two closely related plants through selective breeding, for example, resistance to a particular pest or disease or tolerance to climatic conditions. Pollen with the genes for a desired trait is transferred from plants of one crop variety to the flowers of another variety with other desirable traits. Eventually, through careful selection of offspring, the desired trait will appear in a new variety of plants. Traditional plant breeding has produced numerous highly successful new varieties of crops over the centuries (Fig. 20.1).

20.2.2 Classical Breeding with Induced Mutation

Mutations are natural events which occur randomly and induce changes in the genetic makeup of a plant. Sometimes, the mutation can result in beneficial trait. Plant breeders used radiation (X-rays, gamma rays) or chemical (ethyl methanesulfonate) mutagens to induce changes in the sequence of the base pairs of DNA, which provides biochemical instructions for the development of plants (Fig. 20.2). Resultant plants may possess new and desirable characteristics through this modification of their genetic material. During this process, plant breeders must grow and evaluate each plant from each seed produced [39, 58].

The mutation breeding, since it was started 75 years before, has almost 3200 varieties of different crop plants (IAEA database). The important crops that were improved to possess agronomically desirable characteristics include rice (824), barley (312), wheat (274), maize (96), common bean (57), tomato (20), potato (16), sugarcane (13), soybean (2), and other crops as grapefruit, lettuce, and many fruits (FAO/IAEA, 2008) [16]. But as the understanding of genetics developed, newer technologies came into being, and so were the changes in the way the problems were addressed earlier. Also, the dependence of the conventional breeding on screening, evaluation, and selection was another constraint.

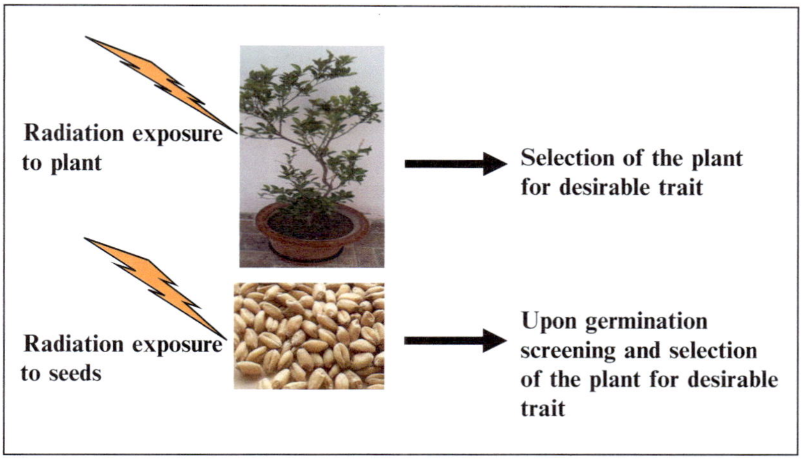

Fig. 20.2 Mutation breeding: the plant and the seeds are exposed to gamma rays. The gamma rays can induce mutations in the plants and the seeds. The seeds, after germination, are screened for desirable trait. The plants are also screened for mutation-induced desirable traits

Open pollinated (OP) seeds: These are the seeds which are produced when the crop is openly, naturally, and randomly pollinated by winds, insects, and birds resulting in a naturally diverse population. After seeds are obtained by open pollination, they can be selected for desirable traits and can be reused for planting in the next season.

Hybrid seeds: A hybrid is produced by crossing between two genetically dissimilar parents. Pollen from male parent (pollen parent) can pollinate, fertilize, and set seeds in female (seed parent) to produce F1 hybrid seeds (Fig. 20.3). For hybrid production, cross between two parents is important. The process is difficult in self-pollinated but easier in cross-pollinated varieties. The plants have adopted various mechanisms to generate diversity by cross-pollination like unisexual flowers (dicliny), time of anther dehiscence and stigma receptivity which are different (dichogamy), non-compatibility with self (self-incompatibility), and spatial separation of anther and stigma (herkogamy).

Pure line: Pure line is a breed or strain of animals or plants that maintains a high degree of consistency in certain characters because of inbreeding for generations.

(continued)

The raising of pure line involves the selection of lines in the existing germplasm which express the desired characteristics (resistance to pest and diseases, early maturity, yield). As these traits may not be present in only one line, thus selected lines are bred together by hand.

In self-pollinated plants, flowers are emasculated by removing the anthers or the male part of the flower by hand and are pollinated by pollen from another line. The female parent is usually the line that possesses the desired agronomic trait while the male parent is the donor of the new trait. F1 (first filial generation) offsprings are planted and selfed, as well as the F2 generation. Breeders then select in the F3 and F4 generation the lines which exhibit their desired agronomic characteristics and the added trait. Testing for resistances to pests and abiotic stresses are conducted also at this time. Lines with desired traits and are rated intermediate to resistant/tolerant to the pests and abiotic stresses are selected and selfed in two to three more generations. Lines which do not lose the new traits and are stable are termed pure lines (Fig. 20.3).

(continued)

Recombinant inbred lines (RILs) are a collection of strains that can be used to map quantitative trait loci. Parent strains are crossed to create recombinants that are then *inbred* to isogenicity, resulting in a permanent resource for trait mapping and analysis. RILs are useful for preliminary mapping of any trait that differs between the parental strains used to generate the population. In RILs the same mapping population can be maintained and used repeatedly to map all kinds of different traits. They can also reveal multiple loci contributing to any trait of interest. The downside is that they are less statistically powerful for analyzing effects of any one particular locus, because each RIL also harbors potentially confounding background genetic variation.

Near isogenic lines (*NILs*): NILs are created by crossing the phenotype of interest with a standard line. The offsprings of F1 are selfed to produce F2 generation. Then F2 progeny with target trait is selected and crossed with the standard line (the recurrent parent) (Fig. 20.3). This process is repeated for many generations. NILs are most useful when any one candidate gene or one particular locus of interest needs to be analyzed. They allow the measurement of the effect of allelic variation at that locus only, while eliminating background genetic variation.

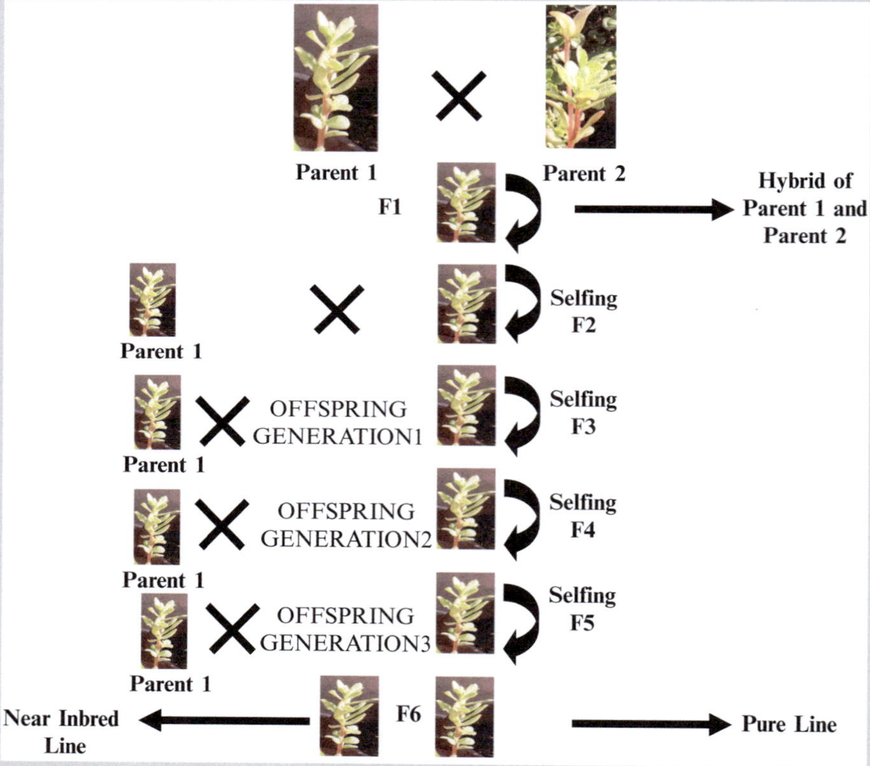

Fig. 20.3 The figure shows the cross of parent 1 and parent 2 to give F1 hybrid. Selfing of these hybrids for 5–6 generations results in the production of pure line. For creation of near inbred lines (NILs), the F1 is selfed, and after appropriate selection, F2 is crossed with recurrent parent for many generations, resulting in the production of NILs

20.2.3 Hybrid Varieties and Their Applications

The hybrids are developed by the breeding of two pure lines with complementing traits from diverse parents. The offsprings thus obtained (F1 generation) are tested for hybrid vigor (agronomically, yield-wise best offspring) and selected [1]. After this technology the yield parameters were largely increased in major crops including rice, corn, wheat, cotton, and other crops including many vegetables.

In the USA, corn yield from 26 bu/acre had increased to 159 bu/acre in 2012 (with fertilizers and pesticides). In China production increased from 140 million tons in 1978 to 188 million tons in 1990.

During that time, there were enormous benefits of conventional breeding. It produced a vast number of varieties and hybrids that contributed immensely to higher grain yield, stability of harvests, and farm income. Despite the successes of the green revolution, the challenges were still there in the form of rapid mushrooming populations, decline in agricultural resources such as land and water, and the apparent plateauing of the yield curve of the staple crops, changing demographics, and inadequate poverty intervention programs which have eroded many of the gains of the green revolution. The green revolution resulted in increased crop management productivity by controlling tillage, water use, fertilization, weed and pest, and harvesting [13]. Important examples during the green revolution were the high-yielding semidwarf cultivars of cereals and the hybrid rice developed in the 1970s.

20.2.4 Limitations of Conventional Plant Breeding

Although conventional plant breeding revolutionized agricultural practices, it had a few limitations:

- Breeding was only possible in sexually compatible plants, limiting the addition of new and useful traits.
- In traditional breeding, crosses are often made in a relatively uncontrolled manner. The breeder chooses the parents to cross, but at the genetic level, the results are unpredictable. DNA from the parents recombines randomly, and desirable traits such as pest resistance may be bundled with undesirable traits, such as lower yield or poor quality.
- A great deal of effort is required to separate undesirable from desirable traits, and this is not always economically practical.
- Traditional breeding programs are time consuming, often taking decades to produce new viable crop varieties, and labor intensive.
- Many potential benefits are lost along the way, as plants that fail to demonstrate the introduced characteristics are discarded. Traditional plant breeding takes on average 12–15 years to produce a new crop variety.

However, for the genetic improvement of food crops to continue at a pace sufficient to meet the needs of the present and future population on earth, both conventional technology and biotechnology are needed [42]. Thus, new crop improvement technologies should be developed and utilized. Biotechnology can help meet our future needs for food and fiber. The adoption of the transgenic crops has been one of the most rapid cases of technology diffusion in the history of agriculture [59]. Between 1996 and 1999, the area planted commercially with transgenic crops has increased from 1.7 to 39.9 million ha. In the last 20 years, biotechnology has developed invaluable new scientific methodologies and products, which need active financial and organizational support to bring them to fruition. So far, biotechnology has had the greatest impact in medicine and public health. However, there are a number of fascinating developments approaching commercial applications in agriculture [29].

20.3 Transgenics in Crop Improvement

The transgenic plants are created by genetic manipulation of the plant by the incorporation of desired gene (refer to Chap. 5 for transgenic plant production) in the host plant. The host plant is referred as "genetically engineered or genetically modified (GM)" plant [1].

Here, a brief comparison is done between conventional breeding and genetic engineering (Fig. 20.4):

- In conventional breeding, it was impossible to manipulate individual genes due to lack of natural introgressable genes between species; thus, the trait transfer was a limitation which can be efficiently addressed by genetic engineering. In genetic engineering, the gene of any organism including humans can be cloned and expressed in plants.
- In conventional breeding, obtaining any particular gene combination, from millions of crosses, is a remote possibility, whereas genetic manipulation allows the direct transfer of one or just a few genes, between either closely or distantly related organisms.
- Conventional breeding can result in transfer of undesirable genes along with desirable genes, or while one desirable gene is gained, another is lost because the genes of both parents are mixed together and reassorted more or less randomly in the offspring. Genetic engineering techniques can add or switch off function of the particular gene without altering other genes in host organism.
- Due to human involvement, the crops can be produced by using "conventional" and "modern" biotechnological approach. Conventional approaches require crosses and selections, for example, development of salt-tolerant crop requires the cross of sensitive variety with resistant variety, followed by selection (backcrossing if required), whereas modern techniques simply require to insert the target gene in the host plant (resultant plant is called genetically modified organism (GMO) or genetically modified food (GMF) for human consumption variety) [52]. By whatever means it was prepared, the resultant plant would have resistance to salt.
- Conventional breeding cannot modify the present genetic makeup of the plant, whereas in genetic engineering techniques, plants may be modified by removing or switching off the particular gene and genetic control (promoters) elements.

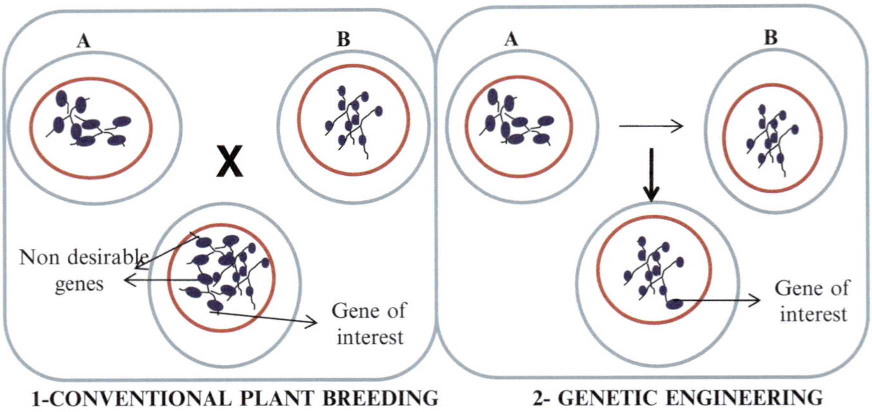

Fig. 20.4 The figure shows cross between two varieties *A* and *B* where a gene from *A* is desirable in *B*. Represents conventional breeding where cross results in transfer of desirable character along with other genes. This shows genetic engineering, where only gene of interest is transferred to *B*

However, the intervention of genetic engineering is required for the following conditions:

1. The gene of interest is not present in the germplasm of the crop.
2. The improvement of trait is difficult by conventional breeding methods.
3. The incorporation of that trait is very difficult or takes a very long time to introduce and/or improve such trait in the crop by conventional breeding methods.

However, genetic engineering techniques are multidisciplinary and required coordinated involvement of tools and elements of conventional breeding techniques, bioinformatics, biochemistry, molecular genetics, and molecular biology [61].

Genetic engineering has helped in the improvement of crops. The area which is now planted with GM crops is consistently increasing. The website http://www.isaaa.org has the updated record of GM plantations across the world, the hectarage planted, the benefits derived from the biotech crops, farmer accounts of planting biotech crops, as well as future prospects and directions of the technology [1].

So far, 27 transgenic crops which are planted commercially are alfalfa, Argentine canola, bean, carnation, chicory, cotton, creeping bent grass, eggplant, flax, maize, melon, papaya, petunia, plum, Polish canola, poplar, potato, rice, rose, soybean, squash, sugar beet, sugarcane, sweet pepper, tobacco, tomato, and wheat.

With genetic engineering, more than one trait can be incorporated into a plant, and these are called stacked traits. These are currently corn, cotton, and soybean crops with both herbicide and insect-resistant traits. Transgenic crops with combined traits are also available commercially such as the herbicide-tolerant and insect-resistant maize and cotton.

Stacking different genes for one trait makes the crop more durable to resist the pest/disease and tolerate more herbicides. Another strategy to improve sustainability of the technology is the use of the refuge. Technology developers have studied effective refuge systems for specific transformation event. These are discussed to farmers extensively for proper implementation and are monitored regularly to observe any resistant insects or weeds.

Refuge Area

To avoid insects from developing resistance for *Bacillus thuringiensis* (Bt), the crops are grown with Bt crops along with some refuge area supply of source of wild-type insects. The transgenic crops as Bt are planted with rows of non-Bt crops separated either by row or in the form of linear block or border. The idea is that all insects would not develop resistance as there would be more chances of mating between resistant and nonresistant insect resulting in progeny which is nonresistant. Thus, growing transgenic Bt crops on 80 % and non-transgenic on at least 20 % area (refuge area) (5 % for corn and 20 % for cotton) can delay the development of resistance in insects.

20.3.1 New and Future Initiatives in Crop Genetic Engineering

Commercial GM crops have delivered benefits in crop production. Many products are in the pipeline which would contribute to food quality. Resistance for pest helps in maintaining clean environment. The plants are also being explored for pharmaceutical production.

Examples of these products include golden rice with higher levels of iron and beta carotene (an important micronutrient which is converted to vitamin A in the body); long life banana that ripens faster on the tree and can therefore be harvested earlier; maize with improved feed value;

delayed ripening papaya; papaya ring spot virus-resistant papaya; tomatoes with high levels of flavonols, which are powerful antioxidants; drought-tolerant maize and wheat; maize with improved phosphorus availability; arsenic-tolerant plants; insect-resistant eggplant and rice; edible vaccines from fruit and vegetables; and low-lignin trees for paper making among others.

The length of time in developing transgenic plant depends upon the gene, crop species, available resources, and regulatory approval. It varies from 6 to 15 years before a new transgenic plant or hybrid is ready for commercial release.

20.4 Genetic Marker-Assisted Breeding

The usage of modern techniques for effective and efficient selection provides better options with less time requirements. The present technology of transgenics and marker-assisted breeding (MAB) gives tremendous opportunities and prospects for breeding [28]. After development of DNA markers and easy and high-throughput ways to analyze them, they have led to the development of molecular breeding (refer to Chaps. 6 and 18 for molecular markers). The marker-assisted breeding, where molecular markers are associated with specific traits, could be used to direct breeding programs [1, 46]. The markers are extensively used and are helpful in germplasm evaluation, genetic mapping, map-based gene discovery, characterization of traits, and crop improvement. This breeding technique offers many advantages over the conventional breeding:

- As it is DNA based, it allows the selection of traits at a very early stage (seedling stage). Thus, genotypes may be screened for selection or rejection by marker-assisted selection (MAS) [1, 6].
- They are selectively neutral, allowing selection under any kind of growth condition. This is useful for trait improvement (which is expressed at particular environmental condition) as resistance to pest, stress tolerance, and so on.
- QTL: QTL is quantitative trait loci which can be typically linked to or contains the single or number of genes that control a particular phenotype. QTL mapping is done by identifying the molecular markers which can affect that particular trait. It basically links certain complex traits to specific regions of chromosomes [55, 56]. This identifies the action, interaction, number, and precise location of these regions; their analysis can be more effective using reliable markers tightly linked to the trait by the use of MAS.
- Heterozygosity of alleles can be determined using codominant markers which is also helpful for disease inheritance and phenotype predictions in a very short time.
- MAS is independent of developmental stage and thus avoids growth time lag and is efficient, cheap, and precise as compared to conventional assays.
  - Molecular markers help in the identification of the gene and are referred as "gene tags" and the process is "gene tagging." The markers are located very close to the desired gene hence they tend to stay together or are linked.
  - For these studies, during mapping of the genome, marker–marker framework map is created, which can specify the position of marker and the position of gene, the tagged markers, and their distance from other known genes. For example, in a population of plants (RIL), one particular marker is found to be associated with flower color. When color is white, the marker used amplifies the gene at lower position (allele is evident as having lower band on the gel), and when the color is pink, the same marker used amplifies the gene at higher position (allele is evident as having higher band on the gel). Now, amplification using the same marker at the seedling stage can predict (1) if band is at lower position, the plant would bear white flowers, and (2) if band is at higher level, the plant would bear pink flowers (Fig. 20.5). Thus, the knowledge of the molecular structure allows the scientists to analyze DNA from just germinated

Fig. 20.5 Marker-assisted selection. The marker is used to amplify genomic DNA. The appearance of these two bands is seen on the gel. The white-colored flowers are associated with a lower band and pink-colored flowers are associated with an upper band

seedlings without waiting for the seedlings to grow into a mature plant [45, 58]. Once analysis is done, it is known whether the seedling contains the target gene or not, based on which the seedling is accepted or discarded.

Nowadays, MAS is a routine step in breeding of most crops where the gene and the markers for a specific trait are known. This technique is being used in the efficient introgression of important genes into various crops including bacterial blight resistance in rice, increased beta carotene content in rice, cassava, and banana, and submergence tolerance in rice [5, 9, 59].

- Random primers like RAPD and AFLP are helpful to predict the genetic profile of a line or variety. The results can then be analyzed for the relatedness of one line to another. The information on genetic diversity of the lines is utilized in selecting for extremely unrelated parents useful for hybrid seed technology.
  - However, MAS suffers a few limitations as it does not give good response in complex traits like yield, protein content, and others, but is often more useful in simply inherited traits.
  - All the markers may not be very useful, but a variety of markers can be checked for better results. Usually, the PCR-based markers are more user-friendly.
  - Marker polymorphism and trait tagging is not universal.
  - Multiple mapping populations are helpful for a better understanding of marker allelic diversity and genetic background effects.
  - In addition, QTL positions and effects also need to be validated and reestimated by breeders in their own germplasm of interest [55].
  - Labor costs are higher in many cases; thus, they can be a supplementary addition to conventional breeding.
  - High costs and technical or equipment demands of MAB will continue to be a major obstacle for its large-scale use in the near future, especially in the developing countries.

20.5 Tissue Culture and Plant Regeneration

Another technology or group of tools and techniques important for plant breeding were development and applications of plant tissue culture. The success of tissue culture was due to "totipo-

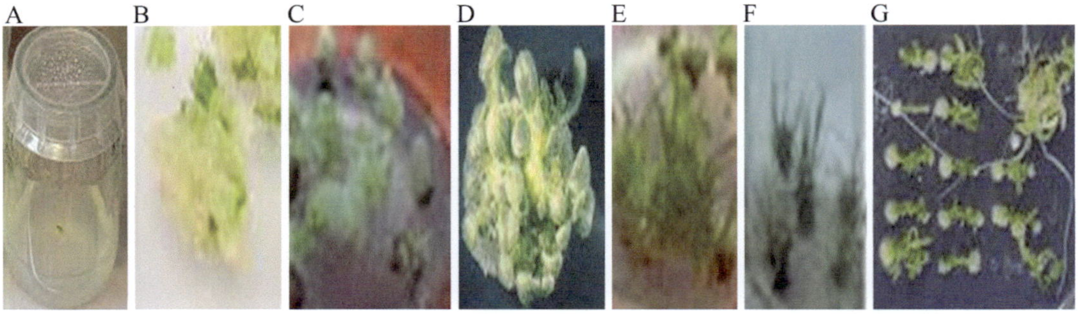

Fig. 20.6 The figure shows the general process of tissue culture. (**a**). Media with explant, (**b**) undifferentiated callus, (**c**) start of differentiation, (**d**) partially differentiated callus, (**e**) shooting and rooting starts, (**f**) and (**g**) buds separated for further subculturing and micropropagation

tency," a very important property possessed by plant cells where they have the ability to regenerate the whole plant.

In tissue culture the normal plant cells, or tissues, or organs or genetically engineered plant cells are cultured aseptically under laboratory conditions on specific defined nutrient media and proper temperature, humidity, and light conditions to create more plants like the parent plant from where the cell is obtained. Under appropriate conditions, an entire plant can be regenerated from a single cell. Gottlieb Haberlandt in 1902 gave birth to the idea of plant tissue culture [54]. Plant tissue culture is a technique that has been around for more than 30 years. There are several types of tissue culture depending on the part of the plant (explant) used and their applications (Fig. 20.6). Let us briefly summarize some applications of tissue culture:

- Tissue culture techniques can produce genetically homogeneous (clones) plants.
- Micropropagation can produce a large number of clones using shoot or meristem culture.
- Homozygous lines may be obtained by haploid culture for studying recessive traits.
- It is useful for inducing somaclonal variations.
- It is useful for producing pathogen and virus-free plants.
- It provides growth of plant cells for production of many important products and secondary metabolites.

20.5.1 Basic Introduction to Cell Culture

Before we proceed with tools and technique of cell culture, let us know some of the terminologies used in plant tissue culture:

- *Aseptic*: Free of any kind of microorganisms or contaminants.
- *Totipotency*: The property of the cell to establish part of the plant or complete plant.
- *Explant*: Small tissue or plant part or cell used to initiate the culture.
- *Plasticity*: Ability of cell to differentiate into any other lineage other than its own under the influence of different growth factors.
- *Dedifferentiation*: When specialized or differentiated cells lose its status and return to a condition where it becomes unspecialized or comes to its primitive state.
- *Redifferentiation*: When dedifferentiated cell again becomes differentiated or specialized.
- *Callus*: The mass of unorganized parenchymal cells obtained after explant culture. The culturing is performed in the dark as light can promote the differentiation.

Requirements The tissue culture work requires a closed culture laboratory where conditions of temperature, moisture, and light can be controlled. The work requires the laminar hood to maintain aseptic conditions during handling of plant material and medium.

Laminar airflow A laminar airflow cabinet has a small air-blowing motor. The air thus passes through a coarse filter. In coarse filter large particles are lost from incoming air. The air then passes through a fine filter known as "high-efficiency particulate air" (HEPA) filter. The HEPA filter removes particles larger than 0.3 mm; thus, after the HEPA filter filters air, the ultra-clean air flows through the working area. The hood has ultraviolet light tube for decontamination of poured media and other plasticware.

Sterilization Sterilization (autoclaving) is a mandatory requirement for making material (tubes, glassware, medium, plasticware) sterile. The heat labile compounds like growth regulators or plant hormones are filter sterilized. For sterilization of tissue, the following reagents 0.3–0.6 % sodium hypochlorite for 15–30 min or 9–10 % calcium hypochlorite for 15–30 min or 10–12 % hydrogen peroxide for 5–15 min or 0.1–1 % mercuric chloride for 2–10 min can be used. The process needs to be controlled as it is dangerous for the plant tissue.

Greenhouse For acclimatization of the plants, a greenhouse is required having facilities for maintaining humidity, misting, light control, and cooling and heating system to provide appropriate growing conditions to the plants.

Media The explants are grown in tissue culture media. There are many types of media with different supplementations. Thus, a medium has to be chosen with best optimizations. The widely used medium is Murashige and Skoog (MS) medium and Gamborg and White's media. The medium consists of inorganic nutrients, macroelements, microelements, and growth regulators (vitamins, growth hormones) and carbon source (sucrose).

The media can have undefined components like coconut milk (CM), casein hydrolysate (CH), corn milk, malt extract (ME), tomato juice (TJ), or yeast extract (YE). The problem with undefined media is variability in the results obtained.

The growth hormones used for appropriate growth of the plant may be:

- *Auxins*: Cause elongation of the stem and internode, tropic movements, and apical dominance; rooting examples are indole-3-acetic acid (IAA), indole-3-butyric acid (IBA), naphthalene acetic acid (NAA), and para-chlorophenoxyacetic acid (2,44,5-T).
- *Cytokinins*: Involved in cell division, shoot differentiation, and modification of apical dominance; examples are benzylaminopurine (BAP), isopentenyl adenine (2-ip), furfuryl aminopurine (kinetin), thidiazuron (TDZ), and zeatin.
- *Gibberellins*: Stimulate normal development of plantlets from in vitro adventitious embryos; example is GA3.
- *Ethylene*: Gaseous plant growth regulator and can have variability (promotory or inhibitory) for the same processes in different systems. It is released in response to stress.
- *Abscisic acid (ABA)*: Promotes morphogenesis; example is ancymidol.

Gelling agents Common gelling agent used is agar, but agarose (β-D(1-3)-galactopyranose and 3,6-anhydro-α-L(1-4) galactopyranose) and gelrite (Merck chemicals) or Phytagel (Sigma) can also be used. For optimum results, balance of correct nutrients, excellent laboratory conditions, good handling techniques, and contamination-free growth are very important.

20.5.2 Culture of Cells

The cells can be separated from the parent tissue by either mechanical method or enzymatic method. The mechanical method involves scrapping of the cells, while the enzymatic method requires disaggregation by the use of enzymes as pectinase or careful use of macerozyme (it degrades middle lamella and weakens cell wall). The culturing of cells can be done by batch culture or suspension culture in fed-batch or continuous mode. The cell culture is useful for production of secondary metabolites [25].

20.5.3 Plant Tissue Culture Techniques

Here some of the plant tissue culture techniques are briefly discussed with their applications in Table 20.1.

Somaclonal Variations

- The genetic variations found in the culture's cells are referred as somaclonal variations and plants derived from such cells are known as "somaclones."
- Variations in tissue or plant originating from gametophytic origin (pollen or ovary culture) are known as "gametoclonal" variations.
- Sometimes, due to variations, some useful traits are observed which have applications in crop improvement like plant disease resistance and quality trait improvement (yield).
- Disease resistance in tomato; a new variety of *Lathyrus sativus Linnaeus* Bio L212 has con-

Table 20.1 The table lists some important plant tissue culture techniques, their basic methodology, principle and potential applications

Technique	Methodology	Principle	Applications
Anther culture	Anthers are placed in culture medium	Pollen inside anther contains haploid set of chromosomes	Successful development of doubled haploid lines of wheat, rice [12], sorghum, barley, etc.
	↓		
Haploid culture	Immature pollen within anther divides	It spontaneously doubles during culture or after colchicine treatment	Salt-tolerant variety of rice PSBRc50 (Bicol) released
	↓		
	Callus is produced	Doubling allows the expression of recessive and suppressed traits, which are masked or undetected otherwise	Eight salt-tolerant and two rain-fed rice released by Philippine Ins.
	↓		
	Healthy calli are used for shoot and root regeneration		
	↓		
	Selection and acclimatization		
Micropropagation	Stage 0: Selection and preparation of mother plant (actively dividing meristematic cells)	Meristem divides faster than disease-causing virus	Clean and uniform planting material is available for oil palm, pine, banana, abaca, date, rubber, field crops, jojoba, pineapple, tomato, cassava, yam, sweet potato, orchids, etc.
	↓		
Production of disease-free high-quality plant	Stage 1: Initiation of culture on nutrient medium	Micropropagated plants establish quickly	Rapid multiplication irrespective of season
	↓		
	Stage 2: Multiple shoot formation from cultured explant	Grow vigorously	Multiplication of sexually derived sterile hybrids
	↓		
(for rapid production of many uniform plants)	Stage 3: Rooting	Have shorter and uniform production cycle	Highly efficient and cost-effective
	↓		
	Stage 4: Acclimatization and transplantation	Have higher yields	Creation of disease-resistant hybrids

(continued)

Table 20.1 (continued)

Technique	Methodology	Principle	Applications
Somatic hybridization	Isolation of protoplasts by degradation of cell wall by enzyme and osmotic pressure which is controlled	Promote fusion between incompatible and two distinct species	Creation of environmental tolerance like salt, frost, etc.
	↓		
Protoplast fusion: two different plant species fuse together to form hybrids	Fusion of different protoplasts (by PEG, high Ca^{++}, or electrofusion)	Protoplasts of desired strains/ species are mixed in almost equal proportion	Cytoplasmic male sterility by using cybrids
	↓		
	Selection of hybrid cells by biochemical, visual, or cytometric or auxotrophic mutant selection (auxotroph which cannot grow on minimal media)	The protop last suspension recovered has (1) unfused protoplasts of the two species/ strains, (2) products of fusion between two or more protoplasts of the same species (homokaryons), and (3) "hybrid" protoplasts produced by fusion between one (or more) protoplast(s) of each of the two species (heterokaryons)	Tomato has developed resistance against TMV, spotted wilt virus, and insect pest
	↓		
	Hybrids can be symmetrical (with chromosomes from both species) or asymmetrical (with chromosomes of only one species) or cybrids (nucleus of one and cytoplasm of both)		

tent of neurotoxin; *Citronella java* Bio-13 is with more oil and citronellol; super tomatoes.

Embryo Rescue

- Immature embryos are cultured in a special medium.
- It prevents abortion of young embryos by supporting germination in culture conditions.
- It is done in breeding between incompatible genomes.
- New rice plant for West Africa was generated by a cross between Asian *Oryza sativa* and African *Oryza glaberrima*. The resultant embryos and thus the plant had yield traits of *sativa* and local adaptation from *glaberrima*.
- It offers resistance for bacterial blight from wild rice (*Oryza longistaminata*) to cultivated variety IR24.
- Salt-tolerant rice was developed by a cross between highly salt-tolerant wild rice *Oryza coarctata* and cultivated rice variety IR56 [47].

20.6 Applications of Agricultural Biotechnology

20.6.1 Applications of Plant Tissue Culture

Various techniques of tissue culture are able to help plant breeders and scientists to induce continuous ease of handling of the plant material and embryos [18] which was not possible before:

- Capable of creating distant hybrids in incompatible crops
- Micropropagation of various plants which otherwise have long dormancy period or are difficult to grow
- Production of pathogen- and virus-free plants
- Multiplication of sexually derived sterile hybrids
- Haploid production and production of double haploids

- Production of stress-tolerant varieties of various species
- Helpful to incorporate the characters of wild varieties (stress and pest tolerance) in the cultivated ones for better adaptability

20.6.2 Production of Biopharmaceuticals

The production of recombinant proteins in plants is advantageous as they have all the properties of higher systems for production of protein. They can perform correct posttranslational modifications and are free from mammalian pathogens, are flexible, and offer low-cost production of high-quality, bioactive recombinant proteins [40]. The success has been achieved in obtaining plants that produce monoclonal antibodies or other therapeutic proteins or that may serve as a source of edible vaccines. Many studies are being conducted for using plants for production of therapeutic agents including hormones (as insulin, somatotropin, erythropoietin), coagulation factors, blood components, interferons, drugs for cancer, infectious agents, respiratory system, infertility etc..

Human glutamic acid decarboxylase This is an enzyme which is localized in pancreatic β-islet cells and the brain. It catalyzes the conversion of glutamate to γ-aminobutyric acid (GABA) and carbon dioxide.

Plants are being tried for its production, and chloroplasts of *Chlamydomonas reinhardtii* transformed with an hGAD65 vector produced the immunoreactive protein as 0.3 % of the total soluble protein (TSP) in the algal cells.

Norwalk virus-like particles (NVLPs) It is a single-stranded, positive-sense virus which belongs to a group of highly infectious viruses that are responsible for more than 95 % of epidemic outbreaks of viral gastroenteritis in adults in developed and developing countries.

Recombinant Norwalk virus coat protein (NVCP) self-assembled into NVLPs accounted for 0.23 % of soluble proteins in transgenic tobacco leaves to 0.37 % of total soluble proteins in transgenic potato tubers (34 μ g/g of tuber weight), although only ~50 % self-assembled into NVLPs. Their oral immunogenicity was demonstrated in mice.

Monoclonal antibody 2G12 The antibody 2G12 can neutralize anti-HIV-1 human IgG1 that recognizes a high-mannose glycan cluster on the surface of the virus glycoprotein 120 (gp120). The broad biological activity of 2G12 allows it to defend against infection either by direct virus neutralization or by combination with other effector cells and complement activation. The antibody was constitutively expressed in the leaves of wild-type *Arabidopsis* plants.

Human Interleukin-6 (hIL-6) hIL-6 is a 26-kDa secreted glycoprotein having diverse physiological functions like the induction of the acute phase response and inflammation, the regulation of the immune response, and the promotion of B-cell differentiation into immunoglobulin-secreting cells. hIL-6 production was reported in transgenic tobacco plants.

Plant-specific glycans may also be desirable; the presence of terminal mannose residues on plant-derived recombinant glucocerebrosidase was shown to increase its uptake by macrophages and thus its efficacy for the treatment of Gaucher disease.

20.6.2.1 Edible Vaccines

Edible vaccines are composed of antigenic proteins (subunit vaccines) and are devoid of pathogenic genes, and thus they cannot establish infection. Conventional subunit vaccines are expensive and technology intensive, need purification, require refrigeration, and produce poor mucosal response.

Edible vaccines hold great promise as they are cost-effective, easy to administer, easy to store, and fail-safe and are currently being developed for a number of human and animal diseases. They exhibit good genetic stability and are heat stable, do not require cold chain maintenance, and can be stored near the site of use, eliminating long-distance transportation.

Edible vaccines activate both mucosal and systemic immunity, as they come in contact with the digestive tract lining. This dual effect would provide first-line defense against pathogens invading through the mucosa, like *Mycobacterium tuberculosis*, and agents causing diarrhea, pneumonia, STDs, and HIV to name a few.

Fear of contamination with animal viruses—like the mad cow disease, which is a threat in vaccines manufactured from cultured mammalian cells—is eliminated, because plant viruses do not infect humans.

Administration of edible vaccines to mothers might be successful in immunizing the fetus in utero by transplacental transfer of maternal antibodies or the infant through breast milk. Edible vaccines are currently being developed for a number of human and animal diseases, including measles, cholera, foot and mouth disease, and hepatitis B, C, and E [33].

Another unique advantage of plant-based systems is the natural "bioencapsulation" provided by edible plant organs when pharmaceuticals are intended for oral administration.

20.6.3 Production of Secondary Metabolites

The secondary metabolites are not directly involved in plant growth and development but are thought to have a major role in the adaptation of plants and are an important source of pharmaceuticals. They are economically important as drugs, dyes and pigments, food additives for flavor, fragrances, and pesticides, etc. Due to their commercial value they are now focused for production. Their in vitro production in suspended plant cell culture from various medicinal plants has been reported. They are produced differentially by taxonomic groups.

These secondary metabolites include alkaloids, glycosides (steroids and phenolics), terpenoids, latex, tannins, etc. Based on their biosynthetic origins, the natural secondary metabolites may be divided into three major groups:

- The terpenoids: All the terpenoids, including primary metabolites and nearly 25,000 secondary compounds, are all derived from the 5-carbon precursor isopentenyl diphosphate (IPP).
- The alkaloids: Nearly 12,000 alkaloids, have one or more nitrogen atoms, and they are biosynthesized from amino acids.
- The phenolics: Approximately 8000 phenolic compounds are formed by either the shikimic acid pathway or the malonate/acetate pathway.

Production of Secondary Metabolites

- They can be synthesized in the roots by the hairy root culture system based on inoculation with *Agrobacterium rhizogenes*. The hairy root phenotype is characterized by fast hormone-independent growth, lack of geotropism, lateral branching, and genetic stability. The secondary metabolites produced by hairy roots arising from the infection of plant material by *A. rhizogenes* are the same as those usually synthesized in intact parent roots, with similar or higher yields.
- They can be produced by cell culture, during their morphological differentiation and maturation during plant growth, some cells produce them. The cell cultures contribute in several ways to the production of natural products. These are:
  (a) A new route of synthesis to establish products, for example, codeine, quinine, pyrethroids
  (b) A route of synthesis to a novel product from plants difficult to grow or establish, as thebaine from *Papaver bracteatum*
  (c) A source of novel chemicals in their own right, for example, rutacultin from culture of *Ruta*
  (d) As biotransformation systems either on their own or as part of a larger chemical process, for example, digoxin synthesis

The advantages are: (1) growth of cells can be controlled, (2) product isolation and purification is easy and economical, (3) production can be enhanced as in vitro systems are under better control, and (4) high-yielding mutant lines can be used. However, natural yield is higher.

There are multiple uses of secondary metabolites obtained from plants; enlisting all of those is not possible; however, some of the uses are mentioned here. They are used as:

- *Antimalarial drugs*—artemisinin (*Artemisia sp.*), quinine (*Cinchona officinalis*)
- *Antitumor or anticancer drugs*—vincristine and vinblastine (*Catharanthus roseus*), taxol (*Taxus* sp.), cephalotaxine and camptothecin (*Camptotheca* sp.)
- *Insecticide*—azadirachtin (*Azadirachta indica*), pyrethrin (*Chrysanthemum cinerariaefolium*)
- *Antibacterial agents*—berberine (*Coptis japonica*)
- Some others are used to cure rheumatic pain (capsaicin from *Capsicum sp.*), hypertension (scopolamine from *Datura stramonium*), cardiovascular disorders (digoxin), and many others.

Shikonin is a dye produced by the cells of *Lithospermum erythrorhizon* on a commercial scale; *Papaver somniferum* is a commercial source of the analgesics (morphine and codeine). L-3, 4-Dihydroxyphenylalanine (L-DOPA) is known as a precursor of alkaloids, betalain, and melanin, obtained from *Vinca faba* and *Baptisia* and being used as a potent drug for Parkinson's disease, a progressive disabling disorder associated with a deficiency of dopamine in the brain. Diosgenin is a precursor of steroidal drugs and is obtained by the use of cell cultures of *Dioscorea deltoidea*.

The applications of some different secondary metabolites are tabulated in Table 20.2.

20.6.4 Production of Stress-Tolerant Crops

Abiotic stresses include high and low temperatures, salinity, drought, flooding, heavy metal stress, and many other environmental factors. They have negative impact of nonliving factors on the living organisms resulting in crop loss worldwide. However some plants are able to tolerate these stresses and their tolerance to these stresses is dependent on the molecular networks involved in stress perception, signaling, and the expression of specific stress-related genes and metabolites.

Compatible solutes (osmoprotectants) as glycine betaine, soluble proline, and mannitol accumulating in the cytoplasm of stressed plants make them more tolerant to stresses. A physiological role of betaine was proposed as it was accumulated in high quantities in plants subjected to osmotic stress. It has been shown to protect enzymes and membranes from cold, heat, salt, and freezing stress [10]. Betaine may also stabilize the photosystem II protein–pigment complex in the presence of high NaCl concentrations.

Polyamines (PAs), widely present in living organisms, are now regarded as a new class of growth substances which includes spermidine (Spd, a triamine), spermine (Spm, a tetramine), and their obligate precursor putrescine (Put, a diamine) which play a pivotal role in the regulation of plant developmental and physiological processes [8]. They play an important role in modulating the defense response of plants to diverse environmental stresses [48], which include metal toxicity [21], oxidative stress [65], drought [62], salinity [14], and chilling stress.

Enhanced plant stress tolerance has been achieved by genetic engineering of the compatible solutes like glycine betaine [24].

20.6.5 Production of Insect-/Pest-Resistant Crops

Bacillus thuringiensis (Bt) are gram-positive spore-forming bacteria with entomopathogenic properties. Bt produce insecticidal proteins during the sporulation phase as parasporal crystals. These crystals are predominantly comprised of one or more proteins (Cry and Cyt toxins), also called δ-endotoxins. Cry proteins are parasporal inclusion (crystal) proteins from *Bacillus thuringiensis* that exhibit experimentally verifiable toxic effect to a target organism or have significant sequence similarity to a known Cry

Table 20.2 The table shows secondary metabolites along with their producers and applications

Producer plant	Secondary metabolites	Applications
Alkaloids: immensely use in medicine		
Papaver somniferum L.	Codeine	Analgesic and is also used in cough syrups
P. bracteatum	Thebaine	It can be converted to codeine
Nicotiana	Nicotine	Interferes with nerve transmission
Coptis japonica	Berberine	Intestinal disorders
Phellodendron amurense		
Duboisia myoporoides	Scopolamine	Anesthetic and antispasmodic
D. leichhardtii	Hyoscyamus	
Atropa		
Digitalis species	Cardenolides	Heart diseases
Lithospermum erythrorhizon	Pyrrolizidine	Antipyretic, reduces inflammation
Anticancer drugs	Vinblastine	Anticancer
Catharanthus roseus	Vincristine	
Camptotheca accuminata	Camptothecin	
Other higher plants	Maytansine	
	Tripdiolide, bruceantin, baccharin	
Terpenoids		
Mentha spicata and related members	Menthol, linalool	Hirsutism, interferes with neurotransmission
Salvia officinalis (sage)	All produce essential oils as menthol, peppermint, limonene, citronella, citrine, camphor, terpineol	Culinary use, local anesthetic
Tanacetum vulgare		Aromatic and insect repellent
Foeniculum vulgare and other members		
Parthenium	Parthenolide	Contact dermatitis
Gossypium	Gossypol	Blocks phosphorylation
Digitalis	Digitogenin	Stimulates heart muscles
All plants	Carotene	Antioxidant
Hevea	Rubber	Tires and rubber items
Spinach	Spinasterol	Interferes with animal hormone action
Artemisia maritima	α-Santonin	Antihelminthic
Eucalyptus	Cineole	Perfume industry and nasal decongestant
Phenolics: key to defense response against biotic and abiotic stresses		
Carrots	Umbelliferone	Blocks cell division
All plants	Caffeic acid	Causes oxidative damage and browning in fruits and wine
All plants	Anthocyanin	Gives color to fruits, antioxidant
Oak hemlock	Gallotannin	

protein. Similarly, Cyt proteins are parasporal inclusion proteins from *Bacillus thuringiensis* that exhibit hemolytic (cytolytic) activity or have obvious sequence similarity to a known Cyt protein (Fig. 20.7).

The first commercialized GM crops in the mid-1990s were that of corn (maize), potato, and cotton plants which expressed genes encoding the entomocidal δ-endotoxin from *Bacillus thuringiensis* (Bt, also known as Cry proteins). Since then 12 transgenic crops (corn, tomato, soybean, cotton, potato, rapeseed (canola), squash, beets, papaya, rice, flax, and chicory) have been approved for commercial production

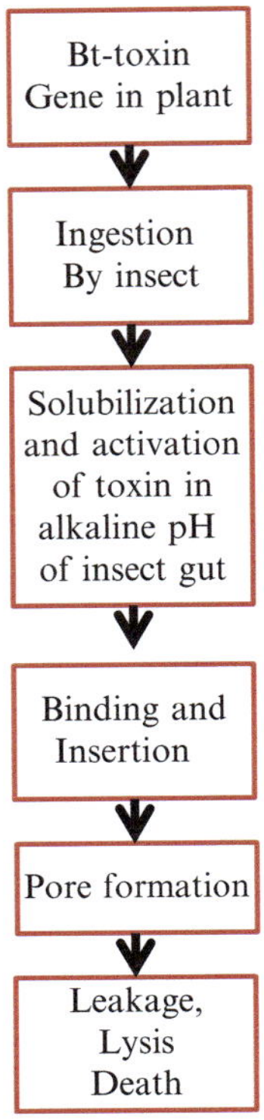

Fig. 20.7 The figure shows the fate of Bt toxin in insect gut leading to its death

in the USA [26, 27]. The most widely grown are "Bt" corn and cotton and glyphosate-resistant soybeans. Bt corn and cotton have had DNA from a naturally occurring insecticidal organism, *Bacillus thuringiensis*, incorporated into their genome; it kills some of the most serious insect pests of these crops (European and southwestern corn borers, and cotton budworms, and bollworms) after they feed on the plant, while beneficial insects are left unaffected.

However, because of the potential for pest populations to evolve resistance, and owing to lack of effective control of homopteran pests, alternative strategies are being developed. Some of these are based on *Bacillus* spp., for example, vegetative insecticidal proteins (VIPs) or other insect pathogens.

Vegetative insecticidal proteins (VIPs) are secretable proteins from *Bacillus thuringiensis* which do not share sequence homology with known Cry proteins and display insecticidal activity against a wide variety of lepidopterans and coleopterans and some sap-sucking insect pests [50, 57]. The members of the Vip3 family characterized to date exhibit activity against lepidopterans, and several of them do not compete with Cry proteins for binding sites. They are classified into two subfamilies (Vip3A and Vip3B), and some are especially toxic for species with little susceptibility to several Cry proteins [15, 35]. All of these features have made VIPs a research target for broadening the host range of *B. thuringiensis*-based biopesticides and for the management of insect resistance to *B. thuringiensis* proteins [44, 53].

Other strategies are based on the use of plant-derived or animal-derived genes, including those from insects, such as those encoding immunosuppressive proteins. More recently, the potential to identify and exploit endogenous resistance genes using functional genomics and the use of RNAi are actively being investigated.

In 2010, 148 million ha of biotech crops were grown in 29 countries, representing 10 % of all 1.5 billion hectares of cropland in the world. The global value of this seed alone was at US $11.2 billion in 2010, with commercial biotech maize,

Case Study

The big companies are continuously working for improvement of traits of crop plants. In this line the US Supreme Court heard arguments filed by Monsanto against 75-year-old Indiana soya-bean farmer Vernon Hugh Bowman, who used the progeny of Monsanto seeds to sow his land for eight seasons. The company says that by not buying seeds for each generation, Bowman violated its patents. Bowman was a regular customer for Monsanto's herbicide-resistant soybeans for his main crop, but bypassed the company by purchasing seeds for a late-season crop from a grain elevator known to contain Monsanto's transgenic seed. In 2007, Monsanto sued him.

The companies were trying to limit the offspring of naturally "self-replicating" technologies by the patents. The patents owned by Monsanto required the insertion of three

(continued)

different genes into the plant genome. Thus through modern biotechnology, it may be possible to develop crops that will not produce viable offspring seeds. Sterile seed technology—dubbed "terminator technology" in the popular press—is one type of gene use restriction technology (GURT) in which seed produced by a crop will not grow.

Thus Monsanto, the agricultural biotechnology giant in St Louis, Missouri, was surprised by the furore that followed when it announced that it might acquire a method for engineering transgenic crops to produce sterile seed, which would force farmers to buy new seeds for each planting. Early patents on "gene use restriction technologies" (GURTs)—later rebranded as "terminator" technology by the media and activists opposed to them—described a genetic modification that switched on production of a toxin that would kill off developing plant embryos. The result: a seed that could be harvested for food but would not produce offspring. The controversial proposal raised concerns that it would make farmers dependent on industry for their livelihood [34].

Monsanto says it is currently not researching the techniques. Monsanto maintained, "We remain committed not to commercialize sterile seed technology in food crops." After consulting with international experts and sharing many of the concerns of small landholder farmers, Monsanto made a commitment in 1999 not to commercialize sterile seed technology in food crops. There are alternatives to making sterile seeds ("Terminator, the sequel").

- One approach would be to switch off the transgene of interest in seeds, so that they could grow into new plants but would not pass on the benefits of the engineered trait.
- Another approach is to place the transgene under the control of a switch that must be activated by a proprietary chemical.

That would give companies control over the engineered trait by forcing buyers to return each year to purchase the chemical (Ledford 2014).

http://www.monsanto.com/newsviews/pages/terminator-seeds.aspx

soybean grain, and cotton valued at approximately US $150 billion per year.

20.6.6 Protease Inhibitors and Pest Resistance

Long back it was noticed that larvae of certain insects were unable to develop normally on soybean products due to the presence of the trypsin inhibitors which were toxic to the larvae of flour beetle, *Tribolium confusum*. The term "protease" includes both "endopeptidases" and "exopeptidases". These protease inhibitor genes can confer resistance to a wide range of insect pests including lepidopterans such as *Heliothis* and *Spodoptera*, coleopterans such as *Diabrotica* and *Anthonomus*, and Orthoptera such as locusts.

Protease inhibitors are a promising complement to Bt toxins for the development of insect-resistant transgenic crops, but their limited specificity against proteolytic enzymes and the ubiquity of protease-dependent processes in living organisms raise questions about their eventual nontarget effects in agroecosystems [17].

Case Study

Transgenic maize (*Zea mays*) engineered to express genes for various insecticidal protein endotoxins (Bt toxins) from the soil bacterium *Bacillus thuringiensis* is one of the most widely grown transgenic crop. The main target species for Bt toxin is the European corn borer (*Ostrinia nubilalis*), which accounts for huge losses of the maize crop (exceed over $1 billion annually in the USA). Bt toxins are selectively toxic to only those insects as lepidopteran larvae that have a gut alkaline enough to activate the Bt protoxin by enzymatic proteolysis. Receptor binding by the C-terminal domain of the active toxin is the major determinant of host specificity by the different Bt toxins.

However, a preliminary study by Losey et al. (1999) [36] raised serious concerns about the ecological safety of Bt maize cultivation to nontarget lepidopterans, in particular the larvae of monarch butterfly (*Danaus plexippus*) [41]. The group of Losey et al. found that Bt corn plants might represent a risk because most hybrids express the Bt toxin in pollen, and corn pollen is dispersed over at least 60 m by wind. Corn pollen is deposited on other plants near cornfields and can be ingested by the nontarget organisms that consume these plants. Based on laboratory assays, the authors concluded that monarch larvae reared on milkweed (*Asclepias syriaca*) leaves dusted with pollen from Bt maize ate less, grew more slowly, and suffered higher mortality than those reared on leaves dusted with nontransformed maize or on leaves without pollen. Losey et al. (1999) examined the effects of only one type of transgenic pollen (Cry1Ab event 176). However these findings were confirmed but only with Cry1Ab event 176. However the fact was that spray of chemical pesticides even had deleterious effects on not only major larvae but also on the environment. Thus effects of

(continued)

Bt maize cultivation on monarch butterfly was negligible, and current evidence suggests that Bt maize is an environmentally safer insect control strategy than conventional chemical spraying. Thus, for every GMO released, the assessment of risk should be carried out on a case-by-case basis with repeated studies by different groups and different geographical locations (Minorsky 2001; Losey et al. 1999).

20.6.7 Production of Herbicide-Resistant Crops

Weeds are important problems which affect crop productivity as they not only compete with crops for water, nutrients, sunlight, and space but also harbor insect and disease pests, clog irrigation and drainage systems, undermine crop quality, and deposit weed seeds into crop harvests. With the advancement of technology, the development of herbicide-tolerant crop by engineering of herbicide glyphosate gene and glufosinate gene has helped in effective control of weeds and improving crop yield.

- Glyphosate-tolerant crops: Glyphosate herbicide kills plants by blocking the 5-enolpyruvy lshikimate-3-phosphate synthase (EPSPS) enzyme which is involved in the biosynthesis of aromatic amino acids, vitamins, and many secondary plant metabolites. Engineering glyphosate-tolerant form of EPSPS can protect crop plant from herbicide.
- Glufosinate-tolerant crops: Glufosinate herbicides contain the active ingredient phosphinothricin, which kills plants by blocking the enzyme responsible for nitrogen metabolism and for detoxifying ammonia, a by-product of plant metabolism. Crops modified to tolerate glufosinate contain a bacterial gene that produces an enzyme that detoxifies phosphinothricin and prevents it from doing damage.
- Crops can be genetically modified to produce new protein that detoxifies the herbicide.

- Producing physical or physiological barriers prevents the entry of the herbicide into the plant.

Glyphosate-resistant soybeans are unharmed by the broad-spectrum herbicide glyphosate, a characteristic that allows farmers to kill yield-reducing weeds in soybean fields without harming the crop. The advantages of herbicide tolerance are (1) excellent weed control, (2) higher crop yields, and (3) reduced numbers of sprays in a season.

Case Study
Glyphosate-resistant crops, also known as "Roundup Ready" (RR), are now important crops in the USA. In 2004, approximately 13 % corn, 85 % soybean, and 60 % of cotton acreage was occupied with RR varieties. Roundup (glyphosate) is a broad-spectrum herbicide that kills a wide range of plants. It is not applied directly to crops. The RR technology incorporates genetic resistance to glyphosate into crop plants by inserting a single bacterial gene that modifies 5-enolpyruvylshikimate-3-phosphate (EPSP) synthase, an enzyme that is essential for plant growth. Glyphosate's mode of action is to inhibit a plant enzyme involved in the synthesis of the aromatic amino acids: tyrosine, tryptophan, and phenylalanine.

If a farmer has to grow on Roundup Ready crops, they would have to use Roundup (from the same company) to treat their fields.

20.6.8 Production of Virus-Resistant Crops

Plants are susceptible to viral, bacterial, and fungal diseases. Much progress has been made in evolving transgenic plants resistant to viruses. For example, expression of a gene that encodes the coat protein of tobacco mosaic virus (TMV) in transgenic tobacco plants has been shown to cause the plants to resist TMV infection. A number of other virus-resistant plant species have been developed including squash and potatoes.

20.6.9 Production of Biofuel and Biodiesel

Nowadays, global demand for energy is being fulfilled by fossil fuels, but they are associated with the greenhouse effect and environmental pollution. Due to fast growth in fossil fuel-requiring human activities worldwide, the oil reserves may also run out after 2050. Therefore, huge efforts are being made in developing biological solutions for CO_2 fixation and reduction technologies and in finding alternative and renewable energy sources. Biodiesel has received significant attention since it is made from non-toxic and biodegradable materials, and its use leads to a huge decrease in the emissions of greenhouse gases (such as CO_2) and air pollutants [2, 43]. However, to produce enough biodiesel from oleaginous crops (such as soybean, palm, and rapeseed) to supply the existing demand, the crop land would be required, thus sending the world in food shortage and supply.

Microalgae are very small in sizes usually measured in micrometers, which normally grow in water bodies or ponds. Microalgae contain more lipids and have the faster growth in nature with excellent CO_2 fixation ability. Marine microalgal strains *Chlorella sorokiniana*, *Nannochloropsis* sp. *F&M-M24*, *Nannochloropsis gaditana*, *D. tertiolecta ATCC30929*, and *C. protothecoides* may serve as economic feedstock for microalgae-based biodiesel production due to their characteristics of high salt tolerance and high lipid content. Production of biofuels and chemicals from renewable feedstocks is necessary to meet the energy demand in a world where petrol fuels are becoming scarce and more expensive.

One of the main problems associated with biofuels is still the production costs, which can be reduced if residues of biofuel production processes are converted into valuable coproducts [64]. Biodiesel is an alternative fuel that reduces net greenhouse effects and its use has become mandatory in many countries. It is mainly obtained by the transesterification of fat and vegetable oils in the presence of a catalyst by a primary alcohol (usually methanol) leading to a fatty acid methyl ester (FAME), which is used as a biofuel. Sunflower, rape, soybean, and palm oils are the main substrates to make biodiesel worldwide; however, there are local variations on which it is the main source. In Brazil, for example, 80 % of the biodiesel produced in 2010 was from soybean oil.

Production of the two main types of residues, pies and crude glycerol, is increasing concurrently with the biodiesel industry. Pies, which are produced by pressing of palms, seeds, and others for oil extraction, are usually used as feed for animals or as fertilizers, consequently adding value to the biodiesel production chain. Crude glycerol, which is derived from the transesterification reaction of fat and vegetable oils (triglycerides) to produce biodiesel, contains methanol, salts, soaps, and water as the main contaminants.

Thus, the development of biorefineries based on crude glycerol is expected to favor the biodiesel industry economy, by reducing costs associated with the disposal of residues and increasing production of value-added chemicals [2]. Biorefineries are based on the integration of the biomass conversion processes to produce power, fuels, and chemicals [2, 4]. In this context, the utilization of glycerol generated in the biodiesel production process offers an excellent opportunity to obtain chemicals by microbial fermentation. Production yields of fuels and chemicals from glycerol as high as 90 % of theoretical maximum have been obtained.

A vast range of fuels and chemicals can potentially be produced by microbial fermentation of glycerol [63]. Several engineered yeasts and bacteria, especially *E. coli* and others from the *Enterobacteriaceae* and *Clostridiaceae* families, have the potential for biorefineries. However process can be further enhanced after engineering for desirable productions. Thus, the researches show strong potential of crude glycerol use for the development of biorefineries by the production of several chemicals like alcohols (1,3-propanediol, 2,3-butanediol, ethanol, butanol), ketone and organic acids (dihydroxyacetone phosphate, glyceric acid, lactic acid, succinic acid, citric acid [49], oxalic acid), polyols (mannitol, erythritol, arabitol), and polyhydroxyalkanoates (PHAs) using different routes and microorganisms.

Polyhydroxyalkanoates (PHAs) production is also being focused due to their potential application as renewable, biodegradable, and biocompatible thermoplastics.

20.6.10 Improved Nutritional Traits by Biotechnological Interventions

Agricultural biotechnology is continuously evolving with its tools and techniques to improve the nutritional quality of the food crops (taste, nutrition, allergen reduction) by either correcting the deficiency or altering the undesirable trait.

Thus, for better health, the macronutrients as protein, carbohydrates, fats, and fibers and micronutrients as vitamins and minerals need to be enhanced, while allergens and antinutrients (phytate which inhibits the biosorption of certain nutrients) need to be controlled or removed. However these require understanding of the metabolic process along with the ways they have evolved during crop domestication for devising more targeted improvement strategies for meeting current requirements [32]. Some GM crops are shown here along with their important trait (Fig. 20.8).

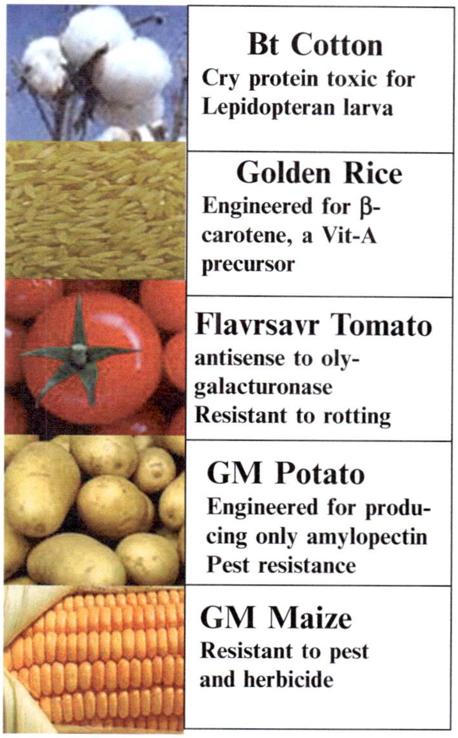

Fig. 20.8 Some publicized genetically modified plants and their traits

Concerns with Transgenics

In 1994, the first GM food to reach the market was the Flavr Savr tomato. It was launched for its flavor and long shelf life. In this the gene of tomato responsible for ripening was blocked so that the protein responsible for ripening was not produced. This tomato had longer shelf life and better flavor. However it could not attract customers.

Other GM crops may come in the near future termed as "pharma crops." The State Food and Agricultural Department was advised by California Rice Commission to allow Ventria Bioscience to grow 50 ha of GM rice. Two types of rice were modified with human genes (1) to prepare human lactoferrin gene (for anemia) and (2) to produce lysozyme (for diarrhea). These were aimed for children under 5 years of

(continued)

age for prevention of anemia [37] and diarrhea [11]. However later on the request was denied to Ventria after the rice growers raised concerns for rejection of their rice by international customers out of fear of contamination. The concerns raised were due to cross-pollination and transfer of foreign genes in other crops.

A nontoxic anthrax vaccine is being developed by INB Biotechnologies (Philadelphia) by raising transgenic petunias. This transgenic petunia would express the new protein and when eaten would promote the development of anti-anthrax antibodies.

20.6.10.1 Protein

Protein energy malnutrition is the most lethal form (Food and Agriculture Organization, 2006) of malnutrition and affects every fourth child worldwide, according to the World Health Organization (2006). The Food and Agriculture Organization estimates that 850 million people worldwide suffer from undernutrition, to which insufficient protein in the diet is a significant contributing factor.

Plant proteins are dietary source of proteins, but they are deficient in certain essential and indispensable amino acids. For example, cereal proteins (maize, wheat (*Triticum aestivum*), rice) tend to be low in lysine (Lys) (1.5–4.5 vs. 5.5 % of WHO recommendation), while legume (soybean, pea [*Pisum sativum*]), root, tuber, and most vegetable proteins are deficient in the sulfur-containing amino acids [methionine (Met) [3] and cysteine (Cys), 1.0–2.0 % vs. 3.5 % of the WHO reference protein]. Thus, improvement in the nutritional qualities of some of the crop plants was attempted, for example:

- The glycinin of soybean was modified by insertion of 4 contiguous methionine residues into the variable regions of glycinin gene and was placed under the under the control of

GluB-1 promoter. After its transformation and expression in rice, it accumulated to 5 % as total protein [30].

- Cysteine and methionine rich albumin was expressed in seed-specific phaseolin promoter into tobacco plant. Accumulation of the sulfur-rich protein in the seeds of tobacco amounted to 3–8 % of the total seed protein, resulting in up to 30 % increase in the content of Met in the seeds.
- This successfully increased the essential amino acid Met content of seeds; expression of the gene with high Cys and Met in lupine led to 40 % increase in the content of sulfur-containing amino acids; successful examples of improving amino acid balance were observed in maize, canola (*Brassica napus*), and soybean.
- Another solution to this was creation of artificial protein with essential amino acids Met, Thr, Lys, and Leu in a stable conformation and expressing it in soybean under a seed-specific promoter and sweet potato (*Ipomoea batatas*). There was two- to fivefold increase in the protein content with essential amino acids in seed and leaves and roots, respectively.

20.6.10.2 Carbohydrates

The carbohydrates like fructans are an important ingredient as they promote healthy colon. High levels of fructan were produced in a transgenic sugar beet (*Beta vulgaris*). Inulins are also derived from fructans; the transgenic potato was able to synthesize the full spectrum of inulin.

20.6.10.3 Fiber

Fibers are present in plant foods and provide bulk in the diet which helps in better digestibility without much calories. But they are poorly metabolized for energy or other nutritional uses. The mutants were generated with significantly lower lignin which leads to softer cell walls compared with the wild type, improving its digestibility.

20.6.10.4 Novel Lipids

Novel lipids are good in food. Transgenic canola accumulated high levels of capric and caprylic acids. The substitution of medium-chain triglycerides (MCTs) for long-chain triglycerides (LCTs) in the diet makes animals gain less weight, store less adipose tissue, and experience an increase in metabolic rate.

Edible oils rich in monounsaturated fatty acids provide improved oil stability, flavor, and nutrition for human and animal consumption. Transgenic soybean with high oleic acid is naturally more resistant to degradation by heat and oxidation and so requires little or no post-refining processing (hydrogenation), depending on the intended vegetable oil application. In 2009, DuPont hopes to introduce soybean oil composed of at least 80 % oleic acid, linolenic acid of about 3 %, and over 20 % less saturated fatty acids than commodity soybean oil.

20.6.10.5 Vitamins and Minerals

Micronutrient malnutrition, the so-called hidden hunger, affects more than half of the world's population, especially women and preschool children in developing countries (United Nations System Standing Committee on Nutrition, 2004). Even mild levels of micronutrient malnutrition may damage cognitive development, lower disease resistance in children, and increase the incidence of childbirth mortality.

Using various approaches, vitamin E levels are being increased in several crops, including soybean, maize, and canola, while rice varieties are being developed with the enhanced vitamin A precursor, β-carotene, to address vitamin A deficiency that leads to macular degeneration and affects development. Likewise iron contents are improved in lettuce (*Lactuca sativa*).

Golden rice was made by cloning the majority of carotenoid biosynthetic enzymes from plants during the 1990s. In golden rice, β-carotene expression was done in the endosperm and had enormous health benefits. A similar method was used by Monsanto to produce β-carotene in canola.

Here some of the important modifications for improved nutritional quality are mentioned in Table 20.3.

Table 20.3 The table lists some modifications tried for improved nutritional quality of the crop plants

S. No.	Nutrient	Plant type	Function enhanced
1. A	Protein quality and levels	Bahia grass	High protein
		Maize	
		Potato	
		Rice	
		Sweet potato	
1. B	Amino acids	Canola	Essential amino acids
		Maize	Lys
		Potato	Lys, Met
		Lupine	Met
		Soybean	Met
		Sorghum	Lys, Trp
			Lys
2.	Oils and fatty acids	Canola	Lauric acid, ω3-FA
		Cotton	Oleic acid, stearic acid
		Linseed	ω3-FA
		Maize	Oil
		Oil palm	Oleic acid or palmitic acid
		Rice	α-Linolenic acid
		Soybean	Oleic acid
3.	Carbohydrates	Chicory	Fructans
		Maize	Fructans
		Potato	Fructans
		Sugar beet	Fructans
		Soybean	Fru, raffinose, stachyose
		Potato	Inulin
		Rice	Amylase
4.	Vitamins and carotenoids	Canola	Vitamin E
		Maize	Vitamin E, vitamin C
		Mustard	β-Carotene
		Potato	β-Carotene, lutein
		Rice	β-Carotene
		Strawberry	Vitamin C
		Tomato	Folate, phytoene, β-carotene, lycopene, provitamin A
5.	Minerals	Alfalfa, maize, soybean, wheat	Phytase increased
		Lettuce, rice, maize	Ferritin

20.6.11 Removal/Reduction of Antinutrients, Allergens, and Toxins

Many secondary metabolites produced by the plants are protective to the plants and are also beneficial for humans and animals. However, some of them are allergens and interfere with the normal process of intake and absorption (antinutrients) and thus are not desirable.

For example, phytate (phosphorous containing) is an antinutrient, as it strongly chelates iron, calcium, zinc, and other divalent mineral ions preventing their absorption. Due to it, animals also excrete high amounts of phosphorous causing water pollution. When low-phytate soybean meal is utilized along with low-phytate maize for animal feeds, the phosphate excretion in swine and poultry manure is halved. A number of groups have added heat- and acid-stable phytase from *Aspergillus fumigatus* to make the phosphate and liberated ions bioavailable in several crops.

To promote the reabsorption of iron, a gene for a metallothionein-like protein has also been engineered.

Other antinutrients that are being examined as possible targets for reduction are trypsin inhibitors, lectins, and several other heat-stable components found in soybeans and other crops.

Researchers have been trying to reduce food allergens (albumins, globulins), malabsorption and food intolerances (gluten), and toxins (glycoalkaloids, cyanogenic glucosides, phytohemagglutinins) in crop plants.

Biotechnology approaches can be employed to downregulate or even eliminate the genes involved in the metabolic pathways for the production, accumulation, and/or activation of these toxins in plants.

20.6.12 Production of Perfumes and Scent

The biotechnological techniques are innovative ways to synthesize flavor and fragrance compounds by microbial fermentation and plant tissue culture.

Microorganisms are also used to produce aroma chemicals; they not only have the capability to enhance the quality of some fragrance or flavor compounds but also are able to produce these by microbial fermentation. Microbes can produce and/or biotransform natural precursors into valuable flavor/fragrance chemicals via microbial metabolic pathways. Microbial biotransformation and biosynthesis of flavor and fragrance chemicals offer the potential benefits of producing optically active isomers which often have marked differences in flavor and fragrance quality and sensory intensity.

20.6.13 Phytoremediation

Phytoremediation uses plants to remove and store or degrade, sequester, or bioaccumulate pollutants from contaminated soil and water. It offers an environmentally friendly, cost-effective, and carbon neutral approach for the cleanup of toxic pollutants from the environment [19, 31, 51]. There have been many studies and reports on the successful use of phytoremediation for the cleanup of sites contaminated with volatile or nonvolatile organic pollutants, heavy metals, radioactive compounds, and pesticides. However, the use of plant-based technologies does have limitations due to the fact that plants are not ideally suited to the breakdown and metabolism of organic pollutants (refer to Chap. 19 for phytoremediation).

20.7 Germplasm Conservation and Cryopreservation

Germplasm refers to the complete genetic material. Germplasm conservation is important when the crop is endangered or has very long dormancy or is under any kind of threat or seedless plant. The conservation of these can be done in the storage banks or with the help of tissue culture in the form of artificial seeds. The meristematic tissue to the plant to be preserved is used as an explant for the culturing. It develops into callus with numerous embryoids (heart shaped structures). These embryoids can be encapsulated in sodium alginate or other encapsulation materials and can be stored for a long period till further requirements in freezing conditions.

The ultrafreezing is done in cryopreservation, that is, freezing at cryogenic temperature of liquid nitrogen (−196 °C). The material to be cryopreserved is treated with cryoprotectant as glycerol or dimethyl sulfoxide (DMSO). It maintains the water balance and avoids damage due to ice crystal formation. At cryogenic temperature, the cells are in a state of "absolute quiescence." Here all the biochemical changes and reactions stop.

20.8 GMOs and Risk Assessment

- All the GMOs need careful evaluation as they can pose risk of toxicity, allergenicity, or antibiotic resistance for human health. The changes induced by transgene may be deleterious as they might affect the host metabolism or can alter the genetic composition (due to random insertion of transgene).
  - Risk assessment of agricultural and food technologies is not a new concept.
  - Innovative agricultural practices were with their own set of potential risks. Be it usage of insecticides or pesticides.
  - The risks associated with GM are similar to those of crop hybridization, the keystone of the first green revolution. Whereas hybridization leads to the transfer of thousands of genes from one plant (often from different species) to another that leads to multiple effects, GM transfers one to a few genes, resulting in more predictable effects. Therefore, GM should result in fewer unintended risks.
- The gene transferred into an organism or the resultant products can actually remain in the environment leading to environmental problems. The intentional release of GMOs into the environment has led to an increased interest in possible interactions that may occur between other organisms in the environment. Such changes can lead to production of new proteins that may be toxic or allergenic or may disrupt or alter metabolic pathways that play a role in making the GMO successful. Accidental crossbreeding between GMO plants and traditional varieties through pollen transfer can contaminate the traditional local varieties with GMO genes resulting in the loss of traditional varieties of the farmers.
  - Gene flow or the movement of pollen from one plant to another is made possible when the parental plants are:
    (a) Flowers at the same time.
    (b) Close enough to allow a vector (insect, wind, or animal) to transfer pollen to receptive females.
    (c) Able to produce pollen that can result in embryos developing into viable seeds and germinating. Successful pollination also depends on the longevity of pollen viability, pollen travel distance, and the mode of pollination the plant has, whether self- or cross-pollinated.
  - Gene flow may present significant economic or environmental risks for either conventionally bred or GE crops on a case-by-case evaluation. Crop-to-wild relative gene flow could result if the plants grow in overlapping regions resulting in new combinations of genes that can improve, harm, or have no effect on the fitness of recipient plants. Genes can also flow from wild relatives to cultivated crops, introducing new traits into the next-generation seeds, but only affect the crop if it is replanted. Planting of GE varieties in areas of genetic diversity of plants needs additional precautions to reduce possible impacts of introgression of GE traits and the potential significant environmental consequences. To minimize this occurrence, planting of GE crops near wild species should be avoided or other technologies could be used to prevent gene(s) from moving to wild varieties. Gene flow could also occur when compatible plants are present within the vicinity. GE varieties like conventional plants can also persist in the environment. Organic farmers should be aware of these occurrences to be able to adopt the necessary precautions of spatial and temporal isolation.
- The potential benefits of planting insect-resistant transgenic crops include decreased insecticide use and reduced crop damage. However, the innate ability of insect populations to rapidly adapt to environmental pressures poses a serious threat to the long-term efficacy of insect resistance. Adaptation by insects and other pests to pest protection mechanisms can have environmental and health impacts.

- Resistance of insects against synthetic insecticides and Bt toxins in sprays occur and this will be true for GE crops. To slow this development in GE crops, several strategies have been developed. First-generation GE crops produced only one Bt toxin in each plant. Planting refuges of non-Bt crops near Bt crops in the field is the primary strategy of delaying insect resistance. This is based on the idea that insects feeding on plants in the refuge are not selected for resistance.
- Insect resistance to Bt toxins is recessive. The heterozygous offsprings produced when homozygous resistant insects mate with susceptible insects are killed by the Bt crops. This high-dose/refuge strategy creates plants that produce Bt toxin concentrations high enough to kill heterozygous insects, making resistance functionally recessive. Insect resistance to Bt toxins can thus be postponed substantially.
- Another approach is called the pyramid or stacking strategy that combines two or more toxins in a single plant, each with different modes of action. An example is Bollgard II cotton producing Cry1Ac and Cry2b, which targets the same pest in two different ways. Other approaches to delaying insect development are (1) mixing seeds of Bt and non-Bt varieties under small-scale experiments, (2) the use of inducible promoter to drive Bt gene expression only during insect attack, and (3) the use of modified toxins to kill resistant insects, as exemplified by the use of modified Bt toxin that will not be affected by the mutations in the midgut cadherins. Cadherins promote toxin oligomerization of Cry1A protein which has alpha helix in the binding site. Modified Cry1A which does not contain the alpha helix is independent of the cadherins and can thus be effective with insects which have developed resistance due to mutated or silenced cadherins. To date, the elapsed time before the first cases of field resistance of insects to Bt crops were reported has been longer than what was predicted under worst-case scenarios, suggesting that management strategies may have delayed resistance development. Despite documented cases of resistance, Bt crops remain useful against most target pests in most regions. As insect resistance to Cry toxins currently deployed in Bt crops increases, other strategies to create GE crops resistant to insects are being developed.

- There is also fear about the development of superweeds, that is, a weed that has acquired the herbicide-tolerant gene due to genetic contamination with a herbicide-tolerant GMO through in-field crossbreeding to related species or through horizontal gene transfer. Loss of biodiversity/reduction of cultivars: there have been concerns about reduction in the genetic diversity in cropping systems by the development and global spread of improved crop varieties to the green revolution.
  - Development of herbicide-tolerant weeds has occurred with both traditionally bred and GE crops. This phenomenon reduces the effectiveness of certain weed control strategies and decreases weed management options. Strategies have been developed to minimize the development of herbicide-tolerant weeds, such as:
  - Use of herbicide-tolerant (HT) cultivars with resistance genes for herbicides with alternative modes of action that can be used in rotation.
  - Use of restriction technologies to prevent gene passage to the next generation through the pollen, that is, transgenes can be targeted to the cytoplasmic organelles, not in the pollen.
  - Use of HT crops with different modes of action or with non-HT crops. A point to consider in using HT crops is that weeds can also escape herbicide treatment based on application rate, weed age and size, spray volume adjuvants used, water quality, and interactions with other herbicides that affect efficacy. Late germination of weeds can also escape herbicide applica-

tion; thus, a second pass of sprays can be done.

- However an important concern is that genetic erosion has occurred as the farmers have replaced the use of traditional varieties with monocultures. This is expected to further intensify as more and more transgenic crops are introduced which bring in considerable economic benefits to the farmers. The relative rate of susceptibility to any unforeseen infections or destructive situations increases when single varieties are used in cropping system in place of multiple varieties. Changes in the soil ecology: many plants leak chemical compounds into the soil through their roots.

Case Study

The various GMO crops are available for farmlands like corn, soybeans, cotton, alfalfa, sugar beets, canola, papaya, and squash. The Rural Advancement Foundation International (RAFI), a Canada-based organization, has identified several technologies that can be used to genetically teach the plants to respond only to certain combinations of agrochemicals, popularly known as "traitor technology."

RAFI has announced that it has uncovered over three dozen new patents describing a wide range of techniques that can be used for the genetic sterilization of plants and seeds. The disclosure follows after controversial patent, the "terminator," generating worldwide protest and debate because it renders farm-saved seed sterile—forcing farmers to return to the commercial seed market every year. The terminator patent is jointly owned by the US Department of Agriculture and a Monsanto subsidiary, Delta and Pine Land Company.

A number of the patents use benign-sounding technical terms such as "controlled gene expression" linked to

(continued)

"inducible promoters" to describe their sterilization techniques. Other patents describe "killer genes" that destroy pollen, or "GRIM proteins" that do the same to invertebrates or even mammalian cells.

Sterile seed technologies are extremely dangerous, "because over 1.4 billion farmers—primarily poor farmers in Africa, Asia, and Latin America—depend on farm-saved seed as their primary seed source. If they can't save seed, they can't continue to adapt crops to their unique farming environments, and that spells disaster for global food security." "These technologies are intended to force farmers to buy seed every season and to take still more crop production control away from farmers." "Genetic seed sterility is not about improving the productivity or quality of crops, it's a quest to increase seed industry profits."

The technology, as in Zeneca's new plant killer patent, is to insert a gene into the plant which produces barnase, a compound that can kill cell. The barnase gene is linked to a cysteine protease promoter (active during germination and growth of the plant). If grown alone, the promoter will induce barnase production and the plant will kill itself during germination or shortly afterward. This can be overcome by another pair of disrupter genes linked to an inducible promoter. The disrupter gene is turned on when the seed or the plant is exposed to a chemical; after activation it either blocks the action of the cysteine protease promoter or stops the barnase-producing gene itself.

Thus, the companies are developing suicide seeds whose genetic traits can be turned on and off by an external chemical "inducer"—mixed with the company's patented agrochemicals. In the not-so-distant future, we may see farmers planting seeds that will develop into productive (but sterile) crops only if sprayed with a carefully

(continued)

prescribed regimen that includes the company's proprietary pesticide, fertilizer, or herbicide. If traitor technologies are developed for commercial sale, "the farmers will be forced to surrender control of their seed supply and the Gene Giants will ultimately dictate what the farmer grows, how to grow it, and where to sell it.

The big companies might argue that engineered seed sterility is highly beneficial to the environment because it will eliminate the problem of horizontal gene transfer—it will prevent cross-pollination and thus the escape of engineered genes from transgenic plants to nearby weeds or wild relatives."

There is concern that transgenic plants could pass genes on to wild plant relatives—thus creating "superweeds" that could wreak havoc on the environment.

The big companies might argue that suicide seeds prevent preharvest crops from sprouting prematurely and that it will decrease the cost of producing hybrid seeds. They can argue that they cannot continue to develop new, more productive varieties for agriculture unless they get a fair return on their investment.

No matter what rationale is used by the Gene Giants to engineer social acceptance of seed sterility, the technology is unacceptable to growing numbers of civil society organizations around the world. These technologies, in case materialized, can capture the farming world which can lead to serious unimaginable disasters.

Kumar S. and Bhat V. traitor technology—A threat to the national food security (www.currentscience.ac.in)

http://nature.berkeley.edu/srr/Alliance/novartis/sterile.htm

20.9 Safety Assessment

GM crops are under strict regulatory bodies to detect any unexpected outcomes. At a very fundamental level, a recent report (Baack and Rieseberg 2007) on genome-wide analyses of introgression from oak (*Quercus* spp.) to fruit flies indicates that a substantial fraction of genomes are malleable. Hybridization gives rapid genomic changes, chromosomal rearrangements, genome expansion, differential expression, and gene silencing (transposable elements). In the context of this sea of malleability, reports have demonstrated that GM crops have a composition more similar to the isogenic parental strain used in their development than to other breeding cultivars of the same genus and species and in some instances even the location in which they are grown, and on occasion the latter "terroir" effect demonstrated greater variation than breeding strategy. As more metabolic modifications are introduced, we must continue to study plant metabolism and the interconnected cellular networks of plant metabolic pathways to increase the likelihood of predicting pleiotropic effects that may occur as a result of the introduced genetic modification.

20.10 Future Prospects

- Research to improve the nutritional quality of plants has historically been limited by a lack of basic knowledge of plant metabolism and the almost insurmountable challenge of resolving complex branches of thousands of metabolic pathways. With the tools now available to us through the fields of genomics and bioinformatics, we have the potential to fish for genes of value across species, phyla, and kingdoms and subsequently to study the expression and interaction of transgenes on tens of thousands of endogenous genes simultaneously by in silico analysis.
- With advances in proteomics, we should also be able to simultaneously quantify the levels

and interactions of many proteins or follow posttranslational alterations that occur.

- With these newly evolving tools, we are beginning to get a handle on the global effects of metabolic engineering on metabolites, enzyme activities, and fluxes.
- Right now, for essential macronutrients and micronutrients that are limiting in various regional diets, the strategies for improvement are clear and the concerns, such as pleiotropic effects and safe upper limits, are easily addressed. However, for many other health-promoting phytochemicals, clear links with health benefits remain to be demonstrated. Such links, if established, will make it possible to identify the precise compound or compounds to target and which crops to modify to achieve the greatest nutritional impact and health benefits.
- The achievement of this aim will be a truly interdisciplinary effort, requiring expertise and input from many disparate fields, ranging from the obvious human physiology and plant research to the less obvious "omics" and analytic fields.
- With these emerging capabilities, the increase in our basic understanding of plant secondary metabolism during the coming decades will be unparalleled and will place plant researchers in the position of being able to modify the nutritional content of major and minor crops to improve many aspects of human and animal health and well-being [38].

20.11 Conclusions

- Many different tools are available for increasing and improving agricultural production. These tools include methods to develop new varieties such as classical breeding and biotechnology. Traditional agricultural approaches are experiencing some resurgence today, with renewed interest in organic agriculture; an approach that does not embrace the use of genetically engineered crops. The role that genetic engineering stands to play in sustainable agricultural development is an interesting topic for the future.
- As with the development of any new technology, there are concerns about associated risks, and agricultural biotechnology is no exception. All crops developed using genetic engineering are subjected to extensive safety testing before being released for commercial use.
- Risk assessments are conducted for these new varieties, and only those that are safe for human use are released. Some concerns arise through people not fully understanding the reporting of risk. Many consider any level of risk unacceptable.
- Extensive risk assessment and safety testing of crops developed through the use of genetic engineering has shown that there are no varieties in use that pose risks to consumers. This is not to say that new varieties should not be carefully examined for safety; each case should be considered on its unique merits.

20.12 Chapter End Summary

- Agricultural biotechnology aims to bring the innovative practices in agriculture which can supply food to world's growing population. Though the conventional breeding and hybridization technology brought green revolution with highly enhanced yield, they had their own limitations.
- In creating hybrids, there were compatibility issues, introgression of selected gene was not possible, and the technology was time consuming. Then came molecular breeding where transgenics were created. In creation of transgenics, it was possible to introgress the human gene in plants.
- The creation of transgenics allowed the major revolution and improved quality traits tremendously in plants. Marker-assisted breeding helped to quickly correlate the results without any time lag or waiting for the seedlings to grow.
- Plant cell and tissue culture has explored the power of totipotency of plant cells and has made possible the haploid culture, micropropagation, embryo culture, and embryo rescue to name a few. Because of these technologies,

we are able to produce biopharmaceuticals, edible vaccines, secondary metabolites, and many other important products.

- The plants are engineered to tolerate abiotic stresses as drought, salt, frost. These stress-tolerant plants can be grown on land which was earlier unsuitable for agriculture.
- The plants are genetically modified for genes against insects and pests, which make them capable of killing their pests. This also reduces load of insecticides and pesticides.
- Nutritionally enriched crops are also an advantage where crops can be more nutritious with minimal toxicity or allergies.
- However there are a few developments like terminator and traitor technologies which might have deleterious effects upon their release.
- Ethical and biosafety issues are high with the technology. There are unknown and unseen fears at the consumer end.
- To sum up, transgenic organisms can offer a range of benefits beyond those that emerged from innovations in traditional agricultural biotechnology.

Multiple Answer-Type Questions

1. Induced mutation resulted in:
   (a) Beneficial mutations
   (b) The crops acquiring new trait
   (c) Harmful mutations
   (d) All of the above
2. Open pollination is:
   (a) Self-pollination
   (b) Pollination by birds and insects
   (c) Pollination in laboratory condition
   (d) All of the above
3. Pure line is obtained by:
   (a) Repeated selfing of first filial generation
   (b) Repeated selfing of parent with recessive characters
   (c) Repeated backcross
   (d) None of the above
4. Near isogenic lines are produced by:
   (a) Repeated selfing of first filial generation
   (b) Repeated backcrossing with one parent
   (c) Repeated test crossing with one parent
   (d) All of the above
5. In hybridization it is possible to:
   (a) Cross two sexually compatible varieties
   (b) Cross two sexually incompatible varieties
   (c) Transfer the gene of interest
   (d) Self-pollinate and fertilize
6. Transgenics are a method of choice when:
   (a) The gene of interest is not present in the germplasm of the crop.
   (b) Trait improvement is difficult with breeding.
   (c) Important traits need to be transferred.
   (d) All of the above.
7. Transfer of stacked trait means:
   (a) Transfer of genes for more than one trait
   (b) Transfer of many copies of genes for a single trait
   (c) Removal of the gene for a particular trait
   (d) All of the above
8. The markers which can predict about heterozygosity of alleles are:
   (a) Dominant markers
   (b) Biochemical markers
   (c) Codominant markers
   (d) Quantitative trait loci
9. Production of identical clones is possible by:
   (a) Genetic engineering
   (b) Conventional breeding
   (c) Marker-assisted breeding
   (d) Tissue culture
10. Dedifferentiation is:
    (a) Differentiation of newly formed callus
    (b) Differentiation of meristematic cells
    (c) When differentiated cells become undifferentiated
    (d) None of the above
11. Sterile hybrids can be multiplied by:
    (a) Haploid culture
    (b) Embryo culture
    (c) Micropropagation
    (d) All of the above
12. Embryo rescue is important when:
    (a) Embryo is from fusion of incompatible gametes.

(b) Embryo is sterile.
(c) Embryo is from fusion of compatible gametes.
(d) None of the above.

13. Golden rice has:
    (a) Herbicide tolerance
    (b) Insecticidal Bt gene
    (c) β-carotene gene
    (d) None of the above
14. Terminator technology is:
    (a) Spraying of terminator as herbicidal agent
    (b) Production of sterile seeds
    (c) Production of transgenics
    (d) All of the above
15. Roundup is:
    (a) A crop variety
    (b) An insecticide
    (c) A herbicide
    (d) None of the above
16. Which of the following statement is true regarding classical plant breeding?
    (a) It can only be done between two sexually compatible plants.
    (b) Many undesirable traits are transferred along with traits of interest.
    (c) Both of the above.
    (d) None of the above.
17. Mutation breeding involves the use of:
    (a) X-rays
    (b) Sodium azide
    (c) EMS
    (d) All of the above
18. Which of the following is not a plant tissue culture medium?
    (a) MS medium
    (b) LB medium
    (c) Gamborg's medium
    (d) White's medium
19. Which of the following technique is used for producing virus-free plants?
    (a) Anther culture
    (b) Embryo culture
    (c) Meristem culture
    (d) Root culture
20. A process in which only adventitious shoot bud initiation occurs:
    (a) Rhizogenesis
    (b) Caulogenesis
    (c) Mutagenesis
    (d) Embryogenesis
21. Which of the following is true regarding suspension culture?
    (a) It consists of only single cells.
    (b) Physiologically and biochemically different cells are used.
    (c) Ideally multiple cells are used.
    (d) None of the above.
22. Cybrids have:
    (a) Nucleus from one parent
    (b) Cytoplasm from one parent
    (c) Nucleus from one and cytoplasm from both
    (d) Both a and b
    (e) Both a and c

Answers

1. (d); 2. (b); 3. (a); 4. (a); 5. (a); 6. (d); 7. (a); 8. (c); 9. (d); 10. (c); 11. (c); 12. (a); 13. (c); 14. (b); 15. (c); 16. (c); 17. (d); 18. (b); 19. (c); 20. (b); 21. (a); 23. (e)

Review Questions

Q1. Enumerate the advantages of plant tissue culture technique over classic plant breeding techniques.
Q2. Write a short note on mutation breeding.
Q3. Explain the role of growth regulators in plant tissue culture.
Q4. Discuss somaclonal variations and their applications in tissue culture.
Q5. Differentiate among the following:
  (i) Cybrid vs. hybrid
  (ii) Anther vs. embryo culture
  (iii) Polymorphic vs. monomorphic marker
Q6. Explain the various tissue culture techniques.
Q7. What are GM crops? How are they prepared and why are they controversial?
Q8. What is marker-assisted breeding?
Q9. What are terminator and traitor technologies?

References

1. Alfonso A (2007) Rice biotechnology. Presentation during PhilRice R&D. Mar 13–15
2. Almeida et al (2012) Biodiesel biorefinery: opportunities and challenges for microbial production of fuels and chemicals from glycerol waste. Biotechnol Biofuels 5:48
3. Altenbach SB, Pearson KW, Meeker G, Staraci LC, Sun SS (1989) Enhancement of the methionine content of seed proteins by the expression of a chimeric gene encoding a methionine-rich protein in transgenic plants. Plant Mol Biol 13:513–522
4. ANP: Agência Nacional do Petróleo, Gás natural e Biocombustíveis. Boletim mensal de biodiesel 2011. http://www.anp.gov.br
5. Baack EJ, Rieseberg LH (2007) A genomic view of introgression and hybrid speciation. Curr Opin Genet Dev 17:513–518
6. Babu R, Nair SK, Prasanna BM, Gupta HS (2004) Integrating marker-assisted selection in crop breeding – prospects and challenges. Curr Sci 87:607–619
7. Borlaug NE (2000) Ending world hunger. The promise of biotechnology and the threat of antiscience zealotry. Plant Physiol 124:487–490
8. Bouchereau A, Aziz A, Larher F, Martin-Tanguy J (1999) Polyamines and environmental challenges: recent development. Plant Sci 140:103–125
9. Collard BCY, Mackill DJ (2008) Marker-assisted selection: an approach for precision plant breeding in the twenty-first century. Phil Trans R Soc B 363:557–572
10. Cuevas JC, Lopez-Cobollo R, Alcazar R, Zarza X, Koncz C, Altabella T et al (2008) Putrescine is involved in Arabidopsis freezing tolerance and cold acclimation by regulating ABA levels in response to low temperature. Plant Physiol 148:1094–1105
11. Dalton R (2004) California edges towards farming drug-producing rice. Nature 428:591
12. Desamero NV (2007) Genetic enhancement of in vitro culture-derived tungro resistant rice breeding lines. Paper presented during the 19th Federation of Crop Science Societies of the Philippines, Development Academy of the Philippines, Tagaytay City. June 13–15
13. DeVries J, Toenniessen G (2001) Securing the harvest: biotechnology, breeding and seed systems for African crops. The Rockefeller Foundation, New York
14. Duan JJ, Li J, Guo SR, Kang YY (2008) Exogenous Spermidine affects polyamine metabolism in salinity-stressed *Cucumis sativus* roots and enhances short-term salinity tolerance. J Plant Physiol 165:1620–1635
15. Estruch JJ et al (1996) Vip3A, a novel Bacillus thuringiensis vegetative insecticidal protein with a wide spectrum of activities against lepidopteran insects. Proc Natl Acad Sci U S A 93:5389–5394
16. FAO/IAEA Symposium on induced mutations (2008) Vienna, Austria. http://www.pub.iaea.org/mtcd/meetings
17. Gatehouse et al (2011) Insect-resistant biotech crops and their impacts on beneficial arthropods. Phil Trans R Soc B 366:1438–1452
18. George EF, Hall MA, De Klerk G-J (eds) (2007) Plant progapagation by tissue culture 3rd edition. Volume 1. Background. Springer
19. Germaine KJ, McGuinness M, Dowling DN (2012) Improving phytoremediation through plant associated bacteria. In: de Bruijn FJ (ed) Molecular ecology of the rhizosphere. Wiley-Blackwell, Hoboken
20. Goldberg I (1994) Functional foods, designer foods, pharmafoods, nutraceuticals. Chapman and Hall, New York
21. Groppa MD, Benavides MP, Tomaro ML (2003) Polyamine metabolism in sunflower and wheat leaf discs under cadmium or copper stress. Plant Sci 161:481–488
22. Hall RD, Brouwer ID, Fitzgerald MA (2008) Plant metabolomics and its potential application f or human nutrition. Plant Physiol 132:162–175
23. Heffner EL, Sorrells ME, Jannink JL (2009) Genomic selection for crop improvement. Crop Sci 49:1–12
24. Huang et al (2000) Genetic engineering of glycinebetaine production toward enhancing stress tolerance in plants: metabolic limitations. Plant Physiol 122(3):747–756, March
25. Hussain et al (2012) Current approaches toward production of secondary plant metabolites. J Pharm Bioallied Sci 4:10–20
26. James C (1999) Global review of commercialized transgenic crops: international service for the acquisition of agri-biotechnology applications briefs No.12 preview. International Service for the Acquisition of Agri-biotechnology Applications, Ithaca
27. James C (2010) Global status of commercialized biotech/GM crops, ISAAA brief no. 42. ISAAA, Ithaca
28. Jiang GL (2013a) Molecular markers and marker-assisted breeding in plants. In: Anderson SB (ed) Plant breeding from laboratories to fields. InTech, Croatia, pp 45–83
29. Jiang GL (2013) Plant marker-assisted breeding and conventional breeding: challenges and perspectives. Adv. Crop Sci Tech. 1–3
30. Katsube T, Kurisaka N, Ogawa M, Maruyama N, Ohtsuka R, Utsumi S, Takaiwa F (1999) Accumulation of soybean glycinin and its assembly with the glutelins in rice. Plant Physiol 120:1063–1073
31. Khan S, Afzal M, Iqbal S, Khan QM (2013) Plant – bacterial partnerships for the remediation of hydrocarbon contaminated soils. Chemosphere 90:1317–1332
32. Kitch L, Koch M, Sithole-Nang I (2002) Crop biotechnology: a working paper for administrators and policy makers in sub-Saharan Africa (FAO)
33. Lal P, Ramachandran VG, Goyal R, Sharma R (2007) Edible vaccines: current status and future. Indian J Med Microbiol 25:93–102
34. Ledford H (2014) Seed-patent case in Supreme Court. Loss of patent control could rekindle 'terminator' technology. Nature 494:289–290

35. Lee MK, Walters FS, Hart H, Palekar N, Chen JS (2003) The mode of action of the Bacillus thuringiensis vegetative insecticidal protein Vip3A differs from that of Cry1Ab delta-endotoxin. Appl Environ Microbiol 69:4648–4657
36. Losey JE, Rayor LS, Carter ME (1999) Transgenic pollen harms monarch larvae. Nature 399:214
37. Lucca P, Hurrell R, Potrykus I (2002) Fighting iron deficiency anemia with iron-rich rice. J Am Coll Nutr 21:184S–190S
38. McGloughlin MN (2008) Nutritionally improved agricultural crops. Plant Physiol 147:939–953
39. McLintock B (1950) The origin and behavior of mutable loci in maize. Proc Natl Acad Sci U S A 36:344–355
40. Merlin et al, (2014) Comparative evaluation of recombinant protein production in different biofactories: the green perspective. BioMed Research International ID:136419
41. Minorsky PV (2001) The Monarch butterfly controversy. Plant Physiol 127:709–710
42. Nakaya A, Isobe SN (2012) Will genomic selection be a practical method for plant breeding? Ann Bot 110:1303–1316
43. O'Connor D (2011) Report T39-T3. Biodiesel GHG emissions, pas, present, and future. A report to IEA Bioenergy Task 39. In Commercializing Liquid Biofuels from Biomass. International Energy Agency (IEA).(www.ieabioenergy.com)
44. Palma et al (2012) Vip3C, a novel class of vegetative insecticidal proteins from Bacillus thuringiensis. Appl Environ Microbiol 78:7163–7165
45. Pray LA (2008) Discovery of DNA structure and function: Watson and Crick. Nature Education Knowledge
46. Ribaut JM, de Vicente MC, Delannay X (2010) Molecular breeding in developing countries: challenges and perspectives. Curr Opin Plant Biol 13:1–6
47. Rice Varieties Adaptable to Abiotic Stress Conditions. (2012) National Seed Industry Council, Department of Agriculture – Bureau of Plant Industry
48. Rider JE, Hacker A, Mackintosh CA, Pegg AE, Woster PM, Casero RA Jr (2007) Spermine and spermidine mediate protection against oxidative damage caused by hydrogen peroxide. Amino Acids 33:231–240
49. Rywinska A, Rymowicz W (2010) High-yield production of citric acid by Yarrowia lipolytica on glycerol in repeated-batch bioreactors. J Ind Microbiol Biotechnol 37:431–435
50. Sattar S, Maiti MK (2011) Molecular characterization of a novel vegetative insecticidal protein from *Bacillus thuringiensis* effective against sap-sucking insect pest. J Microbiol Biotechnol 21:937–946
51. Segura A, Ramos JL (2013) Plant–bacteria interactions in the removal of pollutants. Curr Opin Biotechnol 24:467–473
52. Semagn K, Bjornstad A, Ndjiondjop MN (2006) Progress and prospects of marker assisted backcrossing as a tool in crop breeding programs. Afr J Biotechnol 5:2588–2603
53. Sena JA, Hernández-Rodríguez CS, Ferré J (2009) Interaction of Bacillus thuringiensis Cry1 and Vip3A proteins with Spodoptera frugiperda midgut binding sites. Appl Environ Microbiol 75:2236–2237
54. Thorpe TA (2007) History of plant tissue culture. Mol Biotechnol 37:169–180
55. Wang B, Chee PW (2010) Application of advanced backcross quantitative trait locus (QTL) analysis in crop improvement. J Plant Breed Crop Sci 2:221–232
56. Wang X, Jiang GL, Green M, Scott RA, Hyten DL, Cregan PB (2012) Quantitative trait locus analysis of saturated fatty acids in a population of recombinant inbred lines of soybean. Mol Breed 30:1163–1179
57. Warren GW (1997) Vegetative insecticidal proteins: novel proteins for control of corn pests. In: Carozzi N, Koziel M (eds) Advances in insect control: role of transgenic plants. Taylor and Francis, London, pp 109–121
58. Watson JD et al (1992) Recombinant DNA, 2nd edn. W. H. Freeman, New York
59. Wieczorek A (2003) Use of biotechnology in agriculture—benefits and risks. Biotechnology May 3
60. Wieczorek AM, Wright MG (2012) History of agricultural biotechnology: how crop development has evolved. Nat Educ Knowl 3:9
61. Xu Y (2010) Molecular plant breeding. CAB International Oxfordshire OX10 8DE UK
62. Yamaguchi K et al (2007) Protective role for the polyamine spermine against drought stress in Arabidopsis. Biochem Biophys Res Commun 352:486–490
63. Yazdani SS, Gonzalez R (2007) Anaerobic fermentation of glycerol: a path to economic viability for the biofuels industry. Curr Opin Biotechnol 18:213–219
64. Zhang Y-HP (2011) What is vital (and not vital) to advance economically-competitive biofuels production. Process Biochem 46:2091–2110
65. Zhao H, Yang H (2008) Exogenous polyamines alleviate the lipid peroxidation induced by cadmium chloride stress in *Malus hupehensis* Rehd. Sci Hortic 116:442–447

Some Related Resources

American Association for the Advancement of Science. Annual meeting (2011)

Development_of_the_first_salt-tolerant_rice_cultivar_through_indicaindica_anther_culture

Estruch JJ, et al (1998) Plant pest control. Patent WO 9844137

http://books.google.com/books?hl=en&lr=&id=55X_Wjct7f0C&oi=fnd&pg=PP6&dq=%

http://www.agbioworld.org/biotech-info/articles/biotech-art/biosafety.html

http://www.biotechnology4u.com/plant_biotechnology_applications_cell_tissue_culture.html
http://www.nseedcouncil.bpinsicpvpo.com.ph/downloadables/ccvar2012-2nd.pdf
http://www.researchgate.net/publication/223276104_
http://deskuenvis.nic.in/pdf/chairman-inauguraladd-261107.pdf (Prof. S. Kanniyan)
https://isaaa.org/resources
IRRI (2013) Wild Parent Spawns super salt tolerant rice. http://irri.org/news/media-releases/wild-parent-spawns-super-salt-tolerant-rice Apr 15
Land Grant Universities (2011)
NERC (2011) Can GM crops harm the environment?
West Africa Rice Development Association (WARDA) http://www.warda.cgiar.org

Tissue Engineering and Artificial Organ

21

Abstract

Tissue engineering is an exciting technique, which has the potential to create tissues and organs de novo. Tissue engineering was defined in 1988 as "application of the principles and methods of engineering and life sciences toward fundamental understanding of structure–function relationship in normal and pathological mammalian tissues and the development of biological substitutes for the repair or regeneration of tissue or organ function." It was later summarized as "an interdisciplinary field which involves fundamentals of life sciences, medical sciences, and principles of material sciences, which can provide a functional substitute for damaged or diseased organ restoring, maintaining, or improving tissue function or a whole organ." The existence of tissue engineering dates to the sixteenth century, when complex skin flaps were used to replace the nose. Initially, the field was recognized as a subfield of biomaterials. Most definitions of tissue engineering cover a broad range of applications; in practice, the term is closely associated with applications that repair or replace portions of or whole tissues (i.e., bone, cartilage, blood vessels, bladder, skin, and so on). It has the potential to produce a supply of immunologically tolerant "artificial" organ and tissue substitutes that can grow in the patient.

21.1 Introduction to Tissue Engineering

Tissue engineering is an exciting technique which has the potential to create tissues and organs de novo. Tissue engineering was defined in 1988 as "application of the principles and methods of engineering and life sciences toward fundamental understanding of structure–function relationship in normal and pathological mammalian tissues and the development of biological substitutes for the repair or regeneration of tissue or organ function." It was later summarized as "an interdisciplinary field which involves fundamentals of life sciences, medical sciences, and principles of material sciences, which can provide a functional substitute for damaged or diseased organ restoring, maintaining, or improving tissue function or a whole organ." The existence of tissue engineering dates to the sixteenth century, when complex skin flaps were used to replace the nose. Initially the field was recognized as a sub-

V. Gupta et al., *Basic and Applied Aspects of Biotechnology*, DOI 10.1007/978-981-10-0875-7_21

field of biomaterials. Most definitions of tissue engineering cover a broad range of applications; in practice, the term is closely associated with applications that repair or replace portions of or whole tissues (bone, cartilage, blood vessels, bladder, skin, and so on). It has the potential to produce a supply of immunologically tolerant "artificial" organ and tissue substitutes that can grow in the patient [27].

The term regenerative medicine has also been applied to efforts which are performed for specific biochemical functions using cells within an artificially created support system (e.g., an artificial pancreas or a bioartificial liver), although those involved in regenerative medicine place more emphasis on the use of stem cells to produce tissues.

In tissue engineering technologies, tissue loss or organ failure can be treated either by implantation of an engineered biological substitute or alternatively with ex vivo perfusion systems. The tissue-engineered products may be fully functional at the time of treatment (liver assist devices, encapsulated islets) or have potential to integrate and form the expected functional tissue upon implantation (chondrocytes embedded in a matrix carrier). In some cases, biomaterials are modified to enhance migration and attachment of the specific cell populations, which repair or replace the damaged tissue.

As the tissue-engineered products are a new generation of the medical supplements, thus they do not easily fit in the definition and classification of the regular and traditional product range of the Food and Drug Administration (FDA), for example, live cell-containing devices may be attributes of drugs or biologics and may be subjected to multiple regulatory definitions and classification. Their demand is very high with huge economic potential [34].

21.2 Source of Tissue/Cells

The diseased/affected tissues require readily available tissue which could replace the affected tissue with certain mechanical and structural properties for proper functioning. In majority of the cases, autografts can be used. Autografts/autologous cells are obtained from the same individual to whom they will be reimplanted. Autologous cells have the fewest problems with rejection and pathogen transmission; however, in some cases, they might not be available. For example, in genetic disease and in ill or elderly persons and in patients suffering from severe burns, obtaining suitable autologous cells is not possible. Since autologous cells need to be harvested and cultured from the patient before they can be used, there complications may include infection and donor site morbidity or chronic pain.

Allografts may be obtained from the body of a donor of the same species (Fig. 21.1). Allograft/allogenic cells could provide healthy tissue from another donor but may suffer from limitation of the tissue, possibility of rejection, or disease transmission. While there are some ethical constraints to the use of human cells for in vitro studies, the employment of dermal fibroblasts from human foreskin has been demonstrated to be immunologically safe and thus a viable choice for tissue engineering of skin.

Xenogeneic cells can be obtained from individuals of another species (e.g., pig) (Fig. 21.1). In particular, animal cells have been used quite extensively in experiments aimed at the construction of cardiovascular implants. Syngenic or isogenic cells may be obtained from genetically identical organisms, such as twins, clones, or highly inbred research animal models. The issues with contaminating viruses and bacterial agents are even more significant when xenogeneic cells and/or culture components are used, because potentially infectious xenogeneic agents may be introduced into the human population with this vehicle. Finally, when genetically modified cells are used in a tissue-engineered product, there are additional concerns such as cell transformation by the vector, vector stability, and optimal function of the inserted gene.

21.3 Requirements of Tissue Engineering

The growth of replacement tissue using tissue engineering and regenerative medicine is one method of using technology to create an effec-

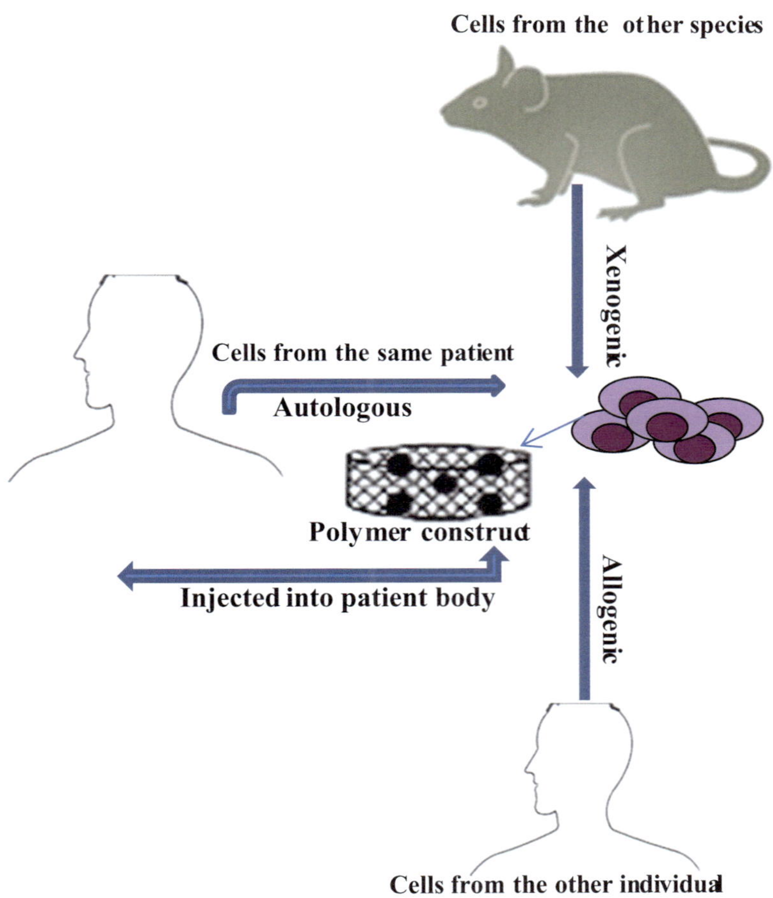

Fig. 21.1 The figure shows the autologous, allogenic, and xenogeneic transplantation of cells for repair and regeneration

tive replacement for these types of graft tissue. It involves the in vitro seeding and attachment of human cells onto a scaffold. These cells then proliferate, migrate, and differentiate into a specific tissue while secreting the extracellular matrix components required for creation of the tissue.

Tissue engineering encompasses three approaches: (1) cells and cell substitutes, (2) factors for tissue induction that can mimic natural physiological conditions, and (3) the seeding of cells onto matrices. The most common tissue engineering approach is the third one, to place cells onto a biomaterial matrix.

The various tissue regeneration strategies can be (1) direct injection of bolus cells into the tissue of interest or the systematic circulation; (2) implantation of cells after they have been combined to form a three-dimensional tissue structure, often within a bioreactor; and (3) scaffold-based delivery of signaling molecules such as low molecular weight drugs, proteins, and oligonucleotides that stimulate cell migration, growth, and differentiation [35].

21.3.1 Cells

The most important aspect of tissue engineering are cells as they have to regenerate and repair the tissue by proliferation and differentiation, perform cell-to-cell signaling, biomolecule production, and formation of extracellular matrix.

They should be easily accessible and capable of proliferation in response to specific growth factors while maintaining their differentiated function.

The functionality of an engineered tissue may be structural (bone, cartilage, and skin) or metabolic (liver, pancreas), or both.

Cells may be a part of an engineered tissue, or alternatively, these cells may be recruited in vivo with the help of biomaterials and/or biomolecules. When selecting the cellular component of an engineered product, it is important to identify appropriate cells and to be able to isolate them from the primary source.

In addition, expansion of these cells without permanently altering the phenotype and function during the expansion phase and without introduction of any adventitious and species-specific bacterial/viral agents poses significant challenges. Stem cells may provide unlimited supply of cells; however, differentiation and regulation of lineage of stem cell is critical for normal tissue development.

Cells are often implanted or "seeded" into scaffold which is capable of supporting three-dimensional tissue formation which not only provides mechanical support but also supply critical nutrients and transport metabolites to and from the developing tissue (Fig. 21.2).

21.3.2 Growth Factors

The critical molecules that drive the differentiation of the cells to a particular lineage are growth factors. They are important signaling molecules that drive cell differentiation during development, and one may achieve tissue regeneration in the adult by enabling control over growth factor delivery. These signaling molecules are broadly grouped into the overlapping categories of mitogens (stimulate cell division), growth factors (originally identified by their proliferation-inducing effects, but have multiple functions), and morphogens (control generation of tissue form). Precise control over the signaling of these factors in a local area may potentially allow control over a regenerative process.

The broad term growth factor is used for all these molecules and they affect cell migration, proliferation, and cellular differentiation. Popular growth factors in tissue regeneration are angiopoietin (Ang), basic fibroblast growth factor (bFGF), bone morphogenetic protein (BMP), epidermal growth factor (EGF), fibroblast growth factor (FGF), hepatocyte growth factor (HGF), insulin-like growth factor (IGF), nerve growth factor (NGF), platelet-derived growth factor (PDGF), transforming growth factor (TFG), vascular endothelial growth factor (VEGF), etc. Along these lines, the molecules such as transforming growth factor, insulin, and dexamethasone have been shown to differentiate the mesenchymal stem cells along the chondrocytic and astrocytic lineages [4]. The various growth factors along with their functions are listed in Table 21.1.

21.3.3 Polymers

Many different materials (natural and synthetic, biodegradable and permanent) have been investigated. For bone tissue engineering, hydroxyapatite and tricalcium phosphate are used, the advantage being their resemblance with natural inorganic components of bone and osteoconductive properties; unfortunately, they lack mechanical properties of the bone. Other synthetic and natural polymers are an attractive alternatives and versatile in their applications to the growth of most tissues (Table 21.2).

21.3.3.1 Synthetic Polymers

The polymers which can be degraded readily in the body are more useful as synthetic polymers. Aliphatic polyesters such as polyglycolic acid (PGA), polylactic acid (PLA), and their copolymers as poly(lactic-co-glycolic acid (PLGA) and polycaprolactone (PCL) are commonly used. Their degradation products (glycolic acid and lactic acid) are present in human body. Polylactic acid (PLA) is a polyester which degrades within the human body to form lactic acid, a naturally occurring chemical which is easily removed from the body.

Degradation mechanism of polyglycolic acid (PGA) and polycaprolactone (PCL) is similar to that of PLA, but they exhibit a faster and a slower rate of degradation, respectively, compared to PLA (Table 21.3).

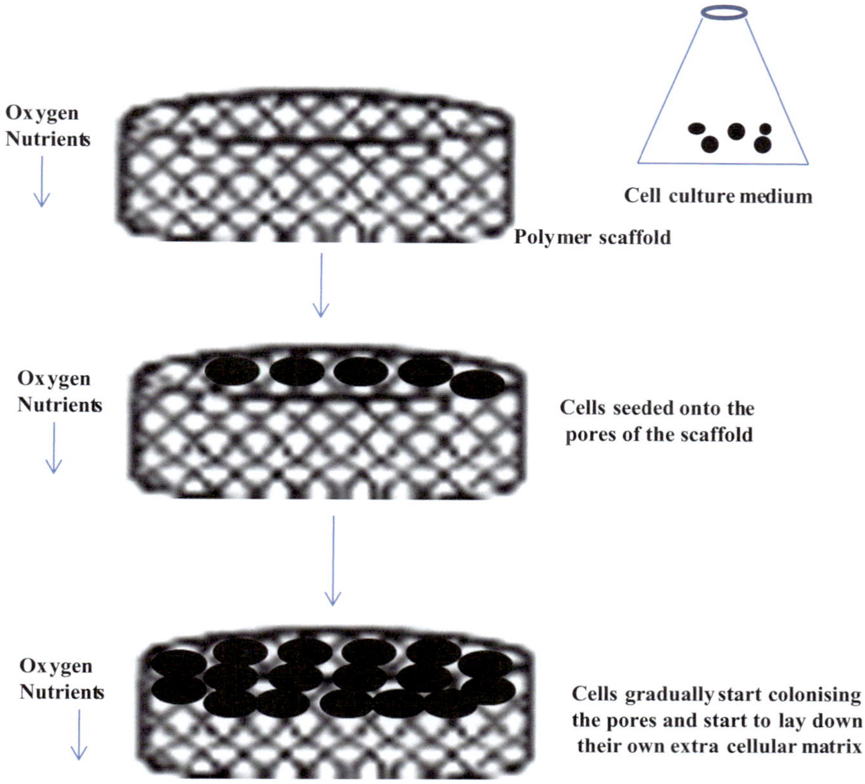

Fig. 21.2 The figure shows the growth of the cells on polymer scaffold

21.3.3.2 Natural Polymers

Natural biological components as protein or carbohydrate have been used for tissue engineering. Naturally occurring biomaterials can be obtained from their natural sources and processed to make porous scaffolds.

These materials can be in their native form, such as extracellular matrix (ECM) from allografts and xenografts, or can be in the form of smaller building blocks, which include but not limited to inorganic ceramics such as calcium phosphates and organic polymers such as proteins, polysaccharides, lipids, and polynucleotides.

Natural biomaterials usually have superb biocompatibility so that cells can attach and grow with excellent viability. Table 21.4 lists some of the natural polymers along with their features.

However, one issue with natural materials is their limited physical and mechanical stability, and therefore, they may not be suitable for some load-bearing applications.

Different derivatives of the extracellular matrix have been studied to evaluate their ability to support cell growth. Proteic materials, such as collagen or fibrin, and polysaccharidic materials, like chitosan or glycosaminoglycans (GAGs), are highly suitable in terms of cell compatibility, but some issues still exists with potential immunogenicity.

Among GAGs, hyaluronic acid, possibly in combination with cross-linking agents (e.g., glutaraldehyde, water -soluble carbodiimide, etc.), is one of the possible choices as scaffold material. Functionalized groups of scaffolds may be useful in the delivery of small molecules (drugs) to specific tissues. The surface properties of the materials used for the scaffold are important for adhesion, migration, and cell differentiation. Ongoing research is focused on tethering growth

Table 21.1 The table shows various growth factors along with their physiological functions and their role in tissue engineering for cell differentiation

Growth factors	Functions	Role in tissue engineering
Angiopoietin-1	Vascular growth factors	Blood vessel, maturation and stability
	Play an important role in embryonic and postnatal angiogenesis	
Angiopoietin-2, VEGF	Vascular endothelial growth factor along with angiopoietin is used	Blood vessel destabilizes, regresses, and disassociates endothelial cells from surrounding tissues
		Blood vessel migration and proliferation
FGF2	Fibroblast growth factor family of GFs, angiogenesis, wound healing, embryonic development	Migration of the blood vessel, bone, skin, spine, and nerve; muscle migration positively affects endothelial cells
PDGF-AB (or PDGF-BB)	Platelet-derived growth factor regulates cell growth and division	Influences development of the muscles, bone, cartilage, skin
		Blood vessel migration and proliferation
Bone morphogenetic proteins	Induce the formation of bone and cartilage	Bone and cartilage differentiation and migration of osteoblasts
	BMP-2 is a very potent GF	
TGF-β	Transforming growth factor beta	Bone and cartilage proliferation and differentiation of bone-forming cells
	It controls proliferation and differentiation of many cell types	
IGF-1	Insulin-like growth factor 1 (IGF-1) also called as somatomedin C	Has proliferative effects on many cells and inhibits cell apoptosis
	Important role in childhood growth	
	Has anabolic effects in adults	
EGF	Epidermal growth factor stimulates cell growth, proliferation, and differentiation	Skin and nerve regulation of epithelial cell growth, proliferation, and differentiation
Erythropoietin (EPO)	Controls erythropoiesis or red blood cell production also called hematopoietin and hemopoietin	Promotes the development of red blood cells
Nerve growth factor (NGF)	Released by neurons	Promotes the nerve cell survival and recovery from ischemia

Table 21.2 The table shows the examples of synthetic and natural polymers

Polymers	Material
Synthetic polymers	Polyglycolic acid (PGA)
	Polylactic acid (PLA)
	Poly(lactic-co-glycolic acid) (PLGA)
	Polycaprolactone
Natural polymers	Proteic collagen or fibrin
	Polysaccharide chitosan or glycosaminoglycans

factors or peptide sequences to the surface of the scaffold to improve adhesion and migration.

The goal of tissue engineering is to surpass the limitations of conventional treatments based on organ transplantation and biomaterial implantation. It has the potential to produce a supply of immunologically tolerant artificial organs and tissue substitute that can grow with the patient. This would also help in permanent solution to the damaged organ or tissue without the need for

Table 21.3 Shows the properties of various synthetic polymers

Polymer	Polymer composition	Properties	Degradation and disadvantage	Clearance time	Commercial usage
Polyester	Polyglycolic acid (PGA) [14]	Rigid thermoplastic material, high crystallinity, not soluble in most organic solvents	Degrades into glycolic acid, may induce tissue damage locally	6–12 months	Resorbable sutures (DEXON, American Cyanamid Company)
	Polylactic acid (PLA) [15]	Semicrystalline solid; more hydrophobic; l-isomer is used; biocompatible	Lactic acid	>24 months	
	d(-)				
	l(+)				
	Racemic (d,l)				
	Poly(lactic-co-glycolic acid)	GA/Lain 70/30 copolymer is used; biocompatible, nontoxic, noninflammatory	Degrades easily into lactic and glycolic acid, local inflammation, adverse response due to release of toxic compounds during tissue repair	5–6 months	Suture material (Vicryl by Ethicon Inc.), bone repair
	Polycaprolactones	Semicrystalline polymer, low melting temperature, nontoxic, and compatible with a range of other polymers	Degradation varies with copolymer composition; no inflammatory response in subcutaneous implantations	2–3 years	Long-term implantable drug delivery systems (MONOCRYL, Ethicon Inc.)
	Polypropylene fumarate [16]	No inflammatory response is observed in subcutaneous implantation	Degrades into propylene glycol, poly(acrylic acid-co-fumaric acid), and fumaric acid		Injectable biodegradable material
Polyanhydrides	Polyanhydrides	Biocompatible and biodegradable, nontoxic, excellent controlled release characteristics, limited mechanical properties	Degrades by hydrolysis of anhydride linkages and has well-defined degradation characteristics	1 year	Drug delivery at known rates

(continued)

Table 21.3 (continued)

Polymer	Polymer composition	Properties	Degradation and disadvantage	Clearance time	Commercial usage
Tyrosine-based polycarbonates	Tyrosine-based polycarbonates [3]	Copolymers with differing and good mechanical properties, degradation rates, and cellular response	Hydrolysis of carbonate yields alcohol and CO_2, so no local inflammation	Slow degradation	Orthopedic applications
Poly(ortho esters)	Poly(ortho esters)	Addition of lactides in polymer can tune degradation rates from 15 to 100 days	Slow degradation		Orthopedic applications
Polyurethane based on LDI and poly(glycolide-co-γ-caprolactone) [9]	Polyurethane based on LDI and poly(glycolide-co-γ-caprolactone)	Good biocompatibility, excellent mechanical properties	Long-term implants Degradation products are lysine and glycolic and caproic acids	1–2 months	Cardiac pacemaker and vascular grafts, scaffold for tissue engineering [19]
Ethyl glycinate polyphosphazenes	Ethyl glycinate polyphosphazenes	Biostable incorporation of specific side groups makes them biodegradable	Phosphates and ammonia from backbone and other products depending on side chain	2–3 months	Polymer supported the growth of osteogenic cell lines

Table 21.4 The table lists the characteristics of natural polymers

Polymer material	Composition	Properties	Scaffolding method	Commercial usage
Collagen [28]–glycosaminoglycan	Collagen–glycosaminoglycan	Controlled porosity, stability, and degradation	Chemical cross-linking by glutaraldehyde	Regeneration of the dermis, skin, sciatic nerve, knee meniscus
Alginic acid	Copolymer of β-D-mannuronic acid (M) and α-L-guluronic acid (G)	Gelation kinetics with $CaSO_4$ is difficult to control	Gel beads encapsulating living cells after cross-linking with calcium sulfate	Cell delivery vehicles
Chitosan [17]	Cationic polymer b-(1–4) Glucosamine and N-acetyl-d-glucosamine	Solubility is difficult to control in common solvents	Freeze-drying or lyophilizing a chitosan gel solution	Hemodialysis membrane, drug delivery system, orthopedic and dental coating material, and artificial skin

supplementary therapies. It is therefore evident, that the choice of scaffold is crucial to enable the cells to behave in the required manner to produce tissues and organs of the desired shape and size.

21.4 Properties of Biomaterial

For medical sutures, collagen and some polyesters were in use as biomaterials since long [30]. New biomaterials have been engineered to have ideal properties: (a) injectability, (b) synthetic manufacture, (c) biocompatibility, (d) non-immunogenicity, (e) transparency, (f) nanoscale fibers, (g) low concentration, and (h) resorption rates. PuraMatrix is one of these new biomimetic scaffold families which is impacting clinical tissue engineering.

The design of the scaffold prior to exposure to cells is of vital importance (Table 21.5):

1. The scaffold must present a surface that promotes cell attachment, growth, and differentiation while providing a porous network for tissue growth. Incorporation of signal peptides into the material has been attempted to effectively mimic the extracellular matrix and induce cell migration
2. A high porosity and an adequate pore size and continuity are necessary to facilitate cell seeding and diffusion throughout the whole structure of both cells and nutrients
3. Its degradation as well as its degradation products should be biocompatible. Biodegradability is important since scaffolds should preferably be absorbed by the surrounding tissues without the necessity of a surgical removal. However, its degradation rate should match the regeneration rate of the tissue and the resulting degradation products should be nontoxic to the host. Premature degradation of the material combined with lack of timely in vivo development of replacement tissue may result in reduced mechanical strength of an engineered tissue over time, which may lead to its failure
4. The scaffold must have the mechanical properties necessary to temporarily offer structural support until the new tissue has formed. Their mechanical properties may generate an inflammatory/immune reaction
5. In addition to consisting of an effective biomaterial, the scaffold must also possess key morphological characteristics. It must be highly porous and offer a suitable path for nutrient transmission and tissue in growth. To achieve these requirements, tissue engineering scaffolds are often designed to mimic the structure of the naturally occurring extracellular matrix (ECM). Vascularization is critical more specifically for three-dimensional

Table 21.5 The table shows the properties of scaffold

S. No.	Property of scaffold	Vital role
1.	Surface property	It should present a surface that promotes cell attachment, growth, and differentiation while providing a porous network for tissue growth
2.	Porosity	A high porosity and an adequate pore size are necessary to facilitate cell seeding and diffusion throughout the whole structure of both cells and nutrients
3.	Biodegradability	Biodegradability is important since scaffolds should preferably be absorbed by the surrounding tissues without the necessity of a surgical removal
4.	Mechanical properties	It must have the mechanical properties necessary to temporarily offer structural support until the new tissue has formed
5.	Non-immunogenic	Their chemical composition should not generate strong inflammatory/immune reaction

engineered tissues greater than 1 mm^3 to meet their nutritional and metabolic requirements. Investigators have incorporated angiogenic factors such as vascular endothelial growth factor (VEGF) [40] and platelet-derived growth factor (PDGF) or their genes into the implants to stimulate angiogenesis in engineered tissues.

21.5 Designing of Scaffolds

21.5.1 Fabrication Methods

A number of different methods has been described in literature for preparing porous structures to be employed as tissue engineering scaffolds. Each of these techniques presents its own advantages, but none is devoid of drawbacks:

1. *Solvent casting and particulate leaching (SCPL)*: This preparation of porous structures with regular porosity with a limited thickness is done. Initially, the polymer is dissolved into a suitable organic solvent (as polylactic acid could be dissolved into dichloromethane), and then the solution is casted into a mold filled with porogen particles. Such porogen can be an inorganic salt like sodium chloride, crystals of saccharose, gelatin spheres, or paraffin spheres. The size of the porogen particles will affect the size of the scaffold pores, while the polymer to porogen ratio is directly correlated to the amount of porosity of the final structure. After the polymer solution has been casted, the solvent is allowed to fully evaporate, and then the composite structure in the mold is immersed in a bath of a liquid suitable for dissolving the porogen. Once the porogen has been fully dissolved, a porous structure is obtained. The disadvantage being the small thickness range and use of organic solvents which must be fully removed to avoid any possible damage to the cells seeded on the scaffold. In addition to pore size, pores interconnectivity in a scaffold is also important for cell migration, cell signaling, and mass transport. While particle leaching techniques can create well-formed macropores, it is difficult to control the degree of interconnectivity.
2. *Phase separation*: A biodegradable synthetic polymer is dissolved in molten phenol or naphthalene, and biologically active molecules such as alkaline phosphatase can be added to the solution. The temperature is then lowered to produce a liquid–liquid phase separation and quenched to form a two-phase solid. The solvent is removed by sublimation to give a porous scaffold with bioactive molecules incorporated in the structure.

3. *Gas foaming*: To overcome the necessity to use organic solvents and solid porogens, a technique using gas as a porogen has been developed. First, disk-shaped structures made of the desired polymer are prepared by means of compression molding using a heated mold. The disks are then placed in a chamber where they are exposed to high-pressure CO_2 for several days. The pressure inside the chamber is gradually restored to atmospheric levels. During this procedure, the pores are formed by the carbon dioxide molecules that abandon the polymer, resulting in a sponge-like structure. The main problems related to such a technique are caused by the excessive heat used during compression molding (which prohibits the incorporation of any temperature labile material into the polymer matrix) and by the fact that the pores do not form an interconnected structure.
4. *Melt molding*: This process involves filling a Teflon mold with PLGA powder and gelatin microspheres, of specific diameter, and then heating the mold above the glass transition temperature of PLGA while applying pressure to the mixture. This treatment causes the PLGA particles to bond together. Once the mold is removed, the gelatin component is leached out by immersing in water and the scaffold is then dried. Scaffolds produced this way assume the shape of the mold.
5. *Emulsification/freeze-drying*: This technique does not require the use of a solid porogen like SCPL. First, a synthetic polymer is dissolved into a suitable solvent (polylactic acid in dichloromethane), then water is added to the polymeric solution, and the two liquids are mixed in order to obtain an emulsion. Before the two phases can separate, the emulsion is casted into a mold and quickly frozen by means of immersion into liquid nitrogen. The frozen emulsion is subsequently freeze-dried to remove the dispersed water and the solvent, thus leaving a solidified, porous polymeric structure. While emulsification and freeze-drying allows a faster preparation if compared to SCPL, since it does not require a time-consuming leaching step, it still requires the use of solvents; moreover, pore size is relatively small and porosity is often irregular. Freeze-drying by itself is also a commonly employed technique for the fabrication of scaffolds. In particular, it is used to prepare collagen sponges: collagen is dissolved into acidic solutions of acetic acid or hydrochloric acid that are casted into a mold, frozen with liquid nitrogen, and then lyophilized.

21.5.1.1 Limitations of Conventional Tissue Engineering Scaffolds

Scaffold is required to provide adequate mechanical support. In the absence of mechanical support, excessive deformation might occur. Designed scaffolds have better performance over conventional scaffolds. Nonbiodegradable synthetic polymers such as polytetrafluoroethylene and polyethylene provide well-defined mechanical and structural properties, but their long-term presence in the body can lead to a chronic inflammatory response, which results in poor tissue quality.

Polylactic acid and polyglycolic acid are biodegradable polymers; however, the degradation of synthetic polymers, both in vitro and in vivo conditions, releases acidic by-products which raise concerns that the scaffold microenvironment may not be ideal for tissue growth. Lactic acid release during degradation reduces the pH, which further accelerates the degradation rate due to autocatalysis, resulting in a highly acidic environment adjacent to the polymer. Such an environment may adversely affect cellular function. Although the degradation of these materials can be partially controlled, however, nonuniform degradation and varying rates in different anatomic locations represent challenges [21–23, 25]. The rate at which degradation occurs has to coincide as much as possible with the rate of tissue formation. This means that while cells are fabricating their own natural matrix structure around themselves, the scaffold should be able to provide structural integrity within the body. Then eventually it should break down leaving the newly formed tissue to take over the mechanical load. The use of polymers has major applications in drug delivery systems, in orthopedic fixation devices (as rods, pins, and screws), and in resorbable sutures.

Cells attached to scaffolds are faced with several weeks of in vitro culturing before the tissue is suitable for implantation. During this period, even small pH changes (6.8–7.5) in the scaffold microenvironment can significantly affect cell growth. Furthermore, particles released during polymer degradation can affect tissue remodeling processes along with eliciting an inflammatory response and inducing tissue resorption in vivo [21, 22, 23].

Moreover, current synthetic polymers do not possess a surface chemistry which is familiar to cells. In in vivo condition, cells thrive on an extracellular matrix made mostly of collagen, elastin, glycoproteins, proteoglycans, laminin, and fibronectin. In contrast, collagen is the major protein constituent of the extracellular matrix and is recognized by cells, and being chemotactic, collagen scaffolds present a more native surface relative to synthetic polymer scaffolds for tissue engineering purposes. However, like other natural polymers, it may elicit an immune response. The antigenicity of collagen can be reduced by treating with pepsin to remove the telopeptide regions or by cross-linking.

Conventional scaffold fabrication techniques are incapable of precisely controlling pore size, pore geometry, spatial distribution of pores, and construction of internal channels within the scaffold. In them, the cells do not necessarily recognize such surfaces, and most importantly cells cannot migrate more than 500 μm from the surface. The lack of oxygen and nutrient supply governs this depth. Solid freeform fabrication (SFF) uses layer-manufacturing strategies to create physical objects directly from computer-generated models. It can improve current scaffold design by controlling scaffold parameters such as pore size, porosity, and pore distribution, as well as incorporating an artificial vascular system, thereby increasing the mass transport of oxygen and nutrients into the interior of the scaffold and supporting cellular growth in that region.

Only thin scaffold cross-sections can be produced due to difficulty in removing salt particles deep in the matrix. For gas foaming, it has been reported that only 10–30 % of the pores were interconnected.

Conventional scaffold fabrication techniques use organic solvents, like chloroform and methylene chloride, to dissolve synthetic polymers at some stage in the process. The presence of residual organic solvent is the most significant problem facing these techniques due to the risks of toxicity and carcinogenicity it poses to cells.

In addition, conventional fabrication techniques produce scaffolds that are foam structures. Cells are then seeded and expected to grow into the scaffold. However, this approach has resulted in the in vitro growth of tissues with cross-sections of less than 500 μm from the external surface.

The pioneering cells cannot migrate deep into the scaffold because of the lack of nutrients and oxygen and insufficient removal of waste products; cell colonization at the scaffold periphery is consuming, or acting as an effective barrier to the diffusion of, oxygen and nutrients into the interior of the scaffold. Furthermore, for bone tissue engineering, the high rates of nutrient and oxygen transfer at the surface of the scaffold promote the mineralization of the scaffold surface, further limiting the mass transfer to the interior of the scaffold.

Thus cells are only able to survive close to the surface. In this connection, it should be noted that no cell, except for chondrocytes, exists further than 25–100 μm away from a blood supply. The low oxygen requirement of cartilage may be the reason why only this tissue has been successfully grown in vitro to thick cross sections usually greater than 1 mm using conventional scaffold fabrication techniques.

Skin is a relatively simple 2D tissue and thus thick cross-sections of tissue are not required, thereby explaining the success of producing this tissue with conventional scaffold fabrication techniques [5]. However, most other 3D tissues require a high oxygen and nutrient concentration. The human body supplies its tissues with adequate concentrations of oxygen and nutrients via blood vessels. Tissue engineering scaffolds should embrace this approach and have some form of an artificial vascular system present within them to increase the mass transport of oxygen and nutrients deep within, and removal of waste products from, the scaffold.

21.5.2 Modern Fabrication Methods

Modern techniques require scaffolds that balance temporary mechanical function with mass transport to aid biological delivery and tissue regeneration. Far from being a passive component, scaffold material and porous architecture design (here architecture refers to features 10–1,000 um in size) play a significant role in tissue regeneration by preserving tissue volume, providing temporary mechanical function, and delivering biofactors. A successful scaffold should balance mechanical function with biofactor delivery, providing a sequential transition in which the regenerated tissue assumes function as the scaffold degrades. This balance often presents a trade-off between a denser scaffold providing better function and a more porous scaffold providing better biofactor delivery:

1. *Nano-fibrous scaffolds*: Diameter of these fibers is smaller than those fabricated by conventional method. A variety of synthetic and natural biomaterials, including poly(lactic-co-glycolic acid) (PLGA), poly(L-lactic acid) (PLLA), polycaprolactone (PCL), poly(ethylene oxide) (PEO), poly(vinyl alcohol) (PVA), gelatin, collagen, silk protein, and fibrinogen, have been used to form nano-fibrous scaffolds. For fabrication of nano-fibrous scaffolds, electrospinning, self-assembly, and phase separation are employed:
   (i) *Electrospinning* uses an electric field to draw a polymer solution from an orifice to a collector, producing polymer fibers with diameters in the range of nanometers to micrometers. Electrospinning typically is used to produce thin two-dimensional (2D) sheets. While three-dimensional (3D) nano-fibrous scaffolds have been fabricated by layering these 2D sheets, it is inherently difficult to create 3D scaffolds with well-defined pore architecture and complex geometries, including interconnected macropore networks. These hydrogel scaffolds have shown superior in vivo toxicology and biocompatibility compared with traditional macroscaffolds and animal-derived materials.
   (ii) *Molecular self-assembly* is a useful approach for fabricating supramolecular architectures. Molecular self-assembly is mediated by noncovalent bonds such as hydrogen bonds, van der Waals interactions, electrostatic interactions, and hydrophobic interactions. Self-assembled molecules are ubiquitous in nature. Biomolecules, such as peptides and proteins, interact and self-organize to form well-defined architectures that are associated with functionality. Under certain conditions, nanostructured fibers can be formed by molecular self-assembly. The fiber diameter created by molecular self-assembly usually is much smaller than those produced using electrospinning. Molecular self-assembly has limited ability to form macropores, which are important for cell accommodation and mass transport. The mechanical properties of self-assembled scaffolds also have to be improved before they can be used in tissue engineering applications.
   (iii) *Thermally induced phase separation (TIPS)*: A novel thermally induced phase separation (TIPS) technique was developed recently to fabricate nanofibers to mimic natural collagen fibers. The TIPS process for nanofiber formation typically includes five steps: polymer dissolution, phase separation and gelation, solvent extraction, freezing, and freeze-drying under vacuum. The fiber network formation depends on the solvent of the polymer solution and the gelation temperature. The fibers formed in this manner have diameters ranging from 50 to 500 nm and have a porosity in excess of 98 %. This

nano-fibrous matrix has a much higher surface-to-volume ratio than those of fibrous nonwoven fabrics fabricated with the textile technology or foam fabricated with other techniques. A distinct advantage of the TIPS technique is that it can be combined with other processing techniques (such as particulate leaching or 3D printing) to design complex 3D structures with well-defined pore morphologies.

To engineer functional tissues and organs successfully, the scaffolds have to be designed to facilitate cell distribution and guide tissue regeneration in three dimensions. Macroscopic pores (>100 μm) in a scaffold play an important role in cell seeding distribution, cell migration throughout the 3D space, and neovascularization after implantation of the scaffold in vivo.

To create a nano-fibrous matrix with a macroscopic pore network, a combination of TIPS and particulate leaching can be used. For surface modification of nano-fibrous scaffolds, it is important to consider the interactions of cells with the scaffold materials as the nature of the scaffold surface can directly affect cellular response. Molecular self-assembly is one of the few methods to create biomaterials with properties similar in scale and chemistry to that of the natural in vivo extracellular matrix (ECM).

2. *Biomimetic polymer/apatite composite scaffolds*

   This method creates composite scaffold structures with high porosity, good interconnectivity, and varying degrees of control over pore size and shape. However, all these methods involve a ceramic content that is largely contained within the bulk of the scaffold material, rather than at the surface. Since all interactions with biological components occur at the pore surface, the nonexposed ceramic is in effect wasted. In order to better utilize the ceramic component, polymer/ceramic composite scaffolds have also been fabricated using a biomimetic approach, where a prefabricated polymer scaffold is soaked in simulated body fluid.

3. *Cell sheet technology*

   Cell sheet technology (CST) is based on the use of thermo-responsive polymers, poly(N-isopropylacrylamide) (PIPAAm). The surface of PIPAAms is formulated in such a way as to make its typical thickness <100 nm. PIPAAm-grafted surface can be prepared and its functionalization are important to be able to harvest a functional cell sheet, to be further transplanted. It can perform three-dimensional reconstruction of a tissue in vitro showing its clinical applications. Animal experimentation of the technology is done for cardiomyopathy, visual acuity, periodonty, esophageal ulcerations, and type I diabetes.

4. *Peptide nanostructures*

   Peptide nanostructures containing bioactive signals offer exciting novel therapies of broad potential impact in regenerative medicine. These nanostructures can be designed through self-assembly strategies and supramolecular chemistry and have the potential to combine bioactivity for multiple targets with biocompatibility. It is also possible to multiplex their functions by using them to deliver proteins, nucleic acids, drugs, and cells. Progress is made in the use of self-assembling peptide amphiphiles toward applications in the regeneration of the central nervous system, vasculature and hard tissue along with the transplant of islets, and the controlled release of nitric oxide to prevent neointimal hyperplasia.

   Nanoparticles: Nanoparticles are defined as being on the scale of less than 100 nm in diameter. They possess a high surface-to-volume ratio, which makes them an effective vehicle for the release of biological factors such as growth factors, drugs, and genes. By incorporating PLGA nanospheres, a type of nanoparticle, into nano-fibrous PLLA scaffolds, recombinant human bone morphogenetic protein (rhBMP-7) was delivered to the region of cell culture, and ectopic bone formation was positively affected.

5. *CAD/CAM Technologies*

Since most of the above described approaches are limited when it comes to the control of porosity and pore size, computer-assisted design and manufacturing techniques have been introduced to tissue engineering. Advances in computer-aided technology and its application with biology, engineering, and information science to tissue engineering have evolved a new field of computer-aided tissue engineering (CATE). This emerging field encompasses computer-aided design (CAD), image processing, manufacturing, and solid freeform fabrication (SFF) for modeling, designing, simulation, and manufacturing of biological tissue and organ substitutes. To achieve hierarchical design, libraries of unit cells may be created (not to be confused with a biological cell) at different physical scales that can be assembled to form scaffold architectures. First, a three-dimensional structure is designed using CAD software or using image-based design approach, and then the scaffold is realized by using ink-jet printing of polymer powders or through fused deposition modeling of a polymer melt. These approaches have been used to optimize functional elastic properties with a constraint on porosity or to maximize permeability with a constraint on desired elastic properties and permeability. This technique is used to design microstructures whose permeability is maximized for cell migration and mass transport, but whose effective linear elastic properties match those of natural bone tissue. The purpose of image-based design is to create the scaffold architecture within any arbitrarily complex 3D anatomic defect. This stage draws heavily on commonly used medical imaging modalities, especially computed tomography (CT) and magnetic resonance imaging (MRI), and directly introduces patient medical information into the scaffold fabrication process. Both CT and MRI produce structured voxel datasets where patient anatomy is defined by density distribution. The anatomic defect shape of interest is isolated from the CT or MRI scan.

21.6 Assembly Methods

One of the continuing, persistent problems with tissue engineering is mass transport limitations. Engineered tissues generally lack vasculature, thus making it difficult for any implanted cells to obtain sufficient oxygen and nutrients to survive, and/or function properly. Self-assembly may play an important role here, both from the perspective of encapsulating cells and proteins, as well as creating scaffolds on the right physical scale for engineered tissue constructs.

It might be possible to print organs, or possibly entire organisms. A recent innovative method of construction uses an ink-jet mechanism to print precise layers of cells in a matrix of thermoreversible gel. Endothelial cells, the cells that line blood vessels, have been printed in a set of stacked rings. When incubated, these are fused into a tube [5].

21.6.1 Tissue Culture

In many cases, creation of functional tissues and biological structures in vitro requires extensive culturing to promote survival, growth, and inducement of functionality. In general, the basic requirements of cells must be maintained in culture, which include oxygen, pH, humidity, temperature, nutrients, and osmotic pressure maintenance. Tissue-engineered cultures also present additional problems in maintaining culture conditions. In standard cell culture, diffusion is often the sole means of nutrient and metabolite transport. However, as a culture becomes larger and more complex, such as the case with engineered organs and whole tissues, other mechanisms must be employed to maintain the culture.

Another problem associated with tissue culture is lack of vascular and nervous system supply in cultured tissue and introduction of the proper factors or stimuli required to induce functionality. In many cases, simple maintenance culture is not sufficient. Growth factors, hormones, specific metabolites or nutrients, and chemical and physical stimuli are sometimes required. For example, certain cells respond to changes in oxygen tension as part of their normal development, such as chondrocytes, which must adapt to low oxygen conditions or hypoxia during skeletal development. Others, such as endothelial cells, respond to shear stress from fluid flow, which is encountered in blood vessels.

21.6.2 Bioreactors

In many cases, bioreactors are employed to maintain specific culture conditions. The devices are diverse, with many purposes built for specific applications. By controlling agitation and rates of flow, the transfer of nutrients, gases, metabolites, and regulatory molecules can be maximized.

21.7 Artificial Organs

Future developments in the field aim toward development of artificial organs that use cells embedded into appropriate support structures [8, 13, 33]. A recent report describes the use of polyurethane foam as a good matrix into which liver cells grew as spheroids, the system showing promise as an artificial liver. Vascular grafts are also in line with attention on taking a scaffold that is a structurally intact xenogeneic vessel, such as a pig aorta, removing all cells and repopulating this with human autologous cells. A recent report showed how this could be done in 2–3 weeks, opening up the way for a good alternative to vascular engineering.

Researchers are excited for developing artificial human thyroid capable of producing T cells. Finally, stem cells and their manipulation for therapeutic purposes will continue to be a major area of development, because of the pluripotency of these cells.

Case Study

Early developments in tissue engineering was in mid-1960s where artificial skin for burn victims was a symptomatic therapy later on synthetic fibers as artificial skin grafts were tried. Actual tissue engineering started in the 1980s, with development of skin equivalent consisting of silicone cover over a sponge of porous collagen cross-linked with chondroitin for treating severe burns in 1981. During that period, several products were launched in the market.

(continued)

Interpore's Pro-Osteon coral-derived bone graft was introduced in 1993. Integra's artificial skin was approved for in vivo use in non-biological tissue regeneration product in 1996. Then, in 1998, Organogenesis first manufactured living human organ product Apligraf (Graftskin) which was a human skin equivalent and was approved for treating venous leg ulcers. Life cell developed tissue grafts for transplantation and the preservation of transfusable blood. Advanced tissue Sciences had its clinical trials for Dermagraft which is an engineered human dermal tissue combined with a synthetic epidermal layer. Dermagraft is a three-dimensional human neonatal dermal fibroblast culture that has been grown on a biodegradable scaffold and cryopreserved. It has been applied to foot ulcers that develop as a side effect of long-term diabetes. In clinical trials, significant healing occurred with this material, especially when the Dermagraft cells were alive and functioning properly.

21.8 Achievements

- Tissue engineering skin substitutes have been approved by the US Food and Drug Administration (US FDA). Some tissue-engineered products, like human foreskin-derived neonatal dermal fibroblasts on polylactide-co-glycolide scaffold which is subsequently degraded in aqueous environment, are

Table 21.6 The table lists the success of tissue engineering products in the market along with their trade names and corporates

S. No	Name of the product	Company	Details in brief
1	Alloderm	LifeCell	Acellular dermal matrix
2	Apligraf	Organogenesis	Living keratinocytes and fibroblast cells for venous leg ulcers and diabetic foot ulcers
3	Carticel	Genzyme Biosurgery	Chondrocytes for cartilage repair
4	Epicel	Genzyme Biosurgery	Living skin equivalent for burn patients
5	Durasis	Cook Surgical Products	Xenogeneic pericardium as dural substitutes
6	Surgisis	Cook Surgical Products	Xenogeneic graft for dermal wounds and reinforcement of weakened tissue
7	Orcel	Ortec	Living skin equivalent for burn patients
8	Dermagraft	Smith & Nephew	Living skin equivalent for full-thickness diabetic foot ulcers

approved by the US Food and Drug Administration (FDA).

- During the 1990s, the tissue engineering field has progressed rapidly, and biological substitutes are in development for several tissues in the body. Tissue-engineered products such as bioartificial skins with viable and nonviable cells (Apligraf from Organogenesis and TranCyte from Advanced Tissue Sciences) and autologous cultured chondrocytes (Carticel from Genzyme Tissue Repairs) have reached the market. Tissue engineering products have made tremendous strides recently (Table 21.6).
- Scientists are now engineering cardiovascular tissues such as heart valves and blood vessels.
- Encapsulated pancreatic islets have been implanted in the patients for the treatment of diabetes and liver assist systems containing encapsulated hepatocytes have been used clinically to provide extracorporeal support to the patients with liver failure.
- A kidney support system with encapsulated urothelial cells is in development for the treatment of patients with kidney failure [29] and a bioartificial bladder [39] has been developed as a replacement engineered organ.
- Additionally, investigators have attempted to engineer the nervous system by encapsulation of genetically modified ciliary neurotrophic factor (CNTF) secreting neural cells to treat amyotrophic lateral sclerosis, and nerve guidance channels have been developed for peripheral nerve regeneration and spinal cord repair [46].
- Ophthalmologic engineering efforts are targeted to develop cornea and retina and lens tissue [20, 26, 38].
- Significant progress has been made in orthopedic tissue engineering for the repair of bones, tendons, cartilage, and ligaments [11]. Another polymer graphene and its derivatives (graphene oxide and reduced graphene oxide) have received increasing attention for biomedical applications as they present remarkable properties such as high surface area, high mechanical strength, and ease of functionalization. These biocompatible carbon-based materials can induce and sustain stem cell growth and differentiation into various lin-

eages. Furthermore, graphene has the ability to promote and enhance osteogenic differentiation making it an interesting material for bone regeneration research [18].

- Landis and coworkers have demonstrated the formation of small phalanges and whole joints using bovine chondrocytes and tenocytes, and bovine periosteum on biodegradable polymer matrices [34, 36].
- Scientists have also attempted to generate dental pulp using dental fibroblasts and synthetic matrix [7].
- Novel approaches have been pursued to develop vascularized network inside a tissue-engineered skin for grafting on deep wounds to improve vascularization of the skin graft [5].

21.9 Industry Challenges

The quality control of the materials used in various surgical applications is a key challenge for the tissue engineering industry. For use the materials need to be produced and cultured under good manufacturing practice (GMP) conditions to meet FDA standards—especially the cells that are grown ex vivo. As a result, the tissue engineering industry is striving to create appropriate quality control standards and evaluate them [2], for example, the autologous cultured chondrocytes for knee damage. Recently, Genzyme Tissue Repair reported on a quality control program for this material that was based on evaluating 303 patients who had received the material. The program was based on the analysis of several quantifiable parameters, including viability and freedom from pathogens; results showed that the materials were, indeed, appropriate for their use and demonstrated one way to follow up with quality control monitoring of tissue-engineered products.

Another important aspect is the basic fundamental understanding of tissue differentiation mechanisms which may be utilized for the development of tissue-engineered products. Obstacles as oxygen diffusion limitations due to absence of intrinsic capillary network are a challenge as maximal thickness achievable by present methods is not more than 150–200 μm.

The development of in situ polymerizable polymers that can function as cell delivery systems are attractive. For orthopedic repair, the scaffold made from synthetic and natural polymers have been extensively investigated [10]. The advantages may be (1) ability to generate desired pore structure and (2) matching size, shape, and mechanical properties for variety of applications. However, the problem with scaffold is that it is difficult for them to take shape to fit cavities or defects with complicated geometries. The major challenge which industry is facing is to develop vasculature in surgical alternatives [37]. For example, grafts in dog model contained tissue-engineered blood vessel without any synthetic or exogenous materials and demonstrated good handling and saturability characteristics.

One product that the tissue engineering industry is pursuing is "bone on demand," to be used in cases where new bone formation is needed, having bone morphogenetic protein (BMP) complex as an important component [35].

Immune rejection needs to be addressed probably by developing "universal donor" concept by masking the proteins of MHC. Availability is another issue, where the product should be available at sophisticated hospitals as well as at the places where appropriate facilities for cell culture and cell proliferation are not available.

Presently many companies are involved in the production of natural tissue supplements with regenerative powers.

21.10 The Future

Success has been obtained in tissues as cartilage [41, 45] and skin. Tissue engineering is emerging as an industry with a huge potential market having major impact on future health-care practice [12, 32]. The biomaterials, scaffolds, artificial organs, and differentiating cells that are combined to create a tissue engineering product address significant medical needs, such as major tissue and organ damage or failure.

Tissue engineering has significant market potential driven in part by positive results regarding specific products and processes in clinical settings. Technical advances in the various components of the industry will contribute to market growth like availability of biomaterials that act as scaffolds for tissue repair and reconstruction or for the deposition of engineered tissues and cells preceding implantation. An increasing amount of research is directed for the properties of these scaffolds with the goal of creating materials that have the desired functional profiles for various applications [42–44].

For the biological component of tissue engineering, rapid advances are being made in identifying new cell types for use in tissue regeneration, like undifferentiated stem cells for their capacity to be transformed into almost any cell type that may be needed, and even fat cells can be directed to produce appropriate tissues [6]. Artificial nerve grafts or nerve guidance channels hold tremendous potential for nerve regeneration [1, 46].

Development of effective transportation devices: Development of temperature-controlled shipping units for refrigerated transportation of highly perishable products as tissue-engineered replacement product or stem cells or organ for transplantation. Kidney perfusion capabilities were present in the first organ transportation device. It had organ preservation solutions which can effectively transport postmortem organs (by expanding the ischemia time for effective organ acceptance) and xenogeneic organs (upon approval for human usage). Future devices would be able to maintain temperature for longer periods (current devices can maintain for two days). A refrigerated transportation device is in preclinical trials for a chemically defined solution (Unisol™) for maintaining cell viability and tissue function.

The future focus would be on the development of the product under safety and efficacy standards, including (a) source of cells and tissues, (b) characterization and testing of the materials, (c) quality assurance and control, (d) preclinical and clinical evaluation, (e) safety issues, and (f) transportation device.

21.11 Chapter End Summary

- Injury or disease can cause tissue loss or end-stage organ failure and transplantation of tissues or organs in these patients is severely limited due to low availability of compatible donors.
- Tissue engineering is an interdisciplinary field which has the potential to provide the biological substitutes. These can improve tissue/organ function by restoring or maintaining tissue.
- In this a temporary scaffold is needed to serve as an adhesive substrate for the implanted cells and a physical support to guide the formation of the new organ.
- The biomaterials can be modified to enhance migration and attachment of the specific cell populations for repair or replacement of the damaged tissue and should facilitate cell adhesion, promoting cell growth and allowing the retention of differentiated cell functions.
- The scaffold should be biocompatible, biodegradable, highly porous with a large surface/volume ratio, mechanically strong, and malleable.
- There are a huge number of scaffolds fabricated by conventional and modern tissue engineering approaches.
- These scaffolds need to be biocompatible, with controlled porosity and good mechanical properties.
- Tissue engineering can provide treatments which may need time with superior performance in terms of effectiveness and minimal side effects, without long-term complications.

Multiple Choice Questions

1. Tissue engineering is:
   (a) Supply of artificial tissues and organs
   (b) Repair of partial or whole tissue
   (c) Supply of protein deficient in the body
   (d) All of the above
2. Autografts are obtained from:
   (a) Same individual in which they need to be implanted
   (b) Sibling of individual in which they need to be implanted
   (c) Unrelated individual
   (d) A different species
3. Syngenic cells are obtained from:
   (a) Same individual in which they need to be implanted

(b) Sibling of individual in which they need to be implanted
(c) Unrelated individual
(d) A different species

4. Polylactic acid is an example of:
(a) Natural polymer
(b) Tissue graft
(c) Synthetic polymer
(d) All of the above

5. An ideal scaffold should have:
(a) Mechanical and immunogenic properties
(b) Biodegradability and toxicity
(c) Mechanical stability, biodegradability, and biocompatibility
(d) Non-immunogenicity and nonbiodegradability

6. In solvent casting and particulate leaching:
1. There is compression molding and exposure to high-pressure CO2.
2. Polymer is dissolved in molten phenol with subsequent lowering of temperature.
3. Polymer in organic solvent is casted into mold filled with porogen particle.
4. Polymer is casted and solvent is fully evaporated.
(a) 1 and 3
(b) 2 and 3
(c) 3 and 4
(d) 1, 3, and 4

7. Limitations of conventional tissue engineering scaffolds are:
(a) Precise control of pore size
(b) Toxicity induced by biodegradation
(c) Immunogenicity
(d) All of the above

8. Surgisis is a tissue-engineered product for:
(a) Dermis reconstitution
(b) Leg ulcers
(c) Xenogeneic graft for dermal wounds
(d) Acellular dermal matrix

Answers

1. (b); 2. (a); 3. (a); 4. (a); 5. (c); 6. (c); 7. (d); 8. (c)

Review Questions

Q1. What is tissue engineering?
Q2. What tissues of the human body can be best treated by tissue engineering?
Q3. What should be the properties of scaffolds?
Q4. What are the limitations of conventional method of tissue engineering?
Q5. Give a brief account of tissue-engineered products in the market.

References

1. Aebischer P, Tan SA, Deglon N, Heyd B, Zurn A, Baetge E, Sagot Y, Kato AJ (1995) Encapsulation of neurotrophic factor-releasing cells for the treatment of neurodegenerative diseases. Restor Neurol Neurosci 8:65–66
2. Ahsan T, Nerem RM (2005) The science, the technology, and the industry. Orthod Craniofacial Res 8:134
3. Andriano KP, Tabata Y, Ikada Y, Heller J (1999) In vitro and In vivo comparison of bulk and surface hydrolysis in absorbable polymer scaffolds for tissue engineering. J Biomed Mater Res (Appl Biomater) 48:602–612
4. Awad HA, Butler DL, Boivin GP, Smith FN, Malaviya P, Huibregtse B, Caplan AI (1999) Autologous mesenchymal stem cell-mediated repair of tendon. Tissue Eng 5:267–277
5. Black AF, Berthod F, L'Heureux N, Germain L, Auger FA (1998) In vitro reconstruction of a human capillary-like network in a tissue-engineered skin equivalent. FASEB J 12:1331–1340
6. Blight AR (1999) Presentation. Pittsburgh: Cambridge HealthTech Institute's 2nd annual tissue engineering/ regenerative healing/ stem cell biology meeting. Oct 3–5.
7. Bohi KS, Shon J, Rutherford B, Mooney DJ (1998) Role of synthetic extracellular matrix in development of engineered dental pulp. J Biomater Sci Polym Ed 9:749–764
8. Borkenhagen M, Stoll RC, Neuenschwander P, Suter UW, Aebischer P (1998) In vivo performance of a new biodegradable polyester urethane system used as a nerve guidance channel. Biomaterials 19:2155–2165
9. Bruin P, Smedinga J, Pennings AJ, Jonkman MF (1990) Biodegradable lysine diisocyanate-based poly(glycolide-co-ε-caprolactone)-urethane network in artificial skin. Biomaterials 11:191–295

10. Burg KJL, Porter S, Kellam JF (2000) Biomaterials development for bone tissue engineering. Biomaterials 21:2347–2359
11. Cao Y, Vacanti JP, Ma X, Paige KT, Upton J, Chowanski Z, Schloo B, Langer R, Vacanti CA (1994) Generation of neo-tendon using synthetic polymers seeded with tenocytes. Transplant Proc 26:3390–3392
12. Chapekar MS (2000) Tissue engineering: challenges and opportunities. J Biomed Mater Res 53:617–620
13. Chen SC, Mullon C, Kahaku E, Watanabe F, Middleton Y, Arkadopoulos N, Demetriou AA (1997) Bioartificial organs: science, medicine, and technology. Ann N Y Acad Sci 831:350–360
14. Chu CC (1981) Hydrolytic degradation of polyglycolic acid: tensile strength and crystallinity study. J Appl Polym Sci 26:1727–1734
15. Cutright D, Beasley J, Perez B (1971) Histologic comparison of polylactic acid sutures. Oral Surg 32:165–173
16. Domb AJ (1989) Poly(propylene glycol fumarate) compositions for biomedical applications. United States Patent 4888(413):1–31
17. Duarte ARC, Mano JF, Reis RL (2010) Preparation of Chitosan scaffolds for tissue engineering using supercritical fluid technology. Mater Sci Forum 636:22–25
18. Dubey et al (2015) Graphene: a versatile carbon-based material for bone tissue engineering. Stem Cells Int. doi:10.1155/2015/804213
19. Fabiani JN, Dreyfus GD, Marchand M, Jourdan J, Aupart M, Latremouille C, Chardigny C, Carpenter AF (1995) The autologous tissue cardiac valve: a new paradigm for heart valve replacement. Ann Thorac Surg 60:189–194
20. Germain L, Auger FA, Grandbois E, Guignard R, Giasson M, Boisjoly H, Guerin SL (1999) Reconstructed human cornea produced in vitro by tissue engineering. Pathobiology 67:140–147
21. Gibbons DF (1992) Tissue response to resorbable synthetic polymers. In: Plank H, Dauner M, Renardy M (eds) Degradation phenomena on polymeric biomaterials. Springer Verlag, New York, pp 97–104
22. Gilding DK, Reed AM (1979) Biodegradable polymers for use in surgery-polyglycolic/poly(acetic acid) homo- and copolymers: 1. Polymer 20:1459–1464
23. Gogolewski S, Pennings AJ (1982) Biodegradable materials of polylactides, 4 Porous biomedical materials based on mixtures of polylactides and polyurethanes. Makromol Chem Rapid Commun 3:839–845
24. Goldstein S (1999) Pittsburgh: Cambridge HealthTech Institute's 2nd annual tissue engineering /regenerative healing/stem cell biology meeting. Oct 3–5.
25. Gunatillake PA, Adhikari R (2003) Biodegradable synthetic polymers for tissue engineering. Eur Cells Mater 5:1–16
26. Hadlock T, Singh S, Vacanti JP, McLaughlin BJ (1999) Ocular cell monolayers cultured on biodegradable substrates. Tissue Eng 5:187–196
27. Kasper DL, Braunwald E, Fauci AS, Hauser SL, Longo DL, Jameson JL (eds) (2008) Harrison's principles of internal medicine, 17th edn. Mc Graw-Hill Medical publishers, New York
28. Huang D, Chang TR, Aggarwal A, Lee RC, Ehrlich HP (1993) Mechanisms and dynamics of mechanical strengthening in ligament-equivalent fibroblast-populated collagen matrices. Ann Biomed Eng 21:289–305
29. Humes HD, Buffington DA, Mackay SM, Funke AJ, Weitzel WF (1999) Replacement of renal function in uremic animals with a tissue-engineered kidney. Nat Biotechnol 17:451–455
30. Huynh T, Abraham G, Murray J, Brockbank K, Hagen PO, Sullivan S (1999) Remodeling of an acellular collagen graft into a physiologically responsive neovessel. Nat Biotechnol 17:1083–1086
31. Isogai N, Landis W, Kim TH, Gerstenfeld LC, Upton J, Vacanti J (1999) Formation of phalanges and small joints by tissue-engineering. J Bone Joint Surg 81A:306–316
32. Kronenthal RL (1975) Biodegradable polymers in medicine and surgery. Polymer Sci Technol 8:119–137
33. Lanza R, Chick WL (1997) Bioartificial organs: science, medicine, and technology. Ann N Y Acad Sci 831:323–331
34. Lavik E, Langer R (2004) Tissue engineering: current state and perspectives. Appl Microbiol Biotechnol 65:1
35. Lee et al (2011) Growth factor delivery-based tissue engineering: general approaches and a review of recent developments. J R Soc Interface 8:153–170
36. Matsushima S, Isogai N, Jacquet R, Lowder E, Tokui T, Landis WJ (2011) The nature and role of periosteum in bone and cartilage regeneration. Cells Tissues Organs 194:320–325
37. Niklason LE, Gao J, Abbott WM, Hirschi KK, Houser S, Marini R, Langer R (1999) Functional arteries grown in vitro. Science 284:489–493
38. Nishi K, Mano C, Ichihara M, Honda T (1998) The inhibition of lens epithelial cell migration by a discontinuous capsular bend created by a band-shaped circular loop or a capsule-bending ring. Ophthalmic Surg Lasers 29:119–125
39. Oberpenning F, Meng J, Yoo JJ, Atala A (1999) De novo reconstitution of a functional mammalian urinary bladder by tissue engineering. Nat Biotechnol 17:149–155
40. Peters MC, Isenberg BC, Rowley JA, Mooney DJ (1998) Release from alginate enhances the biological activity of vascular endothelial growth factor. Biomater Sci Polym Ed 9:1267–1278

41. Schreiber RE et al (1999) Repair of osteochondral defects with allogeneic tissue engineered cartilage implants. Clin Orthop 367S:382–395
42. Shea LD, Smiley E, Bonadio J, Mooney DJ (1999) DNA delivery from polymer matrices for tissue engineering. Nat Biotechnol 17:551–554
43. Sheih SJ, Vacanti JP (2005) State-of-the-art tissue engineering: from tissue engineering to organ building. Surgery 137:1
44. Skalak R, Fox CF (1988) Tissue engineering. Liss, New York, pp 26–29, Granlibakken, Lake Tahoe: Proc workshop
45. Stading M, Langer R (1999) Mechanical shear properties of cell-polymer cartilage constructs. Tissue Eng 5:241–250
46. Xu XM, Zhang SX, Li H, Aebischer P, Bunge MB (1999) Regrowth of axons into the distal spinal cord through a Schwann-cell-seeded mini-channel implanted into hemisected adult rat spinal cord. Eur J Neurosci 11:1723–1740

Some Related Resources

Clinical Tissue Engineering Center
http://www.ncbi.nlm.nih.gov/books/NBK6008/ Kelvin GM Brockbank (2000) Tissue Engineering Constructs and Commercialization.
https://www.coursera.org/course/tissue
Tissue Engineering and Regenerative Medicine International Society
www.liebertpub.com/overview/tissue-engineering
www.tissue-engineering.net

Lifestyle, Stress, and Disorders 22

Abstract

Lifestyle-related disorders are the important risk factors, which are widely distributed in the population and greatly influenced by the lifestyle of the urban people. Lifestyle diseases or diseases of longevity or diseases of civilization are diseases that appear to increase in frequency as countries become more industrialized and people live longer. Some maintain a distinction between diseases of longevity and diseases of civilization. Certain diseases, such as diabetes or asthma, appear at greater rates in young populations living in the "Western" way; their increased incidence is not related to age, so disease of longevity cannot accurately be used interchangeably for all diseases.

In many developed as well as developing countries, people's diet has changed dramatically which generally is with increases in consumption of meat, dairy products, vegetable oils, fruit juice, soft drinks, and alcoholic beverages and decreases in consumption of starchy staple foods such as bread, potatoes, rice, and maize flour. Factors in diet, sedentary lifestyle, and the house are thought to influence susceptibility to the many diseases. These determinants of the lifestyle are expressed through intermediate risk factors, such as raised blood glucose, abnormal blood lipids, high blood pressure, and overweight/obesity.

The other aspects of lifestyle have also changed like reduction in physical activity, mostly living indoors, stress, and consumption of alcohol or smoking. Stress is now being linked to cardiovascular diseases, cancer, respiratory infections, and suppressed immune functions. Due to many of these changes in lifestyle, people are getting prone for many diseases like diabetes and cancer. Studies also suggest that diets and/or lifestyle of different populations might partly determine their rates of cancer, and the basis for this hypothesis was strengthened by results of studies showing that people who migrate from one country to another generally acquire the cancer rates of the new host country, suggesting that environmental rather than genetic factors are the key determinants of the international variation in cancer rates.

V. Gupta et al., *Basic and Applied Aspects of Biotechnology*, DOI 10.1007/978-981-10-0875-7_22

22.1 Introduction

Lifestyle-related disorders are the important risk factors, which are widely distributed in the population and greatly influenced by the lifestyle of the urban people. Lifestyle diseases or diseases of longevity or diseases of civilization are diseases that appear to increase in frequency as countries become more industrialized and people live longer [17]. Some maintain a distinction between diseases of longevity and diseases of civilization. Certain diseases, such as diabetes or asthma, appear at greater rates in young populations living in the "Western" way. Their increased incidence is not related to age, so disease of longevity cannot accurately be used interchangeably for all diseases.

In many developed as well as developing countries, people's habits have changed dramatically which generally are with increases in consumption of meat, dairy products, vegetable oils, fruit juice, soft drinks, and alcoholic beverages and decreases in consumption of starchy staple foods such as bread, potatoes, rice, and maize flour [15]. Factors in diet, sedentary lifestyle, altered and disturbed sleeping habits, and the house are thought to influence susceptibility to many diseases [1, 4, 20]. These determinants of the lifestyle are expressed through intermediate risk factors, such as raised blood glucose, abnormal blood lipids, high blood pressure, and overweight/obesity.

The other aspects of lifestyle have also changed like reduction in physical activity, mostly living indoors, stress, and consumption of alcohol or smoking. Stress is now being linked to cardiovascular diseases, cancer, respiratory infections, and suppressed immune functions. Due to many of these changes in lifestyle, people are getting prone for many diseases like diabetes and cancer. Studies also suggest that diets and/or lifestyle of different populations might partly determine their rates of cancer, and the basis for this hypothesis was strengthened by results of studies showing that people who migrate from one country to another generally acquire the cancer rates of the new host country, suggesting that environmental rather than genetic factors are the key determinants of the international variation in cancer rates [7].

In contrast, individuals in rural areas and developing countries usually had diets that are based on one or two starchy staple foods, with low intakes of animal products, fat, and sugar (Fig. 22.1).

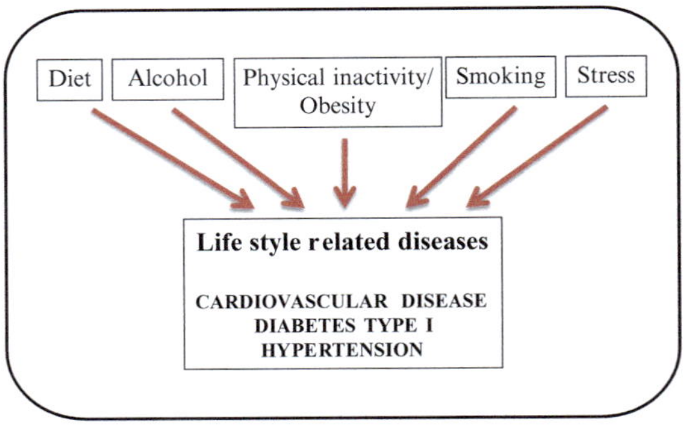

Fig. 22.1 The predominant causes of lifestyle-related diseases. These include diet, alcohol, obesity, smoking, and stress which are leading to cardiovascular diseases, diabetes, and hypertension

Case Studies

The data from Key et al. (2002) authenticates the existence of lifestyle-related diseases. The data shows the international variation in the cancer rate argument that evidences the existence of lifestyle diseases. It was noted in the 1970s that people in many Western countries had diets high in animal products, fat, and sugar and high rates of cancers of the colorectum, breast, prostate, endometrium, and lung. By contrast, individuals in developing countries usually had diets that were based on one or two starchy staple foods, with low intakes of animal products, fat, and sugar and low rates of these cancers.

According to reports of the World Health Organization (WHO) and the World Economic Forum:

(continued)

- India would suffer from an accumulated loss of $236.6 billion by 2015 for unhealthy diet and lifestyle.
- Sixty percent of deaths (35 million) throughout the world in 2005 were due to noncommunicable diseases.
- Major sufferers (80 %) would be in low- and medium-income countries like India, which already is struggling with infectious diseases, malnutrition, and poor maternal and perinatal conditions.

A survey by the Associated Chambers of Commerce and Industry of India (ASSOCHAM) shows that nearly 68 % of women who are working in age range of 21–52 years were afflicted by lifestyle disorders like hypertension, obesity, depression, and diabetes.

Stress due to long working hours in corporates with deadlines also causes anxiety and depression. In India 10 % of the population is hypertensive; 25–30 million are diabetics; 3/1000 are prone for stroke and many die due to heart attacks.

Death statistics shows that in 1900, pneumonia/influenza, tuberculosis, and diarrhea/enteritis were the top three causes of death in the USA. Then, communicable diseases accounted for about 60 % of all deaths. In 1900, lifestyle diseases like heart disease and cancer were ranked #4 and #8, respectively. Since the 1940s, most deaths in the USA have resulted from heart disease, cancer, and other lifestyle diseases. And, by the late 1990s, lifestyle diseases accounted for more than 60 % of all deaths [2, 16, 19].

Drug abuse, especially tobacco smoking and alcohol drinking, as well as the lack of exercise, may increase the risk of certain diseases in later life. Another important factor is the very fast pace of life which is full of hassles, deadlines, frustrations, and demands. For many people, stress is so commonplace that it has become a way of life. Stress is not always bad. In small doses, it can help one to perform under pressure and motivate one to do the best. However, when people are constantly running in emergency mode, the mind and body have to pay the price. When one perceives a threat, the nervous system responds by releasing a flood of stress hormones, including adrenaline and cortisol. These hormones rouse the body for emergency action.

22.2 Stress

Stress is a very broad concept and difficult to define in a concise way. Greek philosopher Hippocrates perhaps was the first to attempt to define the word stress in terms of "balance" which was conceived as an essential state of health and "disharmony" which is manifested as disease when perturbed. Hans Selye in the early twentieth century proposed the general adaptation syndrome which provided the first comprehensive biological theory of stress.

The definition of stress may be the state of threatened balance, equilibrium, or harmony. Those causing threats to homeostasis are called "stressors." Stress can be defined in many different ways; all these definitions, however, share a common component of adaptive physiological responses following challenges to homeostasis. The adaptive reactions to stressors may involve mobilization of a wide variety of physiological responses including the immune response. Stress responses usually include physical perturbations that can encompass either the entire body or specific cellular compartments. "Psychological stress" can be defined as a psychologically perturbing condition occurring in response to adverse external influences capable of affecting mental and physical health. Transportation, fear (that is fright and flight response), overcrowding, and weaning in the form of social reorganization are a few of the important types of psychological stressors identified in the literature.

Short-term stress enhances leukocyte response thereby augmenting innate and adaptive immunity. The set of lifestyle factors like health status, sleep and recovery, nutrition, and psychological issues can influence immune function. These persistent or long-term stressors have been linked to many conditions including immune suppression, disease susceptibility, hypertension, and reproductive dysfunctions [3]. Stress is a major concern because it is ubiquitous, is recurring in nature, and has detrimental effects on health.

It is now established that in approximately 50 % of autoimmunity cases there is a role for "unknown trigger factors." Psychological and physical stress may have a role in the development of autoimmune disease [10, 22]. Many human and animal researches have shown the effect of sundry stressors on immune function [8]. Up to 80 % of patients reported uncommon emotional stress before the onset of the disease in many retrospective studies. Thus, reduction of stress can have a beneficial therapeutic effect in autoimmune disease patients. The health-care providers' and patients' collective efforts to recognize the potential for stress and its impact on autoimmune diseases along with stress management should be considered in a multidimensional treatment approach [10, 22].

Psychological stress has been shown to significantly increase disease susceptibility. Until 20 years ago, researchers investigating the psychological factors contributing to human disease focused primarily on coronary heart disease and cancer and neglected studies on infectious diseases [24]. However, interest in this area started to shift with the reports showing that psychological factors influenced immune function [9]. Furthermore, there was an increasing awareness that stress and other psychological factors played a role in the onset and progression of acquired immunodeficiency syndrome (AIDS).

The studies demonstrated a significant role for psychological stressors in compromising immunity, and thus studies on the effects of stressors in other diseases were initiated [25]. Studies have also linked a variety of psychological stressors with an increased incidence and severity of respiratory infections in humans and animals. Now it is an established fact that psychosocial factors and health-related behaviors were associated with increased susceptibility to colds, which then led to an exacerbation of asthma. It is known that respiratory disease has a huge economic impact in the areas of human health care, animal welfare, and the food industry. A viral challenge study has provided the strongest evidence for a link between stress and susceptibility to the common cold. Social support can often act as a buffer against the effects of stress, as researches showed that social support reduced viral replication rate and increased mucociliary clearance of infection.

Alcohol consumption, personality, and demographic factors were also shown to be important predictors of susceptibility. In contrast, smoking was related to illness severity but not disease susceptibility. These results show that one must consider a range of psychosocial factors, personality traits, demographic factors, and health-related behaviors in studies of individual differences in susceptibility and severity of infections.

22.3 Mechanism of Stress-Induced Infection Susceptibility

While it has been assumed that psychological (psychogenic) and physical (neurogenic) stressors are most closely aligned with depression, the suspicion has arisen that systemic stressors, including immune alterations, may also act in such a provocative capacity. Communication occurs between the immune, endocrine, autonomic, and central nervous systems such that psychological events that affect the central nervous system and then neurochemical processes may affect immune activity. Conversely, immune activation may affect hormonal processes and the activity of neurotransmitters. Thus, by virtue of the neurochemical effects, imparted immune activation may affect behavioral outputs and may even be related to behavioral pathology such as depressive illness.

As compared to lifestyle-related diseases, the communicable diseases have specific causative biologic agents and are cured through the application of specific or a combination of treatment regimens like vaccines and antibiotics. The degenerative diseases are usually caused by a combination of factors that have yet rendered cure or elimination virtually impossible. Individuals are bound to experience any of these diseases to a certain degree after reaching a certain age. The age of onset and rate of progression of these degenerative diseases depend on factors that can be influenced by the socioeconomic status, physical environment, the health-care system, and the person's behaviors and practices. These degenerative diseases share many risk factors that lead to their onset and progress. The major risk factors identified by experts include the lifestyle, which includes the intake of high-salt and high-fat diet, stressful and sedentary lifestyle, and smoking. These factors, together with the interplay of the individual's genetic and physical environment and exposure to critical levels of disease-causing agents and environmental hazards, increase a person's susceptibility to developing degenerative diseases.

Case Studies

In many countries, chronic lifestyle-related diseases as cardiovascular disease, diabetes, and hypertension are on the rise and are causes of millions of deaths each year.

Reports from several places show that the population is on the brink of a major epidemic with disastrous and far-reaching medical, social, and economic consequences [21]: the epidemic of cardiovascular diseases, diabetes mellitus, and hypertension. In China the biggest health problems are strokes, cancer, respiratory diseases, and cardiac diseases.

Various causes for hypertension are high sodium intake, excessive consumption of calories, physical inactivity, excessive consumption of alcohol, low intake of potassium, and excessive smoking. Hypertension subsequently leads to cardiovascular risk. Diabetes mellitus is also prevalent the population and is major cause of blindness in those above 40 years. It is also a major cause of cardiovascular diseases.

(continued)

Thus, Alzheimer's disease, hypertension, heart disease, cancer, arteriosclerosis, chronic liver disease/cirrhosis, diabetes, chronic obstructive pulmonary disease (COPD), nephritis/CRF, and stroke are some common diseases whose incidents are increasing due to the current pattern of life. Apart from diet and sedentary lifestyle, there are many other factors also like all kinds of pollution (sound, dust, fumes, toxic gases, smoke) responsible for augmenting the deteriorating physical conditions. They are also responsible for allergy and respiratory problems. They have insidious onset and may take years to develop. Therefore, it is very important to adopt a lifestyle and habits which go with nature and biological clocks [18]. These diseases in many instances are preventable and the risks can be lowered by taking appropriate precautions.

22.4 Management of Lifestyle-Related Diseases

Once they are acquired, these become lifetime chronic diseases that can be stabilized only through a combination of behavioral, medical, or surgical interventions. The health management paradigm that shifts from directly fighting lifestyle-related diseases to fighting off the risk factors and risk behaviors acquired by the individual has proven to be a difficult challenge to health-care providers. The means and control switches are with the patient, not with the health-care provider [13, 14]. The prerequisite education and training of most health practitioners do not include the development of skills for behavior modification, nor is their school training adequate for the challenging task of counseling patients.

A set of public health measures to address the multiple factors contributing to the development of these diseases had been appropriately identified and is promoted as "healthy lifestyle." The healthy lifestyle approach includes the promotion of proper diet and nutrition, increased physical activity, and smoking prevention and cessation among others. Public health measures should aim to arm every individual with information and motivational tools to promote health, prevent the early onset, and stall the advancement of these diseases. Health sector managers would need more than the seasonal mass media campaigns for people to get rid of old habits and make correct choices for health. The healthy lifestyle campaign should be clear, consistent, and competitive enough to overcome the effects of contradictory information and persuasions from the commercial sector. The most rational and most efficient public health approach to these diseases is prevention.

To prevent and control lifestyle-related diseases, several activities and programs which can reduce the burden of disease has been instituted. The nationwide launch of the enforcement of laws passed on tobacco industry like regulation and republic act on increasing taxation for tobacco and alcohol, respectively, is important. The combined effects of these initiatives, if sustained and pursued by local governments and other stakeholders, are expected to break the alarming trend of life-threatening diseases like heart and vascular diseases, cancer, chronic obstructive pulmonary diseases, and diabetes mellitus.

22.5 Related Diseases

Diseases of the heart, diseases of the vascular system, diabetes mellitus, chronic respiratory diseases, and malignant neoplasms are among the leading causes of death. Cardiovascular diseases (CVDs) can develop at any time throughout the life cycle of individuals. Prevention and control of heart diseases entail application of appropriate measures at different stages of the development of the disease [11, 24]. The first step is the promotion of healthy lifestyle and the application of preventive measures before the illness begins. The second step is protection through early diagnosis and prompt treatment. This is important so that illness may not progress and lead to disability or death. Arteriosclerotic changes in blood vessels may start developing in early adulthood and may progress to hypertension, coronary artery disease, or ischemic heart disease which may result into myocardial infarction or heart attack [11]. The incidences of hypertension in adults above 20 years of age are increasing. Cerebrovascular accident or stroke may happen in uncontrolled hypertension especially among the elderly (Table 22.1).

The development of CVDs is multifarious. Some are acquired and some are inherited. Others are due to environmental causes. Among the most predominant risk factors in the development of CVDs are smoking, physical inactivity, and obesity. Lifestyle modification is necessary to prevent and control the development of CVDs [11, 26]. Prevention and control of heart diseases entail application of appropriate measures at different stages of the development of the disease. Noncompliance to medications and not following the suggested lifestyle modification like low-fat and low-salt diet, mild to moderate exercise, and smoking cessation among persons with CVDs lead to complications and eventual demise from the disease [12, 13].

There are two types of diabetes mellitus (DM): DM type I usually starts early due to lack of insulin, while DM type II usually sets in among the older group secondary to decreased activity of the insulin produced which results in increased sugar level in the blood. The signs and symptoms of DM include polydipsia or frequent intake of water due to thirst, polyphagia or frequent intake of food due to hunger, and polyuria or frequent urination. Other symptoms are blurring of vision and difficult wound healing. Rehabilitation to limit disability and prevent early death is the third step in managing the disease.

Diabetes mellitus can be determined through examination of the blood sugar level and presently there is no known cure for this disease, but the blood sugar level can be controlled by insulin injections. In both types of diabetes the management can start with lifestyle modification.

Table 22.1 Predominant lifestyle diseases, their affects, symptoms, and treatment modalities

Disease	Possible causes	Affects	Symptoms	Treatment
Cardiovascular disease	Inherited and acquired, predominant lifestyle disease caused by smoking	Coronary arteries are at risk for narrowing as cholesterol deposits, called plaques, build up inside the artery, affect the heart muscle or the blood vessels of the heart which can lead to congestive heart failure, and may also cause disturbances in normal heartbeat, called arrhythmia; clot in artery due to rupture of plaque can lead to myocardial infarction	Chest pain, shortness of breath, sweating, nausea, and indigestion	Medical intervention with multistep approach, lifestyle modification, regular exercise, cholesterol reduction, quitting smoking and alcohol consumption, and stress-free life
	High blood pressure (hypertension)			
	High cholesterol			
	Diabetes			
	Family history			
	Peripheral artery disease			
	Obesity			
Hypertension	High cholesterol levels, atherosclerosis leading to high blood pressure	Due to high pressure, the heart has to work harder to pump, leading to organ damage and several illnesses such as heart attack, stroke, heart failure, aneurysm, or renal failure	A blood pressure of 140/90 or above is hypertension	Medical treatment includes several drugs and lifestyle modification by losing weight, quitting smoking, eating a healthful diet, reducing sodium intake, exercising regularly, and limiting alcohol consumption
			Extreme conditions may cause severe headaches	
			Fatigue or confusion	
			Dizziness	
			Nausea	
			Problems with vision	
			Chest pains	
			Breathing problems	
			Irregular heartbeat	
	Predominant lifestyle disease caused by smoking, physical inactivity, obesity, stress, diabetes, high-salt intake, alcohol consumption, and aging		Blood in the urine	

(continued)

Table 22.1 (continued)

Disease	Possible causes	Affects	Symptoms	Treatment
Diabetes	Low insulin secretion by the pancreas resulting in hyperglycemia of blood	Diabetes mellitus is of four kinds	Increased fatigue, polydipsia, polyuria, polyphagia, weight fluctuation	Lifestyle modification with diet modification, reducing obesity, oral hypoglycemic therapy, and insulin treatment
		Type I diabetes or insulin-dependent diabetes mellitus (IDDM), juvenile-onset diabetes, brittle diabetes, or ketosis-prone diabetes (autoimmune)		
		Type II diabetes is also called non-insulin-dependent diabetes mellitus (NIDDM), adult-onset diabetes, ketosis-resistant diabetes, or stable diabetes. Type II often develops in overweight adults		
		Type III is gestational diabetes and occurs in some women during pregnancy		
		Type IV includes other types of diabetes which are linked to disease of the pancreas, hormonal changes, side effects of drugs, or genetic defects	Blurry vision, irritability, infections, and poor wound healing	

(continued)

Table 22.1 (continued)

Disease	Possible causes	Affects	Symptoms	Treatment
Arteriosclerosis	Arterial wall becomes thickened and loses elasticity. Atherosclerosis is the most common and serious vascular disease	Plaques (atheromas) deposited in the walls of arteries are major causes of heart disease, chest pain (angina pectoris), heart attacks, and other disorders of the circulation. In atherosclerosis yellowish plaques of cholesterol, fats, and other remains are deposited in the walls of large- and medium-sized arteries. Atherosclerosis usually occurs with aging. It is linked to overweight, high blood pressure, and diabetes	As the plaque builds up, it can result in serious health problems as heart attack and angina. Symptoms will depend on which part of the body is not receiving enough blood and oxygen due to the narrowing of arteries resulting in symptoms of ischemia like paresthesias, tingling, and numbness	Lifestyle modification by losing weight, quitting smoking, eating a healthful diet, reducing sodium intake, exercising regularly, limiting alcohol consumption, and medical and surgical intervention
Chronic obstructive pulmonary disease (COPD)	A disease characterized by slowly progressing, irreversible airway obstruction	Breathing problems while exercising, difficulty in breathing in or out deeply, and sometimes a long-term cough. The condition may result from chronic bronchitis, emphysema, asthma, or chronic bronchiolitis. Cigarette smoking and air pollution make it worse	Chronic obstruction of the flow of air through the airways and out of the lungs, chronic bronchitis, chronic asthma, and emphysema	COPD has no treatment, but there are lifestyle modifications as quitting smoking, some medicines, vaccines, pulmonary rehabilitation, oxygen therapy, and surgery
Cancer	Diseases characterized by uncontrolled, abnormal growth of cells	Cancer is definitely considered the number one disease of civilization There are more than 150 different kinds of cancer and many different causes	Cachexia (weight loss), fatigue, and fever, depending upon the type of cancer	Medical (chemo-, radio- and immunotherapy) and surgical, healthy lifestyle habits

(continued)

Table 22.1 (continued)

Disease	Possible causes	Affects	Symptoms	Treatment
Alzheimer's disease	Brain disease; breakdown of cells of the brain, age, family history, and genetics play o role in the onset of disease	There is no treatment, but good nutrition may slow the progress of this lifestyle disease, which lasts about 7 years in most people who have it	Memory problems but no single definitive medical test to diagnose the disease, significant memory problems, and thinking deficits	Cholinesterase inhibitors
				Namenda (protects nerve cells from excessive amount of glutamine)
				Treatment for anxiety, depression, and psychosis
				Prescription assistance programs, to avoid the disease; lifestyle modification is required, maintaining a healthy heart and normal blood pressure
				Heart disease, stroke, diabetes Watching the weight, avoiding tobacco and excess alcohol, staying socially connected, and exercising both your body and mind

Patients may develop end-organ damages secondary to the elevated blood sugar level such as nephropathy that may lead to end-stage renal disease, neuropathy that leads to decrease sensation in the limbs, and retinopathy that may lead to blindness. Other complications include hypertension and heart disease. Patients may go into diabetic coma or die because of multi-organ damage in the absence of treatment.

Chronic respiratory diseases (CRDs) include asthma and the chronic obstructive pulmonary diseases (COPDs) such as chronic bronchitis and emphysema. COPDs are adverse results of one or a combination of the following: infections, genetic susceptibility, occupational hazard, or environmental pollutants like smoke. Asthma is due to the person's sensitivity to specific external allergens or from internal non-allergenic factors. Most extrinsic asthma begins in children and starts with signs of atopy such as eczema and allergic rhinitis. Most cases of asthma, especially in adults, are preceded by severe respiratory infections [18]. Allergens that may precipitate asthma attacks include pollen, animal dander, house dust, molds, feather, and cotton tree or kapok.

22.6 Chapter End Summary

- In general the lifestyle of the people is rapidly changing. In this, odd hours of waking up and sleeping cause less sleep or sleeplessness. In addition to it, no exercise, no exposure to sunlight, unhealthy eating habits, eating junk food, many commitments, demands, and stress are added risk factors.
- The result is overweight and more susceptibility to develop lifestyle-related diseases like hypertension, diabetes, arteriosclerosis, heart attack, pulmonary embolism, etc.
- Individuals who have preexisting hypertension or heart diseases are required to give more attention on their lifestyle. Regular medical checkups should be done by the specialist to monitor the changes in blood pressure medication, sugar medication, cholesterol medication, and other factors.
- Diseases of the heart, diseases of the vascular system, diabetes mellitus, chronic respiratory diseases, and malignant neoplasms are among the leading causes of death. These diseases can be inherited, but unfortunately incidences

of these diseases are increasing at a very fast pace which impose a major socioeconomic burden. Prevention of these requires modifications in the lifestyle.

- Prevention and control of these requires application of appropriate measures at different stages of the development of the disease by promotion of healthy lifestyle before the symptoms appear and measures to diagnose the disease early for early medical intervention. This is important so that illness may not progress and lead to disability or death.

Multiple Choice Questions

1. The lifestyle changes in urban population, which are causes of the diseases, are:
   (a) Regular exercise and eating junk food
   (b) Less sleep, eating junk food, and stress
   (c) Stress and regular sleep
   (d) All of the above
2. The most prevalent disease due to lifestyle modification is:
   (a) Cardiovascular disease
   (b) Influenza
   (c) Tetanus
   (d) Smallpox
3. In arteriosclerosis:
   (a) Arteries carry more blood.
   (b) Arteries burst.
   (c) Arteries develop less pressure.
   (d) Cholesterol is deposited and plaques develop in the artery.
4. An example of lifestyle-related disease is:
   (a) Hypertension
   (b) Diabetes
   (c) Atherosclerosis
   (d) All of the above
5. Which of the following comments regarding stress is true?
   (a) Stress is always harmful.
   (b) In small doses it improves performance.
   (c) Immune system performance is improved.
   (d) All of the above.
6. The reason why a person who is only on carbohydrate-rich diet without fat but still becomes obese is because:
   (a) Carbohydrate generates much energy.
   (b) Fat is more in the body by any means.
   (c) Carbohydrates are one of the precursors of fat biosynthesis.
   (d) None of the above.
7. A person on a glucose-rich diet with minimal physical activity may likely develop:
   (a) Cardiovascular disease
   (b) Diabetes
   (c) Obesity
   (d) Hypertension
8. A diabetic patient who is on starvation is likely to develop:
   (a) Kidney problem
   (b) Liver problem
   (c) Heart disease
   (d) Obesity
9. Excessive cholesterol-rich diet results in its deposition in the blood vessel leading to:
   (a) Diabetes
   (b) Obesity
   (c) Nephropathy
   (d) Arteriosclerosis
10. A person with obesity and cardiovascular disease forms a large amount of ketone bodies because of:
    (a) Improper digestion of carbohydrates
    (b) Improper digestion of proteins
    (c) Improper digestion of lipids
    (d) None of the above

Answers

1. (b); 2. (a); 3. (d); 4. (d); 5. (b); 6. (c); 7. (b); 8. (a); 9. (d); 10. (c)

Review Questions

Q1. Why the incidences of lifestyle-related disease are increasing?

Q2. Why cardiovascular disease is coming up as most prevalent lifestyle disease?
Q3. What is stress?
Q4. What is the contribution of stress in initiation of diseases?

References

1. Burkitt DP (1982) Western diseases and their emergence related to diet. S Afr Med J 61:26
2. Death statistics data from National Center for Health Statistics, National Office of Vital Statistics, 1947 for the year 1900 (page 67), for the year 1938 (page 55)
3. Dhabhar FS (2013) Psychological stress and immunoprotection versus immunopathology in the skin. Clin Dermatol 31:18–30
4. Forrester T, Cooper RS, Weatherall D (1998) Emergence of Western diseases in the tropical world: the experience with chronic cardiovascular diseases. Br Med Bull 54:463–473
5. Galobardes B, Costanza MC, Bernstein MS, Delhumeau C, Morabia A (2003) Trends in risk factors for lifestyle-related diseases by socioeconomic position in Geneva, Switzerland, 1993–2000: health inequalities persist. Am J Public Health 93:1302–1309
6. Key TJ, Allen NE, Spencer EA (2002) The effect of diet on risk of cancer. Lancet 14:861–868
7. Lichtenstein P, Holm NV, Verkasalo PK (2000) Environmental and heritable factors in the causation of cancer. N Engl J Med 343:78–85
8. Marshall GD (2011) The adverse effects of psychological stress on immunoregulatory balance: applications to human inflammatory diseases. Immunol Allergy Clin N Am 31:133–140
9. Martínez-Carrillo BE et al (2011) Repeated restraint stress reduces the number of IgA-producing cells in Peyer's patches. Neuroimmunomodulation 18:131–141
10. McCray CJ, Agarwal SK (2011) Stress and autoimmunity. Immunol Allergy Clin N Am 31:1–18
11. McGovern PG, Pankow JS, Shahar E et al (1996) Recent trends in acute coronary heart disease—mortality, morbidity, medical care, and risk factors. The Minnesota Heart Survey Investigators. N Engl J Med 334:884–890
12. Meng L, Maskarinec G, Lee J (1999) Lifestyle factors and chronic diseases: application of a composite risk index. Prev Med 29:296–304
13. Morabia A (1996) From disease surveillance to the surveillance of risk factors. Am J Public Health 86:625–627
14. Murray CJ, Lopez AD (1997) Global mortality, disability, and the contribution of risk factors: global burden of disease study. Lancet 349:1436–1442
15. Painter NS (1982) Diverticular disease of the colon. The first of the Western diseases shown to be due to a deficiency of dietary fibre. S Afr Med J 61:1016–1020
16. Pappas G, Queen S, Hadden W, Fisher G (1993) The increasing disparity in mortality between socioeconomic groups in the United States, 1960 and 1986. N Engl J Med 29:103–109
17. Roelcke V (1995) Between individual therapy and political intervention: campaigns against "diseases of civilization" between 1920 and 1960. Gesundheitswesen 57:443–451
18. Sasanabe R, Watanabe Y, Shiomi T (2012) A sleep disordered breathing as the lifestyle-related diseases. Nihon Rinsho 70:1211–1215
19. Sharma M, Majumdar PK (2009) Occupational lifestyle diseases: an emerging issue. Indian J Occup Environ Med 13:109–112
20. Sobra J, Ceska R (1992) Diseases of civilization from the aspect of evolution of the human diet. Cas Lek Cesk 131:193–197
21. Steyn K, Fourie J, Bradshaw D (1992) The impact of chronic diseases of lifestyle and their major risk factors on mortality in South Africa. S Afr Med J 82:227–231
22. Stojanovich L, Marisavljevich D (2008) Stress as a trigger of autoimmune disease. Autoimmun Rev 7:209–213
23. Thacker SB, Stroup DF, Rothenberg RB (1995) Public health surveillance for chronic conditions: a scientific basis for decisions. Stat Med 14:629–641
24. Vartiainen E, Puska P, Pekkanen J, Tuomilehto J, Jousilahti P (1994) Changes in risk factors explain changes in mortality from ischaemic heart disease in Finland. BMJ 309:23–27
25. Walsh NP, Gleeson M, Pyne DB, Nieman DC, Dhabhar FS, Shephard RJ, Oliver SJ, Bermon S, Kajeniene A (2011) Position statement. Part two: maintaining immune health. Exerc Immunol Rev 17:64–103
26. Zöllner N (1990) The relevance of diet for civilization diseases, especially atherosclerosis. Wien Med Wochenschr Suppl 106(suppl):11–12

Some Related Resources

CDC Scientific Data, Surveillance, Health Statistics, and Laboratory Information
health.india.com › Diseases and Conditions › Diabetes
Morbidity and mortality weekly report
www.bips.uni-bremen.de/en/the…and…/lifestyle-related-disorders.html
www.ncbi.nlm.nih.gov/pubmed

23 Intellectual Property Rights

Abstract

Intellectual property is the creation of the human mind. It is intangible as it lacks the physical component and it cannot be touched. Although it is intangible, it can be bought, sold, or rented. It may be the idea or invention, which might be the outcome of many years of hard work. Other tangible properties like pieces of land and vehicles are given protection by their registration and, thus, when stolen, can be recovered. However, if the intellectual property without any protection is leaked or disclosed, the inventor is at lost as he/she cannot claim that it was his/her original creation. The rights given for the creations of the minds (such as any inventions, artistic works, literary works, designs, names, images, and symbols) are intellectual property rights (IPR). The rights are given to the creator for exclusive use of his creation for a defined period. The various forms of law or modes in which these rights are exercised are patent, copyright, trademark, and trade secret. These let people get financial benefit from their creation, and thus it leads to the use of more innovative ideas by exercising their creativity and innovation.

In India, the intellectual property rights are well established at statutory, judicial, and administrative levels. The agreement establishing the World Trade Organization has been ratified by India, which has an Agreement on Trade-Related Aspects of Intellectual Property Rights (TRIPS) (came into force on 1 January 1995). TRIPS's standard agreements ensure protection and enforcement of intellectual property rights in all the member countries for promoting international trade. Thus, in TRIPS agreement, the member countries have obligations related to provision of the minimum standard of protection in their legal systems and practices.

V. Gupta et al., *Basic and Applied Aspects of Biotechnology*, DOI 10.1007/978-981-10-0875-7_23

23.1 Introduction

Intellectual property is the creation of the human mind. It is intangible as it lacks the physical component and it cannot be touched. Although it is intangible, it can be bought, sold, or rented. It may be the idea or invention, which might be the outcome of many years of hard work. Other tangible properties like pieces of land and vehicles are given protection by their registration and, thus, when stolen, can be recovered. However, if the intellectual property without any protection is leaked or disclosed, the inventor is at lost as he/she cannot claim that it was his/her original creation. The rights given for the creations of the minds (such as any inventions, artistic works, literary works, designs, names, images, and symbols) are intellectual property rights (IPR). The rights are given to creator for exclusive use of his creation for a defined period. The various forms of law or modes in which these rights are exercised are patent, copyright, trademark, and trade secret. These let people get financial benefit from their creation, and thus it leads to the use of more innovative ideas by exercising their creativity and innovation.

In India, the intellectual property rights are well established at statutory, judicial, and administrative levels. The agreement establishing the World Trade Organization has been ratified by India, which has an Agreement on Trade-Related Aspects of Intellectual Property Rights (TRIPS) (came into force on 1 January 1995). TRIPS's standard agreements ensure protection and enforcement of intellectual property rights in all the member countries for promoting international trade. Thus, in TRIPS agreement, the member countries have obligations related to provision of the minimum standard of protection in their legal systems and practices.

Thus, intellectual property law works for:

- Safeguarding creators/producers of intellectual property
- Granting the owner rights for a certain time limit to control the usage of that product
- Giving protection to literary, artistic, and scientific works
- Protecting performances of performing artists, phonograms, and broadcasts
- Protecting all inventions as scientific discoveries and industrial designs
- Protecting trademarks, service marks, commercial names, and designations
- Once the protection is granted by the government office, it is the responsibility of the owner to check the misuse of his intellectual property.

IPR is divided into "copyright" and "industrial property." The fields of literary, artistic, and scientific works would get protection under copyright and "related rights." The rights related to copyright would include the performances of performing artists, phonograms, and broadcasts.

Industrial property includes the inventions, industrial designs, trademarks, service marks, and commercial names and designations.

23.2 Trade-Related International Agreements

23.2.1 General Agreement on Tariffs and Trade (GATT)

The General Agreement on Tariffs and Trade (GATT) was concluded by 23 countries at Geneva in 1947 (came to effect on 1 January 1948, it was an interim arrangement awaiting the formation of United Nations agency to supersede it) aimed at trading between member nations without any discrimination where each member nation opens its market equally to every other. Thus, it included a set of multilateral trade agreements for the abolition of quotas and the reduction of tariff duties among the contracting nations. Its most favored unconditional clause supported that if once a country and its largest trading partner had agreed to reduce the tariff, that reduced tariff would be automatically extended to every other GATT member. The agreements of GATT (1) were the most effective instrument of world trade liberalization and (2) played a major role in the massive expansion of world trade. GATT negotiated specific trade problems affecting particular commodities or trading nations. Major multilateral trade conferences were held periodically to work out tariff reductions (seven rounds were held from 1947 to 1993). It started at Geneva in 1947 (establishing GATT); at Annecy, France, in 1949; at Torquay, England, in 1951; and at Geneva in 1956 and again in 1960–1962, Kennedy Round (1964–1967), the Tokyo Round (1973–1979), and the Uruguay Round (1986–1994), all three held at Geneva.

During the Uruguay Round, the important agreements were as follows:

- Trade liberalization agreements were negotiated.
- Worldwide trade treaty slashed tariffs on industrial goods by an average of 40 %.
- Agricultural subsidies were reduced.
- Groundbreaking new agreements were made on trade in services.
- The treaty created a new and stronger global organization, the World Trade Organization (WTO), to monitor and regulate international trade, thus replacing GATT; however, the WTO adopted its principles and the many trade agreements reached under its auspices [14].

When GATT was being replaced by the World Trade Organization (WTO) in 1995, 125 nations were signatories to its agreements, which had become a code of conduct governing 90 % of world trade.

23.2.2 World Trade Organization (WTO)

The WTO is the only international organization which deals with the global rules of trade between nations. Its main function is to ensure that trade flows as smoothly, predictably, and freely as possible. All the decisions in the WTO are taken by consensus among all member countries and they are ratified by members' parliaments. As trade barriers were lowered, the WTO's system also breaks down other barriers between peoples and nations. Thus, WTO liberalized trade and settled trade disputes with a set of rules governing the trade.

Its main tasks include:

- Administering trade agreements
- Acting as a forum for trade negotiations
- Settling trade disputes
- Reviewing national trade policies
- Assisting developing countries in trade policy issues, through technical assistance and training programs
- Cooperating with other international organizations

The WTO has about 160 members, accounting for about 95 % of world trade. Around 25 others have negotiating membership.

23.2.3 World Intellectual Property Organization (WIPO)

On 14 July 1967, the World Intellectual Property Organization (WIPO) was established in a convention in Stockholm, and it came into power in 1970, was amended in 1979, and included all the rights in intellectual property. In 1974, WIPO (an intergovernmental organization) became one of the specialized agencies of the United Nations system of organizations [15].

Thus, WIPO was established as the global forum for intellectual property services, policy, information, and cooperation to lead the development of a balanced and effective international intellectual property (IP) system that enables innovation and creativity for the benefit of all. It has 188 member states and is a self-funding agency of the United Nations. Thus, it is working to protect the rights of creators and owners of intellectual property (IP) and to make sure that inventors and authors are recognized and rewarded for their original and inventive work [10].

Thus, their main motto is:

> Delivering global services for protecting IP by IP laws; shaping international IP rules according to the changing world; give access to the world's IP information; help-use IP for development so that all members can enjoy the benefits of an effective

and affordable system of IP; (1) the progressive development of international intellectual property law; (2) assistance to developing countries to build intellectual property capacity at national and regional levels and encourage more effective use of IP as tool for economic development; and (3) services to industry and the private sector to facilitate the process of obtaining intellectual property protection in multiple countries to promote wealth creation and economic development.

These all are responsible for safeguards and incentives to further flourish human creativity, extending the boundaries of science and technology and enriching the world of literature and the arts.

The name of the organization, before it became WIPO, was BIRPI, the acronym of the French-language version of the name: United International Bureaux for the Protection of Intellectual Property (in English). In 1960, BIRPI moved from Berne to Geneva [10–13].

WIPO allowed the use of valid international treaties in all state parties that create the facility of a single procedure to apply for patents and register trademarks and industrial designs; the Patent Cooperation Treaty (PCT), the Madrid Agreement and Protocol Concerning the International Registration of Marks, and the Hague Agreement Concerning the International Deposit of Industrial Designs have all given rise to an increased volume of registration activities.

To support WIPO's work, an international treaty came into existence in June 2000, namely, the Patent Law Treaty, whose purpose was to streamline application procedures and to reduce the cost of obtaining simultaneous patent protection in several countries.

23.2.4 Agreement on Trade-Related Aspects of Intellectual Property Rights (TRIPS)

TRIPS is an international agreement administered by the World Trade Organization (WTO) which was negotiated at the Uruguay Round of the General Agreement on Tariffs and Trade (GATT) in 1994. TRIPS sets minimum standards for many forms of intellectual property (IP) regulations as applied to nationals of other WTO members. TRIPS is the trade agreement in relation with the production of and access to drugs, particularly in developing countries; copyright rights (including related rights of performers, producers of sound recordings, and broadcasting organizations); geographical indications, including appellations of origin; industrial designs; integrated circuit layout designs; patents; monopolies for the developers of new plant varieties; trademarks; trade dress; and undisclosed or confidential information.

TRIPS also specifies enforcement procedures, remedies, and dispute resolution procedures. The TRIPS agreement introduced intellectual property law into the international trading system for the first time and remains the most comprehensive international agreement on intellectual property to date. In 2001, developing countries were concerned that developed countries were insisting on an overly narrow reading of TRIPS, which initiated a round of talks that resulted in the Doha Declaration: a WTO statement that clarifies the scope of TRIPS, stating, for example, that TRIPS can and should be interpreted in light of the goal "to promote access to medicines for all."

23.2.5 Paris Convention for Protection of Industrial Property

The Paris Convention applies to industrial property including patents, trademarks, industrial designs, utility models, service marks, trade names, and geographical indications.

Its main points applicable on all industrial properties are:

- National treatment: Each contracting state should grant the same protection to nationals of other contracting states as it grants to its own nationals.
- Right of priority: If an applicant has filed the first application for grant of protection in one contracting state, the applicant within a time limit (12 months for patents and utility models and 6 months for industrial design) can apply for protection in any other contracting state.
- Independence of patents: Patents granted in different contracting states for the same inven-

tion would be independent of each other so the different states may or may not grant patent; however, if the applicant has been granted patent in one contracting state and declined by another, the patent which is granted cannot be refused or terminated.

- Measures must be taken for direct or indirect use of false indication and effective protective measures should be taken against unfair competition.
- The Paris Convention was concluded in 1883; was revised at Brussels in 1900, at Washington in 1911, at the Hague in 1925, at London in 1934, at Lisbon in 1958, and at Stockholm in 1967; and was amended in 1979.

23.2.6 Patent Cooperation Treaty (PCT)

The World Intellectual Property Organization (WIPO) situated in Geneva governs all the activities. It is a multilateral treaty, which came into force in 1978. India was part of it in 1998.

Through PCT, (1) an inventor or patentee of one of the member countries can obtain priority for his invention in all of the member countries, (2) separate filing of applications in each member country is not required, and (3) an inventor can save cost toward fees and attorney charges. The time granted for filing patent application under the PCT is from 20 to 31 months, and search report prepared is beneficial to comment about the novelty of the invention.

These applications are referred to the specially appointed "International Searching Authority" (ISA) under the PCT and result in international search reports.

23.2.7 Budapest Treaty

The treaty was signed in 1973 in Budapest and was later amended in 1980. India joined it in 2001. This is regarding microorganisms. When patents are sought on microorganisms and then any contracting state allows or requires the deposit of microorganisms, in that condition the state should recognize officially approved "international depositary authority" or approved "culture collection" for the deposit of microorganism. The deposit authority can be inside or outside of the contracting state.

This is advantageous for the patentee as, firstly, it saves cost and attorney fees for applying separately in each state; secondly, the deposition of culture is not required in each member state where patent is sought; and, thirdly, the patentee can obtain his culture from the culture collection depositories after that. In India the "Institute of Microbial Technology" (IMTECH), Chandigarh, and "National Centre for Cell Science" (NCCS), Pune, are the international depositary authorities.

23.2.8 Patent Law Treaty (PLT)

It came into effect in 2000 to further simplify the procedures in the national and regional patent applications by making them more generous and user-friendly with exception of filing date requirements and implementation of provision for electronic filing (both hardcopies and softcopies are maintained).

23.2.9 Mailbox Provision

Mailbox provision is used in accepting product patent applications. TRIPS requires that the countries which are not providing product patent (for pharmaceuticals) should adopt some mechanism where they can accept product patent applications with effect from 1 January 1995. These applications can be later on reviewed, once suitable amendments are created in national patent law.

23.2.10 Exclusive Marketing Rights (EMR)

TRIPS recommends that some member countries of WTO, which are not having provisions for granting product patent (pharmaceuticals), should introduce EMR for such products on conditions that (1) patent application has already being filed in any of the member countries after 1995, (2) patent has been granted in any of the member countries, and (3) approval for marketing has been obtained in any of the member countries. EMR is valid for a period of 5 years or until patent laws are exercised in that country.

23.2.11 Union for the Protection of New Varieties of Plants (UPOV)

The Union for the Protection of New Varieties of Plants or Union Pour la Protection des Obtentions Végétales is an international convention with headquarters in Geneva (Switzerland) that is responsible for examination of plant varieties for its merits and justifications for granting protection under UPOV or not. These criteria are distinctness, uniformity, stability, novelty, and appropriate denomination. The convention protects the rights of the breeder or manufacturer to prohibit others from using and resowing the seeds without their consent and royalty payment. India is not a member country.

- The plant for which protection is sought must be distinct from its existing relatives.
- It must be novel, that is, it must not be used or marketed in the country where protection is being applied.
- It should be stable, that is, its traits should remain true to type after repeated rounds of propagation.

23.3 Types of Intellectual Property and Rights

In a broad sense, intellectual property is divided into two streams: (1) industrial property and (2) copyright.

23.3.1 Industrial Property

The Paris Convention for the Protection of Industrial Property has set out the industrial applications as it applies to the commerce and industry, agricultural and extractive industries, and manufactured or natural products, for example, medicines, grain, wines, fruit, tobacco leaf, cattle, mineral waters, beer, flowers, and so on. The important rights granted for these industrial properties include patents (for protection of inventions) and industrial designs (for protection of ornamentation or aesthetic creation of industrial product). Under industrial property, rights are also granted for trademarks, trade names, layout designs of integrated circuits, and geographical indications and protection against unfair competition. The reason for granting the protection is to encourage innovation, with continuous enhancement of quality [10–14].

23.3.1.1 Patents

A patent is an IPR document which is issued by the government office (or regional office for several countries), upon application of the patentee (original creator or inventor who files the application) for his invention in the form of a legal situation in which he gets sole authorization for usage of his invention. Thus, it grants protection in the form of exclusive rights for a particular period (generally 20 years) to the patentee to manufacture, use, sell, and import his invention. Though the patents are granted by the government offices, they do not enforce them. Their inappropriate use has to be checked by patentee only.

- Although patents are a sort of "monopolies," a patent does not allow the inventor to make, use, or sell anything.
- The patent gives rights that persons other than the owner of the patent cannot exploit a patented invention in the country unless the owner agrees to such exploitation.

- Thus, the owner is given a statutory right to prevent others from using or commercially exploiting (making, using, or selling) his invention.
- The patentee has the right to start legal proceedings against any person exploiting the patented invention in the country without his agreement.
- Thus the patent owner has right to derive the material benefits to which he is entitled as a reward for his intellectual effort and work.

Criteria for Patenting

Utility or industrial applicability: Invention stands for a solution to specific problem in the area of technology. It may relate directly to the product or the process to produce that product. It must be useful, that is, can be practically used, not be purely theoretical.

Novelty: The invention must be new (novel). However, novelty cannot be proved but its absence can be proved.

Inventiveness or non-obvious: It must exhibit a sufficient "inventive step" (be non-obvious); it is to ensure that the protection is not being granted to what is already known either as part of prior information or anything which the person of ordinary skill could deduce.

Disclosure of the invention: The invention needs to be sufficiently disclosed in the application such that the invention may be carried out by a person skilled in the art.

Utility Model or Short-Term Patent

In some countries, the inventions are protectable in the name of "utility model" or "innovation patents or utility innovations" or "short-term patent." More than 30 countries have utility models in the laws, some countries as Australia and Malaysia use innovation patents or utility innovations, and some other countries, as Hong Kong, Slovenia, and Ireland, use short-term patent. The requirements and procedure for obtaining protection as compared to patent are less strict and simple in respect of inventive step, the fees are lower, and the duration of protection is shorter, but otherwise the rights under the utility model or short-term patent are similar. These are usually sought when the invention is technically not very complex or for inventions with short commercial life but the invention has to be novel, and this requirement must always be met. The projected duration of protection is usually between 7 and 10 years.

Process and Product Patent

There is obvious difference in the inventions where protection may be granted on the products and inventions, which require protection on the processes. Thus, the creation of a novel or new alloy is a product invention, whereas the invention of a novel pathway or protocol or process of creating the existing or new alloy is a process invention. Likewise, the patents are also referred as a product patent and a process patent, respectively.

Non-applicability of Material for Patents

Thus, patents cannot be applied for materials already existing in nature; scientific theories or mathematical methods; plants and animals other than microorganisms and essentially biological processes for the production of plants and animals, other than non-biological and microbiological processes; schemes, rules, or methods, such as those for doing business, performing purely mental acts, or playing games; methods of treatment for humans or animals; or diagnostic methods practiced on humans or animals (but not products for use in such methods).

Patent Infringement

A patentee acquires the right to decide who shall or who shall not exploit his patented invention or in other words the patentee obtains right to prevent others from commercially exploiting his innovation. That right is valid until the term of the patent.

23.3.1.2 Trademarks

Any mark or sign that can specify the producer company or enterprise is called trademark. Trademark not only is an identification mark of a particular brand but is also capable of distinguishing them from the same product of a different company. Trademarks can be words, letters, numbers, picture, shape, drawings and symbols, smell, logo, colored marks, or a combination of any of these. These are widely used by companies, which enable customers to identify the product.

The countries allow the registration of simpler forms of trademark, like the Coca-Cola bottle, audible signs as sound as the roar of the lion in MGM films, or olfactory signs as smells in perfumes. However many countries have limitations on signs registered as a trademark, they allow signs, which are visually perceptible or represented graphically.

There are other categories of marks also like:

Collective mark: They are generally owned by an association, like the association, which may represent accountants or engineers. Their members use the mark for their identification.

Certification marks: The certification marks are granted when some standards are set for the product; once the product manufactured meets those standards, it is given that mark, showing excellent quality, for example, a Woolmark is given for compliance with defined standards or a Hallmark is given for compliance with set standards. They can be certified and obtained by any manufacturer and are not confined to any membership.

Service marks: This is a trademark which is used in relation with services and is known as service mark. Service marks may be used by airlines, tourist agencies, hotels, restaurants, laundries, car rental agencies, and cleaners.

Thus, the trademark is helpful in distinguishing marked goods or services, their commercial origin, their quality, and their promotion in the market place.

Trademarks are registered and issued by the government authority (in some countries the patent offices handle trademark documents) and are granted for 10 years but can be renewed continuously.

Granting Trademark

Granting trademark requires certain criteria as:

- The trademark should be represented graphically on paper.
- It should be a distinctive mark capable of differentiating goods of other competitors.
- It should be able to relate goods with the owner enterprise.

The Indian Trademark Act, 1999, includes the following conditions for granting trademark:

- Trademarks can include shape of goods and/or packaging and a combination of colors.
- Single registration is allowed for the trademark; registration is also allowed for service marks. The companies can also register collective marks.
- Punishments related to trademark offenses are at par with the Copyright Act, 1957, to promote sale of only original and authorized goods.
- Term of registration of trademark is 10 years, subject to renewal thereafter.

23.3.1.3 Trade Names

To distinguish their goods and services, enterprises may use one or many different trademarks and trade names. Unlike trademarks and service marks, however, trade names distinguish one enterprise from others, quite independently of the goods or services that the enterprise markets or renders. In most countries trade names are registered with government authority. However, the Paris Convention's Article 8 states that without the obligation of filing or registration, a trade name must be protected. The trade name of one company may not be used by other company in any form that is neither as trade name or trade or service mark.

23.3.1.4 Trade Secret

It is an IPR that can be incorporated in service contract agreements with the business employees. Thus, it can remain as trade secret either indefinitely or as long as the secrecy is novel (no body is able to do the same thing independently). In case of trade secret, it is essential that the information has the commercial value and disclosure of such information may result in loss of the commercial value of its products thereby damaging the economic interest of the holder.

23.3.1.5 Industrial Designs

This IPR may include two-dimensional or three-dimensional external shape, configuration, pattern, ornamentation, or composition of lines or colors applied to any article. It may be a creative activity of obtaining formal or ornamental appearance within a reasonable budget for products produced in huge quantity to satisfy and attract the potential customers along with its intended functions.

Industrial designs are relevant to toy company (cartoon characters) and costume jewelry areas (dolls, puppets, brooches, and so on). The design protection is important when the creation of the character is done intentionally for commercial gains. The designs are also subjected for registration for ownership from the date of registration of the design. Their main attributes because of which they are important and need protection are visual appeal, which influences consumers in their preference for one product over another. With equivalent technical performance, customers choose things based on price and aesthetic appeal; therefore, manufacturers protect the distinctive element that can determine market success. The maximum term is from 10 to 25 years, requiring renewal of registration.

Industrial property can usefully be divided into two main areas:

Particular trademark: Protection of unique and distinctive signs

Geographical indications: Identifies the geographical place of origin of good

These signs or geographical indications promote:

- Fair competition
- Informed choices for customers between various goods

23.3.1.6 Geographical Indications

A sign or an indication used and marked on the goods which have known geographical origin bearing the reputation or qualities of their place of origin is termed as "geographical indication." Apart from design law, there is no division of intellectual property law, which protects the field of geographical indications. The geographical origin has importance in agricultural products as their growth is affected by climate and other factors. They are used for a variety of products as olive oil "Tuscany" produced in Italy or cheese "Roquefort" produced in France; Swiss shows geographical indications for products manufactured in Switzerland.

Some other examples for names which are well known and associated by their geographical connections throughout the world are "Champagne," "Cognac," "Chianti," "Pilsen," "Porto," "Sheffield," and "Darjeeling." Products having very specific quality that is exclusively due to the geographical origin of the product have appellations of origin. The agreement for the Protection of Appellations of Origin and their International Registration include tobacco "Habana" produced in the Havana region (Cuba), or spirit "Tequila" produced in Mexico. The indications also cover symbols, which may be capable of indicating the actual origin of goods for indirect geographical indications; the examples are the Eiffel Tower for Paris, the Matterhorn for Switzerland, or the Tower Bridge for London.

These are given protection according to national laws under (1) laws against unfair competition, (2) laws for consumer protection, (3) laws for the protection of certification marks, or (4) special laws for the protection of geographical indications or appellations of origin. Thus, the rights prohibit the usage of geographical indications by unauthorized parties as it may mislead the public and may trigger court injunctions for prevention of unauthorized use along with the payment or fines for damages and imprisonment in serious cases.

23.3.1.7 Integrated Circuits (ICs)

This type of protection intended to be granted to the topography or layout design of integrated circuits is relatively new. Manufacturing of integrated circuits is done in accordance with layout designs or detailed plans. These designs are creation of the human mind, which require inputs in the form of expertise, time, and financial resources. New layout designs are required for (1) reducing the dimensions of existing ICs and (2) improving and increasing their functions. The smaller IC requires less material for its manufacture and less space for perfect fittings. The ICs are used in huge range of products as computers and servers, watches, washing machines, television sets, and cars.

While creating a new layout design, tremendous expertise, time, and cost are incurred, but it is easy to copy them; therefore, they require protection. Layout designs of integrated circuits are not considered to be industrial designs in the sense of the laws providing for the registration of industrial designs. They cannot be granted protection in any of the listed IPs; thus, under WIPO's auspices, the Treaty on Intellectual Property in Respect of Integrated Circuits was adopted on 26 May 1989. However, the treaty was not implemented, but its substantive provisions have been incorporated (by reference) in 1994 in the Agreement on Trade-Related Aspects of Intellectual Property Rights (TRIPS).

23.3.2 Copyright

The rights in the form of copyrights are given to authors of literary and artistic works for a minimum period of 50 years after the death of the author. These include books and other writings, musical compositions, paintings, sculpture, computer programs, and films. Copyright law deals with mass communication-related creativity including all forms and methods of communication with the public, which may be printed (books, paintings, drawings) or exist as sound and television broadcasting, such as films. However, copyright protects only the form of expression, not the idea. Thus, copyright law would protect the creativity in the choice, organization, and arrangement of words and musical notes or shapes and colors and so on.

The copyright law prohibits the usage of works in their original form as expressed by the creator.

23.3.2.1 Copyright-Related Rights

These are also referred as "neighboring" rights and include the rights of performers (e.g., actors, singers, and musicians), producers of phonograms (sound recordings), and broadcasting organizations. The main social purpose of protection of copyright and related rights is to encourage and reward creative work.

Copyrights are subjected to a time frame for a projected duration during which the rights of the owner exist.

23.3.2.2 Piracy and Infringement

Infringement of copyrights occurs when the owner is not given authorization but somebody else gets it without his consent. When unauthorized copying of copyrighted material is done for commercial gains, it is termed as "piracy."

23.4 Intellectual Property Rights: Indian Scenario

In India the patent offices are located at New Delhi, Chennai, Mumbai, and Kolkata (head office) [3]. They all can handle international applications for patenting or applications through PCT. The importance of intellectual property in India is well established at all levels—statutory, administrative, and judicial [5]. India ratified the agreement establishing the World Trade Organization (WTO). For trademarks the offices are in Mumbai (head office), Delhi, Kolkata, Chennai, and Ahmedabad. For geographical indications the office is based at Chennai and design office is based at Kolkata.

23.4.1 Indian Patents Act, 1970

In this, most provisions amended in the 1970 Act were brought into force on 20 April 1972. The Patents Act is applicable to the whole of India [6].

Patent Conditions Patent can only be given if it, the "invention," is a new product or process with inventive step(s), "capable of industrial application"; this relates to the invention that it should be capable of production or used in an industry. Inventive step is non-obviousness for it, that a person of ordinary skills should not be able to deduce it. Patent is allowed in all branches of technology where the product or processes are patentable [6].

In India the patent is granted for 20 years from the date of filing for all types of inventions [8]. The maintenance of patent requires payment of yearly maintenance fees; failure of doing such, the patent would be ceased and therefore can be utilized by a third party without the risk of infringement.

There are two types of documents: (1) provisional specifications and (2) complete specifications, which can be submitted by the patentee [3, 8].

1. *Provisional specification*: Filed to establish priority of invention at conceptual stage, where submission of complete specifications can be submitted within 12 months (extendable by three more months) of filing the provisional application. This can establish earliest ownership of the invention but does not grant any kind of protection.
2. *Complete specifications*: Submission of complete documents with all specifications is necessary to obtain patents.

The general format of the application is:

1. Title of the invention.
2. Field to which the invention belongs.
3. Background of the invention including prior art giving drawbacks of the known inventions and practices.
4. Complete description of the invention along with experimental results.
5. Sketch or drawing which are essential for understanding the invention.
6. Claims, which are statements related to the invention on which legal proprietorship is being sought. Therefore, the claims have to be drafted very carefully.
7. Whether a patent granted in one country enforceable in other countries.

The Indian Patents Act was formalized in 1970 and was amended in 1999, 2002, and 2005. The patent rules were finalized and amended in 2006.

Patents Amendment Act, 1999

- This amendment was applicable retrospectively from 1 January 1995.
- Mailbox was introduced for pharmaceuticals and agrochemicals.
- India can grant exclusive marketing rights (EMR).

Patents Amendment Act, 2000

- Patent term was increased to 20 years.
- Inventive step was included as important criteria for patenting.
- Introduction of mandatory compulsory license provision for food and drugs.

Patents Amendment Act, 2005

- Product patents for food, chemical, and pharmaceutical
- In this, rights for both product and process patent provided
- The term of patent being 20 years
- Substantially reduced timelines
- Fast track mechanism for disposal of appeals
- Provisions made for protection of biodiversity and traditional knowledge
- Publication of applications after 18 months with facility for early publication

23.4.2 Plant Breeders' Rights or Protection of Plant Varieties and Farmers' Rights Act

Agriculture is indispensable for life. To maintain the interest of private companies in agriculture-related research, legislation to protect the interest of the private sector was established. The laws were required which could give incentives to breeders by protection of plant varieties. However, these were deleterious to the free exchange of plant genetic resources. Patents and plant breeders' rights were designed to protect plant varieties, and keeping in view the interest of plant breeders, they allowed their monopoly to gain financial incentives for their invention [2]. India does not permit patents on life forms, but TRIPS agreement forces developing countries to give protection to the rights of plant breeders by implementing patent or plant breeders' rights. It allows developing countries to acquire sui generis system according to their requirements [4]. India, being an agricultural country, could not adopt the international agreement of UPOV in view of farmers and their interest. In India, the bulk of its population is engaged in the agricultural sector, where farmers should exercise their control over plant breeding and seed supply [1].

Indian legislation took issue of plant breeders' rights as the Protection of Plant Varieties and Farmers' Rights Act, passed in 2001 [9]. IPR promotes monopolies, where farmers' rights have no significance. Thus, India's act balances the innovation keeping in view the rights of the farmers [2, 7].

UPOV's 1978 Act was a model for developing countries which had provisions for breeders [4]. According to UPOV's 1978 Act, the farmers could use the protected variety as starting material for further breeding without payment of any royalty (breeder's exemption). UPOV then revised these and now required breeders to pay the royalty for using protected variety.

23.4.2.1 The Protection of Plant Varieties and Farmer's Rights Act, 2001

India always resisted patenting of life forms, however, it gradually adopted Bt cotton. It is always believed that agricultural resources should be freely used and shared. With debates and controversies and keeping farmers' rights in view came the Protection of Plant Varieties and Farmers' Rights Act in 2001.

- The act allowed IPR claim to breeders with novelty, uniformity, stability, and distinct varieties.
- Breeders had the right to produce, sell, market, distribute, and import and export the protected variety.
- The act granted an 18-year term to trees and vines and a 15-year term to extant varieties and other fields.
- The act allowed breeders' exemption (usage of variety by farmers or researcher for breeding).
- There are no rights for any species which involved GURT as terminator or traitor technology.

In the provision, rights were allowed to the farmer in the form of "farmers' rights":

- Farmers could save, use, sow, resow, exchange, sell, or share his farm-saved seeds or products which include seeds obtained from a protected variety (except branded seeds).
- In this act, the farmers are recognized as cultivator, conserver, and breeder.
- The act protects farmers from high claims by seed companies for performance of their registered variety. They need to disclose the performance-related information of their variety under given conditions. Failure of performance entitles farmers to claim compensation from the breeding company.
- Protections are granted for innocent and mistaken infringement by farmers, where if they violate breeders rights, they are not punished as they were not aware of such laws.

India's act is the most advanced, which considers the interest of all, the farmers, researchers, and breeders. India is the only country where the rights for farmers are established and secured to this extent. India being a member of WTO and TRIPS required to comply with TRIPS provisions to provide the protection of plant varieties. It complies with TRIPS provision with its 2001 Act; however, UPOV membership was not granted to India as it did not comply with the strict requirements of TRIPS.

Whatsoever, India has emerged as a champion due to consideration of farmers' rights. It balanced both the rights of breeders and farmers. Its provisions but deviations from UPOV were welcomed by majority. It has voiced the concerns of its people and developing countries. It should take initiatives to work for an unbiased trading system. In time to come we will be able to develop and use the products which are environment friendly as biofertilizers and biopesticides.

23.4.2.2 Animal Breeder Rights

There are IP systems, which protect plant varieties. However, no such regulation or system has ever been in place for animal breeds. Therefore, private researchers would not invest in animal breeding.

Some Terms

Branded seed: Any packaged seed with labeling indicative of protection granted to it under the act.

Benefit-sharing fund: Farmers providing material used in registered varieties would be given recognition and rewards through the National Gene Fund.

Researchers' rights: In this the researchers are free to use protected varieties for conducting laboratory experiments, and the protected variety can be used as an initial source to create other varieties.

Essentially derived variety: Indian legislature prevents the marketing of a newly developed variety if it shows genetic similarity to a protected variety without breeders' consent.

23.4.2.3 Safeguards in the Patent Laws

Safeguards in the patent laws were to ensure availability of drugs with reasonable cost with compulsory license; in case of public health emergency, appropriate provisions were incorporated; and patent may be revoked either on grounds of public interest or on security considerations.

23.4.2.4 Indian Patent Excludes

The Indian Patents Act excludes the patenting of any invention or discovery, which is or may be detrimental to the welfare of any kind of life form. Some of them are biological warfare, terminator gene technology, embryonic stem cells, all GMOs (except microorganisms), or a plant or animal as a whole or any part thereof.

23.5 Chapter End Summary

- Intellectual property is any invention or creation which is the outcome of rational thinking of the mind. The protection granted to these is termed as intellectual property rights. The intellectual property may be industrial property and copyright.
- Industrial property includes patents, trademarks, trade names, and industrial designs, and copyright includes copyrights and copyright-related rights.
- The international agreement, which governs their protection and powers, is TRIPS. Other international agreements, which govern trade and intellectual property-based platforms, are WTO and WIPO, and several conventions are held from time to time to upgrade the trade-related problems.
- These international bodies ensure smooth trading with controlled tariffs between different nations. All the member nations of these bodies have to mold their national rules and regulations as per the guidelines laid down by these bodies to bring them in international harmony.
- In India the intellectual property rights are well established at statutory, judicial, and administrative levels. The agreement establishing the World Trade Organization has been ratified by India. The Indian Patents Act was formulated in 1970, and since then it has undergone amendments in 1999, 2002, and 2005.

- After the rights are granted to the applicant (patentee), it is his responsibility to check the misuse of the protection (infringement). Upon infringement, the legal suit can be filed against the persons misusing the patent.
- These protections are motivation and encouragement for industries and organizations to experiment and innovate for better usability and utility of various compounds.

Multiple Choice Questions

1. Intellectual property refers to:
   (a) Land
   (b) Invention
   (c) Jewelry
   (d) Car
2. Intellectual property rights are:
   (a) Copyright
   (b) Patent
   (c) Trademark
   (d) All of the above
3. The General Agreement on Tariffs and Trade says that among the member nations, once any country has reduced the tariff rates for its most favored nation, its applicability to other member nations would be:
   (a) The tariff rates would be very high for other countries.
   (b) There would be no tariff.
   (c) The same tariff would apply as it is for the most favored nation.
   (d) Tariff depends upon the relation to the country.
4. Which of the following is not an industrial property?
   (a) Patent
   (b) Trademark
   (c) Copyright
   (d) Geographical indications
5. The role of the World Trade Organization is:
   (a) To liberalize international trade
   (b) To perform trade negotiations
   (c) To settle trade disputes
   d) All of the above
6. The previous name of WIPO was:
   (a) GATT
   (b) TRIPS
   (c) WTO
   (d) BIRPI
7. According to the Patent Cooperation Treaty, the outcome of a company seeking patent after submitting patent application through it is:
   (a) It would be immediately granted patent in all member countries.
   (b) It need not submit separate patent application in all the member countries.
   (c) Its patent application would not be valid.
   (d) It can produce and sell products in all member countries.
8. Budapest Treaty is about:
   (a) Genetically engineered plants
   (b) Genetically engineered animals
   (c) Genetically engineered microorganisms
   (d) All of the above
9. Patenting requires the invention to be:
   (a) Non-obvious
   (b) Novel
   (c) Inventive
   (d) All of the above
10. The following are the examples of geographical indications:
    (a) Amul
    (b) Coca-Cola
    (c) Champagne
    (d) All of these
11. Copyright is related to:
    (a) Films and broadcasting
    (b) Mass communication-related publishing
    (c) Novels
    (d) All of the above
12. The patent head office in India is in:
    (a) New Delhi
    (b) Mumbai
    (c) Kolkata
    (d) Chennai

Answers

1. (b); 2. (d); 3. (c); 4. (c); 5. (d); 6. (d); 7. (b); 8. (c); 9. (d); 10. (c); 11. (d); 12. (c)

Review Questions

Q1. What is intellectual property?
Q2. What are intellectual property rights?
Q3. What is the role of TRIPS?

Q4. Why are patents important?

Q5. What do you mean by exclusive marketing rights?

Q6. What is patent infringement?

Q7. Write short notes on:

(a) Trademark

(b) Trade secret

(c) Copyright

(d) Integrated circuits

Q8. What are plant breeders' rights?

References

1. Alam G (2004) State of the Indian farmer- a millennium study, vol 5. Academic Foundation, New Delhi
2. Chand R, Swaminathan MS (2005) India's agricultural challenges: reflection on policy, technology & other issues. Centre for Trade and Development, New Delhi
3. Controller general of patents, designs, and trade marks
4. Copyright Office, Government of India
5. Cullet P (2005) Life patents, plant breeders' rights and farmers' rights, intellectual property protection and sustainable development. LexisNexis, New Delhi
6. http://indiainbusiness.nic.in/newdesign/index.php?param=investment_landing/267/3
7. http://ipindia.nic.in/ipr/patent/patents.htm
8. http://www.farmersrights.org
9. http://www.indianpatents.org.in/faqpat.htm
10. http://www.lawteacher.net/free-law-essays/human-rights/plant-breeders-rights.php
11. http://www.wipo.int/edocs/pubdocs/en/intproperty/895/wipo_pub_895.pdf
12. http://www.wipo.int/export/sites/www/about-ip/en/iprm/pdf/ch2.pdf
13. http://www.wipo.int/export/sites/www/about-ip/en/iprm/pdf/ch5.pdf
14. http://www.wipo.int/treaties/en/text.jsp?file_id=288514
15. https://www.wto.org/english/tratop_e/trips_e/t_agm2_e.htm
16. The World Intellectual Property Organization (WIPO)

Biosafety and Bioethics

24

Abstract

The advancement in technology is likely to tame several life forms present on earth. Microorganisms are posing a big challenge due to difficulties encountered to control the diseases caused by them. Working with deadly disease-causing microorganisms for their characterization, diagnostics or therapeutics and vaccine development purposes are posing increasingly potential biosafety problems for laboratory workers. Thus, an appropriate biosafe working environment may protect workers from laboratory-induced infections.

Biotechnology has the ability to solve the upcoming problems of the world's increasing population. However, there is often reluctance among the public to accept and support biotechnological products in medicine, industry, or agriculture. There are many safety and ethical issues raised for GM crops and human cloning. Raising transgenic animals and plants has fueled ethical concerns, and the scientists have faced a lot of resistance where genetically modified crop plants or reproductive cloning research of human beings is involved. Thus, biosafety and bioethics are continuously being expanded to combine the rationale of ever-increasing scientific knowledge in biotechnology that is often in conflict with the long-standing social and moral value system of our society.

24.1 Introduction

The advancement in technology is likely to tame several life forms present on earth. Microorganisms are posing a big challenge due to difficulties encountered to control the diseases caused by them. Working with deadly disease-causing microorganisms for their characterization, diagnostics or therapeutics and vaccine development purposes are posing increasingly potential biosafety problems for laboratory workers. Thus, an appropriate biosafe working environment may protect workers from laboratory-induced infections.

Biotechnology has the ability to solve the upcoming problems of the world's increasing population. However, there is often reluctance among the public to accept and support biotechnological products in medicine, industry, or agriculture. There are many safety and ethical issues

V. Gupta et al., *Basic and Applied Aspects of Biotechnology*, DOI 10.1007/978-981-10-0875-7_24

raised for GM crops and human cloning. Raising transgenic animals and plants has fueled ethical concerns, and the scientists have faced a lot of resistance where genetically modified crop plants or reproductive cloning research of human beings is involved. Thus, biosafety and bioethics are continuously being expanded to combine the rationale of ever-increasing scientific knowledge in biotechnology that are often in conflict with the long-standing social and moral value system of our society.

However, biotechnological tools have resulted in high-yielding crop plants, more nutritive values of food grains, longer shelf life, and resistance to insects and pests, but the public acceptance to these biotechnological products is rather low. For example, GM foods have low support in Europe and India. With the passing years, there is further decline in public support, probably due to media focus and the visible debates on GM crops for fear of their long-term effects and unknown risks and environmental safety issues. The controversies from nongovernment organizations (NGOs), scientists, and media have led to a negative impact on the GM crops. GM crop usage has been controversial with fears being expressed on *Flavr Savr* tomato and many others. GM food again went into controversy due to labeling issues. The Bt crops were challenged with the development of resistance in insects. So-called terminator technology and a gene use restriction technology (GURT) met with a lot of resistance due to introduction of negative traits of sterility in the seeds and never came to the market.

The development of cloned animals and its impact on other wild-type animals and the environment also triggered many safety and ethical concerns. Whether or not animals should be used in research, their welfare, sufferings, and well-being were debated throughout the world. In India many plants and animals are linked to religious beliefs and are respected and worshipped for their contribution in the life of human beings.

Likewise usage of embryonic stem cells faced controversies and concerns. Protestants agree to have stem cell research under strict public regulations. Many however are opposed to embryonic stem cell research as it destroys the embryo. The debates and issues surrounding the stem cell are all centered on the potential source of these cells. Embryonic stem cell usage either is prohibited or is under strict public regulation. Somatic stem cells and dedifferentiated somatic cells can be used for therapy.

The public support for xenotransplantation was low due to safety issues with fear of spread of unknown contaminants. However, some xenogeneic tissue-engineered products are now available to treat dermal wounds and burn patients. The usage of biological weapons (use of living organism or its product for human killing) in war is presently banned.

24.2 Biosafety and Biorisk

The biotechnology has tremendously evolved including the technical aspects of medical sciences like diagnostics and therapeutics. With all these, we are also witnessing rapid and dangerous adaptations in microorganisms especially for developing resistance to antibiotics. Microbial pathogens are causative agents of many diseases and due to mutation in their genes, they have developed multiple drug resistance. It is gradually becoming difficult to control these multi-drug-resistant infectious pathogenic microorganisms. To search for solutions to overcome MDR of these pathogenic agents, several scientists and medical workers are handling these pathogens. This poses enormous biohazards and raises serious "biosafety" issues, i.e., scientific practices, methods, and use of appropriate equipment in a biosafe environment.

The biosafety aspects have become very important in various conditions and require many precautions in health-care systems as hospitals, diagnostic laboratories, animal care systems, biological laboratories, and so on. The precautions, which can be taken to reduce or nullify the risk, associated with samples by continuously monitoring and recognizing potential hazards, their risk assessment, and preventive measures to avoid exposure which might result in infection. Individual worker should be appropriately trained and must understand the conditions like containment (conditions where infectious agent can be safely manipulated) and good laboratory practices which can minimize exposure to pathogens. [4]

Biorisk Risk is the likelihood of the occurrence of an adverse event, thus biorisk is the likelihood of the occurrence of serious infection due to exposure to pathogenic microorganism or biohazards. Upon exposure there may be mild to severe infections, allergies, or other clinical problems associated with the pathogen. The biorisk can be managed by risk assessment, effective biosafety measures, and biocontainment.

The first report on laboratory-associated infections was published at the beginning of the twentieth century. However, 4,079 infections associated with laboratory were reported with 168 deaths somewhere between 1930 and 1978. The major pathogenic agents responsible for these infections were *Brucella* spp., *Blastomyces dermatitidis*, *Chlamydia psittaci*, *Coccidioides immitis*, *Coxiella burnetii*, *hepatitis B virus* (*HBV*), *Salmonella typhi*, *Francisella tularensis*, *Mycobacterium tuberculosis*, and *Venezuelan equine encephalitis virus* [23, 29, 30, 38].

After this 1978 report, many more laboratories associated infections and deaths were reported [31, 32]. In these cases also the infectious agents were *Arboviruses*, *Brucella* spp., *Coxiella burnetii*, *Cryptosporidium* spp., *hantavirus*, *Mycobacterium tuberculosis*, *HBV*, *Salmonella* spp., *Shigella* spp., and *hepatitis C virus*.

24.2.1 Assessment of Risk

The processes used for risk assessment are (1) identification of the hazardous properties of a familiar infectious agent or material, (2) the activities responsible for pathogen exposure to the person, (3) the chances of the exposure becoming a laboratory-associated infection, and (4) the ultimate consequences of infection.

Risk assessment needs very alert judgments. If risks are underestimated, they might result in serious and adverse consequences. Overestimation of risks can result in undue protective measures, which may be a burden for the laboratory.

Assessment of risk also needs to evaluate the factors responsible for infections i.e., whether it is due to hazardous agents or due to hazards of laboratory procedure, or the working attitude of the laboratory staff. The staff's capability can be improved by appropriate training and inculcating good laboratory practices with training of accidental spillage/inoculations.

24.2.2 Biohazards

Biohazards may be the microbial infectious agent or other biological materials posing a risk for human health, parasites, viruses, prions, or biologically derived toxins, allergens, venoms, or recombinant DNA that can adversely affect human or animal health and environment [24]. The properties of an agent which make it potentially hazardous are [40] (1) infection capability, (2) ability to cause diseases in human or animal host, (3) its minimum infective dose, (4) severity of the disease, (5) availability of vaccine, (6) availability of therapeutic modality to control it, (7) the probable route followed by it for transmission, (8) stability in the environment, and (9) host (animal–human or only human).

The probable routes of pathogen transmission may be:

1. Through exposed part of the body such as the skin, eye, or mucous membrane
2. Through inhalation
3. Through needle or other sharp objects
4. Through accidental ingestion
5. Through bites of mosquito

The World Health Organization (WHO) has recommended the classification based upon the risks and route of transmission of the pathogenic agents in a laboratory environment. According to their disease-causing capacity and preventive measures available, the human pathogenic agents have been divided into four groups (according to WHO and NIH) [44, 45].

Risk groups	Risk group 1	Risk group 2	Risk group 3	Risk group 4
Properties of microorganisms	Microorganisms not causing any disease	Disease-causing microorganisms	Virulent strains	Highly virulent strain
	No history of diseases in healthy adult human, e.g., nonpathogenic *E. coli*	Low pathogenesis	The pathogen causes severe disease	Causes serious disease which may be lethal
		Disease is not very serious	The disease is aggressive	Difficult to manage by therapeutic intervention
		Preventive measures are present, e.g., measles and mumps virus	Therapeutic intervention is feasible	Vaccine not available
			The worker is at high individual risk, e.g., hepatitis B	The pathogen poses high individual and community risk, e.g., *Ebola* virus
Pathogenicity and infectivity	It poses minimal individual or community risk	Minimal individual and community risk	High individual risk	It poses high individual and community risk
	Microorganism does not cause disease in humans and animals	Is capable of causing disease without serious symptoms	Pathogen capable of causing serious infection in humans or animals	Highly contagious infection
		Sometimes may lead to serious infection but manageable by treatment		Causes serious infection in humans and animals
Preventability and treatment	Preventive and therapeutic measures available but usually not required	Preventive measures available	Effective therapeutic interventions available	Effective therapeutic interventions not available
		Spread of infection is limited	Preventive measures available	Effective preventive measures not available
Biosafety levels requirement	BSL-1	BSL-2	BSL-3/BSL-4	Always using BSL-4

24.2.3 Laboratory Biosafety

National biosafety guidelines came into being because of efforts of microbiological and biomedical communities. These involved the recognition of hazards, assessment of risk, and appropriate use of measures including biocontainment to prevent hazards. They promoted safe laboratory practices and usage of safety equipments along with occupational health programs to reduce laboratory-induced infections while handling microorganisms [5, 6]. Preventive measures also require personal protective equipment like gloves, mask, eye protection, laboratory coat, shoe covers, and respiratory protective measures for prevention of exposure.

The biosafety level (BSL) indicates the kinds of precautions of biocontainment for handling dangerous microbial pathogens. The BSL-1 indicates lowest safety measures while BSL-4 indicates highest possible safety measures.

The containment of laboratory refers to:

- Primary and secondary physical containment barriers
- Contained dressing and shower rooms
- Sealed service penetrations
- Specialized doors and entry and exit avenues to prevent cross-contamination
- Specialized air handling systems for contamination control
- Personal protective equipment
- Biosafety cabinets

  Current biosafety and biocontainment practices and procedures are designed to reduce the exposure of laboratory personnel, the public, agriculture, and the environment to potentially hazardous biological agents [2].

Biosafety levels (BSL) are laboratory designations, which are based upon the degree of risk. They are designated as BSL-1, BSL-2, BSL-3, and BSL-4. These safety levels ensure different levels of protection when working with virulent microbial strain. These laboratories are very sophisticated and have very good design, engineering control, and safe work practices.

Their different levels are used depending upon the pathogen properties like vaccine preventability, infectivity and its contagious nature, severity of the diseases, and risk assessment in case of infection. Every level of containment has its own safety aspects:

Biosafety level 1 (*BSL-1*) in this open bench work can be done in basic teaching or research laboratory. Work involves using chemicals and agents, which do not cause any disease in humans as *E. coli* (nonpathogenic strains) and yeast. This facility may not have biosafety hood.

Biosafety level 2 (*BSL-2*) it can handle moderate-risk microorganisms responsible for infection and pose risk for infection via percutaneous or mucous membrane exposure. In these laboratories provisions for biosafety cabinets are present for containment of aerosols. Protective clothing and biohazard sign is required. The diagnostic and research laboratories are equipped with all these facilities for handling pathogens like *mumps virus*; *hepatitis A*, *B*, and *C virus*; *measles virus*; and *Salmonella*.

Biosafety level 3 (*BSL-3*) is used for agents with known risk of aerosol transmission and causing serious and potentially lethal infections. They may be indigenous or exotic in origin. These are special research or diagnostic laboratories and have biosafety cabinets and other facilities for safe handling, special clothings, controlled access, and maintained directional airflow. The organisms handled in BSL-3 are *Leishmania donovani*, *Mycobacterium tuberculosis*, *SARS coronavirus*, *Yersinia pestis* (plague), *West Nile virus* (encephalitis), and *Rickettsia rickettsii* (Rocky Mountain spotted fever).

Biosafety level 4 (*BSL-4*) is required for high-risk exotic agents which pose risk of life-threatening diseases. Their transmission may be through infectious aerosols. The lack of effective treatment against the diseases caused by these pathogens restricts them to be handled only in high containment laboratories. These laboratories handle dangerous pathogen, have highly advanced biosafety cabinets and positive pressure suits, and have double-ended autoclave, filtered air facility. The laboratories have airlock entry and shower exit with special waste disposal mechanisms. The laboratory is safe to work with *Ebola virus* and *Variola virus* (smallpox agent).

24.2.4 High and Maximum Containment

Any kind of activity that includes usage of potential hazardous human pathogens, zoonotic agents (rabies virus, influenza virus, Trypanosomes (sleeping sickness)), toxins, and agricultural threats which may pose danger to human civilization is recommended for use only in high biosafety levels. The various biosafety levels appropriate for these works are biosafety levels 3 and 4, animal facility/vivarium (ABSL-3 and ABSL-4), and biosafety level 3 agricultural facilities (BSL-3-Ag). High containment refers to the highest level of biosafety. Good biosafety and biocontainment practices can help in effective laboratory biosecurity.

24.2.5 The Importance of Biocontainment Laboratories

There is an urgent requirement to provide adequate safety for all life forms in case of natural emergency. Natural emergency can be the outbreak of any disease or other conditions requiring attention due to attempt for spreading bioterrorism. Thus federally funded research programs are being developed to protect all life forms.

In the measures the products were designed and developed which can protect public health, are called medical countermeasures. These include public health and hygiene, diagnostics, vaccines, therapeutics, and development of biosafety level laboratories.

For the purpose of biosafety, the general laboratory workers (students, scientists, and laboratory staff), along with non-laboratory workers as electricians, plumbers, and sweepers, should also be given appropriate training. The trainings are required in the areas of:

- Usage of microbiological techniques
- Safe use of biosafety cabinets
- Medical and laboratory waste management
- Blood/human samples/tissue handling
- Recombinant DNA technology with viruses or bacterial vectors or working with animal or plant systems [5]

Thus briefly biosafety requires:

- Effective policies for biosafety, biocontainment, and biosecurity.
- Appropriate risk identification, assessment, and continuous monitoring for minimal chances of exposure.
- Training all those working in the laboratory regarding safety, responsibilities, handling in accidental situations, and reporting of concerns.
- Both workers and management should work with full cooperation and critically perform self-review.

24.3 Biotechnology and Bioethics

> Every scientific revolution brings with it a host of ethical and social questions. The so-called genetics revolution is no exception, giving rise to a broad international debate on how the undoubted benefits of progress in this area can be reconciled with certain core human values. (UNESCO 2002a: 1)

Present-day biotechnology opens many opportunities in research and development, addressing medical issues and new ways to explore things; improving human health conditions, fight food, and feed problems; and so on. It has greatly influenced medicine and agriculture. It has tremendously affected the thinking process also like consuming vegetables and fruits without pesticides, using biofertilizer in place of chemical fertilizer, using renewable sources of energy, and switching to products which are biodegradable. It has also advocated the sustainability of systems (agricultural, environmental), knowledge sharing, and brought many products in the market. As the techniques involved gene manipulations in many life forms (plants, animals, microbes), thus with these advancements came the concerns in the form of ethical issues giving rise to bioethics.

In our routine life also, we encounter ethical questions in terms of right or wrong or ethical or unethical; thus, in science also things are judged as being ethical or unethical. Bioethics addresses policy and ethical issues arising by researches and products targeted for human applications. Bioethics addresses the ethical issues in all the streams of life sciences like health care, genetics, and medical research by applying the principles of morality and philosophy [37]. Bioethics has evolved from medical ethics and moral philosophy. The ethical concerns for patient well-being were in use in the form of the Hippocratic Oath.

It starts right from the laboratory to industry and government and affects a very wider society. The study of the social and moral responses arising due to scientific invention or experimentation is "bioethics." Thus it led to granting of ethical clearance for any proposed research projects requiring animal or human experimentations. The project is first reviewed by research/institutional ethics committee (REC/IEC). The REC evaluates the risks, benefits, animal sufferings, utility of work and then only grants approval for conducting research projects and proposals.

Nuremberg Code (1947) This code surfaced in the response of human rights abuses that had taken place under the Nazis during World War II. The war captives were subjected to abuses through experiments on human subjects. This propounded the use of informed consent and non-maleficence, beneficence, justice, and autonomy. These provided normative framework used by researchers and medical practitioners.

Some Important Ethical Terms

- *Morality*: It is a system of moral values and conduct, which govern our decision of right and wrong or good and bad.
- *Empathy*: The ability of one to understand and share the feelings of the other.
- *Euthanasia*: It refers to painless killing of any terminally ill patient suffering from painful and incurable disease on demand of the patient and the court. Removing all life support system from a patient in irreversible coma is also termed as euthanasia.
- *Autonomy* is freedom from external control or influences, independence.
- *Justice*: It is the treatment or behavior, which is genuine, right, or just according to the prevailing laws.
- *Equality*: State of being at par or equal in status or ideology or opportunities.
- *Beneficence*: It is related to kindness, good, or charity for good.
- *Non-maleficence*: An act done to avoid harm or any act which would not harm or violate the trust of others. In the case of physicians, it is their act which will not harm the patient.
- *Accountability*: The condition where somebody is held responsible, answerable, or accountable for an act.

24.3.1 Analyzing Ethical Issues

The ethical principles consist of certain virtues such as autonomy, non-maleficence, beneficence, and justice. Now the area has gained so much of momentum that before even small experimental work, one needs to justify its rightness or wrongness. Analyzing the ethical issues before starting any new experimental work [11] would depend upon the assessment of these points:

- *Consequences*: What would be effect of this experiment? Who is going to be benefited? What would be harms associated with this work? Thus weighing the possible outcomes and their effects.
- *Rights and responsibilities*: Are we exploiting somebody else's rights? Do those rights need protection? Who would protect these? These are questions whose justification would have to be weighted.
- *Autonomy*: If one feels it is right, should things move on? Or else someone feels it is wrong, should things stop? Alternatively, who would decide things should be or should not be? These questions would again need to be addressed.
- *Virtue ethics*: Is this the best thing to do? What is good about it?
- *Animal Rights*: One big question is "Are we authorized to use animals the way we want?" Knowing that they too can think, they are aware of family members, they feel pain (at different levels), and they are alive.

  Animals like primates and whales resembling human brain features, family behavior, and some sensitivity like humans should not be manipulated but others should be, why?

  Animals have been in use since long. They are used for manipulation of their genes, as model system for study of human diseases, and as transgenics for production of pharmaceuticals. Is inflicting pain in any form, which may be by creating a disease model or studying mutations in animals, ethical or not?

24.3.2 Ethics and Ethical Theories

- *Environmental Ethics*: For everything we are dependent on the environment. The environment and ecosystem are very nicely balanced. Thus what would be the effect on the ecosystem once genetically modified organisms are introduced? What would be the effect on biodiversity?
- *Religious ethics*: In this the religious commands are being followed. It involves obeying divine commands and wills.
- *Medical Ethics*: One of the oldest bioethics is medical ethics, which is in practice since its introduction as the Hippocratic Oath (500BC). The Hippocratic Oath is concerned about the behavior and relation of physicians with the patient, practicing non-maleficence and beneficence; however, medical ethics is not limited to the Hippocratic Oath [10].
- *Philosophical Ethics*: In this before executing any decision, it requires evaluation of reasoning and facts for ethical questions. Thus, all ethical decisions are taken upon evaluation of rational and logical questioning:
  - It tries to come up with theories explaining viewpoints.
  - Provides guidance to decision making
  - Solves conflicts of ethical decisions.
  - Ethical theories help in guiding ethical questions.

Philosophical ethics can be divided into descriptive ethics and normative ethics.

24.3.2.1 Descriptive Philosophical Ethics

It explains the actual moral viewpoints of the people; thus, it gives an accurate estimation of people's ethics. Opinion polls or questionnaire evaluation are used to build a consensus of the view of the people. However, these are not affected by facts and reasons so they may be right or wrong [17, 20, 21].

24.3.2.2 Normative Philosophical Ethics

It explains the viewpoint, which people should have, and accordingly the actions, which people should undertake (according to what ought to be). It questions people ethics and guides us in difficult situations about what should be done [3, 37].

Some of the important ethical theories are discussed:

1. *Ethical egoism*: This theory states that morality is mere fulfilling our self-interest. It criticizes the views associated with morality that are in the way of personal self-interest. It advocates that the actions are entirely dependent on desire and self-interest of the people.
2. *Utilitarianism* (*or Consequentialism*): British philosopher Jeremy Bentham (1748–1832) developed the utilitarian theory which was later refined by John Stuart Mill (1806–1872). This theory advocates the selection of moral action, which in its consequence brings greatest happiness for the large number of people. It works to decide the result or consequences of good action, which would maximize happiness. Thus it deals with consequences of action, in that happiness is judged because of some decision. However, greatest good for an individual or society, human race, or animals is debated. Their supporters say that the theory aims to do good for the human race as a whole including nonhuman animals. For example, terminally ill patients should be allowed for physician-assisted mercy killing to avoid pain and sufferings. Utilitarians also believe that abortion and infanticide may be possible in case the baby has severe disability. Research on human embryo and genetic enhancement, which can benefit humans, may be made possible.
3. *Deontology* (*Kantian ethics*): This theory, which states that morality arises from inner sense of duty, was influenced by the German philosopher Immanuel Kant (1724–1804). According to this all people have rights and duties and they should follow these. It states that the decision should come from inner sense of duty without considering the consequences of action. Thus according to Kant any act that follows categorical imperative is moral. Kantian theory requires the use of morality based on rationality with respect to persons and human dignity. Kantian views

refrain from hiding the truth from patients who have terminal cancer. Kant's categorical imperative—"Act in such a way that you treat humanity, whether in your own person or in the person of any other, never merely as a means to an end, but always at the same time as an end" (Kant 1785/1968) [19]. It has been used to avoid abuses in research experiments on human subjects.

This theory has been propounded and supported by a number of philosophers [1].

Main features of his theory are as follows:

(a) *Respect*: Dignity of each and every human being should be respected, and one person should not use the other person as a means to fulfill one's own desire.
(b) *Universality*: If something ethically right is applicable to us as an individual, then it applies to others also, making it a universal right. It should be respected and followed by everyone.
(c) *The formula of ends*: Rational agents should be treated as ends.

Virtue Ethics There are certain ethical virtues and any decision or action is considered morally correct when taken considering the virtues. These promote flourishing and well-being. Thus, any action with a right motive based upon good character or virtue is considered morally right, for example, an action where treatment of a patient is sponsored to gain publicity and honor. The action is right but morally it is not good.

Jansen (2000) had highlighted ethical issues with virtue ethics: (1) problem of content, vague virtues are unable to give proper guidance, and (2) problem of pluralism, competing conceptions of the good life complicate a sound solution.

Feminist Bioethics The social and political background of feminist bioethics is feminism and feminist theory with its major social and political goal to end the oppression of women and to empower them to become an equal gender.

24.4 Ethical Issues in Transgenic Animal Production

The use of transgenic animals has sparked significant controversy in many areas. Some groups see the generation of genetically modified organisms as interfering with natural biological states or processes. They feel that these natural biological states have evolved over long periods which should not be altered. A few other groups are concerned with the limitation of modern science to fully comprehend the potential negative ramification and unforeseen possible effects of genetic manipulation.

Despite the importance of transgenic animals in biomedical research, there are some concerns raised about their use in research. Transgenic animals suffer more abnormalities than regular research animals. The introduction of DNA into an animal can be very complex and the possible side effects can be difficult to predict. Possible harm might arise from surgical techniques used to harvest and reimplant embryos, the collection of tissue from the tip of the tail for genotyping, and nonspecific effects caused by damage to genes adjoining the altered area. In addition, reduced fertility and/or oversize fetuses may result from this technology. In most cases the mutations highly affect specific metabolic processes or cell receptors without actually resulting in disease but causing discomfort, pain, or malfunctioning in the animals.

Transgenic animals not expressing foreign DNA or not containing a particular gene modification are destroyed. Because transgenesis is a complex science, it is not 100% efficient. However, new methods are being developed to increase the accuracy in transgenesis. Again, it should be remembered that such genetic alteration could only be attempted if the authorities are persuaded that there is no other way to pursue important research. The potential risks of transgenics to animals, humans, and the environment are too great to justify their use. The Genetic Modification of Organisms regulations and the Environmental Protection Act (1990) address the risks to those working with animals and the impact on the environment of accidental or planned releases.

The intrinsic worth of animals may be devalued and their integrity violated by genetic modification. Transgenic animals have not chosen to have foreign DNA or other genetic modifications. However, this potential "cost" to the animals is routinely assessed under the ethical review of proposed procedures and weighed against the potential benefits. Medical researchers only employ this technology when no alternative research avenue exists but with appropriate breeding facility of animals. As the Royal Society concluded in its 2001 Report "The Use of Genetically Modified Animals", the use of transgenic animals is fundamentally little different from the use of other animals in biomedical research.

However, the technology has opened new frontiers in biomedical applications and has provided new opportunities for exploring the organization, biological pathway, regulation, and pathological function of molecular processes.

Contractarianism: It says that the animals are important as many people feel their importance. Humanity is the virtue of morality. Thus, humans have obligations toward other animals.

Utilitarianism: It says that as animals can also suffer; thus, they are morally relevant. It chooses the action, which aims to give happiness to all. Thus, its goal is to maximize the total sum of happiness in the world and is impartial, that is, the pain and happiness of everyone count.

Animal activists view: According to these whatever is the motif there is no justification of harming animals for human purposes.

Ruthless view: This has no concern about animals, whichever way they are being used.

However, principles governing animal rights are the 3Rs, which mean *replace*, *reduce*, and *refine*. It involves ethical and humane approach while using animals. The alternatives of animals in the research were also explored. The 3Rs are replacement of conscious living animals with nonsentient animals or materials, reduction of the number of animals used in an experiment or procedure, and refinement of the techniques used in order to decrease the incidence or amount of animal pain and distress [34].

24.5 Genetically Modified Crops and Bioethics

Genetically modified crops or GM crops have given new dimensions to agriculture. With the advancements in the technology, the agricultural productivity is increased, the food's nutritive value is enhanced, the requirement of pesticides and insecticides is greatly reduced, and pharmaceutical substances (vaccines) are being produced in the plants.

However, GM foods have become the target of public concern due to unknown and unseen fears of their effects on the ecosystem and risks to human health. The GM foods are also not labeled leading to distrust and fear in the customers regarding the safety of food.

24.5.1 Genetically Modified Organisms and Environment

The ethical issues raised are:

- What about evolution?
- Each living being has evolved to grow in one condition, and if we put human gene or fish's gene in a plant or animal, what would be its effects?
- What would happen if the gene can find an escape route to introduce itself in other members of that species or other species?
- What would happen if the herbicide resistance gene could transfer itself in weeds, generating superweeds?
- If we have all GM crops of any grain or pulses, then occurrence of any disaster would lead to complete loss of the crop, as there is no biodiversity where one is sensitive and the other one is resistant. This would lead to disaster and damage of whole of the crop.

- Would the monogenetic crops, if not able to react appropriately to environmental stress, again face a situation like "Ireland's potato famine"?
- What if GM crops breed with other crops, would the biodiversity be lost?
- What are their risks to other animals like birds, insects, and other higher animals (like monarch butterfly and Bt crops)?

24.5.1.1 Genetically Modified Organisms and Human Health

- There is tremendous fear of GMOs on human health.
- If the GM can induce allergy due to insertion of unrelated gene what would happen, because GM foods are not labeled. For example, a Brazil nut gene was transferred to a soybean variety; this nut gene product was a source of allergy for many individuals. As GM products are not labeled, therefore, the consumers may consume soybean without being aware of nut gene product in them. This might lead to allergy causing major allergic reactions in the consumers. *However resultant modified crop was never released to the public.*
- Human food supply may have GM products as evidenced by the settlement between Syngenta and the US government over the accidental sale of unapproved GM (Bt10) corn seed to farmers.

24.5.1.2 Economic and Societal Issues

- These seeds can only be afforded by big farmers and landowners, thus negatively affecting small-scale farmers.
- The farmers need to buy seeds every year, as they cannot collect, store, and replant these seeds.
- There would be huge differences in the farming practices in one country only, and that gap would be widened much between developed and developing nations [17].
- As GM food is not labeled, thus, public trust and acceptance of GM food would be a big challenge.

24.5.1.3 Advantages of Genetically Modified Crops

- The GM crops have higher productivity as they are resistant to pests; thus, loss due to diseases is prevented.
- They are engineered to survive in high salt and frost, thus utilizing the land, which was unused for agricultural purposes.
- They also help to reduce contaminants as insecticides and pesticides, thus preventing their entry in soil, water, and organic matter.
- All these effects would ultimately affect the environment and consumers.
- Biotechnology can fulfill the increasing requirements of food, with maximum utilization of land.
- Biotechnology can help reduce the environmental contaminants as insecticides and pesticides preventing their entry in food chain.
- It can help maintain clean environment.
- It can enhance nutritive values of the food, for example, golden rice can help prevent blindness due to nutritional deficiency of vitamin A.
- The longer shelf life of fruits and vegetables can decrease their wastage and increase their availability with simple storage conditions.
- The genes for allergins can also be engineered so that the allergins can be effectively removed from the food crops.
- Thus they are beneficial for humans and animals.
- It can also eliminate trans fats, producing healthier foods.
- Nowadays, plants are also being explored and used for production of biopharmaceuticals like vaccines, ScFv, antibodies, and so on. There is reduced risk of adverse reactions in terms of animal pathogen transmission, and the protein undergoes posttranslational modifications.

These genetically modified varieties faced tremendous resistance in Europe and India. Their public acceptance has also met with resistance in terms of fears, controversies, and acceptance. The controversy from NGOs, media, and scientists has led to tremendous negative impacts on the GM crops. The environmental council of the European Union further augmented the controversy by halting regulatory approval of GM crops.

The ethical issues and safety issues in GM cops can be addressed by evaluating their safety and other benefits and risks. The risks associated with humans, animals, and plants should be well considered along with the intention to do good. The consideration of well-being of all (humans, animals, plants, and environment) is very important for the sustainability of life, along with moving forward with positive aspects of technology. The ultimate aim of the technology should be improvement and sustainability in the future of all life forms with minimal adverse effects on our surroundings and other plants and animals.

Concerns

The experiments being done to create transgenics have left the people with many concerns regarding the health and life. The changes which unnatural gene transfer might be doing to the environment and the society? There have been concerns regarding the transfer of animal genes into plants (for vegetarians) and food animals. This has raised not only safety issues but also ethical issues:

1. Impact of breeding of these transgenic animals and plants on other plants and animals when they breed in their natural environment.
2. Potential health hazards which we can face after consumption of GMOs.
3. Risk of transmission or emergence of potential pathogens post gene manipulation or xenotransplantation.
4. Changes in the environment and ecology by unnatural gene transfer and crossing of species boundary.

(continued)

5. Aren't we imparting sufferings for the plants and animals by using these tools and technology on them?
6. Rights about the well-being of other plants and animals.

Thus, there is a requirement of well-defined guidelines when using gene manipulations [39] as well as labeling of the products produced by gene transfer technologies. There is no doubt that apart from genetically modified plants and animals, the gene manipulation techniques are well accepted in production of recombinant pharmaceuticals for production of life-saving drugs. The technique has enormous potential for solving food problems and sustainable agricultural practices and preventing and treating diseases.

There are many concerns regarding reproductive cloning of humans. Thus, the concerns should be there whether the technology is being used for what is worth or we simply want to see the extremes of what technology can do.

Each achievement of science has both negative and positive aspects; nuclear power when used judiciously can help us solve energy problems but can also result in disaster like what happened in Hiroshima and Nagasaki during World War II. The gene manipulation technology is with tremendous potential but should be used just for benefit of mankind, the environment, and animals.

24.6 Bioethics and Reproductive Technology

Assisted reproductive technology (ART) techniques are used in the case of male or female infertility. The technique is detailed in Chap. 15. The process of ART involves ovulation, artificial insemination, in vitro fertilization, and implantation. It may also involve preimplantation genetic diagnosis, gestational surrogacy, gamete donation, and sex selection.

Bioethical issues are important as they are directly related to society:

- The technology enables children to be conceived by couples who have fertility defects, thus it came up as a very good medical intervention for childless couples.
- However it enables children to be conceived who do not have any genetic relationship with either one or both of their parents (using donor sperm or ovum).
- Though preimplantation genetic diagnosis is recommended for medical reasons, it can be misused.
- The male and female homosexual couples, single men, single women, or postmenopausal women would seek the assistance of assisted reproductive technologies.
- As the mother and father would not have child–parent relationship, who would be responsible for the child and his welfare? The child too have right to be with biological father and mother.
- In conditions where both donor sperms and ovum are used for having the child, what would be the legal rights of the child?
- This technology if not judiciously used could create scattered and torn families and disconnect genetic, gestational, and social child–parent relationship.
- What would be done with excess IVF embryos, would they be treated like human embryos or model embryos available for experimentations?

Although many people are in favor of IVF and surrogacy, a survey shows that people are against posthumous sperm procurement or reproductive cloning. The technology is surrounded by many issues of creation and destruction of embryos [13, 33]. Bioethical issues are for seeking justice, non-maleficence, and beneficence of all involved in any issue. Adequate ethical guidelines and moral evaluations of new technologies are essential so that laws can be framed for national regulations and restrictions of unacceptable practices [12].

24.7 Stem Cells and Bioethics

Stem cells for use as therapy have given new dimensions to medicine. Due to capacity of these cells to differentiate into almost all different kinds of cells, they are being looked as potential alternative for the cure of many diseases.

Probably sometime in the future, scientists would be able to generate whole organs by stem cells and tissue engineering to fulfill the increasing demand for organs.

Depending upon their potency and differentiation capabilities, they can be:

- Adult stem cells: These are present in each tissue of the human body, where they help in replenishing cells for regular wear and tear of the cells.
- Embryonic stem cells are located in the human embryo at the blastocyst stage (5–6 days of age). Leftover embryos at this age are often unwanted in assisted reproductive technology, and some parents can donate them for research. In many countries very strict laws are there for their usage. Some of the nations recommend their use when no other option is available.
- Cord blood stem cells are derived from the umbilical cord, which is often still routinely discarded at birth.

The ethical issues raised for stem cell research is the usage and destruction of human embryos for obtaining embryonic stem cell [14]. As the life begins with the embryo, thus destruction of life is immoral; therefore, its usage and the extraction of embryonic stem cells are unethical.

However, using adult stem cells or umbilical cord blood stem cells has been considered ethically superior alternative to the use of embryonic stem cells. Their rejection and uncontrolled growth would be big challenges to fully explore their therapeutic potential.

24.8 Human Cloning and Bioethics

Human cloning is a very controversial issue [22, 43] to discuss and its cloning can be "reproductive" cloning and "therapeutic" or "research" cloning. Though both the terms are not scientifically accurate, they are widely used. In somatic cell nuclear transfer (SCNT), the nucleus from the somatic cell is transferred into the enucleated egg. The resulting embryo can either be implanted into a mother or surrogate mother for development (the cloned sheep Dolly was the outcome of SCNT) or the embryo can be used for research purposes [7]. Implantation of embryo for gestation is referred as reproductive cloning, whereas in therapeutic cloning embryo is harvested for embryonic stem cells [16]. The stem cells obtained from embryo are embryonic stem cells, which can be induced to form different cells [8, 25].

- People and scientific community in favor of human cloning say that it can combat infertility and very useful for therapeutic purposes.
- Many people are against the creation and usage of embryo for research purposes; some are concerned about the risks like ovarian hyperstimulation syndrome in egg donors.
- Many more object to the destruction of embryos as they represent the initial phase of human life thus considered morally equivalent to humans; therefore, they consider therapeutic cloning at par to reproductive cloning.
- Human reproductive cloning is unacceptable and unethical for many scientists and philosophers [9]. Most of the countries have banned all kinds of human cloning.
- In the Universal Declaration on the Human Genome and Human Rights (UDHGHR) by UNESCO in 1997, it says that the "Practices which are contrary to human dignity, such as reproductive cloning of human beings, shall not be permitted."
- The resolution in this line was also passed by the WHO saying that human reproductive cloning is against human dignity and urged member states to prohibit it. Likewise The Council of Europe's Additional Protocol to the Convention on Human Rights and Biomedicine (1998), The Charter of Fundamental Rights of the European Union (2000) prohibits the reproductive cloning of human beings (Article 3).
- The potential and the power of the technology require caution when we ourselves are the objects of biotechnology [35].

24.9 Impact of Biotechnology on Society: Future Prospects

Biotechnology has had a tremendous impact on human race [26]. The technology has potential to change what humans are [28]. Humans may come up as transhumans. Transhumanism is the international movement which is aimed to transform humans by the use of technology to augment brain function and enhance human intellect, physical potential, and psychological capabilities [27]. Their goal is to extend human life, enhance the brain and its capacities, delay aging, and improve physical conditions (so-called superhumans). Attempting technological augmentation on the normal human function [17, 36] can take us far away from our own species to "superhuman" function. Thus, technological interventions can shift us from what we called as *Homo sapiens* to "superhuman" *Homo sapiens technologicus*—a species that uses, fuses, and integrates technology to enhance its own function [46, 47].

We are living in the best of times as exhibited by the fast-paced economic growth of many countries around the world. Various initiatives that were taken in the fields of molecular biology, genetics, and recombinant DNA technology are beginning to show fruits, strengthening the basis of research.

The technological advances have made our life easier. The problems encountered in the form of ethical dilemma need appropriate attention with a view of the interests of society. Our ultimate aim should be to generate thorough knowledge on all aspects of modern biology and garner excellence in the making of an economically, socially, and morally sustainable society [18].

24.10 Chapter End Summary

- Biotechnology has made tremendous progress in providing therapeutics, new ways to diagnose diseases, regenerative medicines, food and feed solutions, food with nutritive values, and other countless achievements.
- Biosafety is a very important aspect as the health-care and laboratory workers are continuously exposed to highly contagious and infectious pathogenic agents. They work with these agents either for the purposes of diagnostics or research.
- Many laboratory-induced infections were reported among the workers and several of them were serious and resulted in death of workers. These laboratory-induced infections were due to organisms in biorisk-3 and biorisk-4 groups.
- Thus, it becomes very important to do appropriate risk assessment, monitoring, and installation of appropriate preventive measure like biosafety levels and biocontainment facility and training to the laboratory workers for prevention.
- Bioethical issues are there with all the major achievements of biotechnology. The issues were being raised because of rightness or wrongness of the experiment, to ensure experiments are conducted in a way that it ensures safety for all, the humans, animals, and the environment.
- Bioethical issues related with genetically modified crops are related to development of resistance in insect-resistant crops and their impact on the environment, animals, and humans. Many ethical issues were raised for genetically modified animals and their sufferings.
- Embryonic stem cell therapy and assisted reproductive technologies along with genetically modified crops faced too much of ethical discussions and resistance. Human reproductive cloning created tremendous furor; thus, human cloning in all forms was banned in most of the countries.
- As with any other technological advancements, each technology is faced with ethical and social issues and so is biotechnology. However, it is important that before we keep on proceeding, we should ask some basic questions to ourselves: (1) Why am I doing this? (2) Would it be beneficial for anyone? (3) Would this disturb the well-being of other animals, humans, or the environment?
- Ultimately technological achievements are to create and garner a society with the well-being of all life forms in harmony with nature.

Multiple Choice Questions

1. Which of the following techniques is facing bioethical issues?
   (a) DNA microarray
   (b) Fluorescence activated cell sorter
   (c) Embryonic stem cell therapy
   (d) All of the above
2. The concerns raised with GM foods are:
   (a) Development of resistance in insects due to insect-resistant crops
   (b) Appropriate labeling of GM foods
   (c) Their interaction with wild-type varieties may lead to problems
   (d) All of the above
3. Biosafety aspects are:
   (a) Assessment of risk while working with pathogenic agent
   (b) Killing the disease-causing microorganism
   (c) Genetically engineering pathogens to reduce their virulence
   (d) None of the above
4. Laboratory-associated infections may be due to:
   (a) *Candida albicans*
   (b) *Escherichia coli*
   (c) *Brucella* spp.
   (d) *Entamoeba histolytica*
5. The microorganism classified in risk group four would be:
   (a) Vaccine preventable
   (b) Preventable with medicines
   (c) Both of the above
   (d) None of the above

6. Which of the following procedures is commonly used for the production of embryonic stem cells?
   (a) Recombinant DNA technology
   (b) Reproductive cloning
   (c) Therapeutic cloning
   (d) In situ hybridization
7. Bioweapons may be:
   (a) Avirulent bacteria
   (b) Cauliflower mosaic virus
   (c) Toxins derived from pathogens
   (d) All of the above
8. The code established for human rights protection in biomedical research after World War II was:
   (a) Nuremberg code
   (b) UNO code
   (c) Hippocrates code
   (d) UNICEF code
9. Which of the following terms is not a part of the three R's of animal ethics?
   (a) Replacement
   (b) Rejuvenation
   (c) Reduction
   (d) Refinement
10. The technique of somatic cell nuclear transfer involves:
    (a) Nuclei from ovum in an enucleated somatic cell
    (b) Nuclei from somatic cell in enucleated ovum
    (c) Nuclei from zygote in an enucleated ovum
    (d) All of the above
11. Transhumans are:
    (a) Humans born by the process of in vitro fertilization
    (b) Usage of humans as bioweapons
    (c) Usage to technology to improve human overall capabilities
    (d) All of the above
12. An example of euthanasia is:
    (a) Understand others feeling
    (b) Removing life support from brain dead patient
    (c) Independence to express oneself
    (d) All of these
13. Beneficence is:
    (a) Related to genuine behavior
    (b) Related to kindness
    (c) Related to accountability
    (d) All of the above
14. Non-maleficence is:
    (a) An act to retain confidence
    (b) An act to avoid harm
    (c) An act to gain trust
    (d) All of the above

Answers

1. (c); 2. (d); 3. (a); 4. (c); 5. (d); 6. (c); 7. (c); 8. (a); 9. (b); 10. (b); 11. (c); 12. (b); 13. (b); 14. (d)

Review Questions

Q1. What is the importance of bioethics?
Q2. Why is biosafety important?
Q3. What do you understand about risk groups?
Q4. What is biocontainment?
Q5. What are biosafety levels?
Q6. Why did Nuremberg code come into being?
Q7. What are ethical issues related with GMOs?
Q8. What is transhuman? Do you think human cloning should be allowed or should not be allowed? Justify.

References

1. Altman M (2011) Kant and applied ethics: the uses and limits of Kant's practical philosophy. Wiley-Blackwell, Malden
2. American Public Health Association. Control of communicable diseases manual. 18th ed. DL Heymann, editor. Washington, DC; Oxford Journals 2005.
3. Brom FWA (2002) Science and Society: Different Bioethical Approaches towards Animal Experimentation. Presentation at a symposium "Use of animals in research: a science-society controversy?" held by the Doerenkamp-Zbinden-Foundation.
4. Centers for Disease Control and Prevention (1998) Fatal Cercopithecine herpesvirus 1 (B virus) infection following a mucocutaneous exposure and interim recommendations for worker protection. MMWR Morb Mortal Wkly Rep 47:1073–1076

5. Chatigny MA, Hatch MT, Wolochow H et al, (1979) Studies on release and survival of biological substances used in recombinant DNA laboratory procedures. National Institutes of Health Recombinant DNA Technical Bulletin
6. DP O'M' n, Scott PA, McAuley A, Walsh-Daneshmandi A, Daly B (2005) Health care rights and responsibilities: a review of the European Charter of Patients' Rights. Irish Patients' Association, Dublin
7. Galton D, Doyal L (1998) "Goodbye Dolly?" The ethics of human cloning. J Med Ethics 24:279
8. Giubilini A, Minerva F (2012) After-birth abortion: why should the baby live? J Med Ethics Published Online
9. Gogarty B (2003) What exactly is an exact copy? And why it matters when trying to ban human reproductive cloning in Australia. J Med Ethics 29:84–89
10. Gordon J-S, Rauprich O, Vollmann J (2011) Applying the four-principle approach. Bioethics 25:293–300
11. Habermas J (2003) The future of human nature. Polity Press, Cambridge
12. Hostiuc S (2013) Conventional vs unconventional assisted reproductive technologies: opinions of young physicians. J Obstet Gynaecol 33:67–70
13. Human genetic data: preliminary study of the IBC on their collection, processing, storage and use, Rumball S and Smith AM, International Bioethics Committee, Paris: SHS-503/01/CIB-8/3 (Rev.2) 2002a
14. Isasi R et al (2004) Legal and ethical approaches to stem cell and cloning research: a comparative analysis of policies in Latin America, Asia, and Africa. J Law Med Ethics 32:626–640
15. Jansen LA (2000) The virtues in their place: virtue ethics in medicine. Theor Med Bioeth 2:261–276
16. Jensen D (2008) Human reproductive cloning and reasons for deprivation. J Med Ethics 34:619–623
17. Jonas H (1984) The imperative of responsibility: in search of an ethics for the technological age. University of Chicago Press, Chicago
18. Jonas H (2004) Ethics and biogenetic art. Soc Res 71(3):569–582
19. Kant I (1968) Grundlegung zur Metaphysik der Sitten (Groundwork for the Metaphysics of Morals) [1785] Akademie-Ausgabe Kant Werke IV. Walter de Gruyter, Berlin
20. Kass L (2002) Life, liberty & the defense of dignity: the challenge for bioethics. Encounter Books, New York
21. Kuhse H, Singer P (2009) A companion to bioethics, 2nd edn. Wiley-Blackwell, Oxford
22. Langlois A (2013) Negotiating bioethics: the governance of UNESCO's bioethics programme. Routledge, London
23. Lennette EH, Koprowski H (1943) Human infection with Venezuelan equine encephalomyelitis virus. JAMA 123:1088–1095
24. Lennette EH, Panel V (1973) Common sense in the laboratory: recommendations and priorities. Biohazards in biological research. Proceedings of a conference held at the Asilomar conference center. Cold Spring Harbor Laboratory, Pacific Grove, p 353
25. Lo B et al (2010) Cloning mice and men: prohibiting the use of iPS cells for human reproductive cloning'. Cell Stem Cell 6:16–20
26. Mathuna DPO (2007) Bioethics and biotechnology. Cytotechnology 53:113–119
27. Mcnamee MJ, Edwards SD (2006) Transhumanism, medical technology and slippery slopes. J Med Ethics 32:513–518
28. Pearson H (2002) Your destiny, from day one. Nature 418:14–15
29. Pike RM (1976) Laboratory-associated infections: summary and analysis of 3921 cases. Health Lab Sci 13:105–114
30. Pike RM (1978) Past and present hazards of working with infectious agents. Arch Pathol Lab Med 102:333–336
31. Pike RM (1979) Laboratory-associated infections: incidence, fatalities, causes, and prevention. Annu Rev Microbiol 33:41–66
32. Pike RM, Sulkin SE, Schulze ML (1965) Continuing importance of laboratory acquired infections. AmJ Publ Health 55:190–199
33. President's Council on Bioethics (2004) Monitoring stem cell research. GPO, Washington, DC
34. Russell WMS, Burch RL (1959) The principles of humane experimental technique. Methuen London, ALTEX (Alternatives to Animal Experimentation) official publication of the Johns Hopkins Center for Alternatives to Animal Testing. ISBN 0900767782
35. Schuurman E (2005) The technological world picture and an ethics of responsibility. Dordt College Press, Sioux Center
36. Schwartz B (2004) The paradox of choice: why more is less. Harper Perennial, New York
37. Singer P (1979) Practical ethics. Cambridge University Press, New York
38. Sulkin SE, Pike RM (1951) Survey of laboratory-acquired infections. AmJ Publ Health 41:769–781
39. The National Institutes of Health (US), Office of Biotechnology Activities (2002). NIH guidelines for research involving recombinant DNA molecules. Bethesda he
40. Tigertt WD, Benenson AS, Gochenour WS (1961) Airborne Q fever. Bacteriol Rev 25:285–293
41. UNESCO, Universal Declaration on the Human Genome and Human Rights. Paris, France 1997
42. UNESCO, Human Genetic Data: preliminary study of the IBC on their collection, processing, storage and use, Paris: International Bioethics Committee (IBC), SHS-503/01/CIB-8/3 (Rev.2) 2002a.
43. WHO, Fifty-first World Health Assembly Agenda Item 20: ethical, scientific and social implications of cloning in human health, World Health Organization, Geneva, WHA51.10, 1998
44. World Health Organisation (2005) World health report, World Health Organization, Geneva
45. World Health Organization (2004) Laboratory biosafety manual, 3rd edn. World Health Organization, Geneva

46. Zehr EP (2011) Inventing iron man: the possibility of a human machine. Johns Hopkins University Press, Baltimore
47. Zehr EP (2015) The potential transformation of our species by neural enhancement. J Mot Behav 47:73–78

Some Related Resources

Adapted from the MMWR 2011; 60 (Suppl): Guidelines for Biosafety Laboratory Competence
European Union, Charter of Fundamental Human Rights 2000
http://biotechlearn.org.nz/themes/bioethics
http://www.beep.ac.uk/content/424.0.html
http://www.bioethics.org.au/Resources
http://www.cdc.gov/mmwr/preview/mmwrhtml/su6002a2.htm
http://www.iep.utm.edu/bioethic/
http://www.iep.utm.edu/bioethic/
http://www.niaid.nih.gov/about/organization/dir/rml/Pages/default.aspx
http://www.phe.gov/Preparedness/legal/boards/biosafety-taskforce/Documents/transfedbiocontainmentrpt092009
http://www.scu.edu/ethics/practicing/focusareas/medical/conference/presentations/genetically-modified-foods.html
http://www.who.int/whr/2005/en/index.html
https://bioethics.georgetown.edu/explore-bioethics/animal-research-ethics/
www.dcu.ie/nursing/patients_rights.shtml. Cited 20 October 2006

Index

V. Gupta et al., *Basic and Applied Aspects of Biotechnology*, DOI 10.1007/978-981-10-0875-7

MIX
Papier aus verantwortungsvollen Quellen
Paper from responsible sources
FSC® C105338

If you have any concerns about our products,
you can contact us on
ProductSafety@springernature.com

In case Publisher is established outside the EU,
the EU authorized representative is:
Springer Nature Customer Service Center GmbH
Europaplatz 3, 69115 Heidelberg, Germany

Printed by Libri Plureos GmbH
in Hamburg, Germany